THE
MATHEMATICAL PAPERS OF
ISAAC NEWTON
VOLUME II
1667-1670

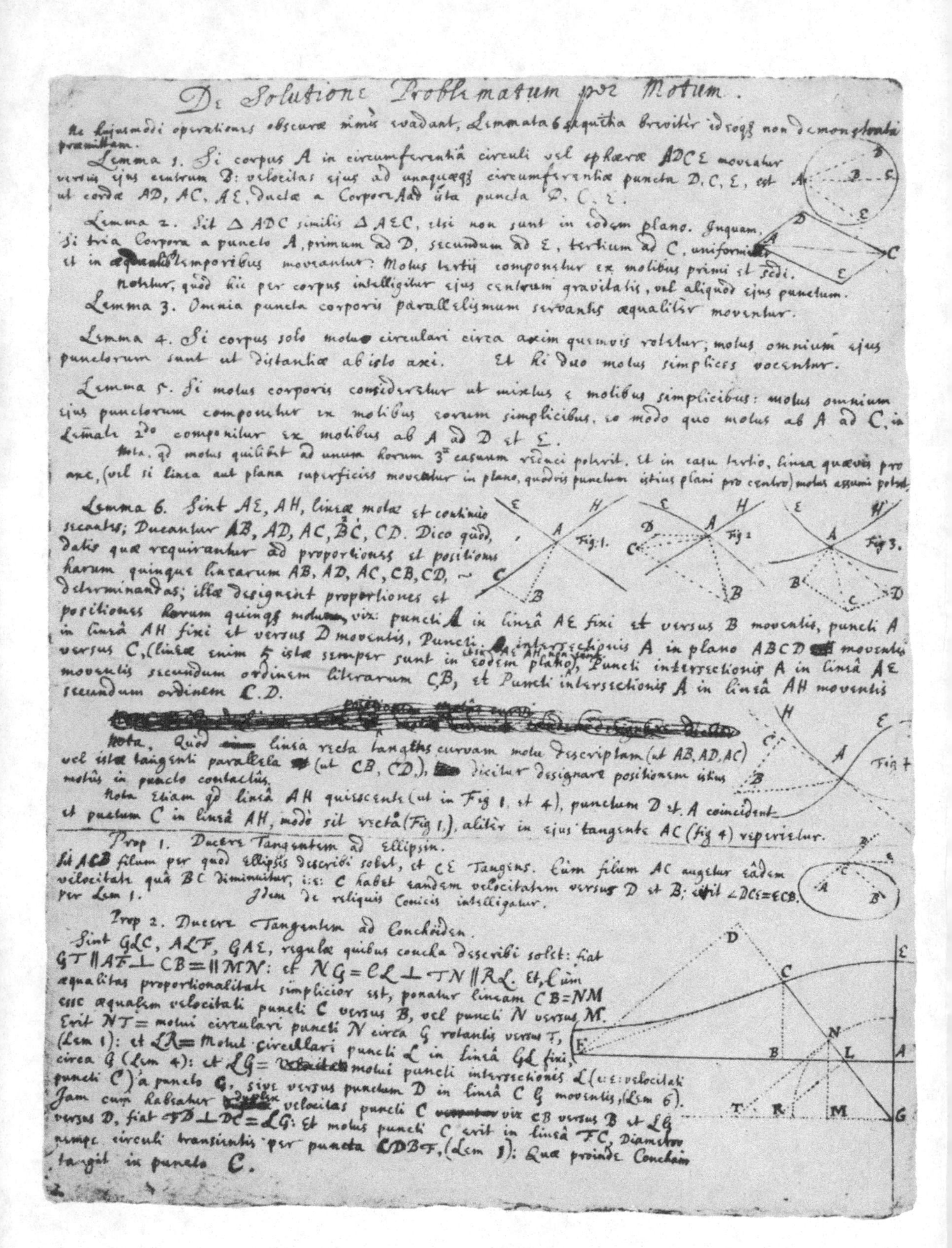

De Solutione Problematum per Motum.

Ne hujusmodi operationes obscuræ nimis evadant, Lemmata 6 sequentia breviter ideoque non demonstrata præmittam.

Lemma 1. Si corpus A in circumferentiâ circuli vel sphæræ ADCE moveatur versus ejus centrum B: velocitas ejus ad unaquæque circumferentiæ puncta D, C, E, est ut cordæ AD, AC, AE, ductæ a Corpore A ad ista puncta D, C, E.

Lemma 2. Sit Δ ADC similis Δ AEC, etsi non sunt in eodem plano. Inquam, Si tria Corpora a puncto A, primum ad D, secundum ad E, tertium ad C, uniformiter et in æqualibus temporibus moveantur: Motus tertij componetur ex motibus primi et sedi.

Notetur, quod hic per corpus intelligitur ejus centrum gravitatis, vel aliquod ejus punctum.

Lemma 3. Omnia puncta corporis parallelismum servantis æqualiter moventur.

Lemma 4. Si corpus solo motu circulari circa axim quemvis rotetur; motus omnium ejus punctorum sunt ut distantiæ ab isto axi. Et hi duo motus simplices vocentur.

Lemma 5. Si motus corporis consideretur ut mixtus e motibus simplicibus: motus omnium ejus punctorum componetur ex motibus eorum simplicibus, eo modo quo motus ab A ad C, in Lemate 2do componitur ex motibus ab A ad D et E.

Nota, quod motus quilibet ad unum horum 3m casuum reduci poterit. Et in casu tertio, linea quævis pro axe, (vel si linea aut plana superficies moveatur in plano, quodvis punctum istius plani pro centro) motus assumi potest.

Lemma 6. Sint AE, AH, lineæ motæ et continuo secantes; Ducantur AB, AD, AC, BC, CD. Dico quod, datis quæ requiruntur ad proportiones et positiones harum quinque linearum AB, AD, AC, CB, CD, determinandas; illæ designent proportiones et positiones horum quinque motuum, viz: puncti A in lineâ AE fixi et versus B moventis, puncti A in lineâ AH fixi et versus D moventis, Puncti intersectionis A in plano ABCD moventis versus C, (lineæ enim istæ semper sunt in eodem plano), Puncti intersectionis A in lineâ AE moventis secundum ordinem literarum CB, et Puncti intersectionis A in lineâ AH moventis secundum ordinem CD.

Nota, Quod linea recta tangens curvam motu descriptam (ut AB, AD, AC) vel istæ tangenti parallela (ut CB, CD), dicitur designare positionem istius motûs in puncto contactûs.

Nota Etiam quod lineâ AH quiescente (ut in Fig 1, et 4), punctum D et A coincident et punctum C in lineâ AH, modò sit recta (Fig 1), aliter in ejus tangente AC (Fig 4) reperietur.

Prop 1. Ducere Tangentem ad Ellipsin.

Sit ACB filum per quod ellipsis describi solet, et CE Tangens. Cùm filum AC augetur eâdem velocitate quâ BC diminuitur, i:e: C habet eandem velocitatem versus D et B; erit ∠DCE=ECB. per Lem 1. Idem de reliquis Conicis intelligatur.

Prop 2. Ducere Tangentem ad Conchoiden.

Sint GLC, ALF, GAE, regulæ quibus concha describi solet: fiat GT∥AF⊥CB=∥MN: et NG=CL⊥TN∥RL. Et, Cùm æqualitas proportionalitate simplicior est, ponatur lineam CB=NM esse æqualem velocitati puncti C versus B, vel puncti N versus M. Erit NT= motui circulari puncti N circa G rotantis versus T, (Lem 1): et LR= Motui circulari puncti L in lineâ GL fixi, circa G (Lem 4): et LG= motui puncti intersectionis L (i:e: velocitati puncti C) a puncto G, sive versus punctum D in lineâ CG moventis, (Lem 6). Jam cum habeatur velocitas puncti C viz CB versus B et LG versus D, fiat FD⊥DC=LG: Et motus puncti C erit in lineâ FC, Diametro nempe circuli transientis per puncta CDBF, (Lem 5): Quæ proinde Concham tanget in puncto C.

The 'solution of problems by motion' (2, 2, §2).

THE MATHEMATICAL PAPERS OF ISAAC NEWTON

VOLUME II

1667-1670

EDITED BY

D. T. WHITESIDE

WITH THE ASSISTANCE IN PUBLICATION OF

M. A. HOSKIN

CAMBRIDGE
AT THE UNIVERSITY PRESS
1968

CAMBRIDGE UNIVERSITY PRESS
Cambridge, New York, Melbourne, Madrid, Cape Town, Singapore, São Paulo

Cambridge University Press
The Edinburgh Building, Cambridge CB2 8RU, UK

Published in the United States of America by Cambridge University Press, New York

www.cambridge.org
Information on this title: www.cambridge.org/9780521058186

First published 1968
This digitally printed version 2008

A catalogue record for this publication is available from the British Library

Library of Congress Catalogue Card Number: 65–11203

ISBN 978-0-521-05818-6 hardback
ISBN 978-0-521-04596-4 paperback
ISBN 978-0-521-72054-0 paperback set (8 volumes)

TO HAROLD HARTLEY

FOR SO MUCH

PREFACE

This volume, the second both in sequence and to appear of eight projected, continues the chronological reproduction of all Newton's mathematical papers now known still to exist, together with appropriate editorial commentary. To what was said in preface to the first volume regarding the aims underlying the present edition I have nothing to add, but I might perhaps remark, that with the exception of the 'De Analysi' (here edited from the original autograph manuscript retained by him), all of Newton's papers now reproduced effectively make their first appearance in print.

For permission to reproduce documents in their custody my collective gratitude must be expressed to the Librarians and Syndics of the Bodleian, Oxford; the Royal Society, London; the Niedersächsische Landesbibliothek, Hanover; and above all the University Library, Cambridge. The efficiency and courtesy of their staffs I again gratefully acknowledge, while to Professor J. E. Hofmann and especially Dr Theo Gerardy personal acknowledgement is due for their effort in making available a transcript and photocopy of Leibniz' notes on the 'De Analysi'. To a private owner my continuing thanks for permitting publication of his Newtonian papers. For financial assistance during the period of preparation of this volume I am largely indebted to the good graces of what is now the Science Research Council, while certain incidental expenses continue to be met by the Royal Society. Let me note in anticipation, however, that the Sloan Foundation, the Leverhulme Trust and the Master and Fellows of Trinity College, Cambridge, have each in their way pledged themselves to be future benefactors. To Sir Harold Hartley my inadequate thanks for his omnipresent support in ways of which only he can know the full detail.

At a technical level my acknowledgement is twofold. To Mr A. Prag, who has both proof-read and indexed this volume, rectifying much that was faulty in the editorial commentary, and to Dr M. A. Hoskin, who has been all—and more—that a helper can be, I cannot begin to express my gratitude. I myself, of course, remain uniquely responsible for the deficiencies, omissions, vagaries and other imperfections of the present volume.

As before, my final word of appreciation goes to the Syndics of the Cambridge University Press for the unsparing efforts of their staff in bestowing logical consistency and typographical beauty on the longhand manuscript which was initially submitted to them.

D.T.W.

1 December 1966

EDITORIAL NOTE

For the principal printing conventions here used and their underlying *raison d'être* the reader is referred to pages x–xiv of the first volume. The Newtonian text reproduced is strictly faithful to the autograph manuscript used, in that contractions and suffixes appear unchanged and grammatical inconsistencies are left unaltered or their editorial correction is indicated. For clarity's sake, however, a very few silent liberties have been taken in italicizing textual subheads and in inserting endpoints to sentences. Otherwise, all insertion in Newton's text is bracketed off in square parentheses and should be accepted as extraneous, though in each instance the reader is invited to consider why insertion was made. In English translation of Latin text we have not 'transliterated' existing English phrases but merely repeat them within square parentheses (as 'editorial' insertion in that translation). As in the first volume, two thick vertical bars in the left-hand margin alongside a piece of text denote that the section so marked off has been cancelled by Newton (being here reprinted for its intrinsic interest). This convention should not be confused with the two faint vertical parallels '||' used within the text (here in **3**, 1, §§1/2) to indicate a page division in the manuscript reproduced: in all cases, in the left-hand margin immediately opposite is inserted a complementary '||[...]', where the square brackets enclose the arabic number of the new page which begins at that point. A few *ad hoc* conventions used in reproducing particular texts are explained in an opening footnote to the piece in question. Universally, the convention 'I, **3**, 4, §3: note (5)' refers back to note (5) of [Volume] I, [Part] **3**, [Section] 4, [Subsection] 3: specifically, to note (5) on page 544 of the first volume. For brevity of quotation portions of the notation are frequently omitted from the left when reference is made to the same volume, part, section and so on: for example in note (121) on page 500 below the full reference is [II,] **1**, 3, §2, while in note (24) on page 9 it should be understood to be [II, **1**,] 3, §1, 2/3.

GENERAL INTRODUCTION

This second volume of Newton's mathematical papers reproduces that portion which, in our estimate, was composed during the years 1667 to 1670. The reason for choosing these chronological bounds is largely one of editorial convenience. On the one hand the October 1666 fluxional tract effectively terminates the thirty-month period of Newton's first creative mathematical researches, and indeed represents his conscious attempt at that time to gather the offshoots of his thoughts on calculus into a collective unity. On the other, the lengthy 1671 tract on fluxions and infinite series (which will appear in the next volume) opens a new cycle of analytical investigation which endured, if somewhat fitfully, till the early 1680's. The dearth of accurate documentary information relating to this period of his development, surely the least known of all the Newtonian dark ages, has not made the task of editing easy. The background to the *De Analysi* is now reasonably well established in consequence of the resurgence of interest in it at the time, forty years after its first circulation, of Newton's dispute with the Leibnizians over calculus priority: even so, its date of composition can only somewhat vaguely be bounded by the appearance of Mercator's *Logarithmotechnia* (in September 1668) and its communication by Barrow to Collins (in early July 1669). On the printing history of Newton's 'Observations' on Kinckhuysen's Dutch *Algebra* we are remarkably well informed, largely thanks to Collins' preservation of his correspondence with Newton and others on the topic. A glimmer of light is shed on his researches into the organic construction of curves and the geometrical construction of equations by Newton's later letters to Collins, particularly that of 20 August 1672. But for those papers which in time of composition precede mid-1669, when Newton's extant correspondence with his contemporaries opens,(1) editorial commentary must inevitably in large part reduce to essentially unsupported circumstantial argument.

Of Newton's life in general during this four-year period we know very little. A century ago Joseph Edleston(2) afforded some insight into his daily routine

(1) The first letter of Newton's now known, the abrupt note to an unknown correspondent reproduced (from Collins' copy for want of an autograph original) in *The Correspondence of Isaac Newton*, 1, 1959: 3–4, is dated late February 1669. We should mention that the often quoted 'Newtonian' advice of '*c.* 1661' to a drink-sodden friend, which exists in two forms in the Pierpont Morgan notebook in New York, is in fact signed by Newton with the initials 'T.N.': in other words, the letter is one form or other of a copybook exercise. Compare the photocopy published by D. E. Smith in W. J. Greenstreet's bicentenary compendium, *Isaac Newton: 1642–1727* (London, 1927): 26.

(2) In notes which he added to his 'Synoptical View of Newton's Life', in his *Correspondence of Sir Isaac Newton and Professor Cotes* (London, 1850): xxi–lxxxi, especially xlii–xliv, with appendices (lxxxii–lxxxv) analysing both Newton's fellow's dividends and his exits and redits.

at Trinity College. This, together with a little additional information to be gleaned from Newton's current items of expenditure as listed in one of his pocket-books[3] and a few off-hand remarks made by him a few years later relating to the sequence of his optical discoveries,[4] is the slender documentary basis on which any reasoned account of his immediate post-graduate years in Cambridge has to be constructed. Enough remains, however, to show up inadequacies in the conventional picture of a nervous, wholly diffident, badly dressed young don interested solely in the flights of his intellect. If only to stress the normality of Newton's social behaviour, at this period at least, and so extirpate any lingering temptation to relate his exceptional intellectual growth to a hypothetical (in fact, non-existent) physical immaturity, we may briefly touch upon some biographical points.

Geographically, at the opening of the year 1667 Newton was in Lincolnshire awaiting the reconvening of his Cambridge college. (Its fellows and students had, we will remember, been 'dismissed' the previous summer because of a renewed outbreak of plague in the town.) With the ending of winter the university sprang back to life and Newton travelled to Cambridge on 22 April. There he remained without break till April 1671 apart from a brief visit 'into ye countrey'—no doubt to visit his mother—over Christmas 1667[5] and two short trips to London in the summer of 1668 and the autumn of 1669,[6] during the latter of which John Collins met him for the first time 'somewhat late upon a Saturday night at his Inne'.[7] Having satisfied his B.A. examiners in January 1665—though he was not, in fact, to pay for his 'Bachelors Act' till some days after his return to Cambridge in April 1667[8]—Newton was now

(3) That now in the Fitzwilliam Museum, Cambridge. The expenses, listed line by line on six and a half unpaginated 16° pages, cover the period 23 May 1665 to 'April' 1669 but are evidently incomplete. The most significant double page is reproduced in photocopy (facing page 52) in the illustrated version of John Taylor's *Catalogue* of the Portsmouth papers auctioned at Sotheby's on 13/14 July 1936. Some additional material relevant to these listed expenses is contained in a stray sheet (sold as part of Lot 201 at the 1936 sale) of about 1667, the present whereabouts of which is not known. All unidentified financial entries in this introduction are taken from the Fitzwilliam pocket-book.

(4) Particularly in his letter to Oldenburg on 5 February 1671/2. Compare note (15) below.

(5) The Fitzwilliam notebook lists his departure from Cambridge on 4 December 1667 and return on the following 12 February, together with an item of five shillings 'For keeping Christmas'.

(6) Specifically, with regard to the first Newton wrote in the Fitzwilliam notebook that 'I went to London on Wednesday Aug 5^t^ & returned to Cambridge on Munday Sept 28, 1668' (and, to be sure, he signed his redit at Trinity the following day) together with 'Spent in my Journey...5. 10. 0. As also 4^li^ 5^s^ more wch my Mother gave mee in ye Country'. His college exit and redit for the latter journey indicate that it was made between 26 November and 8 December 1669.

(7) Collins to James Gregory, 24 December 1670. See **3**, Introduction: note (9) below.

(8) This may mean that Newton did not officially achieve B.A. status till April 1667. Evidently the termination of his undergraduate days was celebrated in traditional fashion for,

in a position to enjoy the first pleasures and privileges of academic position. Certainly, his advancement in college status was rapid enough for any young don determined to make his mark: a minor fellow of Trinity from October 1667, he was quickly promoted to major fellow on 16 March 1668 (duly, and no doubt proudly, paying his shilling for his 'Fellows Key' and being allotted his fellow's quarters).(9) His necessary elevation at university level from pupil status followed speedily in the following July when he was created M.A. (then, as now, largely a matter of satisfying a residential qualification and paying out the requisite sum of money).(10) His election, finally, to the Lucasian Professorship on 29 October 1669, upon Barrow's retirement, gave him security, academic independence and a useful stipend at the expense of only a moderate portion of his time.(11) Socially, it is evident that Newton during his first years as a senior member of Cambridge university made a determined effort to live up to his position. During the two years 1667 and 1668, academic necessities apart, he spent over twenty pounds with his tailor 'Mr Jeffreys' on new clothes(12) and footwear, while for his newly decorated dining room he bought a table and chairs, paying out sixteen shillings for 'A Table cloth' and 'Six Napkins'—for the use of invited guests no doubt. Not a few of his evenings were spent 'wth Mr Lusmore, Hautrey, Salter' and 'other Acquaintance' or with his room-mate Wickins wining and gambling at the local tavern, playing cards—twice losing heavily—or coming out an equally expensive second-best on the college bowling green. The fifty pounds his mother contributed to supplement Newton's income during this period were presumably quickly spent, given away in tips to his college servant Caverly or loaned—at interest

in the Fitzwilliam notebook on the line immediately beneath that where he recorded payment of '0. 17. 6' for his 'Act', he listed a pound spent 'At ye Taverne severall other times &c' together with a further 12*s.* 6*d.* paid out 'on my Couz. Ayscough'.

(9) The fellow's room Newton received in early October 1667 was the celebrated 'Spirituall chamber' (Edleston's *Correspondence* (note (2)): xliii), but it would appear that he himself did not occupy it for in late 1668 he 'Received for Chamber rent 1. 11. 0'. In the spring of the previous year he had, in fact, laid out a considerable sum of money refurbishing the set of rooms he shared with Wickins, putting in new glass, making minor repairs to the fireplace, walls and woodwork, having it repainted, installing new furniture and carpets and, not least, contributing almost a pound for 'My part of a Couch'. He would naturally be reluctant to move out of such comfortable quarters after only six months' tenancy of them.

(10) In fact, 'For my degree to ye Colledg 5. 10. 0. To ye Proctor 2. 0. 0. Expences caused by my Degre 0. 15. 0. 18 yards of Tammy for my M^r^ of Arts Goune 1. 13. 0. Lining 0. 3.6. Making y^t^ & turning my Batchelors Goune 1. 0. 6. A Hood 1. 3. 6.' (Fitzwilliam pocket-book.)

(11) Newton's appointment to the Lucasian Professorship and the concomitant restrictions and responsibilities it imposed upon him raise several interesting questions but we will delay our commentary on these till the next volume. See also Edleston's *Correspondence* (note (2)): xliv.

(12) With his countryman's good sense Newton bought the best cloth, lined 'Woosted Prunella' and 'Stuffe'.

we may be sure—to 'D[r] Wickins' or other acquaintances (such as Perkins, Boucheret or Wadsley) whose names are listed by Newton for loans of between five shillings and two pounds together with a cross to signify repayment. To ensure that his garments were well washed and his rooms clean, warm and well lit he made regular payments to his 'Laundresse' and 'Goodwif Powell', buying 'New Feathers', and a 'Ticken' for his bed and calling frequently on the local chandler for 'coales & sedge'. Altogether the picture we have is of a young man spending his income to the full in the mild pursuit of luxury and pleasure, an acceptably mature version of the cautious, money-wise young puritan who had entered Trinity half a dozen years earlier.

Enough of the worldly face which Newton presented to his fellows at this time. What of the intellectual giant within?

There can be no doubt that Newton's scientific researches proceeded apace during these years, notably in optics but also in chemistry and to some lesser degree—as the extant manuscripts now reproduced themselves attest—in mathematics. The impressive corpus of optical theory and experiment which (in continuation of Isaac Barrow's professorial investigations of the 'genuinæ rationes' of the phenomena of white light(13)) he began to present to his Cambridge audiences from January 1670 was evidently neither conceived nor systematized in one single, brief creative spell, while we have Newton's own confirmation that as early as mid-1668 he was hard at work constructing a 'small prospective', a first version of the portable reflecting telescope of his own design and fabrication which he gave to the Royal Society in 1672.(14) His pocket-book accounts, moreover, list purchases in summer 1668 of a 'Lath & Table' and 'Iron worke for it' together, with 'Drills, Gravers, a Hone & Hammer & a Mandrill' and 'files', all clearly destined for grinding lenses and mirrors. A few months later he listed purchases of 'Glass bubbles 0. 4. 0' and '3 Prismes 0. 3. 0', spending a total of twenty-nine shillings on 'Glasses' both in Cambridge and in London.(15) Newton's practical interest in chemistry,

(13) Published in 1669 (and reissued in 1670 with his *Lectiones Geometricæ*) as his *Lectiones XVIII, Cantabrigiæ in Scholis publicis habitæ; In quibus Opticorum Phænomenωn Genuinæ Rationes investigantur, ac exponuntur.* These eighteen lectures were given on unknown dates in 1667 and 1668 to the university at large.

(14) Newton's Lucasian lectures on optics, later to be published, from a secondary transcript of the version (ULC. Dd. 9.67) deposited in 1674 in the university archives, as his *Lectiones Opticæ, Annis, MDCLXIX, MDCLXX & MDCLXXI. In scholis publicis habitæ* (London, 1729), will be discussed in detail in the third part of the next volume. For Newton's first metal-reflector telescope see his *Correspondence*, **1** (1959): 3–4, 74–7.

(15) Presumably when he wrote to Oldenburg on 6 February 1671/2 (*Correspondence*, **1** (1959): 92–102, especially 95) that he had 'left off' his 'Glass-works' in 1666 he was reporting something less than the truth.

(16) Listed in the Musgrave library roll (R. de Villamil, *Newton the man*, London, 1931: 98) as '6 vols.... 8[vo]'. Curiously, these chemical entries occur a decade earlier than the first dated

inspired no doubt by his reading of Boyle, Hooke and other mechanical philosophers of his period, began about the time of his first visit to London in August 1668. Shortly afterwards he recorded the purchase of two pounds worth of 'Aqua Fortis, sublimate, oyle, perle, fine silver, Antimony, vinegar, Spirit of Wine, White lead, Allome, Niter, Tartar, Salt of Tartar, [Mercury]' (from London, no doubt, since he added 'Carriage of ye oyle 0. 2. 0') and a charge of fifteen shillings for the building of two furnaces (one for 'tin'), acquiring also Ashmole's encyclopedic 'Theatrum Chemicum [Britannicum]'.(16) But apart from a stray reference to his buying 'Gunters book & sector &c' from 'Dr Fox' for five shillings(17) Newton's expense lists are barren of mathematical entries, though, for example, it is certain that he bought the complete set of James Gregory's published works before late 1670.(18) The autograph texts of Newton's mathematical investigations are themselves, however, firm testimony to the volume and quality of his geometrical and fluxional researches during the years 1667 to 1670—if, that is, our present dating of these papers is accurate. Internal evidence convinces us that their posited chronological sequence is indeed correct: the *De Analysi* at one point, for example, borrows its terminology from preceding discussions of the nature of asymptotes,(19) while the concluding tract on the geometrical construction of equations leans heavily in its description of the Wallisian cubic on a prior knowledge of his innovations in the organic construction of curves.(20) For independent external support of our early dating of this group of mathematical papers we can call upon John Collins, who saw a representative selection of Newton's scientific manuscripts some time in the early 1670's and was allowed to make detailed transcripts (now in private possession) of those portions which interested him. Subsequently, towards the end of 1677, when the Oxford Savilian Professor, John Wallis, wrote to him enclosing for comment an early draft of what was to appear in 1685 as his *Treatise of Algebra, both Historical and Practical* and in which bare mention was made of Newton's researches in infinite series, Collins in reply(21)—after an initial rebuke 'in regard you lye under a censure from

passage, an entry on 10 December 1678, in Newton's chemical notebook (ULC. Add. 3975) and we may conjecture that the earliest manuscript record of his researches in that field has been lost. Compare A. R. and M. B. Hall, 'Newton's Chemical Experiments', *Archives internationales d'Histoire des Sciences*, **11** (1958): 113–52, especially 120. (17) I, **3**, 2, §1: note (14).

(18) Newton's library copy of Gregory's *Optica Promota* (London, 1663), sold at the partial auction of his books in the early 1920's, has now vanished but his *Vera Circuli et Hyperbolæ Quadratura* (Padua, 1668 reissue), *Geometriæ Pars Universalis* (Padua, 1668) and *Exercitationes Geometricæ* (London, 1668) are now in Trinity College, Cambridge (NQ. 9.48, bound up with Newton's copy of Nicolaus Mercator's *Logarithmotechnia* (London, 1668) and John Wallis' review of it which appeared in the *Philosophical Transactions* in 1670).

(19) See **2**, 2: note (81) below. (20) Compare **3**, 2, §2: note (121).

(21) ULC. Add. 3977.13. published in the *Correspondence of Isaac Newton*, **2** (1960): 241–3, especially 242–3.

diverse for printing discourses that come to you in private Letters without permission or consent'—went to some trouble in elaborating for Wallis the content of some of the Newtonian papers he had earlier seen:

I...must take liberty to tell you some things concerning your intended Explanation of Mr Newtons Series. If I had been so minded I could about 9 yeares since namely at the beginning[!] of 1669 have imparted to you a full treatise of his(22) of that Argument....

In your narrative you say Mr Newton began to fall into these methods in 1669 or 1670, whereas in the larger Letter he tells you he seemed delighted *hisce* [*in*]*ventis* namely in Calculating Logarithmes and Van Ceulens Numbers in his retiremt from the University in the Plague yeare in 1665,(23) and in 1666 he writt the treatise above mentioned.(24) All the account you can give out of these Letters is but very slender in relation to his performances. He intends a full treatise of Algebra consisting of these Parts according to the best of my apps

1 an Introductory part from Kinckhuysẽ out of low Dutch turned by Mercator into Latin, which he bought(25) and is so excellent, that it comprehends many of Huddens reductions, and those mentioned by Dary at the end of his tract of Interest (26) & some others to which Mr Newton added much of his owne.(27)

2 A discourse about bringing Problemes to an Æquation with a Collection of diverse notable ones.(28)

3 A Treatise about the Construction of Problemes and Æquations which I have seen. All Solid Problems viz those of 4 and 3 Dimensions are solved by ayd of one

(22) The 'De Analysi per æquationes numero terminorum infinitas', reproduced as **2**, 2 below. Collins did not, in fact, see Newton's tract till August 1669.

(23) In his 'larger Letter' (the *epistola posterior* to Oldenburg for Leibniz, 24 October 1676) Newton had remarked of his 1665 retirement to Lincolnshire that 'tunc sanè nimis delectabar inventis hisce'. Collins had communicated the content of the *epistola prior* (of 13 June 1676) to Wallis in September 1676 (*Correspondence of Isaac Newton*, **2** (1920): 101) and his transcript of the second letter was no doubt communicated soon after Oldenburg sent his own copy off to Leibniz. In an omitted portion of the present, undated letter to Wallis, Collins noted that 'Mr Newton last yeare sent up these Letters, you have seene with particular leave upon my importunity to print the same...if I had not [imparted the first Letter to you] I believe you had not seen either to this day'. This is evidently Collins' quick rejoinder to Wallis' letter to him of 2 October 1677 with its assertion that 'I am stil of Opinion yt Mr Newton should perfect his notions, & print them suddenly. These letters, if printed, wil need a little review by himself, for there be some slips in hasty writing them' (*Correspondence*, **2**: 238).

(24) Collins here confuses the 'De Analysi' with the October 1666 fluxional tract, which no doubt he had likewise seen some time before. The Wickins transcript (I, **2**, **7**: note (1)) may indeed have already been in his possession.

(25) Compare **3**, Introduction: note (50).

(26) Michael Dary, *Interest Epitomized, both Compound and Simple.... Whereunto is added, A Short Appendix For the Solutions of Adfected Equations in Numbers by Approachment: Performed by Logarithms* (London, 1677: Newton's copy is now Trinity College. NQ. 16.62). The 'Short Appendix' (pages 32–8) is 'Laid down in Two Methods', the first of which is an adapted form of Stevin's numerical technique, the latter being Dary's own method of iteration applied to the equation $y^8 = 6y^3+200$. Kinckhuysen in his *Algebra* (see **3**, 2, §1: note (108)) 'compre-

Constant Circle (if so desired) supposed to be intersected by Conick Sections, the description whereof is avoyded by helpe of mooveable angles, that give the severall Points of Intersection sought. Other Equations betweene the 5 and 9 degree he performes by ayd of a Cubicall Parabola that being once described in like manner remaines constant, and is to be intersected by a Conick section the description whereof is avoyded as before &c. He hath also diverse tentative Constructions for Cubicks and Biquad^s^ from Plaine Geometry.(29)

4 A Discourse concerning the severall kinds of infinite Series considering which kinds are most convincing and fitt for Calculation, and which for Construction and Demonstration, of this Argum^t^ and of the whole buisinesse of Series he hath written a new and large treatise since that above mentioned,(30) and hath performed aboundantly more than is either mentioned or can be guessed from the Lettters above mentioned.(31)

5 A Treatise *de Locis*.(32)

6 The same(33) applyed to Dioptriques concerning the worth of both which Dr Barrow affirmed he was not only surprized but others would thinke it incredible.(34)

The first three—and the fifth also, if we identify its vague title correctly—of the unpublished Newtonian mathematical tracts here listed by Collins are reproduced for the first time in print in the present volume, while the fourth and portions of the sixth will form the centre-piece of that following. Without further prevarication let us hasten to the rich detail of the papers themselves and appreciate Newton's mathematical genius in its original dress.

hends' only the former. Writing to Newton about July 1675 Collins had already indicated a second point of resemblance between Dary's researches into equations and Kinckhuysen's 'Huddenian' reductions: 'Mr Dary by observing the Complication of the Coefficients hath well performed to this Purpose without the ayd of a Cubick æquation viz. Any Biquadratick æquation being proposed without a Resolvend, to break the same into two rationall quadratick æquations, whose Resolvends shall be what they will happen, and consequenty to give the series of all that shall rationally breake, and withall to breake with one or two Assayes as neare as may be rationally to any Resolvend offered' (*Correspondence*, **1**: 346–7).

(27) Mercator's Latin version of Kinckhuysen's *Algebra Ofte Stelkonst* (Haerlem, 1661) and Newton's 'Observationes' upon it are reproduced below in **3**, 1, §§1/2.

(28) In his letter to Collins on 27 September 1670 Newton spoke of 'having composed somthing pretty largely about reducing problems to an æquation' (*Correspondence*, **1**: 43) but seems merely to refer to his amplified opening to Kinckhuysen's 'Pars Tertia. Quomodo quæstio aliqua ad æquationem redigatur'. Compare **3**, 1, §2: note (110). In October 1676 Leibniz noted of this discourse that 'Collins has not yet seen it' (*Correspondence of Isaac Newton*, **2**: 236).

(29) These 'Problems for construing æquations' are reproduced in **3**, 2, §2 below.

(30) The 'De Analysi', that is: see note (22).

(31) Newton's 1671 tract on fluxions and infinite series (ULC. Add. 3960. 14/4), which will appear in the third volume.

(32) Perhaps the enumeration of cubics reproduced in **1**, 1, §3.

(33) Collins means the 1671 tract described in the fourth part: item 5 is a late insertion.

(34) A clear reference to the later draft (ULC. Dd. 9.67) of Newton's *Lectiones Opticæ* (note (14)), whose section on geometrical dioptrics (Book 1, Part 4) makes extensive use of limit-increment arguments and, on one occasion, a series expansion.

Construct[illegible] Circle (if se desired) appear to be intersected by Conick Sections, the description whereof is avoided by helpe of insc[illegible] angles, that have the severall Points of Intersection in a right line between them and a [illegible] be performed by a[illegible] of a Cubical Parabola that is [illegible] described in like manner [illegible] constant, and [illegible] be intersected by a Conick Section the description whereof [illegible] as before. (3) Each [illegible] Representative Constructions for Cubicks and Equation [illegible] the Circle.

[illegible] concerning [illegible] kinds of [illegible] constitution which [illegible] convenient and fit for Calculation, [illegible] which for Construction, and Demonstration of the [illegible] of the whole [illegible] except the last, which [illegible] that above mentioned[28] hath performed abundantly [illegible] can be guessed from the Letters above mentioned.[29]

5. A Treatise of Lines.[30]

6. The same [illegible] applyed to Dioptricks, concerning the worth of both which D' Barrow affirmed [illegible] was not only surprized but others would think it incredible.[31]

The first three—and the fifth also, if we identify it with... lectures dy... the [illegible] Newtonian mathematical treatises here [illegible] reproduced for the first time in print in the present volume, while the fourth and [illegible] of the sixth will form the centre-piece of the [illegible] further [illegible] to the [illegible] papers themselves and [illegible] Newton's mathematical genius in its original [illegible].

[illegible] only the former. Writing to Newton about July 1675, Collins [illegible] a second point of resemblance between [illegible] into containers and [illegible] Hudde [illegible] by observing the [illegible] of the [illegible] both well performed to this purpose without the aid of a Cubick equation or any Biquadratick equation at the proposed without a Resolvent to break those [illegible] two rational quadratick equations whose Resolvends shall be what they will happen, and consequently to give the [illegible] that shall rationally break... and withall to [illegible] both one [illegible] near as may be rationally to any Resolvend [illegible] (Correspondence, [illegible]).

(27) [illegible] version of [illegible] (Leiden, 16[illegible]) and Newton [illegible] reproduced below in [illegible] 1, § 2.

(28) [illegible] 17 September 1670 Newton [illegible] composed [illegible] about reducing problems to an equation' (Correspondence, 1: 43) but [illegible] merely to [illegible] entitled [illegible] Orie[illegible] [illegible] In October 1676 Collins [illegible] that Collins has not yet seen.

(29) [illegible] Problems [illegible] equations [illegible] 2, § [illegible].

(30) [illegible] 2 [illegible].

(31) [illegible] 3806 [illegible] volume.

(32) [illegible] reproduced in [illegible] 82.

(33) C[illegible] 1671 tract described in [illegible] Lucasian Lectures [illegible] elementary [illegible] (U.L.C. Dd. 9. 67) [illegible]

(34) [illegible] of geometrical Optics (Book [illegible] Part [illegible] of [illegible] comment [illegible] series expansion.

ANALYTICAL TABLE OF CONTENTS

PREFACE *page* vii

EDITORIAL NOTE viii

GENERAL INTRODUCTION ix

LIST OF PLATES xxiii

PART 1
RESEARCHES IN PURE AND ANALYTICAL GEOMETRY
(1667–1668)

INTRODUCTION 3

Preceding researches into co-ordinate transformation (Descartes, Newton), 3. Historical divisions of the conic into component species: no previous known attempt to classify cubics, 4. Newton's analytical subdivision of the general cubic by co-ordinate transform: geometrical interpretation by means of the diametral hyperbola, 5. Later cubic classifications (Newton, Euler): 'Maclaurin's' theorem on cubic tangents, 6. Generalization of conic properties to higher curves, 7. The 'organic' method of constructing new curves from given ones (Schooten), 8. The Newtonian theory and its essential biunivocal structure: Maclaurin's *Geometrica Organica*, 9

1. ANALYSIS OF THE PROPERTIES OF CUBIC CURVES AND THEIR CLASSIFICATION BY SPECIES 10

§1 (ULC. Add. 3961.1: 2r–3r). First attempt to reduce the general cubic by co-ordinate transform. Transformation to new axes in an oblique Cartesian system: analytical theory, 10. The transform of the defining equation of a general cubic curve evaluated as a cubic of 84 terms, 12. Reduction of this to the primary canonical form of cubic, 14. The three particular reduced canonical forms, 16

§2 (ULC. Add. 3961.1: 10r–13r, 32r–34r). Distinction of the primary canonical cubic by 'species' and 'forms'. The first species: construction of its three real asymptotes, 20. The diametral hyperbolas, 22. The six forms of its first 'case', distinguished by the nature of the meets of the cubic with a diametral hyperbola, 24. Diagrams for these, 28. The second case, when the three asymptotes are coincident: its three forms, 30. Illustrative figures, 32

§3 (ULC. Add. 3961.1: 6r–9r, 14r–16r, 22r–30r). First systematic enumeration of cubics and their classification into sixteen species. The four canonical forms of cubic developed as nine 'cases', 38. The first case (general tridiametral), first species: its six forms, 42. General diagrams, 48. Observations on the first species, 52. First case, second species (symmetrical round a diameter): its seven forms, 54. Third species (asymptotes coincident):

PART 2
RESEARCHES IN CALCULUS
(*c.* 1667–*c.* 1670)

PART 3

RESEARCHES IN ALGEBRA AND THE CONSTRUCTION OF EQUATIONS

(*c.* 1670)

Elementary problems: to construct the line (plane) defined by two (three) fixed points, and to draw a circle of given centre and radius, 452. To lay off a given line-segment between two given lines so as to pass through a given point: construction of parallels, normals and other simple problems, 456. Application of these to finding two or three mean proportionals between given line-segments (no proofs given), 460. Similar application to angular sections, 464. [2] Nine problems on the geometrical construction of the (real) roots of quadratics, cubics and quartics. The quadratic constructed as the meet of a circle and straight line, 468. The reduced cubic equation constructed by an Apollonian 'verging' between a straight line and a straight line or circle, 470. The general cubic equation constructed as the meet of a circle and a conic (ellipse is drawn by a trammel): the generalization of this 'in an infinity of ways' by homothety, 468. Similar construction of the general quartic equation (solid problem), 490. Organic construction of the Wallisian cubic by a describing hyperbola (one pole at infinity), with generalization to the oblique case, 498. Its use in constructing higher equations, 504. Conclusion: the preceding cubic and quartic constructions (previously given synthetic proof only) are investigated by analytic means (elimination of one co-ordinate variable between the defining equations of two conics yields a quartic equation whose coefficients are then equated with those of the cubic or quartic to be constructed), 510

INDEX OF NAMES 518

LIST OF PLATES

PART 1

RESEARCHES IN PURE AND ANALYTICAL GEOMETRY (1667-1668)

INTRODUCTION

In this first part of the present volume we reproduce the text of certain geometrical researches carried out by Newton during, we presume,[1] the years 1667 or 1668. Roughly and with some overlap these fall into two definable groups. In those of the first he sought to explore the varied properties of cubics and general algebraic curves analytically and above all by means of co-ordinate transformation, while in the second he attempted to determine which properties and singularities of curves remained invariant under a particular biunivocal point-correspondence (his organic construction) and thereby construct classes of curves by a uniform method from curves of lower algebraic degree. The background to these pieces, unfortunately, is not known. No letter or pertinent biographical fragment from this period now exists nor were these manuscripts ever the subject of a clarifying priority dispute: nowhere have we been able to trace any documentary material, however slight or untrustworthy its nature, which fixes their origin in time or their author's intention in composing them. Such conjectures as we now make are based solely on our assessment of internal and secondary evidence and on our determination to set these papers coherently and consistently within what we understand to be the chronological development of Newton's geometrical thought. In so far as the reader finds our viewpoint unattractive he may discard it at will.

As we have seen[2] Newton had already mastered the elements of Cartesian co-ordinate geometry by the late autumn of 1664. It had then been his especial aim to formulate general analytical tests by which the several diameters, centres and asymptotes of any curve might be determined from its Cartesian defining equation: in the spirit of Descartes' *Geometrie* he had there achieved a limited success by first transforming the co-ordinates so that this defining equation took on a simpler, more amenable form. Over several months this technique of co-ordinate transformation was perfected till in December 1664 he had attained a sophisticated theory of the general analytical transform from one pair of oblique co-ordinates to a second pair defining the same terminal point in a Euclidean plane. At that point, however, he seems temporarily to have laid his researches on one side. In part, no doubt, his energies were devoured by more immediately pressing problems in calculus, optics and astronomy, but it is evident also that for the moment he had run up against a difficulty which was frustrating all contemporary research in analytical geometry—a mere handful of particular curves were yet known and even of

(1) Compare 1, §1, note (1) and 3, §1, note (1) below.
(2) I, 2, 1: passim.

these the elementary properties were largely unexamined, so that any one who sought more general properties of curves had for the most part to fish around in twilight.

From the days of Menæchmus and Euclid, Greek mathematicians, and supremely Apollonius,(3) had dealt fairly exhaustively with conics, reducing them synthetically to their three non-degenerate species and exploring a variety of their metrical and projective properties: more recently it had been Descartes' not inconsiderable achievement in the second book of his *Geometrie*, with some subsequent help from De Beaune and Schooten,(4) to translate the elements of the Greek reduction to those three component species into analytical terms. Of particular curves of higher degree little was known in Greek times, and all such geometrical curves were lumped together without differentiation as the 'linear' class: unique examples of cubics and quartics, the cissoid and conchoid respectively,(5) had indeed been propounded but were studied only because they facilitated the finding of two and three mean proportionals between given quantities. When in the late 1630's Descartes described his trident and folium(6) to his contemporaries he literally tripled the number of known cubics, but over the next quarter century only two more curves of third degree were discussed in print, the cubical and semicubical parabolas, first described by John Wallis and William Neil respectively.(7) Even more importantly for future developments in the theory of higher curves, no one before Newton had attempted any subclassification of cubics into component species on the analogy of the division of the non-degenerate conic into the ellipse, parabola and hyperbola: evidently those, if any, who tried to emulate Descartes by reducing the cubic's general Cartesian defining equation to amenable canonical forms had been unable to make significant progress. In his 1664 researches Newton himself had come face to face with the difficulties of attempting any synthetic classification and to realize the perils which lay waiting when a given cubic was to be traced from even a much simplified defining equation.(8) As autumn passed into early winter in that year he grew progressively more facile in curve sketching, but his researches of that period finally terminate with no sign of a true breakthrough in the problem of subdivision of the algebraic curve of degree

(3) Compare T. L. Heath, *Apollonius of Perga: Treatise on Conic Sections* (Cambridge, $_1$1896) [= $_2$1961]: Introduction: xviff.; and Paul ver Eecke, *Les Coniques d'Apollonius de Perge* (Bruges, $_1$1923) [= Paris, $_2$1963]: Introduction: viiff.

(4) See De Beaune's *Notæ Breves* which Schooten added, together with his own extensive *Commentarii*, to the first Latin edition of Descartes' *Geometrie* in 1649 (I, 1, Introduction, Appendix 2: note (5)).

(5) In neither case, apparently, was the full curve drawn. See T. L. Heath, *Greek Mathematics*, **1** (Oxford, 1921): 264–6 and 238–40 respectively; compare also 1, §3, note (58) below and I, **3**, 3, §3.

(6) Compare I, **3**, 3, §2, note (12) and I, **2**, 1, §5, note (31).

higher than second into species. Some time in 1667 (or perhaps early 1668), we conjecture, Newton returned fresh to the problem after three years, attempted to reduce the general cubic equation in two variables by an appropriate general transformation of co-ordinates and was ultimately successful. The original worksheets in which this important step was made have long since disappeared and already in the first paper we reproduce he is at work strengthening and systematizing the first gains of battle. Even that is soon obsolete and in quick succession two more unfinished drafts are composed, nomenclature is established in a fourth and finally the general subdivision of the cubic is settled in its main structure once and for all.(9)

In essence, Newton shows that by a suitable transformation of the Cartesian co-ordinates x and y the cubic defining equation

$$ay^3+bxy^2+cx^2y+dx^3+ey^2+fxy+gx^2+hy+kx+l=0$$

may, when b is not zero, be taken without loss of generality in the simpler form $bxy^2+dx^3+gx^2+hy+kx+l=0$,(10) but where b is zero then in the form

$$dx^3+ey^2+kx+l=0,$$

unless e is zero when we may take it as $dx^3+fxy+l=0$, or, when e and f are both zero, as $dx^3+hy=0$. The latter three categories, those (in Newton's later terminology) of the divergent parabolas (whose cusped form is Neil's semicubic parabola), the Cartesian trident and the Wallisian cubic respectively, were not unfamiliar to him and, we may suppose, caused him little difficulty, but the niceties of the first, most general reduced form of cubic were still to be explored. The catalyst which enabled Newton to resolve this general case into its elemental components was his introduction of the diametral hyperbola, the locus of the midpoints of all chords joining the two finite meets of the general ordinate with the cubic.(11) Evidently, all finite singularities of the cubic corresponding to the ordinate direction will lie on this hyperbola (which may or may not degenerate into the line-pair of the co-ordinate axes) and by examining the various possibilities he isolates six 'cases' of the general reduced form, further subdividing

(7) See 1, §3, notes (94) and (88) respectively. The latter was, of course, discussed independently by Heuraet in appendix to the second Latin edition of Descartes' *Geometrie* in 1659. (Compare I, 2, 5, §1, note (51).)

(8) Thus, Newton was then unable adequately to describe the curves corresponding to $y^3-ax^2-a^2x = 0$ (I, 2, 1, §5, note (13)) and to $x^3-3xy^2+2xy^2-2ay^2 = 0$ (I, 2, 1, §7, note (13)).

(9) See 1, §§1, 2/3 and 4 following.

(10) Even in Newton's simplified co-ordinate transform, which considers as new variables not the transformed co-ordinates but their respective oblique projections in the directions of the old pair, the transformed cubic (1, §1 below) has 84 terms. We should admire Newton's ability to control such an equation without making simplifying substitutions (as we have ourselves done in 1, §1, note (19)). In terms of mere mental concentration this surely was beyond the capacity of all but a handful of his contemporaries.

(11) See 1, §2, note (13) and, for its generalization, 2, §2, note (56).

the more complex of these into 'species'. Altogether in the present manuscripts, on including the preceding three minor reduced forms, he distinguishes nine cases and sixteen species of the general cubic: these on occasion he further subdivides into 'grades'.(12)

With that classification Newton here remained content, though at one point he inserted in it off-handedly a theorem on cubic asymptotes(13) which, rediscovered and generalized by Colin Maclaurin to general cubic tangents, appeared in print only eighty years later. Nowhere does he hint at a geometrical interpretation of his reduction of the cubic by co-ordinate transform. To look ahead to our seventh volume, however, when Newton revised his present researches in 1695 (expanding them eventually into a linear enumeration of 72 species) he first simplified this reduction by co-ordinate transform and then abandoned it completely for a more empirical one which in effect affords just such a geometrical interpretation.(14) Specifically, elimination of the terms ay^3 and cx^2y from the general defining equation determines that the co-ordinate axes be parallel to a (real) cubic asymptote and the second asymptote of the corresponding diametral hyperbola, while further elimination of the terms ey^2 and fxy restricts the parallelism to being coincidence in either case. Whether consciously intended at this time or not, Newton's further subdivision into species of the reduced canonical form is closely tied to the various possible characters of the infinite branches of the curve, and we need not be surprised that Leonhard Euler repeated that division exactly when he again approached the problem of classifying the cubic in his *Introductio in Analysin Infinitorum* in 1748 by considering the aggregate of possible variances in the nature of its infinite branches.(15)

---

(12) Compare 1, §3, note (1) below. We may note that W. W. Rouse Ball has briefly categorized the content of these manuscripts in his 'On Newton's Classification of Cubic Curves', *Proceedings of the London Mathematical Society*, **22** (1891): 104–43, especially 108–9, 125. Misled, however, by the present physical ordering of the autograph papers on cubics in Cambridge University Library he uncritically accepted that certain manuscripts on discrimination of genera formed part of these early cubic researches: in fact, Newton's handwriting in those is of much later date and, indeed, they make use of technical phraseology invented by him only in 1695. We accordingly assign that year conjecturally to them as their date of composition.

(13) See 1, §3, note (22).

(14) This is, of course, the scheme eventually adopted by Newton in the cubic classification (*Enumeratio Linearum Tertij Ordinis:* §§VIII–XI: 144–5) which he appended to his *Opticks* in 1704. The earlier approach by co-ordinate transformation remained unpublished till Rouse Ball outlined it in 1891 (note (12) above).

(15) Almost exactly, we should say, for Euler's scheme (*Introductio in Analysin Infinitorum* (Lausanne, 1748): *Liber Secundus, Continens Theoriam Linearum Curvarum, una cum appendice de Superficiebus*, Caput IX. 'De Linearum Tertii Ordinis Subdivisione in Species': 114–26, especially 123–6 [= *Opera Omnia* (1) **9** (Geneva, 1945): 122–33, especially 131–3]) interchanges Newton's Species 1 and 2 with 3, 4 and 5, and again Species 8 and 9 with 10 and 11.

With regard to the remaining papers in analytical geometry here reproduced we may be brief. They represent a further stage, presumably sparked off by Newton's success in subdividing the general cubic, in his discovery of general properties of curves analogous to those which hold for conics. In the late autumn of 1664 Newton had already generalized in this way the definitions of ordinate, diameter, axis, vertex and asymptote which hold for a conic.(16) The verbal forms he there gave are here repeated and made more precise while a new list of 'Conick propertys to bee examined in other curves' is jotted down.(17) The fruits of this more systematic search for general curve properties are seen in sequel where, in 'Theor 1', a distinction is now made between the elementary generalization of a conic diameter as the locus of the midpoints of the chord joining the two real meets of the curve with a line of given direction, and the more sophisticated concept of a 'Newtonian' diameter, the locus (provably a straight line) of the arithmetic mean of the intersections of a curve with a line of given direction. (In the cubic case, of course, the diametral hyperbola is an example of the former when the given direction of the general transversal is that of a real asymptote.)(18) The two remaining theorems and the various corollaries are all straightforward deductions from this first theorem, pretty enough but of no great significance. As with the preceding cubic classification Newton revised and extended this paper in 1695 and finally published its gist in 1704.(19)

In pinpointing the origin of the concluding pieces on the 'mechanical' construction of curves we may perhaps be more definite. We saw in our first volume that Newton in his undergraduate annotations made copious notes, perhaps in the late summer of 1664, on Frans van Schooten's *Exercitationes Mathematicæ* (Leyden, 1657) and particularly on its fourth book, *De Organica Conicarum Sectionum in Plano Descriptione Tractatus, Geometris, Opticis; Præsertim verò Gnomo-*

(Compare the 14 species distinguished by Gabriel Cramer in his *Introduction à l'Analyse des Lignes Courbes Algébriques* (Geneva, 1750): 359–69.) Euler, of course, knew only the published Newtonian classification of the *Enumeratio* but he is justly firm that Newton's further subdivision of the sixteen primary species is relatively arbitrary (p. 123 [= 130]): 'Quod vero inter hanc nostram divisionem ac NEWTONIANAM tantum intercedat discrimen, mirum non est; hic enim tantum ex ramorum in infinitum excurrentium indole specierum diversitatem desumsimus, cum NEWTONUS quoque ad statum curvarum in spatio finito spectasset atque ex huius varietate diversas species constituisset. Quanquam autem hæc divisionis ratio arbitraria videtur, tamen NEWTONUS suam tandem rationem sequens multo plures species producere potuisset, cum equidem mea methodo utens neque plures neque pauciores species eruere queam.' In sequel, Euler gives a useful list which collects the 72 varieties of Newton's published enumeration into their respective sixteen primary species.

(16) See especially I, 2, 1, §4: 'How to find y^{e} axes, vertices, Diamiters, Centers, or Asymptotes of any Crooked Line supposing it to have them.'

(17) 2, §§1.2/2.1 below.

(18) See 2, §2.2 below, particularly notes (36) and (56).

(19) *Enumeratio* (note (14)): 140–2 = §§II–IV, VII.

nicis & Mechanicis Utilis.(20) It had there been Schooten's aim to construct any given conic 'organicè [&] uno ductu', that is, systematically by a uniform method and in one continuous, uninterrupted motion.(21) For the individual conic species (ellipse, parabola and hyperbola) he had been lavishly successful, giving a proliferation of simple, elegant constructions of each, the gist of which Newton duly noted: but the final uniform mechanical construction of the general conic eluded him. It was Newton's present triumph to succeed where Schooten had failed: in one of his most brilliant geometrical moods he frames a universal 'organic' construction-method for all conics, including the degenerate line-pair, and immediately generalizes it with a dazzling virtuoso display of talent to describe cubics, quartics and higher curves of defined singularity.(22)

We need not insist on the details of Newton's construction. In outline, two fixed angles rotating each round a fixed pole in their plane determine a one-to-one continuous correspondence between the two meets of their arms (or 'legs' as Newton chooses to call them). If, consequently, one meet is set on a known 'describing' curve (Newton's 'directrix'), the second will trace a 'described' curve (or 'describend') of equal algebraic degree and equivalent singularities. The apparent power of the construction to raise the dimension of the describing curve in drawing the corresponding described one is explained by the former's (and indeed the latter's) ability on occasion to conceal in itself the polar line

(20) See I, 1, 1, §1.3–8: passim; and compare J. E. Hofmann, *Frans van Schooten der Jüngere* (Wiesbaden, 1962): 9–12. For bibliographical details consult I, 1, Introduction, Appendix 2, notes (12) and (13).

(21) As Schooten himself expressed it in the dedication to the book (*Exercitationes Geometricæ*: 295): '[Græci] parti quæ tractat de Conicis Sectionibus, cujus usum cum longè latéque se extendere agnoscerent, maximum studium adhibuerunt.... Quâ quidem re scientia hæc incredibile incrementum cepit, illéque inde magnam gloriam consequutus est, ita ut eo & sequentibus seculis Geometræ magni nomen meruerit. Sed cum, injuriâ temporis, nulla eorum monumenta vel aliorum, qui ipsis posteriores sunt ad nos pervenerint, aut saltem (quod sciam) publicè extent, in quibus modus ipsas Sectiones uno ductu in plano describendi ostendatur; quarum tamen descript[i]o in praxi quotidiana quàm maximè requiritur: Id ipsum materiam mihi subministravit, ut idonea instrumenta excogitarem: quibúsque modis Conicæ lineæ Organicè, & quovis casu in plano describi possent, peculiari tractatu demonstrare conarer.' His preference for the continuous mechanical description of curves over the more mathematical one of construction by points he clarified in the following *Præfatio ad Lectorem* (pp. 301/302): 'Quod autem attinet ad methodum, quâ hæ lineæ in plano per puncta describuntur,...illa quidem, quum ipsæ majori formâ sunt exhibendæ ut in parietibus, caminis, pavimentis, hortis, & quales vulgò in Gnomonica & Mechanica requiruntur, non æquè facili viâ atq; Organicâ succedit. Prout hoc à primo limine dignoscere licet in Sciothericis hoc pacto constructis, si notemus quàm perperàm plerumq; ipsæ designatæ fuerint: siquidem methodus illa sæpiùs ibidem iteratam punctorum inventionem atq; manus exercitatæ solertiam requirit, nec non ut præterea, ad exactā operis exegesin, natura earundem linearum in aperto sit. Quod utiq; in Organica methodo locum non habet: quippe quæ lineas illas, multifariam iteratâ operâ vixdum acquisitas, suâ quasi sponte uno ductu continuo ob oculos ponit. Placuit porrò præ cæteris is modus Organicus, qui è motu implicato originem ducit, eo insuper habito, quo hæ lineæ circino in eum finem extructo describantur.'

taken once or several times. Newton pursues the various possibilities for straight line and conic directrices with his usual care and acuteness in the present manuscripts and we will not duplicate his account. Not untypically these pieces are, worksheets and more finished drafts alike, by and large mere statements of results presented in a logically coherent but unproven deductive sequence. Above all, he gives no clear indication of how he came upon his construction or how he justified the stated dimensionalities of the described curve, which are crucial in his exposition.(23) In matter of notation, finally, there should be little difficulty in understanding Newton's texts, but we may draw the reader's attention especially to the pictographs which he has entered in a mathematical shorthand for multiple points in his worksheets.(24)

As we shall see in future volumes, Newton was to return to his organic construction several times in later years, expanding, systematizing and at length summarizing his early findings. For the moment let us be content to record its first impact on his immediate contemporaries after he communicated a summary of his present results to Collins in August 1672(25) and to note that he allowed two variant accounts of it to appear in print in his lifetime.(26)

(22) Schooten himself remarked in his *Præfatio ad Lectorem* (p. 302) that 'Aliarum...linearum curvarum superioris generis descriptiones quod attinet, eas in medium afferre non fuit nostri instituti: cum maluerimus meritò eximiis Viris D. des Cartes, D. de Fermat,...& D. Robervallo...relinquere.' Of these three mathematicians Descartes alone had independently developed a mechanical construction for curves 'uno ductu continuo', viz. the 'instrument' which he describes in the second book of his *Geometrie* (*Discours de la Methode* (Leyden, 1637): 319ff). This, however, is a poor construction in comparison with Newton's 'organic' one, for from a straight line it constructs a single conic species (the hyperbola) and from a parabola a single cubic (the Cartesian trident). The source of Newton's inspiration should be evident to anyone who compares Schooten's figure on p. 347 of his *Exercitationes* and that of Newton's in his letter to Collins of 20 August 1672 (reproduced in 3, Appendix 2 below). As Colin Maclaurin remarked in his *Geometrica Organica: sive Descriptio Linearum Curvarum Universalis* (London, 1720: *Præfatio:* a[4]^r), 'Nemo...ante Illustr. Newtonum, universalem aggressus est descriptionem organicam Linearum ordinis secundo altioris. Hujus methodus commodissimam suppeditat viam Lineas tertii Ordinis puncto duplice ditatas mechanice describendi; nonnullas etiam Lineas altiorum Ordinum complectitur'.

(23) From his abortive attempt at a Cartesian treatment of the construction (3, Appendix 1 below) it seems evident that his vindication cannot at this period have been analytical and we have conjectured (3, §1, note (18)) that the degree of the described curve was evaluated by considering the possible aggregate of its meets with an arbitrary straight line.

(24) See 3, §1. 2/3 below.

(25) 3, Appendix 2, notes (1) and (12).

(26) See his *Philosophiæ Naturalis Principia Mathematica* (London, 1687): Liber I, Sectio V, Lemma XXI, Props. XXII, XXIII and the Scholium to Prop. XXVII: 77–83, 94–5; and his *Arithmetica Universalis: sive De Compositione et Resolutione Arithmetica Liber* (Cambridge, 1707): Prob. LIII: 207–9. Newton thought highly of Colin Maclaurin's scholarly commentary on the organic construction which appeared in his old age (*Geometrica Organica* (note (22)): especially Pars Prima, Sectio I, 'De Descriptione Curvarum primi Generis seu Linearum Ordinis secundi': 1–11; and Pars Secunda, Sectio I, 'Ubi demonstratur Organica Curvarum descriptio Neutoniana: 79–86.)

1

ANALYSIS OF THE PROPERTIES OF CUBIC CURVES AND THEIR CLASSIFICATION BY SPECIES

[1667 or 1668?][1]

§1. A FIRST ATTEMPT TO REDUCE THE GENERAL CUBIC BY CO-ORDINATE TRANSFORMATION

From the original manuscript in the University Library, Cambridge[2]

ENUMERATIO CURVARUM TRIUM DIMENSIONUM.

Si linea BC concipiatur moveri supra lineam immobilem AB in dato angulo ABC, dum punctum aliquod C, in eâ movens,[3] describat curvam EC:[4] voco BC descriptorem, & AB Basin in quam insistit & ABC angulum describentis. Et nota quod angulus describentis variatus variat quidem magnitudinem & scitum non autem speciem curvæ.[5] Et sumens aliquod punctum A in linea AB fixum et immobile, Equationem, quæ exprimit relationem inter AB & BC, voco naturam curvæ EC.[6]

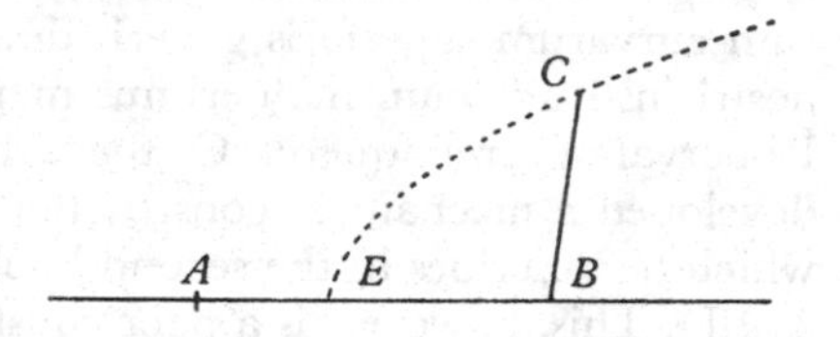

Et quia quævis linea fixa $\alpha\beta$ potest esse Basis & describens βC potest ei insistere in quovis dato angulo: patet naturam ejusdem curvæ posse infinitis modis exprimi. Nostrum est ostendere quo pacto ad simplicissimos deveniatur[7] et inde denumerare & determinare species curvarum.

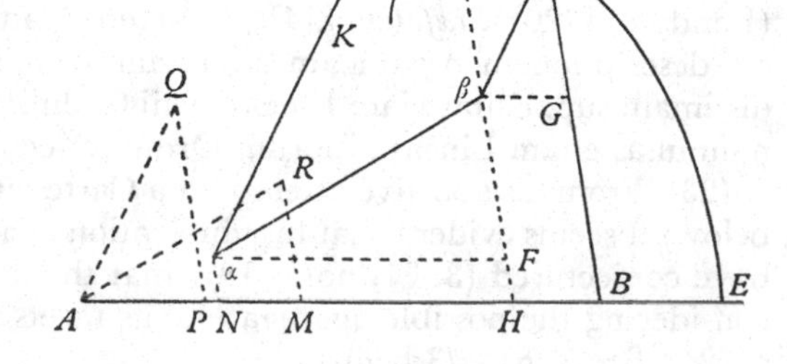

Sit $AB=z$; $BC=v$; & (assumens quaslibet quantitates datas pro a, b, c, &c) sit natura curvæ, $av^3+bzvv+czzv+dz^3+evv+fzv+gzz+hv+kz+l=0$. Quæ omnem curvam trium dimensionum comprehendit. Jam ad arbitrium

(1) We have been unable to glean any information relating to these researches from extant sources, printed or manuscript. Newton, it would seem, left the world wholly in ignorance of them till he chose to reformulate them in the middle 1690's as his *Enumeratio Curvarum Tertij*

Translation

Enumeration of Curves of Three Dimensions

If the line BC be conceived to move on the immovable line AB at a given angle $A\hat{B}C$ while some point C moving[3] in it describes the curve EC, I call BC the describer, AB the base upon which it is ordinate and $A\hat{B}C$ the angle of the describing line.[4] And note that when the angle of the describing line varies it varies also the magnitude and position of the curve, but not its species.[5] And taking some point A in the line AB as fixed and immovable, the equation which expresses the relationship between AB and BC I call the nature of the curve EC.[6]

And since any fixed line $\alpha\beta$ may be the base and the describer βC may be ordinate to it in any given angle, it is evident that the nature of the same single curve may be expressed in an infinity of ways. It is our purpose to show in what manner one may arrive at the simplest[7] and thereby enumerate and determine the species of curves.

Let $AB = z$, $BC = v$, and (assuming any given quantities you wish for a, b, c, ...) take the nature of the curve to be

$$av^3+bzv^2+cz^2v+dz^3+ev^2+fzv+gz^2+hv+kz+l = 0.$$

That comprises every curve of three dimensions. Drawing now $\alpha\beta$ and βC at

---

Ordinis, there suppressing all but an abrupt sketch of his analysis. We base our conjectured dating on our estimation of Newton's handwriting and the relative immaturity of the technical content of the present papers. These present researches are, of course, a great improvement on his first rough investigations into cubics and higher order algebraic curves by analytical techniques in the autumn of 1664 (see I, **2**, 1). That he first began to write out his work in English (note (4) below) is a valuable pointer to the early composition of the piece: little after 1668 he came to manipulate Latin syntax so well that he carried out his researches in that tongue, implicitly rejecting the imperfections of his native English in favour of its well defined, readily available technical vocabulary.

(2) Add. 3961. 1: 2r–3r.

(3) Newton's improvement on his first choice of 'motum' (moved).

(4) Newton first began to write this paragraph in English as 'If upon some immoveable line AB there bee conceived another line BC to move, insisting upon AB wth one end B & in a given angle $A\hat{B}C$ & describing a Curve line wth its other end C', but cancelled his sentence before reaching its main clause and thereafter proceeded in Latin. The suggestion is strong that Newton composed the present draft from a preliminary English worksheet, now lost.

(5) A fundamental observation, here expressed for the first time: as we would say in slightly modified terms, the species (but not the metric) of an algebraic curve is preserved under an affine transformation of co-ordinates.

(6) Terminology apart, a clear and succinct definition of a Cartesian plane co-ordinate system and of what is meant (in Descartes' sense) by the defining equation of a curve.

(7) Newton first wrote more simply 'Nostrum est simplicissimos elligere' (...to select the simplest).

ducta $\alpha\beta$ & βC pro novâ basi & describente; quæro relationem inter $\alpha\beta$ & βC, ut habeam in unâ equatione omnes modos quo natura curvæ possit exprimi per relationem inter quasvis $\alpha\beta$ & βC.

Compleo nempe Parallelogramma βGBH, βGCD, & αNHF; [&] facio triangula APQ, AMR similia triangulis βGC, $\alpha F\beta$ & similiter posita; ita ut PQ & AM sint æquales istæ lineæ quæ supponatur esse[8] unitas.[9] Deinde voco $AP=q$. $MR=p$. $AN=r$. $N\alpha=s$. $\alpha F=x$. & $CG=y$.

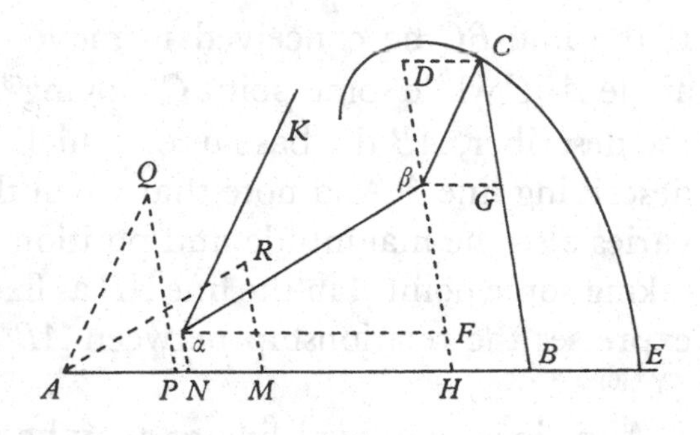

Et inquam $\left\{\begin{array}{l} AM:MR::\alpha F:F\beta. \\ 1\ :\ p\ ::\ x\ :px. \end{array}\right.$ & $\begin{array}{l} GC+\beta F+\alpha N=CB. \\ y+px+\ \ s=v. \end{array}$ Item $\begin{array}{l} PQ:PA::CG:\beta G. \\ 1\ :\ q\ ::\ y\ :qy. \end{array}$ & $\begin{array}{l} G\beta+F\alpha+NA=BA. \\ qy+\ x\ +\ r\ =z. \end{array}$ Et substituens hos valores pro v & z in naturam curvæ av^3+bzvv &c resultabit

$$\left.\begin{array}[t]{r} a \\ +bq \\ +cqq \\ +dq^3 \end{array} y^3 \begin{array}[t]{r} +3ap \\ +b \\ +2bpq \\ +cpqq \\ +2cq \\ +3dqq \end{array} xyy \begin{array}[t]{r} +3app \\ +2bp \\ +bppq \\ +2cpq \\ +c \\ +3dq \end{array} xxy \begin{array}[t]{r} +ap^3 \\ +bpp \\ +cp \\ +d \end{array} x^3 \begin{array}[t]{r} +3as \\ +2bqs \\ +cqqs \\ +br \\ +2cqr \\ +3dqqr \\ +e \\ +fq \\ +gqq \end{array} yy \begin{array}[t]{r} +6aps \\ +2bpqs \\ +2bs \\ +2cqs \\ +2bpr \\ +2cpqr \\ +2cr \\ +6dqr \\ +2ep \\ +f \\ +fpq \\ +2gq \end{array} xy \begin{array}[t]{r} +3apps \\ +2bps \\ +cs \\ +bppr \\ +2cpr \\ +3dr \\ +epp \\ +fp \\ +g \end{array} xx \begin{array}[t]{r} +3ass \\ +bqss \\ +2brs \\ +2cqrs \\ +crr \\ +3dqrr \\ +2es \\ +fqs \\ +fr \\ +2gqr \\ +h \\ +kq \end{array} y \begin{array}[t]{r} +3apss \\ +bss \\ +2bprs \\ +2crs \\ +cprr \\ +3drr \\ +2eps \\ +fs \\ +fpr \\ +2gr \\ +hp \\ +k \end{array} x \begin{array}[t]{r} +as^3 \\ +brss \\ +crrs \\ +dr^3 \\ +ess \\ +frs \\ +grr \\ +hs \\ +kr \\ +l \end{array}\right\}=0.$$

Equatio exprimens relationem inter AF[10] & GC.

Porrò quia ex dato angulo $ABC=APQ=AMR$ dantur triangula APQ AMR & tertium latus AQ & AR. quæ nomino m & n & inquam $\begin{array}{l} QP:QA::CG:C\beta \\ 1\ :\ m\ ::\ y\ :my \end{array}$ & $\begin{array}{l} AM:AR::\alpha F:\alpha\beta. \\ 1\ :\ n\ ::\ x\ :nx. \end{array}$ Itaq; si multiplico y per m & x per n hoc est si in ultimam æquationem substituo $\dfrac{y}{m}$ pro y & $\dfrac{x}{n}$ pro x resultabit æquatio designans relationem inter $\alpha\beta$ & βC nam x designabit $\alpha\beta$ & y designabit βC. Atq; ita, si angulus

(8) 'vocatur' (is called) has been cancelled.

(9) Newton had some difficulty in formulating this sentence. He first wrote, with some cancellation, 'Complens nempe Parallelogramma... centro A & radio quovis AN duco circu-

random for my new base and describing line, I seek a relationship between $\alpha\beta$ and βC such that I may have in one equation all possible ways of expressing the nature of the curve by a relationship between $\alpha\beta$ and βC.

To that end I complete the parallelograms βGBH, βGCD and αNHF, and make the triangles APQ, AMR similar to the triangles βGC, $\alpha F\beta$ and similarly situated with respect to them, such that PQ and AM are equal to the line which may be supposed to be[8] unity[9]. Then I call $AP = q$, $MR = p$, $AN = r$, $N\alpha = s$, $\alpha F = x$ and $CG = y$, and say $AM(1):MR(p) = \alpha F(x):F\beta$, or $F\beta = px$ and

$$GC+\beta F+\alpha N(y+px+s) = CB(v);$$

likewise $PQ(1):PA(q) = CG(y):\beta G$, or $\beta G = qy$ and

$$G\beta+F\alpha+NA(qy+x+r) = BA(z).$$

And by substituting these values for v and z in the nature of the curve there will result

$$\begin{aligned}&(a+bq+cq^2+dq^3)\,y^3+(3ap+b+2bpq+cpq^2+2cq+3dq^2)\,xy^2\\ &\quad+(3ap^2+2bp+bp^2q+2cpq+c+3dq)\,x^2y+(ap^3+bp^2+cp+d)\,x^3\\ &\quad+(3as+2bqs+cq^2s+br+2cqr+3dqqr+e+fq+gq^2)\,y^2\\ &\quad+(6aps+2bpqs+2bs+2cqs+2bpr+2cpqr+2cr+6dqr+2ep+f+fpq+2gq)\,xy\\ &\quad+(3ap^2s+2bps+cs+bp^2r+2cpr+3dr+ep^2+fp+g)\,x^2\\ &\quad+(3as^2+bqs^2+2brs+2cqrs+cr^2+3dqr^2+2es+fqs+fr+2gqr+h+kq)\,y\\ &\quad+(3aps^2+bs^2+2bprs+2crs+cpr^2+3dr^2+2eps+fs+fpr+2gr+hp+k)\,x\\ &\quad+as^3+brs^2+cr^2s+dr^3+es^2+frs+gr^2+hs+kr+l=0,\end{aligned}$$

the equation expressing the relationship between AF[10] and GC.

Further since from the equality of the angles $A\hat{P}Q$ and $A\hat{M}R$ to the given angle $A\hat{B}C$ the triangles APQ and AMR are given, so also are the third sides of each, AQ and AR, which I name m and n. I then say

$$QP(1):QA(m) = CG(y):C\beta,$$

or $C\beta = my$ and again $AM(1):AR(n) = \alpha F(x):\alpha\beta$, or $\alpha\beta = nx$. If therefore I multiply y by m and x by n, that is, if in the last equation I substitute y/m for y and x/n for x, there will result the equation designating the relationship between $\alpha\beta$ and βC, for x will designate $\alpha\beta$ and y βC. And so if the angle $A\hat{B}C$ should be

---

lum MQ & ducta AQ parallela ipsi $C\beta$, & æqualem unitati, cui æqualis sit AM', then he further revised the latter portion as '...duco circulum MQ & facio triangula AQP, AMR similes [sic] triangulis βGC, $\alpha F\beta$ et similiter positos [sic]. Deinde vocato $AM = 1 = AQ$, duco $AQ \parallel \beta C$ & cujuslibet longitudinis'.

(10) Read 'αF'.

ABC penitus determinaretur, invenire licet(11) relationem inter quamlibet $\alpha\beta$ & βC.

Sed hoc non bene convenit nostro proposito; angulus *ABC* non est determinandus. & angulo *ABC* variato variatur relatio inter αF & $\alpha\beta$, *CG* & $C\beta$. & si ex unâ istarum relationum suppositâ quærerem alteram, Æquatio $\left(\begin{smallmatrix} a \\ +bq \end{smallmatrix} y^3 \text{ \&c}\right)$ nimium exc[r]esceret.

Quoniam verò αF & *CG* non variantur ex variato angulo *ABC* & ijs datis datur punctum *C* (nempe sumendo αF vel $NH = x$. & du[c]endo $H\beta \parallel PQ$, & producendo ad *D* ut sit $D\beta = y$. deinde ducendo $\beta C \parallel AQ$ & $DC \parallel AB$) satiùs erit determinare naturam curvæ per relationem quam αF habet ad *CG*; præsertim cùm æquatio exprimens istam relationem, ejusdem sit(12) formæ cum æquatione designans(13) relationem inter $\alpha\beta$ & βC.(14)

Quam Æquationem ut reducam, quæro radicem *q* hujus æquationis

$$dq^3 + cqq + bq + a = 0.$$

(& si duas habet æquales radices pono tertiam radicem pro *q*(15)); (vel si $a = 0$ pono $q = 0$.)(16) Deinde pono $\dfrac{3dqq + cq}{bq + 3a} = p$. $\left(\text{vel si } q = 0 \text{ pono } \dfrac{-c}{2b} = p.\right)$ Præterea posito $3a + 2bq + cqq = \pi\,(= \quad)$(17) & $b + 2cq + 3dqq = \rho\,(= \quad)$(17).

$$+e + fq + gqq = \sigma. \ \& \ ep + \tfrac{1}{2}f + \tfrac{1}{2}fpq + gq = \tau.$$

(11) Newton here added 'propositam' (...the proposed relation...) and then cancelled it.

(12) Following English idiom Newton first wrote the ungrammatical 'erit'.

(13) Read 'designante' correctly.

(14) Presumably dissatisfied with the verbal expression of his argument, in his manuscript Newton has set cancelling brackets round this and the two preceding paragraphs. The argument itself is sound, though we might quibble at his statement that a variation in the angle $A\hat{B}C$ leaves αF and *CG* unchanged. His general intention is to reduce the cubic $C_3(z, v) = 0$ to a simpler form $C_3'(X, Y) = 0$ by suitably transforming the oblique Cartesian co-ordinates $AB = z$, $BC = v$, defining the general point *C* into the simpler pair $\alpha\beta = X$, $\beta C = Y$, say (where the origin *A* is translated to the point (r, s) and the co-ordinate angle $A\hat{B}C$ passes into

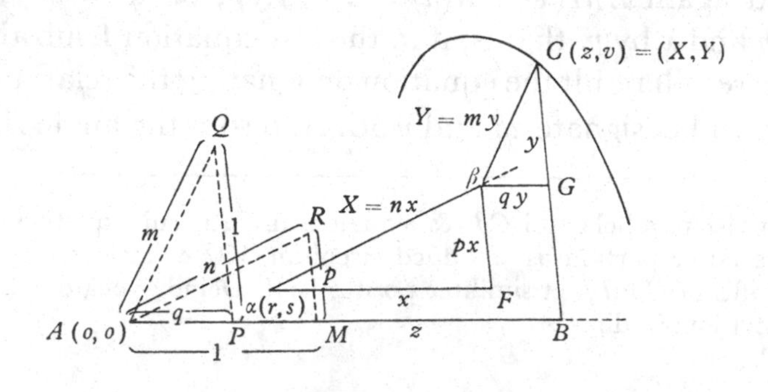

completely determined, this would permit us to find the[11] relationship between any $\alpha\beta$ and βC you wish.

But this is not well suited to our purpose: the angle $A\hat{B}C$ must not be determined and its variation varies the relationship between αF and $\alpha\beta$, and that between CG and $C\beta$; and further if by supposing one of those relationships I should seek the other, the equation $(a+bq\ldots)\,y^3+\ldots$ would grow too big.

Since indeed αF and CG are invariant when the angle $A\hat{B}C$ varies, and when they are given the point C is given also (specifically, by taking αF or $NH = x$ and drawing $H\beta \parallel PQ$, producing it to D so that $D\beta = y$, and then drawing $\beta C \parallel AQ$ and $DC \parallel AB$), it will be well enough to determine the nature of the curve by the relationship which αF has to CG; especially since the equation expressing that relationship will be of the same form with the equation designating the relationship between $\alpha\beta$ and βC.[14]

To reduce that equation I seek the root q of this equation

$$dq^3+cq^2+bq+a=0$$

(and if it has two equal roots I set the third root equal to q,[15] or if $a=0$ I set $q=0$)[16] and then set

$$\frac{3dq^2+cq}{bq+3a}=p \quad \left(\text{or if } q=0 \text{ I set } -\frac{c}{2b}=p\right).$$

Next, having set $3a+2bq+cq^2=\pi(=\quad)$[17] and $b+2cq+3dq^2=\rho(=\quad)$,[17]

$$e+fq+gq^2=\sigma \quad\text{and}\quad ep+\tfrac{1}{2}f+\tfrac{1}{2}fpq+gq=\tau,$$

---

the new angle $\alpha\hat{\beta}C$). Analytically, as he in effect shows, this general transform may be written

$$\begin{cases} z = X/n+qY/m+r \\ v = pX/n+Y/m+s \end{cases}$$

in which the constants m, n, p, q determine the inclinations of AB and BC to $\alpha\beta$ and βC: though Newton fails to note the point, they are in fact connected by

$$q(1+p^2-n^2) = p(1+q^2-m^2) = 2pq\cos A\hat{B}C.$$

We deduce that $\alpha\hat{\beta}C = A\hat{B}C+\beta\hat{\alpha}F+\beta\hat{C}G$, with $\cos\beta\hat{\alpha}F = (1+n^2-q^2)/2n$ and

$$\cos\beta\hat{C}G = (1+m^2-q^2)/2m.$$

Using a trick he had already learnt in late 1664 (compare I, 2, 1, §7) Newton simplifies this transform by considering the oblique projections ($\alpha F = x$, $CG = y$) of $\alpha\beta$, βC in the directions AB, BC respectively, setting $X = nx$, $Y = my$ thereby. As he remarks these 'designations' x of X and y of Y will not detract from the generality of his attempted reduction while affording a substantial reduction in computational complexity: indeed, from Newton's stated designating cubic in x and y we may attain the corresponding $C_3'(X,Y) = 0$ merely by substituting $x = X/n$, $y = Y/m$. This substitution, which replaces the term $x^t y^u$ by $n^{-t}m^{-u}X^tY^u$, evidently does not alter the 'shape' (forma) of the designating cubic, that is, the pattern of the terms X^tY^u which determine the cubic's species.

facio $\left(\dfrac{\tau q+\rho qs}{-\pi p}=\right)r=\dfrac{\rho\sigma q-\pi\tau q}{\pi\pi p-\rho\rho q}$. & $\dfrac{\sigma+\rho r}{-\pi}=s$. (vel si $q=0$ facio $\dfrac{-e}{b}=r$ & $\dfrac{2ce-bf}{2bb}=s$.[18]) & substituendo hos valores pro p, q, r & s, primus tertius quintus & sextus terminus evanescet.[19]

Si ponerem $\dfrac{1}{q}=p$, quatuor primi termini evanescerent, nempe $\alpha\beta$ & βC coinciderent & Curva evaderet recta.[20]

At si $a=0$ licebit quidem ponere $q=\dfrac{c\pm\sqrt{cc-4bd}}{2d}$, (nisi contradicit;)[21] sed præstat ponere $q=0$, $p=\dfrac{-c}{2b}$. $r=\dfrac{-e}{b}$. & $s=\dfrac{2ce-fb}{2bb}$.

Si $dq^3+cqq+bq+a=0$ habet duas æquales radices pone tertiam pro q, & erit $\dfrac{3dqq+cq}{+bq+3a}=p$, una ex æqualibus. &c. & primus 3us 4^tus^ 5^tus^ & 6^tus^ terminus inde evanescet.[22]

Sin tres habet æquales radices, earum unâ pro q posita, sit $\dfrac{f+2gq}{-2e-fq}=p$.[23]

---

(15) In the geometrical equivalent Newton requires that the basic asymptote be unique.

(16) For simplicity's sake.

(17) These bracketed equalities are left blank. Presumably Newton intended to insert the clarifying '$3a+bq+q(b+cq)$' and '$b+cq+q(c+3dq)$' respectively.

(18) At this point the manuscript carries an insertion mark but the referent sheet seems to be lost.

(19) In explanation, if we suppose the designating cubic to be

$$a_1y^3+a_2xy^2+a_3xy^2+a_4x^3+a_5y^2+a_6xy+a_7x^2+a_8y+a_9x+a_{10} = 0$$

and if further we take

$L_1 = 3a+bq,\quad L_2 = b+cq,\quad L_3 = c+3dq,\quad \pi = L_1+qL_2,\quad \rho = L_2+qL_3,\quad \sigma = e+fq+gq^2$

and
$$2\tau = 2ep+f(1+pq)+2gq,$$
then we may find that

$$a_1 = \tfrac{1}{3}(L_1+2qL_2+q^2L_3),\quad a_3 = p^2L_1+2pL_2+L_3,\quad a_5 = s\pi+r\rho+\sigma$$

and
$$a_6 = \frac{2}{q}(pr\pi+qs\rho+q\tau+(qs-r)(pL_1-qL_3)).$$

Hence if $a_1 = a_3 = 0$, then

$$0 = p(L_1+2qL_2+q^2L_3)-q(p^2L_1+2pL_2+L_3) = (1-pq)(pL_1-qL_3).$$

Since $pq = 1$ is a particularizing condition (see following note), we may set $pL_1 = qL_3$, so that
$$a_6 = \frac{2}{q}(pr\pi+qs\rho+q\tau).$$

The remaining conditions $a_5 = a_6 = 0$ then give

$$s = \frac{r\rho+\sigma}{-\pi}\quad\text{and}\quad r = \frac{q(s\rho+\tau)}{-\pi p} = \frac{\pi s+\sigma}{-\rho},\quad\text{or}\quad r = \frac{q(\rho\sigma-\pi\tau)}{\pi^2p-\rho^2q}.$$

I make
$$\left(\frac{\tau q+\rho q s}{-\pi p}=\right) r=\frac{\rho \sigma q-\pi \tau q}{\pi^2 p-\rho^2 q} \quad \text{and} \quad \frac{\sigma+\rho r}{-\pi}=s$$
(or if $q=0$ I make $-e/b=r$ and $(2ce-bf)/2b^2=s$)[18] and on substitution of these values for p, q, r and s the first, third, fifth and sixth terms will vanish.[19]

If I should set $1/q=p$, the first four terms would vanish, indeed $\alpha\beta$ and βC would coincide and the curve would prove to be a straight line.[20]

Should a be zero it is permissible to set $q=([-]c\pm\surd[c^2-4bd])/2d$ (unless it contradicts)[21] but it is still better to set $q=0$, $p=-c/2b$, $r=-e/b$ and $s=(2ce-fb)/2b^2$.

If $dq^3+cq^2+bq+a=0$ has two roots equal set the third equal to q, and then will $(3dq^2+cq)/(bq+3a)=p$ be one of the equal roots and so on, and the first, third, fourth, fifth and sixth terms will vanish thereby.[22]

But if it has three equal roots, on setting one of them equal to q make $(f+2gq)/(2e+fq)=-p$.[23]

Clearly, on reversal these values for q (a root of $0=a+bq+cq^2+dq^3$), $p=qL_3/L_1$, r and s determine that each of a_1, a_3, a_5 and a_6 be zero. When $a=0$ and we take $q=0$ for simplicity, it follows that $a_1=a=0$ (by definition) and we may make $a_3=0$ by taking $p=-c/2b$; also, $a_5=0$ by making $r=e/b$, and thence $a_6=0$ by setting $2b^2s=2ce-bf$.

(20) Newton has set square cancelling brackets round this paragraph, presumably intending to redraft it. Geometrically, if $pq=1$ then $AP:PQ=q:1=1:p=AM:MR$ and so A, Q and R are collinear: hence $\alpha\beta \parallel (AR$ or) $AQ \parallel \beta C$, or $\alpha\beta C$ is a straight line. In more general analytical terms if $pq=1$ then, in the notation of the previous note, $a_2=3p\times a_1$, $a_3=3p^2\times a_1$, $a_4=p^3\times a_1$, $a_6=2p\times a_5$, $a_7=p^2\times a_5$ and $a_9=p\times a_8$, so that the designating cubic becomes $a_1(y+px)^3+a_5(y+px)^2+a_8(y+px)+a_{10}=0$, or the triplet of parallels $y+px=u_i$, where u_i is one of the three roots of $a_1u^3+a_5u^2+a_8u+a_{10}=0$.

(21) We assume it would be 'contradictory' to set q equal to a complex quantity, that is, when $c^2<4bd$; q of course, has to be a root of $a_1=q(b+cq+dq^2)=0$.

(22) With the notation of note (19), the condition for $a+bq+cq^2+dq^3=0$ to have a double root is that $\pi=\rho=0$, or $L_1=-qL_2=q^2L_3$: immediately
$$a_2=p\pi+\rho=0, \quad a_3=(pq-1)^2\times L_3, \quad a_4=ap^3+bp^2+cp+d,$$
$$a_5=s\pi+r\rho+\sigma=\sigma \quad \text{and} \quad a_6=\frac{2}{q}(q\tau+(qs-r)(pL_1-qL_3)).$$
Newton's deductions are curious. If either of a_3 or a_4 is made zero, then $pq=1$ and all of the first four terms are zero (see note (20)); while if a_5 is made zero, $\sigma=0$ and this will usually clash with $a_1=0$. If Newton wishes to retain the simplifying reduction $pL_1=qL_3$ and make $a_6=0$, then $\tau=0$ (see following note).

(23) The condition for $a+bq+cq^2+dq^3=0$ to have a triple root is, in the previous notation,
$$L_1=L_2=L_3=0,$$
or $$\pi=\rho=a_1=a_2=a_3=0, \quad a_4=ap^3+bp^2+cp+d, \quad a_5=\sigma \quad \text{and} \quad a_6=2\tau.$$
The terms a_4 and a_5 cannot be made zero without particularization or possible inconsistency, but Newton removes a_6 by setting $\tau=0$ as his definition of p in terms of q.

§2. CLASSIFICATION OF THE 'FORMS' OF
$bxy^2 = dx^3 + gx^2 + hy + kx + l$

From the original manuscript in the University Library, Cambridge[1]

ENUMERATIO CURVARUM TRIUM DIMENSIONUM.

Si linea BC concipiatur moveri supra lineam immobilem AB in dato angulo ABC, dum punctum aliquod C in ea movens, describat curvam VC: voco BC describentem & AB basin in quam insistit & ABC angulum describentis.

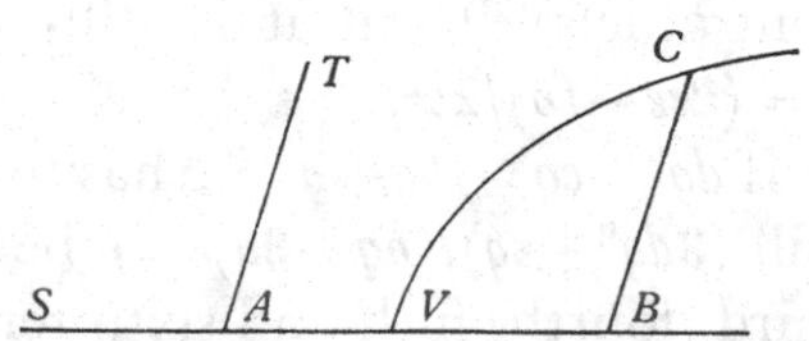

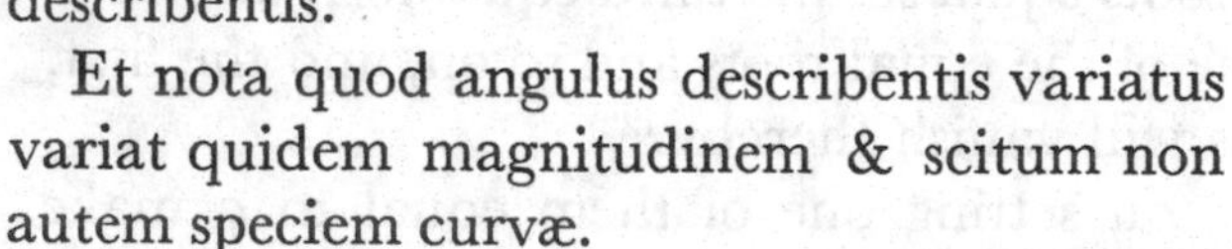

Et nota quod angulus describentis variatus variat quidem magnitudinem & scitum non autem speciem curvæ.

Præterea si in basi AB sumo aliquod punctum A determinatum & immobile & fingam (vel alias acquiram) æquationem quæ exprimet relationem inter AB & BC (hoc est ut quamcunq; longitudinem pro AB sumam, æquatio ista det longitudinem BC) ista æquatio determinabit naturam curvæ VC in quam linea BC semper terminetur sive in quâ omnia puncta C invenientur.[2]

Itaq; vocato $AB = x$, $BC = y$, & assumens pro lubitu quaslibet quantitates pro a, b, c, d, &c vel potiùs fingens assumendas esse: supponem

$$bxyy = dx^3 + gxx + hy + kx + l^{(3)}$$

designare[4] relationem inter AB & BC. sive, radice extractâ,

$$y = \frac{h[\pm]\sqrt{hh + 4blx + 4bkxx + 4bgx^3 + 4bdx^4}}{+2bx}. \text{ sive}$$

$$y = \frac{+h}{2bx} \pm \sqrt{\frac{hh}{4bbxx} + \frac{l}{bx} + \frac{k}{b} + \frac{gx}{b} + \frac{dxx}{b}}.$$

Species prima

Si signa quantitatum b & d sunt eadem, Curva erit Hyperbola trium Asymptotorum, quarum nullæ sunt parallelæ, & per quas terno crurum pari[5] & ad oppositas partes serpit (vid fig 1[)].[6]

Nempe Directrix Adh est una Asymptotorū, & sumptâ $AD = \frac{-g}{2d}$, &

$$\left.\begin{matrix} Ad \\ A\delta \end{matrix}\right\} = \left.\begin{matrix} + \\ - \end{matrix}\right\} \frac{g}{2\sqrt{bd}};$$

(1) Add. 3961.1: 10r–13r with figures on 32r–34r. (The first general diagram has been restored from the revised figure on 3961.1: 22r.) Newton broaches the enumeration of the component species of the first, most general canonical case of the general cubic curve.

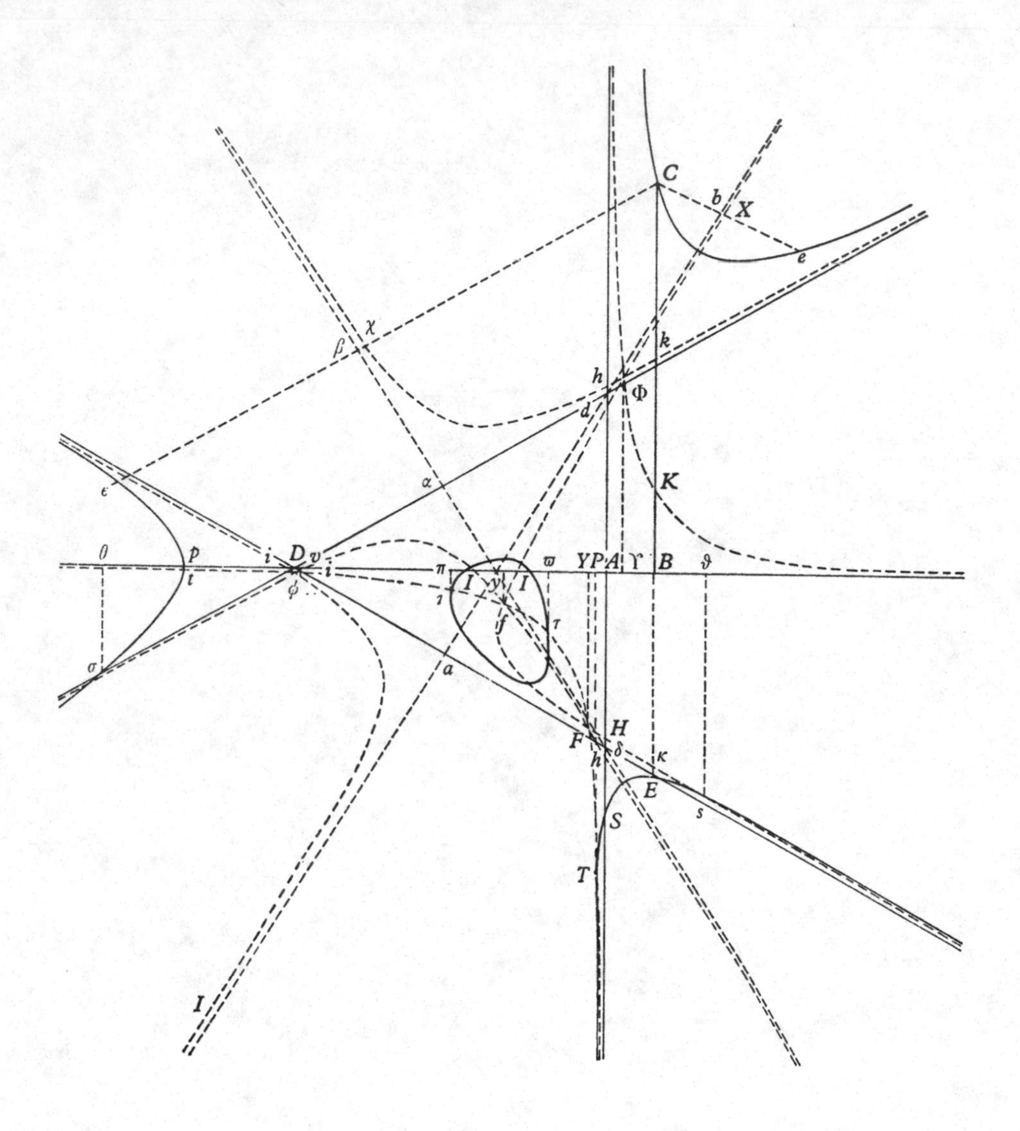

(*facing p.* 19 - *see note* 6)

Translation

Enumeration of Curves of Three Dimensions

If the line BC be conceived to move on the immovable line AB at a given angle $A\hat{B}C$ while some point C moving in it describes the curve VC, I call BC the describing line, AB the base upon which it is ordinate and $A\hat{B}C$ the angle of the describing line.

And note that when the angle of the describing line varies it varies also the magnitude and position of the curve, but not its species.

Further if in the base AB I take some point A as determined and immovable and were to imagine (or otherwise acquire) an equation to express the relationship between AB and BC (that is, such that, whatever length I take for AB, that equation would give the length BC), that equation will determine the nature of the curve VC at which the line BC will always terminate or in which every point C will be found.[(2)]

And so, calling $AB = x$, $BC = y$ and assuming at will any quantities for $a, b, c, d, \ldots$, or rather imagining them to be assumed, let me suppose that $bxy^2 = dx^3 + gx^2 + hy + kx + l$[(3)] designates[(4)] the relationship between AB and BC; that is, extracting the root, that it is

$$y = \frac{h[\pm]\sqrt{(h^2+4blx+4bkx^2+4bgx^3+4bdx^4)}}{2bx},$$

or

$$y = \frac{h}{2bx} \pm \sqrt{\left(\frac{h^2}{4b^2x^2}+\frac{l}{bx}+\frac{k}{b}+\frac{gx}{b}+\frac{dx^2}{b}\right)}.$$

First species

If the signs of the quantities b and d are the same, the curve will be a hyperbola having three asymptotes, none of which are parallel, and round which it crawls in each of its three pairs of infinite branches[(5)] and on opposite sides (see figure 1[(6)]).

Specifically the directrix Adh is one of the asymptotes and, on taking

$$AD = -g/2d \quad \text{and} \quad Ad/A\delta = \pm g/2\sqrt{(bd)}$$

(2) A straight redraft of the opening of the preceding section with some additions.

(3) Newton first wrote '$bxyy + dx^3 + gxx + hy + kx + l = 0$' (forgetting in redraft to cancel the final '$=0$'). As he showed by his method of transforming the co-ordinate axes in the preceding section, this is the most general reduced (canonical) form of the cubic.

(4) 'exponere' (exhibit) is cancelled.

(5) Newton has here cancelled the phrase 'ex adverso latere' (on opposite sides).

(6) We refer the reader to the figure of the general cubic restored from the redraft on 22^r.

ducta[7] Dd, $D\delta$ erunt aliæ duæ. vel si $g=0$; ductâ aliquâ. Describente BC, facio ut sit $b:\sqrt{bd}::AB:BF=Bf$. & per puncta F & f ductæ AF & Af erunt aliæ duæ Asymptoti. Et curva hæc eundem servat respectum ad omnes Asymptotos præcipuè in casu primo secundo & tertio sequenti. in ijs enim[8] secat Asymptoton unamquamq; idq; in unico puncto nempe HAS in S facto $AS=\frac{-l}{h}$. Dd in σ facto $A\theta=\dfrac{4dl+2gh\sqrt{\frac{d}{b}}}{gg-4dk-4dh\sqrt{\frac{d}{b}}}$ & ductâ $\theta\sigma \parallel Ad$: et $D\delta$ in s facto $\dfrac{4bdl-2gh\sqrt{bd}}{bgg-4bdk+4dh\sqrt{bd}}=A\vartheta$; & ducta $\vartheta s \parallel Ad$. Sin aliam ducas lineam alicui Asymptoto parallelam, istā aut in duobus aut in nullo puncto secabit.[9]

Et si Asymptotos Dd in α & $D\delta$ in a biseco & duco lineas $\alpha\delta$, ad; Curva habet eundem respectum ad $\alpha\delta$ & Dd, vel ad ad & $D\delta$ quem habet ad AD & $d\delta$. Nempe, sive referas Curvam ad Basin AD & Directorem Dd sive ad Basin ad & Directorem $D\delta$ sive ad Basin AD & Directorem $d\delta$ Aequatio exprimens naturam Curvæ eandem habebit formam nisi quod aliquando simpliciùs refertur ad unam Asymptoton quam aliam ut statim patebit.[10]

Sic si sumo aliquam AB pro x, & Parallelam Directori Ad duco $BK=\frac{h}{2bx}$. & Asymptotis Ad AB per punctum K describo Hyperbolam conicam ΦK cum suâ Conjugatâ ϕT ista Conica Hyperbola bisecabit omnes lineas CE curvæ interceptas. Pari ratione si sumam $Ah=\dfrac{\sqrt{bk+h\sqrt{bd}}}{b}$, & $AH=\dfrac{\sqrt{bk-h\sqrt{bd}}}{b}$ (vel si radix ista contingit esse imaginaria sumo $A[i]$[11] $=\dfrac{-g\pm\sqrt{gg-3dk-3dh\sqrt{\frac{d}{b}}}}{3d}$. & $AI=\dfrac{-g\pm\sqrt{gg-3dk+3dh\sqrt{\frac{d}{b}}}}{3d}$). & Asymptotis αd, $\alpha\delta$ per punctum h (vel $[i]$[11]) describo Hyperbolam Conicam $\chi h\Phi k$ cum sua Conjugata ϕh. Item Asymptotis

(7) Read 'ductae'.

(8) Newton first wrote 'Præterea hæc curva' (This curve, further).

(9) In the reduced cubic $bxy^2 = dx^3+gx^2+hy+kx+l$ the direction of the infinite branches are given by $bxy^2-dx^3 = 0$, and so the asymptotes are parallel to $x = 0$ and $y = \pm\sqrt{(d/b)}x$. Obviously $x = 0$ is asymptotic. The condition that $y = \pm\sqrt{(d/b)}x+\alpha$ touch the cubic at infinity is that the coefficient of x^2 in the equation which results from substituting this y in the cubic's equation vanishes, or $0 = \pm 2\alpha\sqrt{(bd)}-g$, so that the two oblique asymptotes are

and by drawing Dd, $D\delta$, these will be the other two. Or if $g = 0$, drawing some describing line BC I make $b:\surd bd = AB:BF$ (or Bf); on drawing AF and Af through the points F and f these will be the other two asymptotes. And this curve will keep the same relation to all of its asymptotes, especially in the first, second and third cases following: for in those it cuts each asymptote and then in a unique point, specifically it meets HAS in S on making $AS = -l/h$, Dd in σ on making $A\theta = (4dl + 2gh\surd(d/b))/(g^2 - 4dk - 4dh\surd(d/b))$ and drawing $\theta\sigma \parallel Ad$, and $D\delta$ in s on making $A\vartheta = (4dl - 2gh\surd(d/b))/(g^2 - 4dk + 4dh\surd(d/b))$ and drawing $\vartheta s \parallel Ad$. But should you draw another line parallel to any of the asymptotes, it will meet the curve in two points or none at all.[9]

And if I bisect the asymptotes Dd in α and $D\delta$ in a and draw the lines $\alpha\delta$, ad, the curve will keep the same relation to $\alpha\delta$ and Dd, or to ad and $D\delta$, which it has to AD and $d\delta$. Specifically, whether you are to refer the curve to base AD and director Dd, or to base ad and director $D\delta$, or to base AD and director $d\delta$, the equation expressing the nature of the curve will have the same form except that sometimes it may more simply be referred to one asymptote rather than to another, as will be immediately clear.[10]

Thus if I take some line AB for the x-axis and parallel to the director Ad I draw $BK = h/2bx$ and then with asymptotes Ad, AB through the point K describe the conic hyperbola ΦK with its conjugate branch ϕT, that conic hyperbola will bisect all segments CE intercepted by the curve. Equally, if I were to take $Ah = \surd(bk + h\surd[bd])/b$ and $AH = \surd(bk - h\surd[bd])/b$ (or if that root chances to be imaginary I take $A[i]$[11] $= (-g \pm \surd(g^2 - 3dk - 3dh\surd[d/b]))/3d$ and

$$AI = (-g \pm \surd(g^2 - 3dk + 3dh\surd[d/b]))/3d)$$

and then with asymptotes ad, $\alpha\delta$ through the point h (or $[i]$[11]) describe the conic hyperbola $\chi h\Phi k$ with its conjugate branch ϕh, and likewise with asymptotes ad,

---

$y = \pm\surd(d/b)(x + g/2d)$. The asymptote $x = 0$ meets the cubic such that $hy + l = 0$, or $AS = -l/h$. The asymptotes $y = \pm\surd(d/b)(x + g/2d)$, likewise, meet it such that

$$(g^2/4d \mp h\surd(d/b) - k)x = \pm gh/2\surd(bd) + l;$$

which determines Newton's values for $A\theta$ and $A\vartheta$.

(10) This is evident geometrically but the point may easily be proved analytically and Newton may well have done so. Thus, if we set $\kappa^2 = d/b$, $\lambda = \alpha\delta/\alpha d$ and $\mu = AD/\alpha\delta$ we may transform the co-ordinate axes $AB = x$, $BC = y$ to the new pair $\alpha\beta = X$, $\beta C = Y$, where

$$\begin{cases} x = -\tfrac{1}{2}\mu X + \tfrac{1}{2}\lambda\mu Y - 3g/4d \\ y = \tfrac{3}{2}\kappa\mu X + \tfrac{1}{2}\lambda\kappa Y + 3\kappa g/4d \end{cases}.$$

It follows with some manipulation that the defining equation $bxy^2 = dx^3 + gx^2 + hy + kx + l$ is transformed into a new one $BXY^2 = DX^3 + GX^2 + HY + KX + L$ of the same form. Note too Newton's present use of 'Director' to denote what he has previously called the 'describing' line.

(11) The letter 'i' has been inserted to fill a blank in Newton's text: evidently the corresponding point of his diagram (now lost) was not named through an oversight. Two lines below Newton denoted the point by 'I' but we use the lower-case form to avoid confusion.

ad, *aδ* per punctum *H* (vel *I*) describo Hyperbolam Conicam *XHI* cum sua conjugata [*i*]*φ*: Istæ Hyperbolæ Conicæ bisecabunt omnes rectas *Cχε*, *CXe* curvæ interceptas & parallelas Asymptotis *Dd*, *Dδ*.

Omnes hæ tres Hyperbolæ in eodem puncto Φ, *φ* semper decussant.(12) Nam posito $AB=x$. & *BK*, vel *Bk*, vel $B\kappa=y$. natura Hyperbolæ Φ*K* invenietur $2bxy=h$. Hyperbolæ χΦ*k* invenietur $byy=-2xy\sqrt{bd}+3dxx+2gx+k+h\sqrt{\frac{d}{b}}$. & Hyperbolæ *X*Φ*κ* invenietur $byy=2xy\sqrt{bd}+3dxx+2gx+k-h\sqrt{\frac{d}{b}}$.(13) Si jam in posteriores æquationes substituas $\frac{h}{2bx}$ pro *y*, ad invenienda puncta Φ in quibus aliæ duæ secant Φ*K*; ex utrâq; æquatione proveniet

$$12bdx^4+8bgx^3+4bkxx-hh=0.$$

Adeoq; istæ secant Φ*K* in eodem puncto Φ: positâ nempe $A\Upsilon=x$ radici novissimæ æquationis, erit $\Upsilon\Phi=\frac{h}{2bx}\parallel Ad$.(14)

Et hinc innotescit ratio inveniendi plura curvæ puncta ad eam describendam. nempe si datur aliquod ejus punctum *C*, duco $CK\parallel Ad$ & $C\chi\parallel \alpha d$ & $CX\parallel aD$ & produco eas ad *E*, *ε*, *e* ita ut sit $KE=KC$, $\chi\epsilon=\chi C$ & $Xe=XC$: puncta *E ε* & *e* erunt in curvâ. et eâdem ratione puncta *E*, *ε*, & *e* dant alia.

Punctum *O*, quod est centrum Asymptotorum(15) & trianguli *Ddδ* (positâ nempe $AO=\frac{1}{3}AD$),(16) poterit nominari centrum curvæ; quia unicum est; neque curva ad aliud quodvis punctum eodem modo refertur.(17)

Si vellem cognoscere limites curvæ, istos nempe *T*, *τ*, *7*, *t*, in quibus Describentes *PT*, *ϖτ*, *π7*, *pt* tangunt curvam, hoc est in quibus Hyperbola Conica circa

(12) Newton first wrote, rather more explicitly, 'Et hic notare licet, quod ubi duæ harum Hyperbolarum Conicarum se mutuo secant, in eodem puncto etiam tertiam secabunt' (And here we may note that where two of these conical hyperbolas intersect each other, in the same point(s) also they will intersect the third).

(13) The general ordinate *BC* meets the cubic $bxy^2 = dx^3+gx^2+hy+kx+l$ in two points *C* and *E* (say, $BC = y_1$, $BE = y_2$) such that

$$\left.\begin{matrix} y_1 \\ y_2 \end{matrix}\right\} = \frac{h}{2bx} \pm \frac{1}{2bx}\sqrt{[h^2+4bx(dx^3+gx^2+kx+l)]}.$$

If we denote the ordinate *BK* of the mid-point *K* of *CE* by *Y*, then $Y = \frac{1}{2}(y_1+y_2) = h/2bx$, so that the point *K* is on the hyperbola $0 = 2bxY-h$ (later to be called by Newton the diametral hyperbola). Likewise *Cε* and *Ce* (of Cartesian equation $y = \pm\sqrt{(d/b)}x+m$) meet the cubic such that $(\pm 2b\sqrt{(d/b)}m-g)x^2+(bm^2\mp h\sqrt{(d/b)}-k)x-(hm+l) = 0$, and so, if the ordinate of their mid-point be *X* (the semi-sum of the roots of the preceding quadratic), then $X(\pm 2b\sqrt{(d/b)}m-g) = -bm^2\pm h\sqrt{(d/b)}+k$: hence, on eliminating $m = y\mp\sqrt{(d/b)}X$, we find that the points *χ* and *X* lie on the hyperbolas

$$0 = -by^2\mp 2\sqrt{(bd)}\,Xy+3dX^2+2gX\pm h\sqrt{(d/b)}+k.$$

$a\delta$ through the point H (or I) describe the conic hyperbola XIH with its conjugate branch $[i]\phi$: those conic hyperbolas will bisect all straight lines $C\chi\epsilon$, CXe intercepted by the curve which are parallel to the asymptotes Dd and $D\delta$.

Each of these three hyperbolas will always cross the others in the same points Φ, ϕ. For, on setting $AB = x$ and BK, Bk or $B\kappa = y$, the nature of the hyperbola ΦK will be found to be $2bxy = h$, that of the hyperbola $\chi\Phi k$

$$by^2 = -2xy\surd(bd)+3dx^2+2gx+k+h\surd(d/b),$$

and $by^2 = 2xy\surd(bd)+3dx^2+2gx+k-h\surd(d/b)$ that of the hyperbola $X\Phi\kappa$.(13) If now in the latter equations you substitute $h/2bx$ in place of y in order to find the points Φ in which the other two hyperbolas cut $\Phi\kappa$, from either equation there will come $12bdx^4+8bgx^3+4bkx^2-h^2 = 0$, so that those hyperbolas will cut $\Phi\kappa$ in the same point Φ: specifically, on setting $A\Upsilon = x$ (the root of this most recent equation) $\Upsilon\Phi$ will be equal to $h/2bx$ and parallel to Ad.(14)

And from this may be known a method for finding further points of the curve in order to describe it. Specifically, if some point C of it be given, I draw $CK \parallel Ad$, $C\chi \parallel \alpha d$ and $CX \parallel aD$ producing them to E, ϵ and e so that $KE = KC$, $\chi\epsilon = \chi C$ and $Xe = XC$: then will the points E, ϵ and e be in the curve. And by the same method the points E, ϵ and e will give still others.

The point O, which is the centre of the asymptotes(15) and of the triangle $Dd\delta$ (on setting, that is, $AO = \frac{1}{3}AD$),(16) may be named the curve's centre since it is unique: the curve is not referred in the same manner to any other point.(17)

If I should wish to know the limits of the curve, those points (here T, τ, 7 and t) in which the describing lines PT, $\varpi\tau$, $\pi 7$ and pt are tangent to the curve, that is, in which the conic hyperbola described about the asymptotes AD and Ad cuts

---

We easily show that the respective asymptotes of the hyperbolas (K), (χ) and (X) are

δAd $(x = 0)$, AD $(Y = 0)$; $D\alpha d$ $(y = \surd(d/b)X+g/2\surd(bd))$, $\alpha\delta$ $(-y = 3\surd(d/b)X+g/2\surd(bd))$;

and $Da\delta$ $(-y = \surd(d/b)X+g/2\surd(bd))$, ad $(y = 3\surd(d/b)X+g/2\surd(bd))$.

Further, since $(X)-(\chi) = 2\surd(bd)\times(K)$, the hyperbola (K) is through the meets (F, f, Φ, ϕ) of (χ) and (X). Finally, AH and Ah are given as the meets of Ad $(X = 0)$ with the hyperbolas (χ) and (X), while AI and Ai are their corresponding meets with $AD(y = 0)$.

(14) The meets Φ, ϕ, F, f of the hyperbolas (K), (χ) and (X) are given by

$$0 = 2bxy-h = -by^2 \pm \surd(d/b)\,(2bxy-h)+3dx^2+2gx+k,$$

that is, by the roots of the quartic $0 = -h^2/4bx^2+3dx^2+2gx+k$.

(15) Newton added '& Hyperbolarum Conicarum' (and of the conic hyperbolas) but then cancelled the phrase, rightly so.

(16) Since $d\delta$, δD and Dd are bisected in the points A, a and α respectively, the lines DA, da, $\delta\alpha$ will meet in the centroid (O) of the triangle $Dd\delta$.

(17) This seems an unnatural reason for defining the point O as the centre of the cubic. Newton himself soon abandoned the definition for the more useful and restrictive one which all later writers on cubics have adopted.

Asymptotos AD, & Ad descripta secat curvam: Construo hanc Æquationem $dx^4+gx^3+kxx+lx+\frac{hh}{4b}=0$ (idq; per Parabolam vel alias)(18) & facio ut ejus radices sint AP, $A\varpi$, $A\pi$, Ap (vide figuras(19) primi casus). & duco PT, $\varpi\tau$, $\pi 7$, pt secantes Hyperbolam conicam in T, τ, 7, t, punctis quæsitis. Et harum radicum plurimus est usus in determinando varias curvarum formas; ut postea patebit.

Cas 1. Præterea hæc Species potest in varios Casus distingui. Quorum primus esto, quando nullus terminus in naturâ Curvæ p^r^dictâ(20) deest nisi forsan k vel l vel uterq;, neq; sit $gg-4dk=\pm 4dh\sqrt{\frac{d}{b}}$.(21) Hoc est quando Asymptoti non decussant sese omnes in eodem puncto A: Et crura quæ per Asymptoton quamvis ad diversas partes in infinitum serpunt, jacent ex diverso latere asymptoti. (ut videre est in figuris duodecim(22) Casûs hujus).

Et Figuræ hujus casûs sex præcipue formas induunt prout radices æquationis $dx^4+gx^3+kxx+lx+\frac{hh}{4b}=0$,(23) varijs modis se habent. Nempe

Form 1. Si omnes radices sunt reales, ejusdem signi, & inæquales: figura habet 3 Hyperbolas cum Ellipsi (ut in Fig 1 & 7). Et Ellipsis semper jacet intra Triangulum $Dd\delta$, & etiam inter medios limites τ, 7,(24) in quibus tangitur a lineis $\varpi\tau$, $\pi 7$ quas invenire prius docebas.(25)

Form: 2. Si duæ extremarum radicum (viz duæ maximæ vel duæ minimæ) sunt æquales ($AP=A\varpi$, vel $Ap=A\pi$) Ellipsis et Hyperbolarum una in cuspide junguntur ibi decussantes. (vide fig 2 & 8). Crura nempe Hyperbolæ ST quando $AP=A\varpi$ (ut in fig 2), vel Hyperbolæ σt quando $A\pi=Ap$ (ut in fig 8)(26) sese decussando intra Triangulum $Dd\delta$, in Ellipsin continuantur.

Forma 3. Si tres radices $\left\{\begin{array}{l}AP=A\varpi=A\pi \text{ nihilo propiores}\\ A\varpi=A\pi=Ap \text{ nihilo remotiores}\end{array}\right\}$ sunt æquales Ellipsis

(18) The final term in the quartic (which results from eliminating y between the cubic $bxy^2=dx^3+gx^2+hy+kx+l$ and the diametral hyperbola $2bxy=h$) should read '$-\frac{hh}{4b}$', an error which is carried through the rest of the section. In making the change from his first canonical form $bxy^2+dx^3+gx^2+hy+kx+l=0$ (see note (3)) Newton has not everywhere made appropriate correction. For the Cartesian construction of the quartic by the meet of a circle and parabola see I, 3, 3, §2.

(19) '1, 7, 12 &c' is cancelled.

(20) 'prædictâ'.

(21) This (note (9)) is the condition for one of the meets of the cubic with an asymptote to pass to infinity: in that case the asymptote touches the cubic at infinity, so that the cubic is parabolic.

the curve: I construct this equation $dx^4+gx^3+kx^2+lx+(h^2/4b)=0$ (by a parabola or other means),[18] making its roots AP, $A\varpi$, $A\pi$ and Ap (see the figures[19] of the first case), and then I draw PT, $\varpi\tau$, $\pi 7$ and pt cutting the conic hyperbola in T, τ, 7 and t, the points required. I may add that there is the widest use of these roots in determining the various forms of the curves, as will afterwards be evident.

Case 1. This species, furthermore, can be distinguished into various cases. Let the first of these be when no term in the above-noted nature of the curve is lacking, except perhaps k or l or both, and when g^2-4dk does not equal $\pm 4dh\surd(d/b)$.[21] That is, when the asymptotes do not all cross in the same point A, and the branches which snake round any asymptote into infinity in opposite directions lie on opposite sides of the asymptote (as you may see in the twelve[22] figures of this case).

And the figures of this case take on six especial forms according to the varied mode of the roots of this equation $dx^4+gx^3+kx^2+lx+h^2/4b=0$.[23] Thus

Form 1. If all the roots are real, unequal and of the same sign, the figure has three hyperbolic branches with an oval (as in figures 1 and 7), and the oval lies always within the triangle $Dd\delta$, and indeed between the middle limits τ, 7[24] in which it is touched by the lines $\varpi\tau$, $\pi 7$ which you were previously shown how to find.[25]

Form 2. If the two extreme roots (the two largest, that is, or the two least) are equal ($AP=A\varpi$ or $Ap=A\pi$), the oval and one of the hyperbolic branches are joined in a cusp, crossing each other there (see figures 2 and 8). Precisely, the branches of the hyperbola ST when $AP=A\varpi$ (as in figure 2) or of the hyperbola σt when $A\pi=Ap$ (as in figure 8)[26] by crossing within the triangle $Dd\delta$ are continued into the oval.

Form 3. If three roots ($AP=A\varpi=A\pi$ nearer to zero, or $A\varpi=A\pi=Ap$ more distant from it) are equal, the oval has vanished into the crossing point (T or t)

---

(22) Figures 1–12 below.

(23) See note (18).

(24) Newton has cancelled 'quos invenire per prædictas radices priùs docebatur' (which you were previously shown how to find through the roots of the above-mentioned quartic).

(25) In continuation there is added (and then cancelled) 'In duas etiam æquales partes dividitur ab unaquaq 3ª Hyperbolarum Conicarum, quæ omnes tum in ejus medio tum ab utraq parte inter eam & Hyperbolas decussant. Enim semper decussant inter unumquemq limitem T, τ, 7 & t' (It is also divided into two equal portions by each of the three conic hyperbolas, all of which cross both in its middle and also on its either side between it and the hyperbolic branches. For they always cross each other between each of the limits T, τ, 7 and t). The remark is, of course, just. For 'docebas' read 'doceba[ri]s' or perhaps 'docebam'.

(26) 'extra Asymptotos serpentes &' (snaking outside the asymptotes) is cancelled.

in punctum decussationis $\begin{Bmatrix} T \\ t \end{Bmatrix}$ evanuit, & Crura Hyperbolæ $\begin{Bmatrix} ST \\ \sigma t \end{Bmatrix}$ ibi facientes angulum contactus in cuspidem desinunt, vide fig $\begin{Bmatrix} 3. \\ 9. \end{Bmatrix}$(27)

Forma 4. Si duæ mediæ radices (viz: quarum neutra est omnium maxima vel minima) $A\varpi$ & $A\pi$, sunt æquales: Ellipsis in punctum evanuit.(28) Item si duæ radices sunt imaginariæ & duæ reales & inæquales: Ellipsis nullibi apparet. vide fig 4 & 10. nempe si duæ mediæ radices sunt imaginariæ, Ellipsis imaginaria est; sin duæ extremæ, tum Ellipsis absorbetur in Hyperbolarum unam, & quandoqꝫ videtur quasi angusto Hiatu in eam continuata, ut videre est in fig 10. In hac itáqꝫ forma solæ manent tres Hyperbolæ sine aliqua Ellipsi decussatione vel cuspide quales in formis 1, 2, 3 &(29) proxima sequente accidunt.

Forma 5. Si duæ radices $\frac{AP}{A\varpi}$ sunt æquales & alteræ duæ imaginariæ (fig 5), vel reales cum signis diversis a signis æqualium radicum (fig 11); Tum duæ Hyperbolarum in quartâ formâ sese in cuspide contingunt vel potiùs mutantur in duas alias Hyperbolas sese in isto puncto $\frac{T}{\tau}$ decussantes; quod punctum $\frac{T}{\tau}$ datur ope radicis æqualis $\begin{matrix} A[P], \text{fig } 5 \\ A\varpi, \text{fig } 11. \end{matrix}$ Tertia Hyperbola jacet in angulo D cùm omnes radices sunt reales ut in fig 11; sin duæ sint imaginariæ jacebit in uno angulorum d vel δ ut in fig 5ta. Nempe semper jacet in angulo qui opponitur puncto $\frac{T}{\tau}$.

Forma 6. Si omnes radices sunt imaginariæ (fig 6), vel reales & inæquales quarum duæ sunt affirmativæ & duæ negativæ (fig 12): tum ex Hyperbolis decussantibus fit Hyperbola cum aliâ figurâ quæ intorta circa unam Asymptoton ex adverso ejus latere ad oppositas partes in infinitum procedit; Quam ideò Concham intortam(30) nominabo. Cùm omnes radices sunt reales, habet Directricem Ad pro Asymptoto [(]fig 12[)]: cum vero sunt imaginariæ Asymptoton(31) erit Dd vel $D\delta$ (fig 6); nempe ista cujus sectio (in σ vel s) propior est ad directricem Ad.(32) Intra reliquas duas Asymptotos ad oppositas anguli partes duæ Hyperbolæ jacent.

(27) Newton has cancelled the further sentence, 'In isto etiam cuspide omnes Hyperbolæ Conicæ et sese & curvam contingunt, vel potius secant in angulo contactus' (In that cusp, too, all the conic hyperbolas touch each other and the curve, or rather meet in a contact angle). The angle of 'contact' (or 'horn' angle) is the second-order indefinitely small angle between a continuous curve and a tangent at a point of it. This angle, less than finite but greater than the angle made between two coincident straight lines, was much discussed in the seventeenth century both for its philosophical importance and for its relevance to the concept of the curvature of a curve. In particular John Wallis had, not wholly obtusely, defended its real existence in a tract (*De Angulo Contactus et Semicirculi Disquisitio Geometrica*, published as the first piece in his *Operum Mathematicorum Pars Altera* (Oxford, 1656)) at which we may presume Newton at least glanced in his youth. (Compare I, 1, Introduction, Appendix 2, especially note (20).)

and the branches of the hyperbola (ST or σt) there making a contact angle end in a cusp (see figures 3 and 9).[27]

Form 4. If the two middle roots $A\varpi$ and $A\pi$ (neither of which, that is, are the greatest or least of them all) are equal, the oval has vanished into a point.[28] Likewise if two roots are imaginary and two real and unequal, the oval appears nowhere (see figures 4 and 10). Specifically, if the two middle roots are imaginary the oval is imaginary; but if the two extreme ones are, then the oval is absorbed into one of the hyperbolic branches and is then seen, as it were, continued into it through a narrow cleft, as you can see in figure 10. In this form, consequently, the three hyperbolas alone remain without any of oval, crossing or cusp of the kind which occurs in the first, second, third and [29] next following forms.

Form 5. If two roots AP and $A\varpi$ are equal and the other two imaginary (figure 5) or real and opposite in sign to the equal roots (figure 11) then two of the hyperbolic branches in the fourth form are tangent in a cusp, or, better, are changed into two other hyperbolas crossing one another in the point T (figure 5) or τ (figure 11): that point T or τ is determined by the help of the corresponding equal root AP or $A\varpi$. The third hyperbola lies within the angle D when all the roots are real, as in figure 11; but if two are imaginary it will lie in one of the angles d or δ, as in figure 5: precisely, it will always lie in the angle which is opposite to the point T or τ.

Form 6. If all the roots are imaginary (figure 6), or real and unequal with two positive and two negative (figure 12), then out of the crossing hyperbolic branches there is produced a hyperbola of different shape which is twisted around one of the asymptotes, passing to infinity on opposite sides and in opposite directions: for that reason I call it a twisted conchoid.[30] When all the roots are real it has the directrix Ad for its asymptote (figure 12), when indeed they are imaginary the asymptote will be Dd or $D\delta$ (figure 6): that one, specifically, which it meets (in σ or s) nearer to the directrix Ad.[32] Between the two remaining asymptotes and on opposite sides of the angle they make lie the two hyperbolic branches.

(28) 'Vanished' because a single point has no magnitude (or perhaps, for Newton at this time, an indivisibly small breadth) and so cannot be seen, even though its position may conventionally be marked by a dot. It is interesting that Newton at present does not distinguish the species of cubic which have a real conjugate point from those which have an imaginary oval.

(29) '5' is cancelled.

(30) On the analogy of the quartic curve, the conchoid (mussel-curve), whose properties were familiar to Newton from his reading of Descartes' *Geometrie*. (Compare I, 3, 3, §3.) Later he will use the descriptive adjective 'conchoidalis'.

(31) Read 'Asymptotos'.

(32) This remark is a generalization of a visual property of figure 6 which is not invariably true: for a counter example see Newton's figure 21 of the following section.

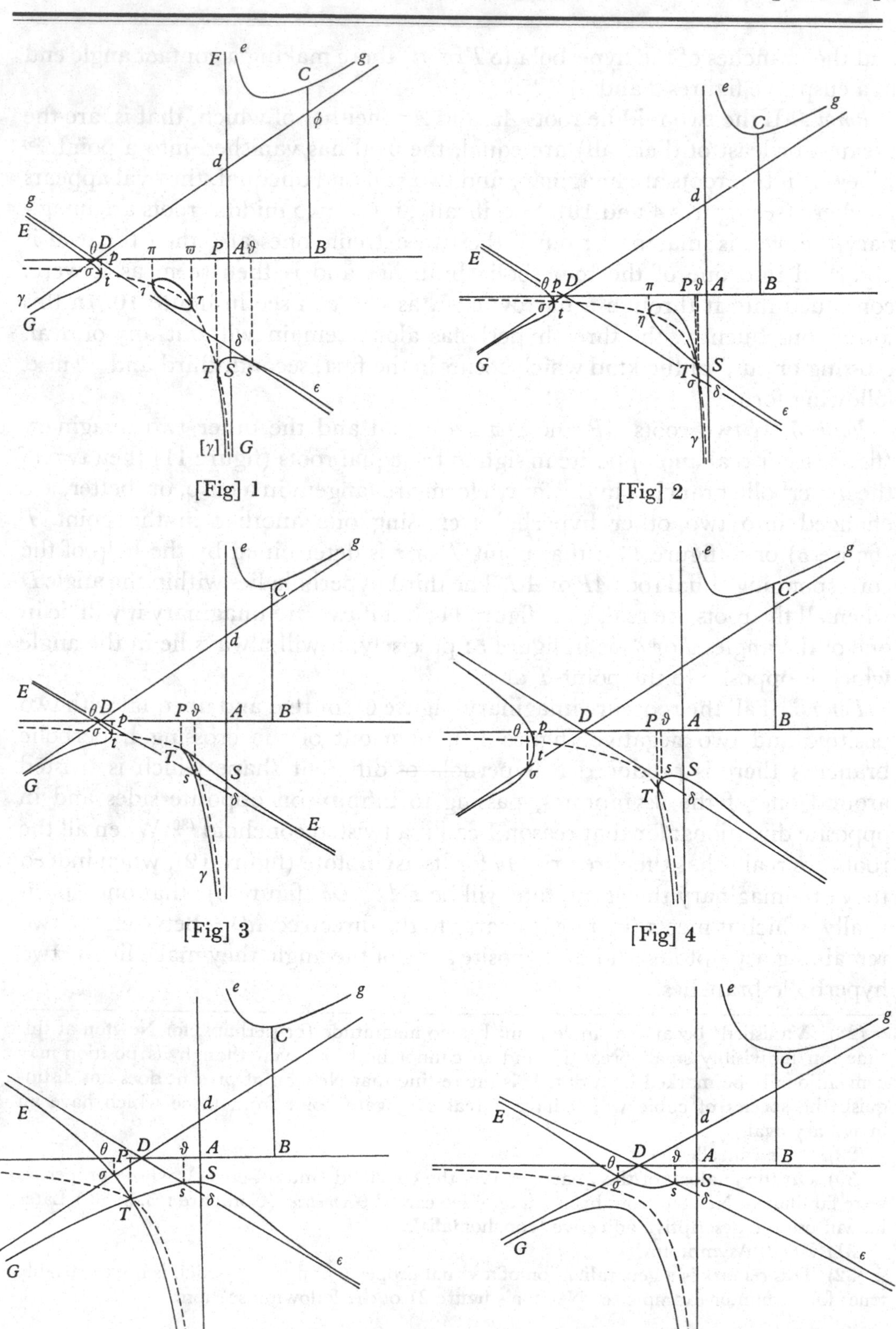

[Fig] 1

[Fig] 2

[Fig] 3

[Fig] 4

[Fig] 5

[Fig] 6

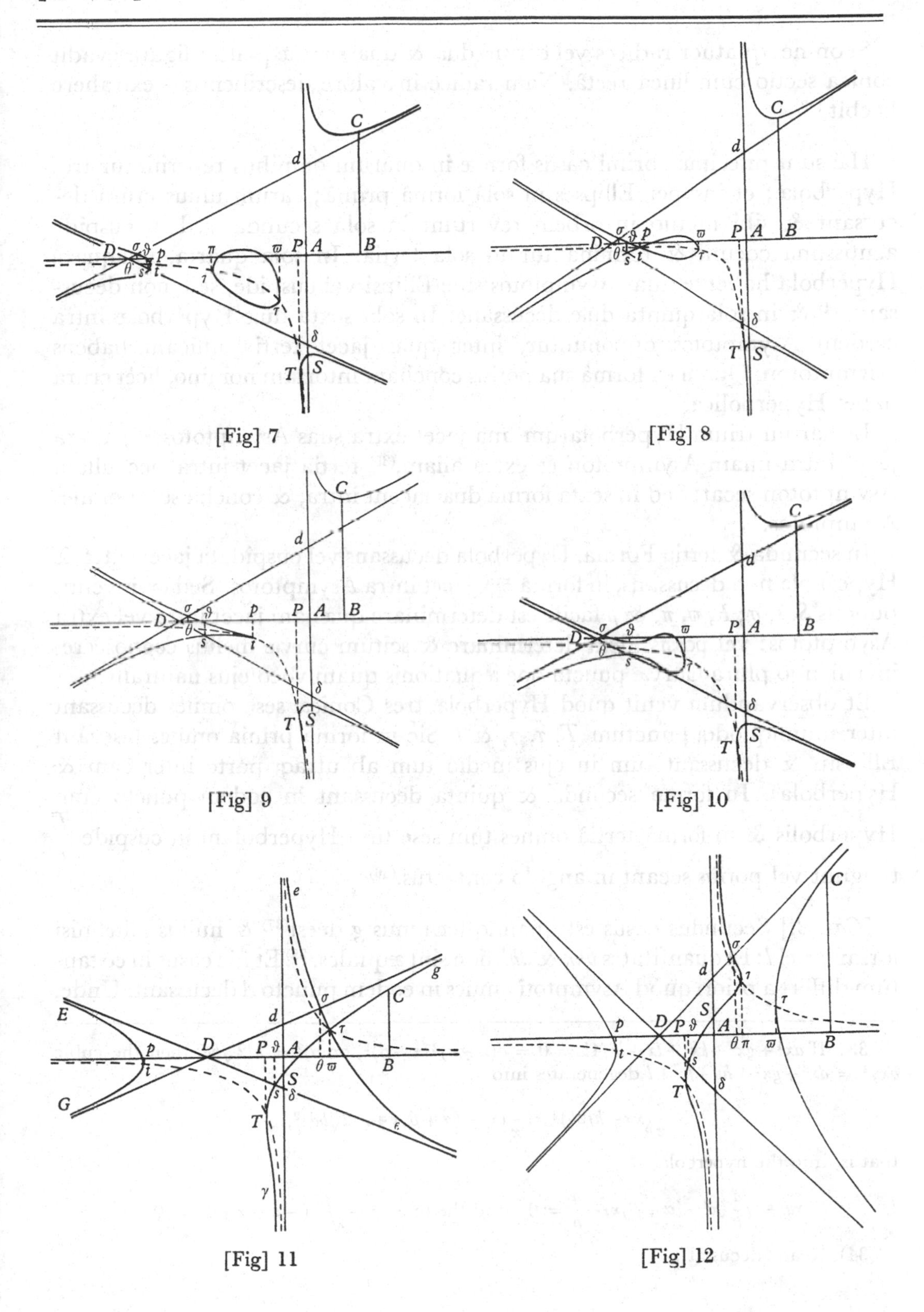

[Fig] 7

[Fig] 8

[Fig] 9

[Fig] 10

[Fig] 11

[Fig] 12

Si omnes quatuor radices vel earum duæ & duæ sunt æquales: figura evadit conica sectio cum lineâ rectâ. Nam radicē in valore describentis y extrahere licebit.[33]

Hæ sunt præcipuæ primi casûs formæ in quarum omnibus reperiuntur tres Hyperbolæ; et insuper Ellipses in solâ formâ primâ; earum unius crura decussant & sibi mutuo in orbem revertunt in solâ secunda, sed in cuspide acutissima coeunt & terminantur in sola tertiâ. In sola quarta, unaquæꝗ Hyperbola habente duas Asymptotos sine Ellipsi vel cuspide, sese non decussant,[34] & in sola quinta duæ decussant; In sola sexta duæ Hyperbolæ intra easdem Asymptotos opponuntur, inter quas jacet tertia unicam habens Asymptoton, Quam ex formâ suâ potiùs concham intortam nomino, licèt crura habet Hyperbolica.

Et harum trium Hyperbolarum una jacet extra suas Asymptotos,[35], altera jacet intra unam Asymptoton & extra aliam,[35] tertia jacet intra nec ullam Asymptoton secat: Sed in sexta forma duæ jacent intra, & concha secat omnes Asymptotos.

In secunda & tertia Forma, Hyperbola decussans vel cuspidata jacet extra & Hyperbola non decussans, in forma 5ta, jacet intra Asymptotos. Sed ex inventis punctis S, s, σ; P, ϖ, π, & p facile est determinare quænam jacet intra vel extra Asymptotos: vel possis idem determinare & scitum curvæ meliùs cognoscere, inveniendo plura Curvæ puncta ope æquationis quam vocò ejus naturam.

Et observandum venit quòd Hyperbolæ tres Conicæ sese omnes decussant inter unumquodꝗ punctum T, τ, 7, & t. Sic in forma prima omnes bisecant Ellipsin & decussant tum in ejus medio tum ab utraꝗ parte inter eam & Hyperbolas. In forma secunda & quinta decussant in eodem puncto cum Hyperbolis & in formâ tertiâ omnes tum sese tum Hyperbolam in cuspide $\genfrac{}{}{0pt}{}{T}{t}$ tangunt vel potiùs secant in angulo contactûs.[36]

[*Cas. 2.*] Secundus casus est quando terminus g deest[37] & nullus alius nisi forsan k vel l: Et quantitates bkk & dhh non sint æquales.[38] Et hic casus in eò tantùm differt a priori quòd Asymptoti omnes in eodem puncto A decussant. Unde,

(33) If $dx^4+gx^3+kx^2+lx+h^2/4b = 0 = d(x-\alpha)^2(x-\beta)^2$, or $\alpha\beta = h/2\sqrt{bd}$, then the cubic $bxy^2 = dx^3+gx^2+hy+kx+l$ degenerates into

$$\frac{b}{x}(xy-h/2b)^2 = \frac{d}{x}(x^2-(\alpha+\beta)x+h/2\sqrt{bd})^2,$$

that is, into the hyperbola

$$xy-\sqrt{\frac{d}{b}}(x^2-(\alpha+\beta)x)-\frac{h}{b} = 0 \quad \text{and the line} \quad y-\sqrt{\frac{d}{b}}(-x+\alpha+\beta) = 0.$$

(34) Read 'decussat'.

If the four roots are all equal to one another, or two by two, the figure will prove to be a conic section together with a straight line. For then the square root in the value of the describing line y can be extracted.(33)

These are the chief forms of the first case. In all of them are found three hyperbolic branches; and ovals too in the first form only: in the second alone the branches of one of them cross and loop back into one another, but they meet in a very sharp cusp in the third only and there terminate. In the fourth alone each hyperbola has two asymptotes without oval, cusp or crossing, but uniquely in the fifth two of them cross. In the sixth alone two hyperbolas are set on opposite sides of the same asymptotes, between which lies the third, having a unique asymptote to itself: though there exist hyperbolic branches, from its shape I prefer to name it the twisted conchoid.

And of these three hyperbolas one lies outside its asymptotes,(35) a second lies within one and without another,(35) the third lies inside without cutting any of the asymptotes at all: but in the sixth form two hyperbolas lie outside their asymptotes and the conchoid cuts them all.

In the second and third forms the crossing or cusped hyperbola lies outside and the non-crossing hyperbola in the fifth form lies inside its asymptotes. But when the points S, s, σ and P, ϖ, π and p are found it is easy to determine which lie inside and which outside their asymptotes: or you might determine that, and better understand the layout of the curve, by finding further points of the curve with the help of the equation which I call its nature.

We should observe too that the three conic hyperbolas cross one another between each of the points T, τ, 7 and t. Thus in the first form they all bisect the oval and cross both in its middle as well as on its either side between it and the hyperbolic branches. In the second and fifth forms they cross in the same point that the hyperbolic branches do, while in the third they all touch each other and the hyperbolic branch in the cusp T or t, or rather they cut one another in a contact angle.(36)

[*Case 2.*] The second case is when the term in g is lacking(37) and no other except perhaps k or l, provided that the quantities bk^2 and dh^2 are not equal.(38) This case differs from the former merely in that all the asymptotes cross in the same point A. Since then the triangle $Dd\delta$ vanishes, the consequence is that the

(35) Newton here has cancelled the clause 'quas/quam ideo secat' (which therefore it cuts).

(36) Following this almost a whole page has been left blank in the manuscript, presumably so that Newton could enter any further general observations which came to mind subsequently.

(37) In that case, $AD = g/2d$ vanishes and so too the triangle $Dd\delta$, so that the three asymptotes meet in a unique point A (coincident with D, d and δ).

(38) The condition of case 1 that $g^2 - 4dk$ be different from $\pm 4dh\surd(d/b)$ when $g = 0$. If $bk^2 = dh^2$ then θ or ϑ (and so σ or s) is at infinity, so that the cubic is parabolic.

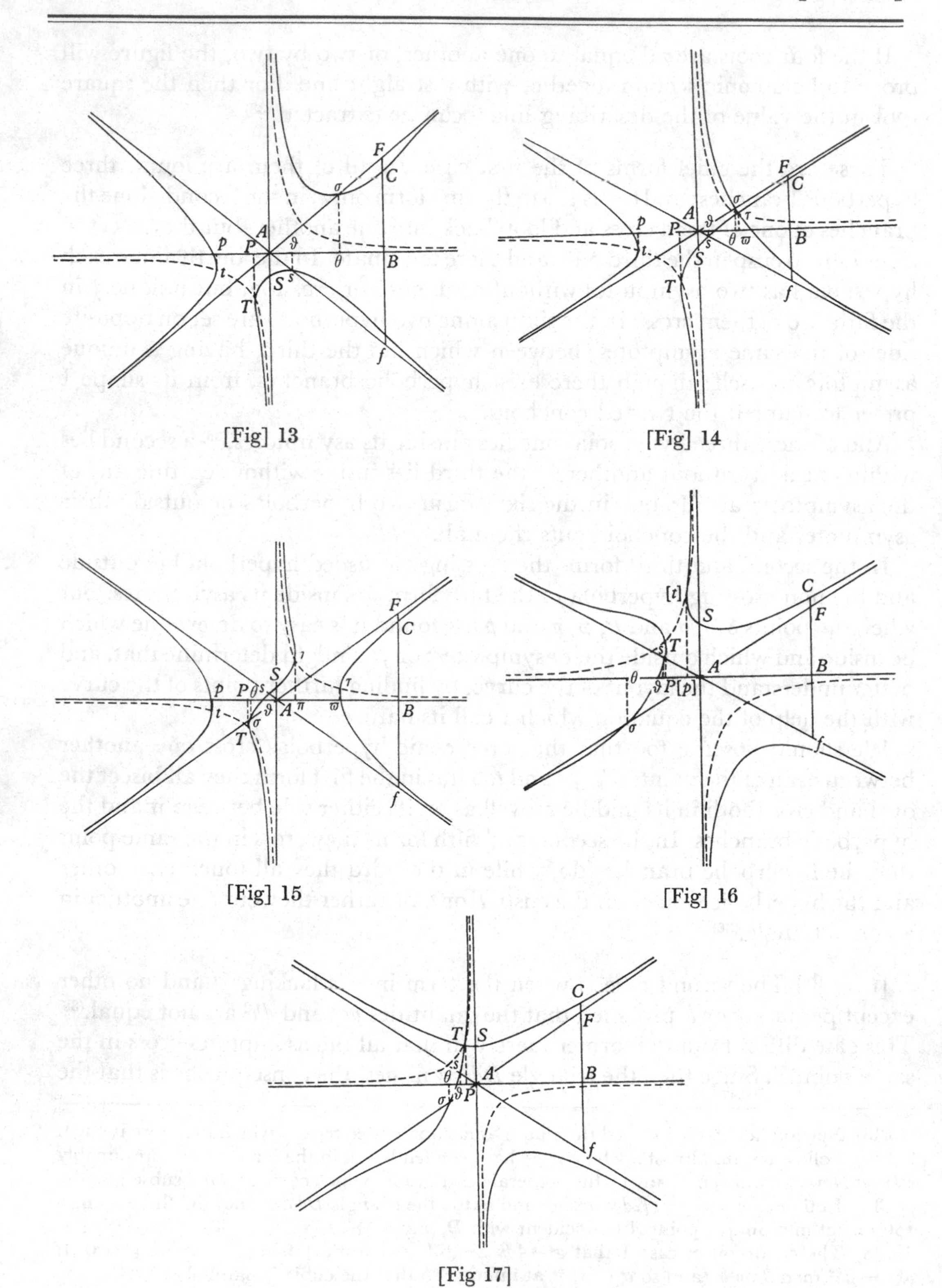

[Fig] 13

[Fig] 14

[Fig] 15

[Fig] 16

[Fig 17]

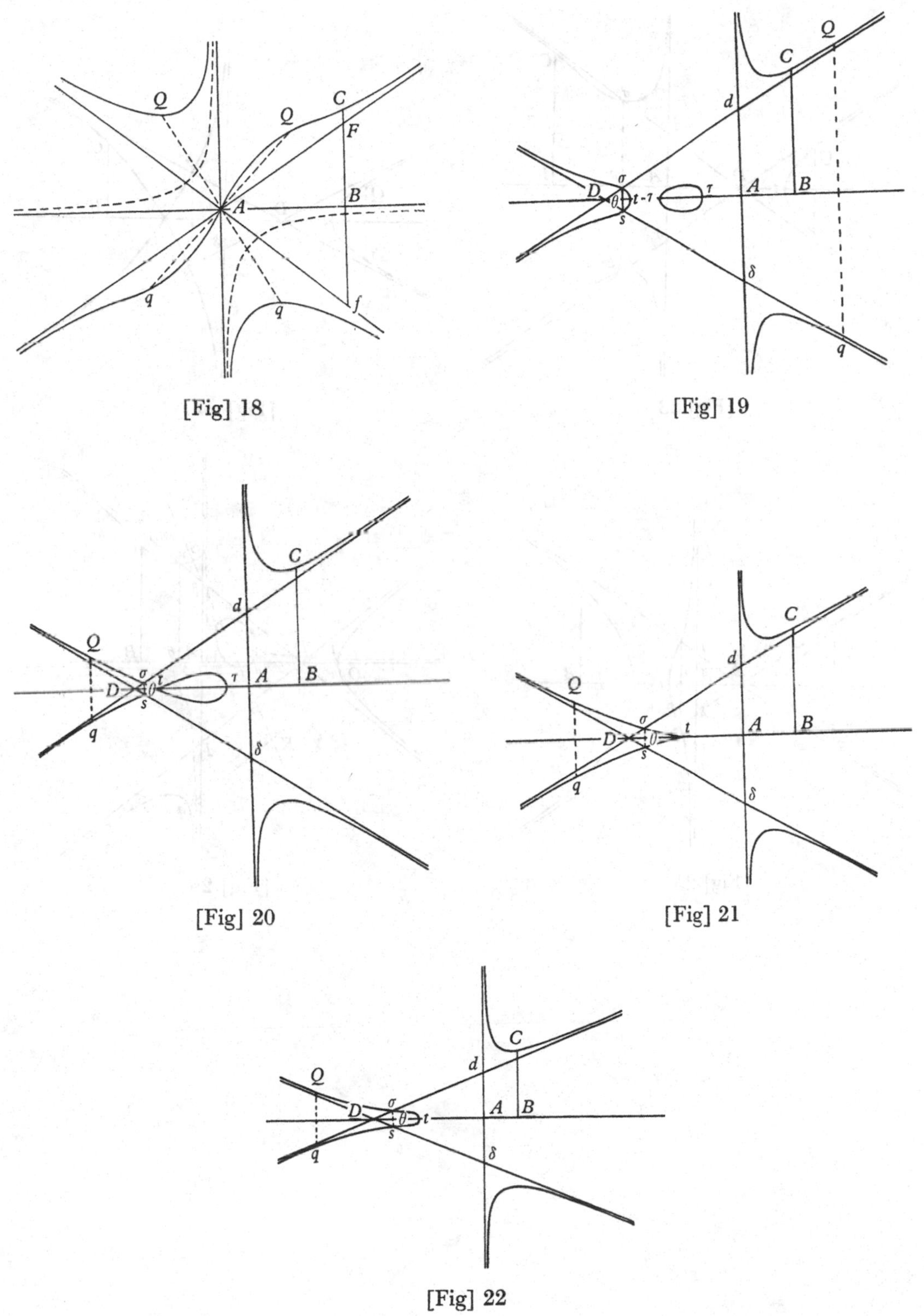

[Fig] 18

[Fig] 19

[Fig] 20

[Fig] 21

[Fig] 22

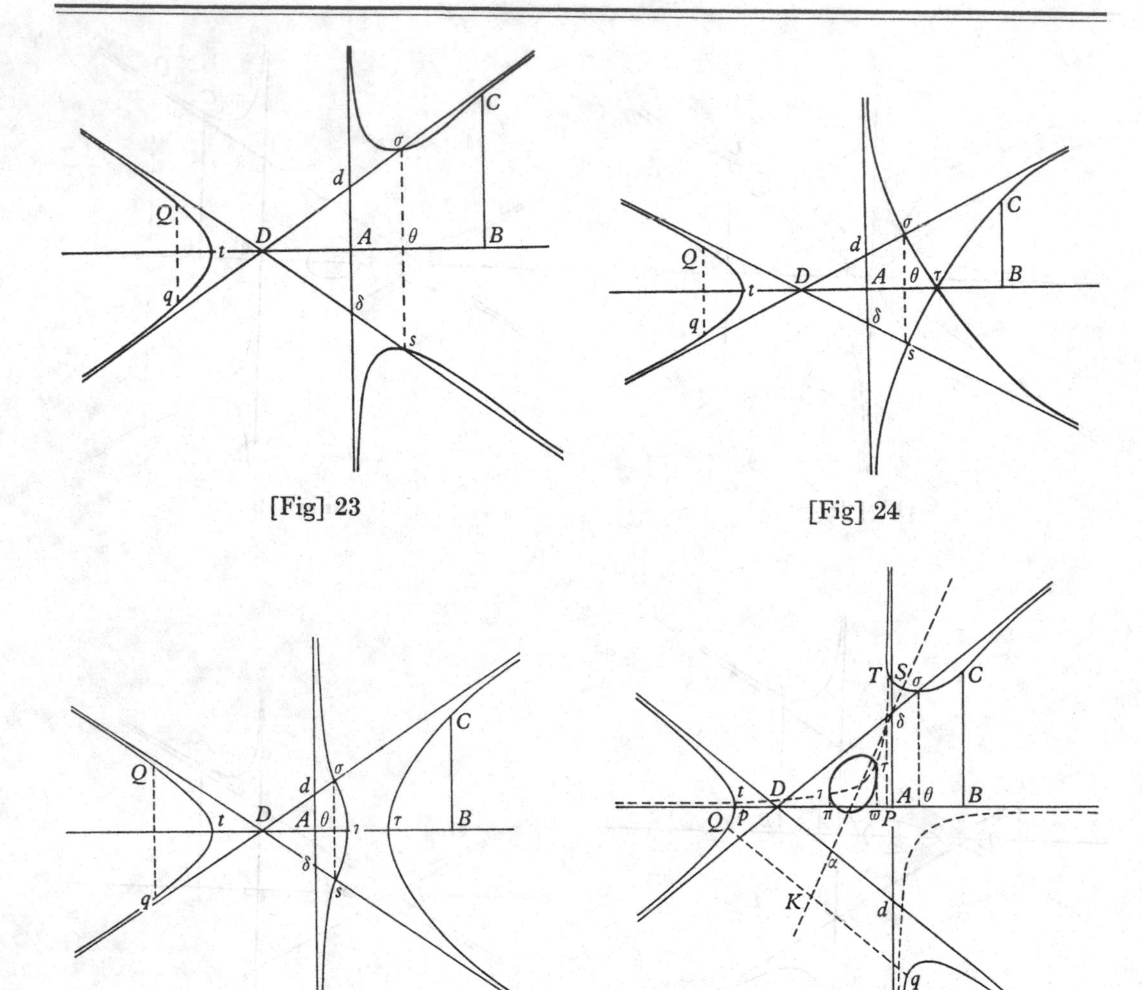

[Fig] 23

[Fig] 24

[Fig] 25

[Fig] 26

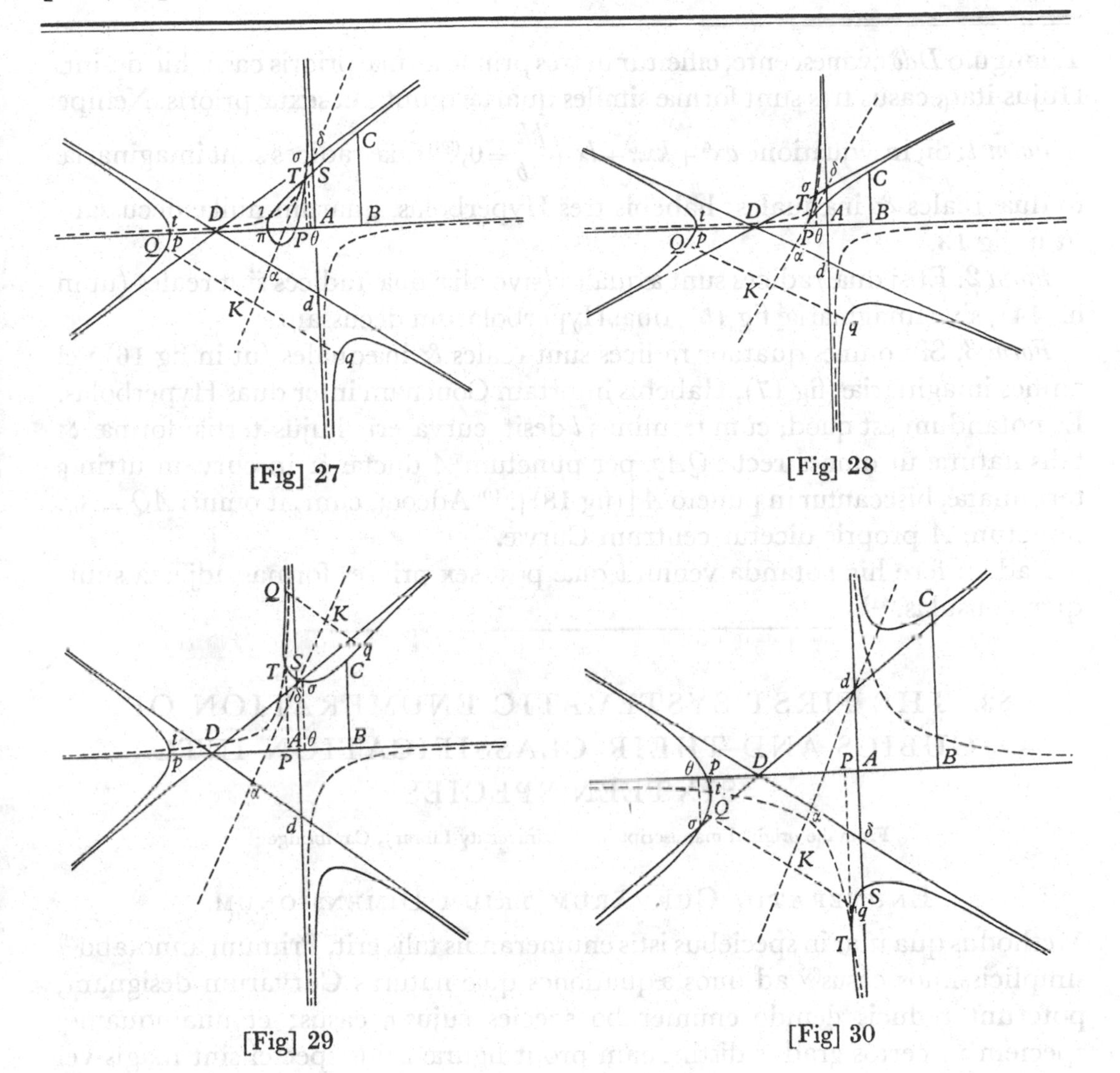

[Fig] 27

[Fig] 28

[Fig] 29

[Fig] 30

Triangulo $Dd\delta$ evanescente, efficitur ut tres primæ formæ prioris casus hic desint. Hujus itaq; casus tres sunt formæ similes quartæ quintæ et sextæ prioris. Nempe

Form 1. Si, in æquatione $dx^4+kxx+lx+\frac{hh}{4b}=0$,[39] duæ radices sunt imaginariæ & duæ reales & inæquales; habebis tres Hyperbolas, quarum nullæ decussant ut in fig 13.

Form 2. Et si duæ radices sunt æquales (sive aliæ duæ radices sint reales (ut in fig 14), sive imaginariæ fig 15), duæ Hyperbolarum decussant.

Form 3. Sin omnes quatuor radices sunt reales & inæquales (ut in fig 16) vel omnes imaginariæ (fig 17), Habebis intortam Concham inter duas Hyperbolas. Et notandum est quòd, cùm terminus l desit, curva erit hujus tertiæ formæ & talis naturæ ut omnes rectæ QAq, per punctum A ductæ & in curvam utrinq; terminatæ, bisecantur in puncto A [(fig 18)].[40] Adeoq;, cùm sit omnis $AQ=Aq$, punctum A propriè dicetur centrum Curvæ.

Eadem fere hic notanda veniunt quæ post sex priores formas adjecta sunt; quæ consulas.[41]

---

§3. THE FIRST SYSTEMATIC ENUMERATION OF CUBICS AND THEIR CLASSIFICATION INTO SIXTEEN SPECIES

From the original manuscript in the University Library, Cambridge[1]

ENUMERATIO CURVARUM TRIUM DIMENSIONUM.

Methodus qua utar in speciebus istis enumerandis talis erit. Primum annotabo[2] simplicissimos casus[3] ad quos æquationes quæ naturas Curvarum designant, poterunt reduci: deinde enumerabo species cujusq; casus; et unamquamq; speciem in certos gradus distinguam prout figuræ istius speciei sint magis vel minùs simplices; et varias etiam describendi formas quas ista species induit: Deniq; ostendam quo pacto quævis æquatio naturam curvæ designans reducatur ad aliquem istorum modorum[4] quorum ope distinxi curvarum species; eodem

---

(39) As in case 1 with g now zero (see note (18)).

(40) For then $dx^4+kx^2+h^2/4b = 0$ or $x^2 = (-k\pm\sqrt{[k^2-dh^2/b]})/2d$, and all four roots will be imaginary when $bk^2 < dh^2$ and not otherwise. Further, any line $y = mx$ through the origin will meet the cubic $bxy^2 = dx^3+hy+kx$ there and in the points $(\pm\mu, \pm m\mu)$ symmetrical with regard to the origin, $\mu^2 = (hm+k)/(bm^2-d)$.

(41) Newton here abandons the present draft though his figures delineate twelve more cubics which he clearly intended to describe. His present results are revised and extended in the tract which follows.

(1) Add. 3961.1: 6^r–9^r, 14^r–16^r and (figures only) 22^r–30^r. This is Newton's first attempt at an exhaustive description of cubics by an analysis of the four standard reduced forms. The first, general canonical form is divided into six 'cases', the first of which is subclassified into

first three forms of the previous case are here missing. Thus, this case has three forms similar to the fourth, fifth and sixth of the preceding. Specifically

Form 1. If in the equation $dx^4+kx^2+lx+h^2/4b=0$(39) two roots are imaginary and two real and unequal, you will have three hyperbolic branches, none of which cross, as in figure 13.

Form 2. And if two roots are equal (whether the other two be real, as in figure 14, or imaginary, figure 15), two of the hyperbolic branches cross.

Form 3. But if all four roots are real and unequal (as in figure 16) or all imaginary (figure 17), you will have a twisted conchoid between two hyperbolic branches. Note that when the term in l is missing, the curve will be of the third form and its nature such that all straight lines QAq drawn through the point A and terminated at the curve on either side will be bisected in the point A [(figure 18)].(40) Hence, since always AQ is equal to Aq, the point A may properly be called the centre of the curve.

Almost the same points to note suggest themselves here as have been added after the six former forms: you should consult them.(41)

Translation

ENUMERATION OF THE CURVES OF THREE DIMENSIONS

The method which I shall use in enumerating those species will be this. First, I will lay down(2) the simplest cases(3) to which the equations designating the natures of the curves may be reduced: then I shall enumerate the species of each case, distinguishing each species into definite grades according as the figures of that species are more or less simple, and also describing the various forms which that species takes on: finally I shall show in what way any equation designating the nature of a curve may be reduced to some one of those modes(4) by whose

---

three component 'species' and the rest into two each, while the remaining three standard forms are each further 'cases' containing a unique species. Each species, finally, is further subclassified into 'forms', each of which may take on one or more of the three 'grades' according as the three real diameters are or are not copunctal or as the cubic has a centre. Altogether species 1 has 6 forms and 3 possible grades, 2 has 7 and 2, 3 has 1 and 2, 4 has 4 and 2, 5 has 5 and 1, 6 has 6 and 1, 7 has 4 and 1, 8 has 2 and 2, 9 has 1 and 1, 10 has 1 and 2, 11 has 1 and 1, 12 has 1 and 1, 13 has 1 and 1, 14 has 4 and 1, 15 has 1 and 1, while species 16 has 1 form and 1 possible grade: in total Newton finds $10+11+2+5+5+6+4+3+1+2+1+1+1+4+1+1$, that is, 58 distinct types of cubic.

(2) A first cancelled continuation is 'quot præcipuè modis æquatio, quæ naturam Curvæ designat, poterit exprimi' (in how many special ways the equation designating the nature of the curve might possibly be expressed).

(3) 'modos' (modes) is cancelled here, and several times below. Evidently Newton chose to alter his nomenclature after he began the present tract.

(4) For consistency read 'casuum' (cases): see note (3) above.

opere docens quomodo cognoscatur cujusnam speciei, gradus & formæ sit curva quævis proposita, & demonstrans me nullam speciem omisisse.(5)

De 9(6) Æquationum casibus

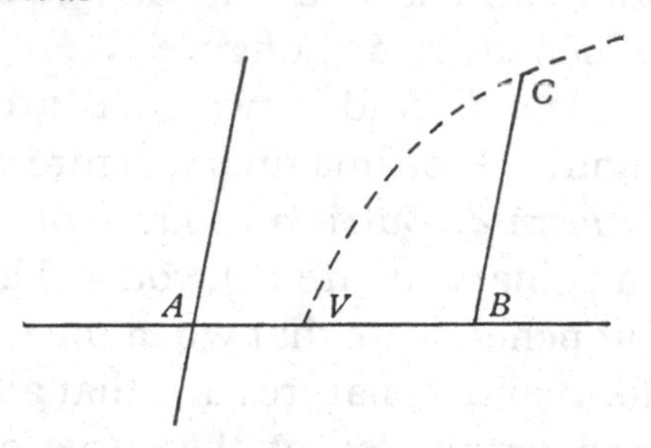

Si linea BC concipiatur moveri supra lineam immobilem AB in dato angulo ABC dum punctum aliquod C in ea movens describat curvam VC: assumens punctum aliquod A in linea AB fixum, voco AB Basin sive x; & BC Basi Applicatam sive y; et ABC angulum Applicationis. Et assumens ad arbitrium quaslibet quantitates pro $a, b, c, d, e, f, g, h, k, l$; vel potiùs fingens assumendas esse, suppono Æquationem $ay^3.\ bxyy.\ cxxy.\ dx^3.\ eyy.\ fxy.\ gxx.\ hy.\ kx.\ l=0$(7) designare naturam Curvæ. Ista enim (ponendo quantitates a, b, c, d &c esse pro lubitu affirmativas vel negativas vel nihilo æquales) omnes hujus generis species in se comprehendit & designat.

Sed ista Æquatio semper poterit reduci ad aliquem horum novem casuum,(8)

1. $bxyy=dx^3+gxx+hy+kx+l.$
2. $bxyy=-dx^3+gxx+hy+kx+l.$
3. $bxyy=gxx+hy+kx+l.$
4. $bxyy=hy+kx+l.$
5. $bxyy=hy-kx+l.$
6. $bxyy=hy+l.$
7. $eyy=dx^3+kx+l.$
8. $fxy=dx^3+l.$
9. $hy=dx^3.$

ut postmodum ostendetur.(9) Ubi notandum est quòd in primo casu Termini b & d debent esse affirmativi, in secundo casu terminus b debet esse affirmativus & d negativus, in quarto $+b$ & $+k$, in quinto $+b$ & $-k$;(10) sed reliqui termini in

(5) This secondary aim is not implemented in the present manuscript. As we shall see (note (13) below) the reduction of the general cubic to its four canonical forms by axis-transformation on the lines of §1 is not difficult. Presumably Newton here intended to classify the species, grade and form of any given cubic by reducing it to standard form and then operating on the reduced equation as in the present scheme. (A quarter of a century later Newton outlined a procedure for discriminating the species—but not the forms—of cubics without reducing to standard form, viz. by considering, as Euler was later to do, the infinite portions of the curve: these manuscripts on discrimination of 'genera' will be reproduced in our seventh volume.) The final phrase is a little suspect. In his published *Enumeratio* of 1704, which distinguishes pure cubics from those with a real conjugate point, Newton found only 72 of the possible 78 varieties separable under his classification.

(6) Originally 'sex' (six). The fourth, fifth and sixth cases below are, in fact, a late addition to a first division into cases 1–3 and 7–9 only.

help I previously distinguished the species of the curve, explaining on the same occasion how the species, grade and form of any proposed curve may be established and proving that I have not omitted any species.(5)

The 9(6) cases of equations

If the line BC be conceived to move on the immovable line AB at a given angle $A\hat{B}C$ while some point C moving in it describes the curve VC, assuming some point A fixed in the line AB (or x) the base, BC (or y) the ordinate to the base and $A\hat{B}C$ the angle of application. Assuming, further, any quantities at will for $a, b, c, d, e, f, g, h, k$ and l, or rather imagining them to be assumed, I suppose that the equation

$$ay^3+bxy^2+cx^2y+dx^3+ey^2+fxy+gx^2+hy+kx+l=0$$

designates the nature of the curve. For, on setting the quantities $a, b, c, d, \ldots$ positive, negative or zero as you please, that comprehends and designates all species of this genus.

But that equation may always be reduced to some one of these nine cases:

1. $bxy^2 = dx^3+gx^2+hy+kx+l.$
2. $bxy^2 = -dx^3+gx^2+hy+kx+l.$
3. $bxy^2 = gx^2+hy+kx+l.$
4. $bxy^2 = hy+kx+l.$
5. $bxy^2 = hy-kx+l.$
6. $bxy^2 = hy+l.$
7. $ey^2 = dx^3+kx+l.$
8. $fxy = dx^3+l.$
9. $hy = dx^3.$

as will afterwards be shown.(9) Here we should note that in the first case the terms in b and d must be positive, and in the second case the term b positive and d negative, while in the fourth we should have $+b$ and $+k$, in the fifth $+b$ and $-k$;(10) but that the remaining terms in any of the cases may be positive,

(7) That is, in modern notation,

$$ay^3 \pm bxy^2 \pm cx^2y \pm dx^3 \pm ey^2 \pm fxy \pm gx^2 \pm hy \pm kx \pm l = 0.$$

The nomenclature is Descartes' (*Geometrie*: Book 3, passim).

(8) Newton first wrote 'sex modorum' (six modes): compare notes (3) and (6).

(9) See note (5).

(10) The last two phrases are a hasty addition inserted with the extra three cases 4–6 (see note (6)) but, as it happens, are logically more acceptable than the more studied earlier pair. Newton intended to assert that in the reduced form $bxy^2 = dx^3+gx^2+hy+kx+l$ (with d and g zero in the latter two cases) we must take b and d or k both positive in cases 1 and 4, but b positive and d or k negative in cases 2 and 5.

quovis casu possunt esse vel affirmativi vel negativi vel etiam nihilo æquales (modò curva non inde evadat Conica Sectio).[11] Ego tamen ijs omnibus signa affirmativa præfixi ne novi characteres pro ambiguis eorum signis adhibiti sequentem calculum perturbarent. Observabis itaq quòd quoties contingit terminum aliquem esse negativum (excepto d in secundo & k in quinto casu), ejus signum in sequenti calculo ubiq erit mutandum.

Ratio hujus numeri Casuum talis est. Natura Curvarum præcipuè dependet ab ijs terminis in quibus x & y sunt plurimarum dimensionum: Termini autem ay^3 & $cxxy$ semper possunt tolli (ut postea docetur); quibus sublatis, soli manent $bxyy$ & dx^3 in quibus incognitæ quantitates sunt plurimarum dimensionum.[12] Quorum signa vel sunt eadem ut in primo casu, vel diversa ut in secundo, vel contingit terminum d deesse ut in tertio, quarto, quinto & sexto, vel contingit terminum b deesse ut in hâc Æquatione dx^3. eyy. fxy. gxx. hy. kx. $l=0$. Quæ, si terminus e non desit potest reduci ad septimum casum; seu e desit & f non desit, potest reduci ad octavum; & ad nonum cùm uterq e & f desunt.[13]

De Primo Casu.

In primo Casu $bxyy=dx^3+gxx+hy+kx+l$[14] Curva erit Hyperbola trium Asymptotorum quarum nullæ sunt parallelæ, & per quarum singulas duo curvæ crura ad oppositas partes in infinitum serpunt. Nempe Applicata δAd est una Asymptotorum; &, si sumatur $AD=\frac{-g}{2d}$, & $Ad=A\delta=\frac{g}{2\sqrt{bd}}$, ductæ Dd, & $D\delta$ erunt aliæ duæ. Vide figuras 1, 2, 3, 4, &c.[15] Vel si contingit terminum g deesse; ductâ aliquâ Applicatâ BC, facio ut sit $b:\sqrt{bd}::AB:BF=Bf$; et per puncta F & f[16] ductæ AF, & Af, erunt aliæ duæ Asymptoti. Vide fig. [2].

(11) Compare note (33) of the preceding section.

(12) Newton omits to note that in cases 1–6 he has removed the terms ey^2 and fxy also.

(13) The details of this reduction are not indicated in the present manuscript (note (5) above) but the procedure of reducing by a general transformation of axes outlined in §1 may easily be extended to cover the present generalization. In that section Newton showed that the general cubic may be reduced to the standard form $bxy^2 = dx^3+gx^2+hy+kx+l$: cases 1–3 follow on taking d positive, negative or zero, and cases 4–6 are the degenerate ones where d and g are both zero and k is taken positive, negative or zero. In the terminology of §1, note (19) the coefficient a_2 is $p(L_1+qL_2)+L_2+qL_3$ and the conditions $a_1 = a_2 = a_3 = 0$ yield $pL_1 = qL_3$ with $p^2L_1L_2+2pL_1L_3+L_2L_3 = 0$. If we use Newton's terminology the cubic in this case has the form $dx^3+ey^2+fxy+gx^2+hy+kx+l = 0$: this we may transform by using the Cartesian axis-change (cf. Descartes, *Geometrie*, 1637: Book 2, 327 ff. = *Geometria*, ${}_2$1659: 27 ff.) $x, y \Rightarrow x', y'$ defined by $x = X+r$, $y = pX+Y+s$ where we, with Newton, use the oblique projections $X = mx'$, $Y = ny'$ of x', y' in the directions of x and y for simplicity. The transformed cubic is then $dX^3+eY^2+FXY+GX^2+HY+KX+L = 0$, in which $F = 2ep+f$,

$$G = 3dr+ep^2+fp+g, \quad H = 2es+fr+h, \quad K = 3dr^2+2eps+fs+fpr+2qr+hp+k$$

negative or zero (provided that the curve does not there come out to be a conic).(11) I, however, have prefixed positive signs to all those lest fresh symbols denoting their ambiguity of sign confuse the following computation. You will observe, therefore, that whenever some term happens to be negative (except for d in the second case and k in the fourth) you must change its sign correspondingly everywhere in the following computation.

The reason for this number of cases is this. The nature of curves depends especially on those terms in its equation in which x and y are of most dimensions. Here the terms ay^3 and cx^2y can always be removed (as we shall subsequently show), and if you do so there remain only bxy^2 and dx^3 in which the unknowns are of most dimensions.(12) Their signs will either be the same (as in case 1) or opposite (as in case 2), or the term in d may chance to be missing (as in cases 3, 4, 5 and 6), or the term in b may be lacking, as in this equation

$$dx^3 \pm ey^2 \pm fxy \pm gx^2 \pm hy \pm kx \pm l = 0.$$

If the term in e is present this may be reduced to case 7; if e is lacking but not f it may be reduced to case 8, and when both e and f are lacking to case 9.(13)

The First Case

In the first case $bxy^2 = dx^3 + gx^2 + hy + kx + l$ the curve will be hyperbolic with three asymptotes, none of which are parallel and along every two of which the branches of the curve snake out to infinity in opposite directions. Specifically, the ordinate δAd is one of the asymptotes, and if we take $AD = -g/2d$, $Ad = A\delta = g/2\sqrt{bd}$ and draw Dd and $D\delta$ these will be the other two. See figures 1, 2, 3, 4(15) and so on. Or if the term in g happens to be missing, drawing some ordinate BC I make $b:\sqrt{bd} = AB:BF$ or Bf and on drawing AF and Af through the points F and f(16) they will be the other two asymptotes. See figure [2].

---

and $L = dr^3 + es^2 + frs + gr^2 + hs + kr + l$. The divergent parabolas of case 7 follow when $e \neq 0$ on setting $p = -f/2e$, $r = (f^2 - 4eg)/12de$ and $s = (4efg - 12deh - f)/24de^2$. When $e = 0$, the cubic is the Cartesian trident and the reduced form of case 8 is obtained by setting $r = -h/f$, $p = (3dh - fg)/f^2$ and $s = (2fgh - f^2k - 3dh^2)/f^3$. When, finally, $e = f = 0$ (with p free) we obtain the Wallis cubic $0 = dX^3 + HY$ on making $r = -g/3d$, $p = (g^2 - 3dk)/3dh$ and $s = (-2g^3 + 9dgk - 27d^2l)/27d^2h$.

(14) We omit a superfluous '= 0.' in the manuscript.

(15) Of species 1, that is. Figure 2 is presumably here included in error.

(16) In Newton's figure 2, in fact, the points 'F' and 'f' are entered as 'g' and 'γ'. The phrase has been taken over unchanged from §2: compare note (9) and figures 13–18 of that section.

De Speciebus in hoc Casu

Hic casus tres præcipuè Curvarum species exhibet. Primæ speciei curva nullam habet Diametrum (nisi Hyperbola Conica pro Diametro habeatur);[17] curva secundæ habet unicam; & tertiæ tres. In primâ specie duo Curvæ crura circa eandem quamvis Asymptoton, jacent ex diverso ejus latere; In secundâ, duo crura jacent ex eodem latere & reliqua quatuor ex diverso; In tertiâ, omnia crura jacent ex eodem latere suarum Asymptotorum, In prima specie Curva secat unamquamqꝫ Asymptoton singulas[18] in unico puncto; in secundâ, secat tantū duas; In tertiâ, nullam. Vide Fig[uras.]

Si neqꝫ terminus h desit neqꝫ sit $gg-4dk=\pm dh\sqrt{\frac{d}{b}}$,[19] Curva erit primæ speciei; sin eorum alterum accidit, erit secundæ; Et tertiæ, si utrumqꝫ.

De Prima Specie

Hujus speciei Curva eundem ferè respectum ad omnes Asymptotos servat.[20] Ducamus enim $C\gamma XBcg$, $\kappa l\beta\xi GC$ and $LCbxk\lambda$ parallelas Asymptotis GL, $\gamma\lambda$, & gl, & secantes Asymptotos in g, γ, G, l, λ, L & Curvam in C, k, κ, c: & bisecentur $g\gamma$, Gl, λL; cC, $C\kappa$, Ck in punctis B, β, b; X, ξ, x. Dico

Primò, si ducantur DAB, $\delta\alpha\beta$, dab (vel AB, $A\beta$, Ab cùm puncta D, d & δ coincidunt in A, fig 2) quòd Æquatio quæ designat relationem inter Basin $\alpha\beta$ & ei Applicatam βC, vel inter Basin ab & ei Applicatam bC, habebit formam primi casus æquè ac Æquatio quæ designat relationem inter Basin AB & ei applicatā BC. Adeoqꝫ parum interest ad quam Basin & applicatam AB & BC vel $\alpha\beta$ & βC vel ab & bC curva referatur.

(17) This diametral hyperbola is, of course, curved and non-degenerate in species 1: in species 2 it reduces to the line-pair of the linear diameter AD and the prime asymptote $d\delta$.

(18) Read 'singulam'.

(19) Newton intends '$bgg-4dk=\pm 4dh\sqrt{\frac{d}{b}}$', the condition that the diametral hyperbola $0=-by^2\mp 2\surd(bd)xy+3dx^2+2gx\pm h\surd(d/b)+k$ degenerate into the line-pair

$$0=(y\pm 3\surd(d/b)x\mp \tfrac{1}{2}g\surd(b/d))\,(y\mp \surd(d/b)x\pm \tfrac{1}{2}g\surd(b/d)):$$

compare note (40) below. Likewise $h=0$ reduces the hyperbola to the pair of lines AD $(y=0)$ and $d\delta$ $(x=0)$. It will be evident (compare §2, note (9) and correspondingly below) that these conditions determine that the meets S, s and σ respectively pass into infinity, or that a diametral cubic is met by the corresponding asymptote in three points at infinity, that is, in one of its points of inflexion.

(20) A first, cancelled continuation here reads: 'Sic quia Basis DA bisecat $d\delta$ in A, si biseco Dd in α & $D\delta$ in a & duco $\delta\alpha\beta$, & dab & applico $\beta\kappa$ Parallelam Asymptoto $\lambda\gamma$, & bc parallelam asymptoto lg: Equatio quæ designat relationem inter Basin $\alpha\beta$ & ei Applicatam

The Species in this Case

This case reveals three species of curve in particular. The first species' curve has no diameter (unless we assume the conic hyperbola to be a diameter);[17] that of the second has one only; that of the third three. In the first species two branches of the curve lie around the same asymptote but on opposite sides, whichever it is; in the second, two branches lie on the same side and the remaining four on opposite ones; in the third, all the branches lie on the same side of their asymptotes. In the first species the curve cuts each asymptote once only in a unique point; in the second, it cuts but two of them; in the third, none at all. See the figures.

If neither the term in h is lacking nor is g^2-4dk equal to $\pm dh\surd(d/b)$,[19] the curve will be of the first species; if only the one of those circumstances occurs, it will be of the second; but will be of the third if both are true.

The First Species

The curve of this species keeps almost the same relationship with each of its asymptotes.[20] For let us draw $C\gamma XBcg$, $\kappa l\beta\xi GC$ and $LCbxk\lambda$ parallel to the asymptotes GL, $\gamma\lambda$ and gl, meeting those asymptotes in g, γ, G, l, λ, L and the curve in C, k, κ, c, and let $g\gamma$, Gl, λL and cC, $C\kappa$, Ck be bisected in the respective points B, β, b and X, ξ, x. I say that:

First, if DAB, $\delta\alpha\beta$ and dab are drawn (or AB, $A\beta$ and Ab when the points D, d and δ are coincident in A, figure 2), the equation designating the relationship between the base $\alpha\beta$ and its ordinate βC, or between the base ab and its ordinate bC, will take on the form of case 1 in line with the equation designating the relationship between the base AB and its ordinate BC. It matters little therefore to which base and ordinate, AB and BC, $\alpha\beta$ and βC, or ab and bC, the curve is referred.

$\beta\kappa$, vel inter Basin ab & ei Applicatam bc, habebit formam primi casus æque ac æquatio quæ designat relationem in[ter] Basin AB et ei Applicatā BC. Adeoq; parùm interest an referas curvam ad Basin AB, $\alpha\beta$, vel ab' (Thus since the base DA bisects $d\delta$ in A, if on bisecting Dd in α and $D\delta$ in a I draw $\delta\alpha\beta$ and dab and apply $\beta\kappa$ parallel to the asymptote $\lambda\gamma$, bc parallel to the asymptote lg, the equation designating the relation between the base $\alpha\beta$ and its ordinate $\beta\kappa$ or between the base ab and its ordinate bc will have the form of the first case in line with the equation designating the relation between the base AB and its ordinate BC). Compare §2, note (10). An added note clarifies what Newton will call the second grade of the species: 'Nota q^d, Si fortè puncta D, δ, & d in A coincidunt duco GI & λL parallelas Asymptotis $\gamma\lambda$ and gl et secans Asymptotos GL, gl, $\gamma\lambda$ in punctis G, l; λ & L et eàs biseco in β & b & duco Bases $A\beta$, Ab &c. vide fig [2]' (Note that if the points, D, δ and d chance to coincide I draw GI and λL parallel to the asymptotes $\gamma\lambda$ and gl meeting the asymptotes GL, gl, $\gamma\lambda$ in the points G, l, λ and L, and bisecting them in β and b I draw the bases $A\beta$, Ab.... See figure [2]).

Dico secundo quod punctum X semper cadet in Hyperbola Conica ΦX cujus Asymptoti sunt AG & AB; & punctum ξ in Hyperbola Conica $\xi h\Phi\chi$ cujus Asymptoti sunt $\alpha\gamma$ & $\alpha\beta$, & punctum x in Hyperbola Conica $KxI\phi$ cujus Asymptoti sunt ag, ab. Nempe posita $AB=x$; & BX, vel $B\chi$ vel $BK=y$: Natura Hyperbolæ Conicæ ΦX erit $2bxy=h$: Hyperbolæ Conicæ $\xi h\Phi\chi$ erit

$$byy=-2xy\sqrt{bd}+3dxx+2gx+k+\frac{h}{b}\sqrt{bd}:$$

Et Hyperbolæ Conicæ $KxI\phi$ erit $byy=2xy\sqrt{bd}+3dxx+2gx+k-\frac{h}{b}\sqrt{bd}$. Unde facile describuntur.

Et hinc duo lucrantur; Primò quoniam Hyperbola Conica in infinitum serpit inter unumquodqꝫ crus hujus Curvæ & suam Asymptoton, adéoqꝫ crura utriúsqꝫ Hyperbolæ jacent ex eodem latere Asymptoti: Inde si cognoscis ex quo latere (sive in quo angulo) Asymptotorum, Crura Hyperbolarum Conicarum jacent; cognosces etiam ex quo latere suarum Asymptotorum Crura hujus Curvæ jacent. Facilè autem cognosces in quibus Asymptotorum angulis Hyperbolæ Conicæ jacent quærendo unicum cujusqꝫ punctum: Nempe, posito $Ah=\frac{\sqrt{bk+h\sqrt{bd}}}{b}$ & $AH=\frac{\sqrt{bk-h\sqrt{bd}}}{b}$; vel (cum istæ radices sunt imaginariæ) $Ai=\frac{-g\pm\sqrt{gg-3dk-\frac{3h}{b}\sqrt{bd}}}{3d}$ & $AI=\frac{-g\pm\sqrt{gg-3dk+\frac{3h}{b}\sqrt{bd}}}{3d}$: punctum h vel i cadet in Hyperbola Conica $\xi h\Phi\chi$ & punctum H vel I in Hyperbola $KxI\phi$. &c

Secundo. Descriptis istis Hyperbolis conicis cum suis Conjugatis per puncta h vel i, & H vel I, earum ope facilè invenias plurima hujus Curvæ puncta ad eam describendam. Nempe si datur aliquod ejus punctum C duco $CX\parallel GL$, $C\xi\parallel\gamma\lambda$, & $C\kappa\parallel gl$ & produco eas ad ϵ, κ, & e;(21) puncta ϵ, κ & e(21) erunt in Curva. Et eadem ratione puncta ϵ, κ, & e(21) dant alia. Sic si quæris puncta S σ & s in quibus curva secat suas Asymptotos GL, $\lambda\gamma$ & lg (ponendo nempe $AS=\frac{-l}{h}$; & $A\theta=\frac{4dl+2gh\sqrt{\frac{d}{b}}}{gg-4dk-4dh\sqrt{\frac{d}{b}}}$, & $A\vartheta=\frac{4bdl-2gh\sqrt{bd}}{bgg-4bdk+4dh\sqrt{bd}}$; & ducendo $\theta\sigma$ & ϑs parallelas Asymptoto GL): ista puncta S, σ, & s dant alia. Nota quod ista puncta S, s et σ jaceant in directum.(22)

(21) In his figure Newton has entered the points 'e' and 'ϵ' as 'k' and 'c' without here remembering to alter his text correspondingly.

(22) This is almost wholly a straightforward revision of corresponding passages in the preceding section (§2). The last sentence, added abruptly at a later stage in a rough scrawl,

Secondly, the point X will fall always in the conic hyperbola ΦX whose asymptotes are AG and AB, the point ξ in the conic hyperbola $\xi h\Phi\chi$ whose asymptotes are $\alpha\gamma$ and $\alpha\beta$, and the point x in the conic hyperbola $KxI\phi$ whose asymptotes are ag and ab. Specifically, on setting $AB = x$ and BX, $B\chi$ or $BK = y$, the nature of the conic hyperbola ΦX will be $2bxy = h$, that of $\xi h\Phi\chi$

$$by^2 = -2\surd(bd)\, xy + 3dx^2 + 2gx + k + h\surd(bd)/b,$$

and $by^2 = 2\surd(bd)\, xy + 3dx^2 + 2gx + k - h\surd(bd)/b$ that of the conic hyperbola $KxI\phi$. They are easily described in consequence.

From this two further points may be gained: First, since a conic hyperbola snakes into infinity between each branch of this curve and its corresponding asymptote, and therefore the branches of each hyperbola lie on the same side of the asymptote, if then you know on what side of the asymptotes (or in which of their angles) lie the branches of the conic hyperbolas, you will know also on what side of their asymptotes lie the infinite branches of this curve. You will easily know, indeed, in which asymptote angles the conic hyperbolas lie by seeking a single point of each: precisely, on setting $Ah = (\surd[bk + h\surd bd])/b$ and $AH = (\surd[bk - h\surd bd])/b$, or (when those roots are imaginary)

$$Ai = (-g \pm \surd[g^2 - 3dk - 3h\surd(bd)/b])/3d$$

and
$$AI = (-g \pm \surd[g^2 - 3dk + 3h\surd(bd)/b])/3d,$$

the point h or i will fall in the conic hyperbola $\xi h\Phi\chi$ and the point H or I in the hyperbola $KxI\phi$, and so on.

Secondly, having described those conic hyperbolas with their conjugate branches through the points h or i and H or I, with their help you may easily find further points of this curve in order to describe it. Precisely, if some point C of it is given, I draw $CX \parallel GL$, $C\xi \parallel \gamma\lambda$ and $C\kappa \parallel gl$, producing them to ϵ, κ and e:(21) these points ϵ, κ and e(21) will then be in the curve. And by the same method the points ϵ, κ and e(21) give still others. Thus, if you seek the points S, σ and s in which the curve cuts its asymptotes GL, $\lambda\gamma$ and lg (by setting, that is, $AS = -l/h$, $A\theta = (4dl + 2gh\surd[d/b])/(g^2 - 4dk - 4dh\surd[d/b])$ and

$$A\vartheta = (4bdl - 2gh\surd[bd])/(bg^2 - 4bdk + 4dh\surd[bd]),$$

and drawing $\theta\sigma$ and ϑs parallel to the asymptote GL), those points S, σ and s give others. Note that those points S, s and σ should lie in a straight line.(22)

---

is new. In proof, the condition for S, s and σ to be in line is that, with respect to the Menelaus triangle $Dd\delta$, $dS \times \delta s \times D\sigma = \delta S \times Ds \times d\sigma$, that is equivalently

$$(AS + A\delta) \times A\vartheta \times D\theta = (AS - A\delta) \times D\vartheta \times A\theta:$$

this is evidently satisfied by the computed values $AS = -l/h$, $AD = -g/2d$, $A\delta = -\frac{1}{2}g/\surd(bd)$ and $A\vartheta$, $A\theta = (4dl \pm 2gh\surd(d/b))/(g^2 - 4dk \mp 4dh\surd(d/b))$. The theorem is a particular case (in which the transversal is the line at infinity) of a result formulated by Colin Maclaurin

Et hic de ordine literarum in Schematibus notandum est quod istam Asymptoton quæ lineis Applicatis AB[23] parallela est designo literis $LGhdHA\delta S$: et Hyperbolam Conicam (cum sua conjugata) quæ per illam serpit designo literis $X\Phi\phi FfT\tau\tau t$. Istam[24] Asymptoton quæ secat tum Hyperbolam Conicam ΦX tum suam Conjugatam designo literis $\lambda\gamma d\alpha D\sigma$: et Hyperbolam Conicam (cum sua conjugata) quæ per eam serpit, litteris $\xi h\chi\Phi\phi Ffi$. Deniq; istam tertiam Asymptoton cum sua conjugata[25] quæ vel neutram vel alteram & non utramq; Hyperbolam Conicam $X\Phi$ & ejus Conjugatam secat designo literis $lDa\delta gs$ et Hyperbolam Conicam cum sua Conjugata quæ per eam serpit, literis $xKH\Phi\phi FfI$. Reliquarum literarum ordo ex supradictis facile percipietur.

De Gradibus hujus Speciei.

Præterea figuras hujus speciei in tres præcipue gradus distinguo, Primus & maximè compositus gradus esto quando Asymptoti sese decussant in diversis punctis D d & δ (fig 1, 3, 4, 5, 6, 7 &[c]): Secundus quando omnes in eodem puncto A decussant (fig 19, 20, 21, 22[)] & Tertius quando insuper A est Centrum Curvæ (fig 22) hoc est cum omnes rectæ QAq per punctum A ductæ & in curvam utrinq; terminatæ bisecantur in puncto A. Si terminus g desit & terminus l non desit erit secundi; & tertij cum uterq; g & l desunt.[26]

De Sex ejus Formis.

Jam ut cognoscantur variæ formæ quas hæc species induit, Consideranda est quod istæ formæ variæ sunt prout Curva variè limitatur reflectendo vel decussando. Ad invenienda autem puncta T, τ, τ, t in quibus Curva reflectit vel decussat (hoc est in quibus secat Hyperbolam Conicam $X\Phi$ vel ejus conjugatam), quæro maximos vel minimos valores quantitatis x & prodit Aequatio $dx^4+gx^3+kxx+lx+\frac{hh}{4b}=0$:[27] Cujus radices pono esse AP, $A\varpi$, $A\pi$, & Ap & duco Applicatas PT, $\varpi\tau$, $\pi\tau$, pt, secantes Hyperbolam conicam $X\Phi$ (vel ejus conjugatam) in punctis T, τ, τ, & t. (hoc est, facio ut sit $AP:\frac{h}{2b}::1:PT$. &c.)

Sed cognitis radicibus hujus Aequationis $dx^4+gx^3+kxx+lx+\frac{hh}{4b}=0$[27]

(*A Treatise of Algebra in Three Parts* (London, 1748): *Appendix: De Linearum Geometricarum Proprietatibus generalibus Tractatus:* §57 = Sectio III, 'De Lineis tertii Ordinis', Prop. VI): the tangents at three collinear points of a cubic meet the cubic again in points which are collinear. Maclaurin's theorem, of course, follows from Newton's immediately by projection.

(23) Read 'BC'.

(24) 'Item' (likewise) is cancelled.

Here, too, concerning the order of the letters in my figures you should note that the asymptote which is parallel to the ordinates AB[23] I designate by the letters $LGhdHA\delta S$, and the conic hyperbola (with its conjugate branch) which snakes along it by the letters $X\Phi\phi Ff T\tau 7t$. That[24] asymptote which cuts both the conic hyperbola ΦX and its conjugate branch I designate by the letters $\lambda\gamma d\alpha D\sigma$, and the conic hyperbola (with its conjugate branch) which snakes along it by the letters $\xi h\chi\Phi\phi Ffi$. That third asymptote, finally, which cuts neither the conic hyperbola $X\Phi$ or its conjugate branch, or just one and not both, I designate by the letters $lDa\delta gs$, and the conic hyperbola which together with its conjugate branch snakes along it by the letters $xKH\Phi\phi FfI$.

The order of the remaining letters will easily be perceived from what has been said above.

The Grades of this Species

Moreover I distinguish the figures of this species into three grades in particular. Let the first and most complex grade be when the asymptotes cross in separate points D, d and δ (figures 1, 3, 4, 5, 6, 7, ...), the second when they all cross in the same point A (figures 19, 20, 21, 22) and the third when in addition A is the centre of the curve (figure 22), that is, when every straight line QAq drawn through the point A and terminated on either side at the curve is bisected in the point A. If the term in g is missing but not that in l the cubic will be of second grade; but of the third when both g and l are missing.[26]

Its Six Forms

Now to know the varied forms which this species takes on, we must consider that those forms are varied according as the curve is variously limited by being bent back or crossing itself. To find the points T, τ, 7 and t, however, in which the curve is bent back or crossed (that is, in which it cuts the conic hyperbola $X\Phi$ or its conjugate branch) I seek the greatest or least values of the quantity x and I turn up the equation $dx^4+gx^3+kx^2+lx+h^2/4b=0$.[27] The roots of this I suppose to be AP, $A\varpi$, $A\pi$ and Ap and then draw the ordinates PT, $\varpi\tau$, $\pi 7$ and pt meeting the conic hyperbola $X\Phi$ (or its conjugate branch) in the points T, τ, 7 and t: that is, I make $AP:h/2b=1:PT$, and so on.

But when you know the roots of this equation

$$dx^4+gx^3+kx^2+lx+h^2/4b=0^{(27)}$$

(25) This superfluous phrase should presumably be cancelled and we omit it in our translation.

(26) The condition that D, d and δ coincide in A is that $AD=-g/2d$ be zero: for the cubic $bxy^2=dx^3+gx^2+hy+kx+l$ to have a centre, clearly at (0, 0), all even terms (here gx^2 and l) must be zero.

(27) See our remarks on §2, note (18) above.

Figuræ primæ Speciei (*Figures for species* 1).

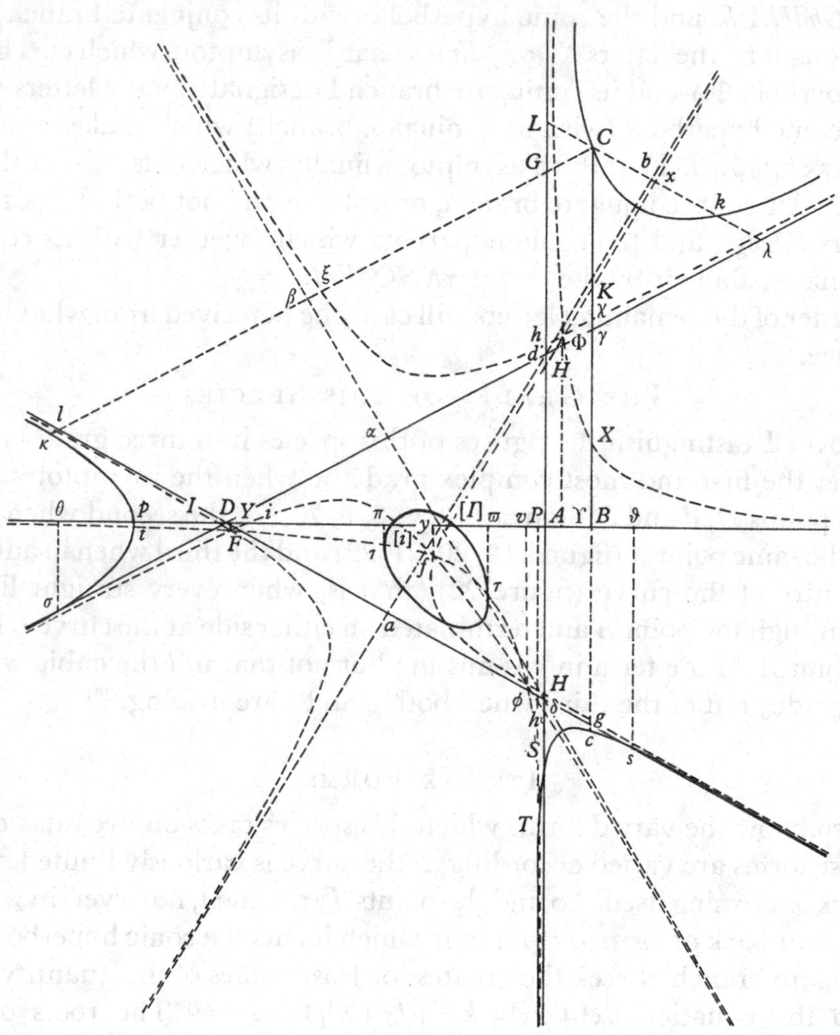

Fig 1

Plate I. The general tridiametral cubic (I, I, §3).

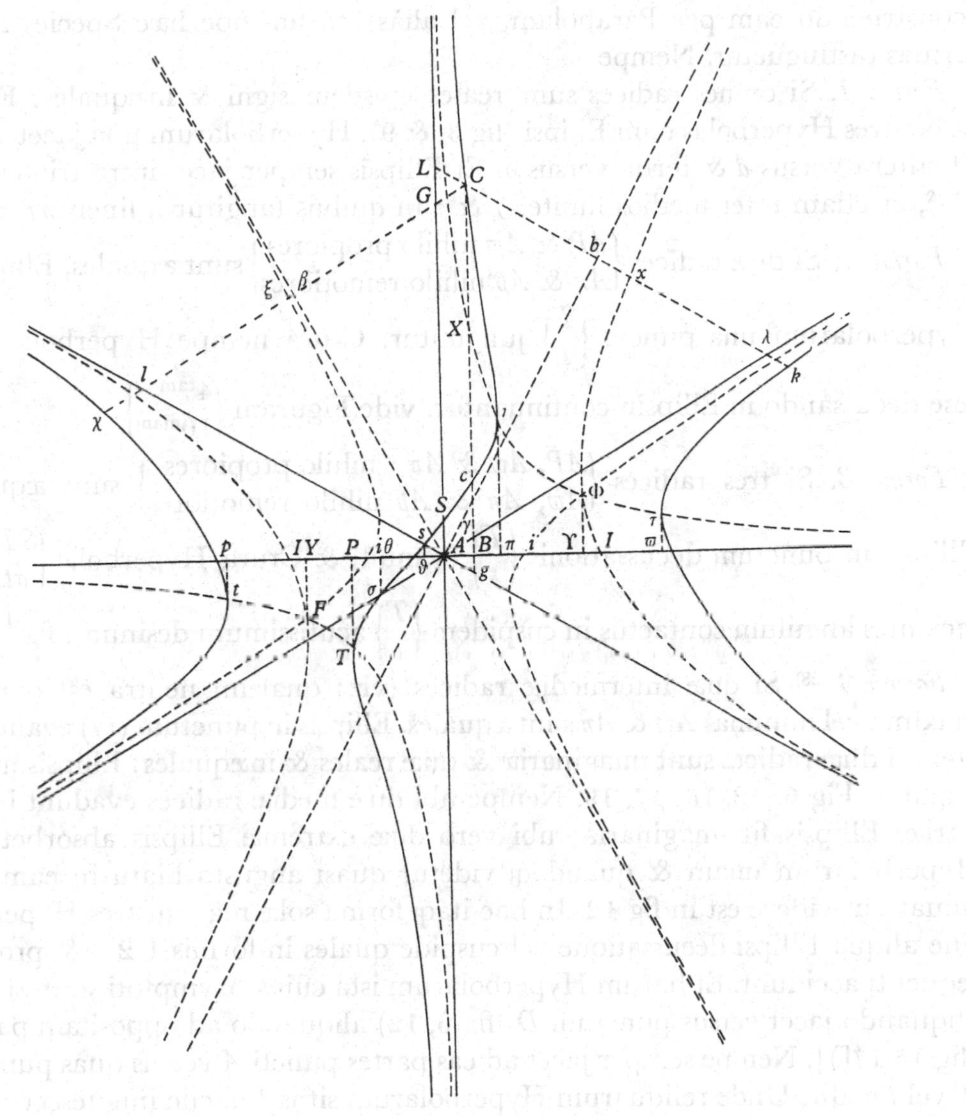

Fig 2

(construendo eam per Parabolam, vel aliàs) earum ope hæc Species in sex formas distinguetur. Nempe

Forma 1. Si omnes radices sunt reales ejusdem signi & inæquales; Figura habet tres Hyperbolas cum Ellipsi (fig 3, & 9). Hyperbolarum una jacet versus D, altera versus d & tertia versus δ. Et Ellipsis semper jacet intra triangulum $Dd\delta$, et etiam inter medios limites 7 & τ in quibus tangitur a lineis $\pi 7$, $\varpi\tau$.

Forma 2. Si duæ radices $\left\{\begin{matrix} AP \text{ \& } A\varpi \text{ nihilo propiores} \\ A\pi \text{ \& } Ap \text{ nihilo remotiores} \end{matrix}\right\}$ sunt æquales, Ellipsis et Hyperbolarum una puncto $\left\{\begin{matrix} T \\ t \end{matrix}\right\}$ junguntur. Crura nempe Hyperbolæ $\left\{\begin{matrix} STs \\ \sigma ts \end{matrix}\right\}$ sese decussando in Ellipsin continuantur. vide Figuram $\left\{\begin{matrix} 4^{\text{tam}} \\ 10^{\text{mam}} \end{matrix}\right\}$.

Forma 3. Si tres radices $\left\{\begin{matrix} AP, A\varpi \text{ \& } A\pi \text{ nihilo propiores} \\ A\varpi, A\pi \text{ \& } Ap \text{ nihilo remotiores} \end{matrix}\right\}$ sunt æquales; Ellipsis in punctum decussationis $\left\{\begin{matrix} T \\ t \end{matrix}\right\}$ evanuit, & Crura Hyperbolæ $\left\{\begin{matrix} STs \\ \sigma ts \end{matrix}\right\}$ ibi facientes angulum contactûs in cuspidem $\left\{\begin{matrix} T \\ t \end{matrix}\right\}$ acutissimum desinunt. fig $\left\{\begin{matrix} 5 \\ 11 \end{matrix}\right\}$.

Forma 4.(28) Si duæ intermediæ radices (viz: quarum neutra est omnium maxima vel minima) $A\varpi$ & $A\pi$ sunt æquales, Ellipsis in punctum $(\tau 7)$ evanuit.(29) Item si duæ radices sunt imaginariæ & duæ reales & inæquales: Ellipsis nullibi apparet. Fig 6, 12, 15, 17, 19. Nempe ubi duæ mediæ radices evadunt imaginariæ, Ellipsis fit imaginaria; ubi vero duæ extremæ Ellipsis absorbetur in Hyperbolarum unam & quandoqȝ videtur quasi angusto hiatu in eam continuata ut videre est in fig 12. In hac itáqȝ forma solæ manent tres Hyperbolæ sine aliqua Ellipsi decussatione vel cuspide quales in formis 1 2 3 & proxima sequenti accidunt. Et harum Hyperbolarum ista cujus Asymptoti sunt $\gamma\lambda$ & gl, aliquando jacet versus punctum D (fig 6, 12) aliquando ad oppositam partem (fig 15 17[)]. Nempe semper jacet ad eas partes puncti A versus quas punctum T vel t cadit. Unde reliquarum Hyperbolarum situs(30) facile innotescet.

Forma 5. Si duæ radices AP vel(31) $A\varpi$ sunt æquales & alteræ duæ imaginariæ (fig 7 16 & 20) vel reales cum signis diversis a signis æqualium radicum (fig 13, 18) tum duæ Hyperbolarum in quartâ formâ in puncto T vel τ junguntur; vel

(28) In looser imitation of the corresponding form of §2 Newton's first version of this paragraph read 'Si duæ radices sunt inæquales & alteræ duæ vel imaginariæ vel æquales & intermediæ inter radices inæquales (i.e. nec majores nec minores utraqȝ inæqualium) tum solæ manent tres Hyperbolæ sine aliqua Ellipsi decussatione vel cuspide quales in formis 1, 2, 3 & proxima sequente accidunt. Nempe si duæ radices intermediæ $A\varpi$ & $A\pi$ sunt æquales Elipsis in punctum $(\tau 7)$ evanuit: Et si sunt imaginariæ Ellipsis est imaginaria' (If two roots are unequal and the other two either imaginary or equal and intermediate between the unequal ones, that is, neither greater than the one nor less than the other of them, then there remain three hyperbolic branches alone without any oval, crossing or cusp of the kind which

(by constructing it through a parabola or otherwise) you will with their aid distinguish this species into six forms. Specifically:

Form 1. If all the roots are real, unequal and of the same sign, the figure has three hyperbolic branches with an oval (figures 3 and 9). One of the hyperbolas lies facing D, a second facing d and the third facing δ, while the oval lies always within the triangle $Dd\delta$ and also between the middle limits τ and τ in which it is touched by the lines $\pi\tau$, $\varpi\tau$.

Form 2. If two roots, AP and $A\varpi$ nearer to zero or $A\pi$ and Ap more distant from it, are equal, the oval and one of the hyperbolic branches are joined in the points T or t respectively. In other words, the branches of the hyperbola STs or σts are by crossing continued into the oval. See figure 4 or 10.

Form 3. If three roots, AP, $A\varpi$ and $A\pi$ nearer to zero or $A\varpi$, $A\pi$ and Ap more distant from it, are equal, the oval has vanished into the crossing point T or t and the branches of the hyperbola STs or σts there making a contact angle with each other end in an exceedingly sharp cusp. Figures 5 and 11.

Form 4.[28] If the two intermediate roots (those, namely, neither of which is greatest or least of them all) $A\varpi$ and $A\pi$ are equal, the oval has vanished into the [conjugate] point τ/τ.[29] Likewise if two roots are imaginary and two real and unequal, the oval nowhere appears. Figures 6, 12, 15, 17, 19. Precisely, when the two middle roots turn out to be imaginary, the oval becomes imaginary; when, indeed, this happens to the two extreme ones the oval is absorbed into one of the hyperbolic branches and is then seen, as it were, continued into it through a narrow cleft: this you can see in figure 10. In this form, consequently, the three hyperbolas alone remain without any of the oval, crossing or cusp of the kind which occur in the first, second, third and next following forms. And of these hyperbolas that whose asymptotes are $\gamma\lambda$ and gl sometimes lies facing the point D (figures 6 and 12), sometimes in the opposite direction (figures 15 and 17): precisely it lies in that direction from the point A in which the point T or t falls. From this the position of the remaining hyperbolic branches will easily come to be known.

Form 5. If two roots AP or[31] $A\varpi$ are equal and the other two imaginary (figures 7, 16 and 20) or real and opposite in sign to the equal roots (figures 13 and 18), then the two hyperbolic branches in form 4 are joined in the point T

---

occur in forms 1, 2, 3 and that next following. Specifically, if the two intermediate roots $A\varpi$ and $A\pi$ are unequal, the point has vanished into the single point τ, τ: whereas if they are imaginary, so also is the oval).

(29) Newton refers to no figure but presumably intends a reference to figure 6 with the conjugate point τ/τ added.

(30) Here, as several times in the preceding pages, Newton first preferred the etymologically less accurate variant 'scitus'.

(31) Read '&' (and).

potius mutantur in duas alias Hyperbolas sese in isto puncto decussantes. Tertia Hyperbola quæ non decussatur, semper jacet in isto angulo Asymptotorum qui maxime opponitur isto puncto decussationis T vel τ.

Forma 6. Si radices sunt vel omnes imaginariæ (fig 8, 21.) vel omnes reales & inæquales quarum duæ sunt affirmativæ & duæ negativæ (fig 14)(32) Tum ex Hyperbolis decussantibus fit Hyperbola cum alia figurâ quæ intorta circa unam Asymptoton, ex adverso ejus latere ad utramꝗ partem in infinitum procedit: Quam ideò Concham Intortam nominabo. Cum omnes radices sunt reales GL est Asymptoton(33) Conchæ, & gl cùm sunt imaginariæ (fig 8, 14, 21).(34)

Præterea notandum venit quod

1. Si omnes quatuor radices vel earum duæ & duæ sunt æquales: Figura evadit Conica sectio(35) cum lineâ rectâ. Nam radicem in valore Applicatæ y extrahere licebit. Unde hæc curva in duobus punctis decussare nequit.(36)

2. Primus Gradus hujus speciei induit omnes sex formas; secundus quartam quintam & sextam; & tertius solam sextam.(37)

3. Trium Hyperbolarum in quatuor primis formis una jacet extra suas Asymptotos quas ideo secat, altera jacet intra unam Asymptoton & extra aliam quam ideo secat; tertia jacet intra nec ullam Asymptoton secat.(38) In quinta formâ una decussantium Hyperbolarum jacet uno crure intra Asymptotos altero extra, & reliquæ duæ jacent intra. In sexta forma duæ jacent intra easdem Asymptotos ad oppositas partes Conchæ. Unde postquam inv[en]eris in quo latere Asymptotorum Crura Curvæ jacent, facile cognoscere quænam crura ad Hyperbolam constituendam sibi junguntur, et situm Curvæ præter propter determinare. Sin velis ejus situm exactiùs cognoscere opus erit ut eam describas.

4. Hyperbolæ tres conicæ (vide fig 1 & 2) sese omnes decussant inter unumquodꝗ punctum $T\tau 7$ & t, idꝗ in eodem puncto Φ, ϕ, F, & f. Sic in forma prima omnes Ellipsin bisecantes decussant tum in ejus medio, tum ab utraꝗ parte inter eam & Hyperbolas. In forma secunda & quinta decussant in eodem puncto cum Hyperbolis; & in forma tertiâ omnes tum sese tum Hyperbolam cuspidatam in cuspide T vel t tangunt, vel potius secant in angulo contactûs. Si vellem invenire ista puncta $F f \Phi$ & ϕ facio ut radices hujus æquationis $3dx^4+2gx^3+kxx-\frac{hh}{4b}=0$(39) sint AY, Ay, $A\Upsilon$, Av, & duco YF, yf, $\Upsilon\Phi$, $v\phi$ parallelas Asymptoto GL & secantes Hyperbolam Conicam ΦX in punctis F, f, Φ & ϕ.

(32) And '22' also, in fact. (33) Read 'Asymptotos'.

(34) This is not significantly variant from the corresponding portion in §2 above. For explanation of some minor points see notes (27), (28) and (30) of that section.

(35) More exactly, 'hyperbola': compare §2, note (33) above.

(36) Alternatively, if a cubic has two distinct points where it crosses itself then the straight line through those points will meet it in four points: which is impossible except in the degenerate case where the cubic breaks into a conic and the line itself.

or τ; or, better, they are changed into two others crossing in that point. The third, uncrossed hyperbola lies always in that asymptote angle which is most extremely opposed to the crossing point T or τ.

Form 6. If the roots are all imaginary (figures 8 and 21) or all real and unequal with two positive and two negative (figure 14)(32) then out of the crossing hyperbolic branches there is born a hyperbola of different shape which is twisted around one of the asymptotes, passing to infinity on opposite sides and in either direction: for that reason I call it a twisted conchoid. When all the roots are real GL is the conchoid's asymptote, gl when they are all imaginary (figures 8, 14, 21).(34)

Furthermore you should note that:

1. If the four roots are all equal to one another, or in pairs, the figure will prove to be a conic(35) together with a straight line: for then the square root in the value of the ordinate y can be extracted. This curve, in consequence, cannot cross itself in two points.(36)

2. The first grade of this species takes on all six forms, the second the fourth, fifth and sixth, the third the sixth alone.(37)

3. Of the three hyperbolic branches in the four first forms one lies outside its asymptotes, both of which consequently cut it; a second lies within one asymptote and without another, which consequently cuts it; the third lies wholly within and is cut by neither asymptote.(38) In the fifth form one of the crossing hyperbolas lies with one branch within, the other without the asymptotes while the two remaining lie wholly within. In the sixth form two lie within the same asymptotes on opposite sides of the conchoid. Accordingly as soon as you have found on which flank of the asymptotes the branches of the curve lie, you should easily know which branches are joined to constitute each hyperbola and roughly determine the situation of the curve. But if you wish to know its situation more exactly you will need to describe it.

4. Each of the three conic hyperbolas (see figures 1 and 2) cross one another between the points T, τ, 7 and t severally and then in the same points Φ, ϕ, F and f. Thus in the first form they all bisect the oval, crossing both in its mid-point as well as on either hand between it and the hyperbolic branches. In the second and fifth forms they cross in the same point as the hyperbolic branches do, while in the third they all touch each other and the cusped hyperbola in the cusp T or t, or rather they cut one another in a contact angle. Should I wish to find those points F, f, Φ and ϕ I make the roots of this equation $3dx^4+2gx^3+kx^2-h^2/4b=0$ equal to AY, Ay, $A\Upsilon$ and Av and then draw YF, yf, $\Upsilon\Phi$ and $v\phi$ parallel to the asymptote GL and meeting the conic hyperbola ΦX in the points F, f, Φ and ϕ.

(37) That is, Newton distinguishes altogether 10 types of cubic in his first species.
(38) Newton, of course, speaks merely of finite intersections.
(39) See §2, note (14).

De Secunda Specie

Dictum fuit Curvam esse secundæ speciei cum vel terminus h desit, vel quantitates $gg-4dk$ & $\pm\sqrt{\frac{hhd^3}{b}}$(40) sint æquales: sed quoniam ostendam quo pacto terminus h tollatur quando accidunt quantitates $gg-4dk$ & $\pm\frac{dh}{b}\sqrt{bd}$(40) esse æquales; itaꝗ hic tantum considerabo istum casum in quo terminus h desit.(41)

Et hæc Species differt a priori in eo quod per absentiam termini h effectum est ut Hyperbola Conica ΦX evanescat in Basin AB, reliquis Hyperbolis Conicis $\xi h\chi$, xHK manentibus ut priùs, (vide fig 1 & 2 primæ speciei). Adeoꝗ hæc Curva habet Basin AB pro Diametro suâ et puncta T, τ, 7, & t in Diametro AB cadentia sunt ejus vertices. vide figuras hujus Speciei.

De septem ejus Formis

Præterea Hæc Species septem formis variatur;

1ma. Si omnes radices At, $A7$, & $A\tau$ hujus Æquationis $dx^3+gxx+kx+l=0$, sunt reales, ejusdem signi & inæquales: tum dantur tres Hyperbolæ cum Ellipsi $\tau 7$ quæ jacet intra triangulum $Dd\delta$. Et Hyperbola versus D jacet extra Asymptotos suas $\gamma\lambda$ & gl[,] Hyperbolæ verò versus d & δ jacent intra. Fig 1 hujus speciei.

Forma 2. Si duæ radicum nihilo remotiores At & $A7$ sunt æquales, Crura Hyperbolæ jacentis versus D sese decussantia in Ellipsin continuantur. fig 2.

Forma 3. Sin istæ tres radices sint æquales Hyperbola ista fit cuspidata sine aliquâ Ellipsi. fig 3.

Forma 4. Si duæ radices nihilo propiores $A7$ & $A\tau$ sint æquales, Ellipsis in punctū $\tau 7$ evanuit. Item si duæ radices sint imaginariæ solæ manent tres Hyperbolæ sine aliqua Ellipsi Decussatione vel Cuspide. Et hinc duæ oriuntur formæ, nempe quære puncta s & σ ubi curva hæc secat Asymptotos ponendo $A\theta=\frac{4dl}{gg-4dk}$(42) & ducendo ordinatim Applicatam $\sigma\theta s$. Si puncta θ & t cadunt ad easdem partes puncti A, Hyperbolæ versus d & δ cadunt intra Asymptotos suas & tertia Hyperbola versus t cadit extra. Fig 4, 8 & 12.

Forma 5. Sin puncta θ & t cadunt ad diversas partes puncti A; Tum Hyperbola

(40) Read '$bgg-4dk$ & $\pm 4dh\sqrt{\frac{d}{b}}$', and compare note (19) above. These three conditions, of course, determine singly that one of the diametral hyperbolas shall degenerate into a line-pair.

(41) Specifically, the reduced cubic whose defining equation is $bxy^2 = dx^3+gx^2+kx+l$. The equivalence of the other two conditions to $h = 0$ follows by transforming the axes AB,

The Second Species

It has been said that the curve is of the second species when either the term in h is lacking or the quantities $[b]\,g^2-4dk$ and $\pm[4]\,dh\surd(d/b)$ are equal, but, since I will show in what way the term in h may be removed when the quantities $[b]\,g^2-4dk$ and $\pm[4]\,dh\surd(d/b)$ chance to be equal, I will merely consider here, therefore, the case in which the term in h is lacking.(41)

This species differs from the former in that through the absence of the term in h the conic hyperbola ΦX is made to vanish into the base AB, while the remaining conic hyperbolas $\xi h\chi$, xHK stay as before (see figures 1 and 2 of the first species). In consequence, this curve has its base AB for its diameter and the points T, τ, 7 and t falling in its diameter AB are its vertices. See the figures of this species.

Its seven Forms

This species, furthermore, is diversified in seven forms:

[*Form*] *1.* If all the roots At, $A7$ and $A\tau$ of this equation $dx^3+gx^2+kx+l=0$ are real, unequal and of the same sign, then three hyperbolic branches are determined together with the oval $\tau 7$ which lies within the triangle $Dd\delta$. Also the hyperbola facing D lies without its asymptotes $\gamma\lambda$ and gl, while the hyperbolas facing d and δ lie within. Figure 1 of this species.

Form 2. If the two roots more distant from zero, At and $A7$, are equal, the branches of the hyperbola lying facing D on crossing are continued into the oval. Figure 2.

Form 3. But if those three roots are equal that hyperbola becomes cusped and is without any oval. Figure 3.

Form 4. If the two roots nearer to zero, $A7$ and $A\tau$, are equal, the oval has vanished into the conjugate point τ, 7. Likewise if two roots should be imaginary only the three hyperbolic branches remain without any oval, crossing or cusp. From this there arise two forms. Precisely, seek the points s and σ where this curve cuts its asymptotes by setting $A\theta = 4dl/(g^2-4dk)$(42) and drawing the ordinate $\sigma\theta s$. If the points θ and t fall on the same side of the point A, the hyperbolas facing d and δ fall within their asymptotes and the third hyperbola facing t falls without. Figures 4, 8 and 12.

Form 5. But if the points θ and t fall on opposite sides of the point A, then the

---

BC into ab, bC or into $\alpha\beta$, βC: this, in effect, merely changes one diametral hyperbola for another. As we have said (note (5)) Newton does not in the present manuscript fulfil his declared intention of discussing the transformation of a cubic by changing the axes.

(42) The particular case of the equivalent result in the preceding species when $h=0$.

versus t jacet intra Asymptotos $\gamma\lambda$ & gl & reliquæ Hyperbolæ jacent uno crure intra Asymptoton GL, altero extra $\gamma\lambda$ & gl. Fig 5, 9 & 13.

Forma 6. Si duæ radices $A\tau$ & $A\jmath$ sunt æquales cum signis diversis a signo tertiæ radicis At duæ Hyperbolæ decussabunt in puncto τ jacentes intra suas Asymptotos & tertia Hyperbola jacet versus t intra Asymptotos $\gamma\lambda$ & gl. Fig 6, 10, 14.

Forma 7. Si duæ radices sunt inæquales cum signis diversis a signo tertiae radicis tum habemus Concham quæ in utramq; partem Asymptoti GL & ex eodem ejus latere in infinitum serpit. Medium punctum $\jmath$ est vertex conchæ & extrema puncta t & τ sunt vertices duarum Hyperbolarum quæ sibi oppositæ jacent ex utraq; parte Conchæ intra Asymptotos $\gamma\lambda$ & gl. Fig 7, 11 & 15.

De ejus Gradibus

Deniq; hæc Species duos præcipue gradus admittit: Primus est cum Asymptoti faciunt triangulum $Dd\delta$, & secundus cùm omnes decussant in eodem puncto A. Primus est omnium formarum (vide fig 1, 2, 3, 4, 5, 6, 7, 8, 9, 10 & 11), & secundus tantùm quatuor ultimarum. (fig 12, 13, 14, & 15). Cum terminus g deest, curva erit secundi gradûs,(43) aliàs primi. Et nota quod terminus k deesse nequit.(44)

De Tertia Specie

Si terminus h desit et insuper quantitates gg & $4dk$ sunt æquales vel desunt, tum omnes Hyperbolæ Conicæ $X\Phi$, $\xi\lambda$, xK, evanescunt in suas Asymptotos (vide fig 1 & 2 primæ Speciei) Adeoq; hæc curva habet tres Diametros AB, $\alpha\beta$ & ab. Et punctum o ubi hæ tres Diametri sese secant erit Centrum figuræ.(45) Nam ab aliquo puncto Curvæ C duc $Ck \parallel lg$, $kQ \parallel GL$, $Qc \parallel \lambda\gamma$, $cq \parallel lg$, $q\kappa \parallel LG$ secantes curvam in k, Q, c, q, κ & junge κC: ista κC erit parallela Asymptoto $\lambda\gamma$,(46) et punctum o erit centrum gravitatis(47) tum rectilineæ figuræ $CkQcq\kappa$, tum Recti=Curvilineæ Figuræ $CekQ\epsilon cqt\kappa C$, tum trium Curvæ portionum simul sumptarum Cek, & Qec, & $qt\kappa$. Vide fig 1 hujus speciei.

Si jungas Cc, kq, & κQ istæ etiam lineæ erunt Asymptotis GL, $\gamma\lambda$ & gl parallelæ. Et hinc facile patet quod datum aliquod curvæ punctum C dat alia quinq; & non plura puncta k, Q, c, q, & κ ad eam describendam; ducendo nempe $Cb \parallel lg$,

(43) For then $AD = -g/2d$ vanishes and the triangle $Dd\delta$ collapses into A.

(44) If $g = h = k = 0$ then the resulting cubic ($bxy^2 = dx^3 + l$) has also $bg^2 - 4dk$ and $4dh\surd(d/b)$ each equal to zero and so to each other, or it is of species 3 (tridiametral). In sum, Newton identifies 11 distinct types of cubic as belonging to his second species.

(45) Compare §2, note (17).

(46) Since $Cb = bk \parallel lg$ we may say that C, k are oblique images in the line aod parallel to lg: similarly C and c, Q and k, c and q, q and κ are oblique images in AoD, AoD, aod and $\alpha o\delta$

hyperbola facing t lies within the asymptotes $\gamma\lambda$ and gl and the remaining hyperbolas lie with one branch within the asymptote GL and the other outside $\gamma\lambda$ and gl. Figures 5, 9 and 13.

Form 6. If the two roots $A\tau$ and $A\jmath$ are equal and opposite in sign to the third root At, two of the hyperbolas will cross in the point τ and lie within their asymptotes while the third lies within the asymptotes $\gamma\lambda$ and gl facing t. Figures 6, 10 and 14.

Form 7. If two of the roots are unequal and opposite in sign to the third, we have a Conchoid which snakes into infinity along the asymptote GL in either direction and on the same side. The middle point $\jmath$ is the conchoid's vertex and the extreme points t and τ are the vertices of two hyperbolic branches which lie opposite to each other on either flank of the conchoid and within the asymptotes $\gamma\lambda$ and gl. Figures 7, 11 and 15.

Its Grades

This species, finally, admits two grades in particular: a first when the asymptotes form the triangle $Dd\delta$, and the second when they all cross in the same point A. The first is common to all the forms (see figures 1 to 11), the second characterizes merely the last four (figures 12 to 15). When the term in g is missing the curve will be of the second grade,(43) otherwise of the first: but note that then the term in k must be present.(44)

The Third Species

If the term in h is lacking and moreover the quantities [b] g^2 and $4dk$ are equal or both lacking, then each of the conic hyperbolas $X\Phi$, $\xi\lambda$ and xK vanishes into its asymptotes (see figures 1 and 2 of species 1) so that this curve has three diameters AB, $\alpha\beta$ and ab and the point o where these three diameters meet will be the centre of the figure.(45) For from any point C of the curve draw $Ck \parallel lg$, $kQ \parallel GL$, $Qc \parallel \lambda\gamma$, $cq \parallel lg$ and $q\kappa \parallel LG$ meeting the curve in k, Q, c, q, κ, and join κC: that line κC will be parallel to the asymptote $\lambda\gamma$(46) and the point o will be the gravity centre(47) equally of the rectilinear figure $CkQcq\kappa$, of the recti-/curvi-linear figure $CekQ\epsilon cqt\kappa C$ and of the three portions of the curve collectively Cek, $Q\epsilon c$ and $qt\kappa$. See figure 1 of this species.

Should you join Cc, kq and κQ, those lines also will be parallel to the asymptotes GL, $\gamma\lambda$ and gl. From this it is readily evident that any given point C of the curve determines a further five points k, Q, c, q and κ and no more to describe

parallel to the corresponding asymptote, so that C and κ are oblique images in $\alpha o\delta$ parallel to $\lambda\gamma$, or $C\beta = \beta\kappa \parallel \lambda\gamma$.

(47) For, since AD, ad and $\alpha\delta$ are medians, o is the centroid of $\Delta Dd\delta$, while the remaining figures are obliquely symmetrical round each of AD, ad and $\alpha\delta$.

$CB \parallel LG$, & $C\beta \parallel \lambda\gamma$ & sumendo $bk = Cb$, $\beta\kappa = C\beta$, & $Bc = CB$ &c: (vel sumendo $\lambda k = LC$, $gc = \gamma C$, & $l\kappa = GC$). Fig 1.

Hujus Speciei unica est forma: nempe habet tres Hyperbolas intra suas Asymptotos jacentes sine aliqua Ellipsi, decussatione vel Cuspide. Fig 1 2 & 3.

Duo vero sunt gradus. Primus est cùm Asymptoti faciunt Triangulum $Dd\delta$ (ut in fig 1 & 2), secundus cùm decussant omnes in puncto A (ut in fig 3). Cùm termini g & k desunt erit secundi gradus, alias erit primi.(48)

De Secundo Casu.

In secundo Casu, ubi habetur $bxyy = -dx^3 + gxx + hy + kx + l$, Curva habet unicam Asymptoton nempe Applicatam(49) AG, circa quam Concha utroꝗ crure in infinitum serpit. Ellipsis etiam quandoꝗ apparet, quandoꝗ in Concham absorbetur, & quandoꝗ imaginaria est.(50)

De Quarta et Quinta Specie

In hoc Casu duæ sunt Species quarum prima (id est Specierum quarta) nullam habet Diametrum, sed omnes lineæ Cc Asymptoto AG parallelæ et in Curvam terminatæ, bisecantur ab Hyperbola Conica XT, cujus Asymptoti sunt AG & AB & potestas $AB \times BX$(51)$= \frac{h}{2b}$ (vide fig: 1 & 2 hujus speciei): At in secundâ (specierū quintâ) istæ lineæ Cc bisecantur a Basi AB quæ proinde erit Diameter figuræ. In quarta specie Concha est intorta & secat Asymptoton in S posito $AS = \frac{-l}{h}$.(52) In quinta non intorquetur nec Asymptoton secat. Si terminus h desit(53) curva erit quintæ Speciei, alias erit Quartæ.

De Formis Quartae Specie

Hujus Speciei quatuor sunt Formæ. Nempe

1. Si omnes radices $A\pi$, AP, Ap, & $A\varpi$ hujus æquationis

$$dx^4 = gx^3 + kxx + lx + \frac{hh}{4b},^{(54)}$$

sunt reales & inæquales habes Concham intortam cum Ellipsi. fig 1.

(48) Both grades, of course, satisfy [b] $g^2 = 4dk$ so that either g and k are both non-zero (grade 1) or both zero (grade 2). In the latter case, where the cubic has the defining equation $bxy^2 = dx^3 + l$ (compare note (44) above), $AD = -g/2d$ is zero and so both g and k also. Altogether Newton here distinguishes but two types of cubic.

(49) Newton has previously called this y-axis the 'Directrix' (in §2).

(50) This last sentence is a late addition. In general the directions of the asymptotes are determined by $0 = bxy^2 + dx^3$, so that one asymptote, AG ($x = 0$), is real while the others are imaginary with slopes $\pm i\sqrt{(d/b)}$.

it; namely, by drawing $Cb \parallel lg$, $CB \parallel LG$ and $C\beta \parallel \lambda\gamma$ and then taking $bk = Cb$, $\beta\kappa = C\beta$ and $Bc = CB$, and so on (or by taking $\lambda k = LC$, $gc = \gamma C$ and $l\kappa = GC$). Figure 1.

There is a single form of this species: it has, specifically, three hyperbolic branches lying within their asymptotes without any oval, crossing or cusp. Figures 1, 2, 3.

But there are, indeed, two grades: a first when the asymptotes form the triangle $Dd\delta$ (as in figures 1 and 2), the second when they all cross in the point A (as in figure 3). When the terms in g and k are lacking the curve is of the second grade, otherwise of the first.(48)

The Second Case

In the second case, where we have $bxy^2 = -dx^3 + gx^2 + hy + kx + l$, the curve has a unique asymptote, namely the ordinate(49) AG, around which the conchoid snakes in either branch to infinity. Also, the oval sometimes appears, is sometimes absorbed into the conchoid and is sometimes imaginary.(50)

The fourth and fifth Species

In this case there are two species: the first of these (that is, species 4) has no diameter, but every line Cc parallel to the asymptote AG and terminated at the curve is bisected by the conic hyperbola XT whose asymptotes are AG and AB and whose power $AB \times BX$(51) equals $h/2b$ (see figures 1 and 2 of this species): in the second, however, (species 5) those lines Cc are bisected by the base AB, which consequently will be the figure's diameter. In the fourth species the conchoid is twisted and cuts the asymptote in S on setting $AS = -l/h$:(52) in the fifth it is not twisted nor does it cut the asymptote. If the term in h is lacking(53) the curve will be of the fifth species, otherwise of the fourth.

The Forms of the fourth Species

There are four forms of this species, specifically:

[*Form*] *1.* If all the roots $A\pi$, AP, Ap and $A\varpi$ of this equation

$$dx^4 = gx^3 + kx^2 + lx + h^2/4b \text{(54)}$$

are real and unequal, you have a twisted conchoid with an oval. Figure 1.

---

(51) That is, the product xy.

(52) When $x = 0$, the corresponding ordinate is determined by $hy + l = 0$.

(53) That is, the point S passes to infinity on AG and the cubic assumes the form

$$by^2 = -dx^2 + gx + k + l/x,$$

which shows that AB $(y = 0)$ is its diameter.

(54) We may write the cubic as $x^{-1}[b(xy - h/2b)^2 - d(x-\alpha)(x-\beta)(x-\gamma)(x-\delta)] = 0$, where α, β, γ and δ are the roots of $0 = -dx^4 + gx^3 + hx^2 + lx + h^2/4b$. (Compare §2, note (18).)

Forma 2. Si duæ radices intermediæ AP & Ap sint æquales, Ellipsis et Concha in puncto T junguntur decussantes. fig 2.

Forma 3. Si tres radices sunt æquales Concha in puncto T acutissime cuspidata est. fig 3.

Forma 4. Si duæ radices extremæ Ap & $A\varpi$ sunt æquales Ellipsis in punctum evanuit. Item si duæ quævis radices sint imaginariæ sola manet concha sine aliqua Ellipsi decussatione vel cuspide. Fig. 4 & 5. Nempe si extremæ radices sunt imaginariæ Ellipsis imaginaria est; sin duæ intermediæ, tum Ellipsis in Concham absorbetur & aliquando apparet quasi angusto hiatu in eam continuata, ut in fig 4.

De ejus Gradibus

Praeterea hanc Speciem in duos gradus distinguo; primus esto qui nullum centrum habet, & secundus qui habet punctum A pro centro: Nempe cùm omnes rectæ qAQ per punctum A ductæ & in curvam terminatæ bisecantur in A. fig 5. Primus gradus est omnium formarum, secundus vero solius quartæ. Si termini g & l desint Curva erit secundi gradus, aliàs erit primi.(55)

De Formis Quintæ Speciei(56)

Hujus Speciei quinq; sunt formæ.

1. Si duæ radices $A\tau$ At hujus æquationis $dx^3 = gxx + kx + l$(57) sunt inæquales cum signis diversis a signo tertiæ radicis AT; tum Ellipsis & Concha jacent ex diverso latere Asymptoti. Fig 1.

Forma 2. Sin omnes radices habent eadem signa, tum jacent ex eodem latere Asymptoti. fig 2.

For: 3. Si duæ radices sunt æquales, & tertia AT est ejusdem signi & nihilo remotior, tum Ellipsis et Concha in puncto T junguntur ibi decussantes. Fig 3.

For 4. Si tres radices AT, At, & $A\tau$ sunt æquales, Concha in puncto T cuspidata est estq; Cissois Dioclëa.(58) fig 4.

For 5. Si duæ radices At; $A\tau$ sunt æquales & tertia AT vel diversi signi, vel ejusdem signi & nihilo propior; tum Ellipsis in punctum evanuit. Item si duæ radices sint imaginariæ sola manet Concha sine aliqua Ellipsi decussatione vel Cuspide. Fig 5, & 6.(59)

---

(55) The condition for a curve to have a centre in the present sense is that all even terms (here gx^2 and l) in its defining equation must vanish. In all, Newton distinguishes 5 types of cubic in his fourth species.

(56) That is, $bxy^2 = -dx^3 + gx^2 + kx + l$: here the diametral hyperbola XT collapses into the line-pair formed by its real asymptote and diameter.

(57) Compare note (54): since $\alpha\beta\gamma\delta = -h^2/4bd$, when $h = 0$ one of the quartic's roots will be zero and Newton here considers the cubic formed by the other three.

Form 2. If the two intermediate roots AP and Ap should be equal, the oval and conchoid cross and join in the point T. Figure 2.

Form 3. If the three roots are equal the conchoid is very sharply cusped in the point T. Figure 3.

Form 4. If the two extreme roots Ap and $A\varpi$ are equal, the oval has vanished into a conjugate point. Likewise if any two roots are imaginary there remains only the conchoid without any oval, crossing or cusp. Figures 4 and 5. Precisely, if the extreme roots are imaginary the oval is imaginary; but if the two intermediate ones, then the oval is absorbed into the conchoid and at times appears as though continued into it through a narrow cleft, as in figure 4.

Its Grades

I further distinguish this species into two grades. Let the first be that having no centre, and the second that having the point A for its centre: namely, when every straight line qAQ drawn through the point A and terminated at the curve is bisected in A (figure 5). The first characterizes all the forms, but the second only the fourth. If the terms in g and l are lacking the curve will be of the second grade, otherwise it will be of the first.(55)

The Forms of the fifth Species(56)

There are five forms of this species:

[*Form*] *1.* If two roots $A\tau$, At of this equation $dx^3 = gx^2+kx+l$(57) are unequal and opposite in sign to the third, AT, then the oval and conchoid lie on opposite sides of the asymptote. Figure 1.

Form 2. But if all the roots have the same sign, then they both lie on the same side of the asymptote. Figure 2.

Form 3. If two roots are equal and the third, AT, is of the same sign and farther from zero, then the oval and conchoid cross and join in the point T. Figure 3.

Form 4. If the three roots AT, At and $A\tau$ are equal, the conchoid is cusped at the point T: it is indeed the cissoid of Diocles.(58) Figure 4.

Form 5. If two roots, At and $A\tau$, are equal and the third, AT, either opposite in sign or of the same sign and nearer to zero, then the oval has vanished into a conjugate point. Likewise, if two roots are imaginary there remains only the conchoid without any oval, crossing or cusp. Figures 5 and 6.(59)

(58) These three words are a later addition: in the notation of note (54) the cubic of this form reduces to $bxy^2 = -d(x-\alpha)^3$, that is, on writing $\alpha - x = X$ and $by^2 = dY^2$, to the more familiar analytical defining equation $(\alpha - X)Y^2 = X^3$ of Diocles' 'ivy-leaf' curve.

(59) Since this species has a unique grade, five types of cubic are distinguished.

De Tertio Casu.

In tertio Casu ubi habetur $bxyy = gxx + hy + kx + l$, Figura habet quatuor infinita Crura: quorum duo sunt Hyperbolicæ naturæ & habent Applicatam(60) AG pro Asymptoto: reliqua vero duo sunt Parabolicæ naturæ, et nullam habent Asymptoton(61) & tendunt ad easdem partes(62) & ex altera parte Asymptoti Ellipsis etiam aliquando apparet. Præterea si linea quævis Rr Basi AB parallela & in Curvam terminata in R & r, bisecetur in χ, istud punctum χ semper cadet in Parabola Conica χI; cujus natura est $byy = 2gx + k$, posito $AB = x$ & $B\chi = y$.(63) Vide fig 1 & 2 sextæ Speciei.

De Sexta et Septima Specie

Et hic casus duas Species exhibet quarum prima (id est Species Septima(64)) nullam habet Diametrum sed omnes rectæ Cc bisecantur ab Hyperbola Conica ut priùs: at secunda (id est Species Septima) habet Basin AB pro Diametro. In primâ [curva] secatur Asymptoto in S, posito $AS = \frac{-l}{h}$,(65) & Crura Hyperbolica jacent ex diverso latere Asymptoti, in secundâ jacent ex eodem latere & non secant Asymptoton. Si terminus h desit figura erit secundæ Specierum, alias erit primæ.(66)

De Formis Sextæ Speciei(67)

Hujus Speciei sex sunt formæ.

1. Si tres radices AP, $A\varpi$, $A\pi$ hujus æquationis $gx^3 + kxx + lx + \frac{hh}{4b} = 0$(68) sint inæquales & ejusdem signi: Figura constat ex Ellipsi et alijs duabus curvis quæ partim Hyperbolicæ sunt & partim Parabolicæ. Nempe crura Parabolica junguntur cruribus Hyperbolicis sibi proximis. Fig 1.

Forma 2. Si duæ radices sint æquales & tertia $A\pi$ ejusdem signi & nihilo remotior, Ellipsis & una dictarum figurarum in puncto T junguntur ibi decussando. fig 2.

For 3. Si tres radices sint æquales ista figura in puncto T cuspidata est, & ellipsis deest. fig 3.

---

(60) Compare note (49).

(61) The remainder of this sentence is a late addition.

(62) The nature of the infinite branches of the cubic is determined by $bxy^2 = gx^2$: in consequence, there is a unique finite asymptote AG ($x = 0$) and an asymptotic parabola, $by^2 = gx$, which together with the cubic itself is touched by the line at infinity parallel to AG (or the parabolic points will be at infinity in the direction AB).

(63) $B\chi = y$ meets the cubic $(gx^2 + (k - by^2)x + (hy + l) = 0)$ such that

$$AB = x = \tfrac{1}{2}(x_1 + x_2) = -(k - by^2)/2g.$$

(64) Read 'Sexta' (sixth).

THE THIRD CASE

In the third case, where we have $bxy^2 = gx^2 + hy + kx + l$, the figure has four infinite branches: two of these are hyperbolic in nature and have the ordinate(60) AG for asymptote, while the remaining two are parabolic in nature, having no asymptote and extending in the same direction,(62) and in addition the oval sometimes appears on the further side of the asymptote. Moreover, if any line Rr parallel to the base AB and terminated at the curve in R and r be bisected in χ, that point χ will fall always in the conic parabola χI, whose nature, on setting $AB = x$ and $B\chi = y$, is $by^2 = 2gx + k$.(63) See figures 1 and 2 of species 6.

The sixth and seventh Species

This case, too, manifests two species: the first of these (that is, the [sixth]) has no diameter but all straight lines Cc are bisected by the conical hyperbola as before, but the second (that is, the seventh species) has the base AB for its diameter. In the first the curve meets its asymptote in S, on setting $AS = -l/h$,(65) and the hyperbolic branches lie on opposite sides of the asymptote; in the second they lie on its same side and do not cut it. If the term in h is lacking the figure will be of the second of the species, otherwise it will be of the first.(66)

The Forms of the sixth Species(67)

There are six forms of this species:

[*Form*] *1.* If the three roots AP, $A\varpi$ and $A\pi$ of this equation

$$gx^3 + kx^2 + lx + h^2/4b = 0^{(68)}$$

be unequal but of the same sign, the figure will consist of an oval and two other curves which are partly hyperbolic and partly parabolic: precisely, the parabolic branches are joined to the nearest hyperbolic ones. Figure 1.

Form 2. If two roots be equal and the third, $A\pi$, of the same sign and more distant from zero, the oval and one of the above-named figures cross and join in the point T. Figure 2.

Form 3. If all three roots be equal that figure is cusped in the point T and the oval is missing. Figure 3.

(65) See note (52).

(66) Compare note (53) when $d = 0$.

(67) Newton made extensive preliminary drafts of the cubic figures for the present enumeration. Many of these (whose originals are now in private possession) are little variant from the revised drawings here reproduced but occasionally there are some major discrepancies. To illustrate the latter we reproduce below, in appendix, his preliminary sketches for cubics in the present (sixth) species. Most notably, in his revision Newton has considerably simplified his first drawings but in compensation expanded their number from three to seven.

(68) Compare note (27) when $d = 0$.

For 4. Si duæ radices $A\pi$ & $A\varpi$ sint æquales et tertia AP ejusdem signi & nihilo propior, Ellipsis in punctum evanuit. Item si duæ radices sunt imaginariæ solæ manent istæ curvæ sine aliqua Ellipsi decussatione vel cuspide. fig 4 & 5.

For 5. Si duæ radices $A\pi$ & $A\varpi$ sint æquales & tertia diversi signi, istæ curvæ junguntur in puncto τ ibi decussantes; vel potiùs mutantur in alias curvas Parhyperbolicas,(69) nempe Crura Parabolica sese decussantia junguntur cruribus Hyperbolicis sibi oppositis. Fig 6.

For 6. Si duæ radices $A\varpi$ & $A\pi$ sint inæquales & tertia diversi signi, Figura erit Concha intorta cum Parabolâ. Fig 7.

De Formis Septimæ Speciei

Septima Species habet sex(70) formas.

1. Si radices $A\gamma$, $A\tau$ hujus Æquationis $gxx+kx+l=0$(71) non sint ejusdem signi habes Concham (non intortam) cum Parabolâ, quæ jacent ex diverso latere Asymptoti AG. Fig 1.

Forma 2. Si sint ejusdem signi Concha et Parabola jacent ex eodem latere Asympt: fig 2.

For 3. Si sint æquales Concha & Parabola junguntur in puncto τ ibi decussantes. fig 3. Sive Crura Parabolica, sese decussantia, junguntur cruribus Hyperbolicis sibi oppositis.

For 4. Si sint imaginariæ, tum Crura Parabolica junguntur cruribus Hyperbolicis sibi proximis & constituunt duas figuras partim Parabolicas partim Hyperbolicas, quæ sese non decussant. Fig. 4.(72)

De Quarto Casu.

In quarto casu, ubi habetur $bxyy=hy+kx+l$,(73) Figura habet tres Asymptotos, quarum duæ sunt parallelæ. Nempe Applicata(74) AG est una Asymptotos: &, posito $Ad=A\delta=\sqrt{\frac{k}{h}}$, lineæ dg & $\delta\gamma$ Basi AB parallelæ erunt aliæ duæ. Et per unamquodqꝫ Asymptoton duo Curvæ Crura ad oppositas partes in infinitum serpunt: Quorum duo jacent inter parallelas Asymptotos & semper sibimet junguntur; reliquorum vero quatuor crurum (quæ non jacent inter parallelas Asympt[ot]os) ista duo & duo sibimet junguntur quæ jacent ad easdem partes parallelorum(75) Asymptotorum. Vide figuras harum specierum

(69) Those 'partly hyperbolic and partly parabolic' of form 1.

(70) Read 'quatuor' (four). Newton is perhaps thinking of the six forms of species 6.

(71) Compare the analogous equation of case 1, species 2 above: here $d = 0$.

(72) Since each of species 6 (6 forms) and species 7 (4 forms) has only one grade, Newton distinguishes altogether 10 types of cubic in the two species.

Form 4. If two roots $A\pi$ and $A\varpi$ be equal and the third AP of the same sign and closer to zero, the oval has vanished into a conjugate point. Likewise, if two roots are imaginary there remain only those curves without any oval, crossing or cusp. Figures 4 and 5.

Form 5. If two roots $A\pi$ and $A\varpi$ be equal and the third of opposite sign those curves are joined in the point τ and cross there; or rather they are changed into other, parhyperbolic(69) curves: precisely, the parabolic branches on crossing are joined to the opposite hyperbolic ones. Figure 6.

Form 6. If two roots $A\varpi$ and $A\pi$ be unequal and the third of opposite sign, the figure will be a twisted conchoid with a parabolic branch. Figure 7.

The Forms of the seventh Species

The seventh species has six(70) forms:

[*Form*] *1.* If the roots $A\gamma$, $A\tau$ of this equation $gx^2+kx+l=0$(71) be not of the same sign, you have an (untwisted) conchoid with a parabolic branch lying on opposite sides of the asymptote AG. Figure 1.

Form 2. If they be of the same sign, the conchoid and the parabolic branch lie on the same side of the asymptote. Figure 2.

Form 3. If they be equal, the conchoid and the parabolic branch are joined in the point τ, crossing there. Figure 3. In other words, the parabolic branches on crossing are joined to the opposite hyperbolic ones.

Form 4. If they be imaginary, then the parabolic branches are joined to the nearest hyperbolic ones and constitute two figures partly parabolic, partly hyperbolic which do not cross. Figure 4.(72)

The Fourth Case

In the fourth case, where we have $bxy^2=hy+kx+l$,(73) the figure has three asymptotes, two of which are parallel: precisely, the ordinate(74) AG is one asymptote and, if you set $Ad=A\delta=\sqrt{(k/b)}$, the lines dg and $\delta\gamma$ parallel to the base AB will be the other two. Also, along each asymptote two branches of the curve snake into infinity in opposite directions: two of these lie between the parallel asymptotes and are always joined together, while as for the remaining branches (not lying between the parallel asymptotes) those lying on the same sides of the parallel asymptotes are joined in pairs. See the figures of these species.

(73) That is, $y=(-h\pm\sqrt{[h^2+4bx(kx+l)]})/2bx$. The infinite branches are determined in direction by $bxy^2=0$, so that there are three asymptotes: $x=0$, and the parallel lines $y=\pm\sqrt{(k/b)}$, that is, the y-axis AG and the parallels dg and $\delta\gamma$.

(74) See note (49).

(75) Read 'parallelarum'.

De Octava & Nona Specie

Præterea hic casus speciem dabit Octavam et nonam. Octavæ nulla est Diameter, sed Basis AB est Diameter Nonæ. In Octavâ, crura circa Asymptoton AG, jacent ex diverso ejus latere; in Nonâ jacent ex eodem, idq versus ista crura quibus junguntur; in utráq, Crura circa parallelas Asymptotos jacent ex diverso earum latere. In Octavâ, curva secat Asymptoton AG in S, posito $AS = \frac{-l}{h}$; in Nonâ non omnino secat; & in utráq non secat Asymptotos parallelas $\delta\gamma$ & dg. Si terminus h desit Curva erit Nonæ speciei, alias erit Octavæ.

De Formis Octavæ Speciei

Octavæ Speciei duæ sunt formæ.

1. Si radices AT, At hujus æquationis $kxx + lx + \frac{hh}{4b}$(76) $[=0]$ sunt reales, Tum Hyperbola interior tota jacet ad easdem partes Asymptoti AG: Et crura(77) Hyperbolarū exteriorum quæ jacent circa parallelas Asymptotos Ag & $A\gamma$, tendunt ad alteras partes. Fig 1.

Forma 2. Sin istæ radices sint imaginariæ, tum Hyperbolæ exteriores jacent ex diverso latere Asymptoti AG; & crura tertiæ figuræ interioris tendunt ad oppositas partes, quam ideò Hyperbolam intortam nominabo. Fig 2 & 3.

Nota. Si radices istæ essent æquales Figura evaderet Conica Sectio.(78)

De Gradibus Octavæ Speciei

Præterea hanc speciem in duos gradus distinguo. Primi nullum est centrum, sed punctum A est centrum secundi, nempe in hoc graduo(79) omnes rectæ QAq per punctum A ductæ & in Curvam terminatæ bisecantur in A (fig 3). Primus gradus est utriusq formæ, secundum tantum secundæ.(80) Si terminus l desit curva erit secundi gradus, alias erit primi.(81)

De Formâ Nonæ Speciei(82)

Hujus s[p]eciei unica est forma: nempe Hyperbolæ exteriores jacent intra suas Asymptotos et ad easdem partes Asymptoti AG; & figura inter

(76) That is, $b(xy - h/2b)^2$.

(77) Newton first continued with 'tertiæ figuræ jacentis intra parallelas Asymptotos' (of the third figure lying within the parallel asymptotes): to correspond, he had first supposed in the previous line that there were two inner hyperbolic branches (compare form 2 below).

(78) More precisely, 'hyperbola' (compare note (35) above). The quadratic

$$kx^2 + lx + h^2/4b = 0$$

The eighth and ninth Species

This case, moreover, gives species 8 and 9. There is no diameter of the eighth, but the base AB is the diameter of the ninth. In the eighth the branches round the asymptote AG lie on its opposite sides, in the ninth they lie on the same side and then facing the branches to which they are joined: in both, the branches round the parallel asymptotes lie on their opposite sides. In the eighth the curve cuts the asymptote AG in S, on setting $AS = -l/h$; in the ninth it does not cut it at all; and in neither does it cut the parallel asymptotes $\delta\gamma$ and dg. If the term in h is lacking the curve will be of the ninth species, otherwise it will be of the eighth.

The Forms of the eighth Species

There are two forms of the eighth species:

[*Form*] *1.* If the roots AT and At of this equation $kx^2+lx+h^2/4b[=0]$(76) are real, then the inner hyperbolic branch lies wholly on the same flank of the asymptote AG, while the branches(77) of the outer hyperbolas lying around the asymptotes Ag and $A\gamma$ extend in the other direction. Figure 1.

Form 2. But if those roots be imaginary, then the outer hyperbolas lie on opposite sides of the asymptote AG and the branches of the third, inner figure extend in opposite directions: for that reason I will call it a twisted hyperbola. Figures 2 and 3.

Note. Should those roots be equal, the figure will turn out to be a conic.(78)

The Grades of the eighth Species

This species, moreover, I distinguish into two grades. Of the first there is no centre, but the point A is the centre of the second: precisely, in this latter grade all straight lines QAq drawn through the point A and terminated at the curve are bisected in A (figure 3). The first grade characterizes both forms, the second merely the second.(80) If the term in l is missing the curve will be of the second grade, otherwise it will be of the first.

The Form of the ninth Species(82)

There is a unique form of this species: specifically, here the outer hyperbolic branches lie within their asymptotes and on the same flank of the asymptote AG,

has equal roots when $h^2k = bl^2$, and under that condition the cubic $0 = bxy^2-hy-kx-l$ degenerates into the pair formed by one of the parallel asymptotes ($y = \pm\sqrt{[k/b]}$) and the hyperbola $0 = xy\pm\sqrt{[k/b]}x-h/b$ whose asymptotes are the remaining two cubic ones ($x = 0$, $y = \mp\sqrt{[k/b]}$).

(79) Read 'gradu'.

(80) In all, then, Newton distinguishes 3 types of cubic in his eighth species.

(81) Compare note (55) when $g = 0$.

(82) That is, $bxy^2 = kx+l$, which has one form of a single grade and so is a unique type.

Asymptotos parallelas, tota jacet ad alteras partes Asymptoti *AG*. Vide Figuram hujus.

De Quinto Casu.

In Casu quinto ubi habetur $bxyy = hy - kx + l$, Figura habet unicam Asymptoton(83) *AG* circa quam Concha jacet.

De Specie Decimâ et Undecimâ

Si terminus *h* non desit Concha erit intorta sine diametro (fig 1 & 2): Et hæc esto species decima cujus duo sunt gradus; primi nullum est centrum (fig 1), sed punctum *A* est centrum secundi (Fig 2); in secundo terminus [*l*] desit, in primo non.

Sin terminus *h* desit; Concha non erit intorta, & habet Basin *AB* pro Diametro. Et hæc esto Species Undecima. Vide figuram hujus speciei.(84)

De Sexto Casu.

In Casu sexto ubi habetur $bxyy = hy + l$, Curva habet duas Asymptotos Basin *AB* & Applicatam *AG*;(85) circa quas duæ Hyperbolæ jacent ad diversas partes Asymptoti *AB*, sed duo crura circa Asymptoton *AB* tendunt ad easdem partes.

De Specie Duodecima et Decima Tertia

Si terminus *h* non desit Curva non habet diametrum & crura circa Asymptoton *AG* jacent ex diverso ejus latere; Quæ Species esto Duodecima. vide fig.

Sin terminus *h* desit, Asymptoton(86) *AB* est Diameter curvæ, & crura circa *AG* jacent ex eodem ejus latere, nempe ex isto versus reliqua crura quibus junguntur: Et hæc esto Species Decima tertia. vide fig.(87)

De Septimo Casu, *et Specie Decima Quarta*

In septimo Casu ubi habetur $eyy = dx^3 + kx + l$, Figura(88) erit Parabola cujus crura jacent ex eodem latere lineæ Applicatæ *AG* & tendunt ad oppositas(89)

---

(83) Here the asymptotes are (compare note (78) when *k* is negative) $x = 0$ and $y = \pm i\sqrt{(k/b)}$, only the first of which is real.

(84) Much as before, species 10 is a central cubic (grade 2) when $l = 0$, while $h = 0$ determines (species 11) that *AB* ($y = 0$) be a diameter. Altogether 3 types of cubic are distinguished in the two species of case 5.

(85) Evidently the infinite branches of this cubic are determined by $bxy^2 = 0$, so that the three asymptotes are the (doubled) base *AB* ($y = 0$) and the 'directrix' *AG* ($x = 0$).

(86) Read 'Asymptotos'.

(87) The two species here, each of a unique form taking on a single grade, yield only one distinct type of cubic each.

(88) One of the class of diverging cubics, here described for the first time generally, though its fourth form ($dy^2 = x^3$) had been discussed extensively in print by both William Neil and Henrik van Heuraet and was, indeed, the first algebraic curve to be rectified, as Newton well knew (compare I, 2, 7: Problem 9, Example 1). See John Wallis, *Tractatus Duo* (I, 1: Appendix 2,

while the figure between the parallel asymptotes lies wholly on its other flank. See the figure for this.

The Fifth Case

In the fifth case, where we have $bxy^2 = hy - kx + l$, the figure has a unique asymptote[83] AG around which the conchoid lies.

The tenth and eleventh Species

If the term in h be not missing, the conchoid will be twisted and without diameter (figures 1 and 2): let this be the tenth species. It takes on two grades: of the first there is no centre (figure 1) but the point A is the centre of the second (figure 2). In the second the term [l] should be missing, in the first not.

But if the term in h be missing, the conchoid will be untwisted, having the base AB for diameter. Let this be the eleventh species. See the figure for this species.[84]

The Sixth Case

In the sixth case, where we have $bxy^2 = hy + l$, the curve has two asymptotes, the base AB and the ordinate AG.[85] Around these lie two hyperbolic branches on opposite sides of the asymptote AB, but the two infinite branches round the asymptote AB extend in the same direction.

The twelfth and thirteenth Species

If the term in h is not lacking, the curve does not have a diameter and the infinite branches round the asymptote AG lie on its opposite sides. Let this be the twelfth species. See the figure.

But if the term in h be lacking, the asymptote AB is the curve's diameter and the branches round AG lie on its same side; namely, on that facing the remaining branches to which they are joined. And let this be the thirteenth species. See the figure.[87]

The Seventh Case *and fourteenth Species*

In the seventh case, where we have $ey^2 = dx^3 + kx + l$, the figure[88] will be a parabola whose infinite branches lie on the same side of the ordinate AG and

note (22)), (Oxford, 1659): 90ff.; and van Heuraet, *Epistola de Transmutatione Curvarum Linearum in Rectas* in Descartes' *Geometria*, ₂1659: 517–20, especially 519. As we shall see in our seventh volume, this class of cubic held particular interest for Newton for its separate forms, each projectively distinct, generate all the other cubics as their optical 'shadows': specifically, any cubic (including the divergent parabolas themselves) may be projected into the corresponding cubic of this class by making the tangent at a point of inflexion the line at infinity (a defining property of the general divergent parabola). Clearly, in forms 1 and 3 two finite inflexion points lie in a line parallel to AG and so are collinear with the third, which is at infinity in the vertical direction. (It is an immediate corollary, perhaps later known to Newton, that in all cubics having three real inflexion points these are in line since they are so in the corresponding divergent parabolas into which each may be projected. Historically, this was first stated

partes. Basis AB est ejus Diameter, sed non habet Asymptoton;(90) Ellipsis etiam aliquando apparet, nempe

Forma 1. Si omnes radices $A\tau$, AT, & At hujus Æquationis $dx^3+kx+l=0$ sunt reales figura constat ex Ellipsi & Parabola quæ Campanæ fere similis est. fig 1 & 2.

Forma 2. Si duæ radices AT & At sunt æquales et ejusdem signi cum termino d, Ellipsis et Parabola in puncto T junguntur, nempe Crura Parabolica sese decussando in Ellipsin continuantur. Fig 3.

Forma 3. Si duæ radices $A\tau$ & AT sunt æquales & signi diversi a signo termini d, tum pro Ellipsi habemus unicum punctum T. Item si duæ quævis radices sunt imaginariæ sola manet Parabola sine aliqua Ellipsi, puncto, decussatione vel cuspide. fig 4 & 5.

Forma 4. Si termini k & l desunt Parabola erit Cuspidata in puncto A. fig 6.(91)

De Octavo Casu *& Specie Decima Quinta*

In casu octavo ubi habetur $fxy=dx^3+l$, Figura(92) habet quatuor crura infinita quorum duo sunt Hyperbolica circa Asymptoton AG & duo Parabolica: Et horum tria tendunt ad easdem partes, quartum (Hyperbolicum) ad oppositas. Ista duo crura (Parabolicum & Hyperbolicum) sibi junguntur quæ jacent ex eodem latere lineæ AG, & ista quæ ex altero.(93) Vide Figuram.

De Nono Casu, *& Specie Decima Sexta*

Deniq; si Casus sit $hy=dx^3$, Figura erit Parabola cujus duo crura ab invicem reflexa tendunt ex diverso latere lineæ AG ad oppositas partes lineæ AB; quam ideo Parabolam intortam(94) nomino. Et punctum A est ejus centrum. Vide figuram.(95)

---

explicitly by Colin Maclaurin in his *De Linearum Geometricarum Proprietatibus generalibus* (note (22)): §68 = Section III, Prop. X.)

(89) 'duas' (two) is cancelled.

(90) We would say, rather, that it was at infinity, parallel to AG.

(91) This case includes four forms of a single grade each, or there are 4 distinct types in it.

(92) This is the cubic (the Cartesian 'trident', as Newton was later to name it) by which Descartes in his *Geometrie*, employing its more manageable equivalent defining equation $dxy = (b-y)(y^2-cd)$, constructed the general sextic as its meets with a defined circle (see I, 3, 3, §2: 495).

(93) The trident's infinite branches are determined by $fxy = dx^3$: of these the line AG ($x = 0$) is its unique real aysmptote and the asymptotic parabola $fy = dx^2$ determines a pair of parabolic points at infinity on AG, so that there the trident crosses itself and is touched by the line at infinity.

extend in opposite[89] directions. The base AB is its diameter but it has no asymptote.[90] An oval also appears sometimes. In detail:

Form 1. If all the roots $A\tau$, AT and At of this equation $dx^3+kx+l=0$ are real, the figure consists of an oval and a parabolic curve which is almost bell-like. Figures 1 and 2.

Form 2. If two roots AT and At are equal and of the same sign as the term in d, the oval and parabola are joined in the point T: precisely, the parabolic branches on crossing are continued into the oval. Figure 3.

Form 3. If two roots $A\tau$ and AT are equal and opposite in sign to the term in d, then instead of an oval we have the single point T. Likewise, if any two roots are imaginary, there remains the parabola alone without any oval, conjugate point, crossing or cusp. Figures 4 and 5.

Form 4. If the terms in k and l are missing, the parabola will be cusped in the point A. Figure 6.[91]

THE EIGHTH CASE *and fifteenth Species*

In the eighth case, where we have $fxy = dx^3+l$, the figure[92] has four infinite branches, two of which are hyperbolic round the asymptote AG and two parabolic: three of these extend in the same direction, the fourth (an hyperbolic one) in the opposite. Those two branches (one parabolic, one hyperbolic) lying on the same side of the line AG are joined, and those lying on the other.[93] See the figure.

THE NINTH CASE *and sixteenth Species*

If, finally, the case be $hy = dx^3$, the figure will be a parabolic curve whose two branches bent back one from the other extend on either side of the line AG in opposite directions from the line AB: for that reason I call it a twisted parabola.[94] Also, the point A is its centre. See the figure.[95]

---

(94) Newton earlier (compare I, 3, §2: note (30); and also 3, 2, §2: note (120) below) called this 'a Parabola of ye 2^d kind': later, in recognition of its inventor (John Wallis, *De Sectionibus Conicis* (I, 1, Introduction: Appendix 2, note (20)): 106: Prop. 46) he will call it a 'Wallis' cubic. Its most important geometrical properties are outlined by him in 3, 2, §2 below. It has a unique real asymptote at infinity parallel to AG and this meets the cubic in a cusp (also at infinity). Since its defining equation contains no even terms it has an anti-symmetric centre at the origin A.

(95) With this final unique type of cubic Newton has succeeded in enumerating altogether 58 distinct kinds. (Compare note (1) and also notes (36), (44), (48), (55), (60), (72), (80), (82), (84), (89), (91) and (93).)

Figuræ primæ Speciei (*Figures for species* I).

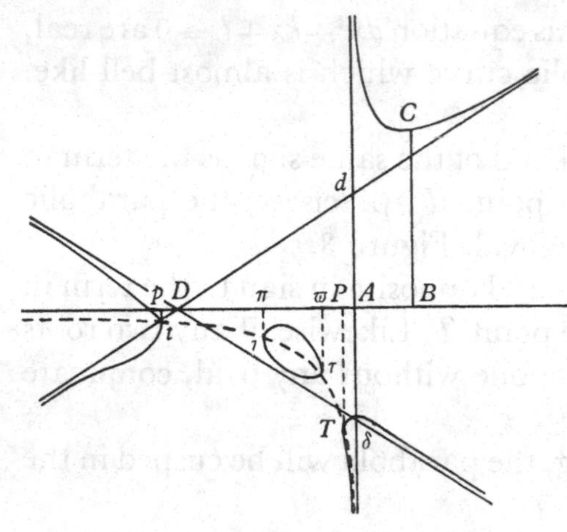

Fig 3

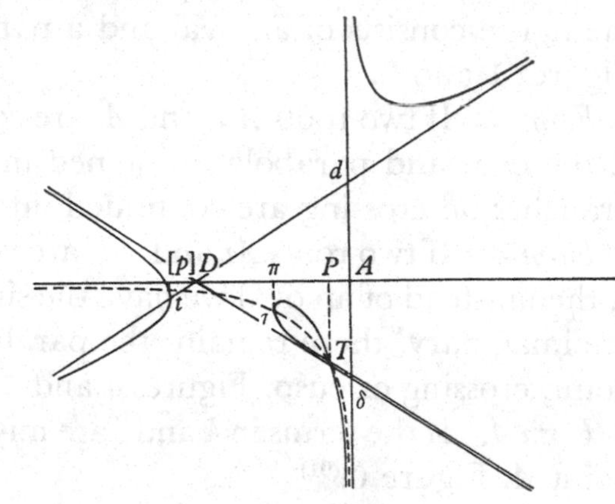

[Fig] 4

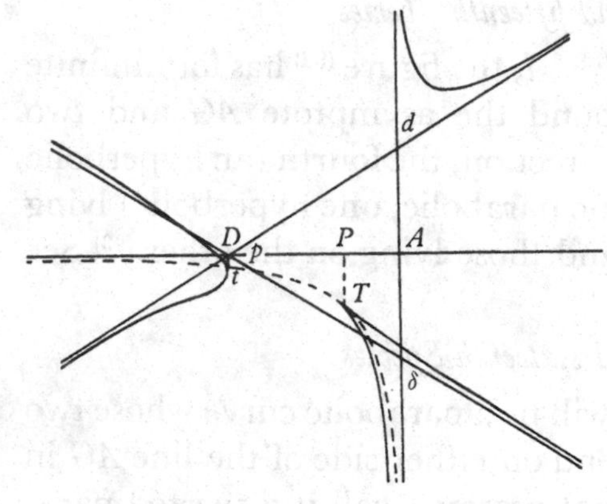

[Fig] 5

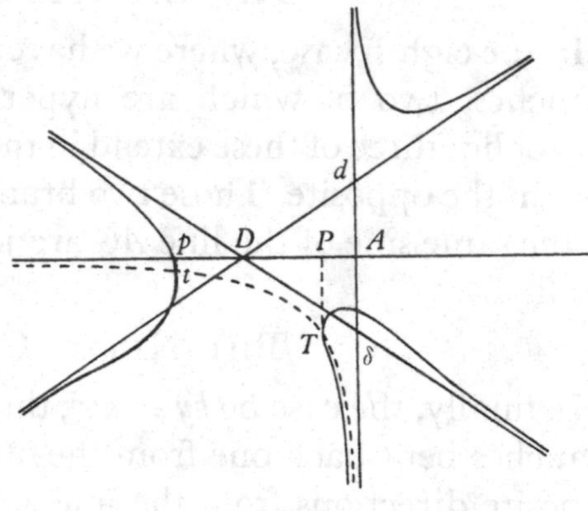

[Fig] 6

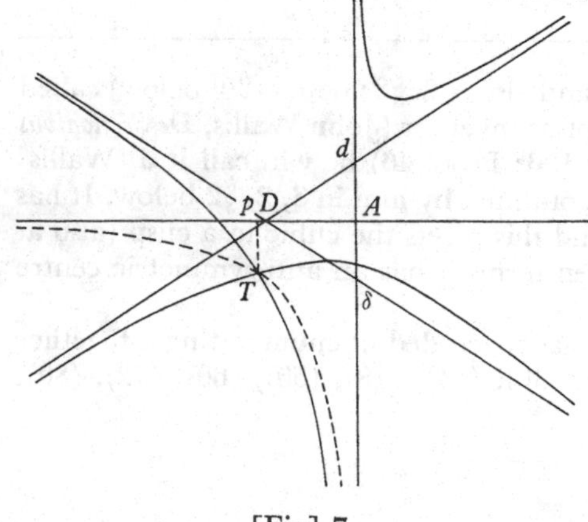

[Fig] 7

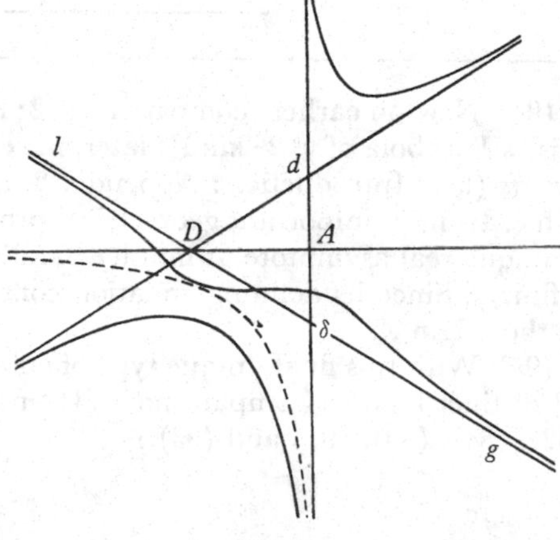

[Fig] 8

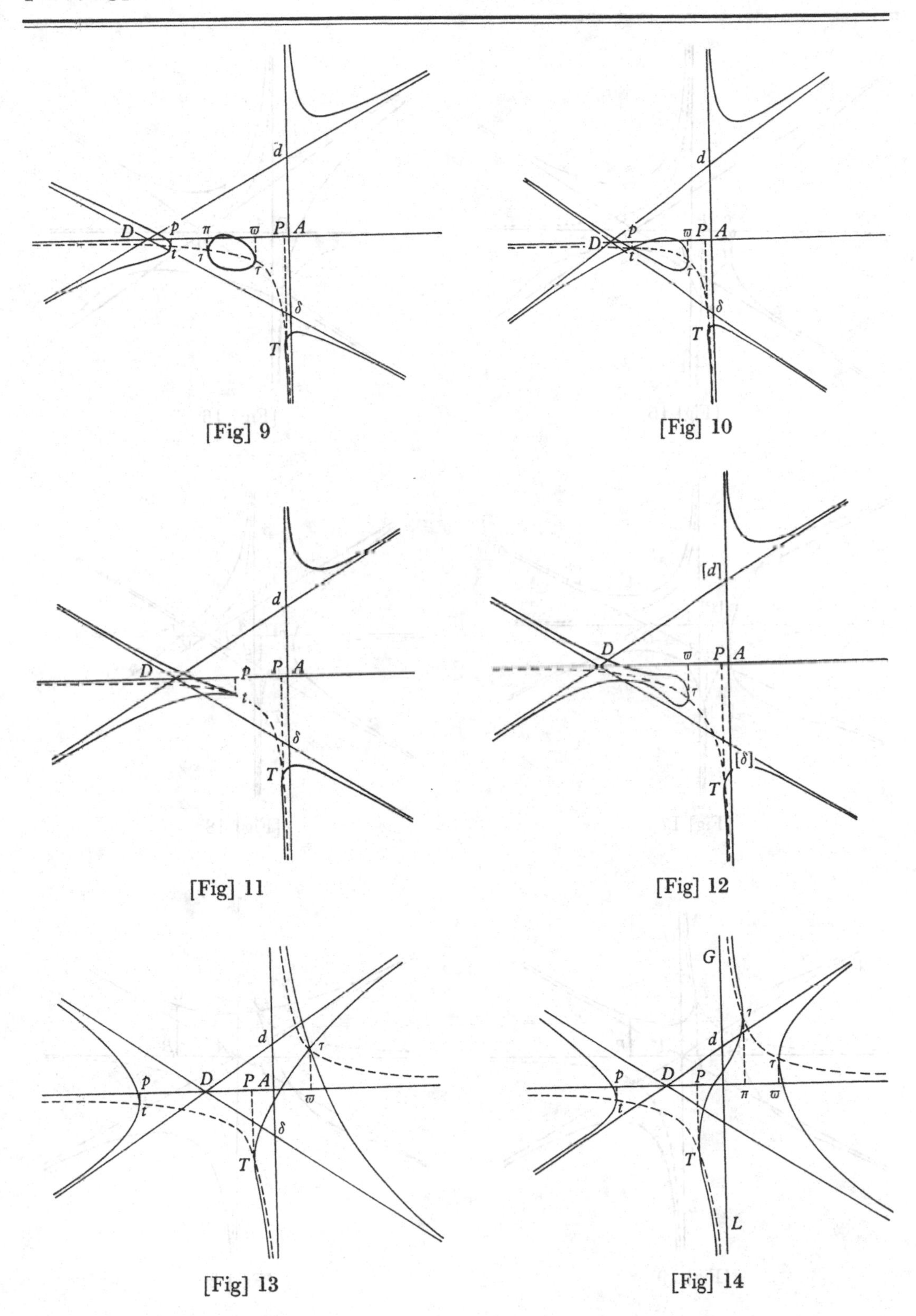

[Fig] 9

[Fig] 10

[Fig] 11

[Fig] 12

[Fig] 13

[Fig] 14

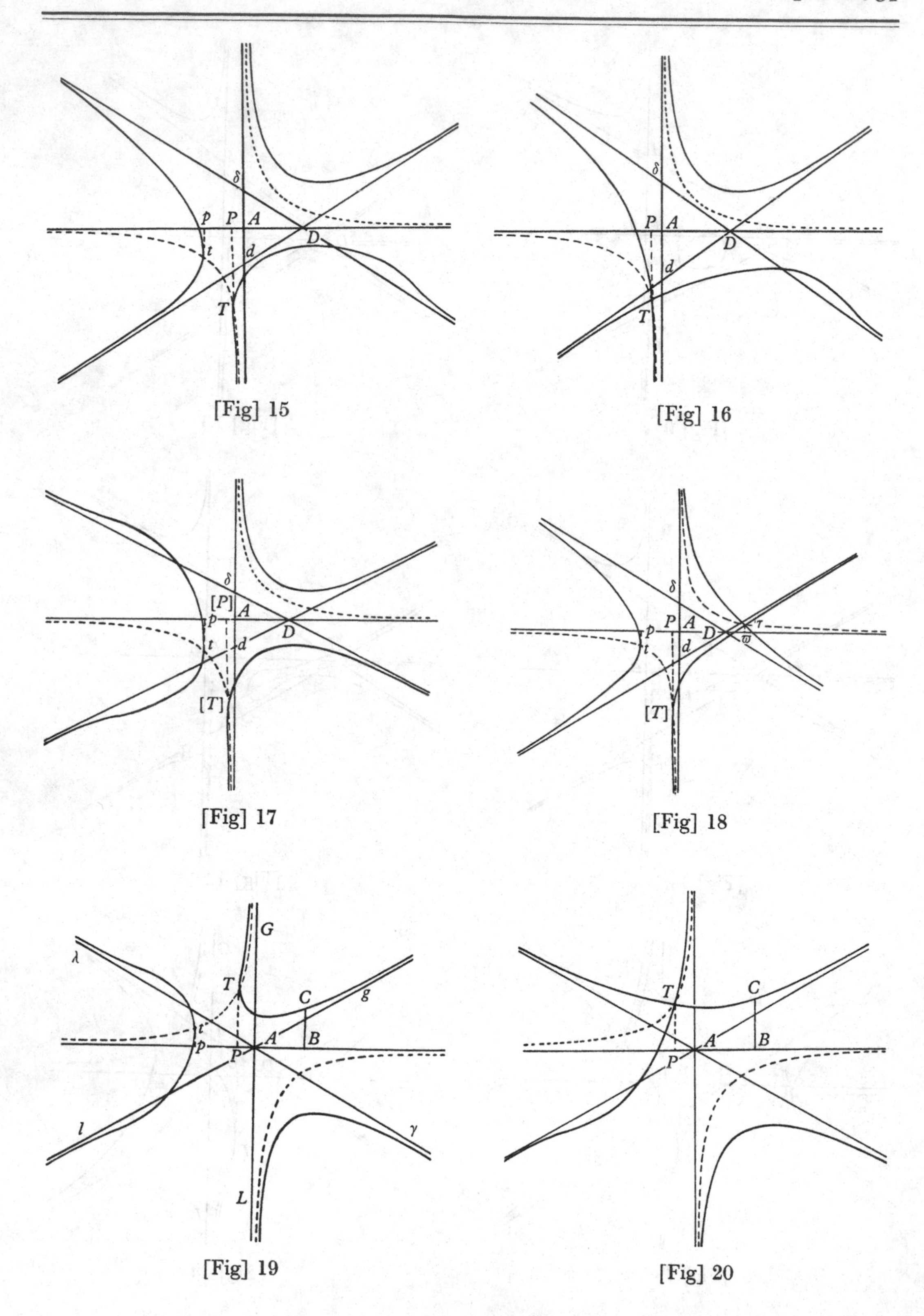

[Fig] 15

[Fig] 16

[Fig] 17

[Fig] 18

[Fig] 19

[Fig] 20

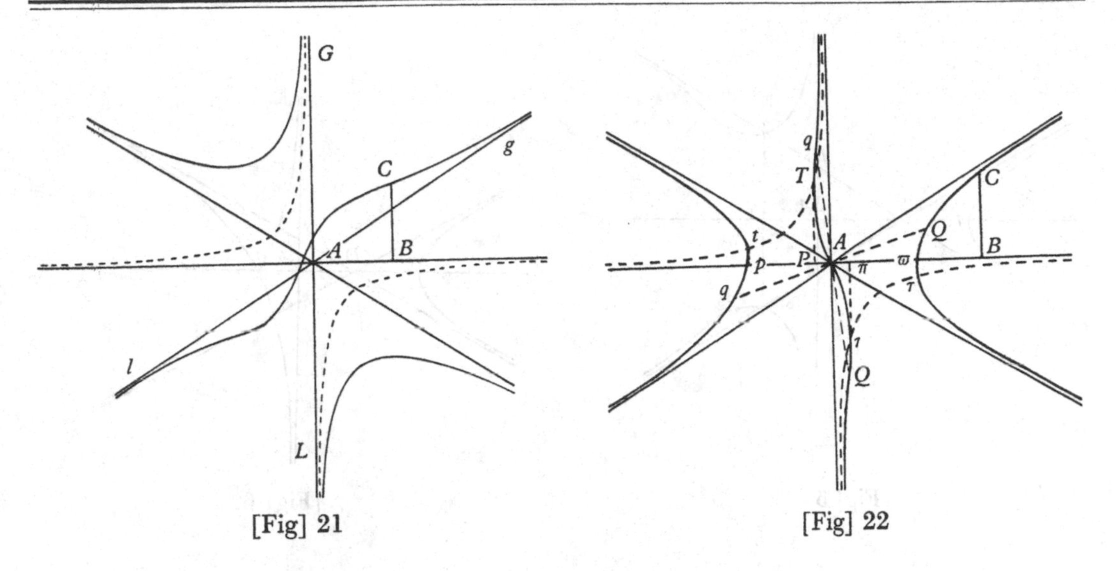

[Fig] 21 [Fig] 22

Figuræ secundæ Speciei (*Figures for species 2*).

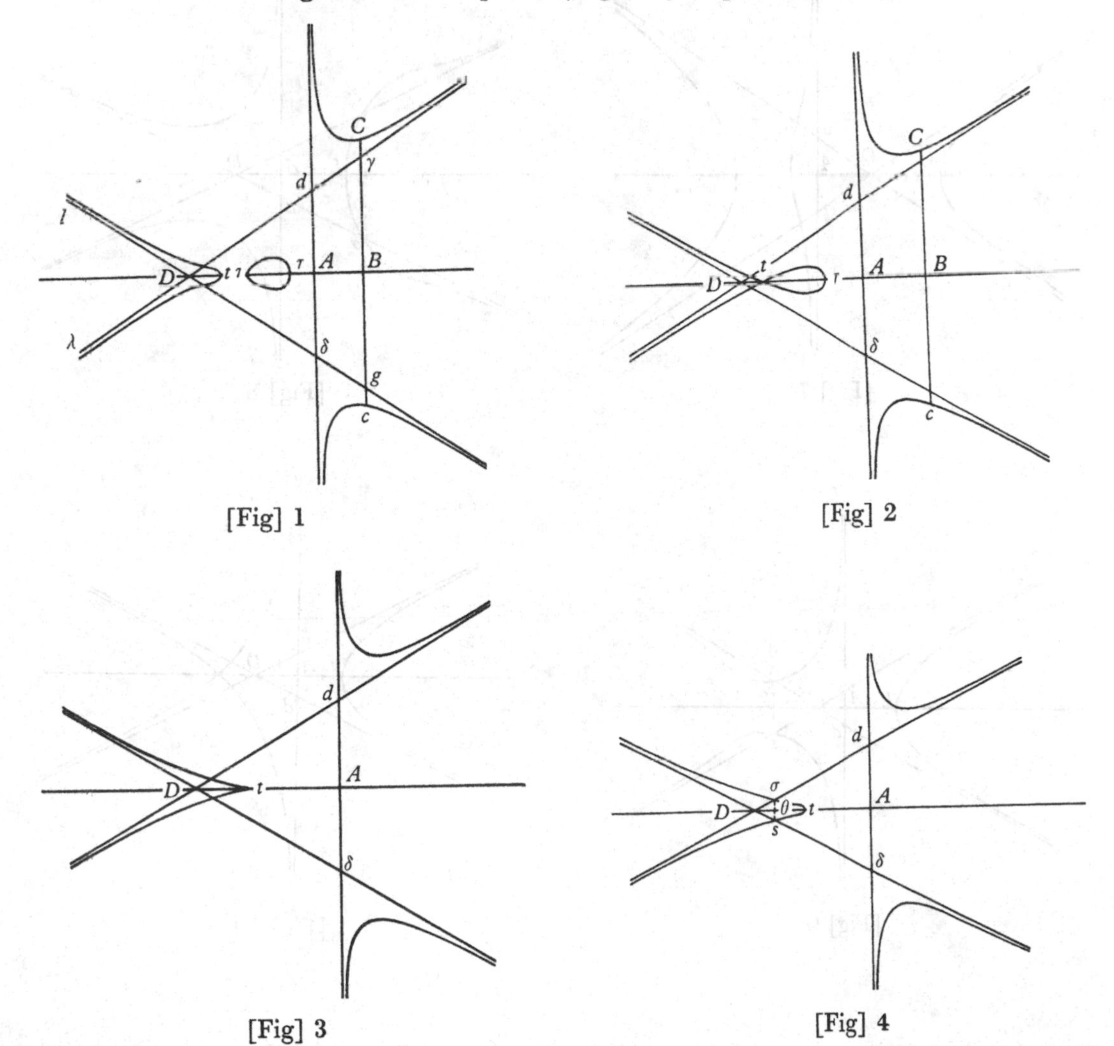

[Fig] 1 [Fig] 2

[Fig] 3 [Fig] 4

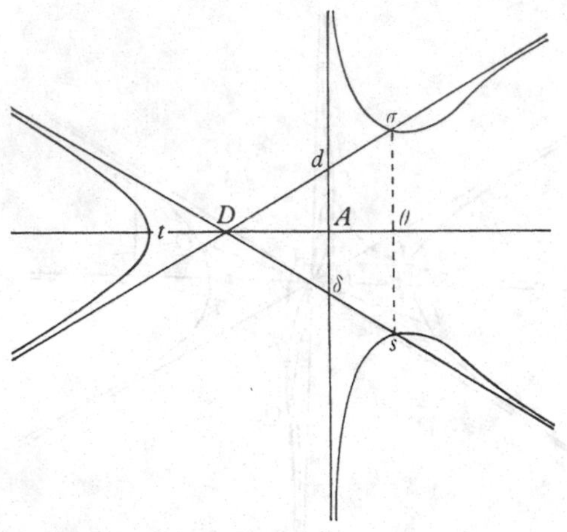

[Fig] 5

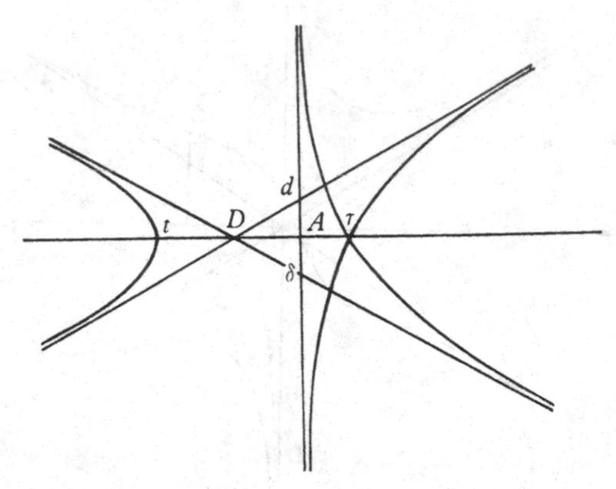

[Fig] 6

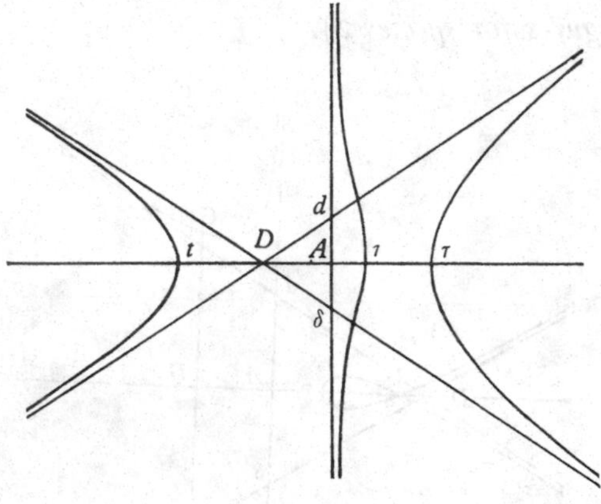

[Fig] 7

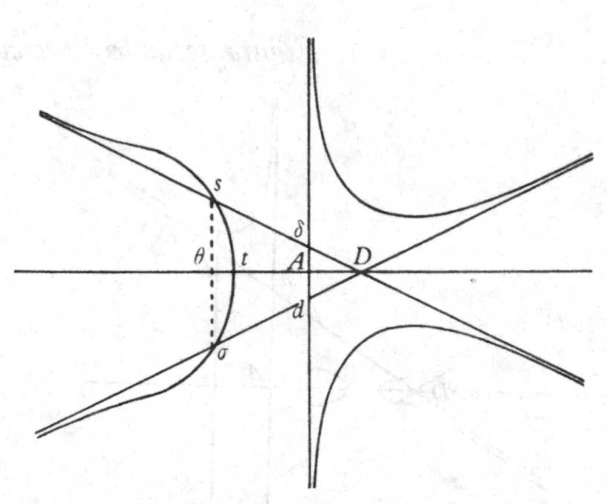

[Fig] 8

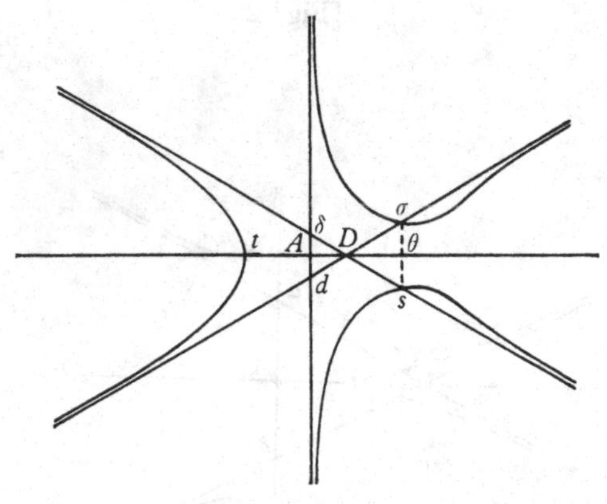

[Fig] 9

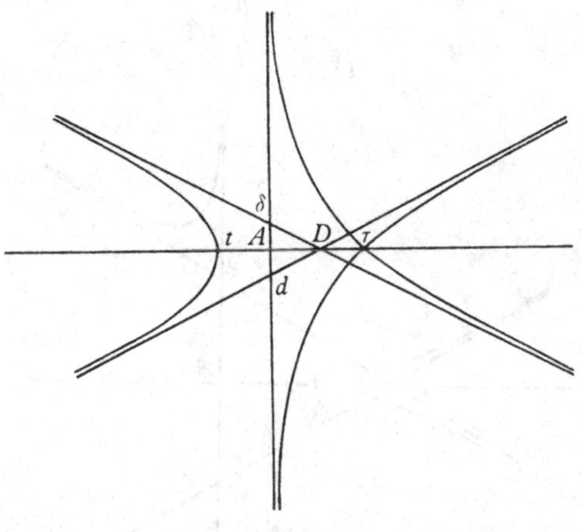

[Fig] 10

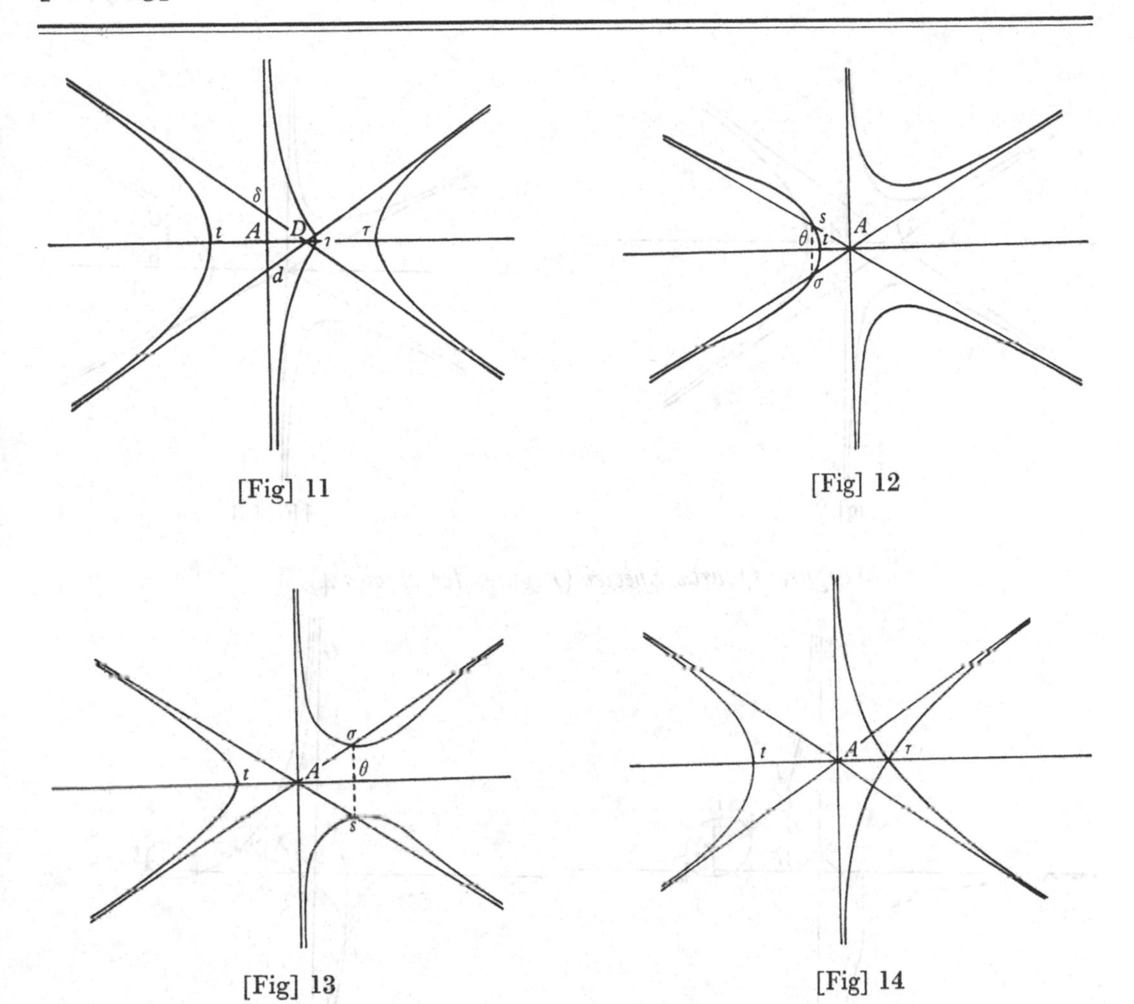

[Fig] 11

[Fig] 12

[Fig] 13

[Fig] 14

Figuræ tertiæ Speciei (*Figures for species* 3).

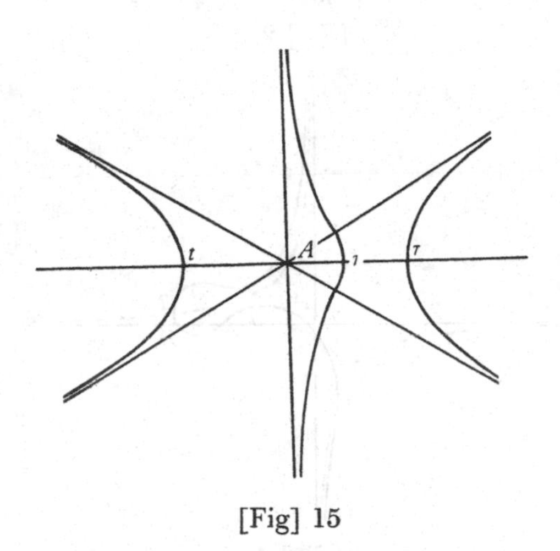

[Fig] 15

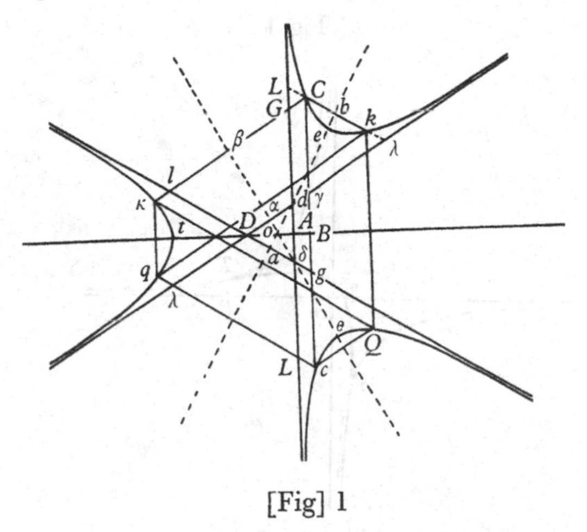

[Fig] 1

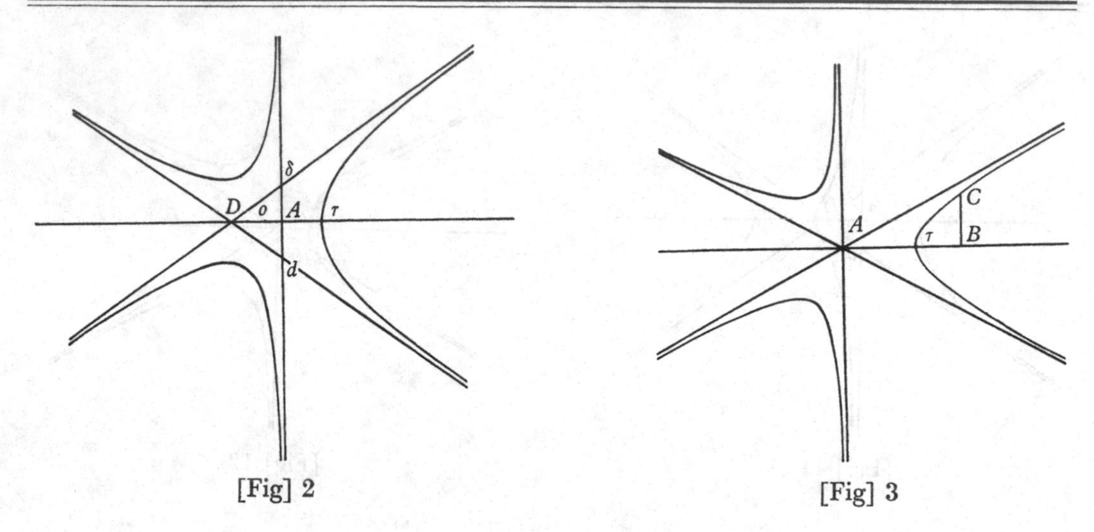

[Fig] 2

[Fig] 3

Figuræ Quartæ Speciei (*Figures for species* 4).

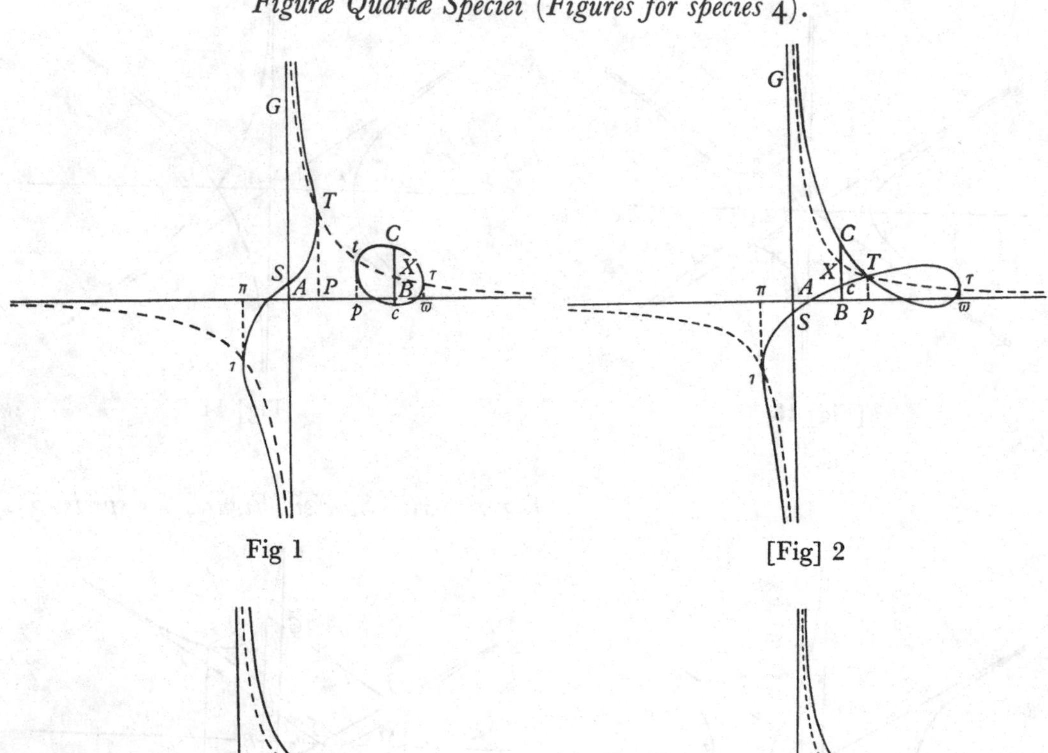

Fig 1

[Fig] 2

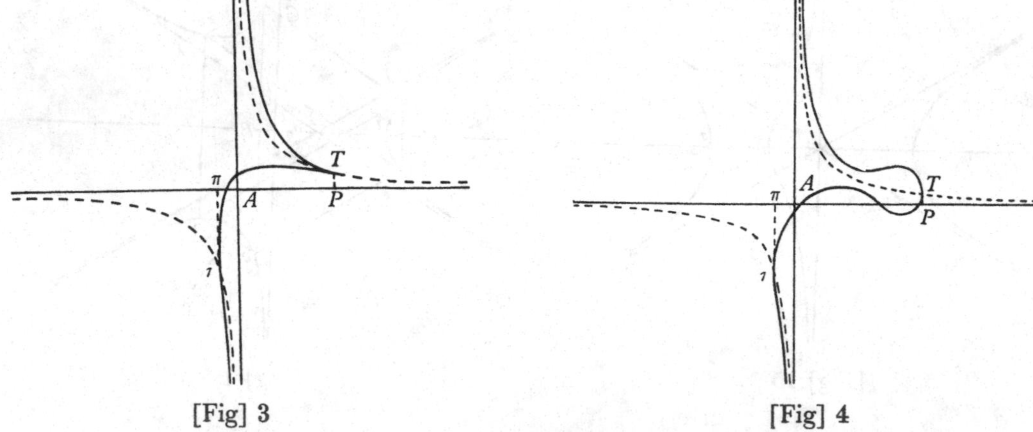

[Fig] 3

[Fig] 4

Figuræ Quintæ Speciei (*Figures for species* 5).

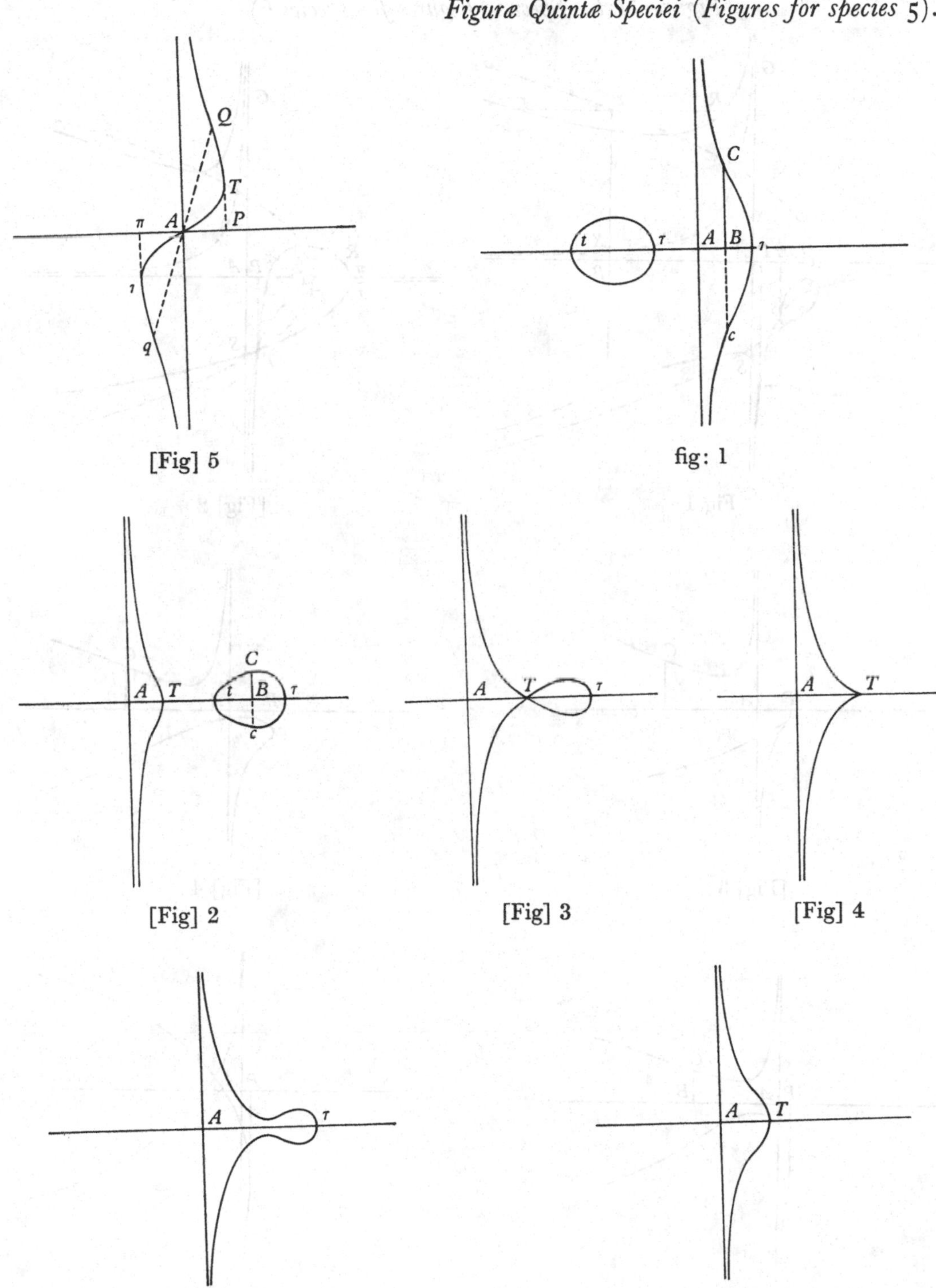

[Fig] 5 fig: 1

[Fig] 2 [Fig] 3 [Fig] 4

[Fig] 5 [Fig] 6

Figuræ Sextiæ Speciei (*Figures for species* 6).

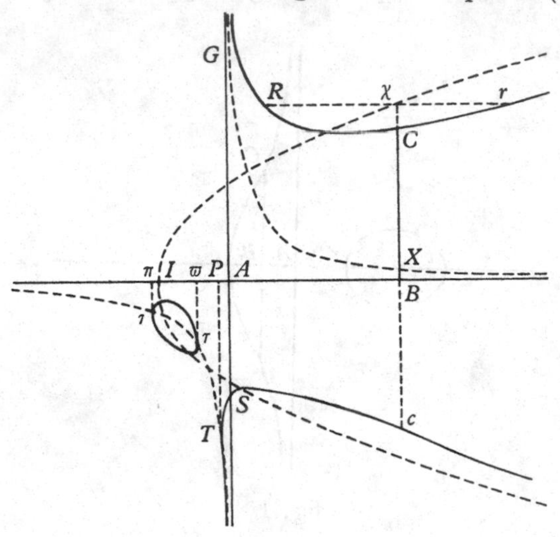

Fig 1

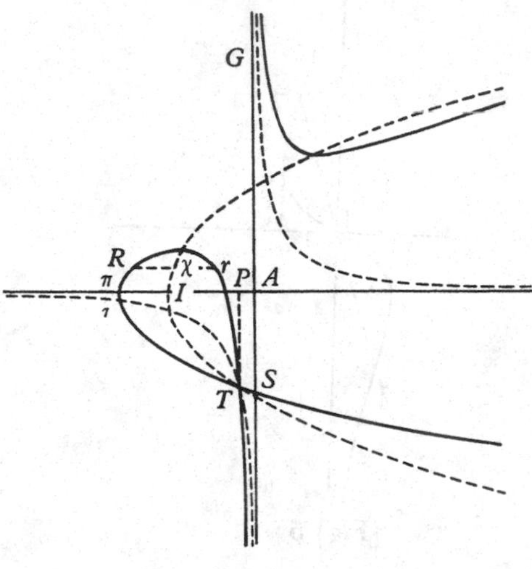

[Fig] 2

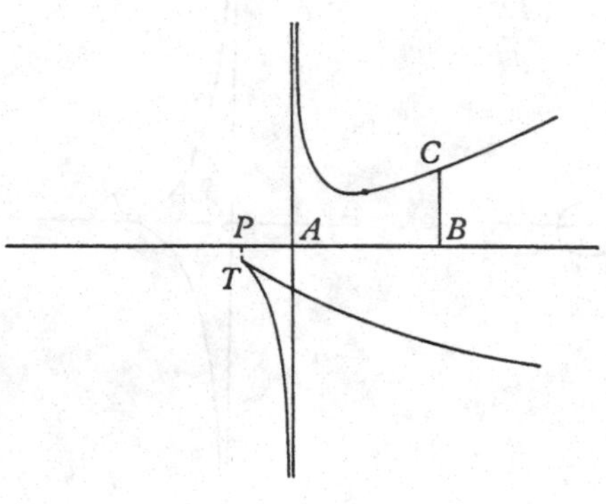

[Fig] 3

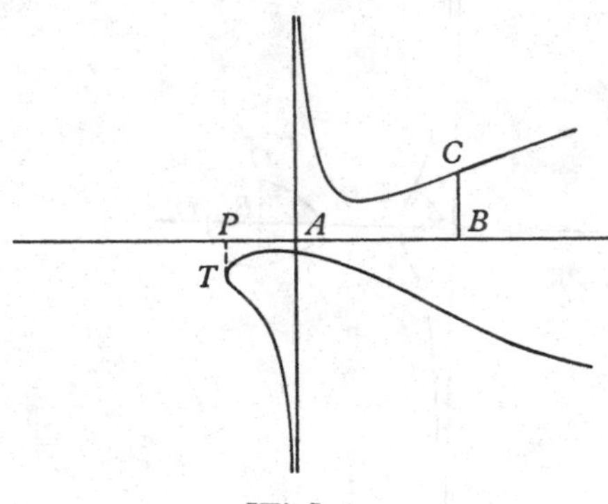

[Fig] 4

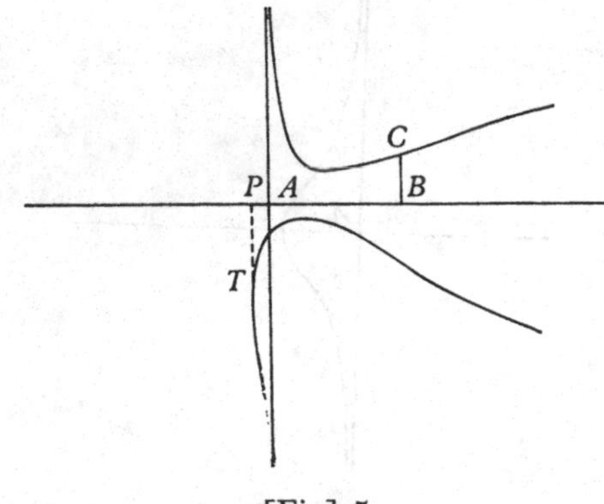

[Fig] 5

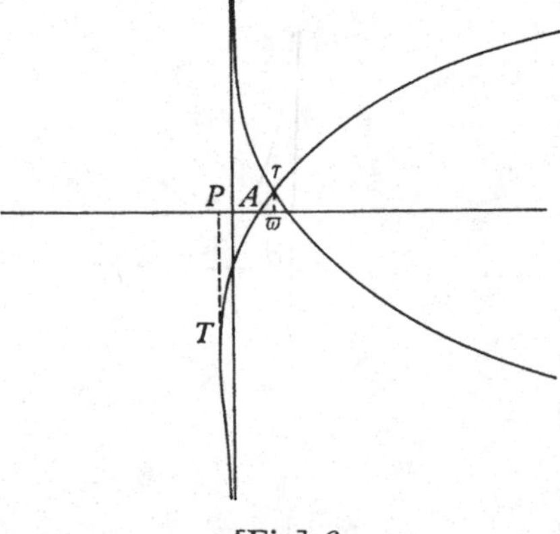

[Fig] 6

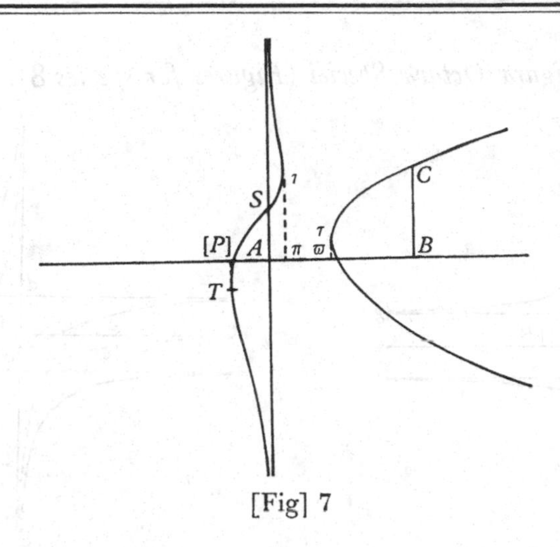

[Fig] 7

Figuræ Septimæ Speciei (*Figures for species* 7).

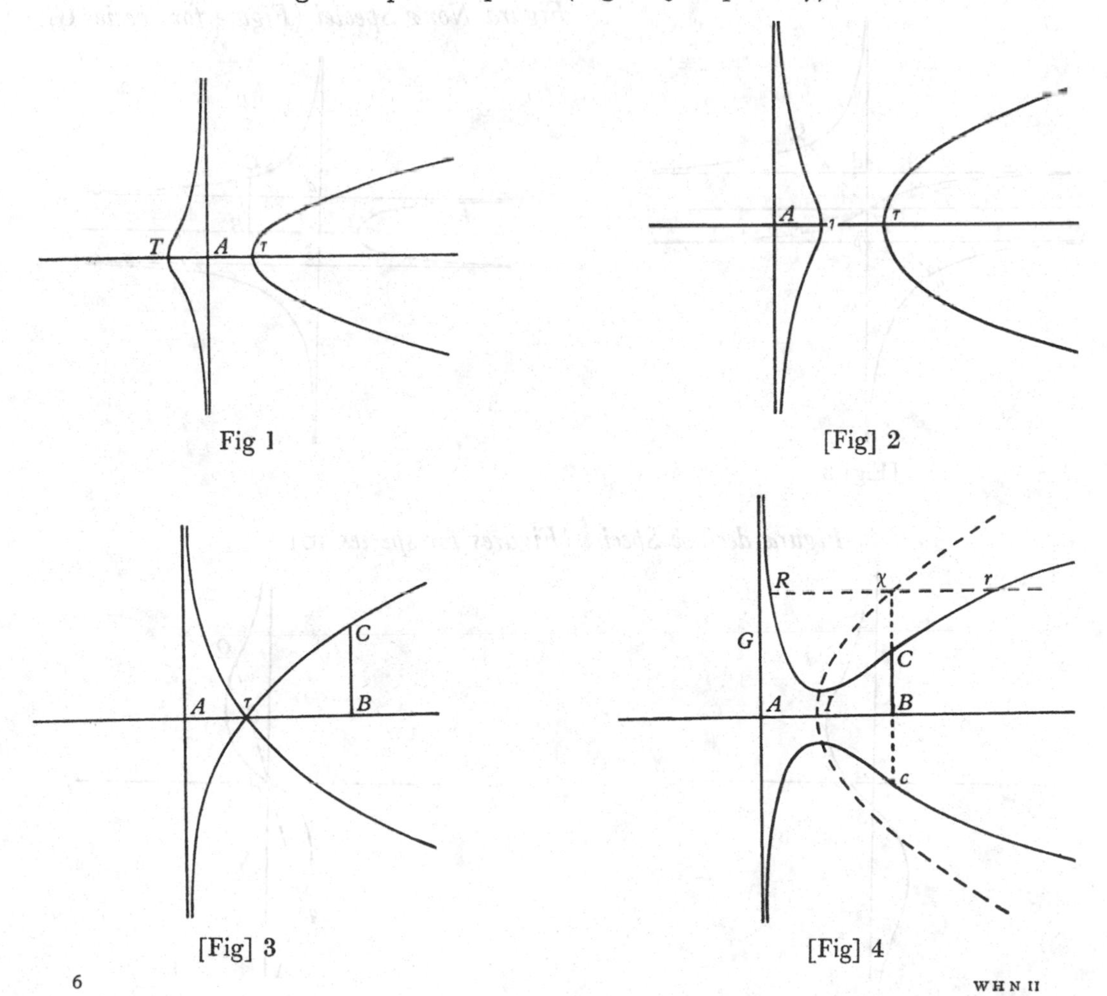

Fig 1

[Fig] 2

[Fig] 3

[Fig] 4

Figuræ Octavæ Speciei (*Figures for species* 8).

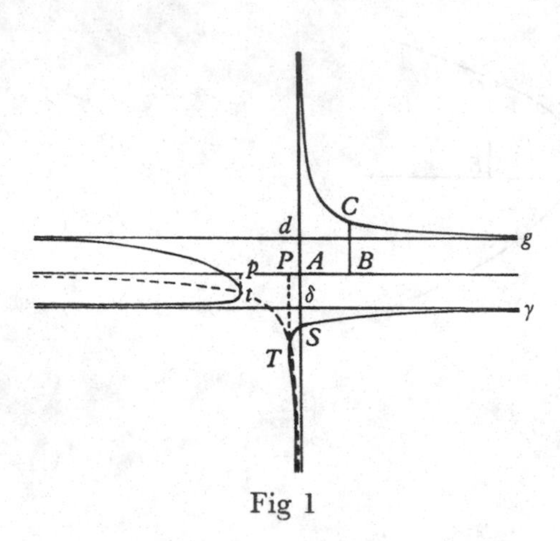

Fig 1

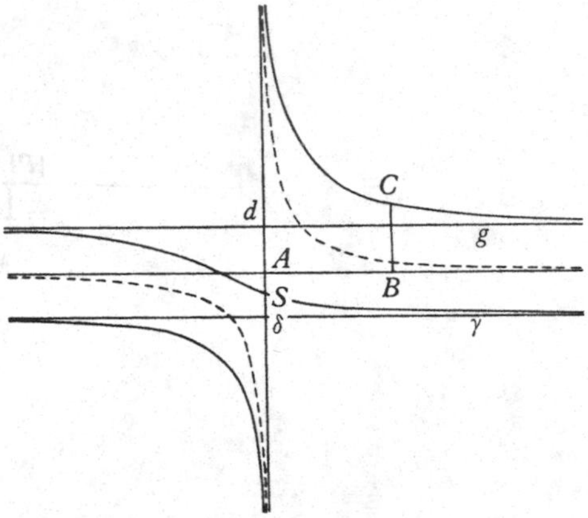

[Fig] 2

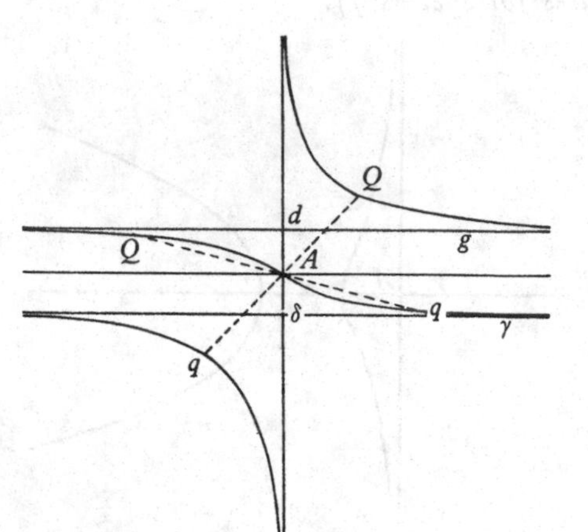

[Fig] 3

Figura Nonæ Speciei (*Figure for species* 9).

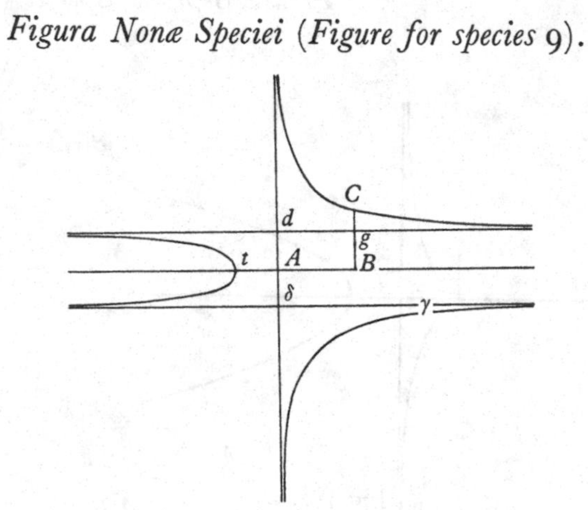

Figuræ decimæ Speciei (*Figures for species* 10).

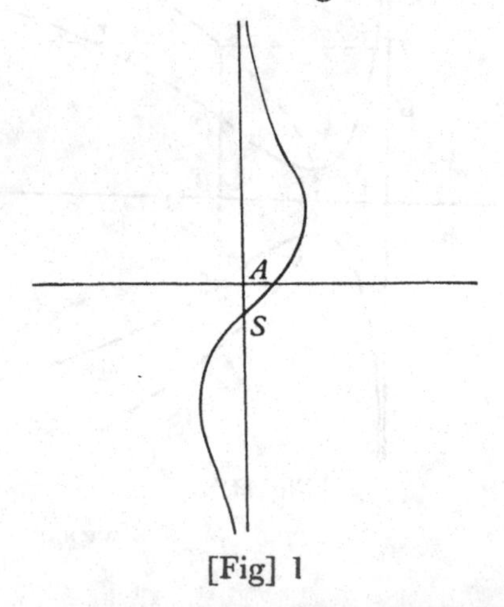

[Fig] 1

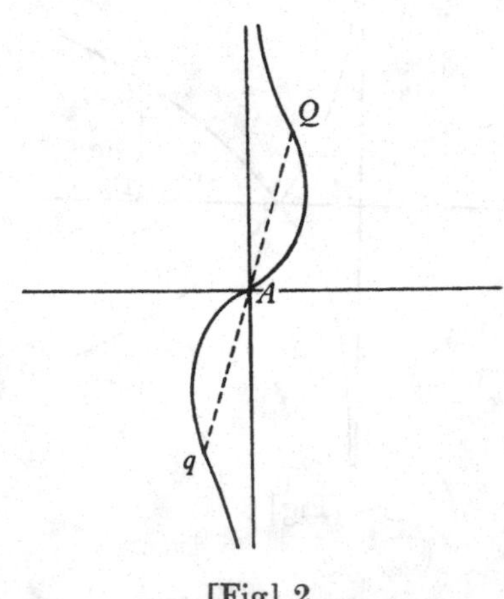

[Fig] 2

Figura undecimæ speciei
(*Figure for species* 11).

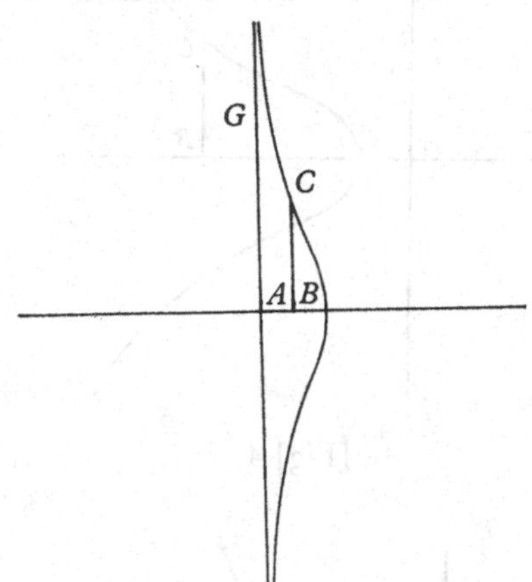

Figura Speciei Duodecimæ
(*Figure for species* 12).

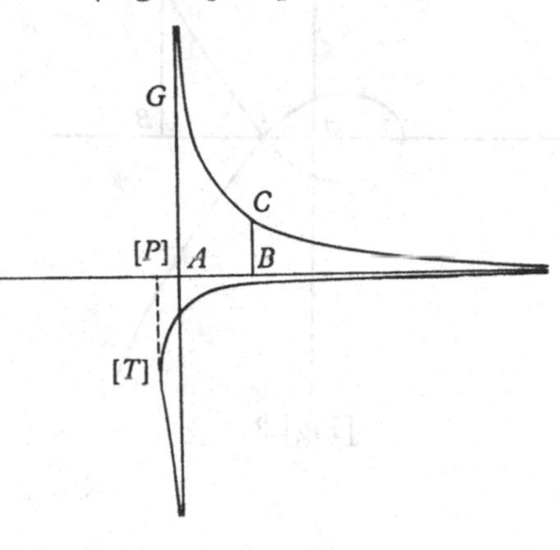

Figura Speciei decimæ tertiæ (*Figure for species* 13).

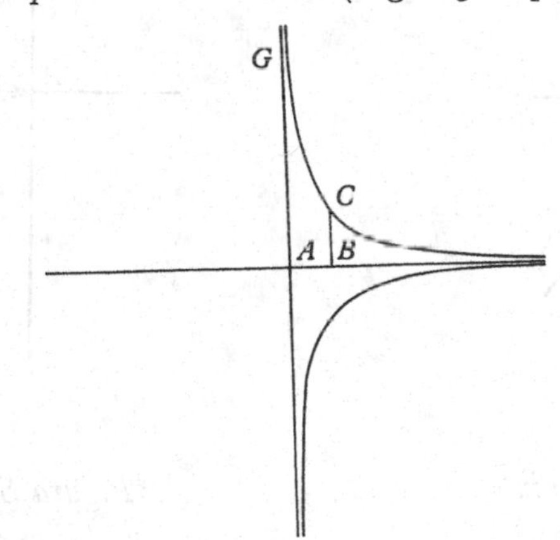

Figuræ Speciei decimæ Quartæ (*Figures for species* 14).

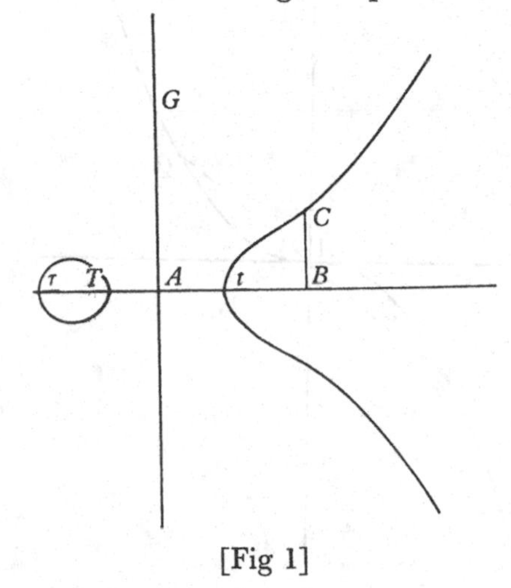

[Fig 1]

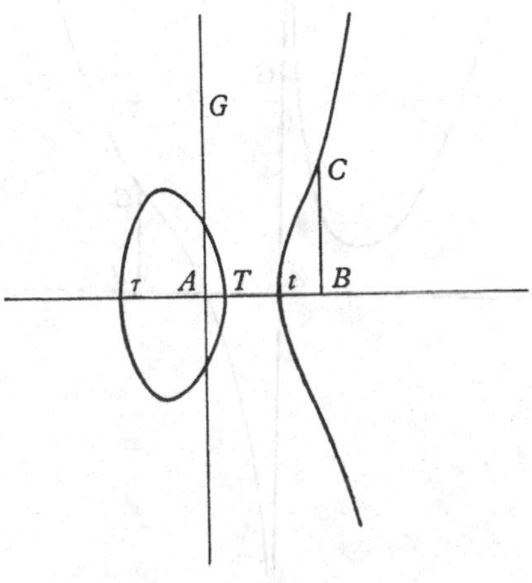

[Fig] 2

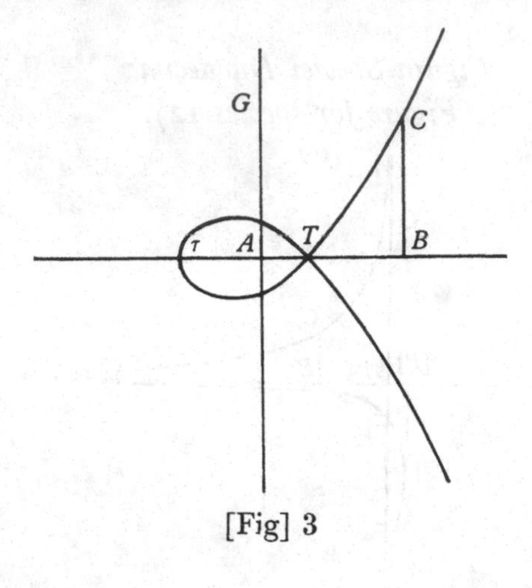

[Fig] 3

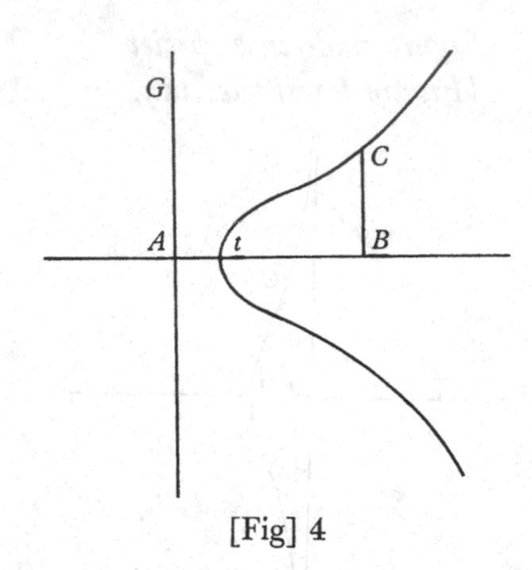

[Fig] 4

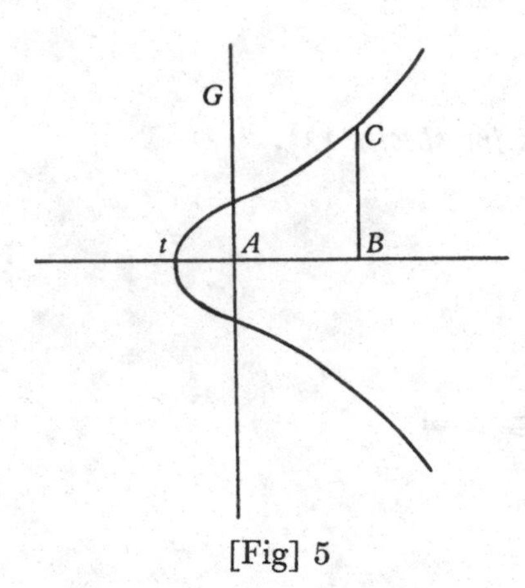

[Fig] 5

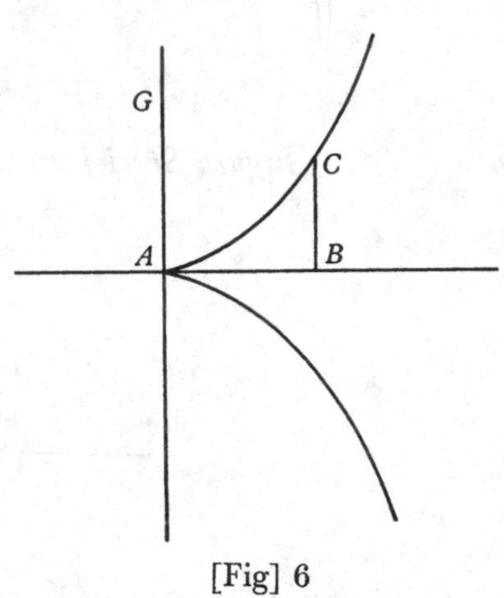

[Fig] 6

Figura Speciei decimæ quintæ
(*Figure for species* 15).

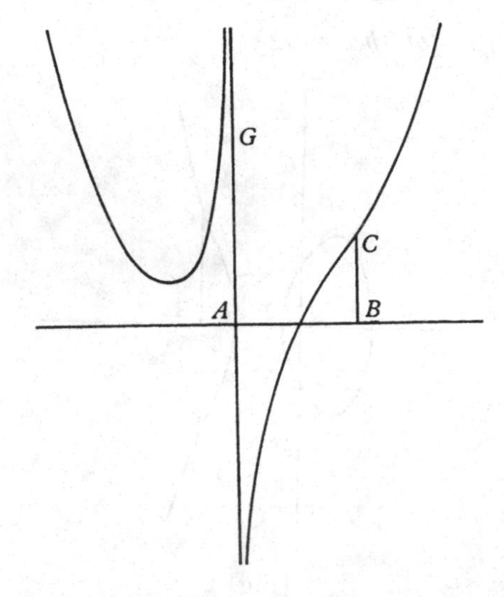

Figura Speciei decimæ Sextæ
(*Figure for species* 16).

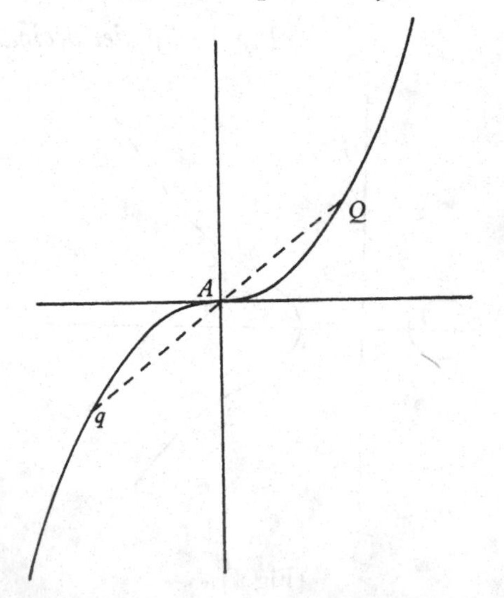

APPENDIX [See §3, note (67).]

Figuræ Sextæ Speciei (*Figures for species* 6) [draft]

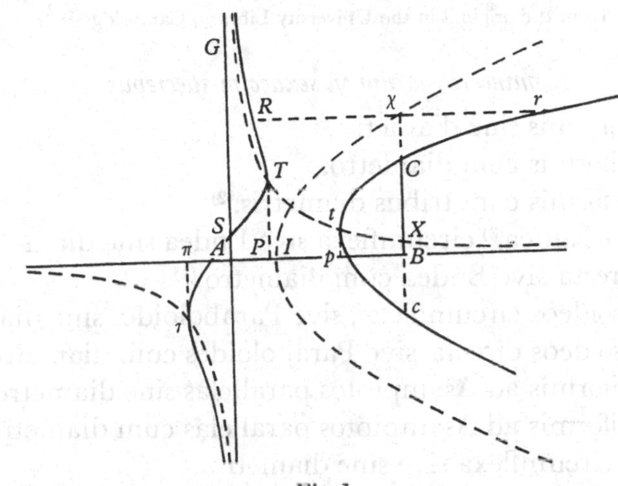

Fig 1

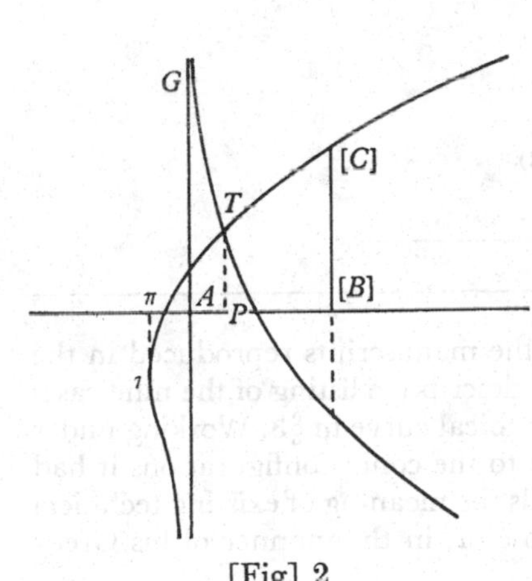

[Fig] 2

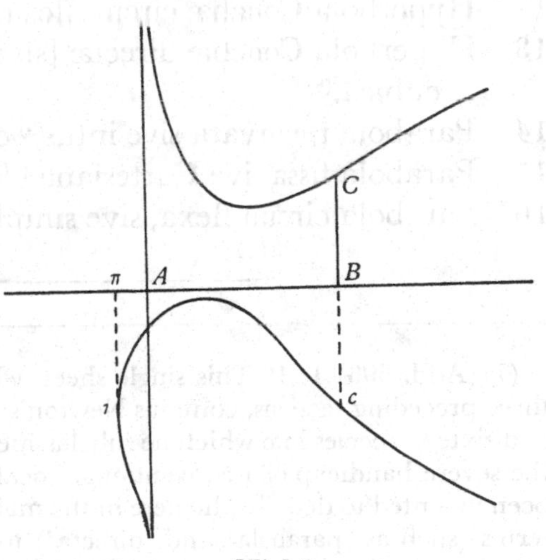

[Fig] 3

§4. BREAKDOWN OF THE CUBIC BY CASE AND SPECIES

From the original in the University Library, Cambridge(1)

[1] *Nomina curvarum in sexdecim speciebus.*

1 Hyperbola triformis sine diametro.
2 Hyperbola triformis cum diametro.
3 Hyperbola triformis cum tribus diametris.(2)
4 Cissoidalis sive Endea(3) circumflexa sive Endea sine diametro.
5 Cissoidalis directa sive Endea cum diametro.(4)
6 Parabola Cissoideos circumflexæ, sive Paraboloides sine diametro.
7 Parabola Cissoideos directæ sive Paraboloides cum diametro.(5)
8 Hyperbola triformis ad Asymptotos parallelas sine diametro.
9 Hyperbola triformis ad Asymptotos parallelas cum diametro.(6)
10 Conchoidalis circumflexa sive sine diametro.
11 Conchoidalis directa sive cum diametro.(7)
12 Hyperbola Conchæ circumflexæ, sive Crux Hyperbolica circumflexa.
13 Hyperbola Conchæ directæ [sive] Crux Hyperbolica inflexa, sive simplex cubica.(8)
14 Parabola recurvata sive intus(9) convexa.
15 Parabola fissa sive Cartesiana.(10)
16 Parabola circumflexa, sive simplex cubica.(11)

---

(1) Add. 3961.1: 1r. This single sheet which envelops the manuscripts reproduced in the three preceding sections, contains Newton's first essays at a descriptive listing of the nine cases and sixteen species into which he subclassified the general cubical curve in §3. Working under the severe handicap of a conventional vocabulary tailored to the conic configurations it had been invented to describe, he here in the main either extends the meaning of existing technical terms (such as 'parabola' and 'directa') to suit his purpose or, in the manner of his Greek predecessors, has recourse to suitable biological analogy.

(2) These three species, which together make up Newton's first case

$$bxy^2 = dx^3 + gx^2 + hy + kx + l$$

will later (see our volume 7) be called by him 'redundant' (excessive) hyperbolas.

(3) We have been unable to satisfy ourselves uniquely on the meaning of 'endeal'. It was our first impulse, one not wholly disowned in retrospect, to suppose it an adjectival variant of 'endiva', that is, *cichoria endivia*, the common variety of which has a curled or frizzled leaf not unlike that of the ivy-plant (to which 'Endea' is here juxtaposed): on that interpretation we should perhaps render it as 'chicory-leaf'. Most, however, may prefer to think of 'endea' as the feminine Latinized participle form of the Greek verb ἐνδέω, taking it, in other words, as a strict equivalent of 'defectiva': compare note (4).

Translation

[1] *The denominations of [cubic] curves in their sixteen species*(1)

1 The three-limbed adiametral hyperbola.
2 The three-limbed diametral hyperbola.
3 The three-limbed tridiametral hyperbola.(2)
4 The round-bent ivy-leaf or endeal(3) or the adiametral endeal.
5 The straight ivy-leaf or diametral endeal.(4)
6 The parabolic form of the round-bent ivy-leaf or the adiametral paraboloid.
7 The parabolic form of the straight ivy-leaf or the diametral paraboloid.(5)
8 The three-limbed adiametral hyperbola with parallel asymptotes.
9 The three-limbed diametral hyperbola with parallel asymptotes.(6)
10 The around-bent or adiametral mussel-shell.
11 The straight or diametral mussel-shell.(7)
12 The hyperbolic form of the around-bent mussel or around-bent hyperbolic cross.
13 The hyperbolic form of the straight mussel [or] inbent hyperbolic cross or simple cubical hyperbola.(8)
14 The backwardly curved or inwardly(9) convex parabola.
15 The cleft or Cartesian parabola.(10)
16 The around-bent or simple cubical parabola.(11)

---

(4) The two species which compose case 2 ($bxy^2 = -dx^3+gx^2+hy+kx+l$) and which Newton will later name collectively the class of 'defective' hyperbolas.

(5) The two species of 'parabolic hyperbolas' which together form case 3, viz.

$$bxy^2 = gx^2+hy+kx+l.$$

(6) These two species, the two varieties of Newton's case 4 ($bxy^2 = hy+kx+l$), will subsequently be denominated 'hyperbolisms'.

(7) The two component species of Newton's case 5 ($bxy^2 = hy-kx+l$), the class of 'elliptical hyperbolisms'.

(8) The phrase 'Crux Hyperbolica inflexa' is a late insertion which grammatically destroys the dependence of 'cubica' on its noun 'Hyperbola'. Species 12 and 13 are the two divisions of case 6 ($bxy^2 = hy+l$), the class of 'parabolic hyperbolisms'.

(9) Newton first wrote 'versus axem' (towards the axis). The present species is his seventh division ($ey^2 = dx^3+kx+l$) of the general cubic, the class of 'divergent' parabolas in his subsequent revised description.

(10) That is, case 8 ($fxy = dx^3+l$), the Cartesian (or parabolic) 'trident' in Newton's later standardized nomenclature. See §2, note (92).

(11) The final case ($hy = dx^3$) of the 'Wallisian' parabola (§2, note (94)).

[2] [*Nomina curvarum in novem casibus*][12]

1 Hyperbola triformis 1 sine diametro, 2 cum diametro, 3 cum tribus diametris.
2 Cissoi[d]alis[13] vel Endea 1 sine diametro 2 cum diametro.
3 Parabola 1 sine diametro 2 cum diametro.
4 Hyperbola Hyperbolæ 1 sine diam 2 cum diam.
5 Hyperbola Ellipseos 1 sine diam 2 cum diam.
6 Hyperbola Parabolæ 1 sine diam 2 cum diam.
7 Parabola reflexa.
8 Parabola fissa[14] sive Cartesiana.
9 Parabola circumflexa sive simplex cubica.[15]

(12) The original lacks any title and our suggested heading is modelled on that of [1]. The revised list which follows gains on that of the preceding in coherence but is less immediately evocative of the curves' geometrical forms.

[2] *[The denominations of cubics in their nine cases]*[12]

1 The three-limbed hyperbola: 1 adiametral, 2 diametral, 3 tridiametral.
2 The ivy-leaf or endeal: 1 adiametral, 2 diametral.
3 The parabolic cubic: 1 adiametral, 2 diametral.
4 The hyperbolic hyperbola: 1 adiametral, 2 diametral.
5 The elliptical hyperbola: 1 adiametral, 2 diametral.
6 The parabolic hyperbola: 1 adiametral, 2 diametral.
7 The back-bent parabola.
8 The cleft[14] or Cartesian parabola.
9 The round-bent or simple cubical parabola.[15]

---

(13) Newton wrote 'Cissoilalis'.
(14) 'inflexa' (inwardly bent) is cancelled.
(15) See §2: passim and notes (2) and (4)–(11) above for the subdivision of these nine cases into the previous sixteen cubic species.

2

RESEARCHES INTO THE GENERAL PROPERTIES OF CURVES

[1667 or 1668?][1]

§1. PRELIMINARY OUTLINES OF THE INVESTIGATION[2]

From the original in private possession

[1] *Some Generall Theorems.*

Eidem puncto (nisi polari[3]) non esse plures tangentes.

Diameter ad duas parallelas est diameter ad omnes, et ad nullas alias.[4]

Cuiqꝫ curvæ plures esse diametros axes sed non semper conjugatas.[5]

Cuiqꝫ Asymptoto parem esse numerum crurum.[6]

Curvas non differre specie propter angulos insistentium solùm mutatos.[7]

Superficies inter insistentes, sive tangentibus asymptotis arcubus subtensis diametro contiguæ esse hinc inde æquales.[8] Hinc a terminis ejusdem lunulæ ductis utcunqꝫ parallelis lunulæ interceptæ habebunt eandem diff[erentiam] vel summam. Idem de lineis inter curvam et Asymptoton intercept[i]s.[9]

---

(1) These present researches are evidently closely connected with the cubic investigations of the preceding division (compare (note (7) below): indeed, we would guess that they were intended to introduce the systematic subdivision of cubics which he evolved there. Since their handwriting and style are closely similar we attach here the same tentatively conjectured composition date as before.

(2) These are entered on two sides of the same folded scrap of paper. Newton here roughly enunciates some general properties of curves and then indicates his basic method of arriving at further general theorems of this kind: since any generalized statement on (algebraic) curves must hold true of conics in particular, we may conversely deduce many of those general theorems by seeking to universalize appropriate 'conick propertys', a selection of which he lists. We are strongly reminded of a much later paper of Euler's, written without knowledge of Newton's present researches of course, in which he tries to differentiate between those properties of conics which are not true of higher curves and those which are not restricted to conics: in particular, in independent rediscovery of many of Newton's unpublished findings Euler defines general classes of curves (exemplified in the case of cubics by the tridiametral

Translation

[1] [*Some Generall Theorems*]

At the same point (unless it is polar)(3) there cannot be several tangents.

A diameter to two parallel lines is a diameter to all lines parallel to them, and to no others.(4)

For each curve there are several diameters, axes but not always conjugate branches.(5)

For each asymptote there are an even number of corresponding infinite branches.(6)

Curves do not alter their species through having merely the angles of inclination of their ordinates changed.(7)

The surfaces cut off between ordinates to tangents, asymptotes or arcs subtended by an adjoining diameter are equal on its either side.(8) Consequently, the lunules intercepted by parallels drawn arbitrarily from the boundaries of the same lunule will have the same sum or difference. The same goes for lines intercepted between the curve and an asymptote.(9)

---

hyperbola classified by Newton here as his third species and later, in his published *Enumeratio*, as species 23) which have simple linear diameters. (See Leonhard Euler, 'Sur quelques Proprietés des Sections Coniques qui conviennent à une Infinité d'autres Lignes courbes', *Mémoires de l'Académie des Sciences de Berlin*, **1** (1745): 71–98 [= *Opera Omnia*, Series I, Vol. 27 (*Commentationes Geometricæ*, Vol. 2, ed. A. Speiser), Zurich, 1954: 51–77], especially §15ff.)

(3) Newton might better have written 'multiplici' (multiple). He clarified the matter in §2 below, where he noted that 'Polus est punctum curvæ duplicatum' (a pole is a double point of a curve): see note (13) of that section.

(4) Newton may intend his 'parallelæ' to be a line n-ple of parallels. In context, however, it would appear that the parallels intersect a general algebraic curve, in which case the Newtonian diameter conjugate to any two of the resulting 'parallelæ [chordæ]' is simultaneously a diameter to all.

(5) In the classical, Apollonian sense in which a conic hyperbola has a primary branch and a conjugate one.

(6) Newton makes the reasonable assumption, true for algebraic curves, that the general curve is continuous at infinity: indeed, without it for him the asymptotic tangent would not be defined.

(7) A generalization of his observation, there restricted to cubics, in 1, §1 (compare note (5)).

(8) This sentence, heavily revised with many illegible cancellations, is abrupt to the point of obscurity. We may guess that Newton is groping after the first theorem of §2, [2] below. Since the 'insistent' lines delimiting the surfaces are ordinate to the same diameter they are, of course, parallel.

(9) We are equally confused in our efforts to give an exact interpretation of this. Presumably (as in Theorems 1 and 2 of §2, [2] below) the parallels are ordinate to a diameter which divides the 'lunules', or better lunule segments, cut off by the parallels between two branches of the same curve (or between two corresponding asymptotes).

Si tangens inter punctū ubi secat curvam(10) trisecetur, ordinata diameter transibit per punctū contactui propius.(11)

Quomodo Axes per Asymptoto[s] inveniendi sunt.(12)

---

[2] *Conick propertys to bee examined in other curves.*(13)

$FV=V\psi$. $HE=LM$. $EA=AL$. $S\beta=\beta P$. $SR=QP$. $R\beta=\beta Q$.(14) $\Delta\delta C\pi=\Delta FC\psi$.(15) $CV=CN$.(16) $FV^q=HL\times LM$.(17) $XY^q=RP\times PQ$.(18) $CF\times FE=CH\times HI$.(19)

$$PDS.IDK::ZBO.EBL::XC^q.CV^q::Z\mu^q.X\mu\times\mu Y.^{(20)}$$

$CA.CX::CX.CG$.(21) $\sigma\tau$ tangit in N sicut $F\psi$ in V.(22) Axis bisecat angulum $FC\psi$.(23) XY ordinatur ad NV.(24) Sunt et curvarum foci $\rho\rho$, a quibus ductæ lineæ habent datam summam vel diff:(25) secant arcus in = (26) angulis.(27)

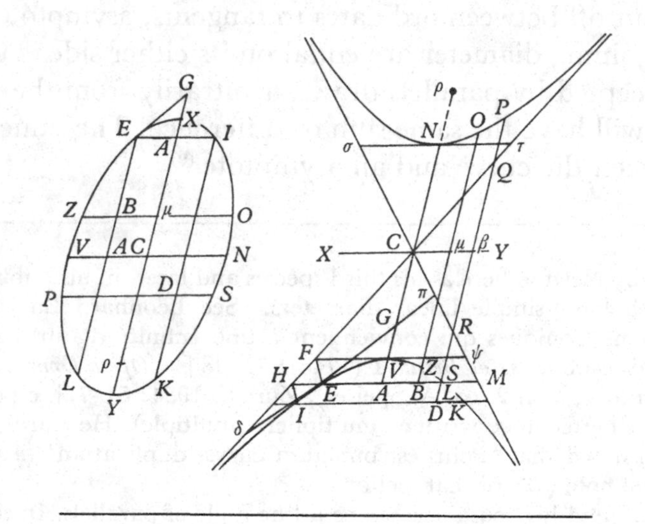

---

(10) That is, understanding 'et punctum contactūs' (and the point of contact).

(11) The remark is true, of course, only for cubics: in higher curves the tangent at a point will not meet its curve again in a unique intersection. In proof, the tangent-point represents a double meet of the tangent-line with the curve and hence the first trisection point will lie on the diameter defined by lines parallel to the tangent (in the sense of §2, [2], Theorem 3 below).

(12) See §2, [2], Theorem 2, Corollary 2 below.

(13) Virtually all the following conic theorems are to be found in Apollonius' *Conics*. Note that two points in Newton's left-hand figure are marked '*A*'.

(14) Apollonius II, 8 and 16: note that the chords *EL* and *PS* are conjugate. For the generalized property see §2, [2], Theorem 1 below.

(15) Apollonius III, 43. It is hard to see how this property can be generalized.

(16) Compare Apollonius I, passim. *VN*, the diameter conjugate to *XY*, is bisected by the conic centre *C*. For the extension to cubics compare **1**, §2: note (17); for the generalization see §2, [2], Theorem 1 below.

If the tangent between the point where it cuts the curve[10] be trisected, the ordinate diameter will pass through that point (of trisection) which is nearer to the point of contact.[11]

How the axes are to be found by means of the asymptotes.[12]

---

[2] [*Conick propertys to bee examined in other curves*]

$FV = V\psi$. $HE = LM$. $EA = AL$. $S\beta = \beta P$. $SR = QP$. $R\beta = \beta Q$.[14]

$\Delta\delta C\pi = \Delta FC\psi$.[15] $CV = CN$.[16] $FV^2 = HL \times LM$.[17]

$XY^2 = RP \times PQ$.[18] $CF \times FE = CH \times HI$.[19]

$PD \times DS : ID \times DK = ZB \times BO : EB \times BL = XC^2 : CV^2 = Z\mu^2 : X\mu \times \mu Y$.[20]

$CA : CX = CX : CG$.[21] $\sigma\tau$ is tangent at N in the same way as $F\psi$ is at V.[22] The axis bisects the angle $FC\psi$.[23] XY is conjugate to NV.[24] There are, too, foci ρ, ρ in these curves: lines drawn from them have a given sum or difference,[25] and meet the arcs in equal angles.[27]

---

(17) Apollonius II, 10.

(18) Read '$CV^q = RP \times PQ$' (Apollonius II, 11). (XY is, in fact, equal to $F\psi$.)

(19) This implies that HI and FE are each parallel to CM (Apollonius II, 12). We reproduce Newton's figure without emendation.

(20) The chord-intercept theorem (Apollonius III, 16ff.) which is virtually a semi-projective definition of the general conic. As we will see in our seventh volume, Newton later found the generalization to higher-order curves which we know as 'Newton's' of this theorem on the constant ratio of the rectangles of the intercepts made by two lines from an arbitrary point each fixed in direction.

(21) The pole-polar relationship for conics (Apollonius III, 30–34).

(22) In fact, $F\psi$, XY and $\sigma\tau$ will be equal and parallel to each other, so that the mid-point C of XY will be the centre of the parallelogram $F\psi\tau\sigma$ as well as of the hyperbola.

(23) The 'axis' (either of the two principal diameters) has not been drawn in either of Newton's figures. The remark is general for any curve which has a main diameter since the asymptotes will be symmetrically placed round it.

(24) For XY is parallel to EL, $F\psi$ and so on, and is the diameter conjugate to NV.

(25) Apollonius III, 51 and 52. (The sum will of course, be constant in the ellipse, the difference in the hyperbola.) There is no full generalization of this to higher curves, for the basic condition for a bipolar definition of a curve is that it be symmetrically situated round the line joining the foci. In general, too, bipolar co-ordinates do not lend themselves easily to the investigation of algebraic curves of odd degree.

(26) Read 'æqualibus'.

(27) That is, the focal lines make equal angles with the instantaneous direction of the curve (or the tangent at the point): compare Apollonius III, 48. There seems no fruitful generalization to high-order curves.

§2. A REVISED ACCOUNT OF THE GENERAL PROPERTIES OF CURVES[1]

From the original in private possession

[1] *Definitions.*[2]

Crus Hyperbolicum est quod recta ad infinitam distantiam possit tangere. Recta ista ad infinitam distantiam tangens dicitur Asymptoton ejus.

Crus Parabolicum est quod recta non potest tangere ad infinitam distantiam.[3]

Basis est recta quævis cui aliæ in dato angulo insistentes, ad cu[r]vam hinc inde terminantur.[4] Et lineæ sic a Basi ad curvam protensæ dicantur[5] insistentes.[6]

Si insistentes ad idem quodlibet punctum Basis aut earum summæ hinc inde sint æquales Basis dicitur diameter & insistentes ordinatim applicatæ.[7] Bases& insistentes, ut et dia[metri] et ordin[atim applicatæ] dicuntur rectang[ulæ] aut obliquā[8] prout ejusmodi angulos conficiunt. Et licet aliqui[9] ex ordinatis sint imag:[9] diameter tamē dicetur siquidem impos[s]ib[il]ium summa nunquam est imposs:[10]

Vertex est punctum curvæ quam tangit insistens.[11]

Axis est diameter cui ordinatim applicatæ ad angulos rectos insistunt. Axis primi generis est quæ bisecat applic[atas] &c.[12]

Nodus curvæ est punctum ubi plures ejus partes conveniunt. quale est polus conchæ cum &c.[13]

---

(1) The scattered thoughts and tentative generalizations of the preceding section are here worked into a more systematic form. It was, without doubt, Newton's intention to set this as a general introduction to his enumeration of cubics in 1, §3 above, and the final sentences of [1] may represent a first sketch of a proposed revision of the piece. However, in a not unusual sequence of events, he first grew dissatisfied with the logical structure of his exposition, breaking for a moment into English with suggestions for revising it, and then abandoned his present researches, not to return to them for almost thirty years.

(2) Newton begins by giving general definitions of terms which he has used in his cubic enumeration in an unconventional significance.

(3) Newton, of course, supposes that the line itself is not at infinity and so cannot accept that a parabolic branch is touched by the line at infinity.

(4) 'protenduntur' (extended to) was first written.

(5) Probably we should read the indicative 'dicuntur' in line with the first paragraph.

(6) Literally, 'standing-ons', but we use Newton's later preferred technical term.

(7) As the numerous cancellations in his manuscript reveal, Newton found great difficulty in adequately formulating this definition of the diameter of a general curve. With some crossing out he first wrote that 'Diameter est linea quæpiam quæ secat omnes lineas alicui datæ || ita ut ejusdem cujusvis parallelæ summæ partiū a diametro ad curvas hinc inde ductarum sint æquales' (A diameter is any line which cuts all lines parallel to some given one

Translation

[1] [*Definitions*][(2)]

A hyperbolic branch is one which a straight line may touch at an infinite distance. That line touching at an infinite distance is called its asymptote.

A parabolic branch is one which a straight line cannot touch at an infinite distance.[(3)]

A base is any straight line from which any others applied in a given angle are terminated on either side at[(4)] the curve. And lines extended in this manner from a base to the curve are to be called[(5)] applicates.[(6)]

If the applicates at the same arbitrary point of the base or their aggregates are, on either side, equal, a base is called a diameter and its applicates ordinates.[(7)] Bases and applicates, and likewise diameters and ordinates, are called rectangular or obliquangular according as they make angles with each other of that kind. And even though some of the ordinates be imaginary, nevertheless a diameter will still be so called inasmuch as the sum of (conjugate) impossibles is never itself impossible.[(10)]

A vertex is a point of the curve to which an applicate is tangent.[(11)]

An axis is a diameter to which the ordinates are applied at right angles. An axis of the first kind is that which bisects the ordinates to it, and so on.[(12)]

A node of a curve is a point where several of its parts meet together. Such as the pole of a conchoid when....[(13)]

---

such that the sums of the parts of each one of these parallels taken from the diameter on either side to the branches of the curve are in each equal): in subsequent revision, cancelled for the final version reproduced in the text, this became 'Diameter est recta quævis cui aliæ omnes rectæ ad idem quadlibet punctum in dato angulo insistentes, et ad curvam protensæ summam hinc inde æqualem conficiunt' (A diameter is any straight line for which all other straight lines applied in a given angle to the same arbitrary point of it and extended to meet the curve together yield the same aggregate on either side).

(8) That is 'obliquangulæ'.

(9) Read '...licet aliquæ...sint imaginariæ'.

(10) The sum of two conjugate complex quantities is always real.

(11) The vertex, that is, corresponding to the base to which the 'insistentes' are applied: a generalization of the vertex of a conic, which is the contact-point of a tangent parallel to the diameter conjugate to that through it. We should perhaps read 'ordinatim applicata' for 'insistens'.

(12) More generally, we assume, an axis of nth kind is a Newtonian diameter of nth order whose ordinates are at right angles to it.

(13) When, that is, the describing *regula* of the conchoid is greater than the distance of the pole from the base-line. (Compare the first figure of I, **3**, 3, §3, note (3).) In an uncompleted preceding paragraph Newton had originally introduced the concept of a 'pole' with the definition 'Polus dicitur ubi duæ [curvæ conveniunt]' (That point is called a pole in which two branches of the curve meet).

Punctum Ellipticum est punctum seorsim existens in quod Ellipsis[14] evanuit. Quale est polus Conchæ cum [&c].[15]

Cuspis est ubi curva acutâ[16] vertice terminatur. Quale est polus Conchæ [cum &c].[17]

Et omnia hæc puncta cu[r]varum poli dicantur propterea q^{d} &c.[18] Possunt etiam dici puncta duplicata, triplicata, &c ubi duæ, tres &c curvæ conveniunt &c. Hoc est per quod[19] linea rotata curvam (extra id) in tot punctis secat vel secare imagi[n]andū [est] demptis 2, 3, 4 &c quot curva habet dimensiones.

Curvæ genere differunt quæ differunt dimensionibus.

Specie differunt quæ si sensim mutantur non possint in se invicem transmigrare nisi earum pars aliqua in infinitum abeat vel ab infinito emergat:[20] Subdifferunt quæ differunt polis. Novem sunt species[21] tertij generis[:] Hyperbola triformis, Hyperbola elliptica, Hyperbola Parabolica in primo ordine;[22] Hyperbola triformis, elliptica, Parabolica transversè[23] in secundo ordine; parabola Elliptica, hyperbolica et reflexa in tertio.[24] Parabola est quæ fit ex asymptotorum parallelismo, vel transitu Ellipsis ad Hyperbolam.[25]

Gradu differunt quos[26] in specie differre plerumqȝ posui, subdifferunt quas posui differ[r]e.

Forma differunt quæ varias formas induunt.

Mechanice differunt pro more describendi.[27]

Cissois est 4ma[28] 4ta $5^{\text{tæ}}$ speciei.[29] Alia etiam huic affinis est species 13 cas $6^{\text{t[i]}}$[30]

---

(14) Newton here carelessly wrote 'Ellipsius' in his manuscript.

(15) When (compare Newton's figure in I, 3, 3, §3) the describing *regula* is less than the distance of the pole from the base-line, in which case the pole becomes a conjugate double point of the conchoid (and the only one in which its 'ellipse' is not nodate or a cusp).

(16) Read 'acuto'.

(17) When the *regula* is equal to the distance of the pole from the base-line. (See the second figure of I, **3**, 3, §3, note (3).)

(18) Presumably, 'quod duæ curvæ ibi conveniunt' (seeing that in each case two branches of the curve meet) in line with the following.

(19) Read 'quæ', and correspondlingly later in the line for 'id' read 'ea'.

(20) Newton first wrote 'quæ differunt quoad crura [sunt] Hyperbolica Parabolica vel in Ellipsin terminata' (which differ in so far as their branches are hyperbolic, parabolic or terminated in an oval). A 'species' is, thus, differentiated by the nature of its infinite branches.

(21) A difference in nomenclature from that used in his preceding cubic enumerations, where he uses 'casus' (case) in this sense. This variance reappears (but not consistently) in the following text, where too 'ordo' (rank) rather than 'species' marks the sixteen main subdivisions of the cubic.

(22) A second use of the word (in conflict with that outlined in the preceding note) to indicate a first, most general tripartite division of the cubic: for the sake of distinction we translate it by 'order'.

(23) Newton first distinguished this 'order' as that of the 'Hyperbola parallelosymptotos', that is 'parallelas Asymptotos habens' (with parallel asymptotes).

An ovate point is a point existing on its own into which an oval has vanished. Such as the pole of a conchoid when....[15]

A cusp is where a curve is terminated in a sharp vertex. Such as the pole of a conchoid when....[17]

And all these points of a curve might be called poles seeing that....[18] They may also be called doubled, tripled,...points when two, three, ...curves meet together, and so on. That is, a rotating line through them meets the curve, or is to be conceived to meet it, in as many (further) points as the curve has dimensions less 2, 3, 4,... correspondingly.

Curves differ in kind which differ in dimensions.

Those differ in species which, if they are sensibly transformed, cannot pass back into themselves in return unless some portion of them go off into infinity or emerge therefrom:[20] they are further differentiated by the differing nature of their poles. There are nine species of the third kind: the three-limbed, elliptical and parabolic hyperbolas in the first order;[22] the transversely three-limbed, elliptical and parabolic hyperbolas in the second order; and the elliptical, hyperbolic and back-bent parabolas in the third.[24] A parabola is what is produced when asymptotes are parallel or occurs in passing from the ellipse to the hyperbola.[25]

Those differ in grade which for the most part I have supposed differing in species: those are further differentiated which I have supposed differing.

Those differ in form which take on various forms.

They differ mechanically according to the manner by which they are described.[27]

The cissoid is the fourth form of the fifth species.[29] A second curve, also, related to this is species 13 of case 6.[30]

(24) These nine 'species' are, it would seem, respectively identical with the nine 'cases' into which he separated the cubic in 1, §4, [2] above.

(25) A postulate of continuity between cubic species which is a generalization of the equivalent 'Keplerian' principle for conics, by which the conic parabola is conceived to be a mean between the Apollonian ellipse and hyperbola. Analytical aspects of the postulate are further explored in the concluding paragraph. (26) Read 'quas'.

(27) Newton, of course, at this time had the general problem of the mechanical construction of curves very much in the forefront of his mind: the care he devoted to exploring the niceties of the 'organic description' of a curve in terms of one of less complexity will be evident from the manuscripts which are reproduced in **1**, 3 below. A decade later he gathered his thoughts on the matter in his concluding Lucasian lectures on arithmetic and algebra. (See his *Arithmetica Universalis* (London, 1707): Appendix, *Æquationum Constructio linearis*, 279–83. This we will reproduce from the original manuscript in our fourth volume.)

(28) Read 'For^ma^'! It is a further proof of Newton's still thinking in English rather than in equivalent Latin word-forms that he should unconsciously and phonetically render the Latin 'For-' by the English 'Four-' and then set it down numerically thus.

(29) Compare 1, §3, note (58). Newton jumps back to the nomenclature for cubics which he has established in his first cubic enumerations.

(30) The particular case of the cissoid when the cusp is at infinity. Compare 1, §3 above.

In prima specie tres sunt ordines,(31) primi ordinis tres gradus & sex formæ. Liquet quomodo prima species(31) transit in secundam mediante 3^a i e d in $-d$ mediante 0, quomodo tertia transit in 4^{tam} et quintam, quarta in 5^{tam} mediante sextâ, tres primæ in 7^{am} 7^a in 8^{vam} et $8^{[a]}$ in $9^{[am]}$.(32)

---

[2] *Theor 1.*(33) Si duæ quævis rectæ BC & bc sibi parallelæ ducantur secantes quamcunq; curvam in tot punctis (C, D, E; c, d, e &c) quot curva habet dimensiones, & in istis rectis sumantur puncta B & b, ita ut summæ distantiarum inter istud punctum & puncta intersectionis sint utrinq; æquales, ($BC = BD + BE$ &c) & per ista 2 puncta (B & b) ducatur recta (Bb): Dico quod ista recta Bb eodem modo dividet aliam quamvis lineam $\chi\beta$ prioribus parallelam hoc est summa distantiarum inter istam lineam Bb & puncta curvæ intersectæ ex una parte(34) erit æqualis summæ distantiarum ex altera ($\beta\delta + \beta\epsilon = \beta\chi$ [&c]) hoc est si ista recta $Bb\beta$ pro basi habeatur & $\beta\chi$, $\beta\delta$, $\beta\epsilon$ [&c] pro Applicatis,(35) Applicatæ ex una parte erunt æquales Applicatis ex altera.(36)

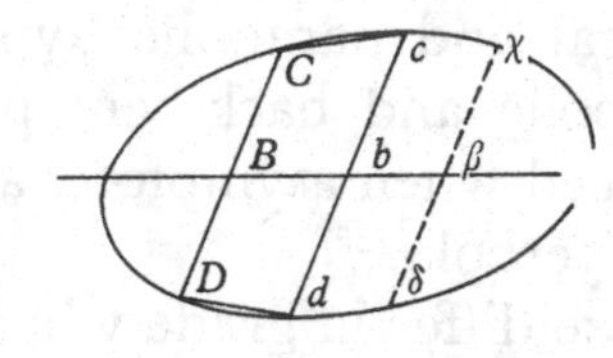

[Fig 1]

Exemplum 1. Si curva sit duûm(37) Dimensionum duco parallelas CD, cd secans curvam in 2 punctis C & c [D & d]; & eas biseco in B & b & duco Bb. ista Bb bisecabit omnes rectas $\chi\delta$ prioribus parallelas.

Exemplum 2. Si Curva sit 3 dimens: duco paralelas CDE & cde, secantes curvam in punctis C, D, E; c, d, e. & si D sit medium punctum(38) sumo $EB = \frac{ED + EC}{3}$(39) (vel facio $CP = EB$,(40) & $DB = \frac{1}{3}DP$) hoc est facio ut $BD + BE$ quæ cadunt ad easdem partes sint

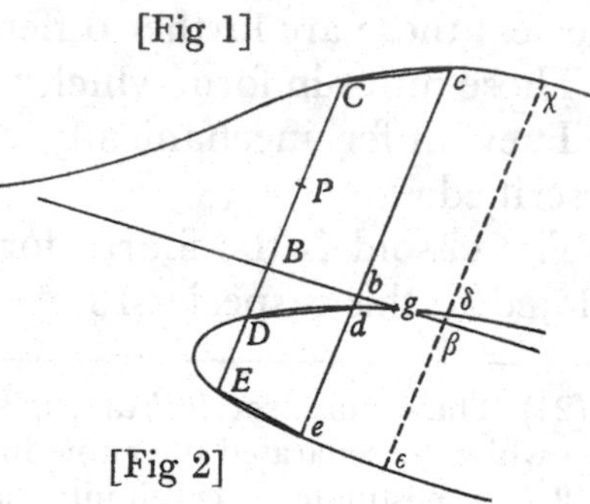

[Fig 2]

---

(31) See note (21) above.

(32) 'Species' 1–3 ($bxy^2 = dx^3 + gx^2 + hy + kx + l$) are distinguished only by having d greater than, less than, or equal to zero. Similarly, 'species' 4 is the particular case where $g = 0$ of 'species' 3, while 'species' 4–6 are differentiated by having k greater than, less than, or equal to zero. When $b = 0$ 'species' 1–3 may be reduced to 'species' 7

$$ey^2 = dx^3 + kx + l,$$

and this in turn when $e = 0$ to 'species' 8 ($fxy = dx^3 + l$) or, when $f = 0$ in the latter case, to 'species' 9 ($hy = dx^3$). All this merely restates the basic division of the cubic into nine 'cases' expounded in 1, §3.

(33) Newton asserts the existence of a general 'Newtonian' diameter for an algebraic curve.

(34) 'versus unam partem' (facing one way) is cancelled.

(35) That is, the 'insistentes' of [1] preceding.

(36) Newton gives no proof, but the following restoration would seem to be structurally

In the first species there are three ranks,(31) and of the first rank three grades and six forms.

It is evident how the first species passes into the second via the third, that is, as d passes into $-d$ via 0; how the third passes into the fourth and fifth, the fourth into the fifth via the sixth, and the first three into the seventh, the seventh into the eighth and the eighth into the ninth.(32)

---

[2]

Theorem 1.(33) If any two straight lines BC and bc are drawn parallel to each other meeting any curve in as many points (C, D, E and c, d, e, and so on) as the curve has dimensions, and in those lines are taken the points B and b such that the aggregates of the distances between each of those points and the intersection points on its either side be equal ($BC = BD+BE$, and so on) and through those two points B and b be drawn the straight line Bb, then I say that that line Bb will equivalently divide any other line $\chi\beta$ parallel to the first two: that is, the aggregate of the distances between that line Bb and the points where the curve is intersected on one flank will be equal to the aggregate of the distances on the other ($\beta\delta+\beta\epsilon = \beta\chi$, and so on), or, in other words, if that straight line $Bb\beta$ be supposed the base and $\beta\chi$, $\beta\delta$, $\beta\epsilon$ and so on its ordinates,(35) then the ordinates on one side will equal the ordinates on the other.(36)

Example 1. If the curve be of second degree, I draw the parallels CD, cd meeting the curve (each) in two points C and c, D and d. These I bisect in B and b and then draw Bb. That line Bb will bisect all straight lines $\chi\delta$ parallel to the first two.

Example 2. If the curve be of third degree, I draw the parallels CDE and cde meeting the curve in the points C, D, E and c, d, e, and if D be the middlemost point(38) I take $EB = \frac{1}{3}(ED+EC)$(39) (or make $CP = EB$(40) and $DB = \frac{1}{3}DP$):

---

unique. If we consider Newton's following second example of the general cubic, whose defining equation we may write (compare 1, §1 above)

$$ay^3+(bx+e)y^2+(cx^2+fx+h)y+dx^3+gx^2+kx+l = 0$$

where the x-co-ordinate is measured along Bb to β and $\beta\chi$ is the corresponding applicate, that applicate will meet the cubic in the three points δ, ϵ and χ such that $\beta\delta$, $\beta\epsilon$ and $\beta\chi$ are the roots of the cubic in y which results from setting x equal to the (instantaneously given) abscissa terminated in β, say $\underline{x}$: that is, $\beta\delta+\beta\epsilon+\beta\chi = -(b\underline{x}+e)/a$. Newton's choice of the points B and b, say $(x', 0)$ and $(x'', 0)$ determines that the corresponding applicates at those points satisfy $y_1'+y_2'+y_3' = -(bx'+e)/a = 0$ and $y_1''+y_2''+y_3''' = -(bx''+e)/a = 0$, so that $b\underline{x}+e \equiv 0$, or $b = e = 0$: that is, for all points β, $\beta\delta+\beta\epsilon+\beta\chi = 0$ (where, of course, due attention must be paid to the sign of the various directed segments). The argument is clearly general.

(37) That is, 'duarum'.

(38) Newton has here cancelled the clarifying phrase '& $DC \sqsubset DE$; sumo versus C $DB = \dfrac{DC-DE}{3}$ vel' (and $DC > DE$, towards C I take $DB = \frac{1}{3}(DC-DE)$ or).

(39) $(EB-ED)+EB = (EC-EB)$ since $DB+EB = BC$.

(40) Read 'ED' since $EB-ED = \frac{1}{3}(EC-ED-PC)$.

æquales ipsi BC quæ cadit ad alteras. Eadem ratione invenio punctum b. Deinde duco rectam Bb, & inquam quod ista recta eodem modo dividet omnes lineas $\chi\delta\epsilon$ prioribus (CDE & cde) parallelas; hoc est ut lineæ $\beta\delta+\beta\chi$ ex una parte, & $\beta\chi$(41) ex altera sint æquales.

Et in Curvis 4 dimensionum eadem est ratio.

Cor 1. Unde(42) superficies ex utraqȝ parte lineæ $B\beta$ erunt æquales. nempe $B\beta\chi C=B\beta\delta D$ in fig 1; & $BbcC=BbeE+BbdD$ vel $b\beta\chi c+g\delta\beta=be\epsilon\beta+bdg$ [in] fig 2.(43)

Cor 2. Hinc si(44) duæ quævis parallelæ ducantur secantes curvam quamvis geometricam in tot punctis C, D, E; c, d, e quot curva habet dimensiones; & arcubus in istas parallelas terminatis Cc, Dd, Ee, ducantur subtensæ: (Cc, Dd, Ee;) summa superficierum istis arcubus & subtensis comprehensarum & jacentium ad easdem partes subtensarum æquatur summæ jacentium ad alteras partes.(45) Sic fig 1 $Cc=Dd$ & fig 2 $Cc+Dd=Ee$. et fig 3(46)

$$Cc+Dp=pd+Ee. \text{ \&c.}$$

Cor 3.(47) Hinc in Curvis ubi Diameter semper est eadem(48) (ut in $ax=by^3$. $axx=by^4+cyy+dy+e$. $axy=by^4+cyy+dy+e$. $ax=by^4+c+dx$. $ax=by^4+cyy$. &c) si duæ quævis parallelæ ducantur secantes curvam in tot punctis quot curva habet dimension[e]s: superficies inter diametrum & Curvam ex una parte equātur superficiebus ex altera.

---

(41) Read '$\beta\epsilon$'.

(42) Newton has cancelled 'in curvis duar^m dimēns' (in curves of second degree).

(43) This is, indeed, an immediate consequence of Cavalieri's principle, the first part of which says that if a set of parallels meet two curves such that the segments cut off by each are in a fixed proportion, then the areas under each curve between any two of the parallels is also in that fixed proportion. (See Bonaventura Cavalieri, *Geometria Indivisibilibus Continuorum Nova Quadam Ratione Promota* (Bologna, $_1$1635 ($_2$1653)): Liber 6 = *Exercitationes Geometricæ Sex* (Bologna, 1647): Liber 1.) Cavalieri's ideas on this point had been widely disseminated, particularly by Grégoire de Saint-Vincent and, in England, by John Wallis in his *Arithmetica Infinitorum* (I, 1, Introduction, Appendix 2, note (20): especially *Pars Prima*, Prop. I ff.).

(44) 'in quacunqȝ Curva Geometrica' (in any geometrical [sc. algebraic] curve) is cancelled.

(45) This follows since the areas of the trapezia $BbcC$, $BbdD$, $BbeE$ and so on are equal to zero (if we suppose that those below Bb have negative area). In proof of the cubic case, evidently a model one, and in the notation of note (36) we may show that the areas of these trapezia are respectively $k(y_1'+y_1'')$, $k(y_2'+y_2'')$ and $k(y_3'+y_3'')$, where $k=\frac{1}{2}Bb\times\sin C\hat{B}b$: immediately their sum is zero since $y_1'+y_2'+y_3'=y_1''+y_2''+y_3''=0$.

(46) The manuscript has no 'fig 3' but Newton intends presumably a central cubic (we show one of his species 8 in I, §3, see opposite page) met by parallels CDE, cde situated equally on either side of the centre p. (The centrality condition is necessary since the corresponding diameter Bb must pass through the meet of Dd and the curve.)

(47) This corollary, a later addition, has been written in on the verso of the preceding over the trio of preliminary diagrams for species 6 of I, §3 above, which are there reproduced in appendix.

that is, I make $BD+BE$, falling on the same side, equal to BC, falling on the other. By the same method I find the point b. Then I draw the straight line Bb, and assert that that straight line will divide equivalently all lines $\chi\delta\epsilon$ parallel to the first (CDE and cde): that is, such that the lines $\beta\delta+\beta\chi$ on one hand and $\beta\chi$[41] on the other be equal.

And in curves of fourth degree the method is the same.

Corollary 1. It follows[42] that the surfaces on either side of the line $B\beta$ will be equal. Thus $B\beta\chi C = B\beta\delta D$ in figure 1, and $BbcC = BbeE+BbdD$ or again $b\beta\chi c+g\delta\beta = be\epsilon\beta+bdg$ in figure 2.[43]

Corollary 2. Hence if [44] any two parallel lines be drawn meeting any algebraic curve in as many points $C, D, E, \ldots$ and $c, d, e, \ldots$ as the curve has dimensions, and in the arcs $Cc, Dd, Ee, \ldots$ bounded by those parallels there be drawn the chords $(Cc, Dd, Ee, \ldots)$, then the aggregate of the surfaces comprehended by those arcs and chords and lying on the same flank of the chords is equal to the aggregate of those lying on the other flank.[45] Thus in figure 1 $(Cc) = (Dd)$, in figure 2 $(Cc)+(Dd) = (Ee)$, and in figure 3[46]

$$(Cc)+(Dp) = (pd)+(Ee).$$

And so on.

Corollary 3.[47] Hence in curves where the diameter is invariant[48] (as in $ax = by^3$, $ax^2 = by^4+cy^2+dy+e$, $axy = by^4+cy^2+dy+e$, $ax = by^4+c+dx$, $ax = by^4+cy^2$, and so on), if any two parallels be drawn meeting the curve in as many points as it has dimensions, the surfaces between the diameter and the curve on the one hand are equal to those on the other.

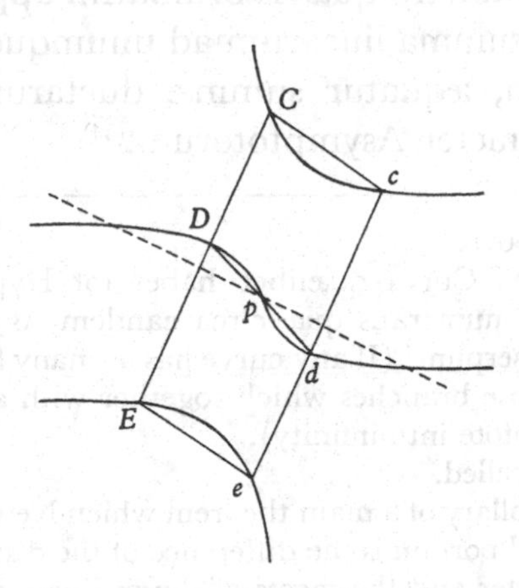

(48) The general nth degree algebraic equation $\alpha y^n+(\beta x+\gamma)y^{n-1}+\ldots = 0$ has the Newtonian diameter $n\alpha y+\beta x+\gamma = 0$ corresponding to parallels in the direction of its y-axis. For this to be independent of the slope of the co-ordinate axes, here evidently a sufficient condition for invariance of the diameter, we require $\beta = \gamma = 0$, so that the diameter must be the axis of the abscissas ($y = 0$) and the curve's defining equation must lack all terms in y^{n-1}.

Schol. Unde linea $Bb\beta$ erit una ex Diametris Curvæ & $\beta\chi$ ei ordinatim applicatur. Nempe si Curva sit duarum dimensionum linea Bb dicitur diameter inter duas superficies (viz $B\beta\delta D$ & $B\beta\chi C$ &c) vel Diameter secundi generis(49) vel simpliciter Diameter. Sin Curva sit trium Dimensionū Diameter erit inter tres superficies vel tertij generis(49) &c. Et ex dictis patet quod linea quævis alicui Diametro ordinatim applicatur, & quo parto Diameter cujusvis ordinatim applicatæ positione datæ inveniatur; (modo linea aliqua isti ordinatim applicatæ parallela potest secare curvam in tot punctis quot curva habet dimensiones).

Theor 2. Si Curva habet tot Asymptotos quot habet dimensiones, ista Asymptoto numerata pro unica [circa] quam non plura quam unum curvæ crus serpunt ad easdem partes[,] pro duabus numerata circa quam duo crura serpunt ad easdē partes, & ista pro tribus circa quam tres &c.(50) Tum si ducatur linea quævis secans omnes Asymptotos & curvam in tot punctis quot Curva habet dimensiones Dico quod summa(51) partium hujus rectæ interceptarum inter unumquodꝗ istorum Curvæ punctarū & Asymptoton sibi propriam & jacentium ad easdem Asymptotorum partes æquat summæ jacentium ad alteras.(52)

Corol 1. Hinc si duæ quævis parallelæ ducantur secantes omnes Asymptotos & Curvam in tot punctis quot Curva habet dimensiones: tum summa istarum superficierum comprehensarum ab arcubus curvæ, suis Asymptotis & lineis parallelis & jacentium ad easdem partes Asymptotorum æquantur summæ jacentium ad alteras.(53)

Cor 2. Hinc si cuiꝗ Diametro quævis ordinatim applicata producatur dum secat omnes Asymptotos summa linearum ad unumquodꝗ Asymptoton versus easdem partes ductarum, æquatur summæ ductarum ad alteras. Hoc est Diameter Curvæ est Diameter Asymptotorum.(54)

---

(49) Compare note (12) above.

(50) Newton first wrote 'Si Curva quælibet habet tot Hyperbolica crura quot habet dimensiones, ijs cruribus non numeratis quæ circa eandem Asymptoton cum aliquo crure numerato ad oppositas partes serpunt' (If any curve has as many hyperbolic branches as it has dimensions, not reckoning those branches which together with an already reckoned branch snake around the same asymptote into infinity).

(51) 'omnium' (all) is cancelled.

(52) This is, in effect, a corollary of a main theorem which Newton presents as his following 'Cor 2', since each intercepted portion is the difference of the distance (in the direction of the ordinates) between the diameter and the meets of the ordinate with the several branches of the curve or the corresponding asymptotes, while of each of these the aggregate is zero (by Theorem 1 and Corollary 2 below respectively).

(53) Each surface is the difference of two surfaces, the aggregate of one of which is zero by Theorem 1, Corollary 1, and that of the second likewise by an analogous argument applied to Corollary 2 of the present theorem.

Scholium. The line $Bb\beta$, consequently, will be one of the diameters of the curve and $\beta\chi$ is ordinate to it. Precisely, if the curve be of second degree the line Bb is called a diameter between two surfaces (namely, $B\beta\delta D$ and $B\beta\chi C$, and so on) or a diameter of the second kind(49) or simply a diameter. But if the curve be of third degree the diameter will be between three surfaces or of the third kind.(49) And so on. From what has been said, again, it is evident that any line you will is ordinate to some diameter and also how the diameter corresponding to any ordinate given in position may be found (providing only that some line parallel to that ordinate can cut the curve in as many points as it has dimensions).

Theorem 2. If a curve has as many asymptotes as it has dimensions, with that asymptote around which no more than one branch of the curve snakes in the same direction counting as one only, that counting as two round which two branches snake in the same direction, that as three round which three, and so on:(50) then, if any line be drawn meeting all the asymptotes and the curve in as many points as the curve has dimensions, I say that the aggregate of(51) the parts of this straight line intercepted between each of those points of the curve and its related asymptote and lying on the same flank of the asymptotes is equal to the aggregate of those lying on the other.(52)

Corollary 1. Hence if any two parallels be drawn cutting all the asymptotes and the curve in as many points as the curve has dimensions, then the aggregate of those surfaces comprehended by the arcs of the curve, their asymptotes and the parallel lines and lying on the same flank of the asymptotes is equal to the aggregate of those lying on the other.(53)

Corollary 2. Hence if any ordinate at random to each diameter be produced till it cuts all the asymptotes, the aggregate of the lines drawn facing the same way to every asymptote is equal to the aggregate of those drawn the other way. That is, the diameter of the curve is the diameter of the asymptotes.(54)

---

(54) This, the fundamental theorem of the present part, follows readily if we view the line n-ple of the asymptotes of an n-degree curve as itself an n-degree curve which coincides with it in its infinite branches. For, since the asymptotic line n-ple touches its curve in n points at infinity, its defining equation can differ from the latter's only in terms of dimension $n-2$ and less. But just these identical terms in each of dimensions n and $n-1$, say

$$\sum_i (a_i x^i y^{n-i}) + \sum_j (b_j x^j y^{n-1-j}),$$

are necessary and sufficient to define the Newtonian diameter in each which corresponds to parallels taken in the direction of the ordinates, and consequently a curve and its asymptotic line n-ple share the same diameter. (Compare, for example, George Salmon, *A Treatise on the Higher Plane Curves* (London, $_2$1879): 39–41, 112–15.) For Newton's further generalization below see note (61).

Theor 3.(55) Si ducantur plures lineæ parallelæ quæ secant curvam in tot punctis dempto uno quot curva habet dimensiones & quærantur puncta (ut priùs) per quæ diameter transire debet ista puncta cadunt in Conica Sectione (Hyperbola vel Parabola[)], & aliquando in linea recta quæ ideò ut plurimum erit Diameter simplicioris generis quam isti curvæ propria est.(56)

☞ Sed ut cognoscas an ista linea sit Hyperbola, Parabola vel linea recta sciendum est quod in hoc casu prædictæ lineæ sibi parallelæ sunt etiam parallelæ vel Alicui curvæ Asymptoto vel 2 cruribus ejus Parabolicis: Si sint parallelæ alicui Asymptoto & duo Curvæ Crura jacent ex diverso ejus latere,(57) tum habes Hyperbolam Conicam pro diametro, sin jacent ex eodem latere tum linea recta erit ejus diameter. Sed cum istæ lineæ sunt parallelæ 2 Cruribus Curvæ Parabolicis(58) si ista Crura jacent ex diverso latere Parallelarum tum loco Diametri habes Parabolam Conicam: sin jacent ex eodem latere tum habes lineam rectam.(59)

(60)This is best made a Theoreme & theoreme 3 a corollary.

Tangents being drawn to ye curve at ye ends of ye ordinatim applic[atæ] to ye same point of the Diameter, viz soe many as the curve hath dimensions[,] if they bee crossed by any line parallel to the said ordinatim applicatæ the parts of y^t^ line on one side of the Diameter is equall to the parts on the other.(61)

The like of Assymptotes for they are tangents at an infinite distance.

Hence the equality of the parts of any line, space &c twixt ye assymptotes & the curve, and of all such parallell lines, spaces &c twixt ye said tangents & curve.(62)

---

(55) Newton here first continued with 'Ordinatim Applicatæ de quibus locuti fuimus non debe[n]t esse alicui Asymptoto vel cruri Parabolico parallelæ. Quoties accidit quod lineæ [istæ sunt parallelæ]' (The ordinates of which we have spoken must not be parallel to any asymptote or the direction of a parabolic branch. As often as [those] lines [are parallel]...).

(56) As his cancellation indicates (note (55)), the set of parallels must meet the curve in one further point at infinity, and so the ordinates must be parallel to an asymptote or the direction of a parabolic branch. Hence, in the general defining equation

$$\alpha y^n + (\beta x + \gamma) y^{n-1} + (\delta x^2 + \epsilon x + \zeta) y^{n-2} \ldots = 0$$

we must have $\alpha = 0$ in this case, and consequently the Newtonian diameter of the curve corresponding to parallels taken in the direction of the y-axis will be of the form

$$(n-1)(\beta x + \gamma) y + \delta x^2 + \epsilon x + \zeta + \lambda x + \mu y + \nu = 0,$$

a hyperbola which may degenerate into a line-pair or (when $\beta = 0$) reduce to a parabola with its infinite branches parallel to $x = 0$. In general a curve of nth degree may have up to n of these diameters 'of n-th kind', as Newton calls them, nor does their real existence depend on the corresponding asymptote being real (as we may clearly see in the case of the conic ellipse).

(57) 'ad oppositas partes' (in opposite directions) is cancelled.

(58) A revised variant of the phrase 'Parabolicæ naturæ' (of parabolic character) which Newton first wrote here.

(59) These remarks are immediate generalizations of the results Newton obtained in his researches into cubics in 1, §§2/3 above.

Theorem 3.(55) If several parallel lines are drawn to meet the curve in as many points less one as the curve has dimensions and the points through which the diameter shall pass be sought (as before), those points fall in a conic (a hyperbola or a parabola) and on occasion in a straight line: that, therefore, will for the most part be a diameter of simpler kind than is proper to that curve.(56)

☞ But to know whether that line be a hyperbola, a parabola or a straight line you must understand that in this case the above-mentioned set of parallel lines is also parallel either to some asymptote of the curve or to a double parabolic branch. If those lines are parallel to some asymptote and two branches of the curve lie on its opposite sides,(57) then you have a conic hyperbola for diameter, while if they lie on the same side then a straight line will be the diameter. But when those lines are parallel to a double parabolic(58) branch if those branches lie on opposite sides of the parallels then in place of a diameter you have a conic parabola, while if they lie on the same side you have again a straight line.(59)

(60)[This is best made a Theoreme & theoreme 3 a corollary.

Tangents being drawn to ye curve at ye ends of ye ordinatim applicatæ to ye same point of the Diameter, viz soe many as the curve hath dimensions, if they bee crossed by any line parallel to the said ordinatim applicatæ the parts of y^t^ line on one side of the Diameter is equall to the parts on the other.(61)

The like of Assymptotes for they are tangents at an infinite distance.

Hence the equality of the parts of any line, space &c twixt ye assymptotes & the curve, and of all such parallell lines, spaces &c twixt ye said tangents & curve.](62)

---

(60) These concluding remarks which follow have been added in a rough, hasty scrawl which is extremely difficult to read. As always at this period, when rushed Newton abandons Latin for English.

(61) This hitherto unpublished theorem of Newton's is, of course, the generalization of the preceding theorem when the line at infinity passes into a finite parallel to the direction of the ordinates. Not untypically he fails to outline his proof but we may conjecture the following 'fluxional' deduction as mirroring his approach. Using the case of the general cubic as a model for a general proof we may as before (note (36)) deduce that the Newtonian diameter corresponding to the ordinates of the cubic $ay^3+(cx^2+fx+h)y+dx^3+gx^2+kx+l=0$ is Bb $(y=0)$: that is, to each point B $(x, 0)$ there correspond ordinates BC, BD, BE (y_1, y_2, y_3) whose aggregate $\sum_i (y_i)$ is zero. It follows, on differentiating, that $\sum_i (dy_i/dx) = 0$ and consequently that $b\chi+b\delta+b\epsilon = \sum_{1\leqslant i\leqslant 3} (y_i+Bb\times[dy_i/dx]) = 0$, Newton's result.

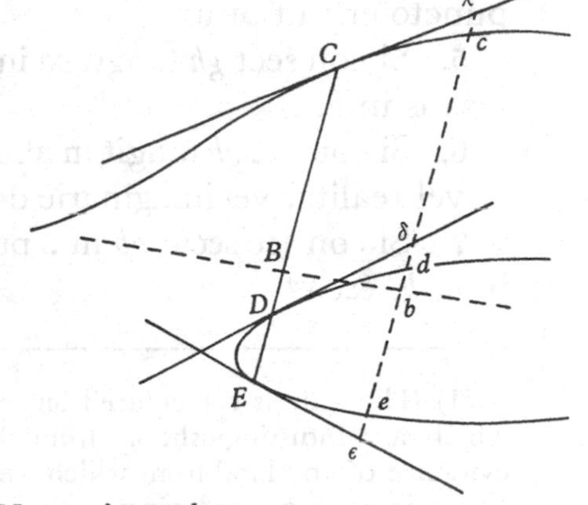

(62) These corollaries are immediate deductions from the main theorem on the style of those elaborated in sequel to the preceding Theorem 3 (the particular case, or better, as Newton would have it, the corollary, of the present one when CDE passes into infinity).

3

RESEARCHES IN THE ORGANIC CONSTRUCTION OF CURVES

[1667 or 1668?][(1)]

§1. PRELIMINARY INVESTIGATIONS[(2)]

From the original worksheets in private possession

[1]

[1.] Si *gh* sit linea recta ipsi *ab* parallela vel inclinata punctum *c* describet Conicam Sectionem[(3)] quæ lineam [*ab*][(4)] secat in punctis [*a*] & [*b*]:[(4)] vel lineam rectam.[(5)]

2. Si *gh* sit conica Sectio secans in *a* & *b* erit *de* conica sectio secans in *a* & *b*. vel linea recta secans in *k*.[(5)]

3. Si *gh* sit con: sect: secans *ab* in uno punctorum *a* erit *de* curva trium dimensionū secans in *b* & (vere vel imaginarie) decussans in *a*. & aliquando erit Con Sect: secans *ab* in *a* & alio puncto[(6)] vel tangens in *a*.

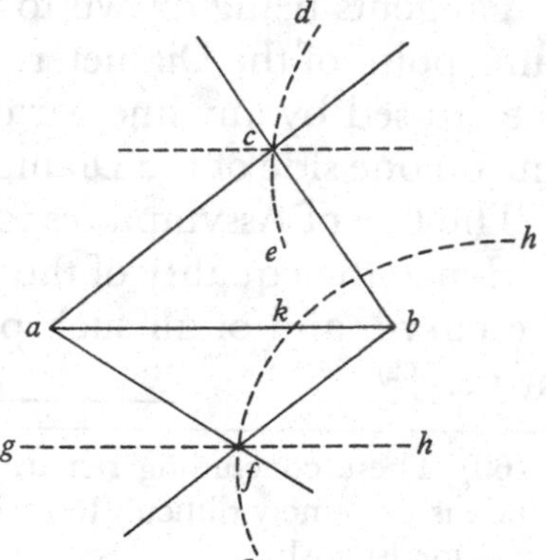

4. Si *gh* sit con sect: secans in *a* & nullo alio puncto erit ut prius.

5. Si con sect *gh* tangit *ab* in *a*, *ed* erit 3ū dimens, secans in *b* vel con sec non secans in *b*.

6. Si con sec *gh* tangit in alio puncto *k* erit curva *ed* 4ū dimens in *c* cuspidata et vel realiter vel imaginariè decussans vel cuspidata in *a* & *b*.[(7)]

7. Si con sec secat *ab* in 2 punctis extra *a* & *b* curva erit 4 dimens decussans in *a*, *b*, &c.[(8)]

---

(1) The date is conjectured on the basis of Newton's handwriting and composition style, which are indistinguishable from those of the preceding sections. (We lack any external evidence of any kind from which we could determine the date of composition.) On not very strong internal grounds we guess it to be slightly later than the researches into cubics and higher curves reproduced above. Precisely, the idea of an organic construction which dominates these present papers is (see the introduction) derived from Schooten's *Exercitationes Geometricæ* and is clearly an after-effect of his deep reading (I, 1, 1) of that work in late 1664, while the

Translation

[1]

[1.] If *gh* be a straight line parallel to *ab* or inclined to it, the point *c* will describe a conic[3] which meets the line *ab* in the points *a* and *b*: or it may itself be a straight line.[5]

2. If *gh* be a conic through *a* and *b*, then will *de* be a conic through *a* and *b*, or perhaps a straight line meeting *ab* in *k*.[5]

3. If *gh* be a conic meeting *ab* in one of those points *a*, then will *de* be a curve of third degree through *b* and with a (real or imaginary) node at *a*. On occasion it will be a conic meeting *ab* in *a* and a second point,[6] or perhaps tangent at *a*.

4. If *gh* be a conic through *a* and no other point of *ab*, the same holds true.

5. If the conic *gh* touches *ab* in *a*, then will *ed* be of third degree and through *b*, or sometimes a conic not through *b*.

6. If the conic *gh* touches it in some other point *k*, then will *ed* be a curve of fourth degree cusped in one of its points *c* and nodate (really or imaginarily so) or cusped at *a* and *b*.[7]

7. If the conic cuts *ab* in two points other than *a* or *b*, the curve will be of fourth degree nodate in *a* and *b*. And so on.[8]

---

writing of the present pieces is considerably more mature than that found in documents composed during Newton's undergraduate years. A certain easy use of technical terms, moreover, which he invented (apparently *ad hoc*) in his researches into cubics is here apparent: we may, for example, point in the following pages to frequent occurrences of 'cuspis', 'punctum decussationis', 'punctum multiplex' and other words and phrases which seem taken over from the manuscripts reproduced in preceding sections in this part. Newton's hand, finally, (with little doubt) is earlier and his Latin idiom less secure than the equivalents in the *De Analysi* (2, 2 below) of early 1669 and subsequent mathematical pieces.

(2) In the few, heavily cancelled sheets here reproduced Newton outlines the fundamental structure of his organic construction, listing (in the first instance with sketchy proof, but later merely as an indigestible sequence of merely enunciated theorems) its application to the description of curves of second, third, fourth and higher degrees by means of corresponding curves of lower or equal degree.

(3) 'parabolam' (a parabola) is cancelled.

(4) Newton here carelessly wrote the equivalent upper case letters.

(5) That is, when the conic hyperbola degenerates into the pair made by the polar line *ab* and the described line: in this case *gh* will be through the pole *a* or *b* or their perpendicular bisector.

(6) The less accurate phrasing 'in unico puncto' (in a unique point) is cancelled.

(7) Newton has cancelled a tentative continuation 'sed potest esse 3^{u} dimens̄, tangens in *b*' (but it can also be of third degree, tangent at *b*).

(8) The preceding statements are presumably to be justified by an elemental Newtonian procedure: since a straight line will, in general, meet an algebraic curve of *n*th degree in *n* points, we may determine the degree and singularities of any curve by considering the number and nature of the meets of a straight line with the described curve. (That line will, of course, be described in the organic construction by some conic through the poles *a* and *b*, and the meets of this conic with the describing curve (*f*) will mirror the meets of the described curve (*c*)

Cum *c* & *f* non sunt in *ab*. tum

1. *ac* & *af* secant suas curvas in eodem numero punctorum, extra *a*.(9)
2. simul tangunt suas curvas.
3. simul secant in puncto decussationis.
4. simul describunt cuspidem.
5. simul describunt unicum punctū.(10)
6. simul tangendo, suas curvas secant.(11)

Cum *f* (vel *c*) est in *ab*, sed in neutro punctorū *a* & *b*, tum

1. *ac* & *bc* secant suam curvam in uno puncto extra *a* & *b*.
2. in isto puncto *c* tot curvæ decussant quot punctis *gh* secat *ab*; extra *a* & *b*.
3. Et tot cuspidantur quot punctis *gh* tangit *ab*.

---

with the straight line described from the conic.) Newton's preceding results and following observations are immediate consequences of this. In effect, the apparatus of the organic construction sets up a one-to-one continuous correspondence between the points (*f*) and (*c*) of

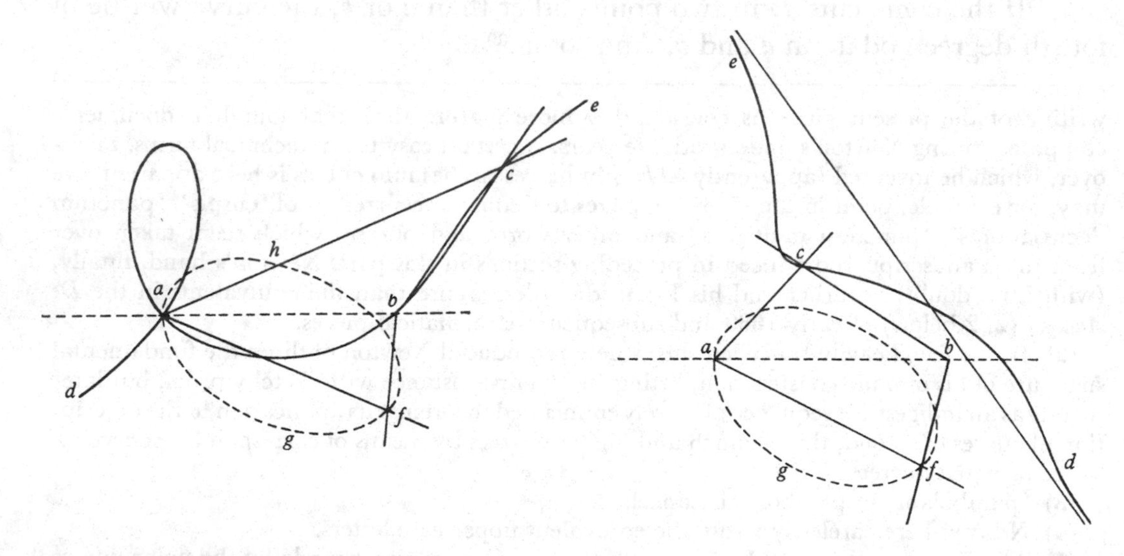

the describing and described curves which preserves topological configurations, tangency properties and, in general, the algebraic degree of the curves thus transmuted: evidently the described curve passes always through the poles *a* and *b* (when *bc* and *ac* respectively coincide with *ab*). Apparent exceptions to this rule are subsumed if we see that the polar line may count as part of the describing locus: once, that is, when the describing curve (*f*) passes through one of *a* or *b*, and twice when it is through neither. In these cases the described curve (*c*) will have appropriate multiple points corresponding to points *a* and *b* in the 'extended' describing curve. In the present case the describing curve (*f*) is a conic (which in 1 degenerates into the line-pair $ab \times de$). When this conic is through both *a* and *b* (1 and 2) the described locus is a conic (which may degenerate into a line-pair). When the conic is through *a* (3 and 4), the polar line *ab* has to be counted once as a part of the describing locus (through *b*) and the

When the points c and f are not in ab, then

1. ac and af meet their respective curves in the same number of points, excluding a.(9)
2. They are simultaneously tangent to their respective curves.
3. They meet them simultaneously in a node.
4. They simultaneously describe a cusp.
5. They simultaneously describe an isolated point (conjugate point).(10)
6. They simultaneously meet their respective curves in points of tangency.(11)

When f (or c) is in ab but coincides with neither of the points a and b, then

1. ac and bc meet their respective curve in one point other than a and b.
2. In that point, c, as many curved branches are nodate as there are meets, other than a and b, of gh with ab.
3. And as many are cusped as there are points of tangency of gh with ab.

---

described locus will be a cubic with a double point in a. (The latter will be a node, 'vere decussans', or a conjugate point, 'imaginarie decussans', according as, when bc coincides with a, the arm bf meets the conic in real points or not: in the latter case the points corresponding must be a conjugate complex pair, so that their imaginary component is zero. When the arm bf is tangent to the conic the oval has 'vanished', as Newton would say, and we have a cusp at a in the described cubic.) In much the same way, when the conic is through neither a nor b we must add the doubled line-pair ab to the describing locus (which will then, in effect, have double points at a and b) and the described curve (c) will be a quartic locus with similar double points at a and b: that at a, for example, will be a node, cusp or conjugate point according as, when the arm bc coincides with ab, the other arm bf meets the conic in distinct real, coincident or (conjugate) complex points. In addition the point in the described locus which is constructed from the meet of the conic with the polar line ab will be unique and hence a third double point (nodal, cusped or conjugately doubled according as the conic meets ab in distinct real, coincident or complex points).

(9) When the arm af (and so also ac) is fixed, the rotating angle $f\hat{b}c$ determines corresponding points in af and ac uniquely. Hence each meet (singular or multiple) of af with the describing curve (f) will be uniquely correspondent with the meet (similarly singular or multiple) of ac with the described curve (c).

(10) By the preceding, if af meets (f) in two coincident points, then ac meets (c) in two coincident points correspondingly. Likewise three coincident points in (f) correspond uniquely with three in (c), and so on. Since (note (8)) tangency properties are preserved in the organic construction, nodes, cusps and conjugate points in the describing curve (f) are uniquely correspondent with similar multiple points in the described curve.

(11) We suppose Newton intends by this rather mysterious phrase that when two describing curves, say (f_1) and (f_2), are tangent, then the corresponding described curves, (c_1) and (c_2), are tangent in the point corresponding to the contact point of the two describing curves. Newton has started the next line with a '7.' but a blank follows, and the figure itself seems to be lightly cancelled.

4. et tot tanguntur a *bc* quot punctis *gh* tanguntur ab *ab* in *a*.[12]

5. ad reliqua puncta in quibus secant suam curvam ductæ rectæ erunt tangentes ad alteram curvam in *a* vel *b*. Nempe[13]

Cum *f* cadit in *a* tum *bc* secat *ed* in

1. tot punctis alijs, quot curvæ *gh* decussant[14] in *a*.
2. & *a* tangit in tot punctis quot curvæ reflectunt in *a*.
3. & tot curvæ in *bc* cuspidantur quot cuspidantur in *a* modo non fit paralleliter ad *ab*.[15]

---

[2][16]

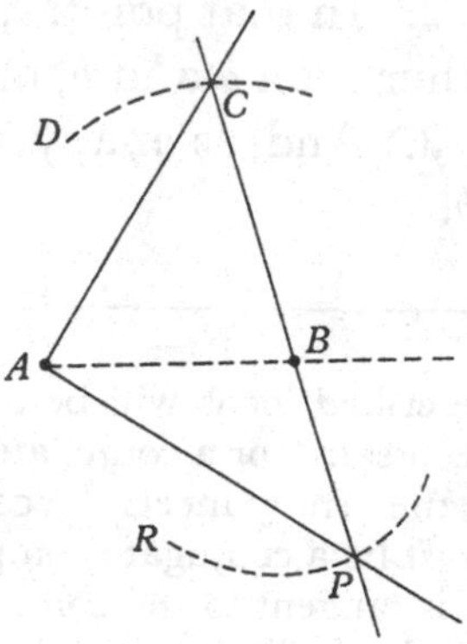

1. Si *PR* sit linea recta *DC* est Con Sect transiens per *A* & *B*.
2. Si *PR* est Con sect transiens per *A* & *B*, *DC* erit recta vel Con sect transiens per *A* & *B*.
3. Si *PR* est Con sect transiens per *A* & non per *B*,[17] *DC* erit curva trium dimensionum[18] habens punctum duplex[19] in *A* & transiens per *B*.
4. Si *PR* est Con sect transiens per *B* & non per *A* *DC* erit 3um dimensionū habens duplex[19] punctum in *B*.

[20] (Nota. In casu 3 & 4 si *AB* tangit Directricem,[21] *DC* habebit punctum cuspidatum[22] in *A* vel *B*.)

6. Si *PR* est Con: sect: transiens per neutrum *A* vel *B*, *DC* erit 4^{or} dimensionum habens 2 puncta multiplicia *A* & *B*. (viz punctum fit ex conjunctione 2^{arum} . &c)[23]

[fig 1]

---

(12) This case differs from the preceding in that the describing curve (*f*) meets the polar line *ab* in real points. To each of these meets (other than *a* or *b* itself) there corresponds a unique point in the described curve (*c*), and this will be multiply nodate or cusped according as those points are distinct or cluster in coincident points (two for a tangent, three for an inflexion point, and so on).

(13) The phrase is left unfinished. The truth of the observation is an immediate corollary of the property of the organic construction that it preserves tangency properties.

(14) 'secant' (meet) is cancelled.

(15) Presumably, that is, when *ab* is tangent to the cusp at *a*, in which case the point corresponding in (*c*) will have not a cusp but a point of inflexion at which *bc* is tangent. Conversely, if the describing curve (*f*) is touched by *ab* in a point of inflexion at *a*, the described curve (*c*) will have an additional cusp at the point corresponding to *a* and the arm *bc* will be tangent to (*c*) at that point.

(16) Newton reviews and extends his preceding notes, now considering higher-order singularities in the curves described by the organic construction. Where previously he had been able to restrict his attention, in the cubics and quartics generated from a conic describing curve, to double points (nodate, cusped or conjugate), he now realizes the necessity for treating

4. And as many are tangent to *bc* as there are points in which *gh* is tangent to *ab* at *a*.[12]

5. Lines drawn to the remaining points in which they meet their curve will be tangent to the second curve in *a* or *b*. Namely[13]

When *f* falls in *a*, then *bc* meets *ed* in

1. as many further points as there are curved branches *gh* nodate[14] in *a*.

2. And *a* touches it in as many points as the curved branches are (times) bent back at *a*.

3. And as many curved branches are cusped in *bc* as are cusped at *a*, provided that does not occur in a direction parallel to *ab*.[15]

---

[2][16]

1. If *PR* be a straight line, *DC* is a conic passing through *A* and *B*.

2. If *PR* is a conic passing through *A* and *B*, *DC* will be a straight line or a conic passing through *A* and *B*.

3. If *PR* is a conic passing through *A* but not *B*,[17] *DC* will be a curve of third degree[18] having a double[81] point at *A* and passing through *B*.

4. If *PR* is a conic passing through *B* but not *A*, *DC* will be a curve of third degree having a double[19] point at *B*.

[20] (Note. In cases 3 and 4 if *AB* touches the directrix,[21] *DC* will have a cusped[22] point at *A* or *B*.)

6. If *PR* is a conic passing through neither of *A* or *B*, *DC* will be a quartic having two multiple points, one each at *A* and *B*. (The double point , namely, results from the coincidence of the two branches , and so on.)[23]

---

higher-order point multiplicities and for the purpose invents an efficient pictographic notation for denoting their topological structure. In figure 1 note that he has taken the angle $P\hat{B}C$ a straight line for simplicity.

(17) The later addition 'vel tangens *AB* in *A*' (or tangent to *AB* in *A*) is cancelled.

(18) The restrictive first version 'decussans' (nodate) of the following phrase has been cancelled.

(19) Newton first wrote 'singulare' (singular) here.

(20) The following 'Nota' was originally paragraph '5.' in the list (and unbracketed). In the sequel Newton has failed to renumber the paragraphs accordingly: thus the following paragraph '6.' should read '5.' and each following paragraph number is correspondingly in error.

(21) The 'directrix' is, of course, the describing curve (*P*) of the organic construction, and is retained with that significance in the following sections.

(22) 'triplex' (triple) is cancelled. The cusp is a triple point in the present connotation only when (compare note (15)) the polar line *ab* is tangent to the curve there.

(23) The significance of Newton's pictographs here (and below) should be evident. These first five paragraphs, of course, summarize the results enunciated in [1].

7. Si *PR* est 3ū dimens transiens per *A* & *B* habens punctum multiplex in *A*, et tangens *AP* in *A* (fig 2)[24] *DC* est Con sect transiens per *A* & non per *B*.[25]

8. Sin secat *AP*, *DC* est 3um dimens habens Punct: Multiplic: in *A* & transiens per *B*.[26]

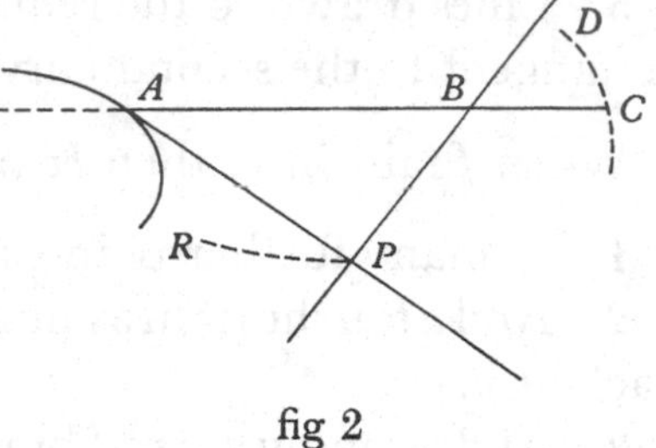

fig 2

9. At si *PR* habet Punc Mult in *B* & tangit *AP* in *A*, *DC* est Con Sect transiens per *B* et non per *A*.

10. Sin secat *AP* in *A* *DC* est 3um Dim habens Pu: Mul: in *B* & transiens per *A*.[27]

11. Si *PR* 3ū Dim: transiens per *A* & *B* habet P: M: extra *AB* in *AP* (fig 2) *DC* erit quatuor Dimens habens punctum multiplex)(in *B* & >(in *A*.

12. Si *PR* habet PM etiam extra *AP* (fig 2) & tangit *AP* in *A*, *DC* erit 3^{D} transiens per *A* & *B* & alicubi habet P: M:

13. Sin secat *AP* tum *DC* erit 4^{D} habens tres P: M: >(viz in *A* & *B*, & in alio puncto.[28]

14. Si *PR* est 3^{D} transiens per *A* et non per *B*, habens P: M: in *A* & tangens *AP* in *A*, *DC* erit 3ū D: habens P: M: in *A*.

15. Sin secat *AP* tum *DC* est 4^{D} transiens per *B* & habens punctum triplex ⋇ in *A*.[29]

16. Si habet P: M: in *B* extra *A*[30] & tangit *AP* in *A* tum *DC* est 4^{D} transiens per *B* & habens PM)(in *A*.

17. Sin secat *AP* in *A*, tum *DC* est 5^{D} habens PM >(in *B* & Punctum Triplex ⋇ in *A*.[31]

---

(24) Figure 2 is the particular case of the organic construction in which the meet *C* of the described curve is made to correspond with some point *P* on the describing curve. The advantage of this choice is that the points of *AP* determine corresponding directions through *B*, while *P* itself corresponds to the meets of the described curve (*C*) with *AB*.

(25) In this case (compare figure 1) the angle $P\hat{B}C$ is straight, while the inclination of the tangent at *A* to the polar line *AB* determines the angle $C\hat{A}P$. Clearly, the point *A* itself determines the line *AB* as part of the (cubic) described locus, so that this must be a conic with the straight line *AB*. Since the describing cubic meets *AP* only at *A* (for *AP* is tangent there in a double point) none of its other points may correspond with *B*, so that the conic is through *A* alone.

(26) Here the described cubic has a point other than *A* which corresponds to a point in *AB*, while, since *BP* meets it in two distinct points, to these will correspond a double point at *A*. Evidently the described cubic cannot degenerate.

(27) These two observations are complements of the preceding (7 and 8). The multiple point at *B* on the describing cubic corresponds with a similar one there on the described cubic while the latter is evidently also through *A*. When (in 9) *P* coincides with *A* the described cubic will degenerate into the line *AB* and a conic through *B*. (Figure 2 is, of course, understood.)

7. If *PR* is of third degree, passing through *A* and *B* with a multiple point at *A* and tangent to *AP* at *A* (figure 2),[24] *DC* is a conic through *A* and not *B*.[25]

8. But if it intersects *AP*, *DC* is of third degree with a multiple point at *A* and passes through B.[26]

9. If, however, *PR* has a multiple point at *B* and touches *AP* in *A*, *DC* is a conic through *B* but not *A*.

10. But if it intersects *AP* in *A*, *DC* is a cubic having a multiple point at *B* and passing through *A*.[27]

11. If the cubic *PR* through *A* and *B* has a multiple point outside *AB* in *AP* (figure 2), *DC* will be a quartic with multiple points at *B* and at *A*.

12. If, in addition, *PR* has a multiple point outside *AP* (figure 2) and is tangent to *AP* at *A*, *DC* will be a cubic through *A* and *B* and with a multiple point elsewhere.

13. But if it intersects *AP*, then *DC* will be a quartic with three multiple points, that is, double points at each of *A* and *B* and a third point.[28]

14. If *PR* is a cubic through *A* but not *B* with a multiple point at *A* and tangent to *AP* at *A*, *DC* will be itself of third degree with a multiple point at *A*.

15. But if it intersects *AP*, then *DC* is a quartic through *B* having a triple point at *A*.[29]

16. If it has a multiple point other than at *A*[30] and touches *AP* in *A*, then *DC* is a quartic through *B* and having a multiple point at *A*.

17. But if it intersects *AP* in *A*, then *DC* is a quintic with a multiple point at *B* and a triple point at *A*.[31]

---

(28) When (11) the describing cubic has a double point in *AP* this must be at *P* (since otherwise *AP* would meet the cubic in four points): in consequence the described curve has double points at *A* and *B*, and is therefore a quartic. But when (13) the double point is outside *AP*, the two meets each of *BP* and *AP* with the cubic other than *B* and *A* will determine double points at each of *A* and *B* in the described curve (a quartic therefore) while the cubic's double point will correspond with a third in the latter: however (12) when *P* coincides with *A*, the quartic will degenerate into the line *AB* and a cubic, the three double points being their meets.

(29) In (15) the three meets of *BP* with the cubic determine a triple point at *A* and the described curve (corresponding with the pair of the describing cubic and the line *AB*) will be a quartic through *B*. But when (14) *P* coincides with *A* the quartic will degenerate into a cubic and the line *AB*.

(30) That is, the describing cubic has its double point not at a pole of the organic construction. (The phrase 'in *B*' is extraneous and not reproduced in our translation.)

(31) The meets of *AP* and *BP* with the describing cubic other than *A* (two and three respectively) define (17) a triple point at *A* and a double point at *B* in the described quintic (which corresponds with the describing cubic and the polar line *AB* taken twice). Degeneracy occurs (16) when *P* coincides with *A*.

18. Si habet PM in *AP*, *DC* est 4^{D} transiens per *B* & habens P: M: χ in *A*.[32]

19. Si habet PM extra *AB* & *AP* & tangit *AP* in *A*, tum *DC* est 4^{dim} transiens per *B* & habens [2] PM χ in *A* & PM χ in alio loco.

20. Sin secat *AP* in *A*, *DC* est 5^{D} habens PM χ in *B* & Punct^{m} 3^{plex} ✶ in *A*, & PM in alio loco.[33]

21. Si *PR* est 3^{D} transiens per *B* et non per *A*, habens PM in *B*; *DC* est 4^{D} transiens per *A* et habens PT[34] in *B*.

22. Si habet PM in *AB* extra *B* *DC* est 5^{D} habens PM χ in *A*, & PT[34] ✶ in *B*.

23. Si $\overline{\text{het}}$[35] PM in *BP*: *DC* est 5^{D} $\overline{\text{hens}}$[36] PM χ in *A* & PT in *B*.

24. Si $\overline{\text{het}}$ PM in *AP*.[37]

25. Si $\overline{\text{het}}$ PM extra *BA* & *BP* *DC* est 5^{D} $\overline{\text{hens}}$ PM in *A* PT in *B* & PM in alio puncto.[38]

26. Si *PR* est 3^{D} per neutrum *A* vel *B* transiens $\overline{\text{hens}}$ PM in *AB* &c: *DC* est 6 Dimens.[39]

[27.] Si *RP* est 3^{D}, forte non habens PM transiens per *A* & *B*, & tangens *AP* in *A*: *DC* est $3^{D:}$ &c.

[28.] Sin non tangit *AP* in *A*; *DC* est 4 $\overline{\text{Dimens}}$ habens PM χ in *A* & *B*.[40]

[29.] Si *RP* transit per *A* et non per *B* & tangit *AP* in *A*: *DC* est 4^{Di} transiens per *B* & habens PM χ in *A*.

[30.] Sin secat *AP* in *A* tum *DC* est 5^{D} Habens Punct^{m} duplex χ in *B* & triplex ✶ in *A*.[41]

---

(32) In this case the multiple point will be at *P*, so that the two distinct meets of *BP* with the describing cubic determine correspondingly a double point at *A*.

(33) These are virtually the same as observations (16) and (17) above.

(34) In analogy with 'PM' (for 'punctum multiplex') read 'punctum triplex'.

(35) Read 'habet', and in the sequel similarly.

(36) Read 'habens' here and subsequently.

(37) Newton has rightly left this unfinished since the describing cubic is not through *A* and so cannot have a double point there!

(38) As before, the multiplicity of the points at *A* and *B* is determined by the (distinct) meets of *BP* and *AP* with the describing cubic, here not counting *B* itself. In (21) *AP* meets the cubic in three distinct points, so that *B* is a triple point on the described curve, while *BP* meets it only in *P*, or the described curve is not multiple at *A*: the latter, in consequence, is a quartic. In (22) *B* is not a double point of the describing cubic, and therefore the intersections of *BP* with it are increased by one. Likewise in (23), when the double point is in *BP* (at *P*, that is). Finally, (25) has an external double point in its describing cubic which corre-

18. If it has a multiple point in *AP*, *DC* is a quartic through *B* and with a multiple point $\times$ at *A*.(32)

19. If it has a multiple point outside both *AB* and *AP* and touches *AP* in *A*, then *DC* is of fourth degree through *B*, having two multiple points, one $\times$ at *A* and a second $\times$ elsewhere.

20. But if it intersects *AP* in *A*, *DC* is of fifth degree with a multiple point $\times$ at *B*, a triple point $\ast$ at *A* and a multiple point elsewhere.(33)

21. If *PR* is a cubic through *B* but not *A* with a multiple point at *B*, *DC* is of fourth degree through *A* with a triple point at *B*.

22. If it has a multiple point in *AB* other than *B*, *DC* is of fifth degree with a multiple point $\times$ at *A* and a triple point $\ast$ at *B*.

23. If it has a multiple point in *BP*, *DC* is of fifth degree with a multiple point $\times$ at *A* and a triple point at *B*.

24. If it has a multiple point in *AP*.(37)

25. If it has a multiple point outside *BA* and *BP*, *DC* is a quintic with a multiple point at *A*, a triple point at *B* and a further multiple point elsewhere.(38)

26. If *PR* is a cubic through neither of *A* or *B* with a multiple point in *AB*, and so on, *DC* is a sextic.(39)

[27.] If *RP* is a cubic, with perhaps no multiple point at all, through *A* and *B* and tangent to *AP* in *A*, *DC* is a cubic, and so on.

[28.] But if it is not tangent to *AP* in *A*, *DC* is a quartic with multiple points $\times$ at *A* and *B*.(40)

[29.] If *RP* passes through *A* but not *B* and is tangent to *AP* in *A*, *DC* is of fourth degree through *B* and with a multiple point $\times$ at *A*.

[30.] But if it intersects *AP* in *A*, then *DC* is of fifth degree with a double point $\times$ at *B* and a triple point $\ast$ at *A*.(41)

---

sponds with one in the described curve, the intersections of *BP* with the cubic (other than *B* itself) remaining two, those of *AP* three.

(39) Each of *AP* and *BP* meet the describing cubic in three distinct points, so that *A* and *B* are both triple points on the described curve, which is therefore a sextic.

(40) As ever, figure 2 is understood. In (28) *AP* and *BP* meet the describing cubic in two distinct points (other than the poles *A* and *B*), and in consequence *A* and *B* will each be a double point on the described curve, which is therefore a quartic since *AB* meets it in four points. When (27) *P* and *A* coincide, the described quartic degenerates into a cubic together with the polar line *AB*.

(41) In (30) *AP* meets the describing cubic in two points other than *A*, while *BP* meets it in three (all distinct in general). Correspondingly, the described curve has a triple point at *A* and a double one at *B*, and is hence a quintic. In (29) *A* and *P* coincide, and the described quintic degenerates into *AB* and a quartic.

[31.] Si *RP* transit per *B* et non per *A*, *CD* est 5^{D} hens 2P(42) in *A* & 3P(43) .(44)

[32.] Si *RP* transit nec per *B* nec *A*: *CD* est 6^{D} habens 3P(43) in *A* et 3P(43) in B.(45)

---

[3](46)

[1.] Si [*R*]*P*(47) est 3^{D} non habens PM transiens per *A* et *B* fig 1 *DC* erit ejusdem generis.

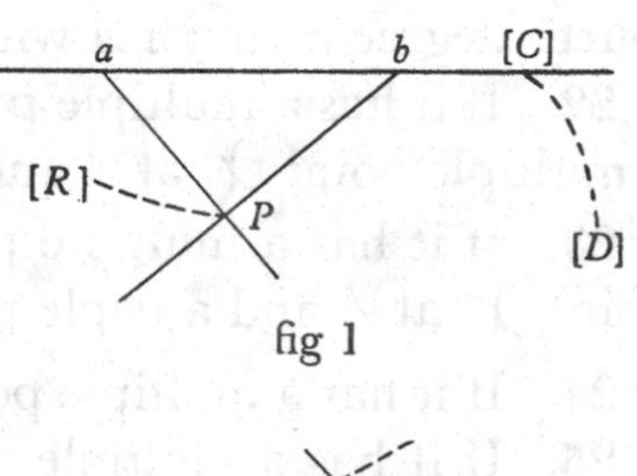

fig 1

[2.] Sin ut in fig 2; *DC* erit 4^{D} habens PM in *A* & *B*.(48)

[3.] Si transit per *A* et non per *B* fig 1 PM(49) est 4^{D} hens PM in *A* et in alio puncto et transiens per *B*.

[4.] Sin ut in fig 2 tum *DC* est 5^{D} hens PM in *B*, PT in *A* & PM in alio puncto.(50)

[5.] Si [*R*]*P*(47) transit per neutrum *A* vel *B* fig 1, *DC* est 5 Dim hens in *A* et *B* & in alio puncto.

[6.] Sin ut in fig 2, tum *DC* est 6^{D} hens PM in *A* et *B* et(51) alio puncto.(52)

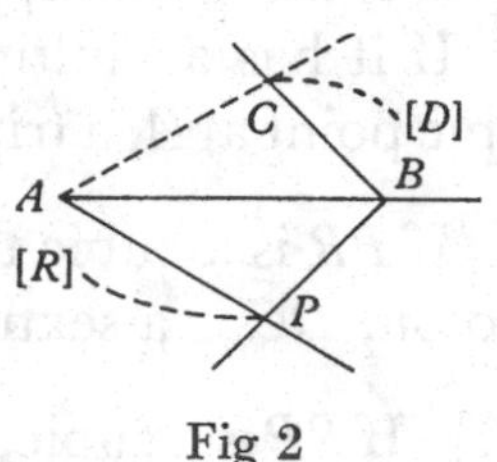

Fig 2

---

(42) Read 'duplex punctum'.

(43) Read 'triplex punctum'.

(44) This is identical with (30) except that *A* and *B* are interchanged.

(45) Here *AP* and *BP* meet the describing cubic in three distinct points each, so that the described curve has triple points at *A* and *B* and is consequently a sextic. (The qualitative distinction Newton makes between these points does not appear always to hold, and indeed he himself elides it in his revision in [3] below.)

(46) Newton revises observations (27)–(32) of [2] preceding. In figure 1 accompanying the line *CD* has been added to the manuscript figure in clarification. Likewise, in figure 2 the arm *AC* and the described and describing curves *PR* and *CD* have been restored.

(47) Newton's figure 1 has two points '*A*' (='*a*'). In line with the text and other figures we have replaced that on the describing cubic by '*R*', and at these two points amended the text to correspond.

(48) These are the restatements of (27) and (28) in [2] above when the angle $P\hat{B}C$ in the organic construction is no longer straight. The two cases of the preceding figure 2, when *P* does and does not coincide with *A*, now become those equivalently when some point *P* of

[31.] If *RP* passes through *B* but not *A*, *CD* is a quintic with a double point at *A* and a triple point.(44)

[32.] If *RP* passes through neither *B* nor *A*, *CD* is a sextic with a triple point at *A* and a second triple point at *B*.(45)

---

[3](46)

[1.] If [*R*]*P*(47) is a cubic without multiple point passing through *A* and *B* (figure 1), *DC* will be of the same kind.

[2.] But if as in figure 2, *DC* will be a quartic with multiple points at *A* and *B*.(48)

[3.] If it passes through *A* but not *B* (figure 1), [*DC*] is a quartic with multiple points at *A* and a further point and passing through *B*.

[4.] But if as in figure 2, then *DC* is a quintic with a multiple point at *B*, a triple point at *A* and a multiple point in a further point.(50)

[5.] If [*R*]*P*(47) passes through neither of *A* or *B* (figure 1), *DC* is a quintic with a double point at each of *A* and *B* and a triple point elsewhere.

[6.] But if as in figure 2, then *DC* is a sextic with a multiple point at *A* and *B* and (51) at a further point.(52)

---

the describing curve (here a cubic) does and does not correspond with the meet *C* of the described curve (a quartic) with the polar line *AB*: in the former (illustrated in the present figure 1) the point *P* of the cubic corresponds with the whole of the line *AB*, and in consequence the described quartic (figure 2) breaks into the pair of a cubic and that line *AB*.

(49) Read '*DC*'.

(50) These two enunciations, (3) and (4) (the cases of figures 1 and 2 respectively), are the generalizations of (29) and (30) in [2] above when the angle $P\hat{B}C$ is no longer straight. Much as before (compare note (48)) (3) is the degenerate case of the quintic described in (4) where *P* itself corresponds to the polar line *AB* and the remaining points of the describing cubic *PR* generate the complementary quartic of the described curve.

(51) Insert 'duplex' (a double point). Since this third multiple point corresponds with a point in the describing cubic not in *AP* or *BP*, it will be a (real or conjugate) double point.

(52) These two final cases are the degenerate and general equivalents respectively of (32) in [2] when $P\hat{B}C$ is no longer a straight line: in the former, *P* itself corresponds with the polar line *AB* while the remaining points of the describing cubic generate the companion (quintic) locus which, together with *AB*, makes up the described sextic curve.

§2. THE FIRST EXTENDED ACCOUNT OF THE CONSTRUCTION(1)

From the original manuscript in private possession

Prob 1. Describere Con sect transeuntem per 5 data puncta a b c d e.(2)

(3)Duæ regulæ similes isti *HGF* sunt fabricandæ ita ut earum crura possunt inclinari, pro lubitu in quovis dato angulo. Et in angulari puncto *F* debet esse cuspis calibeus(4) circa quam regula rotetur, dum iste cuspis alicui dato puncto infigitur. Nempe clavus calibeus quo crura Normæ junguntur possit altero extremo exacui; altero striari,(5) cui nux adaptata arctiùs (prout opus est) constringat crura normæ in angulo dato. Et si harum regularum centra *F* & *K* duobus punctis imponantur,(6) circa quæ tanquam polos regulæ rotentur, dum intersectio *S* duorū crurum in data linea quavis recta vel curva *MN* moveatur, intersectio *R* duorum crurum reliquorum describet novam lineam *PQ*. Et crura *FS*, *KS* voco dirigentia; eorum intersectionem *S*, punctum dirigens; & lineam *MN* in qua istud punctum movetur, Directricem: reliqua crura *FR*, *KR* voco describentia; & earum intersectio[ne]m,(7) punctum describens; & Curvam *PQ* quam describunt, lineam descriptam, vel describendam: angulos deniq; *RFS*, *RKS*, in quibus crura regularum figuntur voco Mobiles.

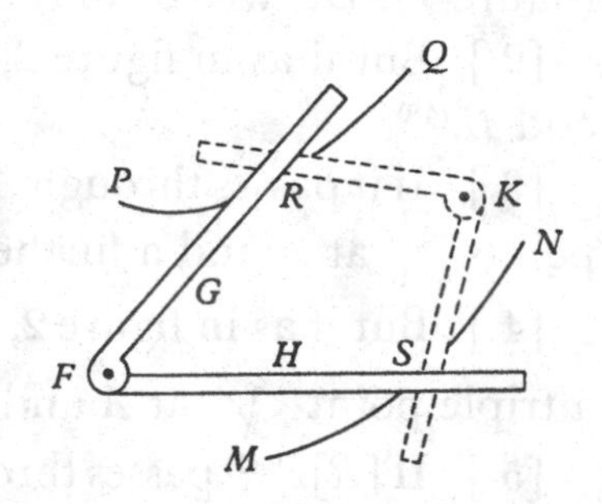

Probl: Solut°. Assumens duo quælibet ex punctis *d* & *e*, pro polis circa quos regulæ sunt rotandæ, istis punctis impono centra regularum. Et crurum alterū applico ad lineam (*de*) Alterum ad tertium aliquod datorum punctorum (*a*)(8) & figo(9) crura regularum in istis angulis *ade*, *aed*: & *da ea* erunt crura describentia & crura *dg*, *ef* dirigentia & *ade*, *aed* anguli mobiles in statione prima.

Deinde circumactis regulis dum crura describentia transeunt per quartum punctorum *b*, noto punctum *p* ubi crura dirigentia decussant: & eodem modo noto punctum *q* ubi crura dirigentia decussant cum crura describentia appli-

---

(1) This tentative, heavily cancelled piece is written in Newton's minute hand on three sides of a single folded sheet, the last of which contains in addition the outline-scheme of §1, [1] above. The fourth side bears the analytical calculations which we reproduce in §4 below, together with some unlettered rough sketches which generalize the present Problem 11 into Lemmas 11, 15 and 17 of §3 following.

(2) As we shall see in the 'solutio' below, this problem is resolved by reversing the first theorem of §1, [1] above. Two of the points (*d*, *e*) are taken for poles, a third (*a*) is made to correspond with the polar line (*de*) so that the described conic breaks into a line-pair (*de* and *pq*), while the remaining two points (*b* and *c*) define the described line (*pq*) uniquely under these limitations. On reversal the describing conic and described line-pair (*de* and *pq*) swop their respective rôles.

Translation

Problem 1. To describe a conic passing through the five given points a, b, c, d and e.[2]

[3]Two rules similar to this one, HFG, are to be manufactured so that their legs may be inclined at will in any given angle. And in the angular point F there should be a steel pin around which the rule may rotate while the pin is fixed on some given point. To be sure, the steel nail by which the legs of each sector are joined might be finely sharpened at one end, and on the other threaded to take a nut more or less tightly (as the need arises) which will clamp the legs of the sector in a given angle. And if the centres of these rules be set on[6] the two points F and K, around which as poles the rules rotate while the intersection S of two of the legs move in some given straight or curved line MN, the intersection R of the two remaining legs will describe the fresh line PQ. The legs FS and KS I call the directing ones; their intersection S the directing point; and the line MN in which that point moves the directrix: the other two legs FR and KR I call the describing ones; their intersection[7] [R] the describing point; and the curve PQ which they describe the described line or describend: lastly, the angles $R\hat{F}S$ and $R\hat{K}S$ in which the legs of the rules are fixed I call mobile.

To the problem's solution. Assuming any two at random, d and e, of the points for the poles around which the rules are to be rotated, on those points I set the centres of the rules. One (each) of the legs I apply to the line de, the second to some third point, a, of the given ones[8] and fix the legs of the rules in those angles $a\hat{d}e$, $a\hat{e}d$: da and ea will then be the describing legs, the legs dg and ef the directing ones and $a\hat{d}e$, $a\hat{e}d$ the mobile angles in this first position.

Then, revolving the rules until the describing legs pass through a fourth, b, of the points, I mark the point p where the directing legs cross: and in the same manner I mark the point q in which the directing legs cross when the describing

(3) In translating the technical terms, largely borrowed from the smithy, which occur in the following paragraph we have been guided by the English equivalents which Newton himself used in expounding the organic construction to John Collins on 20 August 1672. (See Appendix 2 to the present group of papers.) Thus we render 'norma' as 'sector'.

(4) A variant on the more usual forms 'chalybeius' or 'chalybeus'.

(5) That is, 'hollowed' to take a screw-nut.

(6) 'infigantur' (infixed) is cancelled.

(7) The clarifying phrase 'R quæ novam lineam describit voco' (R, which describes this new line, I call) is cancelled.

(8) In a first, cancelled continuation Newton wrote 'ita ut unius regulæ crura contineant angulum *ade*, & crura alterius angulum *aed*, vel *adf* ejus complementum ad duos rectos' (so that the legs of one rule may contain the angle $a\hat{d}e$, those of the other the angle $a\hat{e}d$, or (in the former case) $a\hat{d}f$ its supplement to two right angles).

(9) The more descriptive verb 'compingo' (I lock) is cancelled.

cantur ad quintum punctum *c*. Junctis[10] *p*, *q*: Ista linea recta *pq* erit Directrix. Jam si istæ regulæ ita rotentur circa polos *d* & *e* ut eorum crura dirigentia semper decussent in linea Directrice *PQ*[11] intersectio reliquorum crurum describet Conicam sectionem per data quinqꝫ puncta transeuntem.

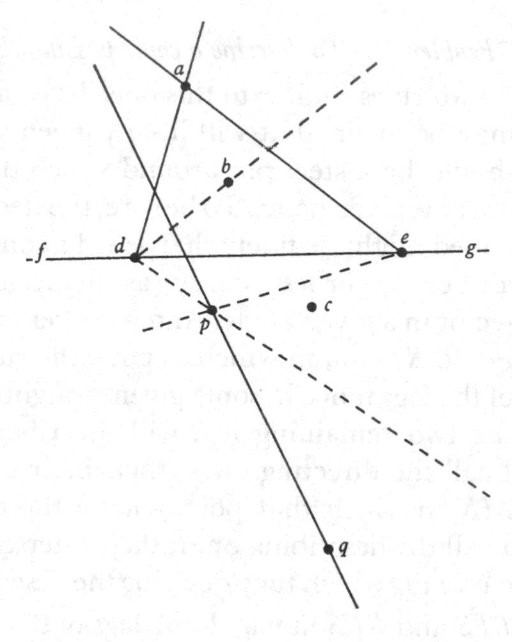

Nota 1. Cum duo crura (dirigentia in quærendis punctis *p* & *q*, vel describentia in describendo curvam) non decussant sed decussarent si ad istas partes versus centrum *F* producerentur; tum pro angulo *ade*, vel *aed* (vel utroqꝫ si opus est) sumendum est *adf* vel *aeg* complementum ad duos rectos. Et si punctum aliquod datum infinitè distat tum crura dirigentia ponantur parallela versus istud punctum &c: sin contingit crura describentia esse parallela tum punctum decussationis quæsitum infinite distat &c & linea *pq* ad eas partes tendet.[12]

2. Postquam uno modo Curvam descripseris, possis alijs modis tentare num exacte descripta sit nempe figendo regulas in alijs angulis (*ade*, *bed* &c) vel sumendo alia puncta pro polis regularum.

3. Cum problema aliquod solvendum est per intersectionem conicæ sectionis cum alia data curva sufficit istas minimas portiones describere quæ dictam curvam secant:[13] Imo ne quidem opus erit ut istæ describentur,[14] sed tantum ut annotentur puncta in quibus conica sectio datam curvam secaret modò describeretur.[15]

4. Hæc et sequentes descripsiones[16] plurimum habent usum in locis solidis[17] determinandis, &c. Nempe ex datis tantū 5 punctis &c[18] sine aliquo calculo præmisso, licet ignorātur vertex, axis, Diametri, centrum et species curvæ modo datur genus [viz] quod sit conica sectio possis tamen curvam describere.[19]

(10) A first harsh assonance with a following word 'punctis' (points) has been avoided by its cancellation.

(11) Newton intends the lower case form '*pq*', of course.

(12) This final sentence is a later addition in the manuscript.

(13) Newton first phrased his thoughts rather differently, writing '...per descriptionem conicæ sectionis, non opus erit amplius describere quam sufficit istas curvæ [portiones]' (through the description of a conic, there will be no need to describe a larger portion of the curve than is requisite).

(14) Read 'describantur'.

legs are applied to the fifth point *c*. Join(10) *p* and *q*: that straight line *pq* will be the directrix. Now if those rules be so rotated round the poles *d* and *e* that their directing legs cross always in the directrix-line *PQ*,(11) the intersection of the remaining legs will describe the conic passing through the five given points.

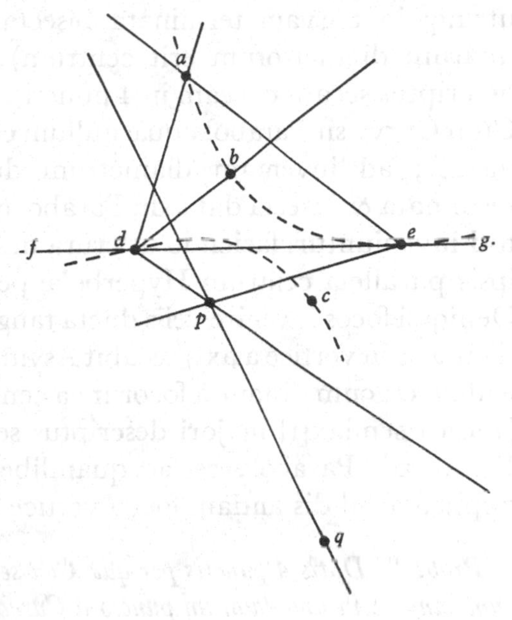

Note 1. When two legs (directing ones when points *p* and *q* are to be sought, or describing ones when the curve is to be described) do not cross but would do so if produced in the direction of the centre *f*, then for the angle $a\hat{d}e$ or $a\hat{e}d$ (or both if need be) you must take its supplement $a\hat{d}f$ or $a\hat{e}g$ to two right angles. And if some given point be infinitely distant, then the directing legs are to be set parallel in the direction of that point, and so on: but if the describing legs chance to be parallel, then the crossing-point you seek is infinitely distant, and so on, and the line *pq* will tend in that direction.(12)

2. After you have described the curve one way, you may test in others whether the description is accurate: precisely, by fixing the rules in other angles (say $a\hat{d}e$, $b\hat{e}d$ and so on) or by taking other points for the poles of the rules.

3. When some problem is to be resolved by the meet of a conic with another given curve, it will be sufficient to describe those minute portions which meet the aforesaid curve: indeed, there will be no need even for those to be described, but merely that the points in which the conic would intersect the curve were it described be noted.(15)

4. This and the following descriptions are of the greatest use in determining solid loci,(17) and so on. Precisely, given merely five points and so on,(18) without any preparatory calculation or knowing the vertex, axis, diameters, centre and species of the curve, provided only that its kind is given (that it is a conic) you should even so be able to describe it. And once you have described the curve it

(15) Presumably by adjusting the organic construction instrument so that in each case a point of the intersected curve be made to correspond with a point of the straight line *pq*.

(16) Read 'descriptiones'.

(17) That is, non-degenerate conic loci.

(18) Newton explained his meaning in the cancelled phrase '& quod curva sit Con sect' (and that the curve be a conic).

(19) 'ac determinare' (and determine) is cancelled.

Et Cur[v]a descripta facile erit reliqua determinare: nempe duæ parallelæ in curvam terminatæ bisectæ dant Diametrum cui applicantur, cujus portio utrinq; in curvam terminata bisecta, dat Centrum Curvæ (vel intersectio duarum diametrorum erit centrum); Quo centro & quovis radio Circulus descriptus secans curvam in 4 punctis, dat axes conjugatos & vertices curvæ.(20) Cùm Curva sit Parabola quæ nullum centrum habet (vel cujus centrum infinite distat); ad inventam diametrum ducta perpendicularis in curva utrinq; terminata & bisecta dat axin Parabolæ. Cùm Curva sit Hyperbola ejus Asymptoti inveniuntur, faciendo ut crura describentia sint sibimet parallela, & linea ipsis parallela centrum Hyperbolæ pertra[n]siens erit una Asymptotorum.(21) Deniq; si focos invenire velis ducta tangens ad verticem Hyperbolæ (viz perpendicularis in vertice a axi) secabit Asymptoton in puncto cujus distantia a centro eadem est cum distantia focorum a centro:(22) & circulus termino minoris axis & [radio] semiax[i] majori descriptus secabit axin majorem in focis Ellipsis:(23) Et, ut axis Parabolæ est ad quamlibet sibi Applicatam ita quarta pars istius applicatæ ad distantiam foci a vertice Parabolæ.(24)

Prob: 2. Datis 4 punctis per quæ Con sec debet transire, una cum linea TV quæ debet eam tangere in uno datorum puncto d Curvam describere.

Solut. Sumpto puncto contactûs *d* cum alio puncto *c* pro polis, applica regulas ad tertium punctum *b* & fige in angulis *bdc*, *bcd*: dein regulæ ad quartum punctum *a* convolutæ dabunt punctum *p* ut priùs, Deniq; applica crus describens regulæ *bdc* ad tangentem *TV* & crus describens regulæ *bcd* ad punctum *d* & intersectio crurum dirigentium dabit punctum *q*. Ducta *pq* erit directrix in qua dum crura dirigentia decussant, reliquorum crurum decussatio describet curvam quæsitam.(25)

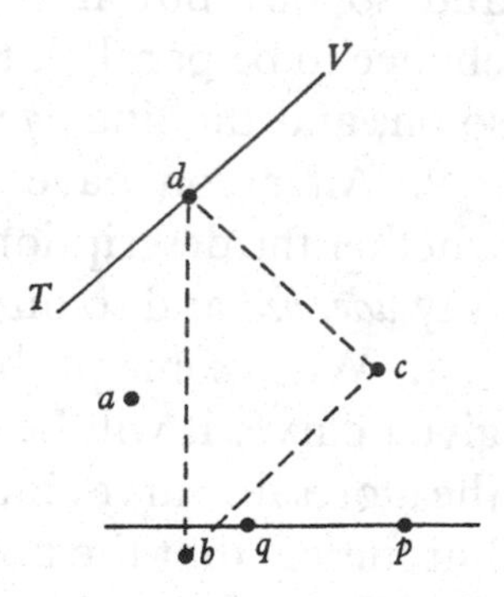

Vel sic,(26) puncto contactûs *a* cum alio puncto *b* pro polis sumto, & ducto *ab*,

(20) Specifically, the main diameter of a central conic will bisect the directions of these four meets from the centre, and the vertices will be the intersections of these main diameters with the conic. This and the preceding were, of course, standard results in Newton's day.

(21) Since the two poles are finite, when the two describing legs meet at the infinite points on a hyperbola they will be parallel and their directions will be parallel to those of the asymptotes (through the hyperbola's centre).

(22) Compare I, 1, 1, §1: 5f. In analytical terms, if we refer the hyperbola to perpendicular Cartesian axes with origin at its centre and $y = 0$ its main diameter, the result is an immediate corollary of reducing the focal-distance property $\sqrt{[(x+ae)^2+y^2]} - \sqrt{[(x-ae)^2+y^2]} = 2a$ to the more easily recognizable defining equation $(e^2-1)x^2-y^2 = a^2(e^2-1)$, in which a is the semi-major axis of the hyperbola and $\pm ae$ its focal distances from the centre.

(23) This standard property of the ellipse is the corollary of its focal-distance property when the general point is, at the ends of its minor axis, equidistant from the foci.

will be easy to determine remaining matters: thus, two parallels terminated at the curve yield by their bisection the diameter to which they are applied, and bisection of that portion of this terminated on either hand at the curve yields the curve's centre (alternatively, the intersection of two diameters will be that centre), while the circle described with this point as its centre and with arbitrary radius which meets the curve in four points yields the conjugate axes and the vertices of the curve.(20) When the curve is a parabola having no centre (or whose centre is infinitely distant), when you have found a diameter draw a perpendicular to it terminated on either hand at the curve: this when bisected yields the parabola's axis. When the curve is a hyperbola its asymptotes are found by making the describing legs parallel to each other, in which case the line parallel to them through the hyperbola's centre will be one of the asymptotes.(21) Should you wish, finally, to find the foci, in the hyperbola draw a tangent at its vertex (that is, a perpendicular at the vertex to the axis) and it will meet an asymptote in a point whose distance from the centre is the same as that of the foci:(22) in the ellipse a circle described on the end of the minor axis with radius equal to the semi-major axis will meet the major axis in the foci:(23) and in the parabola, the distance of the focus from the vertex is a fourth proportional to an arbitrary segment of the parabola's axis, its corresponding ordinate and a quarter of that ordinate.(24)

Problem 2. Given four points through which a conic shall pass together with the line TV which shall be tangent to it at one of those given points, d, to describe the curve.

Solution. Assuming the contact-point d along with any other point c for the poles, apply the rules to a third point b and fix them in the angles $b\hat{d}c$, $b\hat{c}d$; next, when the rules are revolved round to the fourth point a they will yield the point p as before; finally, apply the describing leg of the rule bdc to the tangent TV and that of the rule bcd to the point d and the meet of the directing legs will give the point q. pq, when drawn, will be the directrix, and while the directing legs cross on it, the cross-point of the remaining legs will describe the curve sought.(25)

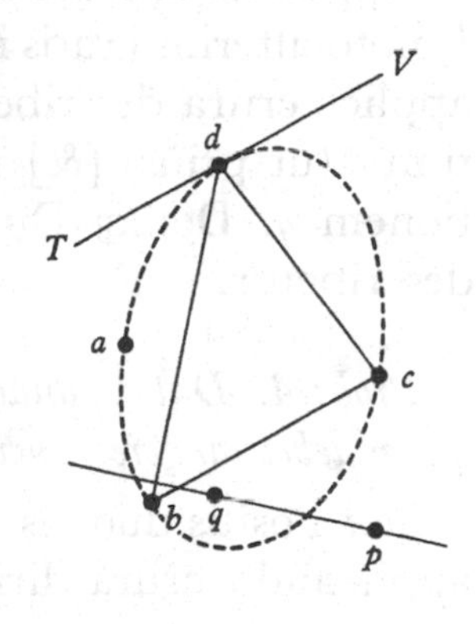

Or thus,(26) taking the contact-point a along with another of the points b for

(24) In analytical terms $x:y = \frac{1}{4}y:f$ (or $y^2 = 4fx$) where x and y are perpendicular coordinates of the general parabolic point and f its focal distance from the vertex (one-fourth its latus rectum).

(25) Newton's trick is to set d as one of the poles, so that when db meets the conic in a second point there the instantaneous direction of the arm ('leg') db will be that of the tangent TV.

(26) Note that point a is now that at which TV is to be tangent.

figo crura regulæ *a* in angulo *baT* vel *baV*, & erit *ab* ejus crus dirigens; sed crura regulæ *b* in directum pono ut unam rectam constituant; deinde regula *b* & crus describens regulæ *a* applicentur ad punctum tertium *d* & nota punctum *p* ubi regula *b* secat crus dirigens, et eodem modo regulis ad quartum punctum [*c*] applicatis notabis intersectionem *q*. Deniq; ducta *pq* erit directrix in qua dum punctum dirigens (i:e decussatio quam regula *b* facit cum crure dirigenti regulæ *a*) movetur, altera regularum decussatio describet quæsitam curvam.(27)

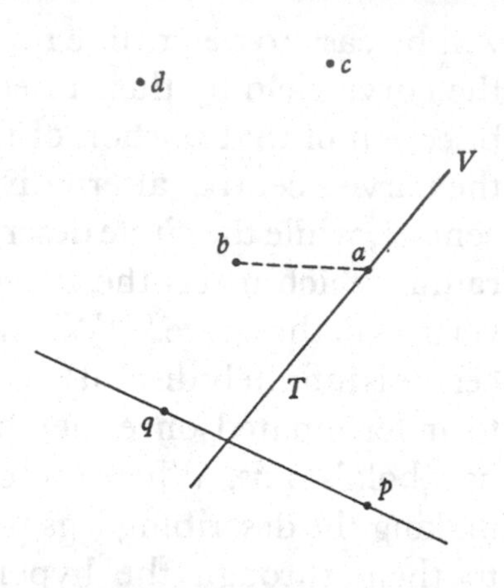

Cor.(28) Hinc si linea *ab* est(29) perpendicularis ad curvam in puncto *a* tum angulus mobilis erit rectus & descriptio erit simplicissima quæ per duos diversos motus fieri poterit.

Prob 3. Datis punctis a & b in quibus Conica sectio debet tangere duas datas rectas aT, bT, una cum tertio puncto c per quod transire debet, Curvam describere.

Sol:(30) Sumptis *a* & *b* pro polis figo crura regulæ *a* in angulo *baT* vel *baV*, posito quod *ab* sit crus dirigens; sed crura regulæ *b* in directum pono. Deinde regulâ *b* ad tangentem *bT* applicata & crure describenti ad punctum *b*, noto alterius cruris intersectionem *p*. Item applico crura describentia ad tertium punctum *c* (ut priùs) [&] noto alteram decussationem *q*. Deniq; Ducta *pq* erit directrix cujus ope curva quæsita ut prius describetur.

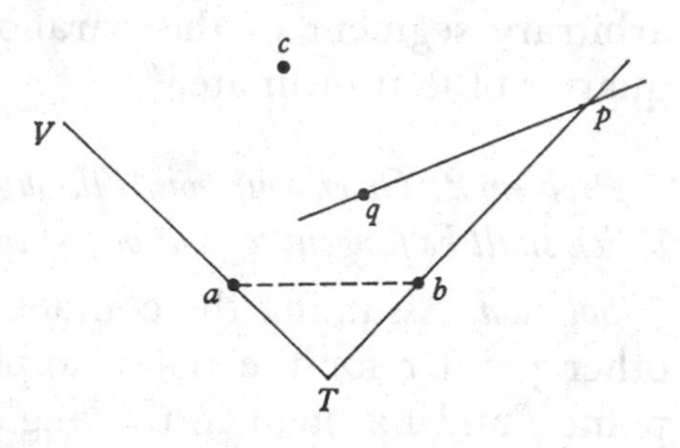

Prob: 4. Datis quatuor punctis per quæ conica sectio debet transire una cum linea recta quam debet tangere: Curvam describere.

Sol: Positis duobus punctis *a*, *b* pro polis, Figo regulas in angulis *cab*, *cba*. Et applicando crura dirigentia ad quartum punctum *d* invenio punctum *p* ut

(27) Much as before (note (25)) the contact-point *a* is made one of the poles and the tangent *TV* is introduced as the 'leg' through *a* which meets the conic in a second point there. Since now the 'crus dirigens' and the 'crus describens' through *b* are the same straight line, Newton denotes them both indifferently by 'regula *b*' (the rule through *b*).

(28) This is a late insertion, here introduced from the bottom of his manuscript page according to Newton's direction.

(29) A first continuation, 'vertex principalis curvæ' (the principal vertex of the curve), is cancelled.

poles, I fix the legs of the rule through a at the angle $b\hat{a}T$ or $b\hat{a}V$ and then will ab be its directing leg; but the legs of the rule through b I set so as to form a single straight line. Next let the rule through b and the describing leg of the rule through a be applied to the third point d and mark the point p where the rule through b intersects the directing leg, and in the same way when the rules are applied to the fourth point [c] you will mark the meet q. Finally pq when drawn will be the directrix and while the directing point (that is, the cross-point which the rule through b makes with the directing leg of the rule through a) moves in it the second crossing of the rules will describe the curve sought.[27]

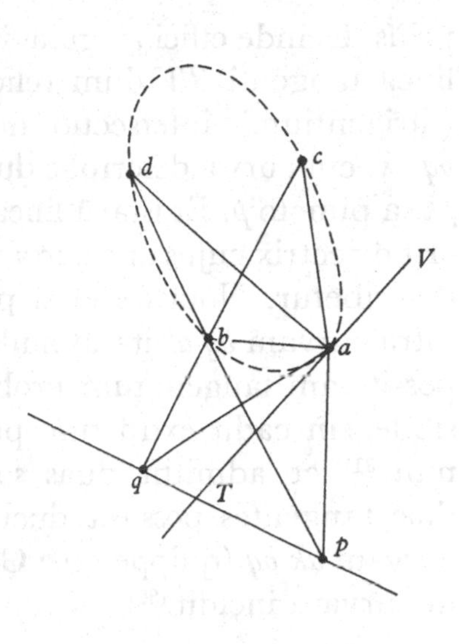

Corollary.[28] Hence, if the line ab is normal to the curve at the point a then the mobile angle will be right and the description will then be the simplest which can result from two different motions.

Problem 3. Given the points a and b in which a conic shall be tangent to the two given straight lines aT, bT together with the third point c through which it shall pass, to describe the curve.

Solution.[30] Taking a and b for poles I fix the legs of the rule through a in the angle $b\hat{a}T$ or $b\hat{a}V$ (on the supposition that ab is the directing leg) but the legs of the rule through b I set in a straight line. Then, applying the rule through b to the tangent bT and the describing leg to the point b, I mark the intersection p of the other pair of legs. In like manner I apply the describing legs to the third point c (as before) and mark the second crossing-point q. Finally, pq when drawn will be the directrix and by its aid the curve sought will be described as before.

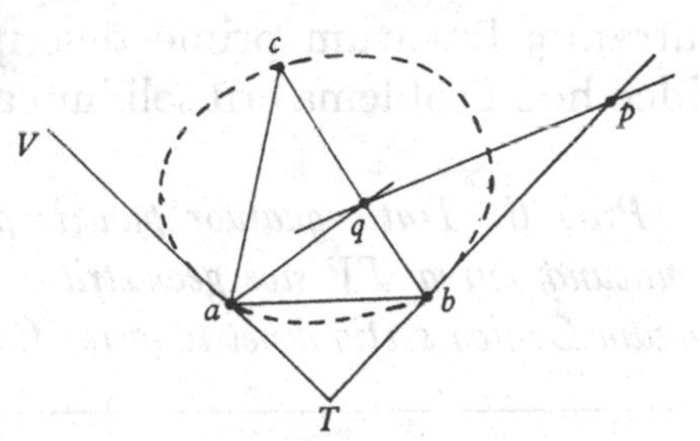

Problem 4. Given four points through which a conic shall pass together with a straight line it shall touch, to describe the curve.

Solution. Setting the two points a and b for the poles, I fix the rules in the angles $c\hat{a}b$ and $c\hat{b}a$, and by applying the directing legs to the fourth point d I find the

(30) The trick which resolved the preceding problem is now applied twice, with each of the contact-points a and b now made a pole.

priùs. Deinde efficio regulas ita rotari ut crura describentia semper decussent in linea tangenti *TV* dum reliquorum crurum (dirigentium) intersectio describet curvam *bq ak*. cui curvæ descriptæ duco tangentes *pq*, *pk*, a puncto *p*. Et istarū lineām *pq*, *pk* utravis erit directrix cujus ope curva quæsita ut prius describetur. Nota quod si punctum *p* cadit intra curvam *bq ak* ita ut nulla linea a *p* ducta possit eam tangere tum problema est impossibile, sin cadit extra tum problema est planum(31) et admittit duas solutiones nempe duæ tangentes possunt duci a puncto *p* ad curvam *ak bq* (quippe quæ Conica sectio est) excepto tantùm cum punctum *p* in curvam incidit.(32)

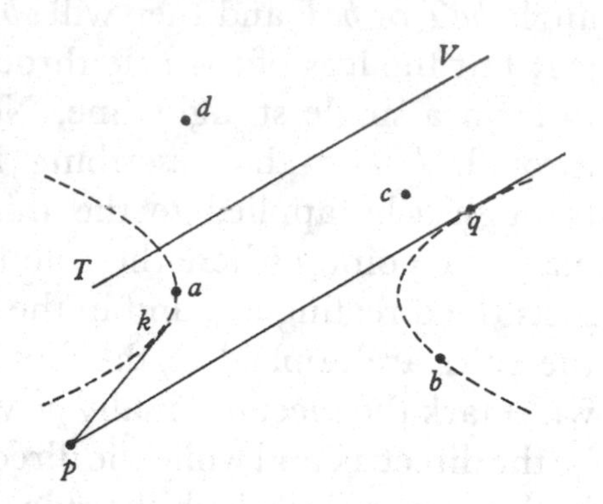

Prob. 5. Datis tribus punctis per quæ conica sectio debet transire una cum duabus rectis quas debet tangere: curvam describere.

Sol: Regulis ad tria data puncta ut priùs applicatis, facio ut eorum crura describentia circumvoluta sese semper decussent in datis tangentibus dum reliquorū crurū intersectio describat alias duas curvas. Deniqȝ si ducatur recta tangens utramqȝ descriptam curvam ista recta erit directrix cujus ope Conica sectio quæsita est describenda.(33)

Nota quoniam quatuor (ut plurimum)(34) rectæ possunt duci quæ tangen[t] utramqȝ linearum primo descriptarum (quippe quæ sunt conicæ sectiones) ideo hoc Problema erit solidum admittens quadruplicem solutionem.(35)

Prob 6. Datis quatuor punctis per quæ conica sectio debet transire, unâ cum alia quacunqȝ curva TV sive geometricâ sive Mechanicâ sive ad arbitrium vel casu descripta, quam Conica sectio debet tangere; Conicam Sectionem describere.

---

(31) That is, constructible as the meet of a straight line with a circle (or *locus planus*).

(32) By means of the organic construction Newton maps the conic *abcd* and its tangent *TV* into a corresponding straight line (*pk* or *pq*) and the derived conic (*bqak*) respectively. Since the former are tangent, so too must the latter be, or in other words *pk* (or *pq*) must be tangent to the conic *bqak* as Newton presupposes in his construction. Our figure is an accurate reproduction of Newton's, in which unfortunately the distribution of the points *a*, *b*, *c* and *d* round the tangent *TV* does not yield a real solution to the problem: precisely, the line-pair $ab \times TV$ does not correspond accurately with the conic *bqak*. However, Newton's argument should be clear without our appending a metrically correct drawing for the case of real solution to the problem.

(33) Much as in the preceding problem Newton reduces the present case of conic tangency to a corresponding but simpler one of straight line tangency by his organic construction. In this correspondence the tangent conic corresponds with an (undetermined) straight line which

point p (which corresponds) as before. Then I make the rules revolve so that the describing legs shall always cross in the tangent line TV while the intersection of the remaining (directing) legs describes the curve $bq\,ak$. To this described curve I draw the tangents pq and pk from the point p, and then either of those lines pq or pk will be the directrix by whose aid the curve sought will be described as before. Note that if the point p falls within the curve $bq\,ak$, so that no (real) line may be drawn tangent to it from p, then the problem is impossible, but if it falls outside it then the problem is plane[31] and admits two solutions: precisely, two tangents can be drawn from the point p to the curve $ak\,bq$ (since it is a conic) excepting only when the point p falls on the curve.[32]

Problem 5. Given three points through which a conic shall pass together with two straight lines it shall touch, to describe the curve.

Solution. Applying the rules as before to the three given points, I make their describing legs cross always, as they revolve, on the given tangents while the intersection of their two remaining legs describes two further curves. If, finally, a straight line be drawn tangent to both these described curves, that line will be the directrix by whose aid the conic sought is to be described.[33]

Note that since four straight lines (at most)[34] can be drawn to touch both the curves first described (since they are conics) this problem will consequently be solid, admitting a four-fold solution.[35]

Problem 6. Given four points through which the conic shall pass, together with any other curve TV, algebraic or transcendental or described arbitrarily or randomly, which shall touch the conic, to describe that conic.

---

is fixed by its touching the two new conics which, in the organic transformation, are the corresponding curves of the two given tangents.

(34) More accurately, there will be four, two or no solutions according as the described conics are distinct and do not contain each other, or as they intersect, or as one contains the other wholly. In addition, two or more of these solutions may coincide when the described conics are tangent to each other.

(35) We should perhaps remark that the duals of Problems 4 and 5, which hold true, yield the same number (two and four) of general solutions. That of the former is not amenable to the organic transformation (which requires two points given on the conic to serve as the poles), while the latter is not conveniently tractable and the number of its solutions not immediately apparent. In this latter case, specifically, when the two given points on the conic are set as poles of the organic construction the conic corresponds with the pair of the polar line and an undetermined straight line while the three given tangent lines correspond with a trio of conics each through the two poles. Existence of any solution depends on our proving that these three new conics share a common tangent, not an easy matter, while enumeration of the possible solutions requires that we show these conics share up to a maximum of four such common tangents (each of which corresponds with a conic which answers the conditions of the problem). Two decades later Newton was able to resolve both these dual cases by optical projection. (See our sixth volume and his *Principia* (London, 1687): Liber I, Props. XXV and XXVI: 87–9.)

Sol. Eodem fere modo cum quarto problemate solvitur. Nam regulæ *b*[*a*] ita sunt movendæ ut earum crura describentia semper decussent in data Curvâ *TV* dum reliquorū crurum [intersectiones] describent novam curvam; ad quam a puncto *p* ducta tangens *pq* erit directrix, cujus ope Conica sectio quæsita est describenda.(36)

Prob 7. Datis tribus punctis per quæ con sec debet transire una cum 2 quibuscunq curvis quas con sect debet tangere; Con sect describere.

[*Solutio.*] Eodem fere modo cum 5° solvitur.(37)

Prob 8. Prædicta problemata solvere cùm conica sectio non debet tangere sed secare datas lineas in datis angulis.

Solut°. In prob 2 & 3 datis angulis intersectionis dantur tangentes, sed in prob 4, 5, 6, & 7 quæstio est difficilior. nempe descriptis lineis *ak qb* ope datarum secantium (ut priùs ope tangentium), in prob 4 & 6 a puncto *p* ducenda est recta *pq* secans curvam *ak qb* in dato angulo(38) & ista recta erit Directrix cujus ope con sectio est describenda. & hoc prob est solidum(39) cum data secans est linea recta. Sed in prob 5 & 7 ducenda est recta *pq* secan[s] utramq curvam in datis angulis,(38) & ista recta erit directrix quæsita. Et Hoc Prob: admittit octo solutiones(40) cum datæ secantes sunt lineæ rectæ.

Prob 9.(41) *Datis 7 punctis a, b, c, d, e, f, g per quæ curva trium dimensionū transire debet & quod in unico istorum punctorū a curva debet vel decussare vel Cuspidari vel habere punctū singulare pro Ellipsi, Curvam describere.*(42)

Vel Facilius sic(43) Solut. Sumpto *a* cum alio puncto quovis *b* pro Polis: applico regulas *ad* tertium aliquod punctum *c*, & eas figo in angulis *cab*, *cba*. Deinde

(36) As Newton says, this is an insignificant generalization of Problem 4, with the tangent conic reduced to a tangent line corresponding to it in the organic construction.

(37) The similar generalization of Problem 5 in which the key point remains the reduction of the tangent conic to a corresponding straight line by an appropriate organic construction.

(38) Newton's implicit assumption here that the angle of intersection of two describing curves is also that of the respectively correspondent ones described by an organic construction, the poles of which are two points on the first pair of curves, is in general fallacious and invalidates the remainder of the present paragraph. (For a counter-example we may take the simplest case in which two straight lines passing one each through the two poles have corresponding organically described curves which are themselves straight lines one each through the poles but which do not in general intersect at the same angle as the first pair.) Newton subsequently realized his error (compare §3: note (28) below).

(39) That is, not resolvable by any equation constructible by the meets of circles.

(40) At most. No solution, for example, is in general possible when the conics are two concentric circles.

(41) In a first, cancelled Problem 9 Newton intended to apply the organic construction to further conic problems, for he began its enunciation 'Dato axe, verticibus, et alio Conicæ

Solution. This is resolved in almost the same manner as the fourth problem. For the rules through *b* and *a* are to be moved so that their describing legs cross always in the given curve *TV* while [the meets of] the remaining legs describe a fresh curve. If to this from the point *p* you draw a tangent *pq*, that will be the directrix by whose aid the conic sought is to be described.[36]

Problem 7. Given three points through which a conic shall pass together with any two curves which it shall touch, to describe the conic.

[*Solution.*] This is resolved in almost the same manner as the fifth.[37]

Problem 8. To resolve the above-mentioned problems when the conic no longer touches but shall intersect given curves in given angles.

Solution. In problems 2 and 3 from the given intersection angles the corresponding tangents are given, but in problems 4, 5, 6 and 7 the question is more difficult. Specifically, in problems 4 and 6 when you have described the curves *ak qb* with the aid of the given intersecting curves (as before you did with that of the tangent curves) from the point *p* there has to be drawn a straight line *pq* meeting the curve *ak qb* at a given angle[38] and that line will be the directrix by whose aid the conic is to be described. This problem is solid[39] when the given secant curve is a straight line. But in problems 5 and 7 there has to be drawn a straight line *pq* meeting both curves in given angles,[38] and then that line will be the required directrix. This problem admits of eight solutions[40] when the given secant curves are straight lines.

Problem 9.[41] *Given the seven points a, b, c, d, e, f and g through which a curve of third degree shall pass and that in one only of those points, a, it shall either cross itself, be cusped or have a conjugate point instead of an oval, to describe the curve.*[42]

Even more easily thus[43] is it resolved. Taking the point *a* together with any second one, *b*, for poles I apply the rules to some third point *c* and fix them in the

sectionis puncto: Curva poterit describi si' (Given the axis, vertices and a further point of a conic, we might be able to describe it if) and then broke off.

(42) This is the problem referred to by Newton in his *Epistola Posterior* of 24 October 1676 to Oldenburg for Leibniz. See *The Correspondence of Isaac Newton*, 2 (Cambridge, 1960): 119, l. 6. In the accompanying note (41) on page 155, where H. W. Turnbull refers mistakenly to a redraft of the present problem in Newton's 'De Modo describendi Conicas sectiones' (§3 following), he is, we should note, unnecessarily confused in his rendering of 'punctū singulare pro Ellipsi': the tangents at a conjugate point are not imaginary but real (if not uniquely defined).

(43) This phrase is a late addition and seems to imply the past existence of a more contrived alternative version, perhaps that of Problem 3 of §3 following. (Compare note (55) of that section.)

applicando crura describentia ad unumquodq; punctorum quatuor reliquorum *d*, *e*, *f*, *g*, decussatio crurum dirigentium dabit alia quatuor puncta *p*, *q*, *r*, *s*, per quæ, & per quintum punctum [*a*] describo conicam sectionem (per Prob 1). Et ista Conica sectio erit directrix[44] cujus ope curva quæsita describetur.[45]

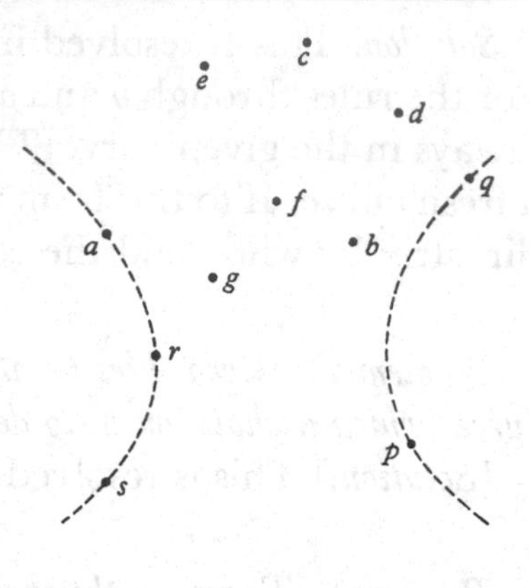

Prob 10. Datis sex punctis a b c d e f per quæ curva trium dimensionum debet transire, tangens datam lineam TV in uno istorum punctorum b, & vel decussans vel cuspidata vel habens punctum singulare in alio punctorum a[; *curvam describere*].

Solut. Sumptis *a* & *b* pro polis, figo crura regulæ *b* in angulo *abT* vel *abV*, sed pono crura regulæ *a* in directum. Deinde applicando crura describentia ad quatuor reliqua puncta *c*, *d*, *e* & *f* Crura dirigentia decussabunt in alijs quatuor punctis *p*, *q*, *r*, & *s*,[46] per quæ & per punctum *a* Conicam Sectionem describe[47] (per Prob 1), & Ista Conica sectio erit Directrix cujus ope Cu[r]va quæsita est describenda.[48]

Nota Si quæsita curva decussaret in puncto contactûs *b*, tum Conica Sectio (quæ directrix est) transiret per punctū *b* & non per punctum *a*; et Reliqua sunt ut priùs.

Quo modo ex alijs datis (sicut in Prob 3, 4, 5 6 7 & 8 in Conicis Sectionibus effectum est) Hujusmodi Curvæ describi queant ex dictis satis patebit; sed plenioris illustrationis grā addo duas sequentes descriptiones.

Prob 11. Datis tangentibus TV, WY ad punctum decussationis a unà cum Asymptoto MN & alijs 2 punctis b, c per quæ curva 3ū dimens debet transire; curvam describere.

Sol: Sumpto *a* cum alio puncto *b* pro polis figo regulam *a* in angulo *Yab* (vel *Vab* &c) & pono regulam *b* in directum.[49] Præterea applico crus describens regulæ *a* ad alteram tangentem *TV* (vel *WY*) & describo rectam *aZ* cui crus

(44) A first continuation 'in qua si Crura dirigentia semper decu[ssant]' (and if the directing legs always cross in it) is cancelled.

(45) The point *c* of the cubic is made to correspond with the polar line *ab*, so that its organically described equivalent breaks into that line *ab* and a conic through the cubic's double point (and organic pole) *a*.

(46) Not shown by Newton in his diagram.

(47) 'describes' (perhaps intended for the hortatory subjunctive 'describas') is cancelled.

(48) The corollary of the preceding problem in which one of the two coincident points at *b* on the cubic is made a pole of the organic construction, while the other is made to corre-

angles $c\hat{a}b$ and $c\hat{b}a$. Then by application of the describing legs to each of the four remaining points d, e, f and g the crossing of the directing legs will give four further points p, q, r and s, and through these and the fifth point a I describe a conic (by Problem 1). That conic will be the directrix(44) by whose aid the curve sought will be described.(45)

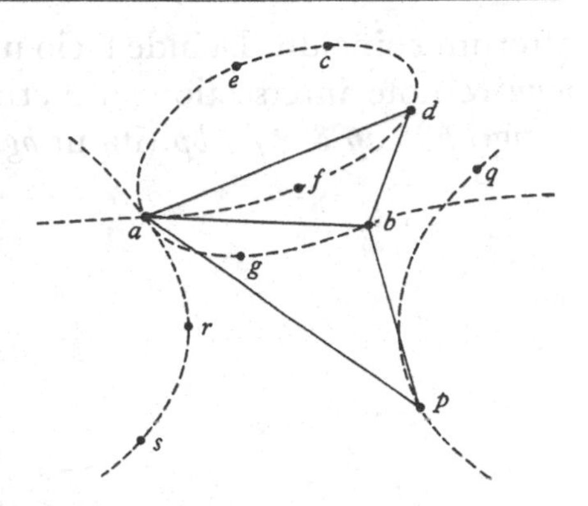

Problem 10. Given six points a, b, c, d, e and f through which a curve of third degree shall pass and also be tangent to the given line TV at one of those points, b, while having a node, cusp or conjugate point at a second, a, to describe the curve.

Solution. Taking a and b for the poles I fix the legs of the rule through b in the angle $a\hat{b}T$ or $a\hat{b}V$, but set the legs of the rule through a in a straight line. Then, on applying the describing legs to the four remaining points c, d, e and f, the directing legs will cross in the four further points p, q, r and s:(46) through these and the point a describe a conic (by Problem 1) and it will be the directrix with whose aid the curve sought is to be described.(48)

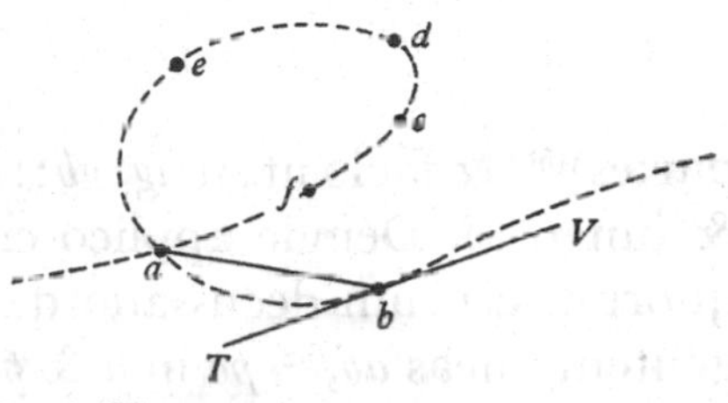

Note. If the curve sought should cross in the contact-point b, then the conic (the directrix, that is) would pass through the point b and not the point a, but the rest would be as before.

How from any given properties curves of this kind may (as has been accomplished in the case of conics in Problems 3–8) be described will be evident enough from what has been said. However, for the sake of fuller illustration I add the following two descriptions.

Problem 11. Given the tangents TV, WY at a nodal point a together with the asymptote mn and two other points b, c through which a curve of third degree shall pass, to describe it.

Solution. Taking the point a along with any other, b, for poles I fix the rule through a in the angle Yab (or Vab, and so on) and set the rule through b in a straight line.(49) Further, I apply the describing leg of the rule through a to one of the tangents TV (or WY) and describe the straight line aZ which is coincident

---

spond with the polar line b: in other words, it is the limit-case of the preceding in which points b and c coincide and g is renamed c. Since point b has to correspond with the line ab, the angle of the legs at a must be straight.

(49) This is obligatory since the rule through b meets the cubic at a in a double point, one of which has to correspond with the polar line through a and b.

alterum coincidit. Deinde facio ut crura describentia sint parallela Asymptoto *mgn* & noto intersectionem p crurum dirigentium. Dein ductis *agb*, *ap* & *bp*, errigo $p\beta \perp ap$ & $p\gamma \perp bp$ (ita ut *bgn* & $bp\gamma$; *agn* & $ap\beta$ sint ad easdem angulares

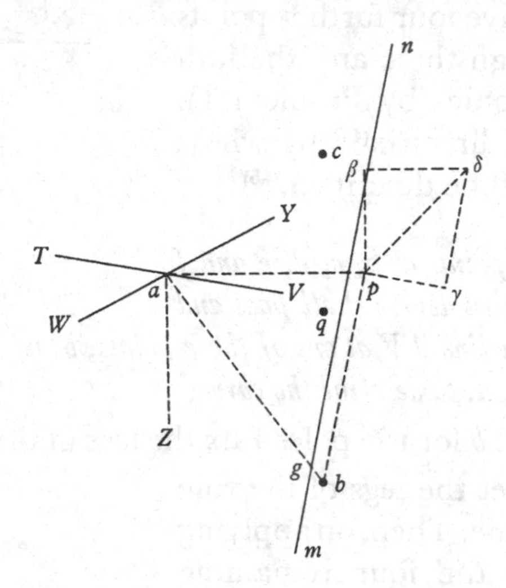

partes)[50] & facio ut sit $ag:gb::p\beta:p\gamma$.[51] & Errigo $\beta\delta \perp p\beta$ & $\gamma\delta \perp p\gamma$ secantes in δ & jungo $p\delta$. Deinde applico crura describentia ad datum punctum c & reliquorum crurum decussatio dat punctum q. Deniq; describo Conic Sect tangentem lineas *ad*,[52] $p\delta$ in a & p & transeuntem per punctum q. & iste Con sec erit Directrix cujus ope curva quæsita describetur.[53]

Nota[54]

---

(50) That is, the pairs of angles are alike clockwise or anticlockwise directed.

(51) Read '$ag:gb = ap \times p\gamma : bp \times p\beta$' and see note (53). In his redraft in §3 following Newton most curiously has the generalization of this correct and its present particularization (with n at infinity) again wrong. (Compare §3: notes (25) and (32).)

(52) Read 'aZ'. In a first version of his figure Newton set a point d (on the cubic) near to the point Z.

(53) In proof we may argue rather more generally. If we suppose that the general point n and the second point r indefinitely near to it correspond in the construction with the point p and d' respectively on the described curve, then $r\widehat{a}n = d'\widehat{a}p$ and $r\widehat{b}n = d'\widehat{b}p$. Hence, if we take nk and nl normal to ar and br, pb' and pc' normal to ad' and bd', we deduce immediately that the triangles akn, $ab'p$ and bln, $bc'p$ are similar pairs, so that $an:kn = ap:b'p$ and $bn:ln = bp:c'p$. In the limit, consequently, as r coincides with n (and therefore d' with p) $an/bn = \lim_{r\to n} (kn/ln) = ap/bp : \lim_{d'\to p} (b'p/c'p)$. When, finally, gn is an asymptote of the cubic, gn will be tangent (at n) at infinity, and in this limiting case $kn = Kn$, $ln = Ln$ (both fixed) while the ratio of an and bn becomes unity, so that with Newton we may write

$$Kn:nL = ag:gb = ap/bp:p\beta/p\gamma,$$

with the second leg. Then I make the describing legs parallel to the asymptote *mgn* and mark the intersection *p* of the directing legs. Next, drawing *agb*, *ap* and *bp*, I erect $p\beta$ normal to *ap* and $p\gamma$ to *bp* (such that *bgn* and $bp\gamma$ together with *agn*

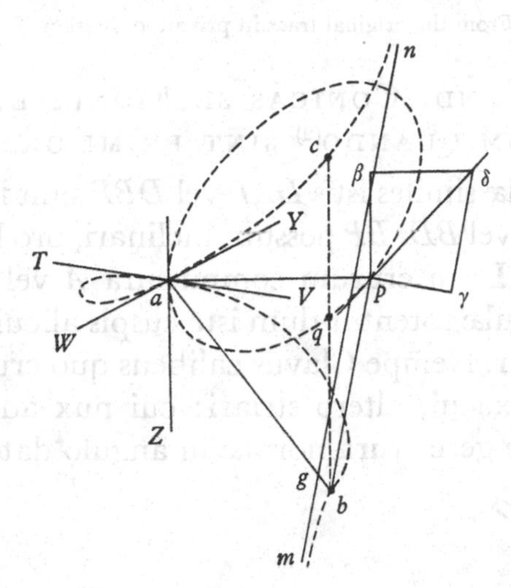

and $ap\beta$ lie in the same angular direction)(50) and make $ag:gb = p\beta:p\gamma$,(51) erecting $\beta\delta$ normal to $p\beta$ and $\gamma\delta$ to $p\gamma$ till they meet in δ and joining $p\delta$. Then I apply the describing legs to the given point *c* and the cross of the remaining legs determines the point *q*. Finally I describe the conic tangent to the lines *a*[*Z*] and $p\delta$ at *a* and *p* and passing through the point *q*: that conic will be the directrix by whose aid the curve sought will be described.(53)

Note that(54) ___

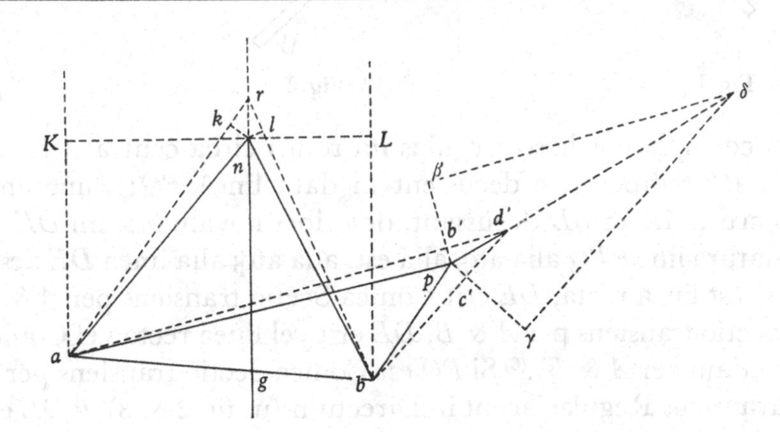

where $p\beta/p\gamma = \lim_{d'\to p} (b'p/c'p)$ and the tangent-direction $p\delta$ of the described curve at *p* is determined by setting δ as the meet of the perpendiculars at β and γ respectively to $p\beta$ and $p\gamma$.

(54) Newton here breaks off without adding any clarification of his intended note. A further line 'Prob 12.' is cancelled.

§3. THE DESCRIPTION OF CONICS AND CUBICS WITH A DOUBLE POINT(1)

From the original tract in private possession

DE MODO DESCRIBENDI CONICAS SECTIONES ET CURVAS TRIUM DIMENSIONUM QUANDO(2) SINT PRIMI GRADÛS. &C.(3)

Lemma 1.(4) Duæ regulæ similes istis *DAP* vel *DBP* sunt fabricandæ (fig. 1) ita ut earum crura *AD*, *AP* vel *BD*, *BP* possunt inclinari, pro lubitu, in quovis dato angulo *DAP* vel *DBP*. Et in crurum commissurâ *A* vel *B*, debet esse cuspis calibeus circa quem regulæ rotentur dum iste cuspis alicui dato puncto *A* vel *B* tanquam centro infigitur. Nempe Clavus calibeus quo crura normæ junguntur possit altero extremo exacui, altero striari; cui nux adaptata possit arctius (prout opus est) constringere crura normæ in angulo dato *DAP*, vel *DBP*.

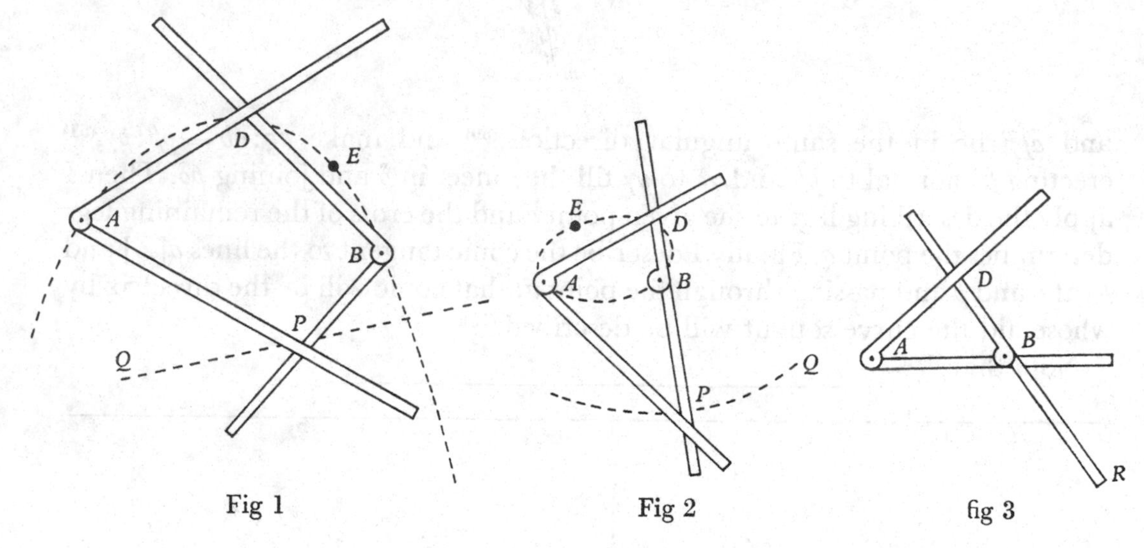

Fig 1 Fig 2 fig 3

2. Jam si concipiamus hasce regulas ita rotari circa centra *A* & *B*, ut duo crura *AP* & *BP* semper sese decussent in data lineâ *PQ*; Punctum *D*, ubi reliqua duo crura *AD* & *BD* decussant, describet novam lineam *DE*. Et prout scitus(5) vel natura lineæ *PQ* alia atq; alia est, alia atq; alia linea *DE* describetur. Nempe si *PQ* est linea recta, *DE* erit Conica Sectio transiens per *A* & *B*; si *PQ* est Conica Sectio transiens per *A* & *B*, *DE* erit vel linea recta vel Conica Sectio (tra[n]siens etiam per *A* & *B*).(6) Si *PQ* est Conica Sectio transiens per *A* & non per *B* & crura unius Regulæ jacent in Directum (ut fig 2 & 3),(7) *DE* erit curva

(1) The following tract is carefully written, with a few minor cancellations and reorderings, on five pages (numbered by Newton (1)–(5)) of a folded, quartered sheet. It is clearly at least the opening of a mature version of the worksheets reproduced in §§1 and 2 above, though towards its close a few new points are made. Its history after its composition by Newton is not

Translation

THE MANNER OF DESCRIBING CONICS AND CUBICS WHEN[2] THEY ARE OF THE FIRST GRADE, AND SO ON[3]

Lemma 1. Two rules similar to these, *DAP* and *DBP* (figure 1), are to be manufactured so that their legs *AD*, *AP* or *BD*, *BP* can be inclined to each other, at will, in any given angle $D\hat{A}P$ or $D\hat{B}P$. And at the junctures *A* or *B* there should be a steel pin-point around which the rules may be rotated while the pin is fixed on some given point *A* or *B* as its centre. To be sure, the steel nail by which the legs of a sector are joined might be finely sharpened at one end, and on the other threaded to take a nut more or less tightly (as the need arises) which will clamp the legs of the sector in the given angle $D\hat{A}P$ or $D\hat{B}P$.

2. Now if we conceive these rules to be rotated round the centres *A* and *B* such that the two legs *AP* and *BP* always cross one another on the given line *PQ*, then the point *D* in which the remaining two legs *AD* and *BD* cross will describe the fresh line *DE*. And according as the situation or nature of the line *PQ* varies from one place to another, so will a correspondingly varying line *DE* be described. Precisely, if *PQ* is a straight line, *DE* will be a conic passing through *A* and *B*; if *PQ* is a conic through *A* and *B*, then *DE* will be either a straight line or a conic (also passing through *A* and *B*).[6] If *PQ* is a conic passing through *A* but not *B* and the legs of one rule lie in a straight line (as in figures 2 and 3),[7]

---

known with certainty, but about 1710 it passed into the hands of William Jones, then gathering material for a second volume on the lines of his 1711 edition of several of Newton's mathematical papers (*Analysis per Quantitatum Series*...) and it is now to be found in a private collection of Jones' manuscripts and letters. Problaby for most of Newton's life it lay unopened among his parcelled early papers.

(2) 'modo' (provided that) is cancelled.

(3) Newton intends 'gradus' (grade) in a different sense from that in which he has used it in the cubic enumeration (1, §3 above). Here the first grade of cubics must be those with a double point (nodal, cuspidal or conjugate), for it is impossible to generate the general acnodal cubic from a conic by the organic construction.

(4) The following paragraphs, and particularly this first lemma, are of course little variant revisions of the introductory remarks to Problem 1 of §2 preceding, where they are nominally restricted to the conic case. For the sake of a logically coherent development Newton has drawn all the general observations of the previous paper into a lengthy generalized discussion of the organic construction, setting their application to the construction of conics and cubics as a short collection of problems in appendix to it.

(5) Compare 1, §3: note (30) above.

(6) According, that is, as, when the legs *AD* and *BD* coincide in the polar line *AB*, the corresponding point *P* does or does not lie on the describing conic. In the former case the described conic (*D*) degenerates into the pair of *AB* and some second straight line.

(7) In the manuscript these are drawn overleaf on Newton's page 2.

trium dimensionum,[8] sin crura utriusq; regulæ ad invicem inclinantur[9] (ut fig 1) curva DE erit[10] Conica Sectio. Si PQ est Conica Sectio transiens per neutrum punctorum A vel B[11]

3. Et distinctionis grā vocabo lineam PQ Directricem; P punctum dirigens; AP & BP crura dirigentia; AD & BD crura describentia, D punctum describens; DE lineam describendam; DAP & DBP angulos mobiles; A & B centra vel polos circa quos moventur.

4. In Descriptionibus hisce tres præcipuè sunt operationes præter Delineationem Describendæ; Nempe Inventio 1 Polorum, 2 angulorum mobilium, 3 Lineæ Directricis. Et prout Data ea, quibus curva est describenda, sunt varia; istarum rerum inventio variatur.

[1] *De Polis*

5. Duo ex datis punctis per quæ curva debet transire pro polis sunt eligenda. Nempe debes[12] ista pro Polis primùm eligere in quibus Curva describenda decussat vel cuspidatur vel habet singulare punctum, deinde ista in quibus duæ rectæ tangunt Curvam describendam, modò talia puncta dantur. Sed si talia non dantur tum elige duo ex punctis per quæ curva est transitura.

Nota 1 Quod[13] iste ordo non est semper necessarius sed tamen est usui maxime accommodatus.

Nota 2 Quod semper necesse est ut duo puncta dentur per quæ curva describenda transibit, & quæ pro Polis eligantur.

Nota 3 Quod punctum infinite distans vicem Poli subeat; et tum crura istius regulæ debent coincidere, & motu parallelo non autem circulari movere. Ut cum Asymptoton[14] Hyperbolæ vel Diamete[r] Parabolæ dantur, tum una regula possit moveri isti Asymptoto vel Diametro parallela. Nempe quia circumferē[tia cujus centrum infinitè dis]tat est linea recta.[15]

---

(8) It is understood that the straight rule pass through B, that pole which is not on the describing conic.

(9) That is, if neither of the angles $D\hat{A}P$, $D\hat{B}P$ is straight.

(10) Read 'potest esse' (can be), viz. if, when the describing legs DA, DB coincide with AB, the corresponding point P is on the describing conic, then the described cubic will degenerate into the pair of the polar line AB and a conic through A.

(11) Newton breaks off in mid-sentence, leaving in the manuscript a small gap (never filled) for its future completion. We may restore its continuation as 'curva DE erit curva quatuor dimensionum, & aliquando trium' (the curve DE will be of fourth degree, but sometimes of the third), since the described quartic degenerates into the pair of a cubic and the polar line AB when a point P of the describing conic corresponds with the line AB (namely, when the describing legs DA and DB coincide with it).

DE will be a curve of third degree, but if the legs of either rule are inclined to each other[9] (as in figure 1), the curve DE will be[10] a conic. If PQ is a conic passing through neither of the points A and B[11]

3. And for distinction's sake I will call the line PQ the directrix; P the directing point; AP and BP the directing legs; AD and BD the describing legs; D the describing point; DE the line to be described (describend); $D\hat{A}P$ and $D\hat{B}P$ the mobile angles; and A and B the centres or poles round which they move.

4. In descriptions of this nature there are, besides the delineation of the describend curve, three especial operations: precisely, the finding of (1) the poles, (2) the mobile angles, and (3) the directrix line. And according as the given conditions under which the curve has to be described are varied, so also the detail of their contrivance varies.

(1) *The Poles*

5. Two of the given points through which the curve must pass are to be chosen as poles. Specifically, you should[12] in the first instance choose those as poles in which the curve to be described either crosses itself, is cusped or has a conjugate point, and then those in which two straight lines touch the curve to be described, provided that such points are given. But if they are not, then choose two of the points through which the curve is to be made to pass.

Note, first, that[13] that order is not always necessary but even so it is the most appropriate in practice.

Secondly, that it is always necessary for two points to be given through which the curve to be described shall pass and which are to be chosen as the poles.

Thirdly, that an infinitely distant point may replace a pole: in that case the legs of the corresponding rule must coincide and move parallel to itself, not circularly. So when an asymptote of a hyperbola or any diameter of a parabola are given, then one of the rules may move parallel to that asymptote or diameter (since, that is, a circle perimeter whose centre is infinitely distant is a straight line).[15]

---

(12) In a first amplification of this introductory phrase Newton wrote 'Quælibet duo possis eligere (præcipuè in Conicis Sectionibus describendis), sed præstat' (You may choose any two (especially when conics are to be described) but it is better to).

(13) The less emphatic 'Unde' (Hence) is cancelled.

(14) Newton's variant of 'Asymptotos'.

(15) We have made suitable restoration of a portion of this sentence which has crumbled away at the bottom of Newton's first manuscript page. A rotating motion whose rest-centre passes to infinity becomes, of course, one of translation.

[2] *De Angulis Mobilibus*

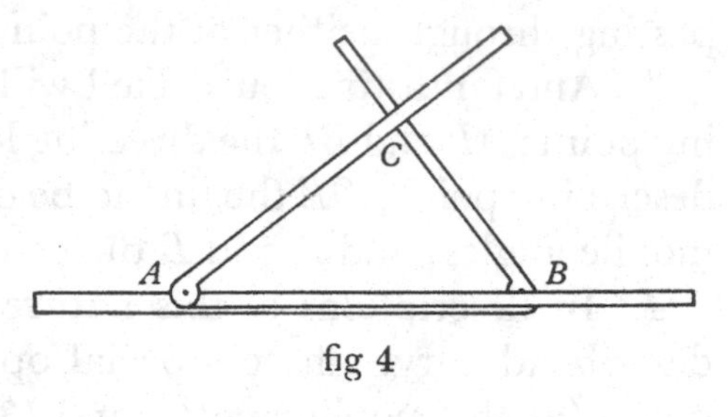

fig 4

6. Postquam articuli regularum sunt istis polis *A* & *B* impositi; anguli mobiles possunt ex varijs datis varie determinari. Nempe 1 si datur recta $\beta\gamma$ (fig 6) quæ debet tangere Describendam in polo *A*; tum junctis *AB*, applico normam *DAP* ad angulum γAB, ita ut crus dirigens *AP* coincidat cum linea *AB* & crus describens cum $A\gamma$, & figo crura in isto angulo $BA\gamma$. Sed crura alterius normæ *DBP* pono in directum,(16) ut in fig 2 & 3.

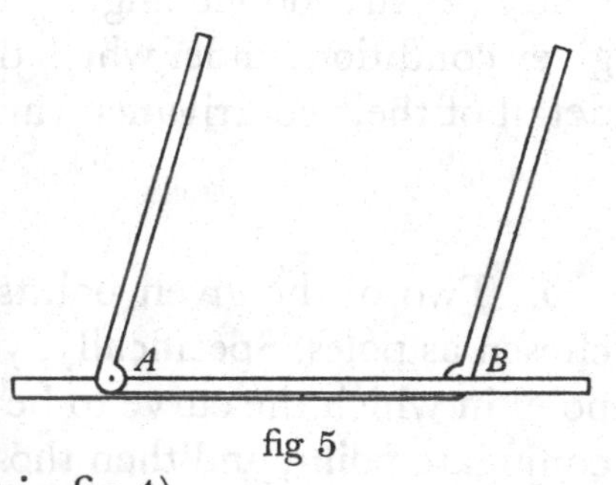

fig 5

7. Sed 2do, si anguli mobiles ex tertio dato puncto *C* (fig 7), per quod Describenda transibit, sunt determinandi; Jungo *AB*, *AC*, *BC* & figo crura regularum in angulis *CAB CBA*, applicando nempe crura dirigentia ad lineam *AB* & crura describentia ad lineas *AC BC* (ut fit in fig 4).

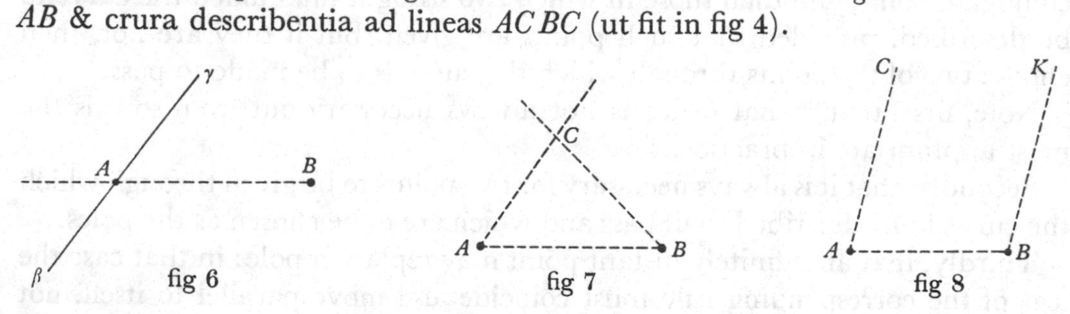

fig 6 fig 7 fig 8

8. Tertio si datur punctum infinite distans versus *C* (fig 8), hoc est, si datur quòd linea *AC* tendit ad eas partes versus quas aliquod crus curvæ Describendæ tendit (linea nempe *AC* existente parallelâ alicui Asymptoto Hyperbolæ vel Diametro Parabolæ &c)(17) duc $BK \parallel AC$ & fige crura regularum in angulis *CAB*, *KBA*, applicando crura dirigentia ad lineam *AB* & Crura describentia ad lineas *AC*, & *BK* (ut in fig 5); & Isti *CAB*, *KBA* erunt anguli mobiles.(18)

Nota 1. quod si data tangens $\beta\gamma$ vel datum punctum *C* coincidit lineæ *AB*, tum anguli mobiles sunt aliunde determinandi.(19)

2. Si datur aliquod tertium punctum *C* in quo Describenda decussat vel cuspidatur vel habet punctum singulare, anguli mobiles sunt inde determinabiles.(20)

(16) Newton has cancelled the explanatory clause 'ita ut ad invicem non inclinati constituant unam lineam rectam' (so that, no longer inclined one to the other, they form a single straight line).

(2) *The Mobile Angles*

6. After the joints of the rules are set on the poles A and B, the mobile angles may be determined in various ways from the various given conditions. Precisely, if, in the first place, there is given a straight line $\beta\gamma$ (figure 6) which must touch the describend in the pole A, then joining AB I apply the sector DAP to the angle γAB so that the directing leg AP coincides with the line AB and its describing leg with $A\gamma$ and then fix the legs in that angle $B\hat{A}\gamma$. But the legs of the second sector DBP I set in a straight line,[16] as in figures 2 and 3.

7. But, secondly, if the mobile angles are to be determined from a third given point C (figure 7) through which the describend shall pass, I join AB, AC and BC and fix the legs of the rules in the angles $C\hat{A}B$ and $C\hat{B}A$, by applying, that is, the directing legs to the line AB and the describing legs to the lines AC and BC (as is done in figure 4).

8. Thirdly, if there is given a point infinitely distant in the direction of C (figure 8), that is, if it is given that the line AC tends in that direction towards which some leg of the curve to be described tends (namely, when the line AC comes to be parallel to some asymptote of a hyperbola or to the diameter of a parabola, and so on)[17] draw BK parallel to AC and fix the legs of the rules in the angles $C\hat{A}B$ and $K\hat{B}A$ by applying the directing legs to the line AB and the describing ones to the lines AC and BK (as in figure 5): those angles $C\hat{A}B$ and $K\hat{B}A$ will then be the mobile angles.[18]

Note (1) That if the given tangent $\beta\gamma$ or the given point C coincides with the line AB, then the mobile angles will have to be determined by other means.[19]

(2) If there be given some third point C in which the describend crosses itself, is cusped or has a conjugate point, the mobile angles are determinable from that.[20]

---

(17) 'tum possis inde determinare angulos mobiles. Nempe' (then you should be able from that to determine the mobile angles. Precisely) is cancelled.

(18) Though not here explicitly stated the whole purpose of the organic construction for Newton is that it is a viable technique for drawing certain types of algebraic (and indeed any arbitrarily defined, smoothly continuous) curve from a 'directrix' of a lower order, or more precisely from the pair of that directrix and the polar line AB counted one or more times. That this latter degeneration occur it is necessary that some point P in the described curve correspond in the construction with that line AB, and this is what Newton is concerned to do in this present section of his paper, attempting further simplification (where possible) by making one of the pivoting angles straight or taking advantage of that which obtains when one of the given points C passes into infinity (as in 8).

(19) In this case (compare previous note) the described curve, too, is degenerate and of the same degree as the describing curve. That, of course, does not interest Newton.

(20) This second note, squashed in almost illegibly between two lines of text themselves narrowly separated, is a late addition. What may be intended as a '3' occurs here in the margin, but no third note which corresponds exists anywhere in the manuscript.

[3] *De Directrice*

9. Ex tribus datis determinavimus Polos et angulos mobiles, manet ut ex reliquis determinemus Directricem. Et Primò si datur quod Curva Describenda transibit per punctum D (fig 9) facio regulas gyrari dum crura describentia decussent in isto puncto D, & punctum P ubi reliqua crura decussant erit in Directrice (fig 1 & 2). Et si forte regulæ decussent in uno polorum, ut in B (fig 3) tum crus dirigens BR istius regulæ DBR tanget Directricem in eodem Polo B.[21]

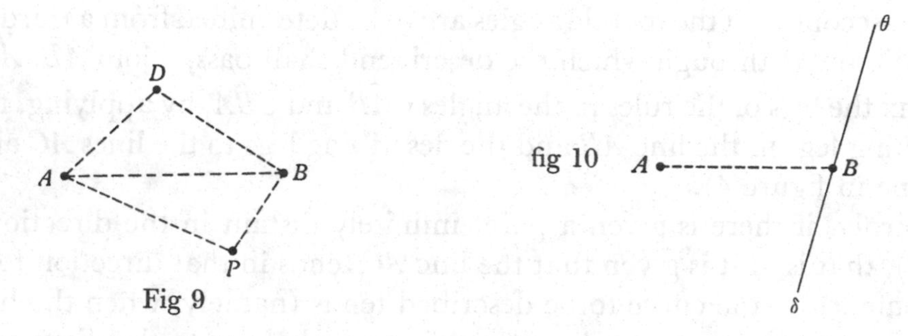

Fig 9 fig 10

10. Secundo si datur quòd Describenda tangit rectam $\delta\theta$ in polo B (Fig: 10) Tum applico crus describens regulæ B ad tangentem $\delta\theta$ (fig 11) & crus describens regulæ A ad punctum B & punctum P ubi reliqua crura decussant erit in Directrice.[22]

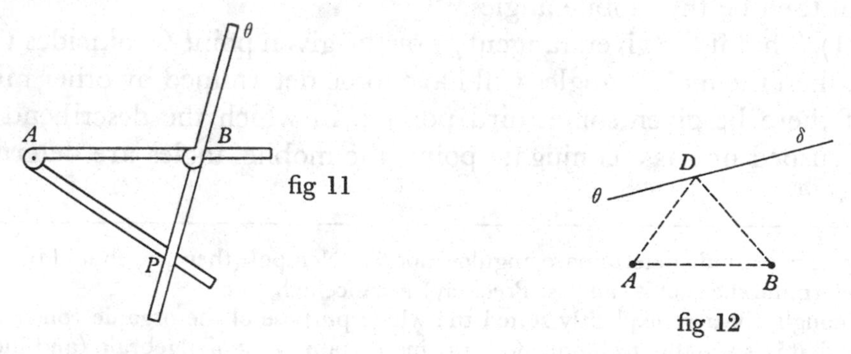

fig 11 fig 12

11. 3° si datur quòd recta $\theta\delta$ tangit Describendam in alio dato puncto D (fig 12); tum applico crura describentia ad istum punctum D & noto punctum P ubi reliqua crura decussant (fig 2). Deinde facio regulas ita moveri ut punctum D ubi crura describentia decussant semper maneat in tangenti $\theta\delta$ dum intersectio reliquorum crurum describit lineam PR: Et ista linea PR tanget Directricem in puncto P.[23]

Vel sic: Puncto P ut priùs invento, Duco AD, BD, AP, BP, & sumens aliquod punctum δ in tangente $\theta\delta$, demitto $\delta\lambda \perp AD$, & $\delta\mu \perp BD$. Deinde erigo $P\tau \perp AP$ & $P\pi \perp BP$ & facio ut sit $AD:\delta\lambda::AP:P\tau$. & ut $BD:\delta\mu::BP:P\pi$. Deniq ductis

(3) *The Directrix*

9. From three given elements we have determined the poles and the mobile angles: it remains for us to determine the directrix from the remainder. And, first, if it is given that the curve to be described shall pass through the point D (figure 9), I cause the rules to revolve until the describing legs cross in the point D and the point P where the remaining legs cross will be on the directrix (figures 1 and 2). And if the rules should chance to cross in one of the poles, in B say (figure 3), then the directing leg BR of the rule DBR will touch the directrix in the same pole B.(21)

10. Secondly, if it is given that the describend touches the straight line $\delta\theta$ in the pole B (figure 10), then I apply the describing leg of the rule through B to the tangent $\delta\theta$ (figure 11) and the describing leg of the rule through A to the point B, and the point P where the remaining legs cross will be on the directrix.(22)

11. Thirdly, if it is given that the straight line $\theta\delta$ touches the describend in some given point D (figure 12), then I apply the describing legs to that point D and mark the point P in which the remaining legs cross (figure 2). Then I make the legs move so that the point D in which the describing legs cross stays always in the tangent $\theta\delta$ while the meet of the remaining legs describes the line PR. That line PR will then touch the directrix in the point P.(23)

Or thus. When the point P has been found as before, I draw AD, BD, AP and BP and then, taking some point δ in the tangent $\theta\delta$, I drop the perpendiculars $\delta\lambda$ to AD and $\delta\mu$ to BD. Next I erect $P\tau$ perpendicular to AP and $P\pi$ to BP and make $AD:\delta\lambda = AP:P\tau$ and $BD:\delta\mu = BP:P\pi$. Finally, drawing πR parallel to

(21) This sentence was added two pages further on with the command 'ad Lemma 9 adjiciatur', which we here obey. In the case where the describing legs meet in a point D coincident with the pole B, the leg BD will touch the described curve at B and so, correspondingly, the directing leg through B will touch the directrix since the organic construction preserves tangency.

(22) Here, much as before, the described curve (the 'describend') will have two points coincident with $\delta\theta$ at B, one of which will, on reversal, yield a corresponding point P on the describing curve.

(23) This is an immediate corollary of the fact that the organic construction preserves tangency between curves. (Compare note (21).) The 'line' PR corresponding to $\theta\delta$ will, in fact, be a conic passing through the poles A and B since the polar line AB is not an invariant under the organic correspondence (unless $\theta\delta$ be coincident with it, in which case PR will shrink into a unique 'punctum Ellipticum' at the point P which corresponds with AB). Compare paragraph 12 below.

$\pi R \parallel BP$ & $\tau R \parallel AP$[24] secantibus in R, jungo PR; et ista PR tanget directricem in puncto P.[25] (vide fig 13[)].

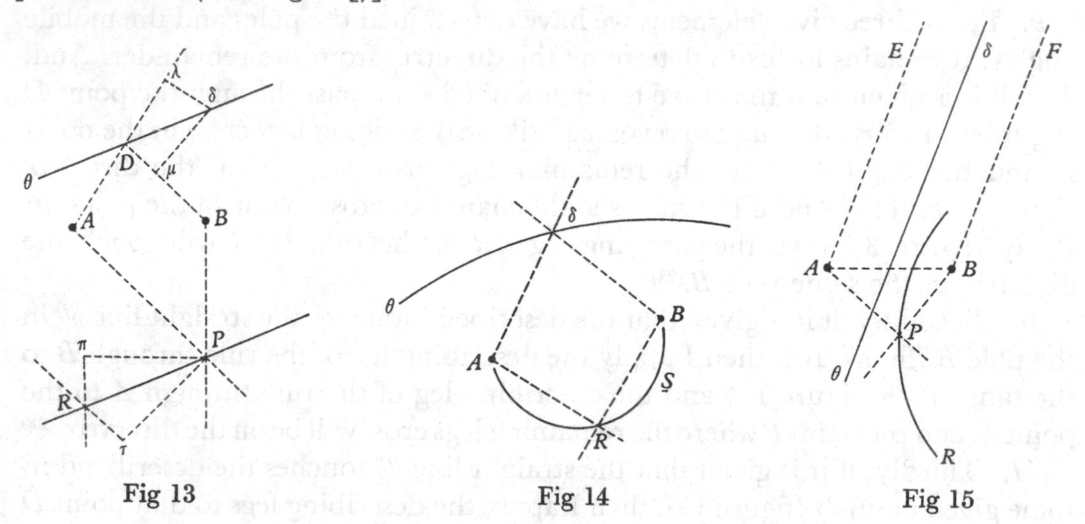

Fig 13 Fig 14 Fig 15

12. 4^{tò} si datur quod linea $\theta\delta$ (sive sit recta sive curva sive Geometrica sive Mechanica & casu ducta) debet tangere curvam describendam sed punctum contactus non datur: tum facio ut, regulis gyrantibus, earum crura describentia semper decussent in linea tangenti $\theta\delta$ dum reliquarum[26] crurum decussatio describet lineam RS; et ista linea RS alicubi[27] tanget directricem. [fig 14.]

13. Quintò,[28] si datur quòd Describenda in infinitum abit ad eas partes versus quas data linea $\theta\delta$ tendit (i: e: quòd linea $\theta\delta$ sit parallela vel Asymptoto Hyperbolæ vel Diametro Parabolæ): tum facio ut crura describentia AE & BF (fig 15) sint parallela isti datæ lineæ $\theta\delta$ & istud punctum P, ubi crura dirigentia decussant, erit in Directrice.[29]

(24) It would perhaps have been more helpful to the reader if Newton had written the equivalent '$\pi R \perp P\pi$ & $\tau R \perp P\tau$'.

(25) This is the generalization of Problem 11 of §2 above. (Compare note (53) of that section.) In proof, if we suppose that d is a near point to D in $\theta\delta$ with r corresponding to it in the describing curve (P), then $D\hat{A}d = P\hat{A}r$ and $D\hat{B}d = P\hat{B}r$. Hence, if we take dl, dm, rp and rt perpendicular to AD, BD, $P\pi$ and $P\tau$, we have two pairs of similar triangles Adl, APt and Bdm, BPp in the limit as d passes into D. In that limit, therefore, $Al:dl = AP:Pt$ and $Bm:dm = BP:Pp$, which is Newton's result since then $dl:dm = \delta\lambda:\delta\mu$, $Pt:Pp = P\tau:P\pi$ and the points l and m both merge with D. (See figure on opposite page.)

(26) Read 'reliquorum'.

(27) Specifically, in the point corresponding in the organic construction to the contact point of $\theta\delta$ and the described curve.

(28) In an attempted generalization of Lemma 12 Newton first wrote in a wholly different Lemma 13: '5to si datur quod dicta linea $\theta\delta$ non tanget, sed in dato angulo secet describendam: tum (ut priūs) facio punctum describens moveri in linea $\theta\delta$ dum punctum dirigens describit lineam RS; & linea RS secabit Directricem in eodem angulo in quo linea $\theta\delta$ secat

BP and τR parallel to *AP*[24] meeting in *R*, I join *PR*: that line *PR* will then touch the directrix in the point *P*.[25] (See figure 13.)

12. Fourthly, if it is given that the line $\theta\delta$ (whether straight or a curve be it algebraic or transcendental and drawn haphazardly) must touch the curve to be described but now the contact point is not given, then I make the describing legs of the rules, as they revolve, cross always in the tangent line $\theta\delta$ while the cross of the remaining legs describes the line *RS*: that line *RS* will then touch the directrix somewhere[27] (figure 14).

13. Fifthly, if it is given that the describend passes into infinity in that direction towards which the given line $\theta\delta$ tends (that is, that the line $\theta\delta$ be parallel to an asymptote of a hyperbola or to any diameter of a parabola), then I make the describing legs *AE* and *BF* (figure 15) parallel to the given line $\theta\delta$: that point *P* in which the directing legs cross will be in the directrix.[29]

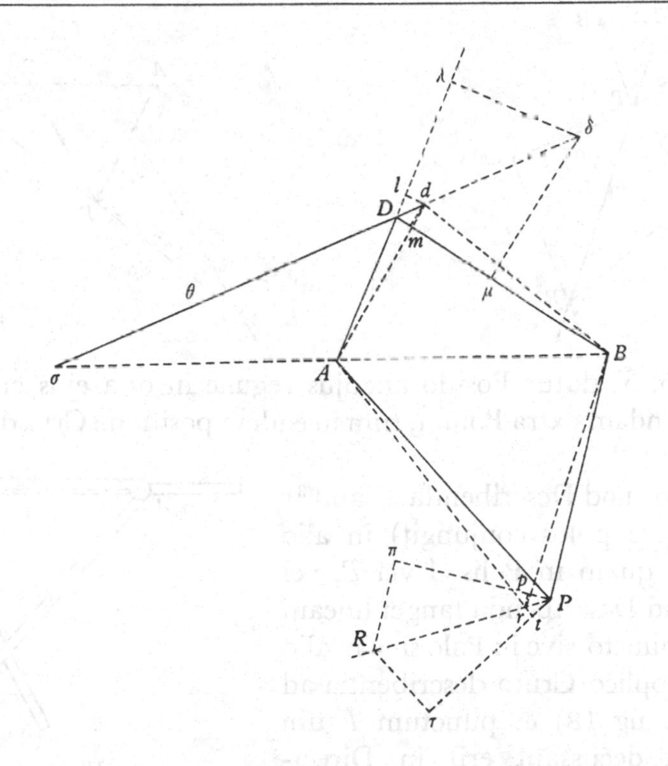

Describendam' (Fifthly, if it is given that the said line $\theta\delta$ does not touch the describend but intersects it in a given angle, then (as before) I make the describing point move in the line $\theta\delta$ while the directing point describes the line *RS*, and this line *RS* will intersect the directrix in the same (!) angle in which the line $\theta\delta$ intersects the describend). Newton realized the fallacy in this argument (compare §2, note (38)) only after he had finished writing up all his lemmas for, as his manuscript shows, he had to revise the numbering of all those subsequent to the present one, decreasing each by one.

(29) The particular case of Lemma 9 when *D* passes into infinity.

14. Sextò, si datur quòd $\theta\delta$ est Asymptoton(30) Describendæ, (i.e. linea quæ tangit curvam in puncto infinite distanti) fig 15 tum, invento puncto P (ut in Lemmate 13), facio ut punctum describens moveatur utrinq; in infinitum in Asymptoto $\theta\delta$ dum punctum dirigens P describit lineam PR; & ista PR tanget directricem in puncto P.(31)

Vel sic: Invento puncto P ut priùs, duco $A\sigma B$, AP, & BP, & erigo $P\tau \perp AP$ & $P\pi \perp BP$ (ad easdem partes angulares cum $A\sigma\theta$, $B\sigma\theta$; fig 16 & 17) & facio ut sit $A\sigma : B\sigma$(32)$:: P\tau : P\pi$. Deinde erectis πR & τR secantibus in R, duco PR & ista recta PR tanget Directricem in puncto P.

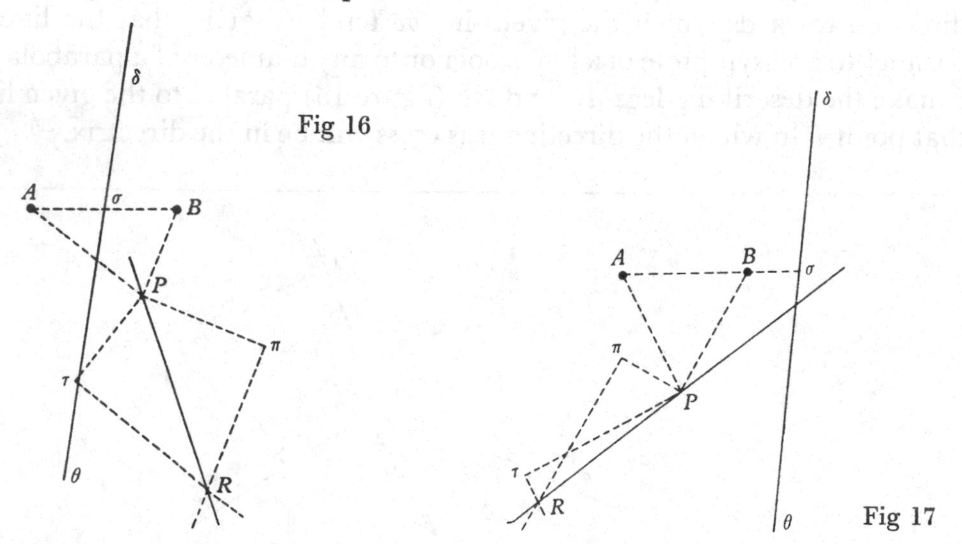

Fig 16 Fig 17

15. Septimò. Si datur Positio alicujus regulæ in qua ejus crus describens tangit Describendam extra Polum, tum in eadem positione Crus dirigens tanget Directricem.(33)

16. 8^{vò}. Dato quod Describenda secabit(34) lineam AB (quæ polos conjungit) in alio aliquo puncto quam in Polis A vel B, vel etiam dato quod Describenda tanget lineam AB in quovis puncto sive in Polo sive in alio puncto: tum applico Crura describentia ad lineam AB (in fig 18) & punctum P ubi reliqua Crura decussant erit in Directrice.(35)

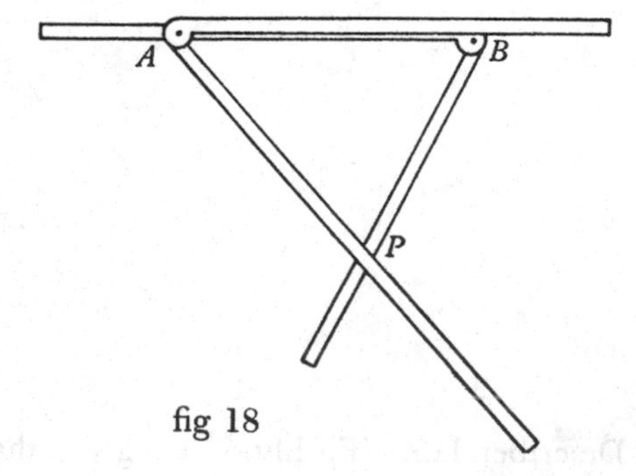

fig 18

17. Nono, si Datur punctum σ ubi Describenda secat vel tangit lineam AB extra polos A & B (fig 19); tum, invento puncto P ut in priori Lemmate, Jungo

(30) We would expect 'Asymptotos'. Newton's variant nominative appears several times below.

(31) This is an immediate corollary to Lemma 11 when D is at infinity.

14. Sixthly, if it is given that $\theta\delta$ is an asymptote of the describend (that is, a line which touches the curve in an infinitely distant point) (figure 15), then, having found the point P as in Lemma 13, I make the describing point move either way into infinity on the asymptote $\theta\delta$ while the directing point P describes the line PR: that line PR will touch the directrix in the point P.(31)

Or thus: having found the point P as before, I draw $A\sigma B$, AP and BP and erect $P\tau$ perpendicular to AP and $P\pi$ to BP (in the same angular direction as $A\sigma\theta$ and $B\sigma\theta$, figures 16 and 17), making $A\sigma:B\sigma$(32) $= P\tau:P\pi$. Next, having erected πR and τR meeting in R, I draw PR: that straight line PR will then touch the directrix in the point P.

15. Seventhly, if there be given the position of some rule, in which its describing leg touches the describend other than at a pole, then in the same position the directing leg will touch the directrix.(33)

16. Eighthly, given that the describend will intersect(34) the line AB (joining the poles) in some point other than the poles A and B, or even given that the describend will touch the line AB in any point at all, pole or otherwise, then I apply the describing legs to the line AB (in figure 18) and the point P in which the remaining legs cross will be on the directrix.(35)

17. Ninthly, if there is given the point σ in which the describend intersects or touches the line AB other than in the poles A and B (figure 19), then, having

(32) Read '$BP \times A\sigma:AP \times B\sigma$'. Newton repeats the incorrect result of Problem 11 of the preceding §2 rather than making the appropriate deduction from Lemma 11 when D passes into infinity. (In that case, the limit of AD/BD is unity and that of $\delta\lambda/\delta\mu$ is $A\sigma/B\sigma$, so that $1:A\sigma/B\sigma = AP/BP:P\tau/P\pi$.) Note that πR and τR are normal to $P\pi$ and $P\tau$.

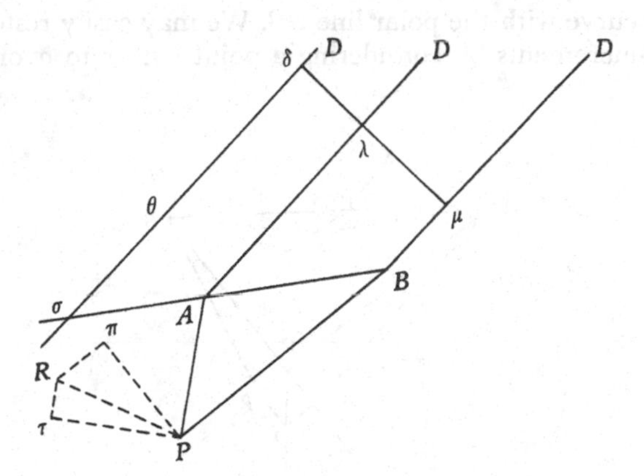

(33) Compare note (23). The lemma is set on the next page of the manuscript but is placed here in agreement with Newton's numbering of it.

(34) 'vel tanget' (or touch) is cancelled.

(35) Evidently, any meet of the described curve with AB (whether a second point coincident with either of the poles A, B or not) will correspond with P.

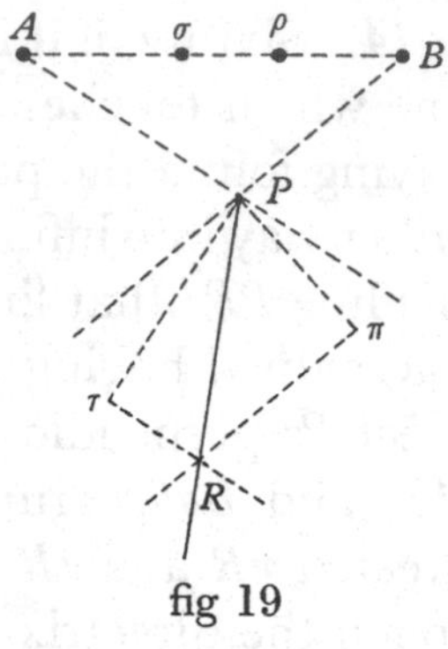

fig 19

AP & BP et errigo $P\tau \perp AP$ & $P\pi \perp BP$ ad partes ab $A\sigma$ & $B\sigma$ diversas: et facio ut sit $\frac{AP \times B\sigma}{A\sigma}:BP::P\tau:P\pi$. Et erectis $\tau R \perp P\tau$ & $\pi R \perp P\pi$ secantibus in R, ducta PR tanget Directricem in P.(36)

18. Decimò, si datur quod Describenda in duobus punctis σ & ρ secat lineam AB tum Directrix in puncto P decussabit: Sin Describenda tangit lineam AB in puncto σ extra polos A & B, tum Directrix in puncto P erit cuspidata. Et e contra, Directrix secans lineam AB bis vel eam tangens efficit ut Describenda decusset vel sit cuspidata.(37)

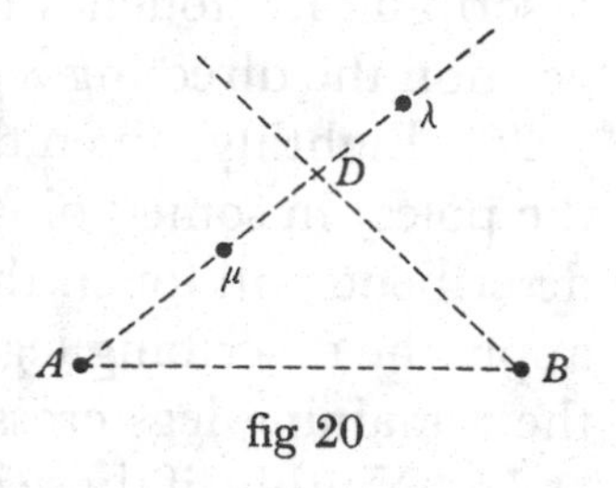

fig 20

19. Undecimò. Cùm crura dirigentia ad lineam AB applicantur (ut in fig 4 & 5) cruribus describentibus in puncto D decussantibus; si Describenda secat crus aliquod describens AD in duobus punctis μ, λ (fig 20) extra punctum D & polos, tum Directrix decussabit in polo opposito B. Sin Describenda tangit crus Describens extra puncta A, B & D tum Directrix erit Cuspidata in Polo opposito B. Et e contra.(38)

20. Duodecimò. Si crura describentia applicentur ad aliquod punctum extra lineas AD, & BD, ubi Describenda decussat vel cuspidata est vel habet punctum singulare, tum Directrix decussabit vel erit cuspidata vel habebit punctū singulare in isto puncto ubi crura Dirigentia decussant. Et e contra.(39)

(36) The complementary case to Lemma 11 in which D is now not at infinity but the meet of the described curve with the polar line AB. We may easily restore Newton's proof as an argument in limit-increments by considering a point s near to σ on the described curve

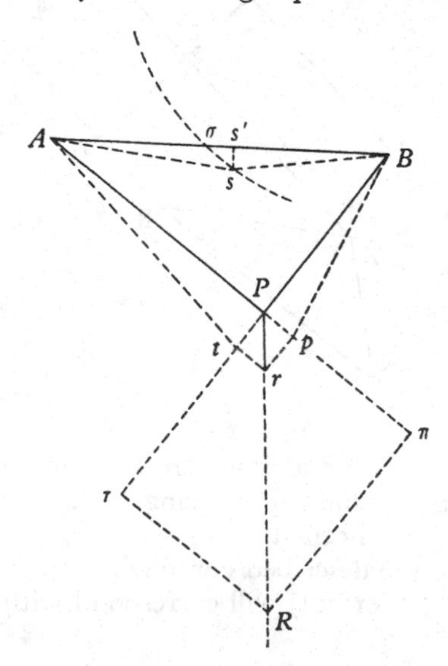

found the point P as in the preceding Lemma, I join AP and BP and erect $P\tau$ and $P\pi$ perpendicular to AP and BP in opposite directions to $A\sigma$ and $B\sigma$: I then make $AP \times B\sigma/A\sigma : BP = P\tau : P\pi$ and, erecting the normals τR and πR to $P\tau$ and $P\pi$ respectively meeting in R, when I draw PR it will touch the directrix in P.(36)

18. Tenthly, if it is given that the describend meets the line AB in two points σ and ρ, then the directrix will cross itself at the point P: but if the describend touches the line AB in the point σ other than at the poles A and B, then the directrix will be cusped at the point P. And, conversely, the result of the directrix intersecting the line AB twice or touching it is to make the describend cross itself or be cusped.(37)

19. In the eleventh place, when the directing legs are applied to the line AB (as in figures 4 and 5) with the describing legs crossing at the point D, if the describend intersects some describing leg AD in two points μ and λ (figure 20) other than at the point D and the poles, then the directrix will cross itself in the opposite pole B. But if the describend touches the describing leg other than in the points A, B and D, then the directrix will be cusped in the opposite pole B. And conversely.(38)

20. Twelfthly, if the describing legs be applied to some point (not in the lines AD and BD) in which the describend crosses itself, is cusped or has a conjugate point, then the directrix will cross itself, be cusped or have a conjugate point at the point in which the directing legs cross. And conversely.(39)

---

and its corresponding point r (near to P) on the describing curve. If ss', rp and rt are normals to AB, $P\pi$ and $P\tau$ respectively, then, since $B\hat{A}s = P\hat{A}t$ and $A\hat{B}s = P\hat{B}p$, the pairs of triangles sAs', tAP and sBs', pBP will be similar in the limit as s passes into σ. In that limit, therefore, $AP:As' = Pt:s's$ and $BP:Bs' = Pp:s's$, so that $AP/BP:As'/Bs' = Pt:Pp$, and Newton's argument is an immediate corollary since $Pt:Pp = P\tau:P\pi$ while s' coincides with σ.

(37) In this case each of the points σ and ρ will correspond with P, so that the describing curve must have a double point there. Since Newton presupposes that the described curve is continuous, this cannot be a conjugate point but will, in general, be a 'punctum decussationis' or node. When, in particular, the described curve touches AB at σ the points σ and ρ are there coincident. In this case, since the described curve in the neighbourhood of σ lies wholly on one side of the polar line AB, the describing curve will emerge from P in the same direction that it enters, so that the nodal oval has shrunk into a cusp.

(38) Clearly, λ and μ correspond to a double point of the directrix at B whether they are distinct or (as in the case of tangency) themselves coincident.

(39) In general, multiple points on the directrix and describend must correspond in the organic construction.

Nota Si quando accidit in præcedentibus operationibus quòd Crura dirigentia non decussant, ad punctum *P* determinandum: sed tamen decussarent modò unum vel utrumq; ad eas partes versus polum produceretur. Tum pro angulo mobili unius (fig 21) vel utriusq; regulæ (fig 22), sumendum est ejus complementum ad duos rectos. Quod idem de Cruribus describentibus, inter Curvam(40) per sequentia problemata delineandam, observabis.

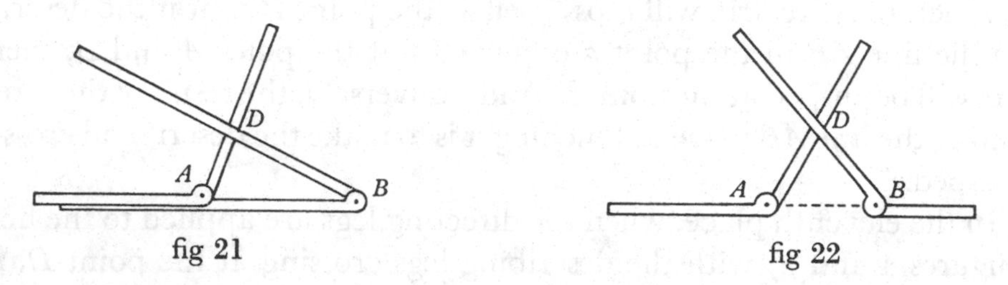

fig 21 fig 22

Jam hisce regulis postquam inveneris quæ ad Directricem determinandam sufficiant superest ut ostendam quo pacto directrix cum Curvâ Describenda delineantur.

Prob 1. Descriptio Directricis & Describendæ cùm sit Conica Sectio.

Si Describenda sit Conica Sectio Directrix erit(41) linea recta. Jam ad Conicam Sectionem quinq; requiruntur, (ut quinq; puncta per quæ curva transibit(42) vel quatuor puncta & linea tangens,(43) vel 3 puncta & 2 tangentes,(44) vel 4 puncta & linea Asymptoto parallela, vel 3 puncta & Asymptoton, vel unum punctum & duo Asymptoti(45) &c:). Et istorum duo dant polos regularum (Lem 5), unum dat angulos mobiles (per Lem 6, 7, 8), & duo reliqua dant Directricem;(46) nempe vel dant duo puncta per quæ directrix transibit (per Lem 9, 10, 13, 16) & curvam quam directrix debet tangere (per Lem 12); vel dant lineam rectam quæ debet tangere Directricem (per Lem 11, 12, 14), vel potius quæ directrici coincidet, quandoquidem Directrix est etiam linea recta.(47) Jam postquam directrix inventa est Describenda facilimè delineatur per intersectionem crurum describentium dum intersectio crurum dirigentium in Directrice mo[ve]tur.

Prob. 2. Delineatio Curvæ trium Dimensionum cujus punctum principale (*i:e: punctum in quo decussat vel cuspidatur vel habet punctū singulare*) *datur.*

Si(48) datur linea tangens describendam in dato puncto tum pro hâc curvâ describendâ Directrix erit Conica Sectio transiens per istum polum qui est principale punctum Describendæ. Jam ad hanc curvam determinandam(49) sex

(40) That is, the describend: 'describendam' is, in fact, cancelled here.
(41) More strictly, 'esse potest' (can be). See note (10).
(42) Compare Problem 1 of §2 above.

Note in the preceding operations whenever, in determining P, the directing legs do not happen to cross but would do so, however, if only one or both were to be produced on the further side of the pole, then for the mobile angle of that one (figure 21) or of both (figure 22) you should take its supplement to two right angles. You will observe the same rule for the describing legs when, in the course of the following problems, the curve[40] is to be drawn.

Now, after you have found by these rules what suffices to determine the directrix, it remains for me to illustrate how the directrix along with the describend curve are to be drawn.

Problem 1. The description of the directrix and the describend when the latter is a conic.

Should the describend be a conic, the directrix will be[41] a straight line. Now for a conic five elements are required (five points through which the curve shall pass,[42] for example, or four points and a tangent line,[43] or three points and two tangents,[44] or four points and a line parallel to an asymptote, or three points and an asymptote, or a single point and two asymptotes,[45] and so on). Of those two give the poles of the rules (by Lemma 5), one the mobile angles (by Lemmas 6, 7 and 8) and the remaining two the directrix:[46] precisely, they give either two points through which the directrix shall pass (by Lemmas 9, 10, 13 and 16) and the curve which the directrix must touch (by Lemma 12) or a straight line which must touch the directrix (by Lemmas 11, 12 and 14), or rather coincide with it since the directrix is here a straight line also.[47] Thereafter, as soon as the directrix has been found, the describend is very easily drawn through the intersection of the describing legs while that of the directing legs moves on the directrix.

Problem 2. The drawing of a curve of third degree whose principal point (that is, the point in which it crosses itself, is cusped or has a conjugate point) is given.

If[48] there is given a line tangent to the describend at a given point, then the directrix for describing the curve will be a conic passing through that pole which is the principal point of the describend. Now to determine this curve there are

(43) Compare §2: Problems 2 and 4.

(44) Compare §2: Problems 3 and 5.

(45) On the lines of Problem 11 of §2 and the preceding when a tangent point passes to infinity.

(46) Strictly, that straight line which, together with the polar line, is the degenerate conic corresponding to the described conic (through the poles). Compare note (41).

(47) Compare note (6).

(48) Compare Problems 6 and 11 of §2 above.

(49) 'describendam' (describe) is cancelled to avoid confusion with the defined meaning of a 'curva describenda'.

requiruntur præter punctum principale (ut[50] quinq; puncta & linea tangens vel quinq; puncta & linea Asymptoto parallela, vel 4 puncta & Asymptoton vel tres Asymptoti &c: Asymptoton pro duobus habeo, quoniam dat punctum infinite distans & insuper Cu[r]vam in isto puncto tangit).[51] Et ex istis septem, punctum principale cum alio puncto dant Polos, tertium datum nempe tangens in uno polorum dat angulos mobiles (Lem 6) & reliqua quatuor dant quatuor conditiones quarum ope Directrix delineanda est, per Problema precedens; Dato etiam (pro quinta conditione)[52] quod Directrix per punctum principale describendæ transibit. Deniq; punctum describens delineabit Describendam, ut priùs, dum punctum Dirigens in Directrice movet.

Sin[53] tangens non datur Prob[54] perficitur per Directricem quæ sit Conica Sectio per neutrum polum transiens.[55] Nempe pro Polis elligo duo puncta a puncto principali diversa et determino angulos mobiles per punctum principale (lem 7, 8), & reliqua quatuor Data dant quatuor conditiones pro Directrice describenda (per Lem 9 10 &c); quintā conditionem per Lemma 16 invenies, nam Describenda alicubi secabit lineam AB extra Polos A & B. Deniq; Conic: S[e]ctio secundum istas quinq; conditiones descripta erit Directrix, cujus ope Describendam, ut priùs, delineabis.

Prob 3. Delineatio Curvæ quatuor dimensionum cujus tria puncta principalia dantur (i:e: puncta in quibus vel decussat vel habet punctum singulare, vel cuspidata est).

Solut: Hujus etiam[56] Describendæ directrix est Conica Sectio quæ per neutrum polum transit. Et ad hanc curvam determinandam octo[57] requiruntur quorum tria puncta principalia dant polos & angulos mobiles, reliqua quinq; dant Directricem (per Lem 9, 10, 11, 12, 13, 14, 15. & Prob 1). Unde Describenda ut priùs delineatur.[58]

---

(50) Newton has cancelled the phrase 'sex puncta per quæ curva trāsibit vel' (six points through which the curve shall pass), choosing instead to devote a whole following paragraph to this special case.

(51) A further continuation, 'vel etiam quia ad Asymptoton determinandū utpote rectam lineam, duæ conditiones requiruntur' (or also since to determine an asymptote, it being a straight line, two conditions are required), is justly cancelled.

(52) Nine conditions, in fact, are necessary (though, if we remember Cramer's rule, not always sufficient) to determine a particular cubic. This fifth double-point condition accounts not for one but two of these.

(53) Compare note (50) and Problem 9 of §2 above. This paragraph is inserted here from the bottom of the manuscript page following Newton's direction 'Prob 2dum'.

(54) 'aliàs sic' (otherwise thus) is cancelled.

(55) With typical ingenuity Newton constructs his double-pointed cubic as a degenerate case of a quartic whose other component curve is the polar line: in effect, the described curve has (compare Theorem 6 of §1. 2 above) three distinct double points, two at the poles. (The cubic's double point corresponds, of course, to the two meets of the describing conic with the polar line.) See also §2: note (43)

required six elements in addition to the principal point ([50]five points and a tangent line, for example, or five points and a line parallel to an asymptote, or four points and an asymptote, or three asymptotes, and so on: I count an asymptote as two since it gives an infinitely distant point and, in addition, touches the curve at that point).[51] Of those seven elements, the principal point along with a second one yields the poles, a third given element (say the tangent at one of the poles) the mobile angles (by Lemma 6) and the remaining four the four conditions by whose aid the directrix is, according to the preceding problem, to be drawn: it is also given (as a fifth condition)[52] that the directrix shall pass through the principal point of the describend. The describing point, finally, will, as before, draw the describend while the directing point moves on the directrix.

But if[53] a tangent is not given the problem is[54] accomplished through a directrix which is to be a conic passing through neither pole. Precisely, for the poles I choose two points distinct from the principal point and determine the mobile angles by means of the principal point (by Lemmas 7 and 8), and then the remaining four given elements determine four conditions for describing the directrix (by Lemmas 9, 10 and the rest): a fifth condition you will find by Lemma 16, for the describend will intersect the line AB somewhere other than at the poles A and B. The conic, finally, described according to those five conditions will be the directrix, and by its aid you will, as before, draw the curve to be described.

Problem 3. The drawing of a curve of fourth degree of which three principal points (that is, points in which it crosses itself, is cusped or has a conjugate point) are given.

Solution. Of this describend also[56] the directrix is a conic which passes through neither pole. To determine this curve eight[57] elements are required: of these the three principal points yield the poles and mobile angles, the remaining five the directrix (by Lemmas 9–15 and Problem 1). In consequence, the describend is drawn as before.[58]

---

(56) This is the case of the second paragraph of Problem 2 when the described quartic no longer degenerates. Its must have three double ('singular') points because (compare §1. 2; Theorem 6) it must correspond with the describing conic and the doubled polar line taken together as a pair: two of the latter's real points (those at the poles) are always double while the third double point of the described quartic corresponds to the meets of the describing conic with the polar line (and will be a node, a cusp or a conjugate point according as those meets are real and distinct, real and coincident or at a conjugate point in the polar line).

(57) Fourteen conditions, in fact, are necessary to determine a particular quartic. That the describing (directrix) conic passes through three 'singular' points adds three further pairs.

(58) Newton here breaks off but it is very possible that he intended to add further problems in illustration on the remaining blank pages of his tract.

APPENDIX 1

AN ATTEMPT AT AN ANALYTICAL APPROACH TO THE ORGANIC CONSTRUCTION.(1)

[1667 or 1668?](2)

From the original worksheet in private possession

[1]

[Pone] $ad=a$. $de=y$. $af=x$. $an=b$. $nm=c$. $dr=d$. $rs=e$. $dh=v$. $ak=z$. [erit]

$a:v::b:\frac{bv}{a}=np$.(3) $mp=$(4)$\frac{ac+bv}{a}$. $mq=\frac{y\sqrt{bb+cc}}{\sqrt{aa+yy}}$.(5) &

$$a[:]\sqrt{a^2+v^2}::\frac{y\sqrt{bb+cc}}{\sqrt{aa+yy}}:\frac{ac+bv}{a}.\text{(6) [hoc est]}$$

$$\overline{ac+bv}\sqrt{aa+yy}=\sqrt{aa+vv}\times y\sqrt{bb+cc}.\ \text{[vel]}$$

$$\left.\begin{matrix} aabb \\ \mp aacc^{(7)} \\ \mp bbvv^{(7)} \\ +ccvv \\ -2abcv \end{matrix}\right. yy=aa\times\overline{ac+bv}\}^2.^{(8)}\ \&\ aby-cvy=aac+abv.$$

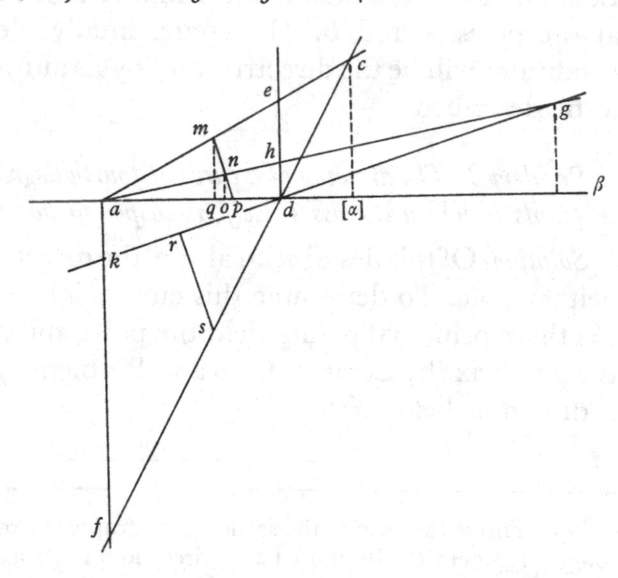

(1) It will be obvious that all of Newton's preceding researches into the organic construction have been conducted without recourse to analytical geometry. Later, as we know, when he compiled his Lucasian lectures on algebra he gave a fully analytical, immediately generalizable proof that, where the describing (directrix) curve is a straight line, then the described curve (describend) is a general conic through the poles of the organic construction's rotating angles. (See his *Arithmetica Universalis: sive de Compositione et Resolutione Arithmetica Liber* (Cambridge,

[2] est $aby-cvy=aac+abv$. [& simili ratione] $adx-ezx=aae+adz$.(9)

[hoc est] $\dfrac{aby-aac}{ab+cy}=v$. [&] $\dfrac{adx-aae}{ex+ad}=z$.

[sit natura curvæ g] $\begin{matrix} fvv+gvz+hzz \\ +kv\ \ +lz \\ +m \end{matrix}=0.$

[erit valoribus pro v & z substitutis]

$$f\times\overline{aabby^2-2a^3bc+a^4cc}\times\overline{eexx+2adex+aadd},$$

$$+g\times\overline{aby-aac}\times\overline{adx-aae}\times\overline{ab+cy}\times\overline{ad+ex},$$

$$+h\times\overline{dax-aae}\times\overline{adx-aae}\times\overline{ab+cy}\times\overline{ab+cy},$$

$$+k\times\overline{aby-aac}\times\overline{ab+cy}\times\overline{ex+ad}\}^2$$

$$+l\times\overline{adx-aae}\times\overline{ex+ad}\times\overline{ab+cy}\}^2+m\times\overline{ab+cy}\}^2\times\overline{ad+ex}\}^2=0.$$

[......](10)

1707: Problem 53 = London, $_2$1722: Problem 48). We will reproduce these lectures from the original autograph manuscript in our fifth volume.) As we have shown in detail elsewhere ('Patterns of Mathematical Thought in the Later Seventeenth Century', *Archive for History of Exact Sciences*, 1 (1961): 307 ff.), the basis of his reduction there is the conscious realization of the fundamental one-to-one continuous correspondence set up in the organic construction between the points of the describing and described curves. In this present first crude attempt at an analytical approach little of that insight is apparent but we think it of sufficient interest to append here, if only as an illustration of how his first brute-force attack on the problem ended in a tangle of irrelevant complexities.

(2) The present calculations are, along with some sketches in revision of the diagram accompanying Problem 11 of those researches, entered on the fourth side of a folded sheet from which we have produced §2 above. (Compare note (1) of that section.) In agreement with the date assigned to §2 we conjecture the present date wholly on our estimate of Newton's handwriting.

(3) That is, $ad:dh = an:np$. We conclude that $a\hat{n}p$ is right.

(4) '$mn+np=$'.

(5) That is, $de\times am/ae$. In consequence mq and ed are to be taken perpendicular to ad.

(6) $ad:ah = mq:mp$. For since $a\hat{q}m = a\hat{n}m = \frac{1}{2}\pi$, $p\hat{m}q = h\hat{a}d$.

(7) Newton's convention that two terms opposite in sign cancel each other.

(8) In modern terms, $(ab-cv)^2y^2 = a^2(ac+bv)^2$.

(9) Newton merely permutes d, e, z and x for b, c, v and y, keeping a fixed.

(10) In the next line Newton began to order the preceding equation by powers of x and y (that is, in the form $Ax^2y^2+Bxy^2+Cy^2+Dx^2y+Exy+Fy+Gx^2+Hx+I=0$) but abandoned the calculation after entering only the partial coefficients deriving from fv^2 and gvz.

[3] [priùs fuit] $\dfrac{aby-aac}{ab+cy}=v=\dfrac{p}{q}$. [&] $\dfrac{adx-aae}{ad+ex}=z=\dfrac{r}{s}$.(11)

[sit natura curvæ g] $fzzvv+gzvv+hvv+kzzv+lzv+mzz+nv+Pz+Q=0$.
[valoribus pro v & z substitutis erit](12)

$$\begin{matrix} fpprr+gpprs+hppss+npqss+Qqqss \\ +kpqrr+lpqrs+mqqrr+Pqqrs \end{matrix}[=0.] \quad \text{[hoc est]}$$

$$\begin{array}{lll} fa^8ccee-2fa^7bcee & -2fa^7ccde & \\ & y & x \\ -ga^7ccde+2ga^6bcde & -ga^6ccee & \\ +ha^6ccdd-2ha^5bcdd & +2ha^5ccde & \\ +ka^7bcee+ka^6bbee & +ga^6ccdd & \&c\ [=0.] \quad \text{sed } Q=0. \\ +la^6bcde-la^5bbde & +2ka^6bcde & \\ +ma^6bbee-ka^6ccee & +la^5bcdd & \\ -na^5bcdd+la^5ccde & -la^5bcdd & \\ -pa^5bbde+2ma^5bcee & +2ma^5bbde & \\ +qa^4bbdd+na^4bbdd & -2na^4bcde & \\ -na^4ccdd & -pa^4bbee & \\ -2pa^4bcde & +pa^4bbdd & \\ +2qq^3bcde & +2qq^3bbde & \end{array}$$

Unde

(1), $-kabbce+lbbcd+2mb^3e-kac^3e+lc^3d+2mbcce-\dfrac{ncbbdd}{ae}-\dfrac{nc^3dd}{ae}-\dfrac{2pb^3d}{a}$
$-\dfrac{2pbccd}{a}+2q\dfrac{b^3dd}{aae}+2q\dfrac{bccdd}{aae}=0.$

&(13) $-kace+lcd+2mbe-\dfrac{ncdd}{ae}-\dfrac{2pbd}{a}+\dfrac{2qbdde}{aae}=0.$

& (2), $-gacdde+2hcd^3+2hcdee+lbdde+lb^3-gace^3$ [&c] $=0.$

&(14) $-gace+lbe+2hcd-\dfrac{pbbe}{ac}-\dfrac{2nbd}{a}+\dfrac{2qbbd}{aac}$ [=0.]

& (3) $+gabcdd+gacbe^2-kabbde-kadc^2e+lc^2dd-lbbe^2-2hbc\dfrac{d^3}{e}$
$-2hbcde+2mbcde+2m\dfrac{b^3de}{c}$ [=0.]

(11) That is, Newton takes $aby-aac=p$, $ab+cy=q$, $adx-aae=r$ and $ad+ex=s$.
(12) An unfinished (and inessential) intervening calculation is omitted.
(13) Dividing through by b^2+c^2.
(14) Dividing through by d^2+e^2.
(15) Newton intends that the bracketed terms following be divided by b^2+c^2.

vel (2')

$$bb+cc)^{(15)}\begin{cases}2fa^3c^3ee-2gaac^3de+2hac^3dd-kaabccee+albc^2de-nbccdd\\+2fa^2bbcee-2gaabbcde+2habbcdd-kaab^3ee+lab^3de-nb^3dd\end{cases}=0$$

$$=2fa^3cee-2gaacde+2hacdd-kaabee+labde-nbdd.$$

Sed prius fuit $-kaacee+lacde+2mabee-ncdd-2pbde=0$

$$=-gaacce+labce+2haccd-pbbe-2nbcd.^{(16)}$$

Explanation. Newton is trying to find the analytical equivalent, in terms of suitably defined co-ordinate systems, of the correspondence between points c and g which arises when the fixed 'mobile' angles $c\hat{a}g$ and $c\hat{d}g$ rotate round the poles a and d of an organic construction system. For the constants of the system he fixes the mobile angles as $\tan^{-1} c/b$ and $\tan^{-1} e/d$ and takes the polar distance $ad = a$. To represent the general points c and g, however, he invents a curious co-ordinate system: the point c is determined by the meet e of ac with the normal and that of dc with the normal at d; and similarly the point g is determined by the corresponding meets h and k of ga and gd with these normals. Thus, if with Newton we set $af = x$ and $de = y$ (paying due consideration to sign) we may represent c by the ordered pair of co-ordinate line-lengths (x, y): similarly, where we set $dh = v$ and $ak = z$, c is (v, z). In [1] Newton shows that the analytical equivalent of the one-to-one continuous correspondence between the points c and g is $\begin{Bmatrix}v = a(by-ac)/(cy+ab)\\ z = a(dx-ae)/(ex+ad)\end{Bmatrix}$. In [2] he assumes that the curve (g) is defined by the quadratic equation $fv^2+gvz+hz^2+kv+lz+m = 0$ and seeks to determine the corresponding defining equation (of the form

$$Ax^2y^2+Bxy^2+Cy^2+Dx^2y+Exy+Fy+Gx^2+Hx+I = 0)$$

for the curve (c) but breaks off. In [3] he assumes a quartic form for the defining equation of the locus g, namely

$$fv^2z^2+gv^2z+hv^2+kvz^2+lvz+nv+mz^2+Pz+Q = 0.$$

(Halfway through his calculation he realizes that the polar line ad must be taken into the equation: for that the sufficient condition is that v and z vanish together, or that $Q = 0$.) The corresponding equation for the curve c, say

$$0 = a^6\alpha+a^5(\beta y+\gamma x)+\ldots,$$

he now calculates, determining the first three coefficients (α, β and γ) of the constant terms and those in x and y. Finally, he seeks to equate each of α, β and γ to zero (why?) and calculates the derived algebraic forms

$$(1) = (2b\alpha+c\beta)/e(b^2+c^2),\quad (2) = (2d\alpha+e\beta)/c(d^2+e^2),\quad (3) = (cd\beta-be\gamma)/ce$$

---

(16) The manuscript breaks off here.

and $(2') = 2ac\alpha - ab^2\beta$, equating them all to zero and in particular juxtaposing $ae\times(1)$ and $ac\times(2)$. He then breaks off.

This present co-ordinate system of Newton's is ingenious but not ultimately very rewarding. Nor does it always uniquely define a point in the plane: in particular, it cannot in general distinguish between points in the polar line *ad* while the poles *a* and *d* themselves are quite generally represented by

$$(0, y) \equiv (0, v) \text{ and } (x, 0) \equiv (z, 0)$$

respectively. We may wonder why Newton does not seem to have considered the same structure in a standard Cartesian form. If, for example, we take *d* as the common origin of the perpendicular co-ordinates $d\alpha = X$, $\alpha c = Y$ and $d\beta = Z$, $\beta g = V$, we deduce immediately that

$$X = ay/(x-y), \quad Y = xy/(x-y), \quad Z = av/(z-v) \quad \text{and} \quad V = zv/(z-v).$$

From this the Cartesian equivalent of the analytical correspondence between the points *g* and *c* is an immediate corollary of Newton's result in [1].

APPENDIX 2

THE ORGANIC CONSTRUCTION AS COMMUNICATED BY NEWTON TO COLLINS IN AUGUST 1672(1)

Extract from the original letter in the University Library, Cambridge(2)

The description of a Conick section wch shall pass through five given points is this. Let the five given points be *A*, *B*, *C*, *D*, & *E* any three of wch as *A*, *B*, & *C* joyn to make a rectilinear triangle *ABC*, to any two angles of wch as *A* & *B* apply two sectors, their poles to ye angular points, & their leggs to the sides of y^e

---

(1) Newton had some time before on one of his visits to London conversed with Collins about mathematics, discussing in particular with him his own researches in Collins' favourite field, the geometrical construction of equations. On 30 July 1672 Collins wrote to him: 'I remember part of your discourse when here, was that you had found out an Instrument that giving 5 points in an Ellipsis or Hyperbola, would describe the Section, whence followed that discourse of the Sector...' (*Correspondence of Isaac Newton*, **1** (Cambridge, 1959): 223). In the present letter, among other things, Newton communicated his apparatus for the numerical construction of equations by 'Gunter's line' (one divided logarithmically; compare I, **3**, **3**, §1: note (6)) and then passed on to the above description of his organic construction of curves. This we reproduce for its intrinsic interest and as a guide to his preferred English equivalents for the Latin technical terms employed in §§2/3 above. Leibniz was shown the letter by Collins in October 1676 but made no recorded comment on this passage (see note (12)).

(2) Add. 3977.10, first reproduced (in full) by H. W. Turnbull in the *Correspondence of Isaac Newton*, **1** (1959): 229–32. However, the accompanying figure (printed on page 231) is not accurately depicted: in the original, in particular, the hand grasps a stylus set at the

triangle. And so dispose them that they may turne freely about their poles *A* & *B* without varying the angles they are thus set at. Which done, apply to y^e other two points *D* & *E* successively their two leggs *PQ* & *RS* wch were before applyed

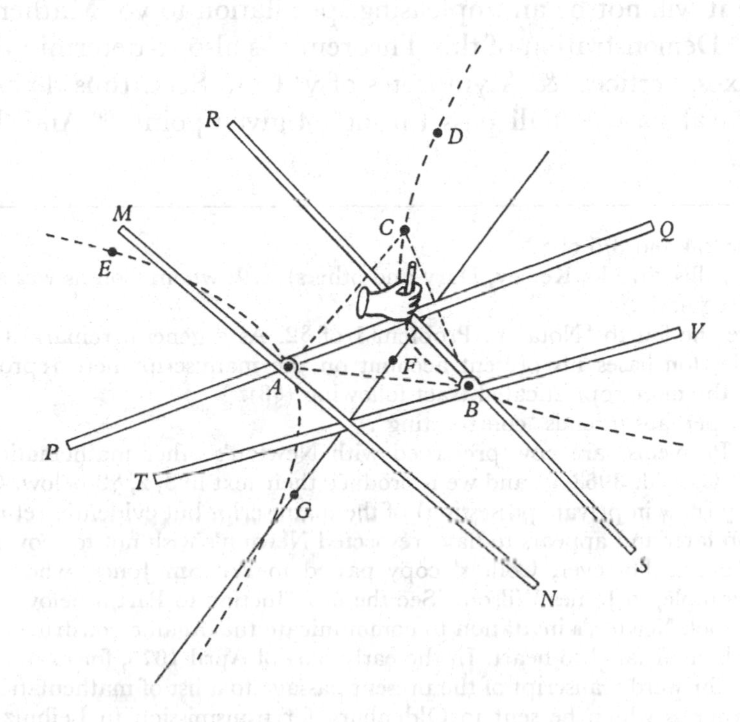

to *C* (wch leggs for distinction sake may be called their describing leggs, & the other two *MN* & *TV* wch were applyed to *AB*, their directing leggs,) & marke the intersections of their directing leggs, wch intersections suppose to be *F* when ye application was made to *D*, & *G* when made to *E*. Draw the right line *FG* & produce it infinitely both ways.(3) And then if you move the rulers in such manner that their directing leggs doe continually intersect one another at the line *GF*, the intersection of their other leggs shall describe the Conic Section wch will pass through all the said five given points.(4)

If three of the given points lye(5) in the same streight line tis impossible for any conick section to pass through them all, And in that case you shall have instead thereof(6) two streight lines.

---

junction of the describing legs and does not merely point in its direction, while (as in §§2/3 above) the two sets of moving angles are intended to be the inner edges of the rulers and not their axes.

(3) Compare the concluding 'Nota' to the Lemmas 'De Directrice' in §3 above.

(4) This is, of course, Problem 1 of §2.

(5) 'all' is cancelled.

(6) This revises the first cancelled phrase 'of wch curve line'.

Much after the same manner a Con. Sect. may be described w[ch] shall pass through 4 given points & touch a given line, or pass through 3 given points & touch two given lines, whether those lines be right or curved. &c[(7)]

I presume it will not be an unpleasing speculation to yo[r] Mathematicians[(8)] to find out y[e] Demonstration of this Theorem. As also to determin the centers, diameters, axes, vertices, & Asymptotes of y[e] Con. Sect. thus described or to describe a Parabola w[ch] shall pass through 4 given points.[(9)] And there-[fore] I omit them.

---

(7) Problems 2/4 and 3/5 of §2.

(8) Those (Wallis, Strode, Kersey, Dary and others) with whom Collins was at this period corresponding regularly.

(9) Compare the fourth 'Nota' to Problem 1 of §2. As a general remark it would seem evident that Newton bases his present account on the manuscript here reproduced as §2 rather than on the more sophisticated tract following (§3).

(10) Newton perhaps intends 'constructing'.

(11) These 'Problems' are now preserved with Newton's other mathematical papers in Cambridge (ULC. Add. 3963. 9) and we reproduce their text in **3**, 2, §2 below. Collins made a complete copy (now in private possession) of the manuscript but evidently returned the original to Newton later and appears to have respected Newton's wish not to show it to any one else. After his death, however, Collins' copy passed to William Jones, who did reveal its contents, for example, to James Wilson. (See the introduction to Part 3 below.)

(12) Collins took Newton's invitation to communicate the organic construction method to his fellow 'Mathematicians' to heart. In the early part of April 1675, for example, he added an almost word for word transcript of the present passage to a list of mathematical researches currently in progress which he sent to Oldenburg for transmission to Leibniz. (See Royal Society MS. LXXI, No. 36: 6–7. Collins' English was rendered into Latin by Oldenburg and sent to Leibniz in his letter of 12 April: compare C. I. Gerhardt, *Der Briefwechsel von Gottfried Wilhelm Leibniz mit Mathematikern*, **1** (Berlin, 1899): 113–22, especially 120–1.) The next year in London when Leibniz broke his journey back from Paris to Hanover (early October?) he was allowed by Collins to see Newton's original letter and make notes on it. Apparently he was not interested in the present passage for in his 'Excerpta ex Epist. Neutoni 20 Aug. 1672 ad [Collins]' it is not mentioned. (Leibniz' 'Excerpta', now folio 2 of the Leibniz Handschriften, 35 (Mathematik), **8**, No. 19 in the Niedersächsische Landesbibliothek in Hanover, are reproduced in J. E. Hofmann's *Studien zur Vorgeschichte des Prioritätstreites zwischen Leibniz und Newton um die Entdeckung der höheren Analysis. I: Materialien zur ersten mathematischen Schaffensperiode Newtons (1665–1675)* [= *Abhandlungen der Preuss. Akad. der Wissenschaften. Jahrgang 1943. Mathem.-Naturwiss. Klasse Nr. 2*], (Berlin 1943): 80.) When, too, about the beginning of 1676 Collins came to compose a *critique* of the three books of Descartes' *Geometrie* (compare the introduction to the present Part 1), he wrote of the latter's analytical reworking of the Greek 3/4 line-locus that '2. As to his Probleme of 3, 4, 5, 6, 7, or more Lines out of Pappus, and his describing the nature of a Locus, it is the best thing in his Workes worthy the Author and perchaunce, if he had not solved it, it had remained hitherto undone. But since him (though not yet extant) M[r] Newton hath solved the said Probleme with much more ease and variety, namely by 5 points in an Ellipsis or Hyperbola and 4 in a Parabola he describes the figure by ayd of two moveable angles, whilst one paire of Legs moove in a right Line the other intersect in a Conick Section: and through one point lesse than those innumerable of the said figures wille passe. Without which ayd, the actuall description of those Sections, in Constructions for æquations would be unpleasing...' (Royal Society MS. LXXXI, No. 39: 1). In the early summer

I herewith send you a set of Problems for construing[10] æquations wch that I might not forget them I heretofore set down rudely as you'le find. And therefore I think 'em not fit to be seen by any but yor selfe, wch therefore you may return when you have perused them.[11] How ye afforesaid descriptions of ye Conick sections are to be applyed to these constructions I need not tell you.[12]

---

of that year, finally, Oldenburg drew up a Latin letter for Tschirnhaus on the basis of Collins' *critique*: though it is now lost we may infer that Tschirnhaus received it from his reply on 1 September (N.S.) (See the *Report on Leibniz–Newton MSS. in the possession of the Royal Society of London. Drawn up by order of the President and Council of the Society* (London, 1880): 7–9; and the *Correspondence of Isaac Newton*, **2** (1960): 17, note (1).) Collins sent both his *critique* of Descartes and Tschirnhaus' answering letter to John Wallis on 20 September 1677. On the following 8 October Wallis returned, characteristically, that 'Your Considerations on Descartes I like well; & they are but faintly answered by Churnhause. I suspect (as I have formerly intimated) that of such communications (of things not published) they make but ill use beyond sea. And particularly that Comiers New Notion of Two means proportional &c [in his *La Duplication du Cube, La Trisection de l'Angle, et l'Inscription de l'Heptagone Regulier dans le Cercle* (Paris, 1677)] was borrowed from those Papers, where you mention (page 1.) Mr Newtons making use of two movable Angles. And no doubt but many other particulars therein imparted, will shortly be published as French Inventions' (*The Correspondence of Isaac Newton*, **2** (1960): 238). Newton's mathematical papers were read by many, but, in all fairness, the suggestion that Comiers saw them is risible.

PART 2

RESEARCHES IN CALCULUS

(c. 1667–c. 1670)

INTRODUCTION

In comparison with the dense flood of manuscript research which flowed from Newton's pen in the two and a half years from mid-1664 and which he summarized in his October 1666 fluxional tract,(1) his calculus investigations during the years immediately following are relatively spasmodic and jejune. Some aspects of his first researches (particularly in the construction of tangents, normals and inflexion points and the computation of logarithms) were clarified, emended and generalized, but on the whole the very act of writing up those researches as a collective unity seems for several years from 1666 to have acted as a brake on further development of his fluxional ideas. In retrospect the short 1669 compendium *De Analysi per æquationes numero terminorum infinitas*, not in itself of striking originality (whatever its future historical significance in terms of calculus priority was to be) but in large part merely a systematic exposition of earlier binomial researches, dominates his other fluxional works of this period more on account of their relative poverty than by its intrinsic excellence. Its true importance, indeed, is that it introduced Newton's name to John Collins and others outside the tight circle of his Cambridge acquaintances and that it led the way to his composing the comprehensive 1671 tract on fluxions and infinite series(2) in which its revised version was incorporated. The *De Analysi* apart, we have found no explicit mention of any of the papers we reproduce in any contemporary Newtonian document and the details of the history of their composition remain mysterious. What little we have seen reason to conjecture from the internal evidence of their text we have inserted at appropriate places in the notes. For the rest, we may well believe that these minor calculus pieces were for him mathematical *divertissements* on which his mind could relax in an idle hour away from his researches in theoretical and applied optics, his overriding preoccupation at this time.

The short opening tract on *Problems of Curves* presents the construction of tangents, normals and oblique transversals to a general algebraic curve from a given fixed point, both gathering the substance of his previous researches on those topics(3) in logical sequence and adding some new observations on the number of possibilities for each problem to be constructed. Here as always Newton aimed at an economy of technique rather than geometrical elegance, seeking in each case an auxiliary curve of least dimensions to achieve his end. His exposition is equally laconic and the fundamental Problem 1, which fixes

(1) I, 2, 7. Compare the scheme displayed on I: 154.

(2) This work, usually known under one of the non-Newtonian titles of *Methodus fluxionum et serierum infinitarum* or *Geometria Analytica*, we will reproduce in our next volume.

(3) I, 2: passim.

the number of possible meets of two algebraic curves of known degree, is left unproved as 'a principle rather then a probleme'.[(4)] A concluding list of allied problems ('much after the same manner') in the elementary differential geometry of curves suggests that he then intended a more comprehensive calculus tract but that his interest, as so often, was unequal to his design: at any rate, the project was not implemented till he began to compose his major fluxional work, the *Methodus fluxionum*, in 1670. To Newton's tract we have prefaced some minor calculations of his, found on an accompanying worksheet, which seem pertinent to its theme.

The extended logarithmic computations which follow are in large part an accurate revision of a worksheet reproduced in our first volume,[(5)] but now in addition Newton derives the logarithms of certain low primes fundamental in the compilation of a canon by adding and subtracting appropriate multiples of those he has previously calculated outright. Though this approach had been broached long before by Henry Briggs and was repeated with variants by James Gregory in a work published about the time that Newton wrote his present paper, the details of his development of it are essentially different and we may suppose that he came upon it independently.[(6)] The computations on which he spent so much effort and time evidently remained for him a theoretical exercise for he never attempted to reduce any of the natural logarithms he thereby obtained to correspondingly accurate Briggsian equivalents: in later reference,

(4) See 1, §2: note (15) below.

(5) I, **1**, 3, §5. The computations are of $\pm\log(1\pm p)$, $p = 0{\cdot}1, 0{\cdot}2, 0{\cdot}02$ and $0{\cdot}0001$.

(6) See Henry Briggs, *Arithmetica Logarithmica sive Logarithmorum Chiliades Triginta* (London, 1624): Caput IX: 17–19; and James Gregory, *Vera Circuli et Hyperbolæ Quadratura* (Padua, 1667): Prop. XXXIII. 'Propositi cujuscunque numeri logarithmum invenire' [= *Christiani Hugenii Opera Varia*, **1** (Leyden, 1724): 407–62, especially, 452–8]. (Compare Christoph J. Scriba, *James Gregorys frühe Schriften zur Infinitesimalrechnung* (Giessen, 1957): 23–5.) Newton appears never to have read any of Briggs' works, but his bound set of Gregory's mathematical tracts (Trinity College, Cambridge NQ. 9. 48) includes an unmarked copy of the 1668 reissue of the *Quadratura.*

(7) 'Pudet dicere ad quot figurarum loca has computationes otiosus eo tempore perduxi. Nam tunc sanè nimis delectabar inventis hisce' (Newton to Oldenburg for Leibniz, 24 October 1676 = *Correspondence of Isaac Newton*, **2** (1960): 114). Newton refers, of course, more particularly to his preliminary calculations (note (5) above) made during his enforced retirement from Cambridge University in the late summer or early autumn of 1665. In comment upon the utter impracticality of his present expansions to 57 decimal places we may add that Napier in his computation of log (2) and log (10) was content with ten significant figures, while even Gregory in calculating log (10) and log (1·024) thought 25 decimal places more than sufficient. (See John Napier's posthumously published tract (written *c.* 1600?) *Mirifici Logarithmorvm Canonis Constrvctio: Et eorum ad naturales ipsorum numeros habitudines* (Edinburgh, 1619): 32; and James Gregory's *Vera Circuli et Hyperbolæ Quadratura* (note (6)): Prop. XXXII. 'Invenire quadratum æquale spatio hyperbolico...'.)

(8) See I, **2**, 6, §4.2 and compare 2, §2: notes (1) and (12) below.

indeed, to the impractical extent of his present calculations he was suitably apologetic.[7]

With regard to the minor Latin pieces we need say little. An ingenious construction of the tangent at a given point on a conic is, in effect, a particular case of Pascal's theorem on the inscribed conic hexagon, while Newton's following cycloid rectification is a mere reworking of Wren's investigation of a decade before. The *De Solutione Problematum per Motum* is in large part merely the Latin version of a paper 'To resolve Problems by motion...' composed in May 1666, although an elegant construction of the inflexion points in a conchoid (discovered independently of Huygens) is introduced in refinement.[8] The *De Gravitate Conicarum*, finally, is an immature, not very well argued paper on gravitational balance round a linear axis.

Inevitably our attention is drawn to the *De Analysi per æquationes numero terminorum infinitas*, and on its historical background, though guided more by circumstantial evidence than documentary fact, we may be somewhat more expansive.

Conventionally, the tract is presented as Newton's first attempt to display his doctrines to public view but it was surely never that. Examination of the original manuscript reveals that it was written hastily with multiple cancellations, several recastings and, not least, some errors in its detail, all of which Newton could and would have eliminated at leisure. Its theme moreover, the employment of infinite series in elementary geometrical analysis, represented only a small portion of the wealth of his mathematical researches from 1664 onwards. A passage in preface to an intended edition of his *De Quadratura Curvarum* written late in life suggests a more plausible reason for his composing the paper:

In [my letter of 24 Octob. 1676 to Mr Oldenburg] I mentioned..that when Mr Mercators *Logarithmotechnia* came abroad Dr Barrow communicated to Mr Collins a Compendium of my method of series.[9] And this is the Tract entituled *Analysis per series* [sic] *numero terminorum infinitas*. The *Logarithmotechnia* came abroad in September 1668.[10] Mr Collins a few months after sent a copy of it to Dr Barrow who replied that

(9) 'Eo ipso...tempore quo [N. Mercatoris *Logarithmotechnia*] prodit communicatum est ab D. Barrow...ad D. Collinsium, Compendium quoddam methodi...serierum, in quo significaveram areas et longitudines curvarum omnium et solidorum superficies et contenta, ex datis rectis et vice versa ex his datis rectas determinari posse, & methodum ibi indicatam illustraveram diversis seriebus' (*Correspondence of Isaac Newton*, **2** (1960): 114).

(10) As a Latin version of the present passage (now in private possession) makes clear, Newton's authority for the date is not his own memory or personal knowledge but John Collins' letter to Thomas Strode on 26 July 1672, which in the Latin translation reproduced on p. 28 of the *Commercium Epistolicum D. Johannis Collins, et Aliorum de Analysi Promota* (London, 1713) reads: 'Mense Septembri 1668, Mercator Logarithmotechniàm edidit suam, quæ specimen...Methodi...Serierum infinitarum in unica tantum Figura, nempe Quadratura Hyperbolæ continet.'

the Method of Series was invented & made general by me about two years before the publication of the *Logarithmotechnia* & at the same time sent back to Mr Collins the said Tract of *Analysis per series.*(11) This was in July 1669....(12)

When Newton turned to the 'Vera Quadratura Hyperbolæ, & Inventio Summæ Logarithmorum' promised in the title of Nicolaus Mercator's work(13) he must have felt crestfallen. There published for all the world to read was his reduction of $\log(1+a)$ to an infinite series by continued division of $1+a$ into 1 and successive integration of the quotient term by term. Even worse, Mercator offered the example of $\log(1\cdot 1)$ calculated to 44 decimals (and, indeed, checked by further computation of $\frac{1}{2}\log(1\cdot 21)$)(14) and it seemed not unlikely that he had privately calculated in addition many of the logarithms over whose accurate evaluation Newton had spent so many now seemingly wasted hours. Mercator's exposition of the procedure was admittedly cumbrous and inadequate, but if he could employ Newton's reduction to infinite series *dividendo*(15) in a particular case how long would it be before he stumbled on the extraction of roots in such

(11) '...A friend of mine here, that hath a very excellent genius to those things, brought me the other day some papers wherein he hath sett downe methods of calculating the dimensions of magnitudes like that of Mr Mercator concerning the hyperbola, but very generall; as also of resolving æquations; which I suppose will please you' (Barrow to Collins, 20 July 1669); 'I send you the papers of my friend I promised, which I presume will give you much satisfaction; I pray having perused them so much as you thinke good, remand them to me; according to his desire, when I asked him the liberty to impart them to you. and I pray give me notice of your receiving them with your soonest convenience; that I may be satisfyed of their reception; because I am afraid of them; venturing them by the post, that I may not longer delay to corrispond with your desire' (Barrow to Collins, 31 July 1669). (*Correspondence of Isaac Newton*, **1** (1959): 13–14.)

(12) ULC. Add. 3968. 41: 84r, extracted from an abortive revised edition of the *Principia* about late 1714, to which the *De Quadratura* was scheduled to be appended.

(13) *Logarithmo–Technia: sive Methodus construendi Logarithmos nova, accurata, & facilis: Scripto Antehàc Communicata, Anno Sc. 1667. Nonis Augusti: Cui nunc accedit. Vera Quadratura Hyperbolæ, &Inventio summæ Logarithmorum* (London, 1668). Newton's unmarked library copy of this tract is now in Trinity College, Cambridge, bound in with his set of James Gregory's mathematical works (NQ. 9. 48: compare note (6) above).

(14) Prop. XV (pp. 28–30) of the *Logarithmotechnia* gives the laborious reduction of $(1+a)^{-1}$ into the infinite series '$1-a+aa-a^3+a^4$ (&c)'; Prop XVI verbally enunciates the general integration theorem

$$\int_0^a a^p \, . \, dx = \frac{1}{p+1} a^{p+1};$$

while Prop. XVII 'Quadrare Hyperbolam' (pp. 31–3) states the fundamental logarithmic expansion in words and then lays out the first stages of the computation of the two logarithms named. Compare J. E. Hofmann, 'Nicolaus Mercators Logarithmotechnia, *Deutsche Mathematik*, **3** (1938): 446–66, especially 459–64. Hofmann has both devoted a detailed monograph (*Nicolaus Mercator* (*Kauffman*) [= *Akademie der Wissenschaften und der Literatur. Abhandlungen der Math.–Naturw. Klasse. Jahrgang 1950, Nr.* 3 (Wiesbaden, 1950)]) to clarifying Mercator's mathematical genius and has also traced the subsequent history of the 'Mercator' logarithmic expansion in his 'Weiterbildung der logarithmischen Reihe Mercators in England' (*Deutsche Mathematik*, **3** (1938): 598–605; **4** (1939): 556–62; **5** (1940–41): 358–75).

series and indeed upon his cherished binomial expansion? Even in the calm of retrospect in his *Epistola posterior* of 1676 Newton's bewilderment still communicates itself.(16) Such news as filtered through to him in Cambridge during the following winter and spring of 1669 could only have added to his depression, for by then not only Mercator but William Brouncker also now claimed privately that he could, in effect, reduce a binomial square root into an infinite series.(17) Gradually it became clear that Newton could choose either to circulate an account of his own discoveries in infinite series or to relinquish the credit for those discoveries to whoever published the inevitable complement to Mercator's book. As we interpret the meagre evidence, towards the end of June 1669 Newton hurriedly compiled the *De Analysi* in revision and partial amplification of his previous researches into the binomial expansion and the resolution of 'affected' equations, and first submitted it almost at once to Isaac Barrow, his college senior and the then Lucasian Professor of Mathematics, for approval. The latter, or so he wrote, communicated it forthwith to John Collins, praising the depth and generality of its content but concealing the name of its author.(18)

The subsequent history of the tract is well documented and we need not repeat its details.(19) Collins wrote back to Barrow full of enthusiasm for the work and

(15) See 3: note (24) below.

(16) '...ubi prodiit ingeniosa illa N. Mercatoris Logarithmotechnia (quem suppono sua primum invenisse) cœpi ea minùs curare, suspicatus vel eum nosse extractionem radicum æque ac divisionem fractionum, vel alios saltem, divisione patefacta, inventuros reliqua priusquam ego ætatis essem maturæ ad scribendum...' (Newton to Oldenburg for Leibniz, 24 October 1676 = *Correspondence of Isaac Newton*, **2** (1960): 114). The latter phrase echoes his own diffidence at this time as well as his doubts.

(17) 'Mr Mercator hath often, yea & very lately in the presence of the Lord Brereton affirmed with much confidence, that he hath now a series for the circle that shall make the sines of any arch, and the converse, and give the area of any sector segment or zone infinitely true, and affirmed that his methods transcend all that is extant, and contain in them as much as rationally can be wished or hoped for, and that he will publish it ere long' (John Collins to James Gregory, 7 January 1668/9 in (ed. H. W. Turnbull) *James Gregory: Tercentenary Memorial Volume* (London, 1939): 60); 'As for Mercators new series I know not what it is...but ...the Lord Brouncker asserts he can turne the square roote into an infinite Series' (Collins to Gregory, 2 February 1668/9, *ibid.*: 66); 'Mr Mercator...affirmes he hath 3 [series] for the Area of the whole Circle, another for the Segments, a third [for] finding the Sine of any Arch proposed' (Collins to Gregory, 15 March 1668/9, *ibid.*: 71). Though Newton did not yet realize it his most dangerous rival in developing the method of infinite series, James Gregory, was working quietly but industriously away from the publicity of metropolitan life in St Andrews. (Compare *James Gregory*: 347–70, and J. E. Hofmann, 'Über Gregorys systematische Näherungen für den Sektor eines Mittelpunktkegelschnittes', *Centaurus*, **1** (1950): 24–37.)

(18) See note (11) above.

(19) References to the pertinent published documents are gathered by J. E. Hofmann in his *Studien zur Vorgeschichte des Prioritätstreites zwischen Leibniz und Newton um die Entdeckung der höheren Analysis. I. Abhandlung: Materialien zur ersten mathematischen Schaffensperiode Newtons (1665–1675)* [= *Abhandlungen der Preussischen Akademie der Wissenschaften. Jahrgang 1943. Math.-Naturw. Klasse Nr. 2*] (Berlin, 1943): 17–20.

was rewarded by him with the author's identity and an estimate of his mathematical talent and proficiency. In addition, Collins was given permission to show the *De Analysi* to Brouncker, perhaps to gain his support as President of the Royal Society but partially, no doubt, to reveal to him how advanced Newton's techniques were in comparison with his own.[(20)] We do not know if Brouncker saw the tract but Collins, certainly, before he returned the original to Newton, took a complete copy which he was later not loth to show around.[(21)] At various times over the next few years, moreover, he circulated news of Newton's work round his wide ring of correspondents—James Gregory in Scotland,[(22)] de Sluse and Bertet in France and Borelli in Italy as well as his fellow Englishmen Francis Vernon and Thomas Strode.[(23)] The question of publication was, of course, posed but Newton appears to have resisted the combined entreaties of Collins and Barrow to have the *De Analysi* set in appendix to the latter's optical lectures.[(24)] Whether in a time of depression in the London book-trade no other way of publication offered itself or whether, now that he had spiked Mercator's gun, Newton was himself unwilling to have the tract printed in its existing form without extensive revision or whether he could not spare the time from preparing his opening series of lectures as the new Lucasian Professor is not clear, but a year afterwards he was busy incorporating its content into the opening passages of his *Methodus fluxionum* and the opportunity to print it as a piece of current research vanished forever. When it was finally published by William Jones in 1711 in his *Analysis Per Quantitatum Series, Fluxiones, ac Differentias*[(25)] it had already acquired historical status and significance as an early exposition of Newton's fluxional method.

The text itself of *De Analysi* requires little introduction. Formally it is an exposition addressed to an unnamed second person, perhaps Isaac Barrow in particular or maybe the reader in general. In the style of Mercator's work three lemmatical 'Regulæ' cite the elementary rules of integration necessary to apply

(20) 'I am glad my friends paper giveth you so much satisfaction. his name is Mr Newton; a fellow of our College, & very young (being but the second yeere Master of Arts) but of an extraordinary genius & proficiency in these things. you may impart the papers if you please to my Ld Brouncker....' (Barrow to Collins, 20 August 1669 = *Correspondence of Isaac Newton*, **1** (1959): 14–15.)

(21) See **3**: note (2) below. John Wallis was an exception: writing to him in late 1677 Collins noted that 'If I had been so minded I could...at the beginning of 1669 have imparted to you a full treatise of his of [Series] but did not, in regard you lye under a censure from diverse for printing discourses that come to you in private Letters without permission or consent as is said of the parties concerned.' (*Correspondence of Isaac Newton*, **2** (1960), 242.)

(22) 'Mr Newton of Cambridge...(before Mercators *Logarithmotechnia* was extant) invented the same method and applyed it generally to all Curves, and diverse wayes to the Circle' (Collins to Gregory, 25 November 1669); '...and all other Curves that are Geometricall In the sense of Deschartes' (Collins to Gregory, 12 February 1669/70). (*Correspondence of Isaac Newton*, **1** (1959): 15, 26.) See also Collins' letter to Gregory of 24 December 1670.

the following three methods of reducing a given analytical expression in two variables into an explicit infinite series expansion 'dividendo', 'radicem extrahendo' and 'per resolutionem æquationum affectarum': of these the first two are respective equivalents of the binomial expansion for index -1 and $\frac{1}{2}$, while the last is an algebraic extension of a slightly modified form of Viète's technique for resolving numerical equations. In sequel, examples are given of the application of the joint method of fluxions and infinite series in problems of quadrature and rectification, while in conclusion Newton urges that his approach remains analytical in Descartes' sense. A retrospective addendum outlines a proof of Rule 1 and then, perhaps most interestingly of all, attempts to apply the 'Eudoxian' first proposition of Euclid x as a convergence test for the infinite series expansion $y = \lim_{n\to\infty} \sum_{0\leqslant i\leqslant n} (a_i x^i)$ when $|a_i| \leqslant 1$ and $|x| \leqslant \frac{1}{2}$. Whether or not Newton knew of Gregory's work on convergence at this time, we may anticipate our next volume and remark that when he came to revise this second observation for his *Methodus fluxionum* he in fact there used Gregory's word 'convergens' in its now standard significance.(26)

We have mentioned the copy which in 1669 Collins took of Newton's tract. In hindsight perhaps the most interesting of all his contemporaries who were allowed to consult that copy over the next few years was Gottfried Wilhelm Leibniz. The latter purchased Mercator's *Logarithmotechnia* at the time of his first visit to England in the spring of 1673 and was thereby inspired, at least in part, to discover his celebrated 'Quadratura arithmetica circuli'.(27) During the following three years he kept up a lively correspondence with the Royal Society's Secretary, Henry Oldenburg, who among other things passed on to him in Latin version several lengthy communications from Collins. Newton's name came up more than once in that correspondence and Leibniz received, for instance, examples of his work in the resolution of algebraic equations and

(23) Compare the Latin versions of these letters in the *Commercium Epistolicum D. Johannis Collins* (note (10)): 21, 26–7, 27, 27–8 and 28–9.

(24) 'Mr Newton...may send up [his method] to be annexed to Mr Barrowes Lectures'; 'I believe Mr Newton...will give way to have it printed with Mr Barrows Lectures.' (Collins to Gregory, 25 November 1669 and 12 February 1669/70) (note (22).)

(25) See 3, Appendix 2: note (1) below.

(26) Newton, in fact, uses the phrase '...quotientes, ut minùs citò ad justam radicem convergant'. (S. Horsley, *Isaaci Newtoni Opera quæ Exstant Omnia*, 1 (London, 1779): 406, which is, in his edited version, the closing paragraph of the *Geometria Analytica*, Caput 3. 'De Speciosâ Æquationum Resolutione'.) For James Gregory's application of convergence see his *Vera Circuli et Hyperbolæ Quadratura* (note (71)): passim, and D. T. Whiteside, 'Patterns of Mathematical Thought in the later Seventeenth Century', *Archive for History of Exact Sciences*, 1 (1961): 179–388, especially 266–70.

(27) See J. E. Hofmann, *Die Entwicklungsgeschichte der Leibnizschen Mathematik während des Aufenthaltes in Paris (1672–1676)*, (Munich 1949): 32–6.

an outline of his organic description of conics.[28] On 12 April 1675, in particular, Oldenburg communicated Collins' observation:

...after Mercators Logarithmotechnia was extant as soone as it came forth I sent [it] to Dr Barrow, and he observing an infinite Series therein used for the making of the Logarithmes writt back that the said method had been some time before found out by his Successor Mr Newton and generally applyed to all Curves and their portions, as well Geometricall as Mechanicall, and sent up some Specimens thereof....[29] And those Series are easily continued ad infinitum.[30]

When during his second visit to London in mid-October 1676 Leibniz met Collins for the first time he was permitted among other things to make the partial transcript of *De Analysi* which is reproduced as the first appendix to Section 3. Apart from its intrinsic interest it must clear Leibniz of any lingering suspicions still felt by the ardent Newtonian supporter that he made good use of this chance to annex for his own purposes the fluxional method briefly exposed there. On the contrary it is patent that Leibniz was interested only in Newton's series expansions: on one occasion, indeed, he used his own 'long s' notation to clarify for himself Newton's less elegant verbal expression of the 'area' under a curve.[31] Let this be proof of Leibniz' independence of thought.

The two final appendices present Leibniz' review of the *De Analysi* in Jones' 1711 published version together with Newton's hitherto unprinted counter-observations on that recension. In part, of course, each of the two authors there presses his own priority in the invention of algorithmic calculus and their accounts are in consequence to some extent emotionally if not factually biased. But it is equally true that each was trying accurately to define his own position in respect to the other's method and notation and with particular regard to Newton's fluxional arguments in the *De Analysi*. Certainly, one or two points are

(28) Compare **1**, 3, Appendix 2: note (12) above. Leibniz' correspondence with Oldenburg is reproduced with annotations in C. I. Gerhardt, *Der Briefwechsel von Gottfried Wilhelm Leibniz mit Mathematikern*, **1** (Berlin, 1899): 39–255.

(29) In this omitted portion Collins listed the series expansion of $z = \sin^{-1} x$ and that conversely of $x = \sin z$.

(30) Royal Society MS. LXXXI, No. 36: 2r. Oldenburg's Latin version of this passage is printed by Gerhardt in his *Briefwechsel* (note (28)): 114.

(31) See 3, Appendix 1: note (4) below. We merely restate, of course, the previously expressed opinions of the two eminent Leibniz scholars, Gerhardt and Hofmann, who have examined the transcript. (C. I. Gerhardt, 'Leibniz in London', *Sitzungsberichte der Königlich Preussischen Akademie der Wissenschaften zu Berlin. Jahrgang 1891*, x: 163–4: 'In Betreff der Excerpte aus Newton's Abhandlung...ist zu bemerken, dass Leibniz sich für die Behandlung der algebraïschen Ausdrücke durch Division und Wurzelausziehung in Reihen interessirte... Wie zur Quadratur zu gelangen war, war ihm bekannt; er deutet es lediglich durch das Summen—d.i. Integralzeichen an.... Über die Bemerkung Newton's am Schluss der Quadraturen, dass die Probleme der Rectification, über die Bestimmung des Inhalts der Körper, über die Bestimmung des Schwerpunktes, auf dieselbe Weise gelöst werden, so wie über die allge-

made which continue to escape many modern Newtonian commentators and we think these appendices will prove a useful complement to the original text of his tract. To that end, in our edited versions we have sought deliberately to prune all which is not directly relevant to technical content.

meine Anweisung, die Newton dazu giebt, geht Leibniz als etwas ihm Bekanntes hinweg'; J. E. Hofmann, *Entwicklungsgeschichte* (note (27)): 182: 'Leibniz' Auszüge beschränken sich ausschliesslich auf Reihenentwicklungen und die dazugehörigen Allgemeinbemerkungen. Der infinitesimale Inhalt der *Analysis* bleibt völlig unberücksichtigt—offenbar deshalb, weil er für Leibniz nichts Neues zu bieten hatte'.)

1

CURVE PROBLEMS AND FURTHER LOGARITHMIC COMPUTATIONS

[*c.* 1667?][(1)]

§1. 'PROBLEMS OF CURVES': MISCELLANEOUS CALCULATIONS.

From the original manuscript in private possession

[1] [Suppose] $b[c]=x$. $cd=y$. $bf=z$. $ef=v$. $ae=a$. [&] $yy=x$.[(2)] [then is] $2x:\sqrt{x}::x-z:v::2\sqrt{x}:1$.[(3)] [or] $x-z=2v\sqrt{x}$.
[Also tis] $x-z+\sqrt{aa-vv}$.[(4)] [Therefore]

$$2vvx^{(5)}=aa-vv=xx-2zx+z^2.$$

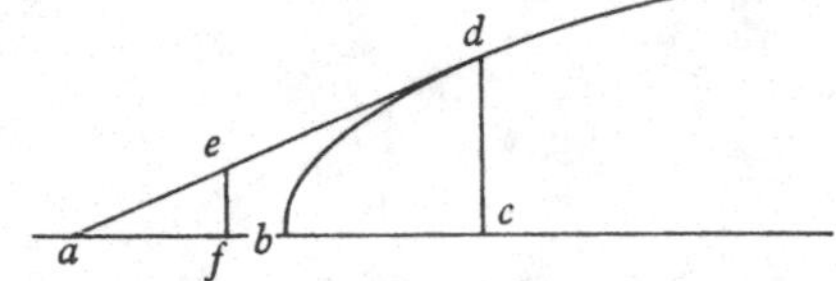

[or] $\sqrt{aa-vv}^{(6)}=\dfrac{aa-vv-2zvv}{2vv}$. [& soe]

$$\frac{aa-vv-2vv\sqrt{aa-vv}}{2vv}=z.^{(7)}$$

---

(1) The date is conjectured solely from our assessment of the handwriting and internal structure of the worksheet we reproduce. These calculations, written on the verso of the sheet which bears the latter portion of the text of **1**, 3, §1.3 above (on Newton's organic construction of curves) is at once closely connected with the geometrical researches we have printed in the previous part and the more coherent text on 'Problems of Curves' which we publish below. ([3], indeed, relates very closely to the cubic enumeration in **1**, 1, §3, while [2] is evidently a draft of the opening of the following section.)

(2) The curve *bd* is consequently a parabola with *bc* as its main diameter.

(3) That is, since $ac = y(dx/dy) = 2x$, $ac:cd = af:fe$ and cd is parallel to fe.

(4) In consequence $\widehat{afe}$, and therefore $\widehat{acd}$, is right.

(5) Read '$4vvx$'. Newton's mistake is left uncorrected in the sequel.

(6) That is, $x-z$.

(7) Where *da* is tangent to the curve whose defining equation in perpendicular co-ordinates is, say, $f(x, y) = 0$, and meets the abscissa in *a*, Newton seeks to define the co-ordinates (z, v) of the point *e* situated in it such that $ae = a$. In his working he supposes that (d) is the parabola $y^2-x = 0$ and determines

$$\frac{dy}{dx} = 2x^{-\frac{1}{2}} = \frac{y-v}{x-z} \quad \text{and} \quad \sqrt{(a^2-v^2)} = y\frac{dx}{dy}-(x-z)$$

[2] [Pone] $ap=a$. $pD=b=rt$. $ab=c$. $bt=d$. $aw=x$. $wv=y$. [tum]

$cq=\frac{y\times Aq}{x}=\frac{b\times\overline{c-Aq}}{d}$.(8) [hoc est]

$\begin{matrix}dy\\+bx\end{matrix}\times Aq=bcx$ [vel] $Aq=\frac{bcx}{dy+bx}$.

[Est ergo] $cq=\frac{bcy}{dy+bx}$. $bq=\frac{cdy}{dy+b[x]}$.

$sc=\frac{bcy-bdy-bbx}{dy+bx}$. [&]

$$sD=\frac{ady+abx-bcx}{dy+bx}.^{(9)}$$

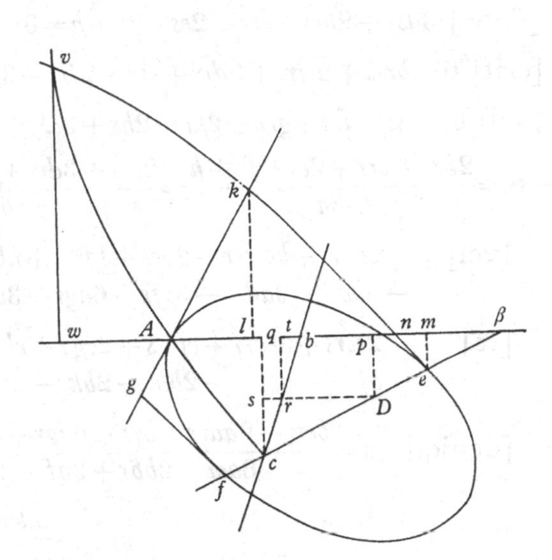

vel posito $c-d=e=At$ vel $c=d+e$. $d=bt$. & $a-c=f=bp$ vel

$$a=c+f=d+e+f.$$

erit $\frac{bey-bbx}{dy+bx}=sc$. & $\frac{ady+bfx}{dy+bx}=sD$.(10) et $\frac{ady+bfx}{ey-bx}=p\beta$.

[Pone] $Am=z$. $me=v$. [erit] $m\beta=a-z+\frac{ady+bfx}{ey-bx}$.(11) [&]

$me=\frac{\overline{a-z}\times\overline{ady+bfx}^{(12)}}{bey-bbx}+b=v$. [posito etiam] $vv+pvz+qzz+rv+sz=0$. [erit]

$mn^{(13)}=\frac{2vv+pvz+rv}{pv+2qz+s}=w$. [&] $w:v::z+x:v+y^{(14)}::2v+pz+r:pv+2qz+s$.(15)

---

as further conditions by which x and y may be eliminated. However, apart from a numerical slip, he mistakenly assumes that $\sqrt{(a^2-v^2)}$ ($=x+z$ correctly in his example) is equal to $x-z$ and the further calculations are consequently vitiated.

(8) That is, $wv\times Aq/Aw=rt\times(Ab-Aq)/bt$. Note Newton's indiscriminate use of upper and lower case forms to indicate the point 'A' in his figure.

(9) Since $bq:cq=bt:rt$, $sc=qc-pD$ and $sD=Ap-Aq$.

(10) That is, $sD\times pD/sc$.

(11) Or, $Ap-Am+p\beta$.

(12) Read '$\frac{\overline{a-z}\times\overline{bey-bbx}}{ady+bfx}$', since $me=-(Ap-Am)\times pD/p\beta+pD$.

(13) That is, $-V\frac{dz}{dv}$.

(14) For, $mn:me=mw:(me+wv)$.

(15) Newton's calculations here break off. Evidently the fixed point c is a pole round which rotates the line cD, meeting the conic Aef in points e and f and the given lines Ab and rb in β and c. (The points g, k, n and v are then determined as the meets of the tangent at A with those at e and f, and of the tangent at e with Ab and Ac respectively.) He then seeks to evaluate the correspondence between $v(x, y)$ and $e(z, v)$, where v is in the fixed line Ac and ve is tangent

[3] $[0=]\ as^3+brss+crrs+dr^3+ess+frs+grr+hs+kr+l.$

[Pone] $3ass+2brs+crr+2es+fr+h=0=bss+2crs+3drr+fs+2gr+k.$

[erit] $0=brss+2crrs+3dr^3+ess+2frs+3grr+2hs+3kr+3l.$

[vel] $0=ess+frs+grr+2kr+2hs+3l.$(16) [Est ergo]

$$-ss=\frac{2brs+crr+2es+fr+h}{3a}=\frac{2crs+3drr+fs+2gr+k}{b}=\frac{frs+grr+2kr+2hs+3l}{e}.^{(17)}$$

[vel] $\begin{array}{l} 2bbrs+bc\ \ rr+2bes+bfr\ \ +hb\ \ =0. \\ -6ac\ \ -9ad\ \ -3afs-6agr-3ak \end{array}$

[&] $\begin{array}{l} 2ecrs+3edrr+ef\ \ s+2egr+ek\ \ =0. \\ -bf\ \ \ -bg\ \ \ -2bh\ \ -2bk\ \ -3bl \end{array}$

[deniqз] $s=\dfrac{bcrr-9adrr+bfr-6agr+bh-3ak}{6acr-2bbr+3af-2be}$

$$=\frac{3derr-bgrr+2egr-2bkr+ek-3bl}{bfr-2ecr+2bh-ef}=\&c:^{(18)}$$

---

to the conic, and the co-ordinates are determined by the perpendicular Cartesian pairs $Aw = x$, $wv = y$; $Am = z$, $me = v$. In analytical terms, he supposes the conic to have the general defining equation

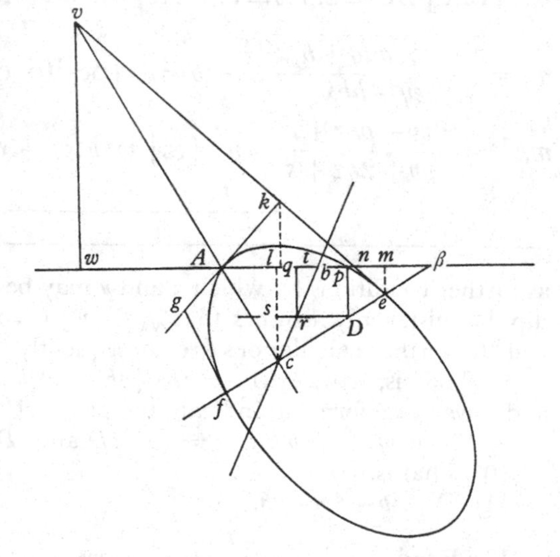

$$v^2+pvz+qz^2+rv+sz = 0$$

(where the constant term is rightly set as zero since the conic is through the origin A) and then finds the correspondence to be determined by

$$\frac{x+z}{y+v} = \frac{2v+pz+r}{pv+2qz+s}\left(=-\frac{dv}{dz}\right)$$

and
$$\frac{z-a}{v-b} = \frac{b(bx-ey)}{bfx+ady}.$$

(Correctly, this last fraction should (note (12)) be inverted.) Obviously, each pair (x, y) determines two pairs (z, v) which correspond, but each (z, v) determines a unique correspondent (x, z): as we would say, the correspondence is (1, 2). This may have been Newton's primary consideration in his present calculation but we cannot be sure. (Note that when c coincides with A, cA becomes the tangent at A and v passes into k. The correspondence between v (that is, k) and e is then (1, 1) since the second tangent kA from k to the conic is fixed.)

(16) If we set $0 = as^3+brs^2+cr^2s+dr^3+es^2+frs+gr^2+hs+kr+l \equiv f(r, s)$, then Newton here determines that $f_s = f_r = 0$ and then evaluates successively $3f-sf_s$ and

$$(3f-sf_s)-rf_r = 3f-(rf_r+sf_s).$$

§2. 'PROBLEMS OF CURVES': THE MAIN DRAFTS.(1)

[*c.* 1667?](2)

From the original manuscript in private possession

[1] *Problems of Curves.*

Prob 1. To find y^e rectangular or given obliquangular diameters[,] *ordinates, either given or their angles given*[,] *Asȳptotes, vertices & centers, nodes* [*&*] *poles of curves.*(4)

1 of the
diamete
2 tange
3 curvit
4 area.(3

---

(17) s^2 is eliminated between $f_s = 0, f_r = 0$ and $3f-(rf_r+sf_s) = 0$.

(18) s, finally, is eliminated between the eliminants of $f_s = f_r = 0$ and of

$$f_r = 3f-(rf_r+sf_s) = 0,$$

leaving a cubic in r. (In a similar way, elimination of r between $f = f_r = f_s = 0$ would yield a cubic in s.) The process will not, of course, be valid when $s = 0$, but in that case s will easily be isolated as a factor of f_r or f_s. In geometrical terms Newton has, in equating the partial derivatives f_r and f_s to zero, determined that the cubic whose Cartesian defining equation is $f(r, s) = 0$ shall have a double point at (r, s): for at a double point (and nowhere else) the tangent-slope $dr/ds = -f_s/f_r$ will not be defined. The conic $3f-(rf_r+sf_s) = 0$ is (see §2, note (19) below) the polar curve of the cubic with respect to the origin, that is, the curve which meets the cubic in the contact-points of all lines drawn through (0, 0) which intersect the cubic in two coincident points. Clearly the cubic's double point will lie on this (and indeed every) polar conic. In further computation he seeks to determine the double point (r, s) of his general cubic by eliminating s between $f_r = f_s = 3f-(rf_r+sf_s) = 0$. (The co-ordinate r will in fact, be a double root of the resultant cubic.) Newton never elaborated this fundamental test for a double (or, more generally, multiple) root but, as we may see by §2 following, he quickly made subtle use of the polar curve in tangent and normal problems.

(1) The manuscript which we reproduce is, in the original, a heavily cancelled single draft but for the purposes of clarity we print it in two parts. [1] is our restoration of the earlier version, and to it we have added Newton's marginal instructions which amend it into the opening of the later version (our [2]). Despite a certain amount of inevitable repetition we think the two versions convey the significant changes in his exposition better than if we attempted to make an accurate reproduction of the several layers of his draft.

(2) Yet again we have been able to find nothing which throws light on the background of this manuscript or determines its composition date accurately. By internal evidence it lies between the October 1666 tract and the 1671 work on fluxions, while the fact that it is written in English suggests the later 1660's as the period during which it was composed. Above all, Problem 1 of [1] seems to suggest a closeness in time to the researches on the properties of curves reproduced above in **1**, 1 and 2, and we accordingly attach the same composition date to it. Much later, about 1708, when William Jones was contemplating an extensive edition of Newton's mathematical papers he made an incomplete and transcriptionally inaccurate copy of the piece, but there is no record of his showing either to anyone else. Intriguingly, Jones' copy is now preserved with Newton's other early fluxional papers in Cambridge University Library (Add. 3958.3: 75^r–76^v) while the original manuscript is in a private collection of Jones' papers but the historical reasons for this seem irretrievable.

(3) Newton's marginal notes suggesting improvements in this first draft. Subsequently the non-fluxional aspects sketched in Problem 1 were discarded, while curvature and area are outlined in the scheme which concludes the revised draft in [2].

(4) Compare I, **2**, 1: passim and **1**, 2 above.

Prob 2. To draw tangents or $\perp^{s}$(5) *to any given point* (*supposing* $VA=x$. $AB=y$.) Put its æquation $=0$. multiply by this progression (6)$0. \frac{y}{x}. \frac{2y}{x}. \frac{3y}{x}$. &c according to yᵉ dimensions of x, for a numerator [yⁿ] chang yᵉ signes & multiply by this $0. \frac{1}{y}. \frac{2}{y}. \frac{3}{y}$ &c for a denominator, wᶜʰ shall make a fraction designing AD(7) to bee taken forwards from A if affirmative, otherwise backwards.(8)

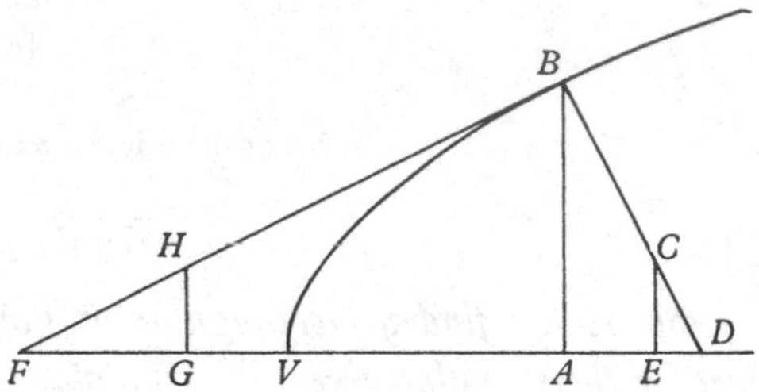

Example. if $a+bxy-xxy+y^3=0$. Then $\dfrac{byy-2xyy}{bx-xx+3yy}=AD$.

Coroll. Hence for AF multiply by yᵉ dimension of y for a numerator & by this $0. \frac{1}{x}. \frac{2}{x}. \frac{3}{x}$ &c according to x for a denomin.(9) Thus results $\dfrac{bx-xx+3yy}{-b+2x}=AF$.

Prob 3. From a point given to draw a tangent to any curve. Suppose $VG=p$. $GH=q$. & $AF=t$. & it is $y-q.\ x+p::y.\ t$. Therefore $ty-tq=xy+py$. If into this you substitute yᵉ valor of t found as before you will have the nature of a curve wᶜʰ described will cut the propounded curve in yᵉ desired tangent points.(10)

Thus if $ax-\frac{a}{b}xx=yy$ then $\dfrac{2ax-\frac{2a}{b}xx}{a-2\frac{a}{b}x}=t$. Note yᵗ the curve to bee described is never more compound then that propounded, but it may bee of an inferior sort. And of this problem there may bee often found particular solutions more

(5) Read 'perpendiculars'.

(6) Newton has cancelled the preceding terms '$\frac{-2y}{x}.\frac{-y}{x}$'.

(7) That is, the subnormal $-ydy/dx$.

(8) More generally, if the Cartesian defining equation of the curve VB be $f(x, y) = 0$, the subnormal $-ydy/dx$ is found as $(y/x)(xf_x)/(1/y)(yf_y)$. (Compare I, 2, 4, §3 and I, 2, 7: Prop. 7.) A first cancelled version of the concluding phrase reads...'to bee taken backwards from A if affirmative, otherwise from it forward'. We may presume that Newton was at first a little confused over the fact that dy/dx and f_x/f_y have opposite signs but his revised version accords with Cartesian convention in measuring positive co-ordinate lengths to the right.

(9) The subtangent $AF = -ydx/dy$ is found correspondingly as $yf_y/(1/x)(xf_x)$.

(10) This is immediate since the slope of the line joining $(-p, q)$ to (x, y) must have the slope of the curve VD ($y/t = dy/dx$) at the latter point. Evidently all such points (x, y) must be on the polar curve $(x+p)/(y-q) = t/y$, and so will be the intersections of that polar curve with VD.

(11) The degree of the polar curve is dealt with more adequately in [2] below.

(12) Newton breaks off in mid-phrase and begins the revised version which we reproduce as [2] following.

(13) 'Given two curves, to find their points of intersection.'

simple then the generall ones as in the conick sections. Nay & in all lines for tangents drawn from any point will all touch in some one & the same ordinatim applicata to the diameter passing through that point.(11)

Prob 4. from a point C, to draw a perpendicular to any curve. Suppose $VE=$(12)

---

[2] *Problems of Curves*

Prob 1. Datis duabus curvis invenire puncta intersectionis.(13) this is rather a principle then a probleme. But rather propounded of y^e^ Algebraicall then geometricall solution & y^t^ is done by eliminating one of the two unknown quantitys out of y^e^ equations. From whence it will appeare y^t^ there are soe many cut points(14) as the rectangle of the curves dimensions.(15)

Prob 2. To draw tangents or $\perp^s$ to any given point (supposing $VA=x$. $AB=y$.) Put its æquation $=0$. multiply by y^e^ dimensions of y for a numerator & by this progression $0.\ \frac{1}{x}.\ \frac{2}{x}.\ \frac{3}{x}$ &c according to x for a denominator, wch shall make a fraction designing AF to bee taken forwards from A if affirmative, otherwise backwards.

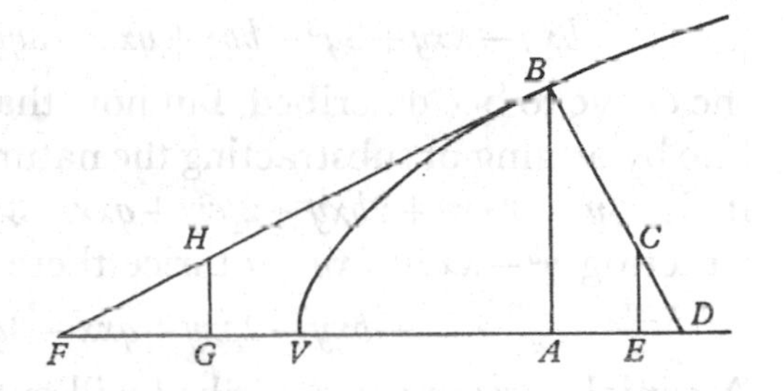

(14) 'roots' is cancelled.

(15) The general proof of this fundamental theorem of algebraic geometry is not so easy as Newton would seem to suggest, nor indeed is it evident analytically that elimination of one of the variables between equations of degrees m and n in the same two variables yields a resultant equation of degree mn in the second. Perhaps the first satisfactory proof was that outlined in 1748 by Leonhard Euler in a paper given to the Berlin Academy ('Démonstration sur le nombre des points ou deux lignes des ordres quelconques peuvent se coupter', *Mémoires de l'Académie des Sciences de Berlin*, **4** (1748): 234–48 = *Opera Omnia* (1), **16**, Lausanne, 1953: 46–59). In his opening paragraph Euler stated the problem fairly: 'La vérité de cette proposition est reconnue de tous les Geometres, quoiqu'on doive avouer, qu'on n'en trouve nulle part une démonstration assés rigoureuse. Il y a des vérités générales que notre esprit est prêt d'embrasser aussitôt qui'l en reconnoit la justesse dans quelques cas particuliers: et c'est parmi cette espece de vérités qu'on peut ranger à bon droit la proposition,...puisqu'on la trouve vraie non seulement dans quelques ou plusieurs cas, mais aussi dans une infinité de cas différens. Cependant on conviendra aisément que toutes ces preuves infinies ne sont pas capable de mettre cette proposition à l'abri de toutes les objections qu'un adversaire peut former, et qu'il faut absolument une démonstration rigoureuse, pour le réduire au silence'. Newton restricted himself, as here, always to the assertion of the general theorem on the analogy of its proven truth for linear, quadratic and cubic cases. (Compare pp. 20–2 of his *Observationes* on Kinckhuysen's *Algebra ofte Stel-konst* (**3**, 1, §2 below): 'Exterminatio quantitatis incognitæ cum plurium in utráq; Æquatione dimensionum existit' = *Arithmetica Universalis* (Cambridge, 1707): 72–5.)

Example. If $a+bxy-xxy+y^3=0$. Then $\dfrac{bx-xx+3yy}{-b+2x}=AF$.

Coroll. Hence for AD multiply by this progression $0.\frac{y}{x}.\frac{2y}{x}.\frac{3y}{x}$ &c according to y^e^ dimensions of x for a numerator [y^n^] chang y^e^ signes & multiply by this $0.\frac{1}{y}.\frac{2}{y}.\frac{3}{y}$ &c for a denominator. Thus results $\dfrac{byy-2xyy}{bx-xx+3yy}=AD$.[16]

Prob 3. From a point given H to draw a tangent HB to any curve. Suppose $VG=p$. $GH=q$. & $AF=t$. & it is $y-q.x+p::y.t$. Therefore $\dfrac{py+xy}{y-q}=t$ or $ty-tq=xy+py$. If into this you substitute y^e^ valor of t found as before you will have the nature of a line wch described will cut the propounded curve in y^e^ desired tangent points.

Example. thus substituting $\dfrac{bx-xx+3yy}{-b+2x}$ for t. there comes

$$bxy-xxy+3y^3-bqx+qxx-3qyy=^{(17)}+2xxy+2pxy-bxy-bpy$$

the curve to bee described. But note that this may bee ever reduced to a simpler line by adding or substracting the nature of y^e^ given line viz: ordering this result it is $3y^3-3xxy+2bxy-2pxy+qxx-3qyy-bqx+bpy=0$. from whence substracting $y^3-xxy+bxy+a$ thrice there rests

$$-bxy-2pxy+qxx-3qyy-bqx+bpy-3a=0.$$

A conick section wch described will cut the curve in y^e^ desired tangent points. Thus if $ax+\frac{b}{a}xx-yy=0$. then $\dfrac{2yy}{a+\frac{2b}{a}x}=t=\dfrac{py+xy}{y-q}$ or

$$2yy-2qy=\frac{2b}{a}xx+ax+\frac{2b}{a}px+ap.$$

---

(16) Compare notes (9) and (8) above.

(17) '$\overline{-b+2x}$ in $\overline{xy+py}$' is cancelled.

(18) That is, in terms of conic geometry, the polar of the point $(-p, q)$ with respect to the conic.

(19) The polar curve of an algebraic curve with respect to the given point $(-p, q)$ is indeed of degree one less than the latter. If we take the defining equation of the curve VD as

$$0=f(x,y)\equiv\sum_{0\leqslant i\leqslant n}(C_i),\quad\text{where}\quad C_i=\sum_{0\leqslant j\leqslant i}(a_{ij}x^jy^{i-j}),$$

then we deduce that

$$\begin{aligned}(x+p)f_x+(y-q)f_y&=\sum_{0\leqslant i\leqslant n}\Big[\sum_{0\leqslant j\leqslant i}\{(j(x+p)/x+(i-j)(y-q)/y)a_{ij}x^jy^{i-j}\}\Big]\\&=\sum_{0\leqslant i\leqslant n}\Big[iC_i+\sum_{0\leqslant j\leqslant i}\{(jpy-(i-j)qx)a_{ij}x^{j-1}y^{i-j-1}\}\Big]\\&=n\sum_{0\leqslant i\leqslant n}(C_i)+\sum_{0\leqslant i\leqslant n-1}(C_i'),\end{aligned}$$

& substracting $ax+\frac{bxx}{a}-yy=0$ twice (viz so oft as y^e curve hath dimensions,) there comes $\frac{2bpx}{a}-ax+2qy+ap=0$ a streight line[(18)] wch drawn will cut the conick in the points desired. And so of the rest where note that this problem in any curve is ever soluble by a line of an inferior degre, & also y^t a line drawn from a given point may touch a curve of 2 dimensions in 1×2 points, of 3 in 2×3 points, of 4 in 3×4 points &c.[(19)] Unlesse i[t] be a polar &c.[(20)]

Prob 3[']. *From a point given C to draw a perpendicular to any curve.* Make $VE=r$, $EC=s$ & $AD=v$. soe is $y-s.\ r-x::y.\ v.$

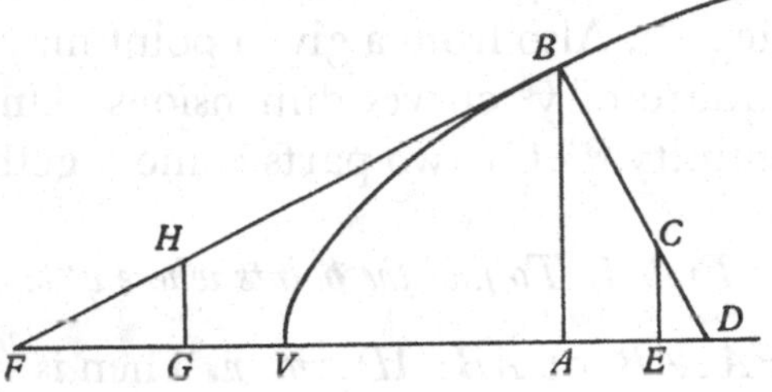

Or $\frac{ry-xy}{y-s}=v$.[(21)] Put this valor of $v=v$[(22)] found as before, & you have a curve wch described will cut the propounded curve in y^e points to wch y^e perpendiculars must bee drawn. But ever try if by meanes of the nature of y^e given curve you can reduce this resulting [æquation] to any simpler forme or degree.[(23)]

Example. if $ax+\frac{b}{a}xx-yy=0$[(24)] then $\frac{1}{2}a+\frac{b}{a}x=v$. Or

$$ry-xy=\frac{1}{2}ay-\frac{1}{2}as+\frac{b}{a}xy-\frac{b}{a}sx.$$

Which appeares not reducible to a simpler forme.[(25)] Yet it may be worth the

---

so that if VD is of degree n, then its polar curves will never be of degree more than $n-1$. Since therefore the polar curve with respect to any given point $(-p, q)$ will meet the parent curve VD (of degree n) in $(n-1)n$ points by Problem 1, from that given point $(n-1)n$ tangents (not all of which need be real) can be drawn to VD and these will meet in the intersections of VD with that polar curve.

(20) That is, when the curve has multiple points. In that case two (or more) coincident points do not determine a tangent direction as the indefinitely small chord which joins them vanishes.

(21) Newton repeats Descartes' notation for the subnormal (*Geometrie*: Book 2).

(22) That is, '*AD*' (in the corollary to Problem 2).

(23) This will not in general be possible, so that the 'normal' curve $(x-r)y+(y-s)v$ (in which $v = y(dy/dx)$) will be of the same degree as the parent curve VD.

(24) A general conic (and a hyperbola when a and b are both positive).

(25) However, when $b=-a$ this polar hyperbola reduces to a straight line: in this case the conic is a circle of centre $(\frac{1}{2}a, 0)$ through the origin and the polar line

$$sx+(\tfrac{1}{2}a-r)y-\tfrac{1}{2}as = 0$$

will pass through that centre and the given point (r, s). In consequence, the two real normals which may be drawn from (r, s) to the circle will coincide with the 'normal' line. For the geometrical discussion of the general conic case see Apollonius, *Conics*, v, 55–63.

while to try if it may bee solved by a circle by assuming $d+ex+fy=xx+yy$(26) & comparing these 3 æquations of y^e given conick, found Hyperbola(27) & assumed circle to find d, e, & f wch if they may bee found by plaine Geometry that circle described will cut the curve in the desired points. This I say might be tryed but it would bee found impossible (unless y^e conick bee a parabola)(28) because there are foure given points through wch ye circle must passe wch make ye prob[lem] contradictious since 3 are enough to determine a circle. This might be done by the Parabola(29) but ye Hyperbola falling in so naturally & being as easily described(30) tis not worth ye while. What is said of ye conicks(31) may be applyd to any other curves. From hence it appeares that this problem is ever soluble by a curve of ye same degree & sometimes perhaps(32) by a curve of an inferiour degree. Also from a given point may be drawne soe many perpendiculars as y^e square of y^e curves dimensions; Unlesse some part extraordinary run out to infinity.(32) Or two parts come together to make it polar.(33)

Prob 4. To find the points where y^e curve hath a given inclination to the basis. Supose $FA.AB$, or $AB.AD::m.n$. Then is $\frac{my}{n}=t$. Put this equall to y^e valor of t found as before & you have a curve wch described will cut the given curve in the desired points. But (if you can) reduce it. thus if $ax+\frac{b}{a}xx-yy=0$. then $\frac{2yy}{a+\frac{2b}{a}x}=t=\frac{my}{n}$. Or $2ny=am+\frac{2mb}{a}x$. The like in other cases. Hence conicks have 1×2 points, those of 3 dimensions(34) 2×3 points[,] of 4 3×4 &c wch satisfy this problem & it is ever soluble by a line of an inferior degree.(35)

(26) The most general defining equation of a circle (in perpendicular Cartesian co-ordinates).

(27) That is, the 'normal' curve $xy(1+[b/a])-[b/a]sx+(\frac{1}{2}a-r)y-\frac{1}{2}as = 0$.

(28) In this case $b = 0$ and one of the normals will lie parallel to the axis meeting the parabola in its infinite point. Since the circle $x^2+y^2-(\frac{1}{2}a+r)x-sy = 0$ may be drawn through the three finite meets of the parabola $y^2-ax = 0$ with its polar hyperbola

$$xy+(\tfrac{1}{2}a-r)y-\tfrac{1}{2}as = 0,$$

it will intersect the parabola in the origin and also in the three finite points to which normals to the parabola may be drawn from the given point (r, s). This construction was first published by Edmond Halley in his *Sereni...De Sectione Cylindri et Coni Libri Duo* (appended to his *Apollonii Pergæi Conicorum Libri Octo* (Oxford, 1710)): Prop. XXXIX, Scholion: 69–70, where he remarks that its justification follows 'non multo opere...ex iis...quæ in *Philosoph. Transact.* Num. 188 [for July/August 1687] tradimus' [: 335–43: 'De Constructione Problematum Solidorum, sive Æquationum tertiæ vel quartæ Potestatis, unica data Parabola ac Circulo efficienda; dissertiatuncula']. See also I, 2, 4, §2.3.

(29) By giving suitable values to the coefficients the parabola $(x+py)^2+qx+ry+s = 0$ may be made to pass through the four meets of the given conic and its polar hyperbola.

Prob 5. From a given point C to draw a line CB wch shall cut any given curve in a given angle. Supose(36) BD y^e perpendicular to the desired point B & from y^e point D raise y^e perp DH & let fall $HF \perp AE$, producing BC & AE to G. & suppose $AD=v$, $VE=r$, $EC=s$, & (y^e angle CBD being given) suppose $BD.DH::m.n.$(37) Now the triangles BAD & DFH being alike(38) tis $m.n::AB.DF::AD.FH$. Therefore $DF=\frac{ny}{m}$ & $FH=\frac{nv}{m}$. Also tis

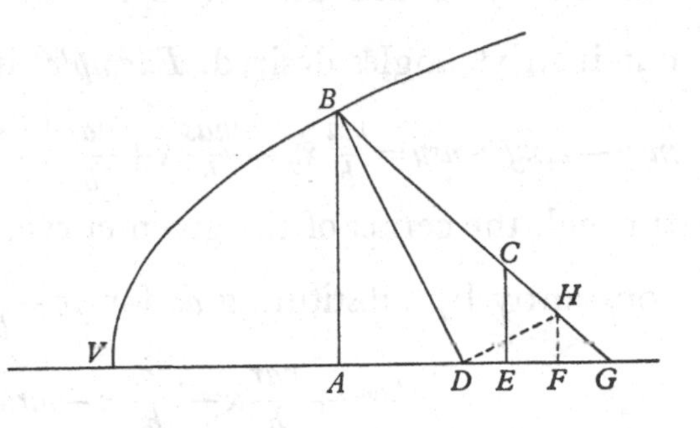

$AB-CE.AE::CE.EG::FH.FG$. Therefore $\frac{rs-sx}{y-s}=EG$ & $\frac{rnv-nvx}{my-ms}=FG$. Lastly since $VE+EG=VG=VA+AD+DF+FG$ there results

$$r+\frac{rs-sx}{y-s}=x+v+\frac{ny}{m}+\frac{rnv-nvx}{my-ms}.$$

Which reduced is $\frac{mry-mxy+nsy-nyy}{my-ms+nr-nx}=v$. this I put æquall to the valor of v

put $mr+ns=$ $nr-ms=$ shorten termes.(

(30) Compare I, 1, 1, §1.5.

(31) The problem of drawing normals from a given point to a central conic was given a general solution by Philippe de la Hire in his *Nouveaux Elemens des Sections Coniques, Les Lieux Geometriques, La Construction ou Affection des Equations* (Paris, 1679: 440–52: *Probleme*): as he said in his introduction of it (p. 440) 'ce probleme est fort facile dans la parabole'. (Compare note (28) above.)

(32) That is, when (compare note (28) in the case of the conical parabola) one or more of the normals from the given point meet the curve at infinity. A curve of lower degree than that of the 'normal' curve may then be drawn to intersect the curve in the finite meets of the normals from the given point with it.

(33) Compare note (20) above. In general, when the parent curve is of nth degree, so also will be the 'normal' curve (note (23)) and their intersection in n^2 points will determine a corresponding number of normals.

(34) 'powers' is cancelled.

(35) If VD has the defining Cartesian equation $f(x, y) = 0$, then by the definition of the intersecting curve

$$\frac{n}{m}=\frac{y}{t}=\frac{dy}{dx}=-\frac{f_x}{f_y},$$

so that its equation is $mf_x+nf_y = 0$: evidently it will be of degree one less than that of VD and so will intersect it in $(n-1)n$ points (where VD is of nth degree).

(36) 'y^e complement of' is cancelled.

(37) In restatement of the problem Newton seeks to draw a line from the given point (r, s) to the curve VB which meets it at the given angle $\tan^{-1} m/n$: as before he takes the subnormal $AD(ydy/dx)$ as v, and assumes $VA = x$ is perpendicular to $AB = y$.

(38) For $B\hat{A}D = D\hat{F}H = B\hat{D}A+F\hat{D}H = \frac{1}{2}\pi$.

(39) Thus, v will now equal $(mx-ny-k)y/(nx-my-l)$.

(found as before)(40) & there results a curve wch described cuts y^e give[n] curve in y^e points desired to wch lines are to bee drawne from C that may cut it in y^e angles desired. *Example.* if $dd+\frac{a}{b}xx-yy=0$ Then is $\frac{a}{b}x=v$. & soe $mry-mxy+nsy=\frac{ma}{b}xy-\frac{mas}{b}x+\frac{nar}{b}x-\frac{na}{b}xx$. W^ch is a conick section passing through the center of the given curve.(41) And it may bee made an Hyperbola constantly by substituting dd for $yy-\frac{a}{b}xx$. For the result will bee

$$dd^{(42)}+\frac{nar}{b}x-\frac{mas}{b}x-mry-nsy+mxy-\frac{ma}{b}xy^{(42)}=0.$$

Hence it appeares that this problem may be ever solved by a curve of the same sort & sometimes perhaps by one of an inferior. And also that from a given point to any curve soe many lines may bee drawne in a given angle as y^e square of its dimensions abating those that are imaginary or coincident to ye pole,(43) or should bee drawn where some part of ye curve vanisheth in infinitum.(44)

Much after the same manner will these be so.(45)

Of ye curvity of Lines.

1. To find their curvity at any given point.(46)
2. To find their points where they have a given curvity.(47)
3. To find their points where they are lesse curve then any circle.(48)
4. To find their points where they are more curve then any circle.(49)
5. To find their points where they are most or least curve.(50)
6. To find the moment of(51) transition from one degree of curvity to another at any point.

(40) By the Corollary to Problem 2.

(41) This is, in fact, a central conic with its centre at the origin $V(0, 0)$, a point clearly on the derived curve since its defining equation lacks a constant term. The accompanying diagram, which makes the curve (B) pass through the origin is a little confusing.

(42) Read 'ndd' and '$+\frac{ma}{b}xy$' respectively.

(43) Compare note (20) above.

(44) Compare note (28). Since $v/y=-f_x/f_y$ where the defining equation of the curve (B) is taken as $f(x, y)=0$, that of the derived curve is $(mx+ny-k)f_x+(nx-my-l)f_y=0$ and it will in general, therefore, be of the same algebraic degree as the parent curve. Accordingly if the parent and derived curves each be of nth degree, each of the n^2 points in which they intersect will yield a solution to the problem.

(45) These appended enunciations of problems evidently elaborate Problems 2–17 of the October 1666 tract (I, 2, 7). In part, as we shall see in our next volume, they are discussed in the later sections of the 1671 tract on analysis (usually known under Jones' title of the *Geometria Analytica*).

(46) Compare I, 2, 7: Problem 2.

7. To find their points where they have a given moment of transition.
8. To find where their transition is infinitely great or none at all.
9. To find where it is most or least.
10. To find ye curve which hath an uniforme transition.(52)

Of ye Areas of lines.(53)

1. To find the line the nature of whose area is given.
2. To find lines at pleasure whose areas are equall or otherwise related to the area of a given line.
3. To find the area of a given line.
4. To examin the ratio of the areas of two propounded lines.

[5.] To find lines whose areas have a give[n] summ rectangle or other relation.

Of the len[g]th of lines.(54)

1. To find lines at pleasure whose lengths may bee compared to the length of a streight line.

[2.] To find lines whose lengths may bee compared to the length of any given line.
[3.] To find the lengths of propounded lines.
[4.] To examin the relation twixt the lengths of two propounded lines.
[5.] To find lines the nature of whose lengths are given.
[6.] To find two lines which have a given summ, rectangle, or other relation expressed by any equation.
[7.] To compare ye area of one line wth ye length of another drawn into an unit.

Of ye axes of gravity.(55)

[1.] To find lines ye nature of whose gravity is given.
[2.] To find lines which æquiponderate others about a given axis or whose weights have any given relation to the weights of others.
[3.] To find ye axes to wch their weights are most simply related.
[4.] To find their weight at any given axis.

(47) Compare I, 2, 7: Problem [2 *bis*].
(48) That is, at an inflexion point: compare I, 2, 7: Problem 3.
(49) That is, at a cusp. Compare I, 2, 7: Problem 2 when the radius of curvature vanishes.
(50) Compare I, 2, 7: Problem 4.
(51) 'proper' was first written for 'moment of'.
(52) These problems (6–10) relating to the 'transition from one degree of curvity to another at any point' are discussed in some detail in Problem 9 of the 1671 tract (note (45)), which deals with the more or less 'inequable' variation in the 'quality' of curvature at a point.
(53) Problems 1, 2 and 3–5 correspond with Problems 5, 6 and 7 of I, 2, 7.
(54) Compare I, 2, 7: Problems 9, 10, 12, 13 and 11.
(55) Compare I, 2, 7: Problems [14_2], [14_1] and 15.

Of centers of gravity.(56)

[1.] To find lines whose centers of gravity have a given nature.
[2.] To find their centers of gravity.
[3.] To find y^e^ axes & centers of gravity of lines as well as superficies.

Of the tangents, curvity, content, & areas of sollids:
Of their axes & centers of gravity.
And
Of their motions about those axes.(57)

---

§3. FURTHER LOGARITHMIC CALCULATIONS.

[*c.* 1667?]

From the original in Newton's Waste Book in the University Library, Cambridge(1)

A Method Whereby to Square lines Mechanichally.

Prop 1.(2) Supposing $ab=x\perp bc=y$. If ye valor of y (in ye Equation expressing ye relation twixt x & y) consist of simple termes, Multiply each terme by x, & divide it by ye number of ye dimensions of x in that terme, & ye quote shall signify ye area acb.(3)

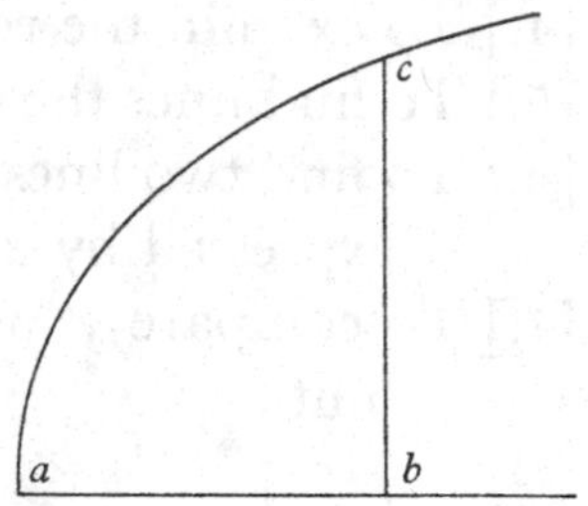

Example. If $ax=yy$. Or $\sqrt{ax}=y$.(4) yn is $\frac{2}{3}\sqrt{ax^3}=\frac{2}{3}x\sqrt{ax}=acb$. Soe if $\frac{xx}{a}=y$. yn is $\frac{x^3}{3a}=acb$. Soe if $a+x+\frac{abb}{xx}+\frac{x^3}{aa}=y$. yn is $ax+\frac{xx}{2}-\frac{abb}{x}+\frac{x^4}{4aa}=acb$. Soe if $\frac{a^4}{x^3}+\sqrt{\frac{a^5}{x^3}}-\sqrt{\frac{x^3}{a}}+\frac{bx^6}{a^6}=y$. yn is $\frac{-a^4}{2xx}-2\sqrt{\frac{a^5}{x}}-\frac{2}{5}\sqrt{\frac{x^5}{a}}+\frac{7bx^7}{a^6}$(5) $=acb$.

(56) Compare I, 2, 7: Problems 16 and 17.

(57) Newton here breaks off. As we have said he will elaborate this outline scheme a few years later in his 1671 fluxional tract (but without incorporating the general properties of tangents and normals in Problems 3–5).

(1) Add. 4004: 80r–81v. It will be clear that this scheme of computation, entered on either side of two folio sheets in the Waste Book and so, by implication, as a finished draft, is the revision and elaboration of I, 1, 3, §5. (Compare note (1) of that section.) We there suggested that the present paper was written in 1676 when Newton came to draft his *epistola posterior* (of 24 October 1676) to Leibniz. Certainly, portions of that letter (see note (12) below) summarize it but we are now persuaded by the closeness of its opening propositions in style to corresponding propositions in the October 1666 tract on fluxions to set the present revised date as its probable year of composition. (Indeed, Newton may at one time have intended to append these logarithmic calculations to that work for he there spoke of evaluating hyperbola-areas 'by a Table of logarithmes as may hereafter appeare'. Compare I, 2, 7: note (15).)

Prop. 2.(2) If any terme in y^e^ valor of y bee a compound ter[me] Reduce it to simple ones by Division or Extraction of Rootes or by Vieta's Method of Resolving Affected Equations, as you would doe in Decimall Numbers,(6) & y^n^ find y^e^ Area by Prop 1st.

Example. If $\frac{aa}{b+x}=y$, bee divided as in decimall fractions it produce[th] $\frac{aa}{b+x}=\frac{aa}{b}-\frac{aax}{bb}+\frac{aaxx}{b^3}-\frac{aax^3}{b^4}+\frac{aax^4}{b^5}-\frac{aax^5}{b^6}+\frac{aax^6}{b^7}-\frac{aax^7}{b^8}$ &c. & by y^e^ 1st Proposition [tis] $\frac{aax}{b}-\frac{aaxx}{2bb}+\frac{aax^3}{3b^3}-\frac{aax^4}{4b^4}+\frac{aax^5}{5b^5}-\frac{aax^6}{6b^6}+\frac{aax^7}{7b^7}-\frac{aax^8}{8b^8}+\frac{aax^9}{9b^9}$ &c$=abc$, y^e^ Hyperbola's Area.

As if $a=1=b=ab=bc$. & $x=0{,}1=be$. The Calculation is as followeth,

[--- --- --- --- ---](7)

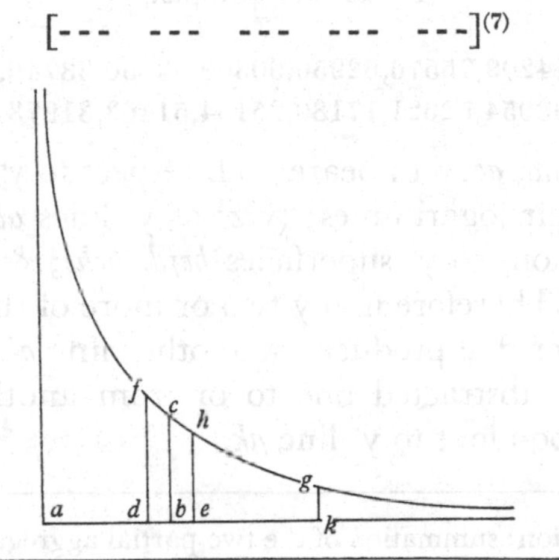

(2) Newton first wrote 'Lemma [1]' and 'Lemma 2^d^' here respectively.

(3) Compare the first part of Proposition 8 in the October 1666 tract (I, 2, 7).

(4) A parabola, of course, as illustrated in the accompanying figure (which repeats the second diagram of I, 2, 7).

(5) Newton forgetfully 'integrates' x^6 as $7x^7$.

(6) Compare I, 2, 7: note (41). Newton, evidently, has not yet attained the simplification of Viète's method which he will expand in his *De Analysi* (§3 below).

(7) These calculations, which repeat the corrected version of those in I, 1, 3, §5: 135–8, contain nothing new and are omitted. In essence where,

$$\alpha(m,n) = \sum_{0\leqslant i\leqslant n}\left(\frac{1}{2i+1}m^{2i+1}\right) \quad\text{and}\quad \beta(m,n) = \sum_{1\leqslant i\leqslant n}\left(\frac{1}{2i}m^{2i}\right),$$

Newton presents in triangular array the computation of

$$\alpha(0{\cdot}1, 26) = 0{\cdot}10033{,}53477{,}31075{,}58063{,}57265{,}52060{,}03894{,}52633{,}62869{,}14595{,}91358{,}63.$$

and

$$\beta(0{\cdot}1, 27) = 0{\cdot}00502{,}51679{,}26750{,}72059{,}17744{,}28779{,}27385{,}30427{,}57503{,}83731{,}49363{,}62.$$

As he would surely realize, the final places of this calculation to $57D$ are not accurate since they depend on how successfully each row rounds off the decimal expansion of its corresponding fraction. Roughly, one would expect the error to be about one-half the number of rows

The summe of these two summes[8] is equall to y[e] area *dbfc*, supposing $ad = 0,9$. And their difference is equall to y[e] area *bche*, supposing $ae = 1,1$. &

That is[9] $ab = 1 = bc \parallel df \parallel he$.

dbfc = [−]0,10536,05156,57826,30122,75009,80839,31279,83061,20372,98327,40725,43.
bche = 0,09531,01798,04324,86004,39521,23280,76509,22206,05365,30864,41991,83.

In like manner If $a = b = 1 = ab = bc$. & $x = 0,2 = be$. The calculation is as followeth

[--- --- --- --- ---][10]

The summe of these two summes is equall to y[e] Area *dbfc*, supposing $ad = 0,8$. And their Difference is equall to y[e] area *bche*, supposing $ae = 1,2$. &

That is[11] $1 = ab = bc \parallel df \parallel he$.

dbfc = −0,22314,35513,14209,75576,62950,90309,83450,33746,01085,54800,72136,12.
bche = 0,18232,15567,93954,62621,17180,25154,51463,31973,89337,91448,69839,22.

Now since the lines *ad*, *ae*, &c: beare such respect to y[e] superficies *bcfd*, *bche*, &c: as numbers to their logarithmes; (viz: as y[e] lines *ad*, *ae*, &c: increase in Geometricall Progression, so y[e] superficies *bcfd*, *bche*; &c: increase in Arithmeticall Progression): Therefore if any two or more of those lines multiplying or dividing one another doe produce some other line *ak*, their correspondent superficies, added or substracted one to or from another shall produce y[e] superficies *bcgk* correspondent to y[t] line *ak*.

---

in the triangular computation: summation of the two partial aggregates would approximately double the error while it would be somewhat reduced in their difference. This is what we find, on average, though some slight arithmetical errors in subsequent computations tend to mask the point.

(8) That is $\alpha(0{\cdot}1, 26)$ and $\beta(0{\cdot}1, 27)$ in the terminology of the previous note.

(9) We have checked this and the subsequent computations against the logarithmic expansions given by W. E. Mansell in his *Tables of Natural and Common Logarithms to 110 Decimals* (Cambridge, 1964). Here we find that the last two places of $(dbfc) = \log(0{\cdot}9)$ should be '64' and the last of $(bche) = \log(1{\cdot}1)$ '5' (with respective errors of 21 and 2).

(10) We omitted this computation in I, 1, 3, §5 (there intolerably vitiated) but we think it here enough to reproduce a facsimile of its correct form in the accompanying Plate II. (Note the corrected row entries for $(1/23)(0{\cdot}2)^{23}$ and $(1/36)(0{\cdot}2)^{36}$ and that the first row in the second array is $\beta(0{\cdot}2, 3)$. It was not, of course, foresight which made Barnabas Smith in 1612 head the next page (81[r]) with the title 'Relapsus'!) Essentially, Newton computes the two partial sums

$$\alpha(0{\cdot}2, 37) = 0{\cdot}20273{,}25540{,}54082{,}19098{,}90065{,}57732{,}17456{,}82859{,}95211{,}73124{,}70987{,}67$$

and

$$\beta(0{\cdot}2, 38) = 0{\cdot}02041{,}09972{,}60127{,}56477{,}72885{,}32577{,}65993{,}50886{,}05873{,}81676{,}01148{,}45.$$

(11) Correctly (note (9)) the last two figures of $(dbfc) = \log(0{\cdot}8)$ should be '71' and those of $(bche) = \log(1{\cdot}2)$ '43' (with respective errors of 59 and 21).

is as followeth

900655773,2174,56,82859,95211,73124,7098767 = summe

0,2027325540,54082,19098,87717775123,47891,61120,82168,25248,~~02~~01
= 32000000000,00000,00000,00000,00000,00000,00000,00000,00000,00000,000000,00
aa^3 = 266,6666666666,66666,66666,66666,66666,66666,66666,66666,66666,666666,66
= 6400000000,00000,00000,00000,00000,00000,00000,00000,00000,00000,000000,00
= 1828571428,57142,85714,28571,42857,14285,71428,57142,85714,28571,428571,28
= 508888888,88888,88888,88888,88888,88888,88888,88888,88888,88888,888888,88
= 130181818,18181,81818,18181,81818,18181,81818,18181,81818,18181,818181,81
630153846,15384,61538,46153,84615,38461,53846,15384,61538,46153,8461,615
21845333,33333,33333,33333,33333,33333,33333,33333,33333,33333,333333
771011,70117,64705,88235,29411,76470,58823,52941,17647,05
27594,10526,31578,94736,84210,52631,57894,73684,210
99864,38095,23809,52380,95238,09523,80952,38095,23
+ 30409,56086,95652,17391,30434,78260,86956,52173
13,42177,28000,00000,00000,00000,00000,00000,00000,00
49710,26962,96296,29629,62962,96296,29629,62
1851,27900,68965,51724,13793,10344,82758,62068,96551
69,27360,60645,16129,03225,80645,16129,03
2,00301,04824,24242,42424,24242,42424,24
98170,08105,14285,71428,57142,85714,2
3771,45603,10054,05405,40540,54054,05
14090,30292,02051,28205,12820,5128
53634,71355,00487,80487,80487,80
2045,60302,84204,65116,27906,97,0744
78,18749,35307,27777,77777,77
2,99441,46458,58042,55319,1489
11488,77455,90180,12244,897959
441,52937,52324,01568,62745
16,99471,55749,83003,77358
655,00903,07084,3578
25,28330,13290,9752
97105,21225,4817
3,78007,05009,07
14640,2730743
507,5921253
22025,96307
8550,117
3325,609
129,37
5,03

+ 364,72226,86956,52173,91304,34782,60869,56521,739

2,02041,09972,60,127,50177,72875,32577,05993,50880,05873,81076,01148,45 = summe
0,02041,09972,60,127,50177,72875,32577,05993,50880,05873,81076,01148,45 = summe $\frac{aa^3}{6} + \frac{aa^6}{66}$
0,02041,00000,66,66666,66666,66666,66666,66666,66666,66666,66666,66666,60
$\frac{aa^8}{8b^8}$ = 320000,00000,00000,00000,00000,00000,00000,00000,00000,00
$\frac{aa^{10}}{10b^{10}}$ = 10240,00000,00000,00000,00000,00000,00000,00000,00000,00
$\frac{aa^{11}}{11b^{11}}$ = 341,33333,33333,33333,33333,33333,33333,33333,33333,33
$\frac{aa^{14}}{14b^{14}}$ = 11702,85714,28571,42857,14285,71428,57142,85714,2
$\frac{aa^{16}}{16b^{16}}$ = 409,60000,00000,00000,00000,00000,00000,00000,00000,0
14,56355,55555,55555,55555,55555,55555,55555,55555,55
524288,00000,00000,00000,00000,00000,00000,00000,00
19065,01818,18181,81818,18181,81818,18181,81818,1
69,90506,66666,66666,66666,66666,66666,66666
2,58111,01538,46153,84615,38461,53846,15384,61
95869,80571,42857,14285,71428,57142,857142
3579,13941,33333,33333,33333,33333,33333,33
134,21772,80000,00000,00000,00000,00000,00
5052,90270,11764,70588,23529,41176,4,70
~~274877906944000000000000000000000~~
190,28743,57777,77777,77777,77777,77777,77
72,33629,13010,52631,57894,73684,2105263
2,74877,90694,40000,00000,00000,00
10471,53931,21523,80952,38095,23
399,82241,01003,63636,36363,63
15,29755,30821,00869,56521,73913,0434
58640,62014,80533,33333,33
2251,79981,36852,48000,00
86,60768,51417,40307,692
3,33599,97239,78145,180
12867,42750,67728,45
496,94892,43995,03
19,21535,84101,14
74382,03255,28
2882,30376,151
111,79844,893
4,34041,03
16865,59
655,88
25,52
99

The summe of these two summes is equall to ye area dbfc, supposing ad = 0,8. And their Difference is equall to ye area ockε, supposing aε = 1,2. + 1 = ab = bc [illegible].

That is

~~dbfc = 0,22314,35513,14209,75576,66060,30770,17388,51206,88042,66974,63440,46.~~

~~bckε = 0,18232,15567,93954,62621,14622,42549,81898,10234,76294,41302,61143,06.~~

dbfc = 0,22314,35513,14209,75576,62950,90309,83450,33746,01085,54800,72136,12

bckε = 0,18232,15567,93954,62621,17180,25154,51463,31973,89337,91448,69839,22

[illegible] since the lines ad, aε, [illegible] beare such respect to ye superficies bcfd, bckε [illegible] as numbers to theire Logarithmes; (viz as ye lines [illegible] increase in [illegible] Geometricall progression, ye superficies bcfd, bckε [illegible] increase in Arithmeticall progression) Therefore [illegible] ~~[illegible]~~ or [illegible] [illegible] ye [illegible] another [illegible] other [illegible] lines corresponding [illegible] added or substracted [illegible] shall produce ye superficies [illegible] corresponding to [illegible]

Soe y[t] since $\frac{1,2\times 1,2}{0,8\times 0,9}=2.\ \frac{1,2\times 1,2\times 1,2}{0,8\times 0,8\times 0,9}=3=\frac{1,2\times 2}{0,8}.$

$$\frac{1,2\times 1,2\times 1,2\times 1,2}{0,8\times 0,8\times 0,8\times 0,9\times 0,9}=5=\frac{2\times 2}{0,8}=\frac{4}{0,8}.$$

$2\times 5=10.\ 10\times 10=100.\ 10\times 100=1000$ &c. $10\times 1,1=11.$ &c.[(12)] The superficies answering to these lines 2. 3. 5. 11. & their products (of one of y[m] multiplying another) may be found. Viz: if $ab=1=bc\perp ab$, &

If y[e] line *ak* is	Then y[e] superficies *bcgk* is[(13)]
2.	0,69314,71805,59945,30941,72321,21458,17656,80755,00134,36025,52539,99.
3.	1,09861,22886,68109,69139,52452,36922,52570,46474,90557,82274,94515,33.
10.	2,30258,50929,94045,68401,79914,54684,36420,76011,01488,62877,29756,09.
100.	4,60517,01859,88091,36803,59829,09368,72841,52022,02977,25754,59512,18.
1000.	6,90775,52789,82137,05205,39743,64053,09262,28033,04465,88631,89268,27.
10000.	9,21034,03719,76182,73607,19658,18737,45683,04044,05954,51509,19024,36.
&c	
11.	2,39789,52727,98370,54406,19435,77965,12929,98217,06853,93741,71747,92.
[&c]	

Having alrcady found y[e] arcas correspondent to y[e] lines 1,1. 0,9. 1,2. 0,8. tis easy by y[e] helpe of these operations to find y[e] areas correspondent to y[e] lines 1,01. 1,001. 1,0001. &c: 0,99. 0,999. 0,9999. &c: 1,02. 1,002. &c: 0,98. 0.998. 0,9998. &c. And since $7=\sqrt{\frac{100\times 0,98}{2}}.\ 17=\frac{100\times 1,02}{6}.$

$$13=\frac{1000\times 1,001}{7\times 11}=\frac{1001}{77}.\quad \&c.^{(14)}$$

Therefore y[e] areas correspondent to y[e] lines 7. 13. 17. &c: are easily found, as followeth. Viz: if $x=0,02$. Then

[--- --- --- --- ---][(15)]

(12) In his *epistola posterior* of 24 October 1676 (*Correspondence of Isaac Newton*, **2** (Cambridge, 1960): 124) Newton employed the revised factorizations $2^{10}=9984\times 1020/9945$, $3^4=8\times 9963/984$ and, assuming log (10) known, $5=10/2$ with $11=99/9$. However, as we shall see in our third volume, in his 1671 fluxional tract he repeated the present reductions.

(13) Correctly (note (9)), the last four figures of log (2) should be '41,21', the last three of log (3) '7,35', the last four of log (10) '60,33' and the last four of log (11) '52,19'. On average, the error in any compounded logarithm will be roughly the sum of the errors for each of its factors.

(14) These factorizations are repeated, in essence, in both the 1671 tract and the *epistola posterior* (note (12)).

(15) This computation we omit since it reproduces the corrected triangular calculation arrays from I, 1, 3, §5: 140–2. In essence Newton computes the two partial sums

$$\alpha(0{\cdot}02,\ 15) = 0{\cdot}02000,22673,06849,58071,70371,83954,64639,04807,62055,62238,59310,49$$

and

$$\beta(0{\cdot}02,\ 16) = 0{\cdot}00020,00400,10669,86769,10081,17069,54599,73717,71328,11118,23899,70.$$

The sum̄e & difference of wch two summes give y^e areas *bcfd*, *bche* as before. That is[16]

If $ad=0,98.$ y^n $bcfd=-0,02020,27073,17519,44840,80453,01024,19238,78525,33383,73356,83210,19.$
$ae=1,02.$ $bche=+0,01980,26272,96179,71302,60290,66885,10039,31089,90727,51120,35410,79.$

And since $7=\sqrt{\dfrac{100\times 0,98}{2}}=\sqrt{\dfrac{98}{2}}=\sqrt{49}.$ & $17=\dfrac{100\times 1,02}{6}$. Therefore

If y^e line *ak* is	The superficies *bcgk* is[17]
7.	1,94591,01490,55313,30510,53527,43443,17972,96370,84729,58186,11881,00.
17.	2,83321,33440,56216,08024,95346,17873,12653,55882,03012,58574,47867,65.

Againe if $x=0,001$. The operation is as followeth.

[--- --- --- --- ---][18]

Therefore[19] $\begin{cases} bcfd=-0,00100,05003,33583,53350,01429,82254,06834,49607,55205,25043,44092,48. \text{ If } ad=0,999. \\ bche=0,00099,95003,33083,53316,68093,98920,53501,14607,55062,39316,65519,96. \text{ If } ae=1,001. \end{cases}$

And since $\dfrac{1000\times 1,001}{77}=13.$ & $\dfrac{1000\times 0,999}{27}=37.$ Therefore

If y^e line *ak* is	the superficies *bcgk* is[20]
13.	2,56494,93574,61536,73605,34874,41565,31860,48052,67944,76020,71159,31.
37.	[3,61091,79126,44224,44436,80956,71031,44716,39000,77587,16763,61629,80.]

But least there may have beene some errors in y^e former calculations it will bee convenient to try if the [q]uantity of some of those areas may bee calculated some other way[.] which may be thus done. viz: since

$$\frac{2\times 2\times 2\times 2\times 2\times 2\times 2\times 2\times 3\times 13}{10000}=0,9984=\frac{256\times 3\times 13}{10000}\text{: Therefore}$$

If y^e line *ad* is	then the area *bcfd* is[21]
0,9984	0,00160,12813,66973,83328,53761,68584,19997,63476,46377,05009,33029,80.

(16) Correctly (note (9)), the last two figures of (*bcfd*) = log (0·98) should be '27' and those of (*bche*) = log (1·02) '86' (with respective errors of 8 and 7).

(17) The final figures of the two following expansions should be '4,59' and '72,97' (with respective errors of 359 and 532). Compare note (13).

(18) In this omitted computation,, repeated from I, 1, 3, §5: 140, Newton finds the two partial sums

$$\alpha(0{\cdot}001, 8) = 0{\cdot}00100,00003,33333,53333,34761,90587,30167,82107,55133,82180,04086,22$$

and

$$\beta(0{\cdot}001, 9) = 0{\cdot}00000,05000,00250,00016,66667,91666,76666,67500,00071,42863,39286,26.$$

(19) Correctly (note (9)), the last two figures of (*bcfd*) = log (0·999) should be '50' and the last of (*bche*) = log (1·001) '7' (with errors of 2 and 1 respectively).

(20) The expansion of log (37), left blank in Newton's manuscript, is inserted (in agreement with his factorization) as 3 log (10) + log (0·999) − 3 log (3). Correctly, the last four figures of log (13) should be '64,19' and those of log (37) '36,45' (with errors of 438 and 665 respectively). Compare note (13).

Which Area may bee otherwise thus found supposing $x=db=-0{,}0016$. Viz:

$\frac{aax}{b}+\frac{aax^3}{3b^3}=$	0,00160,00013,65333,33333,33333,33333,33333,33333,33333,33333,33333,33
	$\frac{aax^5}{5b^5}=$ 2,09715,20000,00000,00000,00000,00000,00000,00000,00
	$\frac{aax^7}{7b^7}=$ 38347,92228,57142,85714,28571,42857,14285,71
	$\frac{aax^9}{9b^9}=$ 7635,49741,51111,11111,11111,11111,11
	$\frac{aax^{11}}{11b^{11}}=$ 1599,28964,04014,54545,45454,54
	346,43074,05669,61230,77
	76,86143,36404,56
	17,36164,14
	3,98
The Sum̄e of $\frac{aax}{b}+\frac{aax^3}{3b^3}$ &c=	0,00160,00013,65335,43048,91681,33197,41816,99469,20181,33677,37988,14.

$\frac{aaxx}{2bb}+\frac{aax^4}{4b^4}+\frac{aax^6}{6b^6}=$	0,00000,12800,01638,40279,62026,66666,66666,66666,66666,66666,66666,66
	$\frac{aax^8}{8b^8}-$ 53,68709,12000,00000,00000,00000,00000,00
	$\frac{aax^{10}}{10b^{10}}=$ 10,99511,62777,60000,00000,00000,00
	2,34562,48059,22133,33333,33
	51469,71002,70913,82
	11529,21504,60
	2623,53
The sum̄e of $\frac{aaxx}{2bb}+\frac{aax^4}{4b^4}$ &c:—	0,00000,12800,01638,40279,62080,35386,78180,64007,26195,71331,95041,94.

The sum̄e of wch two summes[22] is equall to y^e area *bcfd*. Which summe is[23]

| 0,00160,12813,66973,83328,53761,68584,19997,63476,46377,05009,33030,08.

As was found before excepting y^t their difference in y^e two last figures is 28.[24] Which agreement could scarce thus happen in more y^n 50 figures, were not y^e areas, corresponding to y^e lines 2. 3. 5. 7. 13. &c, calculated aright in so many figures.[25]

---

(21) An arithmetical error, consciously contrived or otherwise, makes the following expression much more accurate than it should be. (The calculation of log (0·9984) as

$$8\log(2)+\log(3)+\log(13)-4\log(10)$$

yields a value whose last three figures are '6,29', which differs by 385—and not, as here, by 34—from the correct one.)

(22) These are, in the terminology of note (7) $\alpha(0{\cdot}0016, 9)$ and $\beta(0{\cdot}0016, 10)$ respectively.

(23) The last two figures are, correctly, '14'. Compare notes (9) and (21).

(24) If, however, we eliminate the arithmetical error in the former (note (21)) we find their difference to be 379.

(25) The arithmetical error in the former evaluation of log (0·9984) (note (21)) does little to detract from this justifiably proud assertion: by Newton's estimate the agreement is to

2

MISCELLANEOUS RESEARCHES

[*c.* 1668][1]

§1. CONIC PROPERTIES AND THE RECTIFICATION OF THE CYCLOID.

From the original in the University Library, Cambridge[2]

[1] Fiat $bg \parallel ad$, et ducantur rectæ *ab*, *gd*, *ag*, *bd*: Conicarum portiones *bas*, *gdt*, erunt æquales, item et *gasb*, *bdtg*. (fig 1. 2). Bisectis enim *bg*, *ad* in punctis *r*,

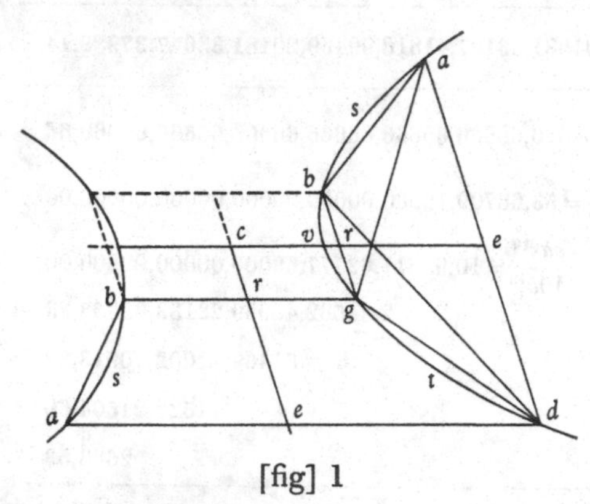

[fig] 1

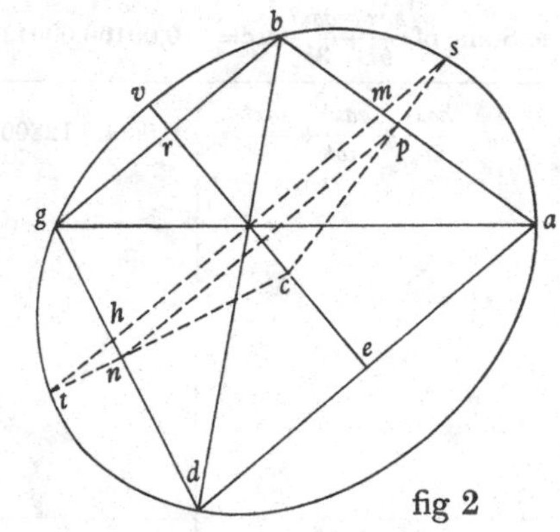

fig 2

et *e*; et ducto *cre*; erunt *brg aed* ordinatim applicatæ ad diametrum *cvre*: proinde portiones $reasb = redtg$, et trapezia $reab = redg$.[3] Ergo eorum differentiæ $bas = gdt$. Deinde $\triangle abg = dgb$ (per 37.1. Elem), Ergo portiones $gasbv = bdtgv$.

Coroll. Hinc pateat modus ducendi tangentes ad Conicas, ignoratis eorum diametris: A dato enim (fig 3) puncto *a*,[4] duc *ac*, *ab*, & ijs parallelas *bd*, *cf*: linea *df* erit parallela tangenti *ah*. Si non, fit $ae \parallel df$, et duc *de*, *ad*, *af*, *cb*; et erit portio

$$ecd = abf = bac = aecd,$$

q^d est impos:[5]

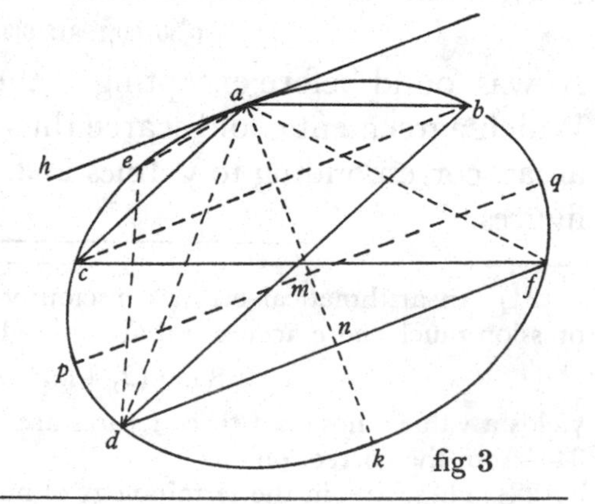

fig 3

55*D*, by the corrected one the two values agree to 54*D*. As we have seen, that agreement is, in fact, a true indication of the correctness of the expansion.

When in October 1676 Newton was engaged in drafting his *epistola posterior* to Leibniz, we may guess that he had his Waste Book open before him at the present page (81^v) for

Translation

[1] Let *bg* be parallel to *ad* and the straight lines *ab*, *gd*, *ag*, *bd* be drawn: the conic segments *bas* and *gdt* will be equal, and likewise *gasb* and *bdtg* (figures 1 and 2). For, on bisecting *bg* and *ad* in the points *r* and *e* and drawing *cre*, *brg* and *aed* will be ordinates to the diameter *cvre*; so that the segments *reasb* and *redtg* and the trapezia *reab* and *redg* are equal,(3) and therefore their differences *bas* and *gdt* also. Further (by *Elements*, I, 37) the triangles *abg* and *dgb* are equal, and consequently so are the segments *gasbv* and *bdtgv*.

Corollary. Hence may appear a method of drawing tangents to conics when their diameters are not known. For, from the given point *a*(4) (figure 3) draw *ac*, *ab* and parallel to them *bd*, *cf*: the line *df* will then be parallel to the tangent *ah*. If not, let *ae* be parallel to *df* and draw *de*, *ad*, *af*, *cb*: then will the segment *ecd*=*abf*=*bac*=*aecd*, which is impossible.(5)

---

immediately below his logarithmic computations he there entered the solution of the fluxional anagram he communicated in that letter:

'Octob. 1676. Memorandum. The letters *6accd æ 13eff 7i 3l 9n 4o 4q rr 4s 8t 12v x* in my second epistle to M. Leibnitz contein this sentence Data æquatione quotcunq; fluentes quantitates involvente, fluxiones invenire: et vice versâ. The other letters in ye same Epistle, viz: *5accd æ 10effh 11i 4l 3m 9n 60qqr 8s 11t 9v 3x*: *11ab 3cdd 10e æ g 10i ll 4m 7n 6y 3p 3q 6r 5s 11t 8vx*, *3ac æ 4egh 5i 4l 4m 5n 8oq 4r 3s 6t 4v*[,] *aad æ 5eiiimmnnooprrr 5sttvv*, express this sentence. Una Methodus consistit in extractione fluentis quantitatis ex æquatione simul involvente fluxionem ejus. Altera tantùm in assumptione seriei pro quantitate qualibet incognita ex qua cætera commodè derivari possunt, et in collatione terminorum homologorum æquationis resultantis ad eruendos terminos assumptæ seriei.' (Compare *The Correspondence of Isaac Newton*, **2** (1960): 115, 129, 153 note (25) and 159 note (72).) In the spring of 1695, along with amanuensis copies of his *epistolæ prior et posterior*, Newton communicated the solutions of his anagrams to John Wallis (compare ULC. Add. 3977.14), and the latter eventually published them in the *Epistolarum Collectio* he added to the third volume of his *Opera Mathematica* (Oxford, 1699). Newton's letter counting here is not good: in the first anagram, for example, there are 9 't''s. (This mistake was rectified by Oldenburg in the copy he made for transmission to Leibniz. See C. I. Gerhardt, *Der Briefwechsel von...Leibniz mit Mathematikern*, **1** (Berlin, 1899): 208.)

(1) Much as before, analysis of the writing of this miscellany of roughly contemporaneous Latin pieces suggests our conjectured date, though evidently §2 (and possibly §3 also) was written after May 1666 (§2, note (1) below).

(2) Add. 3958.3: 73^r. Both the following items were entered in a neat hand on the same side of the same small sheet of paper, and though their content is wholly disparate, we have chosen not to separate them.

(3) Compare **1**, 2, §2.2: Theorem 1, Corollaries 1 and 2.

(4) The point *a* is, of course, always to be taken on the conic.

(5) A pleasant elementary proof by *reductio ad absurdum* of a result we would now deduce as a corollary to Pascal's theorem on the general hexagon inscribed in a conic. (Since *ab* and *bd* are parallel to *fc* and *ca* respectively, the Pascal line of the hexagon (*aabdfc*) which has a vanishingly small side at *a* will be at infinity, and so the tangent at *a* meets *df* on the line at infinity.)

Porro fiat $dn = nf$, et duc ank; fiat $am = mk$, et duc pmq [$\| df$]. Erunt ak, pq diametri.(6)

---

[2] Diametro av describatur circulus $abcdv$. Et centro a, describantur circuli cr, dm, gn, hq. et a punctis c, d, g, h, duc ce, df, gk, hp, perpendiculares ad diametrum av, & duc abe, acf, adk, agp. In triangulis gqh, pqh &c:(7) angulus $hgp = hpq$. & $\angle gqh = \angle pqh =$ recto. Ergo si ponantur trianguli qgh, qph esse infinitè parvi, ut latera hg, hq si[n]t rectæ; erunt similes et æquales, p[r]oinde $gq = pq$. Hinc omnes lineæ gp, dk, cf, be, &[c] sunt duplices linearum gq, dn, cm, br, &c. hoc est longitudo Trochoid:(8) aw est duplex longit:(8) rectæ ag.(9)

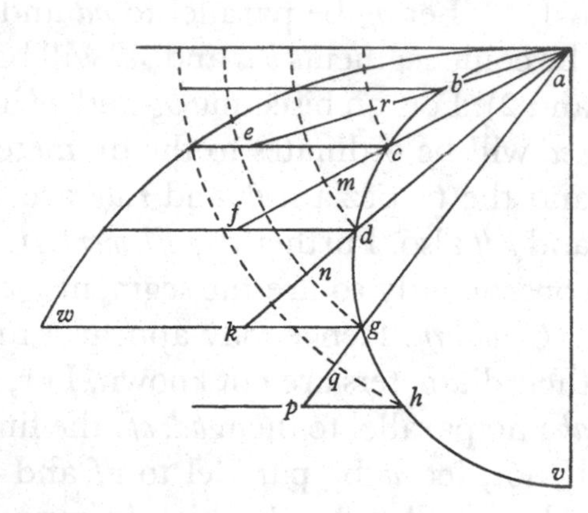

---

(6) The diameter conjugate to the ordinate direction df will bisect all such ordinates and therefore pass through a and n: the diameter conjugate to ak will be the parallel to df which bisects it.

(7) Newton would seem to understand the restriction made in the sequel that the points g and h be indefinitely near to one another. Strictly, when this is not so then the angle $h\hat{p}q$ is equal to $h'\hat{g}p$, where h' is the meet of hp with the tangent to the circle at g. In the limit as h passes into g, of course, the points h and h' coincide.

(8) Read 'Trochoidis' and 'longitudo' respectively.

(9) For ag is parallel to the tangent to the cycloid at w, and correspondingly for the chords ad, ac, ab, ...: hence, in the limit as each of the circle arcs ab, bc, cd, dg, ... becomes vanishingly small, the segments be, cf, dk, ... become equal to corresponding segments of the cycloid arc aw, while their halves br, cm, dn, ... together make up the length of the chord ag. Newton extracts the essence of Christopher Wren's cycloid rectification (1658?), but abandons the latter's turgid exhaustion method of proof. (See John Wallis, *Tractatus Duo* (1, **1**, Introduction, Appendix 2: note (22)) (Oxford, 1659): 62–74; and compare D. T. Whiteside, 'Patterns of Mathematical Thought in the Later Seventeenth Century', *Archive for History of Exact Sciences*, **1** (1961): 179–388, especially 333–5.) His argument, indeed, improves on Wren's by abandoning the latter's unnecessary insistence that the chords ab, ac, ad, ag, ah, ... be taken in geometrical proportion. If we suppose the chord gh to be drawn, with s the meet of wg with ah and w' the meet of the parallel to wg through h with the cycloid, Wren showed essentially that the triangles asg, agh and ahp are similar and (by use of Archimedes' convexity lemmas in the *Sphere and Cylinder*) that the cycloid arc $\widehat{ww'}$ is bounded by the tangent segments $w\omega$ and $w'\omega'$. It follows that $ap^2 : ah^2 = ah^2 : ag^2 = ag^2 : as^2 = ap : ag = ah : as$ and so

$$\left(1+\frac{ah}{ag}\right)(ah-ag) = gp = w\omega > \widehat{ww'} > w'\omega' = hs = \left(1+\frac{ag}{ah}\right)(ah-ag).$$

Newton's basic argument that $\widehat{ww'} = 2(ah-ag)$ follows in the limit as h passes into g, but Wren's further supposition that the chords ab, ac, ad, ag, ah, ... are in continued proportion allowed him to set bounds on the whole cycloid arc $\widehat{aw}$ which, in the limit as the number of

Furthermore, let $dn = nf$ and draw ank, let $am = mk$ and draw pmq [parallel to df] and then ak and pq will be diameters.(6)

---

[2] On the diameter av let there be described the circle $abcdv$. Again, with centre a let there be described the circles cr, dm, gn and hq, from the points c, d, g, h draw ce, df, gk, hp perpendicular to the diameter av, and draw abe, acf, adk, agp. In the triangles gqh and pqh, and so on,(7) the angle $h\hat{g}p = h\hat{p}q$ and $g\hat{q}h = p\hat{q}h$, both right. Therefore, if the triangles qgh and qph be supposed indefinitely small, so that their sides hg and hq be straight, they will be congruent and consequently $gq = pq$. Hence each of the lines gp, dk, cf, be, ... are double those gq, dn, cm, br, ... corresponding. That is, the length of the cycloid aw is double that of the straight line ag.(9)

---

points of division of the corresponding arc $\widehat{ag}$ of the generating circle becomes infinite, both tend to $2ag$. (What his argument gained in elegance thereby, it lost in clarity.)

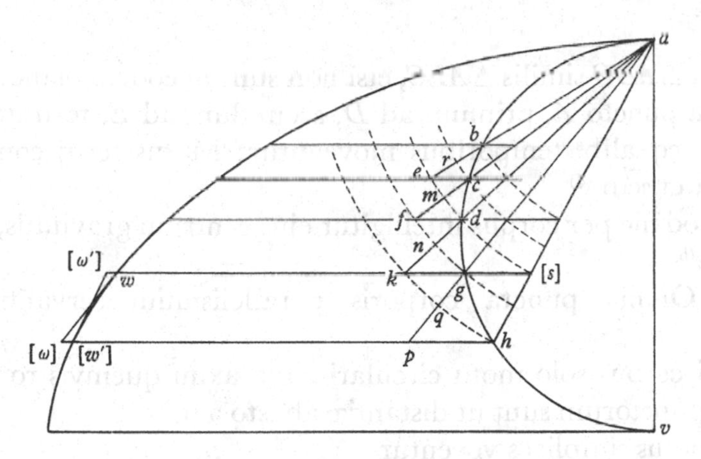

We might note that Christiaan Huygens had independently arrived, more or less, at Newton's simplification in January 1659, provoked by his reading in Pascal's *Historia Cycloidis* of Wren's rectification. (See his *Œuvres Complètes*, **14** (The Hague, 1920): 363–76, especially §2, 364–7. In footnote 9 on p. 365 his editor remarks that 'Il est d'ailleurs curieux de remarquer combien facilement dès ce moment la rectification d'un arc cycloïdal quelconque aurait pu être obtenue; c'est-à-dire, en negligeant, comme Huygens commence à le faire de plus en plus librement, des différences qui disparaissent à la limite': he then makes the crucial addition to Huygens' research of drawing the perpendicular from h to pg which yields Newton's proof.)

§2. 'THE SOLUTION OF PROBLEMS BY MOTION'.

From the original in the University Library, Cambridge[1]

De Solutione Problematum per Motum.

Ne hujusmodi operationes obscuræ nimìs evadant, Lemmata 6 sequētia brevitèr idéoq; non demonstrata præmittam.[2]

Lemma 1. Si corpus *A* in circumferentiâ circuli vel sphæræ *ADCE* moveatur versus ejus centrum *B*: velocitas ejus ad unaquæq; circumferentiæ puncta *D*, *C*, *E*, est ut cordæ[3] *AD*, *AC*, *AE*, ductæ a Corpore *A* ad ista puncta *D*, *C*, *E*.

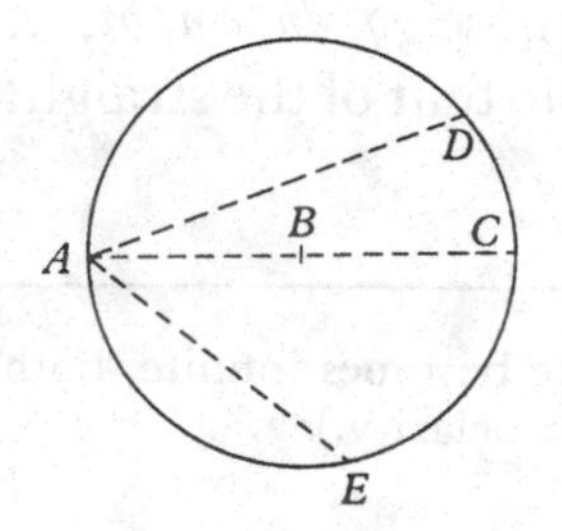

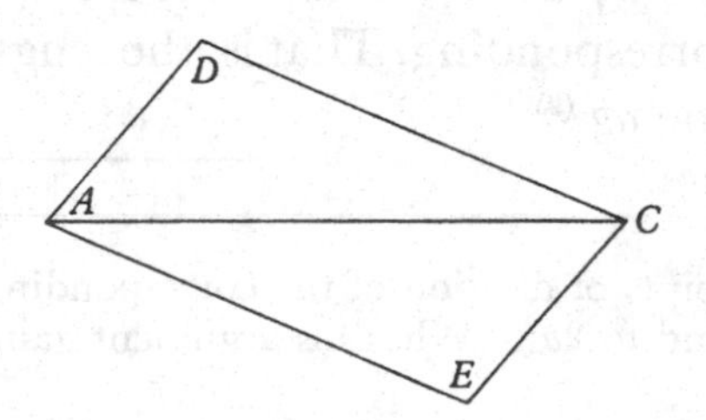

Lemma 2. Sit △*ADC* similis △*AEC*, etsi non sunt in eodem plano. Inquam, Si tria Corpora a puncto *A*, primum ad *D*, secundum ad *E*, tertium ad *C*, uniformiter et in æqualib^s temporibus moveantur: Motus tertij componetur ex motibus primi et scdi.[4]

Notetur, quòd hic per corpus intelligitur ejus centrum gravitatis, vel aliquod ejus punctum.[5]

Lemma 3. Omnia puncta corporis parallelismum servantis æqualiter moventur.

Lemma 4. Si corpus solo motu circulari circa axim quemvis rotetur; motus omnium ejus punctorum sunt ut distantiæ ab isto axi.

Et hi duo motus simplices vocentur.

Lemma 5. Si motus corporis consideretur ut mixtus e motibus simplicibus: motus omnium ejus punctorum componetur ex motibus eorum simplicibus, eo modo quo motus ab *A* ad *C*, in Lem̄ate 2do componitur ex motibus ab *A* ad *D* et *E*.

Nota, q^d motus quilibet ad unum horum 3^ū casuum reduci poterit. Et in casu tertio, linea quævis pro axe, (vel si linea aut plana superficies moveatur in plano, quodvis punctum istius plani pro centro,) motus assumi potest.

Lemma 6. Sint *AE*, *AH*, lineæ motæ et continuò secantes; Ducantur *AB*, *AD*, *AC*, *CB*, *CD*. Dico quòd, datis quæ requirantur ad proportiones et positiones harum quinque linearum *AB*, *AD*, *AC*, *CB*, *CD*, determinandas; illæ designent proportiones et positiones horum quinq; motuum, viz: puncti *A* in lineâ *AE* fixi et versus *B* moventis, puncti *A* in lineâ *AH* fixi et versus *D* moventis, Puncti

Translation

THE SOLUTION OF PROBLEMS BY MOTION

Lest operations of this kind prove too obscure, I shall premise the following six lemmas briefly and accordingly without proof.

Lemma 1. If the body *A* in the circumference of a circle or sphere *ADCE* move towards the latter's centre *B*, its velocity to each point *D*, *C*, *E* of the circumference is as the corresponding chord *AD*, *AC*, *AE* drawn from the body *A* to those points *D*, *C* or *E*.

Lemma 2. Let the triangle *ADC* be congruent to the triangle *CEA*, though not necessarily in the same plane with it. I say: if three bodies move uniformly and in equal times from the point *A*, the first to *D*, the second to *E* and the third to *C*, then the motion of the third will be compounded of those of the first and second.

Let me note that here by a body is understood its centre of gravity or some one point of it.

Lemma 3. All points of a body which keeps parallel to itself move equally.

Lemma 4. If a body rotate by a circular motion alone around any axis, the motions of each of its points are as their distances from that axis.

Let these two motions be called simple.

Lemma 5. If the motion of a body be considered as a mixture of simple motions, then the motion of every one of its points will be compounded of their simple motions in the same way as the motion from *A* to *C* in Lemma 2 is compounded of the motions from *A* to *D* and to *E*.

Note that any motion whatsoever may be reduced to one of these three cases. And that in the third case any line can be assumed for the axis (or, if a line or plane surface move in a plane, any point of that plane for the centre) of motion.

Lemma 6. Let *AE* and *AH* be moving and continuously intersecting lines, and let *AB*, *AD*, *AC*, *CB*, *CD* be drawn. I say that, when conditions requisite to determine the proportions and positions of these five lines *AB*, *AD*, *AC*, *CB* and *CD* are given, they define these five motions: namely, that of the point *A* fixed in the line *AE* and moving towards *B*, that of the point *A* fixed in the line *AH*

(1) Add. 3958.3: 68r–69r. This short Latin piece is evidently a revised version of the English paper 'To resolve Problems by motion...' (I, 2, 6, §4.2) written on 16 May 1666, with an example (on the quadratrix tangent) drawn from the October 1666 tract (I, 2, 7).

(2) The following six lemmas repeat Props. 1–6 of I, 2, 6, §4.2 with little variation.

(3) That is, 'chordæ' more correctly: Newton renders his word 'cordes' of May 1666.

(4) Read 'secundi'.

(5) A slight generalization of his 1666 statement, 'Note yt by a body is meant its centre of gravity'.

intersectionis A in plano $ABCD$ moventis versus C, (lineæ enim 5 istæ semper sunt in eodem plano etsi AE [&] AH non sunt), Puncti intersectionis A in lineâ AE moventis secundum ordinem literarum C, B, et Puncti intersectionis A in lineâ AH moventis secundum ordinem C[,] D.

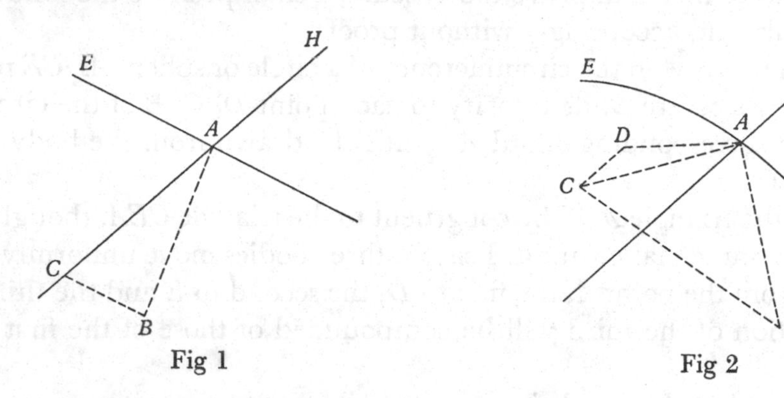

Fig 1 Fig 2

Nota, Quòd linea recta tangens curvam motu descriptam (ut AB, AD, AC) vel istæ tangenti parallela (ut CB, CD), dicitur designare positionem istius motûs in puncto contactûs.

Nota etiam q[d] lineâ AH quiescente (ut in Fig 1, et 4), punctum D et A coincident et pu[n]ctum C in lineâ AH, modò sit recta (Fig 1), alitèr in ejus tangente AC (fig 4) reperietur.

---

(6)*Prop 1. Ducere Tangentem ad Ellipsin.*

Sit ACB filum per quod Ellipsis describi solet,(7) et CE Tangens. Cùm filum AC augetur eâdem velocitate quâ BC diminuitur, i:e: C habet eandem velocitatem versus D et B; erit $\angle DCE = ECB$. Per Lem 1.

Idem de reliquis Conicis intelligatur.

Prop 2. Ducere Tangentem ad Conchoïden.

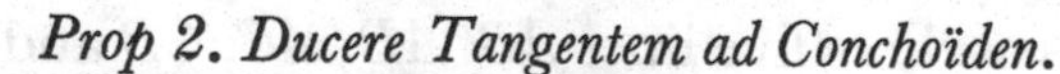

Sint GLC, ALF, GAE, regulæ quibus concha describi solet: fiat $GT \parallel AF \perp CB = \parallel MN$: et $NG = CL \perp TN \parallel RL$. Et cùm æqualitas proportionalitate simplicior est, ponatur lineam $CB = NM$ esse æqualem velocitati puncti C versus B, vel puncti N versus M. Erit $NT =$ motui circulari puncti N circa G rotantis versus T, (Lem 1): et $LR =$ motui circulari puncti

(6) The first three following propositions repeat the first three examples of I, 2, 6, §4.2 though the third is expanded to include its construction by a circle.

(7) That is, in the 'gardener's' construction. (In 1666 Newton wrote simply, 'Suppose y[e] Ellipsis to be described by y[e] thred *acb*.')

and moving towards D, that of the point of intersection A moving in the plane $ABCD$ towards C (for those five lines are always in the same plane even though AE and AH are not), that of the point of intersection A moving in the line AE following the order of the letters C and B, and that of the point of intersection A moving in the line AH following the order of C and D.

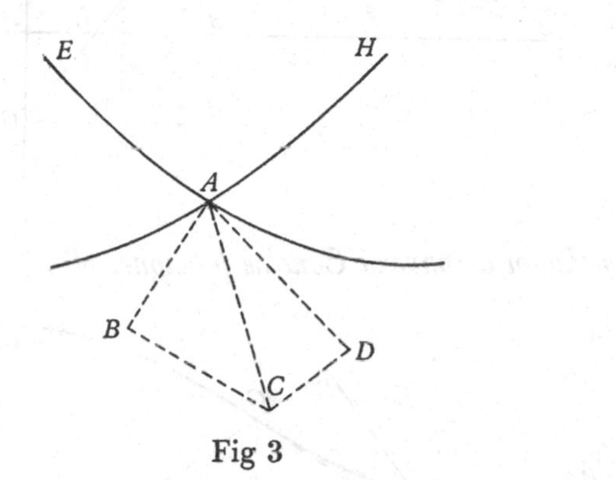

Fig 3

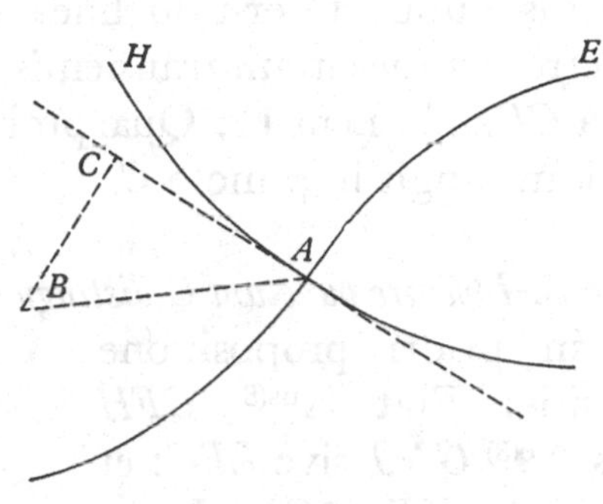

Fig 4

Note that the straight line tangent to the curve described in the motion (as AB, AD, AC) or parallel to that tangent (as CB, CD) is said to define the position of that motion in the point of contact.

Note also that, when the line AH is stationary (as in figures 1 and 4), the points D and A will coincide and that the point C will be found in the line AH should it be straight (figure 1) or else in its tangent AC (figure 4).

[(6)] *Prop. 1. To draw a tangent to the ellipse.*

Let ACB be the thread by which the ellipse is usually described,[(7)] and CE the tangent. Since the thread AC increases at the same velocity as BC diminishes, that is, C has the same velocity towards D and B, the angles $D\hat{C}E$ and $E\hat{C}B$ will be equal (by Lemma 1).

The same is to be understood concerning the other conics.

Prop. 2. To draw a tangent to the conchoid.

Let GLC, ALF and GAE be the rulers by which the conchoid is usually described, and make $GT \parallel AF \perp CB =$ and $\parallel MN$ with $NG = CL \perp TN \parallel RL$. Also, since equality is simpler than proportionality, let the line $CB = NM$ be set equal to the velocity of the point C towards B, or of the point N towards M. Then will NT be equal to the circular motion of the point N rotating round G towards T (Lemma 1); LR to the circular motion of the point L fixed in the line

L in lineâ GL fixi, circa G (Lem 4): et LG = motui puncti intersectionis L (i:e: velocitati puncti C) a puncto G, sive versus punctum D in lineâ CG moventis, (Lem 6). Jam cum habeatur duplex velocitas puncti C viz CB versus B et LG versus D, fiat $FD \perp DC = LG$: Et motus puncti C erit in lineâ FC, Diametro nempe circuli transientis per puncta $CDBF$, (Lem 1): Quæ proinde Concham tangit in puncto C.

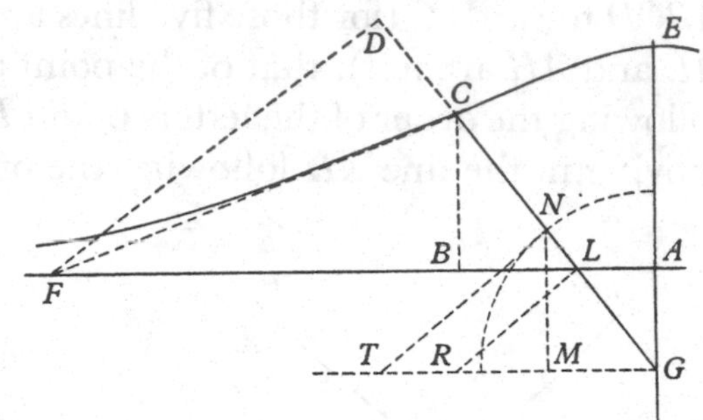

Prop 3. Invenire punctum C distinguens concavam a convexâ Conchæ portione.

Iis in priori propositione suppositis: Fiat $\triangle^{us(8)}$ GFH similis $\triangle^{o(8)}$ GNT sive LBC: et $DF \perp FR \parallel = HK = 2GL$; Jungâtur F, K, et fiat $KP \parallel FD$. Si Linea DF solum motum Parallelum per CD vel FR directum haberet, (quia $CD = GL$) motus omnium ejus punctorum esset FR, (Lem 3): Et si solum motum circularem circa centrū G haberet, motus puncti F, in istâ lineâ DF fixi, esset FH, (Lem 4). At motus istius F ex istis duobus componitur, proinde erit FK, (Lem 5, 2) et motus puncti intersectionis F per lineas AF, DF facti, et in AF moventis, erit FP (Lem 6). Jam si linea CF Concham tangit in puncto quæsito C, facile deprehendatur motum puncti intersectionis F esse nullum; proinde P et F coincidere; sive DF et FK in directum jacere; et $\triangle GDF$, FKH esse similes.

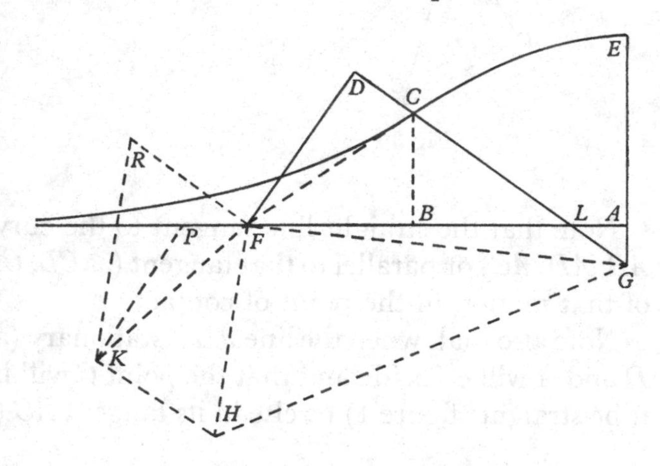

Quæ ut calculo subjiciantur, fiat $AG = b$. $CL = c$. $CB = y$. tum, $BL = \sqrt{cc-yy}$. $2GL = \dfrac{2bc}{y} = HK$. $LD = \dfrac{cb+cy}{y}$. $DF = \dfrac{cb+cy}{\sqrt{cc-yy}}$. Et,

$$\sqrt{cc-yy} : y :: BL : CB :: GF : FH :: DF : KH :: \frac{cb+cy}{\sqrt{cc-yy}} : \frac{2bc}{y} :: by+yy : 2b\sqrt{cc-yy}.$$

Quare, $2bcc - 2byy = byy + y^3$. sive $y^3 + 3byy * - 2bcc = 0$.[9] [Quam] Equatio[nem]

(8) Read 'triangulus' and 'triangulo' respectively.

(9) Compare I, 2, 6, §4.2: note (33) and I, 3, 3, §3: note (2). (Newton's present analysis, indeed, is an improvement on the construction he gave in the latter section.) As we there said,

GL around G (Lemma 4); and LG to the motion of the point of intersection L (that is, to the velocity of the point C) moving in the line CG away from the point G or towards the point D (Lemma 6). Now, since there is had a double velocity of the point C (namely, CB towards B and LG towards D), make $FD \perp DC = LG$ and the motion of the point C will be in the line FC, in the diameter, indeed, of the circle passing through the points C, D, B and F (Lemma 1). That line, in consequence, will touch the conchoid at the point C.

Prop. 3. To find the point C which separates the concave and convex portions of a conchoid.

Taking over the suppositions of the preceding proposition, make the triangle GFH similar to the triangles GNT or LBC, and set $DF \perp FR \parallel = HK = 2GL$; let F and K be joined and make $KP \parallel FD$. Should the line DF have only its parallel motion in a straight line along CD or FR, since $CD = GL$ the motion of all its points would be FR (by Lemma 3): and if it have only its circular motion round the centre G, the motion of the point F, fixed in that line DF, would be FH (Lemma 4). But the motion of that point F is compounded of those two and will consequently be FK (Lemmas 5 and 2), while the motion of the point of intersection F made by the junction of the lines AF and DF and moving in AF will be FP (Lemma 6). Now if the line CF touches the conchoid in the point C which is sought, it may easily be grasped that the motion of the point of intersection F is nil, and consequently that P and F coincide, or DF and FK lie in line and the triangles GDF and FKH are similar.

To reduce this to a calculable form, make $AG = b$, $CL = c$, $CB = y$ and we then have $BL = \surd(c^2-y^2)$, $2GL = 2bc/y = HK$, $LD = c(b+y)/y$, and so

$$DF = c(b+y)/\surd(c^2-y^2)$$

$$\surd(c^2-y^2):y = BL:CB = GF:FH = DF:KH = c(b+y)/\surd(c^2-y^2):2bc/y$$
$$= y(b+y):2b\surd(c^2-y^2).$$

Therefore, $2bc^2-2by^2 = by^2+y^3$, or $y^3+3by^2-2bc^2 = 0$.(9) Resolve this equation

Christiaan Huygens had previously obtained this result on 25 September 1653 (*Œuvres*, **12** (1910): 83–6, especially 83–4: 'In Conchoide Nicomedis invenire flectionis punctum'), evaluating from the Cartesian defining equation of the conchoid, $x^2y^2 = (b+y)^2(c^2-y^2)$, its 'contrary' subtangent $t = y-x(dy/dx)$ as $y(-by^2+c^2y+2bc^2)/(y^3+bc^2)$ and then finding the inflexion by the (insufficient) condition that t be a maximum. (In effect, he determined that $(dt/dy) = 0$, where the derivative was found by a Fermatian adequation: compare Jean Itard, 'Fermat précurseur du calcul différentiel', *Archives internationales d'Histoire des Sciences*, **1** (1947–8): 589–610, especially 606–8: 'Les points d'inflexion'.) The resulting cubic $y^3+3by^2-2bc^2 = 0$ he then constructed in Cartesian manner, finding its roots as the meets of a circle and parabola. (See his letter to Schooten on 23 October 1653, and his *De Circuli Magnitudine Inventa* (Leyden, 1654): Appendix: 210–15. Omission of the extraneous factor $y+b = 0$ is not justified.)

ita resolve.(10) fac $Ao = \frac{8c^4}{27b^3}$.(11) $af = c = AE$. duc *of* et cum diametro *of* describe circulum *fmao*, in quo inscribe $fm = b = AG$. et cum radio *om* fac circulum *mv*, et a pūcto intersectionis *v* duc $vd \perp ad$. erit $3b:2c::VD:Gl$. unde datur punctum *c*.(12)

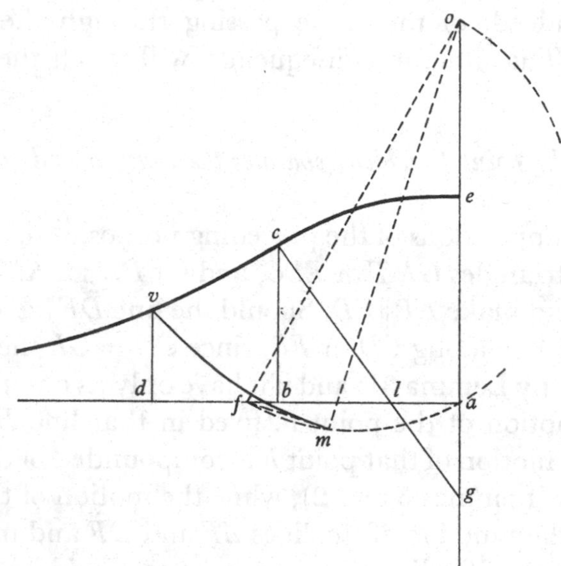

Usus autem hujus methodi (ut intelligo) præcipuus est in lineis Mechanicis, ubi deficit Algebraïca calculatio. Exempli grā, Tangens Quadratrici ita ducetur.

Sint *AG*, *MC* regulæ quibus uniformiter motis Quadratrix describi intelligitur *CB* vocetur motus puncti *C* in lineâ *CM* fixi et versus *B* moventis; tum arcus *GK* erit motus puncti *G* cir[c]a *A* rotantis (sup)(13), et arcus *CL* erit motus puncti *C* in lineâ *AG* fixi et circa [*A*] rotantis, (lem 4). Quare (si fiat $CL = CD \perp AG \parallel DF$, et $BF \parallel CM$) motus puncti intersectionis *C* in plano *AEK* erit *CF* (Lem 6), quæ proinde Quadr[atricem] tangit in *C*.(14)

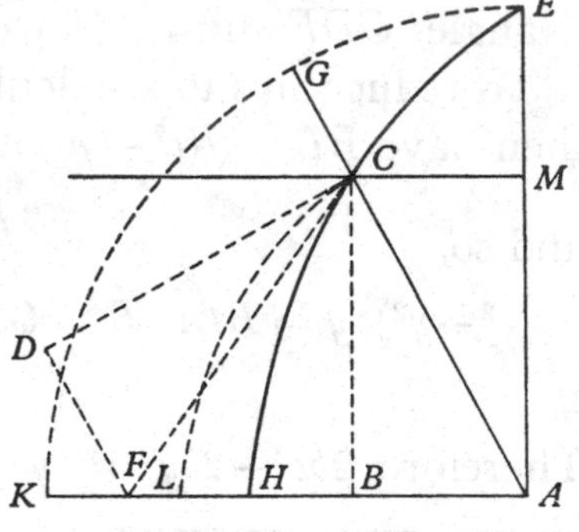

(10) The remainder of this paragraph was added by Newton alongside the previous text and we incorporate it following the implication of his opening phrase 'Equatio $y^3 + 3byy * - 2bcc = 0$, prius inventa, ita resolve'.

(11) Read '$-b + \frac{8c^4}{27b^3}$'.

(12) This construction also, as we now know, was Huygens' independent (and prior) discovery in August 1659, though it appeared in print only in 1910 (*Œuvres*, **12**: 232–7).

in the following way.(10) Make $ao = 8c^4/27b^3$,(11) $af = c = ae$; then draw *of* and with this as diameter describe the circle *fmao*, inscribing in it $fm = b = ag$; and with radius *om* construct the circle *mv*, from the intersection point *v* drawing *vd* perpendicular to *ad*. Then will there be $3b:2c = vd:gl$, and from that is given the point *c*.(12)

But the especial use of this method (as I understand) is in connection with mechanical lines, when algebraic calculation fails. For example, the tangent to the quadratrix will be drawn as follows.

Let *AG* and *MC* be the rulers through whose uniform motion the quadratrix is understood to be described. Also, let *CB* denote the motion of the point *C* fixed in the line *CM* and moving towards *B*: then will the arc *GK* be the motion of the point *G* rotating round *A* (by supposition) and the arc *CL* the motion of the point *C* fixed in the line *AG* and rotating round *A* (Lemma 4). Therefore, if you make $\widehat{CL} = CD \perp AG \parallel DF$ and $BF \parallel CM$, the motion of the point of intersection *C* in the plane *AEK* will be *CF* (Lemma 6), and this in consequence touches the quadratrix at the point *C*.(14)

The basic idea of using the conchoid itself to construct the inflexion condition is due to Henrik van Heuraet, and was announced (compare I, 3, 3, §3: note (4)) by Schooten in his second Latin edition of Descartes' *Geometrie* (*Geometria* (1659): 259–62: *Commentarii in Librum II, O*). Having, as Huygens (note (19)), found the condition $t(y^3+bc^2)+y(by^2-c^2y-2bc^2) = 0$ which determines the contrary subtangent $t = y-x(dy/dx)$, Heuraet then arrived at the condition $3(t+b)y^2-2c^2y-2bc^2 = 0$ which determines that it have a double root in y by Cartesian identification of coefficients with $(t+b)(y-e)^2(y+g) = 0$, and, on eliminating y, at the cubic $t^3+bt^2-\frac{4}{3}c^2t-32b^2c^2-4c^4/27b = 0$: this he constructed as the meet of the conchoid with the circle of radius $\sqrt{[9b^3(\alpha+b)/(8b^2+c^2)+\alpha^2+c^2-b^2]}$, whose centre is in *ae* and distant

$$\alpha = -b+(8b^2+c^2)^2/27b^3$$

from *a*. When Huygens read Heuraet's solution he saw the superiority of using the conchoid itself, rather than an alien parabola, to construct the cubic inflexion condition

$$y^3+3by^2-2bc^2 = 0.$$

On investigation he then found, exactly as Newton in the present paper, that the substitution $y = k^2/z$ reduces the inflexion condition to $z^3-3k^4z/2c^2-k^6/2bc^2 = 0$, while the circle of centre o ($ao = \beta$) and radius $ov = \gamma$ meets the conchoid $v^2z^2 = (b+z)^2(c^2-z^2)$, where $ad = v$ and $dv = z$, in the points v such that $(\gamma+b)z^3-\frac{1}{2}(\gamma^2-\beta^2+b^2-c^2)z^2-bc^2z-\frac{1}{2}b^2c^2 = 0$: on dividing the latter cubic by $\gamma+b$ and identifying coefficients, it follows that

$$\gamma^2-\beta^2+b^2-c^2 = 0,$$

$k^2 = \frac{3}{2}b^2$ and $\gamma+b = 8c^4/27b^3$. Since therefore $gl = cl\times ag/cb = bc/y$, the circle of centre o ($ao = -b+8c^4/27b^3$) and radius $ov = \sqrt{(ao^2+ae^2-ag^2)}$ meets the conchoid in points v such that $vd:gl = 3b:2c$, where gl passes through the inflexion point c.

(13) 'per suppositionem'.

(14) This exemplification of the construction of tangents to non-algebraic curves is taken with little variance from English papers of November 1665 and October 1666 (see I, 2, 6, §2 and I, 2, 7). In our comments on the former (I, 2, 6, §2: note (14)) we saw how growing understanding of the quadratrix' subtangent afforded Newton a fundamental insight into geometrial limit-motion.

§3. 'GRAVITY' IN CONICS.

From the original in the University Library, Cambridge[1]

De Gravitate Conicarum.

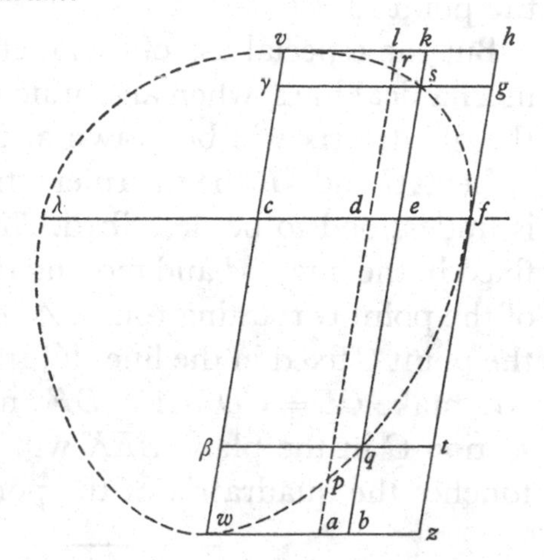

[1][2] Sit λvf Ellipsis, et λwf Parabola. Ita nempe, ut λd vocatâ x; ordinatim applicata sit $dr = \sqrt{rx - \frac{rxx}{q}}$, & $dp = \frac{rx}{2} - \frac{rxx}{2q}$.[3] Fiat $\lambda c = cf = \frac{1}{2}q$. & duc $vcw \parallel hfz \parallel kb \parallel rp$, & $vh \parallel wz \parallel \gamma g \parallel \beta t$, tangentes vel secantes sese aut curvas in punctis $v, l, k, h, r, \gamma, s, g, c, d, e, f, \beta, q, t, p, w, a, b, z$.

Dico quod, posito λf axe gravitatis, pondus parallelogrammi *vkec* est ad pondus portionis *vsec*, sicut parallelogram̄um *wbec* ad portionem *wqec*.

Ponatur enim pondus lineæ *dr* esse $\frac{1}{2} dr \times dr$,[4], sive $\frac{rx}{2} - \frac{rxx}{2q}$, hoc est *dp*. Erit pondus omniū linearū *dr* in superficie *vsec* contentarum, hoc est pondus superficiei *vsec*, æquale superficiei *wqec*.[5] Et eâdem ratione pondus ▱mi[6] *vkec* erit ▱mum[6] *wbec*.[7] Q.E.D.

Coroll̄. Pondus portionis *vks* est *wbq*, et portionis $vs\gamma$ est $wq\beta$, et portionis *esf* est *eqf*, & portionis *fsg* est *fqt* &c: Quæ omnia dat quadratura Parabolæ, est enim $3\beta wq = 2\beta wbq$, &c.[8]

[2] Eâdem ratione gravitas cujuslibet portionis Hyperbolæ cognoscatur, modò axis gravitatis transeat per centrum Hyperbolæ.[9] Et si quævis plana superficies conicis sectionibus ita terminetur, ut omnia conicarum centra sint in eâdem rectâ lineâ: gravitas istius superficiei inveniri potest, positâ istâ rectâ

(1) Add. 3958.3: 74r. This not very important piece applies the method of Problem 15 of the October 1666 tract (I, 2, 7), 'To find ye Gravity of any given plaine in respect of any axis [of gravity], given in position...', to finding the 'pondus' (first moment) of a section of a central conic round a gravity axis which passes through its centre. Newton's original manuscript is written with great care and neatness and, most unusually, the straight lines in the accompanying figures have been drawn with a ruler and not freehand. Evidently, he intended this for restricted circulation in manuscript form if not for publication, but we know nothing of the surrounding circumstances. The immaturity of content and handwriting forcibly indicate an early composition date for the piece, and we suggest 1667 as a year not too far removed from the 1666 tract from which it seemingly depends.

(2) In this first part the axis of gravity is the diameter of an ellipse conjugate to the direction of balance (parallel to *vc*, the ellipse's semi-minor axis).

(3) It follows that the major axis $\lambda f = q$ and that $dr^2 = (r/q)\lambda d \times df = 2dp$.

Translation

The Gravity of Conics

[1][2] Let λvf be an ellipse and λwf a parabola, such indeed that, if λd is called x, their ordinates will be $dr = \surd(rx - rx^2/q)$ and $dp = \frac{1}{2}rx - \frac{1}{2}rx^2/q$.[3] Make $\lambda c = cf = \frac{1}{2}q$ and draw $vcw \parallel hfz \parallel kb \parallel rp$, also $vh \parallel wz \parallel \gamma g \parallel \beta t$ touching or meeting one another or the curves in the points $v, l, k, h, r, \gamma, s, g, c, d, e, f, \beta, q, t, p, w, a, b, z$.

I say that, on setting λf the axis of gravity, the weight of the parallelogram $vkec$ is to the weight of the portion $vsec$ as the parallelogram $wbec$ to the portion $wqec$.

For let the weight of the line dr be set as $\frac{1}{2}dr \times dr$,[4] or $\frac{1}{2}rx - \frac{1}{2}rx^2/q$, that is, dp. Then will the weight of the totality of lines contained in the surface $vsec$, that is, the weight of the surface $vsec$, be equal to the surface $wqec$.[5] And for the same reason the weight of the parallelogram $vkec$ will be the parallelogram $wbec$.[7] As was to be proved.

Corollary. The weight of the portion vks is wbq, that of the portion $vs\gamma$ is $wq\beta$, that of the portion esf is eqf, that of the portion fsg is fqt, and so on. All these are given by the quadrature of the parabola, for $3\beta wq = 2\beta wbq$, and so on.[8]

[2] By the same reasoning the gravity of any portion of a hyperbola may be known, provided the axis of gravity passes through the hyperbola's centre.[9] And if any plane surface be so terminated by conics that all the centres of the conics are in the same straight line, the gravity of that surface can be found if

(4) Strictly, since the gravity of a body acts normally to the gravity axis this should be $\frac{1}{2}dr \times dr \times \sin \widehat{vcf}$, but since Newton intends to measure the 'pondus' by the area under the corresponding parabola λwf, $\sin \widehat{vcf} \times \int (dp) . dx$, he ignores the constant $\sin \widehat{vcf}$ for simplicity.

(5) In more modern notation, the 'pondus' of $(vsec)$ round λf is $\int_{\frac{1}{2}q}^{x} \frac{1}{2}(dr) \times (dr) . dx$, while the surface area of $(wqec)$ is its equal $\int_{\frac{1}{2}q}^{x} (dp) . dx$. (Compare note (3).) The multiplying constant $\sin \widehat{vcf}$ has (note (4)) been ignored.

(6) Read 'parallelogrammi' and 'parallelogrammum' respectively.

(7) This is an immediate deduction from the equality of $\frac{1}{2}cv^2$ and cw (the values of $\frac{1}{2}dr^2$ and dp when d is at c).

(8) A standard property of the parabola, easily proved from first principles in the present case since

$$\int_{\frac{1}{2}q}^{x} (\tfrac{1}{8}qr - y) . dx = \tfrac{1}{3} \times (\tfrac{1}{8}qr - y)(x - \tfrac{1}{2}q) \quad \text{where} \quad y = \tfrac{1}{2}(rx - rx^2/q).$$

(9) The ordinate dr will now be $\surd(rx + rx^2/q)$, of course.

gravitatis axi.(10) Deniqꝫ, centrum gravitatis cujusvis planæ & finitæ superficiei conicis sectionibus ita(11) terminatæ inveniri potest, datâ quantitate istius superficiei, & vice versâ, modò centrum gravitatis axi(12) gravitatis non coïncidat.(13)

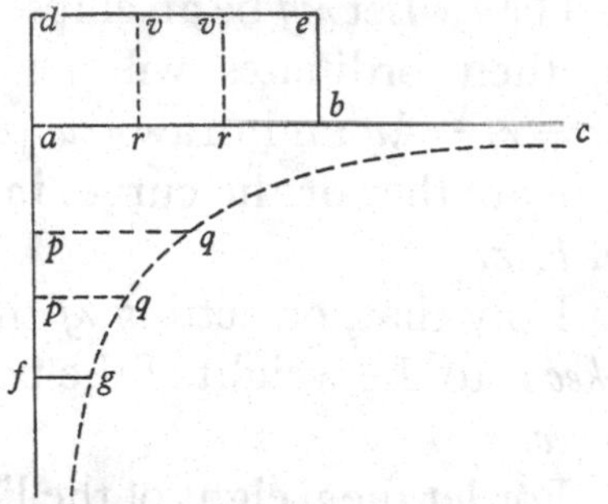

Sint ac, af Asymptota(14) Hyperbolæ gc, in infinitū versus c continuatæ. Duc $de=ab=af$. & $eb=ad=\sqrt{af\times fg}$. & $fg \parallel ac$. Dico q͗d, Parallelogramū ae, & superficies $afgc$ æquiponderant circa axim abc: Etsi superficies $afgc$ versus c sit longitudine & quantitate infinita, & non habet centrum gravitatis.(15)

fiat enim $ar=ap$. & duc $pq \parallel ab$. & $vr \parallel ad$. et sit $pq\times\frac{1}{2}ap$ gravitas lineæ pq, erit $vr\times\frac{1}{2}vr$ gravitas lineæ vr (nam $\angle vrb=\angle apq$).(16) sed $[a]p\times\dfrac{1}{2}pq=\dfrac{af\times fg}{2}=\dfrac{ad\times ad}{2}=vr\times\dfrac{1}{2}v[r]$. æquiponderant ergo lineæ pq, vr. Sed numerus linearum pq in superficie $acgf$ est æqualis numero linearum vr in parallelogrammo ae[,] nam $af=ab$, ergo ae & ag æquiponderant.(17)

---

(10) This is trivial since the conic region so defined may be evaluated as the aggregate of central conics (added or subtracted one from another) whose defining equations are of the form $y^2 = \frac{1}{2}(rx \mp rx^2/q)$.

(11) Presumably, so that each of the conic terminations has its centre in the gravity axis.

(12) Newton wrote 'aixi' here.

you set that straight line as the axis of gravity.(10) Finally, the centre of gravity of any finite plane surface terminated in this fashion(11) by conics can be found when the quantity of that surface is given, and vice versa, provided that the centre of gravity does not coincide with the axis of gravity.(13)

Let *ac* and *af* be the asymptotes of the hyperbola *gc* continued indefinitely in the direction of *c*. Draw $de = ab = af$, $eb = ad = \sqrt{(af \times fg)}$ and *fg* parallel to *ac*. I say that the parallelogram *ae* and the surface *afgc* balance round the axis *abc*, although the surface *afgc* is unbounded in length and quantity in the direction of *c* and has no centre of gravity.(15)

For make $ar = ap$ and draw *pq* parallel to *ab* and *vr* parallel to *ad*. Let also $pq \times \frac{1}{2}ap$ be the gravity of the line *pq*, and then will $vr \times \frac{1}{2}vr$ be the gravity of the line *vr* (for $v\hat{r}b = a\hat{p}q$).(16) But $ap \times \frac{1}{2}pq = \frac{1}{2}af \times fg = \frac{1}{2}ad \times ad = vr \times \frac{1}{2}v[r]$. The lines *pq* and *vr* balance therefore. But the number of lines *pq* in the surface *acgf* is equal to the number of lines *vr* in the parallelogram *ae*, since $af = ab$, and therefore the surfaces *ae* and *ag* balance.(17)

---

(13) This remark seems mysterious unless we restrict its reference to a conic segment lying wholly on one side of a diameter which is its axis of gravity. In that case, of course, for the centre of gravity to coincide with the gravity axis the conic segment must have zero area.

(14) Read 'Asymptoti'.

(15) That is, no finite centre of gravity. Compare Newton's undergraduate notes on Wallis' discussion of the gravity of hyperbola-areas in the latter's *Commercium Epistolicum* (I, 1, 3, §3.4).

(16) Compare note (4) above.

(17) Newton seems to understand an equal division of the equal line-segments *af* and *ab* into 'indivisible' elements. Note with regard to the accompanying diagram that the point *c* is at infinity on both the hyperbola *gq* and the asymptote *ab* (which, of course, passes through the hyperbola's centre *a*).

3

THE 'DE ANALYSI PER ÆQUATIONES INFINITAS'

[June 1669?][1]

From the original in the Library of the Royal Society, London[2]

DE ANALYSI PER ÆQUATIONES NUMERO TERMINORUM INFINITAS.

Methodum generalem quam de curvarum quantitate per infinitam terminorum seriem mensuranda olim[3] excogitaveram, in sequentibus brevitèr explicatam potiùs quàm accuratè demonstratam habes.

Basi AB, curvæ alicujus AD, sit applicata BD perpendicularis: & vocetur $AB=x$, & $BD=y$; & sint a, b, c &c quantitates datæ; & m, n numeri integri. Deinde[4]

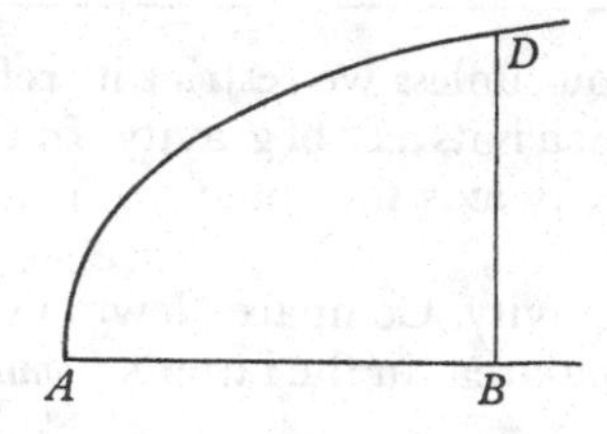

varum
licium
dratura

REG: I. Si $ax^{\frac{m}{n}}=y$, erit $\dfrac{na}{m+n}x^{\frac{m+n}{n}}=$Areæ ABD.[5]

(1) Somewhat incredibly, we have not been able to determine with precision the date when this, perhaps the most celebrated of all Newton's mathematical writings, was composed. In our introduction we have suggested that the piece was written after Nicolaus Mercator published his *Logarithmotechnia* in the autumn of 1668, and we may settle on September of that year as a *terminus ante quem non*. Probably it was composed over a period of a few days in the early summer of 1669: at least, Newton communicated it to his college senior and Lucasian mathematical professor, Isaac Barrow, about the beginning of July 1669 for on the 20th of that month the latter wrote to John Collins that 'A friend of mine here, that hath a very excellent genius to those things, brought me the other day some papers, wherein hc hath sett downe methods of calculating the dimensions of magnitudes like that of Mr Mercator concerning the hyperbola, but very generall; as also of resolving æquations; which I suppose will please you' (*Correspondence of Isaac Newton*, 1 (1959): 13–14). Newton's 'papers' Barrow sent along eleven days later and the copy (see following note) which Collins made at that time before returning the original to Barrow is firm proof that no major addition was made later to its text.

(2) MS LXXXI, No. 2. It would appear that this text was quickly returned to Newton by Collins through Barrow's mediation in 1669, and that it remained in its author's possession during most of his life. However, when in about 1709 William Jones conceived the idea of

Translation

ON ANALYSIS BY EQUATIONS UNLIMITED IN THE NUMBER OF THEIR TERMS

The general method which I had devised some time ago(3) for measuring the quantity of curves by an infinite series of terms you have, in the following, rather briefly explained than narrowly demonstrated.

To the base *AB* of some curve *AD* let the ordinate *BD* be perpendicular and let *AB* be called x and *BD* y. Let again $a, b, c, \ldots$ be given quantities and m, n integers. Then(4)

The Qu ture of simple C

RULE 1. If $ax^{m/n} = y$, then will $[na/(m+n)]\,x^{(m+n)/n}$ equal the area *ABD*.(5)

---

making the selection of Newton's mathematical papers which he finally published as the *Analysis Per Quantitatum Series, Fluxiones, ac Differentias: cum Enumeratione Linearum Tertii Ordinis* (London, 1711), he made a search of a wad of hitherto unexamined papers of Collins which had recently come into his possession and found Collins' copy (made by him in 1669). As Jones says in his *Præfatio*, 'licet neque Auctoris nomen, neque tempus quo scriptus fuerat ullibi comparuerit; multa tamen continere ad D. NEWTONI Methodos spectantia statim agnovi'. Having subsequently, he added, been able to confirm Newton's authorship from Barrow's 1669 correspondence with Collins (which with so many other invaluable items of the latter's hoard of manuscript was in Jones' possession), 'Perspecto jam D. NEWTONUM hujus Tractatus Auctorem esse; ab eo sciscitatus sum num penes se adhuc esset autographum, quod quidem ille exquirens invenit, & mihi tradidit, cum exemplari COLLINSIANO ad verbum usque conveniens'. Jones never returned the 'De Analysi' to Newton but subsequently, with the latter's permission, placed it with Collins' copy in the letters which he placed on deposit in the archives of the Royal Society in 1712 to serve as the documentary foundation for the Society's report on the fluxion priority dispute which ultimately appeared the following March as *Commercium Epistolicum D. Johannis Collins, et Aliorum de Analysi Promota: Jussu Societatis Regiæ In lucem editum.* Jones borrowed his Collins' papers back again from the Society for a period in the late 1730's, but finally returned them in February 1741 'neatly bound in a Book of Guards with proper Titles and References' (*Report on Leibnitz–Newton MSS in the possession of the Royal Society of London* (London, 1880): 21). In that sturdy volume Newton's original has remained undisturbed to the present day in company with Collins' copy, and on its flyleaf may still be read Jones' scrawled identification, 'N° 2. Sent by Dr. Barrow to Mr Collins in a Letter dated July 31. 1669'. We may add that Collins about 1677 made a secondary transcript of his copy for John Wallis who was at that time writing the summary of English algebraical achievement which appeared later as his *Algebra* (London, 1685). At Wallis' death this transcript passed to his fellow Savilian Professor at Oxford, David Gregory, and it is now in the latter's papers in the Library of the University of St Andrews' (MS QA 33 G8 D3: 1–10).

(3) That is, in 1665 and 1666. See I, 1, 3 and 2, 7: passim.

(4) The three 'Regulæ' which follow revise, summarize and exemplify Proposition 8 of the October 1666 fluxional tract (I, 2, 7).

(5) Since *AB* (x) is zero when *B* is at *A*, the lower bound of the integral is zero and Newton correctly evaluates $\int_0^x ax^{m/n}.dx$. In examples 4 and 5 following, however, he avoids the difficulty of having an integrand which is infinite when $x = 0$ by assuming a lower (or rather an upper) bound $A\alpha = \infty$.

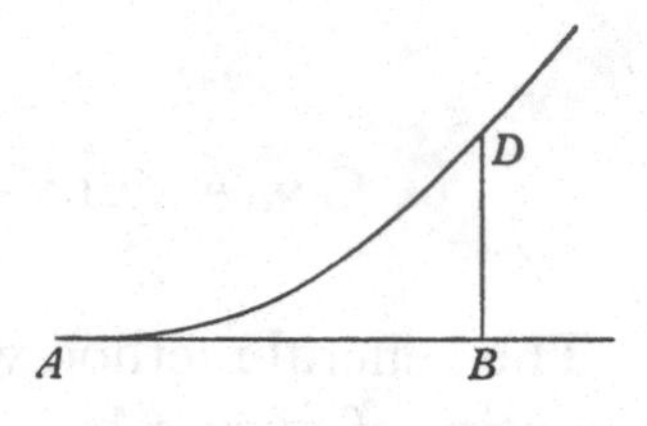

Res exemplo patebit. Exemp 1. Si $x^2(=1\times x^{\frac{2}{1}})=y$; hoc est si $a=1=n$, & $m=2$; erit $\frac{1}{3}x^3=ABD$. Exempl 2. Si $4\sqrt{x}(=4x^{\frac{1}{2}})=y$, erit $\frac{8}{3}x^{\frac{3}{2}}(=\frac{8}{3}\sqrt{x^3})=ABD$. Exemp 3. Si $\sqrt{3:x^5}(=x^{\frac{5}{3}})=y$, erit $\frac{3}{8}x^{\frac{8}{3}}(=\frac{3}{8}\sqrt{3:x^8})=ABD$. Exemp 4. Si $\frac{1}{x^2}(=x^{-2})=y$, id est si $a=1=n$ & $m=-2$, erit $\left(\frac{1}{-1}x^{\frac{-1}{1}}=\right)-x^{-1}\left(=\frac{-1}{x}\right)=\alpha BD$ infinitè versus α protensæ; quam calculus ponit negativam propterea quòd jacet ex altera parte lineæ BD.(6) Exemp: 5. Si

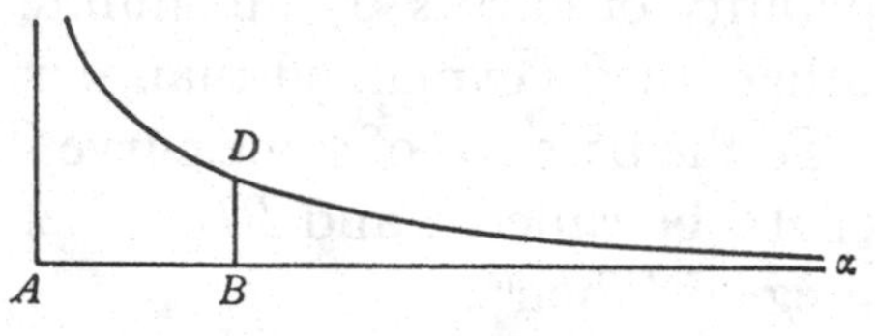

$$\frac{[2}{3\sqrt{x^3}}\left(=\frac{2}{3}x^{\frac{-3}{2}}\right)=y,\text{ erit }\frac{2}{-1}x^{\frac{-1}{2}}=-\frac{2}{\sqrt{x}}=BD\alpha.$$

Exemp 6. Si $\frac{1}{x}(=x^{-1})=y$, erit $\frac{1}{0}x^{\frac{0}{1}}=\frac{1}{0}x^0=\frac{1}{0}\times 1=$ infinitæ, qualis est area Hyperbolæ ex utraq; parte lineæ BD.(7)

[Co]mposi[tor]um(8) ex [sim]plicibus

Reg II. Si valor ipsius y ex pluribus istiusmodi terminis componitur, area etiam componetur ex areis quæ a singulis terminis emanant.

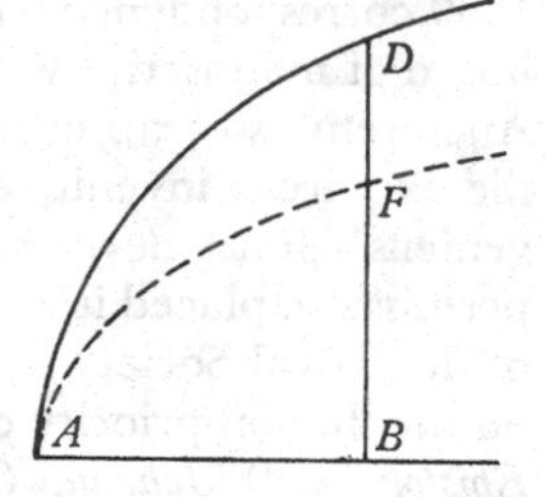

Hujus Exempla prima sunto.(9) Si $x^2+x^{\frac{3}{2}}=y$ erit $\frac{1}{3}x^3+\frac{2}{5}x^{\frac{5}{2}}=ABD$. Etiam si semper sit $x^2=BF$, & $x^{\frac{3}{2}}=FD$; erit ex præcedente Regula $\frac{x^3}{3}=$superficiei AFB descriptæ per lineam BF, & $\frac{2}{5}x^{\frac{5}{2}}=AFD$(10) descriptæ per DF; Quare $\frac{x^3}{3}+\frac{2}{5}x^{\frac{5}{2}}=$totæ(11) ABD. Sic si $x^2-x^{\frac{3}{2}}=y$ erit $\frac{1}{3}x^3-\frac{2}{5}x^{\frac{5}{2}}=ABD$: Et si

$$3x-2x^2+x^3-5x^4=y,$$

erit $\frac{3}{2}x^2-\frac{2}{3}x^3+\frac{1}{4}x^4-x^5=ABD$.

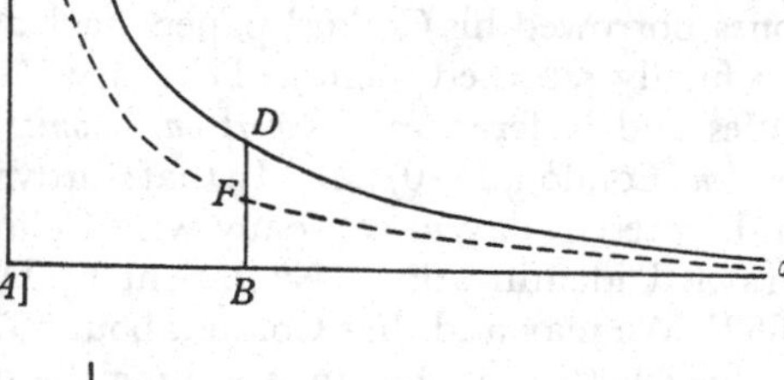

Exempla secunda. Si $x^{-2}+x^{\frac{-3}{2}}=y$, erit $-x^{-1}-2x^{\frac{-1}{2}}=\alpha BD$. Vel si $x^{-2}-x^{\frac{-3}{2}}=y$, erit $-x^{-1}+2x^{\frac{-1}{2}}=\alpha BD$.(12) Quarum signa si mutaveris habebis affirmativum valorem $\left(x^{-1}+2x^{\frac{-1}{2}}\text{ vel }x^{-1}-2x^{\frac{-1}{2}}\right)$ superficiei αBD, modò tota cadat supra Basin $AB\alpha$, sin aliqua pars cadat infra, (quod fit cùm curva decussat suam Basin inter B & α, ut hic vides in δ,) istâ parte a parte superiori subductâ, habebis valorem differentiæ.

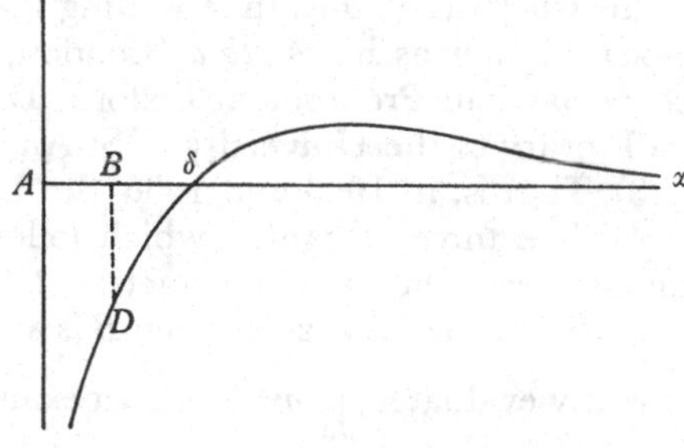

The matter will be evident by example.

Example 1. If $x^2(=1\times x^{\frac{2}{1}})=y$, that is, if $a=n=1$ and $m=2$, then $\frac{1}{3}x^3=ABD$.

Example 2. If $4\surd x(=4x^{\frac{1}{2}})=y$, then $\frac{8}{3}x^{\frac{3}{2}}(=\frac{8}{3}\surd x^3)=ABD$.

Example 3. If $\sqrt[3]{x^5}(=x^{\frac{5}{3}})=y$, then $\frac{3}{8}x^{\frac{8}{3}}(=\frac{3}{8}\sqrt[3]{x^8})=ABD$.

Example 4. If $(1/x^2)\,(=x^{-2})=y$, that is, if $a=n=1$ and $m=-2$, then $([1/-1]\,x^{-\frac{1}{1}}=)-x^{-1}\,(=-[1/x])=\alpha BD$ infinitely extended in the direction of α: the computation sets its sign negative because it lies on the further side of the line BD.(6)

Example 5. If $2/3\surd x^3\,(=\frac{2}{3}x^{-\frac{3}{2}})=y$, then $(2/-1)\,x^{-\frac{1}{2}}=-(2/\surd x)=BD\alpha$.

Example 6. If $(1/x)\,(=x^{-1})=y$, then $(1/0)\,x^{\frac{0}{1}}=(1/0)\,x^0=(1/0)\times 1=\infty$, just as the area of the hyperbola is on each side of the line BD.(7)

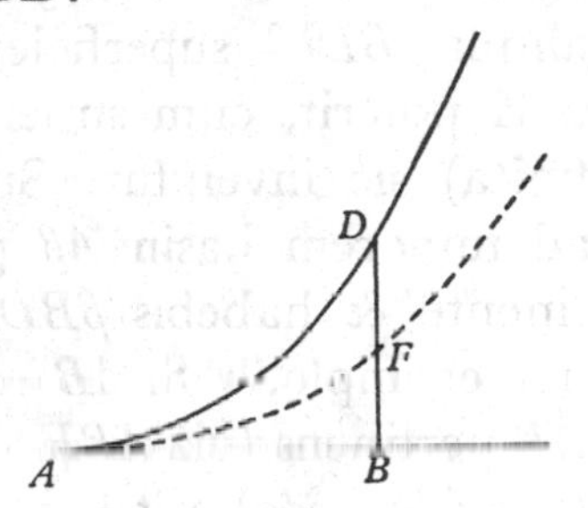

And of those compounded of simple ones(8)

RULE 2. If the value of y is compounded of several terms of that kind the area also will be compounded of the areas which arise separately from each of those terms.

Let its first examples be these.(9) If $x^2+x^{\frac{3}{2}}=y$, then $\frac{1}{3}x^3+\frac{2}{5}x^{\frac{5}{2}}=ABD$. For if there be always $BF=x^2$ and $FD=x^{\frac{3}{2}}$, then by the preceding rule $\frac{1}{3}x^3=$ the surface AFB described by the line BF and $\frac{2}{5}x^{\frac{5}{2}}=AFD$(10) described by DF; and consequently $\frac{1}{3}x^3+\frac{2}{5}x^{\frac{5}{2}}=$ the whole ABD. Thus if $x^2-x^{\frac{3}{2}}=y$, then $\frac{1}{3}x^3-\frac{2}{5}x^{\frac{5}{2}}=ABD$: and if $3x-2x^2+x^3-5x^4=y$, then $\frac{3}{2}x^2-\frac{2}{3}x^3+\frac{1}{4}x^4-x^5=ABD$.

Second examples. If $x^{-2}+x^{-\frac{3}{2}}=y$, then $-x^{-1}-2x^{-\frac{1}{2}}=\alpha BD$. Or if

$$x^{-2}-x^{-\frac{3}{2}}=y, \quad \text{then} \quad -x^{-1}+2x^{-\frac{1}{2}}=\alpha BD.^{(12)}$$

If you change their signs, you will have a positive value ($x^{-1}+2x^{-\frac{1}{2}}$ or $x^{-1}-2x^{-\frac{1}{2}}$) for the surface αBD provided it falls wholly above the base $AB\alpha$; but should any part fall below (which happens when the curve crosses its base between B and α, as you see here in δ), you will have the value of the difference when that part is

(6) What Newton intends, it would seem, is to say that the integration bounds are in effect reversed:

$$\int_x^\infty x^{-2}.\,dx = -\int_\infty^x x^{-2}.\,dx = x^{-1}.$$

An extraneous factor $\frac{2}{3}$ occurring in the following example is omitted by Jones in 1711.

(7) Compare I, 1, 3, §4: note (22), and 2, 7: note (12).

(8) In his 1711 printed version (note (2)) Jones added 'Curvarum Quadratura' here.

(9) Jones in his 1711 version here omitted 'Hujus' and 'sunto', while replacing Newton's incorrectly convex accompanying figure by that which we reproduce in our English version.

(10) Newton has cancelled 'superf[iciei]' (surface).

(11) 'toti' (Jones, 1711).

(12) $\alpha BD = -\int_\infty^x (x^{-2}\pm x^{-\frac{3}{2}}).\,dx = x^{-1}\pm 2x^{-\frac{1}{2}}$.

Earum verò summam si cupis, quære utramqꝫ superficiem seorsim, & adde.[13] Quod idem in reliquis hujus regulæ exemplis notandum volo.

Exempla tertia. Si $x^2+x^{-2}=y$,[14] erit $\frac{1}{3}x^3-x^{-1}=$superficiei descriptæ. Sed hic notandum est quod dictæ superficiei partes sic inventæ jacent ex diverso latere lineæ BD: nempe, posito $x^2=BF$ & $x^{-2}=FD$, erit $\frac{1}{3}x^3=ABF$ superficiei per BF descriptæ, & $-x^{-1}=DF\alpha$ descriptæ per DF. Et hoc semper accidit cum indices $\left(\frac{m+n}{n}\right)$ rationum basis x in valore superficiei quæsitæ sint varijs signis affectæ. In hujusmodi casibus pars aliqua $BD\delta\beta$ superficiei media (quæ sola dari poterit, cùm superficies sit utrinqꝫ infinita) sic invenitur. Subtrahe superficiem ad minorem basin $A\beta$ pertinentem a superficie ad majorem basin AB pertinente & habebis $\beta BD\delta$ superficiem differentiæ basium insistentem. Sic in hoc exemplo,[15] Si $AB=2$ & $A\beta=1$, erit $\beta BD\delta=\frac{17}{6}$.[16] E[te]nim superficies ad AB pertinens (viz $ABF-DF\alpha$) erit $\frac{8}{3}-\frac{1}{2}$, sive $\frac{13}{6}$; et superficies ad $A\beta$ pertinens (viz $A\phi\beta-\delta\phi\alpha$) erit $\frac{1}{3}-1$, sive $-\frac{2}{3}$: et earum differentia (viz

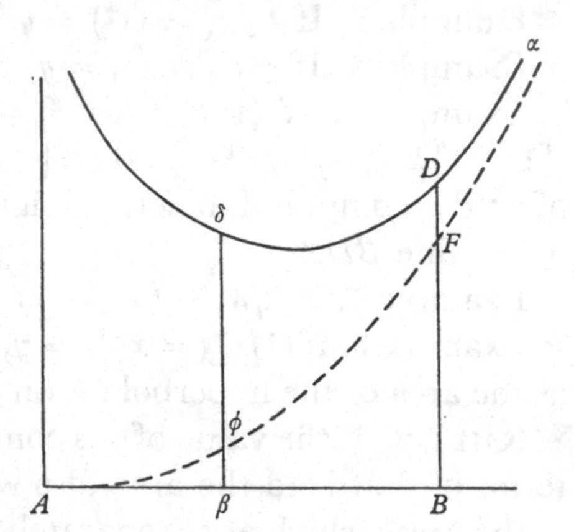

$$ABF-DF\alpha-A\phi\beta+\delta\phi\alpha=\beta BD\delta)$$

erit $\frac{13}{6}+\frac{2}{3}$ sive $\frac{17}{6}$. Eodem modo si $A\beta=1$, & $AB=x$, erit $\beta BD\delta=\frac{2}{3}$[17]$+\frac{1}{3}x^3-x^{-1}$. Sic si $2x^3-3x^5-\frac{2}{3}x^{-4}+x^{-\frac{3}{5}}=y$, & $A\beta=1$; erit

$$\beta BD\delta=\tfrac{1}{2}x^4-\tfrac{1}{2}x^6+\tfrac{2}{9}x^{-3}+\tfrac{5}{2}x^{\frac{2}{5}}-\tfrac{49}{18}.\text{[18]}$$

Deniqꝫ notari poterit quòd si quantitas x^{-1} in valore ipsius y reperiatur, iste terminus (cùm hyperbolicam superficiem generat) seorsim a reliquis considerandus est. Ut si

$$x^2+x^{-3}+x^{-1}=y:\text{[19]}$$

Sit $x^{-1}=BF$, & $x^2+x^{-3}=FD$, ac $A\beta=1$; et erit $\delta\phi FD=\frac{1^{(20)}}{6}+\frac{x^3}{3}-\frac{x^{-2}}{2}$, utpote quæ ex terminis

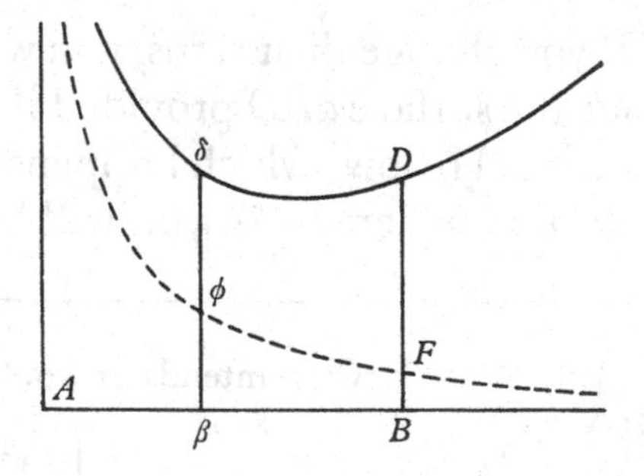

x^2+x^{-3} generatur: quare si reliqua superficies βBFB[21], quæ Hyperbolica est, ex calculo aliquo sit data, dabitur tota $\beta BD\delta$.

t aliarū
ium.[22]

REG III. Sin valor ipsius y vel aliquis ejus terminus sit præcedentibus magis

(13) Since $y = x^{-2}(1\pm\sqrt{x})$, only positive values of x yield real values of y so that the curve takes the two forms sketched by Newton in his figures. Evidently the infinite branches are given by $y = \pm x^{-\frac{3}{2}}$, while $y = x^{-2}(1-\sqrt{x})$ crosses $A\alpha$ when $A\delta = x = 1$.

taken away from the upper part. If indeed you desire their sum, seek each surface separately and add them.[13] I would have you note the same thing in the remaining examples of this rule.

Third examples. If $x^2+x^{-2} = y$,[14] then $\frac{1}{3}x^3-x^{-1} =$ the surface described. But here you should note that the parts of the said surface thus found lie on opposite sides of the line *BD*: precisely, on setting $BF = x^2$ and $FD = x^{-2}$, then $\frac{1}{3}x^3 =$ the surface *ABF* described by *BF* and $-x^{-1} = DF\alpha$ described by *DF*. And this always happens when the indices $(m+n)/n$ of the ratios of the base x in the value of the surface sought are affected with different signs. In cases of this sort any middle portion of the surface $BD\delta\beta$ (and no other can have been given since the surface is infinite on either side) is found in this way. Subtract the surface relating to the lesser base $A\beta$ from the surface relating to the greater base *AB* and you will have the surface $\beta BD\delta$ standing on the difference of those bases. Thus in this example,[15] if $AB = 2$ and $A\beta = 1$, then $\beta BD\delta = \frac{17}{6}$.[16] For the surface relating to *AB* (namely $ABF-DF\alpha$) will be $\frac{8}{3}-\frac{1}{2}$ or $\frac{13}{6}$; and the surface relating to $A\beta$ (namely $A\phi\beta-\delta\phi\alpha$) will be $\frac{1}{3}-1$ or $-\frac{2}{3}$: and so their difference (namely $ABF-DF\alpha-A\phi\beta+\delta\phi\alpha = \beta BD\delta$) will be $\frac{13}{6}+\frac{2}{3}$ or $\frac{17}{6}$. In the same way if $A\beta = 1$ and $AB = x$, then $\beta BD\delta = \frac{2}{3}$[17] $+\frac{1}{3}x^3-x^{-1}$. And thus if

$$2x^3-3x^5-\tfrac{2}{3}x^{-4}+x^{-\frac{3}{5}} = y \quad \text{and} \quad A\beta = 1,$$

then $\beta BD\delta = \frac{1}{2}x^4-\frac{1}{2}x^6+\frac{2}{9}x^{-3}+\frac{5}{2}x^{\frac{2}{5}}-\frac{49}{18}$.[18]

Finally, it might be noted that if the quantity x^{-1} be found in the value of y itself, that term (generating a hyperbolic space) is to be considered on its own apart from the rest. As if $x^2+x^{-3}+x^{-1} = y$:[19] let $BF = x^{-1}$, $FD = x^2+x^{-3}$ and also $A\beta = 1$; and then will $\delta\phi FD = \frac{1}{6}$[20] $+\frac{1}{3}x^3-\frac{1}{2}x^{-2}$, inasmuch as it is produced from the terms x^2+x^{-3}: hence, if the remaining surface $\beta[\phi]FB$, which is hyperbolic, be given by some computational procedure, there will be given the whole $\beta BD\delta$.

RULE 3. But if the value of y or any of its terms be more compounded than the

And of [...] others.[2...]

(14) Evidently the full curve which corresponds to this defining equation will include a mirror-image of $\delta D\alpha$ in the vertical through *A*. No real values of x correspond, of course, to negative values of y.

(15) In his 1711 version Jones here added, in brackets and italicized, the clarifying phrase 'Vide Fig. Præcedentem' (See the preceding figure).

(16) That is, $\frac{1}{3}(2^3-1^3)-(2^{-1}-1)$.

(17) That is, $-(\frac{1}{3}-1)$.

(18) That is, $-(\frac{1}{2}-\frac{1}{2}+\frac{2}{9}+\frac{5}{2})$.

(19) Or $x^3y = 1+x^2+x^5$. Clearly, Newton has sketched the corresponding curve only in the first quadrant. In fact $x^3(y-x^2) = 0$ determines the infinite branches, so that the vertical through *A* meets the curve at infinity in a cusp to which it is tangent, while the remaining infinite portions of the curve are asymptotic to the parabola $y = x^2$.

(20) That is, $-(\frac{1}{3}-\frac{1}{2})$.

(21) Read (Jones' correction in 1711) '$\beta\phi FB$'.

(22) In his 1711 version Jones here added 'Quadratura'.

compositus,[23] in terminos simpliciores reducendus est, operando in literis ad eundem modum quo Arithmetici in numeris decimalibus dividunt, radices extrahunt, vel affectas Æquationes solvunt.

EXEMPLA DIVIDENDO.[24] Sit $\frac{aa}{b+x}=y$, curvâ nempe existente Hyperbolâ: Jam ut æquatio ista a denominatore suo liberetur divisionem sic instituo—

$$b+x\Big)\, aa+0\left(\frac{aa}{b}-\frac{aax}{bb}+\frac{aax^2}{b^3}-\frac{aax^3}{b^4}\ \&c\right.$$

$$aa+\frac{aax}{b}$$

$$0-\frac{aax}{b}+0$$

$$-\frac{aax}{b}-\frac{aax^2}{bb}$$

$$0+\frac{aax^2}{bb}+0$$

$$+\frac{aax^2}{bb}+\frac{aax^3}{b^3}$$

$$0-\frac{aax^3}{b^3}+0$$

$$-\frac{aax^3}{b^3}-\frac{a^2x^4}{b^4}$$

$$0+\frac{aax^4}{b^4}\ \&c.$$

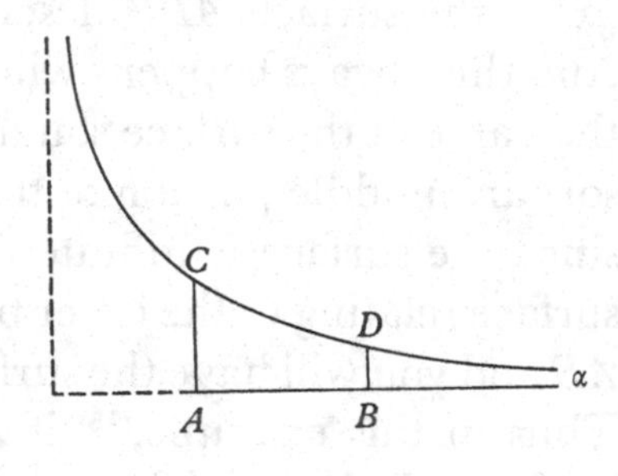

Et sic vice hujus $y=\frac{aa}{b+x}$ nova prodit $y=\frac{aa}{b}-\frac{a^2x}{b^2}+\frac{a^2x^2}{b^3}-\frac{a^2x^3}{b^4}$ &c serie istâc infinitè continuatâ. Adeoq; per Reg 2^{am} erit area

$$ABDC=\frac{a^2x}{b}-\frac{a^2x^2}{2b^2}+\frac{a^2x^3}{3b^3}-\frac{a^2x^4}{4b^4}\ \&c$$

infinitæ etiam seriei, tamen cujus[25] termini pauci initiales erunt in usum aliquem [&] satis exacti cùm[26] x sit aliquoties minor quam b.[27]

Eodem modo si $\frac{1}{1+xx}=y$, dividento prodibit $y=1-xx+x^4-x^6+x^8$ &c: Unde per Reg 2 erit $ABDC$[28] $=x-\frac{x^3}{3}+\frac{x^5}{5}-\frac{x^7}{7}$ &c. Vel si terminus xx ponatur

(23) That is, other than of the form $\sum_i (\pm a_i x^i)$.

(24) Compare I, 1, 3, §5, and 2, §3 above.

(25) In 1711 Jones inverted these two words into a more conventional order.

(26) '...sunt in usum quemvis satis exacti, si modo' (Jones, 1711). Note the change in meaning conveyed by this new phrasing 'sufficiently exact for *any* use'.

foregoing,[23] it must be reduced to simpler terms, by operating in general variables in the same way as arithmeticians in decimal numbers divide, extract roots or solve affected equations.

EXAMPLES BY DIVISION.[24] Let $a^2/(b+x) = y$; whose curve is evidently a hyperbola. Now to free that equation from its denominator I arrange the division as follows:

$$b+x\,\big)\; a^2+0\left(\frac{a^2}{b}-\frac{a^2x}{b^2}+\frac{a^2x^2}{b^3}-\frac{a^2x^3}{b^4}\ldots\right.$$

$$\begin{array}{l}
a^2+\dfrac{a^2x}{b} \\
\hline
0-\dfrac{a^2x}{b}+0 \\
\quad -\dfrac{a^2x}{b}-\dfrac{a^2x^2}{b^2} \\
\hline
\qquad 0+\dfrac{a^2x^2}{b^2}+0 \\
\qquad\quad +\dfrac{a^2x^2}{b^2}+\dfrac{a^2x^3}{b^3} \\
\hline
\qquad\qquad 0-\dfrac{a^2x^3}{b^3}+0 \\
\qquad\qquad\quad -\dfrac{a^2x^3}{b^3}-\dfrac{a^2x^4}{b^4} \\
\hline
\qquad\qquad\qquad 0+\dfrac{a^2x^4}{b^4}\ldots.
\end{array}$$

And thus in place of this equation $y = a^2/(b+x)$ there appears the new one

$$y = \frac{a^2}{b}-\frac{a^2x}{b^2}+\frac{a^2x^2}{b^3}-\frac{a^2x^3}{b^4}\ldots,$$

where the series is continued to infinity. And in consequence by Rule 2 the area *ABDC* will be equal to

$$\frac{a^2x}{b}-\frac{a^2x^2}{2b^2}+\frac{a^2x^3}{3b^3}-\frac{a^2x^4}{4b^4}\ldots,$$

an infinite series also but one whose first few terms will be of some use and sufficiently exact provided x be considerably less than b.[27]

In the same way if $1/(1+x^2) = y$, by division there arises

$$y = 1-x^2+x^4-x^6+x^8\ldots.$$

Hence by Rule 2 there will be $ABDC$[28] $= x-\frac{1}{3}x^3+\frac{1}{5}x^5-\frac{1}{7}x^7\ldots$. Or, if the term

(27) The series for $a^2\log(1+x/b)$ converges, of course, for $-1 < x/b \leqslant 1$.

(28) That is, $\int_0^x (1+x^2)^{-1}.dx = \tan^{-1}x$.

in divisore primus, hoc modo $xx+1)\,1:$[29] prodibit $x^{-2}-x^{-4}+x^{-6}-x^{-8}$ &c pro valore ipsius y. Unde per Reg 2 erit $BD\alpha$[30] $=-x^{-1}+\frac{x^{-3}}{3}-\frac{x^{-5}}{5}+\frac{x^{-7}}{7}$ &c. Priori modo procede cum x sit satis parva, posteriori cùm satis magna supponitur.[31]

Deniq; si $\frac{2x^{\frac{1}{2}}-x^{\frac{3}{2}}}{1+x^{\frac{1}{2}}-3x}=y$, dividendo prodit $2x^{\frac{1}{2}}-2x+7x^{\frac{3}{2}}-13x^2+34x^{\frac{5}{2}}$ &c: Unde erit $ABDC$[32] $=\frac{4}{3}x^{\frac{3}{2}}-x^2+\frac{14}{5}x^{\frac{5}{2}}-\frac{13}{3}x^3$ &c.[33]

EXEMPLA RADICEM EXTRAHENDO.[34] Si $\sqrt{:}\,aa+xx=y$, radicem sic extraho—

$$aa+xx\left(a+\frac{x^2}{2a}-\frac{x^4}{8a^3}+\frac{x^6}{16a^5}-\frac{5x^8}{128a^7}+\frac{7x^{10}}{256a^9}-\frac{21x^{12}}{1024a^{11}}\right.\quad \&\text{c}^{(35)}$$

$$\begin{array}{l}
\underline{aa} \\
0+xx \\
\qquad \underline{xx+\frac{x^4}{4aa}} \\
\qquad 0-\frac{x^4}{4aa} \\
\qquad \underline{-\frac{x^4}{4aa}-\frac{x^6}{8a^4}+\frac{x^8}{64a^6}} \\
\qquad\qquad 0+\frac{x^6}{8a^4}-\frac{x^8}{64a^6} \\
\qquad\qquad \underline{+\frac{x^6}{8a^4}+\frac{x^8}{16a^6}-\frac{x^{10}}{64a^8}+\frac{x^{12}}{256a^{10}}} \\
\qquad\qquad\qquad 0-\frac{5x^8}{64a^6}+\frac{x^{10}}{64a^8}-\frac{x^{12}}{256a^{10}}\quad \&\text{c.}
\end{array}$$

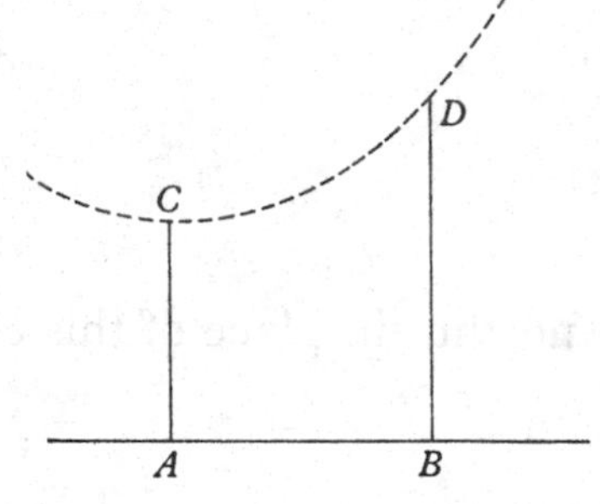

Unde pro[36] $\sqrt{:}\,aa+xx=y$, nova producitur, viz: $y=a+\frac{xx}{2a}-\frac{x^4}{8a^3}$ &c: et area Hyperbolæ quæsita[37]

(29) '$xx+1$),' (Jones, 1711).

(30) That is, $\int_{\infty}^{x}(1+x^2)^{-1}.dx=-\tan^{-1}(x^{-1})$.

(31) In fact, the infinite series for $ABDC$ and $BD\alpha$ are convergent for $|x|\leqslant 1$ and $|x|\geqslant 1$ respectively.

(32) $\int_0^x(2x^{\frac{1}{2}}-x^{\frac{3}{2}})/(1+x^{\frac{1}{2}}-3x).dx$.

(33) In continuation of this section on division Newton began a further paragraph, 'Præstat aliquando partes numeratoris seorsim considerare ad vitandum terminum x^{-1} in quotiente, ut si $\frac{x^2-1}{x^3+x}=y\ldots$' (It is sometimes of advantage to consider the parts of the numerator individually in order to avoid the term x^{-1} in the quotient, as if $(x^2-1)/(x^3+x)=y\ldots$). His cancellation of it seems wise. In particular there seems no way of avoiding the quotient x^{-1} in his example (say, by splitting it into the partial fractions $-x^{-1}+2x/(x^2+1)$) for its indefinite integral is, inevitably, $\log(x+x^{-1})$.

x^2 be set first in the divisor, in this way $x^2+1)\ 1$, there appears

$$x^{-2}-x^{-4}+x^{6}-x^{-8}\ldots$$

for the value of y; and hence by Rule 2 there will be

$$BD\alpha^{(30)} = -x^{-1}+\tfrac{1}{3}x^{-3}-\tfrac{1}{5}x^{-5}+\tfrac{1}{7}x^{-7}\ldots.$$

Proceed by the former way when x is small enough, by the latter when it is taken large enough.(31)

Finally if $\dfrac{2x^{\frac{1}{2}}-x^{\frac{3}{2}}}{1+x^{\frac{1}{2}}-3x} = y$, by division there arises

$$2x^{\frac{1}{2}}-2x+7x^{\frac{3}{2}}-13x^2+34x^{\frac{5}{2}}\ldots:$$

hence there will be $ABDC^{(32)} = \frac{4}{3}x^{\frac{3}{2}}-x^2+\frac{14}{5}x^{\frac{5}{2}}-\frac{13}{3}x^3\ldots.$(33)

EXAMPLES BY ROOT EXTRACTION.(34) If $\surd(a^2+x^2) = y$, I extract its root in the following manner:

$$a^2+x^2\left(a+\frac{x^2}{2a}-\frac{x^4}{8a^3}+\frac{x^6}{16a^5}-\frac{5x^8}{128a^7}+\frac{7x^{10}}{256a^9}-\frac{21x^{12}}{1024a^{11}}\ldots^{(35)}\right.$$

$$\begin{array}{l}
\underline{a^2}\\
0+x^2\\
\qquad \underline{x^2+\dfrac{x^4}{4a^2}}\\
\qquad 0-\dfrac{x^4}{4a^2}\\
\qquad\quad \underline{-\dfrac{x^4}{4a^2}-\dfrac{x^6}{8a^4}+\dfrac{x^8}{64a^6}}\\
\qquad\qquad 0+\dfrac{x^6}{8a^4}-\dfrac{x^8}{64a^6}\\
\qquad\qquad\quad \underline{+\dfrac{x^6}{8a^4}+\dfrac{x^8}{16a^6}-\dfrac{x^{10}}{64a^8}+\dfrac{x^{12}}{256a^{10}}}\\
\qquad\qquad\qquad 0-\dfrac{5x^8}{64a^6}+\dfrac{x^{10}}{64a^8}-\dfrac{x^{12}}{256a^{10}}\ldots.
\end{array}$$

Hence in place of $\surd(a^2+x^2) = y$ there arises a new equation, namely

$$y = a+\frac{x^2}{2a}-\frac{x^4}{8a^3}\ldots,$$

and the hyperbolic area sought(37) will be

(34) Compare Example 2 of the 'mechanicall' resolution of Proposition 8 of the October 1666 tract (I, 2, 7: especially note (43)).

(35) Jones in 1711 omitted the last two terms. The expansion is, of course,

$$\sum_{0\leqslant i\leqslant 6}\binom{\frac{1}{2}}{i}\frac{x^{2i}}{a^{2i-1}}\ \text{'\&c'},\quad \text{where}\quad \binom{\frac{1}{2}}{i} = \frac{1}{2}\times\frac{-1}{4}\times\frac{-3}{6}\times\ldots\times\frac{3-2i}{2i}.$$

(36) Jones here inserted 'Æquatione' in 1711.

(37) $\int_0^x (a^2+x^2)^{\frac{1}{2}}.dx = \frac{1}{2}(a^2\log(x+\surd[a^2+x^2])+x\surd[a^2+x^2])$. Compare I, 1, 3, §4: note (18).

erit $ABDC = ax + \frac{x^3}{6a} - \frac{x^5}{40a^3} + \frac{x^7}{112a^5} - \frac{5x^9}{1152a^7}$ &c.(38)

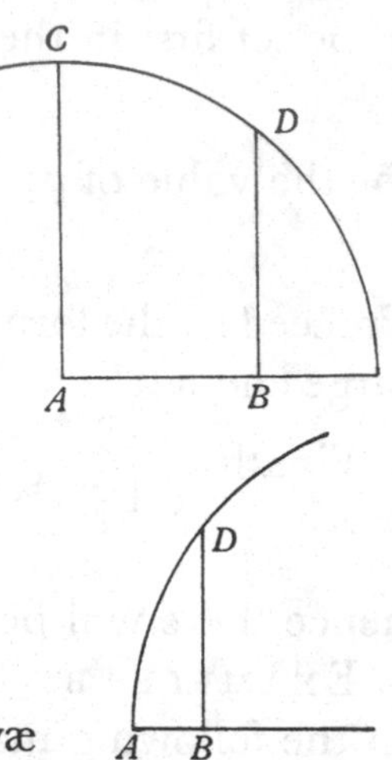

Eodem modo si $\surd : aa - xx = y$ ejus radix erit $a - \frac{x^2}{2a} - \frac{x^4}{8a^3} - \frac{x^6}{16a^5} - \frac{5x^8}{128a^7}$ &c: Adeóq; area circuli quæsita $ABDC$(39) $= ax - \frac{x^3}{6a} - \frac{x^5}{40a^3} - \frac{x^7}{112a^5} - \frac{5x^9}{1152a^7}$ &c.(40) Vel si ponas $\surd : x - xx = y$, erit radix(41) $x^{\frac{1}{2}} - \frac{1}{2}x^{\frac{3}{2}} - \frac{1}{8}x^{\frac{5}{2}} - \frac{1}{16}x^{\frac{7}{2}} - \frac{5}{128}x^{\frac{9}{2}}$ &c et area quæsita

$$ABD^{(42)} = \tfrac{2}{3}x^{\frac{3}{2}} - \tfrac{1}{5}x^{\frac{5}{2}} - \tfrac{1}{28}x^{\frac{7}{2}} - \tfrac{1}{72}x^{\frac{9}{2}} - \tfrac{5}{704}x^{\frac{11}{2}} \text{ \&c:}$$

sive $x^{\frac{1}{2}}$ in $\frac{2}{3}x - \frac{1}{5}x^2 - \frac{1}{28}x^3 - \frac{1}{72}x^4 - \frac{5}{704}x^5$ &c.(43)

Si $\frac{\surd : 1 + ax^2}{\surd : 1 - bx^2} = y$, (cujus quadratura dat longitudinem curvæ Ellipticæ,)(44) extrahendo radicem utramq; prodit

$$\frac{1 + \frac{1}{2}ax^2 - \frac{a^2}{8}x^4 + \frac{a^3}{16}x^6 - \frac{5a^4}{128}x^8}{1 - \frac{1}{2}bx^2 - \frac{bb}{8}x^4 - \frac{b^3}{16}x^6 - \frac{5b^4}{128}x^8}$$

&c. Et dividendo sicut in fractionibus decimalibus, habes

$$1 \begin{array}{l} +\frac{1}{2}b \\ +\frac{1}{2}a \end{array} x^2 \begin{array}{l} +\frac{3}{8}bb \\ +\frac{1}{4}ab \\ -\frac{1}{8}aa \end{array} x^4 \begin{array}{l} +\frac{5}{16}b^3 \\ +\frac{3}{16}abb \\ -\frac{1}{16}aab \\ +\frac{1}{16}a^3 \end{array} x^6 \begin{array}{l} +\frac{35}{128}b^4 \\ +\frac{5}{32}ab^3 \\ -\frac{3}{64}aabb \\ +\frac{1}{32}a^3b \\ -\frac{5}{128}a^4 \end{array} x^8 \quad \text{\&c.}$$

Adeóq; areã quæsitam $x \begin{array}{l} +\frac{1}{6}b \\ +\frac{1}{6}a \end{array} x^3 \begin{array}{l} +\frac{3}{40}b^2 \\ +\frac{1}{20}ab \\ -\frac{1}{40}aa \end{array} x^5$ &c.

Sed observandum est quod operatio non raro abbreviatur per debitam Æquationis præparationem. Ut in allato Exemplo $\frac{\surd : 1 + ax^2}{\surd : 1 - bx^2} = y$ si utramq; partem fractionis per $\surd : 1 - bx^2$ multiplices prodibit $\frac{\sqrt{: 1 \begin{array}{l} +a \\ -b \end{array} x^2 - abx^4}}{1 - bx^2} = y$, & reliquum opus perficitur extrahendo radicem numeratoris tantum & dividendo per denominatorem.

(38) 'Et (per Reg. 2.) Area quæsita *ABDC* erit $= \dots$ &c. Et hæc est Quadratura Hyperbolæ' (Jones, 1711).

(39) $\int_0^x (a^2 - x^2)^{\frac{1}{2}} . dx = \frac{1}{2}(a^2 \sin^{-1}(x/a) + x\sqrt{[a^2 - x^2]})$. Compare I, 1, 3, §4: note (13).

(40) '...Area quæsita *ABDC* erit æqualis ax...&c. Et hæc est Quadratura Circuli' (Jones, 1711).

(41) Jones in his 1711 version here added 'æqualis infinitæ seriei'.

$$ABDC = ax + \frac{x^3}{6a} - \frac{x^5}{40a^3} + \frac{x^7}{112a^5} - \frac{5x^9}{1152a^7} \ldots.^{(38)}$$

In the same way if $\surd(a^2 - x^2) = y$, its root will be

$$a - \frac{x^2}{2a} - \frac{x^4}{8a^3} - \frac{x^6}{16a^5} - \frac{5x^8}{128a^7} \ldots,$$

and consequently the circle area sought will be

$$ABDC^{(39)} = ax - \frac{x^3}{6a} - \frac{x^5}{40a^3} - \frac{x^7}{112a^5} - \frac{5x^9}{1152a^7} \ldots.^{(40)}$$

Or should you set $\surd(x - x^2) = y$ the root will be $x^{\frac{1}{2}} - \frac{1}{2}x^{\frac{3}{2}} - \frac{1}{8}x^{\frac{5}{2}} - \frac{1}{16}x^{\frac{7}{2}} - \frac{5}{128}x^{\frac{9}{2}} \ldots$ and the area sought $ABD^{(42)} = \frac{2}{3}x^{\frac{3}{2}} - \frac{1}{5}x^{\frac{5}{2}} - \frac{1}{28}x^{\frac{7}{2}} - \frac{1}{72}x^{\frac{9}{2}} - \frac{5}{704}x^{\frac{11}{2}} \ldots$, or $x^{\frac{1}{2}}$ multiplied into $\frac{2}{3}x - \frac{1}{5}x^2 - \frac{1}{28}x^3 - \frac{1}{72}x^4 - \frac{5}{704}x^5 \ldots.^{(43)}$

If $\dfrac{\surd(1 + ax^2)}{\surd(1 - bx^2)} = y$ (whose quadrature yields the length of an elliptic arc),(44) then by extraction of both roots there arises

$$\frac{1 + \frac{1}{2}ax^2 - \frac{1}{8}a^2x^4 + \frac{1}{16}a^3x^6 - \frac{5}{128}a^4x^8 \ldots}{1 - \frac{1}{2}bx^2 - \frac{1}{8}b^2x^4 - \frac{1}{16}b^3x^6 - \frac{5}{128}b^4x^8 \ldots}$$

by dividing as is done in decimal fractions you have

$$1 \left.\begin{array}{l} +\frac{1}{2}b \\ +\frac{1}{2}a \end{array}\right\} x^2 \left.\begin{array}{l} +\frac{3}{8}b^2 \\ +\frac{1}{4}ab \\ -\frac{1}{8}a^2 \end{array}\right\} x^4 \left.\begin{array}{l} +\frac{5}{16}b^3 \\ +\frac{3}{16}ab^2 \\ -\frac{1}{16}a^2b \\ +\frac{1}{16}a^3 \end{array}\right\} x^6 \left.\begin{array}{l} +\frac{35}{128}b^4 \\ +\frac{5}{32}ab^3 \\ -\frac{3}{64}a^2b^2 \\ +\frac{1}{32}a^3b \\ -\frac{5}{128}a^4 \end{array}\right\} x^8 \ldots,$$

and consequently the are ι sought

$$x \left.\begin{array}{l} +\frac{1}{6}b \\ +\frac{1}{6}a \end{array}\right\} x^3 \left.\begin{array}{l} +\frac{3}{40}b^2 \\ +\frac{1}{20}ab \\ -\frac{1}{40}a^2 \end{array}\right\} x^5 \ldots.$$

But it should be observed that the operation may not infrequently be shortened by a due preparation of the equation. As in the example adduced $\dfrac{\surd(1 + ax^2)}{\surd(1 - bx^2)} = y$ if you should multiply either part of the fraction by $\surd(1 - bx^2)$ there will arise $\dfrac{\surd(1 + [a - b]\, x^2 - abx^4)}{1 - bx^2} = y$ and the remainder of the work is completed merely by extracting the root of the numerator and dividing by the denominator.

(42) $\int_0^x (x - x^2)^{\frac{1}{2}}.dx = -\frac{1}{8}(\sin^{-1}(1 - 2x) + 2(1 - 2x)\surd[x - x^2])$. In 1711 Jones replaced the following equality sign by 'æqualis erit'.

(43) In his 1711 version Jones here added 'Et hæc est Areæ Circuli Quadratura'.

(44) Precisely, that of the ellipse (counted from the end of its minor axis), whose defining equation in rectangular Cartesian co-ordinates is $bx^2 + (a + b)^{-1}b^2y^2 = 1$.

Ex hisce credo satis patebit modus reducendi quemlibet valorem ipsius y (quibuscunqꝫ radicibus vel denominatoribus sit perplexus, ut hic videre est $x+\frac{\sqrt{x-\sqrt{:1-x^2}}}{\sqrt{3:ax^2+x^3}}-\frac{\sqrt{5:x^3+2x^5-x^{\frac{3}{2}}:}}{\sqrt{3:x+x^2-\sqrt{:2x-x^{\frac{2}{3}}}}}=y$) in series infinitas simplicium terminorum, ex quibus, per Reg 2, quæsita superficies cognoscetur.

meralis tionum ctarum solutio.

Exempla per resolutionem Æquationum affectarum. Quia tota difficultas in Resolutione latet, modū quo ego utor in æquatione numerali primùm illustrabo.(45)

Sit $y^3-2y-5=0$ resolvenda: Et sit 2 numerus qui minùs quàm decimâ sui parte differt a radice quæsitâ. Tum pono $2+p=y$, & substituo hunc sibi valorem in Æquationem; & inde nova prodit $p^3+6p^2+10p-1=0$, cujus radix p exquirenda est ut quotienti addatur: Nempe (neglectis p^3+6p^2 ob parvitatem) $10p-1=0$, sive $p=0,1$ prope veritatē est;(46) itaqꝫ scribo 0, 1 in quotiente, & suppono $0,1+q=p$ & hunc ejus valorem, ut priùs, substituo, unde prodit $q^3+6,3q^2+11,23q+0,061=0$.

$$\left(\begin{array}{r}+2,10000000\\-0,00544853\end{array}\right.$$
$$2,09455147$$

$2+p=y)$	y^3	$+8+12p+6pp+p^3$
	$-2y$	$-4-2p$
	-5	-5
	Summa	$-1+10p+6p^2+p^3$
$0,1+q=p)$	$+p^3$	$+0,001+0,03q+0,3q^2+q^3$
	$+6p^2$	$+0,06\ +1,2\ +6,0$
	$+10p$	$+1,\ +10,$
	-1	$-1,$
	Summa	$+0,061+11,23q+6,3q^2+q^3$
$-0,0054+r=q)$	(47) $6,3q^2$	$+0,000183708-0,06804r+6,3r^2$
	$+11,23q$	$-0,060642\ +11,23$
	$+0,061$	$+0,061$
	Summa	$+0,000541708+11,16196r+6,3rr$
$-0,00004853$		

(45) Newton's method in the numerical example which follows is essentially an improved version of the procedure, expounded by Viète and simplified by Oughtred, on which he made elaborate notes in late 1664 (I, 1, 2, §1: especially note (15)). If we suppose that A is a close approximation to the root $A+E$ of the numerical equation $F(y) = f(y)-N = 0$, Viète had derived a recursive procedure for evaluating the difference E as the limit as $n \to \infty$ of $\sum_{0\leqslant i\leqslant n} (E_i)$, where $E_0 = 0$ and successively $E_{i+1} \approx -F(A+E_i)/g(A+E_i)$, $i = 0, 1, 2, \ldots, n-1$, on setting $g(A+E_i) = F(A+E_i+1)-F(A+E_i)-1$. Newton's improvement is to expand $F(A+E_i+1)$ in powers of the unit as $F(A+E_i)+F'(A+E_i)+\frac{1}{2}F''(A+E_i)+\ldots+1$. By the mean value theorem it follows that $g(A+E_i) \approx F'(A+E_i)$ is the value of the derivative of F when $y = A+E_i$, and this is Newton's approximation: namely, $E_{i+1} \approx -F(A+E_i)/F'(A+E_i)$. His first published use of this iterative process of approximating to the root of the equation $F(y) = 0$ was in his *Principia* in 1687 (Book 1, Prop. 31, Scholium) when he sketched its

From these examples, I believe, will be sufficiently evident the way to reduce any value of the variable y (however complicated with roots and denominators it may be, as here may be seen: $x+\frac{\sqrt{(x-\sqrt{[1-x^2]})}}{\sqrt[3]{(ax^2+x^3)}}-\frac{\sqrt[5]{(x^3+2x^5-x^{\frac{3}{2}})}}{\sqrt[3]{(x+x^2)}-\sqrt{(2x-x^{\frac{2}{3}})}}=y$) into infinite series of simple terms, and from these, by Rule 2, will be known the surface sought.

The numeric resolutio affected equatior

Examples by the resolution of affected Equations. Since the difficulty here lies wholly in the resolution technique, I will first elucidate the method I use in a numerical equation.

Suppose $y^3-2y-5=0$ is to be resolved: and let 2 be the number which differs from the root sought by less than its tenth part. Then I set $2+p=y$ and substitute this value for it in the equation, and in consequence there arises the new equation

$$p^3+6p^2+10p-1=0$$

whose root p must be sought for to be added to the quotient: specifically (when p^3+6p^2 are neglected on account of their smallness) $10p-1=0$ or $p=0{\cdot}1$ very nearly true;(46) and so I write $0{\cdot}1$ in the quotient and suppose $0{\cdot}1+q=p$,

$$\begin{cases}+2{\cdot}10000000\\-0{\cdot}00544853\end{cases}$$
$$2{\cdot}09455147$$

$2+p=y$	y^3	$+8+12p+6p^2+p^3$
	$-2y$	$-4-2p$
	-5	-5
	Sum	$-1+10p+6p^2+p^3$
$0{\cdot}1+q=p$	$+p^3$	$+0{\cdot}001+0{\cdot}03q+0{\cdot}3q^2+q^3$
	$+6p^2$	$+0{\cdot}06\ \ +1{\cdot}2\ \ +6{\cdot}0$
	$+10p$	$+1\ \ +10$
	-1	-1
	Sum	$+0{\cdot}061+11{\cdot}23q+6{\cdot}3q^2+q^3$
$-0{\cdot}0054+r=q$	(47) $6{\cdot}3q^2$	$+0{\cdot}000183708-0{\cdot}06804r+6{\cdot}3r^2$
	$+11{\cdot}23q$	$-0{\cdot}060642\ \ +11{\cdot}23$
	$+0{\cdot}061$	$+0{\cdot}061$
	Sum	$+0{\cdot}000541708+11{\cdot}16196r+6{\cdot}3r^2$
$-0{\cdot}00004853$		

and on substituting this value for it, as before, there arises in consequence

$$q^3+6{\cdot}3q^2+11{\cdot}23q+0{\cdot}061=0.$$

---

application to the Keplerian equation $y-e\sin y-N=0$. The first published systematic discussion of the procedure was that given by Joseph Raphson in his *Analysis Æquationum Universalis, seu ad Æquationes Algebraicas resolvendas Methodus Generalis et Expedita, ex Nova Infinitarum Serierum Methodo Deducta et Demonstrata* (London, $_1$1690, $_2$1697, $_3$1704).

(46) In the notation established previously, $F(y)=y^3-2y-5$, $A=2$, $E_1=p$ and so $p\approx -F(2)/F'(2)$.

(47) A first term '$+q^3$' in the column has been cancelled, presumably in accordance with Newton's observation (note (51)) that $(0{\cdot}0054)^3\approx 0{\cdot}00000016$ is negligible in the present context.

Et cùm $11,23q+0,061[=0]$ ad veritatē[48] prope accedit, sive ferè sit $q=-0,0054$[49] (dividendo nempe donec tot eliciantur figuræ quot locis primæ figuræ hujus & principalis quotientis exclusivè distant,) scribo $-0,0054$ in inferiori parte quotientis, cùm negativa sit. Et supponens $-0,0054+r=q$, hunc ut priùs substituo. Et operationem sic produco quousq; placuerit. Verùm si ad bis tot figuras tantùm quot in quotiente jam reperiuntur, unâ dempta, operam continuare cupio, pro q substituo $-0,0054+r$ in hanc $6,3qq+11,23q+0,061$,[50] primo ejus termino (q^3) propter exilitatem suam neglecto:[51] Et prodit

$$6,3rr+11,16196r+0,000541708=0$$

ferè[,] sive (rejecto $6,3rr$) $r=\dfrac{-0,000541708}{11,16196}=-0,00004853$ ferè, quam scribo in negativa parte quotientis. Denique negativam partem quotientis ab affirmativa subducens, habeo 2,09455147 quotientem quæsitam.[52]

Æquationes plurium dimensionum nihilo seciùs resolvuntur, & operam sub fine, ut hic factum fuit, levabis si primos ejus terminos gradatim omiseris.

Præterea notandum est quòd in hoc exemplo si dubitarem an $0,1=p$ ad veritatem satis accederet, pro $10p-1=0$ finxissem $6pp+10p-1=0$ & ejus radicis[53] primam figuram in quotiente scripsissem. Et secundam vel etiam[54] tertiam quotientis figuram sic explorare convenit ubi in æquatione ista ultimò resultante quadratum coefficientis penultimi termini non sit decies major quàm factus ex ultimo termino ducto in coefficientem termini antepenultimi. Imò laborem plerumq; minues præsertim in æquationibus plurimarum dimensionum, si figuras omnes quotienti addendas dicto modo (hoc est extrahendo minorem radicum ex tribus ultimis terminis æquationis novissimè resultantis)[55] exquiras. Isto enim modo figuras duplo plures in quotiente qualibet vice[56] lucraberis.[57]

Hæc methodus de resolvendis Æquationibus[58] pervulgata an sit nescio,[59]

(48) Here, and several times subsequently, Newton has cancelled the equivalent 'veritati' (which in each case is Jones' reading in his 1711 text).

(49) $q = E_2 \approx -F(2\cdot1)/F'(2\cdot1)$.

(50) 'scilicet' (namely) is cancelled.

(51) Compare note (49).

(52) In conclusion, Newton finds $r = E_3 \approx -F(2\cdot0946)/F'(2\cdot0946)$ and supposes that E is $E_1+E_2+E_3$ to a near enough approximation. In fact, the true value of the root sought is 2·09455148 (to 8D): compare E. T. Whittaker and G. Robinson, *The Calculus of Observations* (London, 1924): 86, note †.

(53) $\frac{1}{6}(-5+\sqrt{31}) \approx 0\cdot0946$, which to 1$D$ is 0·1 as before.

(54) Jones' 1711 version omits 'etiam'.

(55) The less correct participle 'propositæ' (proposed) has been cancelled.

(56) 'qualibet vice Quotienti' (Jones, 1711).

(57) This refinement was to be pursued by Edmond Halley in a paper, 'Methodus Nova, Accurata & Facilis inveniendi Radices Æquationum quarumcunque generaliter, sine prævia Reductione' (*Philosophical Transactions*, **11**, No. 210 (for May 1694): 136–48), later appended

And, since $11{\cdot}23q+0{\cdot}061[=0]$ approaches the truth closely or there is almost $q=-0{\cdot}0054$(49) (by dividing, that is, until as many figures are elicited as the number of places by which the first figures of this and of the principal quotient are distant one from the other), I write $-0{\cdot}0054$ in the lower part of the quotient since it is negative. Again, supposing $-0{\cdot}0054+r=q$, I substitute this as before, and in this way continue the operation as far as I please. But if I desire to continue working merely to twice as many figures, less one, as are now found in the quotient, in place of q in this equation

$$6{\cdot}3q^2+11{\cdot}23q+0{\cdot}061[=0]^{(50)}$$

I substitute $-0{\cdot}0054+r$, neglecting its first term q^3 by reason of its insignificance,(51) and there arises $6{\cdot}3r^2+11{\cdot}16196r+0{\cdot}000541708=0$ nearly, or (when $6{\cdot}3r^2$ is rejected) $r=\dfrac{-0{\cdot}000541708}{11{\cdot}16196}=-0{\cdot}00004853$ nearly. This I write in the negative part of the quotient. Finally, on taking the negative portion of the quotient from the positive part, I have the required quotient $2{\cdot}09455147$.(52)

Equations of more dimensions are resolved not at all differently and you will lighten your labour towards the end if, as was done here, you omit their first terms step by step.

It must be noted, further, that if in this example I had had doubts whether $p=0{\cdot}1$ approached the truth sufficiently, I should have supposed

$$6p^2+10p-1=0$$

in place of $10p-1=0$ and written the root's(53) first figure in the quotient. And it is appropriate to seek out the second or even the third figure of the quotient in this way when in the last resulting equation the square of the coefficient of the penultimate term is not ten times greater than the product of the last term multiplied into the coefficient of the last term but two. Indeed you will for the most part lessen your task, particularly in equations of very many dimensions, if you should search out all the figures to be added to the quotient in the manner I have stated (that is, by extracting the lesser of the roots out of the last three terms of the equation most recently resulting): for in that way you will gain twice as many figures each time in the quotient.(57)

I do not know whether this method of resolving equations is widely known or

by Whiston to the first edition of Newton's *Arithmetica Universalis: sive De Compositione et Resolutione Arithmetica Liber* (Cambridge, 1707): 327–43—and removed by the last from its second (Latin) edition in 1722!

(58) 'resolvendi Æquationes' (Jones, 1711).

(59) In fact, this 'Newton–Raphson' method of resolving numerical equations has a long manuscript history, and its essential structure was known to the fifteenth-century Arabic mathematician al-Kāšī. (The latter used a primitive form of it in root extractions of the form $y^p-N=0$: see B. A. Rosenfeld's and A. P. Yushkevich's 1956 edition of his *Key to Arithmetic*, Book 1.) In western Europe, however, the method was little known. Indeed, we

certè mihi videtur præ reliquis simplex & usui accommodata. Demonstratio ejus ex ipso modo operandi patet, unde cum opus sit in memoriam facilè revocatur. Æquationes in quibus vel aliqui vel nulli termini desint eadem fere facilitate perficit.(60) Et æquatio semper relinquitur cujus radix una cum acquisita quotiente adæquat radicem æquationis(61) primò propositæ: unde examinatio operis hic æque poterit institui ac in reliqua Arithmetica, auferendo nempe quotientem a radice primæ æquationis (sicut Analistis notum est*) ut æquatio ultima vel termini ejus duo tresve ultimi producantur inde. Quicquid laboris hic est, istud in substituendo quantitates unas pro alijs reperietur. Id quod variè possis perficere,(63) at sequentem modū maximè expeditum puto, præsertim cum(64) coefficientes numeri constant ex pluribus figuris. Sit $p+3$ substituenda pro y in hanc $y^4-4y^3+5y^2-12y+17=0$: cum ista potest resolvi in hanc formā $\overline{\overline{\overline{y-4}\times y:+5}\times y:-12}\times y:+17=0$. Æquatio nova sic generabitur $\overline{p-1}\times\overline{p+3}=pp+2p-3$. & $pp+2p+2$ in $p+3=p^3+5p^2+8p+6$. & p^3+5p^2+8p-6 in $p+3=p^4+8p^3+23pp+18p-18$. &

$$p^4+8p^3+23p^2+18p-1=0,$$

quæ quærebatur.(65)

[margin: *Geometr Cartesij(62)]

[margin: Literalis æquationum affectarum resolutio.]

His in numeris sic ostensis: Sit æquatio literalis, $y^3+aay-2a^3+axy-x^3=0$, resolvenda. (66)Primùm inquiro valorem ipsius y cùm x sit nulla, hoc est, elicio radicem hujus æquationis $y^3+aay-2a^3=0$; & invenio esse $+a$.(67) Itaꝗ scribo $+a$ in quotiente(68) & supposito $+a+p=y$, pro y substituo valorem istum,(69) & terminos inde resultantes (p^3+3ap^2+4aap &c) margini appono; ex quibus assumo $+4aap+aax$ ubi p & x seorsim sunt minimarum dimensionum & eas

have been able to trace only a unique occurrence of it which might have been known to Newton: namely, Henry Briggs' use of it in Chapters 4 and 6 of Book 1 of his posthumously published *Trigonometria Britannica* (Gouda, 1633) in resolving the 'trisection' cubic

$$3y-y^3-N=0$$

and the 'quinquisection' quintic $5y-5y^3+y^5-N=0$. (Briggs pursues the former through four stages, the latter through two, and the approach would seem to be his independent discovery.) That Newton does not here know of Briggs' employment of the method seems firm proof that he had not yet read the *Trigonometria*.

(60) 'tractantur' (Jones, 1711).

(61) 'quotientis' (quotient) is cancelled.

(62) This marginal note (not reproduced in Jones' 1711 text) refers presumably to Descartes' [*Geometrie* (Leyden, 1637): 374 =] *Geometria*, ${}_2$1659, Liber III: 71: 'Quomodo augeri vel diminui possint Æquationis radices, ipsis non cognitis'. See also I, **3**, 3, §4.

(63) The equivalent 'perficias' is cancelled, and is also Jones' reading in 1711.

(64) 'ubi' (Jones, 1711). Jones failed also to obey Newton's order to invert the two words following.

(65) Compare I, **3**, 3, §1. Newton's procedure is, of course, equivalent to a contracted division of the equation by $y-3$.

not,[59] but certainly in comparison with others it is both simple and suited to practice. Its proof is evident from the mode of operation itself, and in consequence is easily recalled to mind when needed. It deals efficiently with equations in which either some or no terms are lacking with almost identical ease. Invariably, too, an equation is left whose root together with the quotient already gained equals the root of the equation first proposed: hence a check on the working may be made here equally as well as in other types of arithmetic, namely by taking away the quotient from the root of the first equation (an operation known to analytical mathematicians*) so that the final equation or its two or three final terms may be produced therefrom. What hard work there is here will be found in substituting one group of quantities for another. That you might accomplish by various methods, but I think the following way exceedingly expedite, especially when the numerical coefficients consist of several figures. Suppose $p+3$ has to be substituted for y in this equation

* Desca[rtes] Geomet[rie]

$$y^4-4y^3+5y^2-12y+17=0:$$

since that may be resolved into this form $\{[(y-4)\times y+5]\times y-12\}\times y+17=0$, the new equation may be generated thus:

$$(p-1)(p+3)=p^2+2p-3;\quad (p^2+2p+2)(p+3)=p^3+5p^2+8p+6;$$

$$(p^3+5p^2+8p-6)(p+3)=p^4+8p^3+23p^2+18p-18;\text{ and}$$

$$p^4+8p^3+23p^2+18p-1=0,$$

which was required.[65]

The alg[ebraic] resoluti[on of] affected equatio[ns]

So much for illustration in numerical examples. Suppose now that the algebraic equation $y^3+a^2y-2a^3+axy-x^3=0$ has to be resolved.[66] First I seek out the value of y when x is zero, that is, I elicit the root of this equation

$$y^3+a^2y-2a^3=0,$$

and find it to be $+a$.[67] And so I write $+a$ in the quotient.[68] Again, supposing $y=a+p$, for y I substitute that value and the terms $p^3+3ap^2+4a^2p\ldots$ which thence result I set in the margin. Out of these I take $4a^2p+a^2x$, in which p and x

(66) Understand here, in juxtaposition to the complementary phrasing of the 'Alius modus' which follows, 'Si velis ut valor y tanto magis veritati accedat quanto x sit minor' (If you wish that the value of y should more closely approach the truth the smaller x is).

(67) Newton understandably rejects the conjugate complex roots $\frac{1}{2}a(1\pm i\sqrt{7})$.

(68) Given the 'literal' equation $f(x, y)=0$, Newton seeks to evaluate y as a power series $\phi(x)$ which, it is hoped, is convergent for small x. Evidently the constant term $\phi(0)$ is determined by the condition $f(0, \phi(0))=0$. An equivalent expansion of x as a power series $\phi^{-1}(y)$ will begin from the condition $f(\phi^{-1}(0), 0)=0$, whose real root $\phi^{-1}(0)=-a\sqrt[3]{2}$. Newton will deal with the complications which arise when the given equation $f(x, y)=0$ lacks a constant and other lower-order terms in the revised version of the present passage which he incorporated in his 1671 fluxional tract.

(69) 'supponens $+a+p=y$, substituo pro y valorem ejus' (Jones, 1711).

nihilo ferè æquales suppono, sive $p=\frac{-x}{4}$ ferè, sive $p=\frac{-x}{4}+q$. Et scribens $-\frac{x}{4}$ in quotiente, substituo $\frac{-x}{4}+q$ pro p. Et terminos inde resultantes iterum in margine scribo, ut vides in annexo schemate. Et inde assumo quantitates $+4aaq-\frac{1}{16}axx$, in quibus q & x seorsim sunt minimarum dimensionū & fingo $q=\frac{xx}{64a}$ ferè, sive $q=\frac{+xx}{64a}+r$; & adnectens $\frac{+xx}{64a}$ quotienti, substituo $\frac{xx}{64a}+r$ pro q; & sic procedo quousqꝫ placuerit.(70)

		$a-\frac{x}{4}+\frac{x^2}{64a}+\frac{131x^3}{512a^2}+\frac{509x^4}{16384a^3}$ &c
$+a+p=y.)$	$+y^3$	$+a^3\ +3aap+3app+p^3$
	$+aay$	$+a^3\ +aap$
	$+axy$	$+aax+axp$
	$-2a^3$	$-2a^3$
	$-x^3$	$-x^3$
$-\frac{1}{4}x+q=p.)$	$+p^3$	$-\frac{1}{64}x^3+\frac{3}{16}xxq-\frac{3}{4}xqq+q^3$
	$+3ap^2$	$+\frac{3}{16}ax^2-\frac{3}{2}axq+3aqq$
	$+4aap$	$-aax+4aaq$
	$+axp$	$-\frac{1}{4}axx+axq$
	$+aax$	$+aax$
	$-x^3$	$-x^3$
$+\frac{xx}{64a}+r=q.)$	$+3aqq$	$+\frac{3x^4}{4096a}+\frac{3}{32}xxr+3arr$
	$+4aaq$	$+\frac{1}{16}axx+4aar$
	$-\frac{1}{2}axq$	$-\frac{1}{128}x^3-\frac{1}{2}axr$
	$+\frac{3}{16}xxq$	$+\frac{3x^4}{1024a}+\frac{3}{16}xxr$
	$-\frac{1}{16}axx$	$-\frac{1}{16}axx$
	$-\frac{65}{64}x^3$	$-\frac{65}{64}x^3$

$$+4aa-\frac{1}{2}ax+\frac{9}{32}x^2\Big)\frac{+131}{128}x^3-\frac{15x^4}{4096a}\Big(\frac{+131x^3}{512aa}+\frac{509x^4}{16384a^3}\ \text{[\&c]}.$$

Sin duplo tantùm plures quotienti terminos,(71) uno dempto, jungendos adhuc vellem: primo termino (q^3) æquationis novissimè resultantis misso, & ista etiam parte $\left(\frac{-3}{4}xqq\right)$ secundi ubi x est tot dimensionum quot in penultimo termino

(70) The terms p, q, r, ... will be of respective orders x, x^2, x^3, ... in Newton's scheme and on this basis (in generalization of the preceding resolution of numerical equations where these

separately are of least dimension and suppose them nearly equal to zero, that is, $p=-\frac{1}{4}x$ nearly or $p=-\frac{1}{4}x+q$. And, writing $-\frac{1}{4}x$ in the quotient, I substitute $-\frac{1}{4}x+q$ for p, and the terms thence resulting I again write in the margin, as you see in the appended scheme. From them I take out the quantities $4a^2q-\frac{1}{16}ax^2$, in which q and x separately are of least dimension, and conjecture that $q=x^2/64a$ nearly, or $q=(x^2/64a)+r$: then, appending $+x^2/64a$ to the quotient, I substitute $(x^2/64a)+r$ for q; and so proceed as far as I please.(70)

$$\begin{array}{l r|l}
 & & a-\frac{1}{4}x+\frac{x^2}{64a}+\frac{131x^3}{512a^2}+\frac{509x^4}{16384a^3}\ldots \\
\hline
a+p=y. & y^3 & +a^3\ +3a^2p+3ap^2+p^3 \\
 & a^2y & +a^3\ +a^2p \\
 & +axy & +a^2x+axp \\
 & -2a^3 & -2a^3 \\
 & -x^3 & -x^3 \\
\hline
-\frac{1}{4}x+q=p. & p^3 & -\frac{1}{64}x^3+\frac{3}{16}x^2q-\frac{3}{4}xq^2+q^3 \\
 & +3ap^2 & +\frac{3}{16}ax^2-\frac{3}{2}axq+3aq^2 \\
 & +4a^2p & -a^2x+4a^2q \\
 & +axp & -\frac{1}{4}ax^2+axq \\
 & +a^2x & +a^2x \\
 & -x^3 & -x^3 \\
\hline
+\frac{x^2}{64a}+r=q. & 3aq^2 & +\frac{3x^4}{4096a}+\frac{3}{32}x^2r+3ar^2 \\
 & +4a^2q & +\frac{1}{16}ax^2+4a^2r \\
 & -\frac{1}{2}axq & -\frac{1}{128}x^3-\frac{1}{2}axr \\
 & +\frac{3}{16}x^2q & +\frac{3x^4}{1024a}+\frac{3}{16}x^2r \\
 & -\frac{1}{16}ax^2 & -\frac{1}{16}ax^2 \\
 & -\frac{65}{64}x^3 & -\frac{65}{64}x^3 \\
\hline
\end{array}$$

$$4a^2-\frac{1}{2}ax+\frac{9}{32}x^2\Big)\,\frac{131}{128}x^3-\frac{15x^4}{4096a}\Big(\frac{131x^3}{512a^2}+\frac{509x^4}{16384a^3}[\ldots]$$

But should I wish merely twice as many terms,(71) less one, to be further adjoined to the quotient, I omit the first term (q^3) of the equation which most recently results and also that portion ($-\frac{3}{4}xq^2$) of the second, in which x is of as many

partial quotients are of order 10^{-i}, $i=1, 2, 3, \ldots$) he equates successively at each stage the terms of lowest dimension in x to zero. In the first stage, that is, he sets $4aap+aax=0$ since $p=O(x)$; in the second, $4aaq-\frac{1}{16}axx=0$ since $q=O(x^2)$; and so on correspondingly.

(71) See note (109) below.

quotientis; in reliquos terminos ($3aqq+4aaq$ &c) margini adscriptos ut vides, substituo $\frac{xx}{64a}+r$ pro q. Et ex ultimis duobus terminis

$$\left(\frac{15x^4}{4096a}-\frac{131}{128}x^3\frac{+9}{32}xxr-\frac{1}{2}axr+4aar\right)$$

æquationis inde resultantis, facta divisione $4aa-\frac{1}{2}ax+\frac{9}{32}xx\Big)+\frac{131}{128}x^3-\frac{15x^4}{4096a}$, elicio $\frac{+131x^3}{512aa}+\frac{509x^4}{16384a^3}$ quotienti adnectendos.(72) Deniq; quotiens ista $\left(a-\frac{x}{4}+\frac{xx}{64a}\text{ \&c}\right)$ per Reg 2^{dam} dabit $ax-\frac{xx}{8}+\frac{x^3}{192a}+\frac{131x^4}{2048a^2}+\frac{509x^5}{81920a^3}$ &c pro area quæsita, quæ ad veritatem tanto magis accedit quanto x sit minor.

modus easdem ...lvendi.

Sin velis ut valor areæ tanto magis veritati accedat(73) quanto x sit major, exemplum esto $y^3+axy+xxy-a^3-2x^3=0$; Itaq; hanc resoluturus excerpo terminos $y^3+xxy-2x^3$ in quibus x & y vel seorsim vel simul multiplicatæ sunt & plurimarum & æqualium ubiq; dimensionum.(74) Et ex ijs quasi nihilo æqualibus radicem elicio, quam invenio esse x, & hanc in(75) quotiente scribo. Vel quod eodem recidit, ex y^3+y-2 (unitate pro x substitutâ) radicem (1)(76) extraho,(77) & eam per x multiplico, & factum (x) in quotiente scribo. Deinde pono $x+p=y$ & sic procedo ut in priori exemplo donec habeo quotientem $x-\frac{a}{4}+\frac{aa}{64x}+\frac{131a^3}{512xx}+\frac{509a^4}{16384x^3}$ &c, Adeóq; aream

$$\frac{x^2}{2}-\frac{ax}{4}+\boxed{\frac{aa}{64x}}^{(78)}-\frac{131a^3}{512x}-\frac{509a^4}{32768x^2}\quad[\&c]$$

de qua vide Exempla tertia Reg 2æ. Lucis gratia dedi hoc exemplum in omnibus idem cum priori, modò x & a sibi invicem ibi substituuntur, ut non opus esset aliud resolutionis paradigma(79) hic adjungere.

---

(72) Newton collects all terms of order less than x^5 in his divisor and dividend, so that the quotient is indeed exact to $O(x^5)$. (Note that the term $\frac{9}{32}x^2r$ in the divisor is superfluous since it is of order x^5.)

(73) 'Sin valor...ad veritatem accedere debet' (Jones, 1711).

(74) As before (note (68)) more complicated equations which do not yield to this present simple rule will be dealt with in Newton's 1671 tract by his 'parallelogram'. It is tempting to suppose he has been helped with his study of the behaviour of equations for large x by his work on determining the infinite branches of cubics (I, 1, passim.). Indeed the present rule gives, in simple cases, the directions of the asymptotes to the curve whose defining equation is $f(x, y) = 0$ where previously its counterpart determined that of the tangents at the points where it intersected the y-axis ($x = 0$).

(75) '. Hanc invenio esse x, & in' (Jones, 1711).

(76) Omitting the complex pair $\frac{1}{2}(1\pm i\sqrt{7})$: compare note (67).

(77) 'Radicem extraho quæ his prodit 1' (Jones, 1711).

dimensions as in the penultimate term of the quotient, while I substitute $(x^2/64a)+r$ for q in the remaining terms $(3aq^2+4a^2q\ldots)$ written in the margin, as you see. Again, from the last two terms

$$(\tfrac{15}{4096}x^4/a-\tfrac{131}{128}x^3+\tfrac{9}{32}x^2r-\tfrac{1}{2}axr+4a^2r)$$

of the equation which thence results, on performing the division

$$4a^2-\tfrac{1}{2}ax+\tfrac{9}{32}x^2)\,\tfrac{131}{128}x^3-\tfrac{15}{4096}x^4/a$$

I elicit $+(131x^3/512a^2)+(509x^4/16384a^3)$ to be appended to the quotient.(72) Finally, that quotient $(a-\frac{1}{4}x+[x^2/64a]\ldots)$ will, by Rule 2, yield

$$ax-\frac{1}{8}x^2+\frac{x^3}{192a}+\frac{131x^4}{2048a^2}+\frac{509x^5}{81920a^3}\ldots$$

for the area sought, an expansion which approaches more rapidly to the truth the smaller x is.

But if you wish that the value of the area should approach nearer the truth the greater x is, take this as an example: $y^3+axy+x^2y-a^3-2x^3=0$. Accordingly, ready to resolve this, I take out the terms $y^3+x^2y-2x^3$ in which x and y either separately or multiplied together are of the most and equal dimensions everywhere.(74) From these, set as it were equal to zero, I elicit the root, finding it to be x, and write it in the quotient: or, what comes to the same thing, on substituting unity for x from y^3+y-2 I extract the root (1)(76), multiply it by x and write the product (x) in the quotient. Then I suppose $x+p=y$ and so proceed as in the former example until I have the quotient

Another of resolv[ing] them.

$$x-\frac{1}{4}a+\frac{a^2}{64x}+\frac{131a^3}{512x^2}+\frac{509a^4}{16384x^3}\ldots,$$

so that the area is

$$\frac{1}{2}x^2-\frac{1}{4}ax+\left[\int\frac{a^2}{64x}\cdot dx\right][+]\frac{131a^3}{512x}[+]\frac{509a^4}{32768x^2}\ldots.$$

Relating to this see the third examples of Rule 2. I have given this example, identical in every respect with the former except that there x and a are substituted in each other's place, for clarity's sake so that there would be no need to adjoin here a second pattern for the resolution.

---

(78) That is $\int`\frac{aa}{64x}'.dx$, as we would now write it. This 'square' notation had been used extensively in his manuscript researches in 1665 and 1666, but its present appearance in a paper which achieved a limited private circulation from July 1669 was to be a crucial argument for Newton's priority in inventing the calculus. (Compare Appendix 3.3: note (37) below.) The two minus signs immediately following should be plus.

(79) 'exemplum' (Jones, 1711).

Nota quod area $\left(\frac{x^2}{2}-\frac{ax}{4}+\boxed{\frac{aa}{64x}}\ \&c\right)$ limitatur a curva[80] quæ juxta asymptoton aliquam in infinitum serpit;[81] & termini initiales $\left(x-\frac{a}{4}\right)$ valoris extracti de y, in asymptoton istam semper terminantur: Unde positionem[82] asymptoti facile invenias. Idem semper notandum est cùm area designatur terminis plus plusqȝ divisis per x continuò: præterquam quòd vice asymptoti rectæ quandóqȝ habeatur Parabola Conica vel alia magis composita.[83]

Sed hunc modum missum faciens, utpote particularem quia non applicabilem curvis in orbem ad instar Ellipsium flexis;[84] de altero modo per exemplum $y^3+aay+axy-2a^3-x^3=0$ supra ostenso (scilicet quo dimensiones de[85] x in numeratoribus quotientis perpetuò fiunt plures)[86] annotabo sequentia.

1. Si quando accidit quòd valor ipsius y, cùm x nullū esse fingitur,[87] sit quantitas surda vel penitus ignota, licebit illam litera aliqua designare. Ut in exemplo $y^3+aay+axy-2a^3-x^3=0$, si radix hujus $y^3+aay-2a^3$ fuisset surda vel ignota, finxissem quamlibet (b) pro ea ponendam, et resolutionem ut sequitur perfecissem. Scribens b in quotiente, suppono $b+p=y$, & istum pro y substituo, ut vides; unde nova p^3+3bpp &c resultat, rejectis terminis

$$b^3+aab-2a^3,$$

qui nihilo sunt æquales propterea quod b supponitur radix hujus $y^3+aay-2a^3=0$. Deinde termini $3bbp+aap+abx$

		$\left(b-\frac{abx}{aa+3bb}\right.$ &c
$b+p=y)$	$+y^3$	$b^3+3bbp+3bpp+p^3$
	$+aay$	$+aab+aap$
	$+axy$	$+axb+axp$
	$-2a^3$	$-2a^3$
	$-x^3$	$-x^3$
$\underbrace{\frac{-abx}{aa+3bb}}_{\Vert\ cc}+q=p\Big)$	p^3	$-\frac{a^3b^3x^3}{c^6}$ &c
	$+3bpp$	$+\frac{3a^2b^3x^2}{c^4}-\frac{6ab^2x}{c^2}q$ &c
	$+3bbp$ $+aap$	$\Big\}-[abx+ccq]$
	$+axp$	$-\frac{a^2bx^2}{c^2}+axq$
	$+abx$	$[+abx]$
	$-x^3$	$[-x^3]$ (88)

(80) 'Area autem...terminatur ad Curvam' (Jones, 1711).

(81) The terminology is, of course, taken over from that employed in his cubic researches (1, 1 above).

(82) In 1711 Jones mistranscribed this as 'portionem' and all later editors of the 'De Analysi' have followed him.

(83) As is the case, for example, with species 6 and 7 of the cubic in 1, 1, §3 above. In the present instance the asymptote will be linear ($y = x-\frac{1}{4}a$) since $\lim_{x\to\infty}[(y+\frac{1}{4}a)/x] = 1$.

(84) Evidently such closed curves will have no infinite branches and hence no real asymptotic curves.

(85) 'ipsius' (Jones, 1711).

(86) 'augeantur' (Jones, 1711).

(87) Newton has cancelled the more concrete 'est' (is).

Note that the area $\left(\frac{1}{2}x^2-\frac{1}{4}ax+\left[\int\frac{a^2}{64x}\cdot dx\right]...\right)$ is bounded by a curve which snakes into infinity alongside some asymptote;[81] and that the initial terms $(x-\frac{1}{4}a)$ of the extracted value of y always terminate at that asymptote: you may easily in consequence find the position of that asymptote. The same should always be noted when the area is designated by terms which are continually more and more divided by x, except that in place of a linear asymptote there may sometimes be had one which is a conic parabola or is even more compounded.[83]

But passing over this approach as too restricted, since it is not applicable to curves bent back into closed arcs in the form of ellipses,[84] I will add the following notes on the former method illustrated above by the example

$$y^3+a^2y+axy-2a^3-x^3=0$$

(in which, namely, the dimensions of x in the numerators perpetually become larger).

1. If it happen at any time that the value of y when x is conceived to be[87] zero be a surd quantity or completely unknown, it will be permissible to designate it by some letter. So in the example $y^3+a^2y+axy-2a^3-x^3=0$ if the root of this equation $y^3+a^2y-2a^3$ had been surd or not known, I should have imagined some appropriate letter b to be set in its stead and then completed the resolution as follows. Writing b in the quotient I suppose $b+p=y$ and substitute it for y, as you see: from that results the new equation $p^3+3bp^2...$, where the terms $b^3+a^2b-2a^3$ (equal to zero since b is supposed a root of this equation $y^3+a^2y-2a^3=0$) are rejected. Again, the terms $3b^2p+a^2p+abx$ yield

		$[y-]\ b-\frac{abx}{a^2+3b^2}\cdots$
$b+p=y.$	y^3	$b^3+3b^2p+3bp^2+p^3$
	$+a^2y$	$+a^2b+a^2p$
	$+axy$	$+axb+axp$
	$-2a^3$	$-2a^3$
	$-x^3$	$-x^3$
$\frac{-abx}{c^2}+q=p.$	p^3	$-\frac{a^3b^3x^3}{c^6}\cdots$
[where] $c^2=a^2+3b^2$	$+3bp^2$	$+\frac{3a^2b^3x^2}{c^4}-\frac{6ab^2x}{c^2}q\cdots$
	$\left.\begin{matrix}+3b^2p\\+a^2p\end{matrix}\right\}$	$-[abx\ \ +ccq]$
	$+axp$	$-\frac{a^2bx^2}{c^2}\ \ +axq$
	$+abx$	$[+abx]$
	$-x^3$	$[-x^3]$ (88)

(88) A few terms omitted in the manuscript tabulation have been inserted. Evidently we may round off the procedure at this point by considering terms which are of order less than x^5 and equating them to zero: that is, since q is $O(x^2)$, we may set

$$(3a^2b^3c^{-4}-a^2bc^{-2})x^2-(1+a^3b^3c^{-6})x^3+c^2q+(a-6ab^2c^{-2})xq=0$$

dant $\frac{-abx}{3bb+aa}$ quotienti apponendum & $\frac{-abx}{3bb+aa}+q$ substituendum pro p. &c. Completo opere sumo numerum aliquem pro a, & hanc $y^3+aay-2a^3=0$, sicut de numerali æquatione ostensum supra, resolvo; & radicem ejus pro b substituo.

2. Si dictus valor sit nihil, hoc est si in æquatione resolvenda nullus sit terminus nisi qui per x vel y sit multiplicatus, ut in hac $y^3-axy+x^3=0$; tum terminos $(-axy+x^3)$ seligo in quibus x seorsim & y etiam seorsim si fieri potest,(89) aliàs per x multiplicata, sit minimarum dimensionum. Et illi dant $\frac{+x^2}{a}$ pro primo termino quotientis, & $\frac{x^2}{a}+p$ pro y substituendam.(90) In hâc $y^3-aay+axy-x^3=0$, licebit primum terminum quotientis vel ex $[-]aay-x^3$, vel ex y^3-aay elicere.

3. Si valor iste sit imaginarius ut in hac $y^4+yy-2y+6-xxyy-2x+xx+x^4=0$, augeo vel imminuo quantitatem x donec dictus valor evadat realis. Sic in annexo schemate cum $AC(x)$ nulla est tum $CD(y)$ est imaginaria:(91) Sin minuatur AC per datam AB ut BC fiat x; tum posito quod $BC(x)$ sit nulla, $CD(y)$ erit valore quadruplici (CE, CF, CG & CH) realis;(92) quarum radicum (CE, CF, CG, vel CH) utravis esto(93) primus terminus quotientis, prout superficies $BEDC$, $BFDC$, $BGDC$, vel $BHDC$ desideratur. In alijs etiam casibus, si quando hæsitas, te hoc modo extricabis.

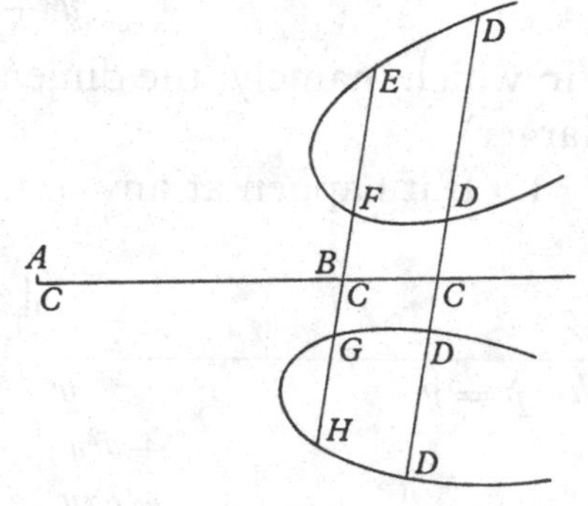

[4.](94) Deniꝗ; si index rationis de(95) x vel y sit fractio, reduco(96) ad inte-

---

and so deduce that

$$q = (a^2bc^{-4}-3a^2b^3c^{-6})x^2+(c^{-2}-a^3bc^{-6}+10a^3b^3c^{-8}-18a^3b^5c^{-10})x^3+O(x^4).$$

(This, of course, reduces to Newton's previous result on replacing b and c by their respective values a and $2a$.) When he came to revise this passage for his 1671 tract Newton himself made a disastrous attempt to round off the expansion, and Jones unwisely incorporated it in his 1711 text. We will return to the point in our next volume (see III, **1**, 2, § 2: note (55)).

(89) The equivalent 'fiat' is cancelled.

(90) Alternatively, we could select the terms y^3-axy and begin the operation from $y = \pm\sqrt{[ax]}$. Newton first began the following sentence with 'Sic' (Thus).

(91) Evidently so since, when x is zero, the equation reduces to $y^4+(y-1)^2+5 = 0$.

(92) For the given equation $x^4-x^2y^2+y^4+x^2+y^2-2x-2y+6 = 0$ this can never be true, unfortunately, and indeed Newton's accompanying figure (whose oblique co-ordinates AC, CD were made rectangular by Jones in his 1711 text) is badly wrong. If we set $s = \sin\frac{1}{2}A\hat{C}D$ and $c = \cos\frac{1}{2}A\hat{C}D$, we may transform Newton's co-ordinates $AC = x$, $CD = y$ to the new, rectangular pair $A\gamma = X$, $\gamma D = Y$ (parallel to the bisectors of $A\hat{C}D$) by taking

$$\begin{Bmatrix} X = s(x+y) \\ Y = c(x-y) \end{Bmatrix}:$$

$-abx/(a^2+3b^2)$ to be appended to the quotient and $-abx/(a^2+3b^2)+q$ to be substituted for p. And so on. When the work is completed, I take some number for a and resolve this $y^3+a^2y-2a^3=0$ as was shown above in the section on numerical equations: its root I then substitute in place of b.

2. If the said value be zero, that is, if in the equation to be resolved there be no term not multiplied by either x or y (as happens in this, $y^3-axy+x^3=0$), then I select the terms $(-axy+x^3)$ in which x on its own and also y on its own if it happens so, or, failing that, multiplied by x, are each of lowest dimensions. These give $+x^2/a$ for the first term of the quotient and $(x^2/a)+p$ to be substituted in place of y.[90] In this case $y^3-a^2y+axy-x^3=0$ it will be permissible to elicit the first term of the quotient either from $-a^2y-x^3$ or from y^3-a^2y.

3. If that value be imaginary, as in this case

$$y^4+(1-x^2)\,y^2-2y+6-2x+x^2+x^4=0,$$

I increase or diminish the quantity x until the said value comes out real. Thus in the accompanying sketch when $AC(x)$ is zero, then $CD(y)$ is imaginary:[91] but if AC be diminished by the given quantity AB so that BC now becomes x, then when $BC(x)$ is supposed zero, $CD(y)$ will be real[92] and fourfold in value (CE, CF, CG and CH). Let any of these roots (CE, CF, CG or CH) be the first term of the quotient according as the surfaces $BEDC$, $BFDC$, $BGDC$ or $BHDC$ are desired. In other cases, too, if at any time you are stuck, you will disentangle yourself in this way.

[4.][94] Finally, if the power index of x or y be fractional, I reduce it to an

---

under this change of co-ordinates Newton's defining equation becomes

$$c^4X^4+14s^2c^2X^2Y^2+s^4Y^4+8s^2c^2(c^2X^2+s^2Y^2)$$
$$-32s^3c^4X+96s^4c^4=0,$$

or

$$sY=\pm c\sqrt{[\pm\sqrt{\{(7X^2+4s^2)^2-(X^4+8s^2X^2-32s^3X+96s^4)\}}-(7X^2+4s^2)]}.$$

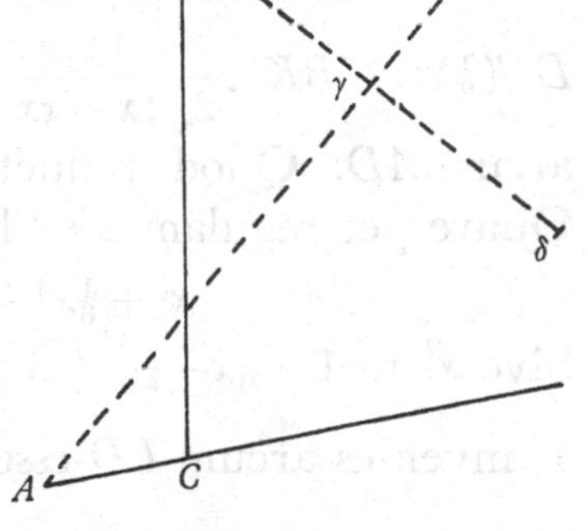

For real values of X, however, the function

$$X^4+8s^2X^2-32s^3X+96s^4$$

is always positive and is a minimum ($\approx 69s^4$) for $X\approx 1\cdot 38s$ (or, more precisely, when X is the real root of

$$X^3+4s^2X-8s^3=0).$$

In consequence, no real values of Y can correspond to any real values of X, and therefore the curve (D) defined by Newton's Cartesian equation is wholly imaginary. In any case, the curve would be symmetrical round the axis $Y=0$ ($x=y$) and not round AC ($y=0$) as shown.

(93) 'quælibet potest esse' (Jones, 1711).

(94) This final observation is a later addition in the margin of the manuscript.

(95) 'potestatis ipsius' (Jones, 1711).

(96) Jones' 1711 text here inserts the clarifying word 'ipsum'.

grum: ut in hoc exem: $y^3-xy^{\frac{1}{2}}+x^{\frac{4}{3}}=0$. posito $y^{\frac{1}{2}}=v$, & $x^{\frac{1}{3}}=z$, resultabit $v^6-z^3v+z^4=0$.[97] cujus radix est $v=z+z^3$ &c sive restituendo $y^{\frac{1}{2}}=x^{\frac{1}{3}}+x$ &c et quadrando $y=x^{\frac{2}{3}}+2x^{\frac{4}{3}}$ &c.

plicatio dictorū reliqua usmodi emata.

Et hæc de areis curvarum investigandis dicta sufficiant. Imò cùm Problemata de curvarum longitudine, de quantitate & superficie solida, deq; centro gravitatis omnia possunt eò tandem reduci ut quæratur quantitas superficiei planæ linea curva terminatæ, non opus est quicquam de ijs adjungere. In istis autem[98] quo ego operor modo dicam brevissimè.

Sit ABD curva quævis, & $AHKB$ rectangulum cujus latus AH vel BK est unitas. Et cogita[99] rectam DBK uniformitèr ab AH motam, areas ABD & AK describere; & quòd $BK(1)$ est[100] momentum quo $AK(x)$, & $BD(y)$ momentum quo ABD gradatim augetur; et quod ex momento BD perpetim dato, possis, per prædictas regulas, aream ABD ipso descriptam investigare, sive cum $AK(x)$ momento 1 descripta conferre.[101] Jam qua ratione superficies ABD ex momento suo perpetim dato per præcedentes regulas elicitur, eâdem quælibet alia quantitas ex momento suo sic dato elicietur. Exemplo res fiet clarior. Sit $ADLE$ circulus

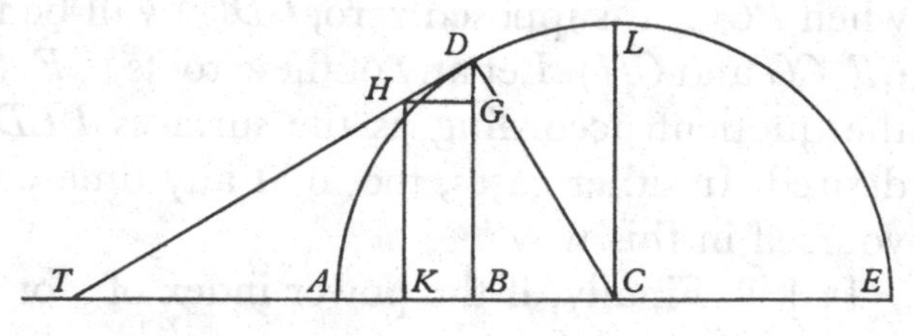

l longitudines curvarū iendas.

cujus arcûs AD longitudo est indaganda. Ducto tangente DHT, & completo indefinitè parvo rectangulo $HGBK$ & posito $AE=1=2AC$:[102] Erit ut BK sive GH momentum Basis AB, ad DH momentum arcus $AD::BT.DT::BD(\surd:x-xx:).DC(\frac{1}{2})::1(BK).\frac{1}{2\surd:x-xx}(DH)$. Adeòq; $\frac{1}{2\surd:x-xx}$ sive $\frac{\surd:x-xx}{2x-[2]xx}$ est momentum arcus AD. Quod reductum fit $\frac{1}{2}x^{-\frac{1}{2}}+\frac{1}{4}x^{\frac{1}{2}}+\frac{3}{16}x^{\frac{3}{2}}+\frac{5}{32}x^{\frac{5}{2}}+\frac{35}{256}x^{\frac{7}{2}}+\frac{63}{512}x^{\frac{9}{2}}$ &c. Quare per regulam 2dam longitudo arcus AD[103] est

$$x^{\frac{1}{2}}+\frac{1}{6}x^{\frac{3}{2}}+\frac{3}{40}x^{\frac{5}{2}}+\frac{5}{112}x^{\frac{7}{2}}+\frac{35}{1152}x^{\frac{9}{2}}+\frac{63}{2816}x^{\frac{11}{2}} \text{ \&c.}$$

Sive $x^{\frac{1}{2}}$ in $1+\frac{1}{6}x+\frac{3}{40}x^2$ &c. Non secus ponendo CB esse x, & radium CA esse 1, invenies arcum LD esse $x+\frac{x^3}{6}+\frac{3}{40}x^5+\frac{5}{112}x^7$ &c.[104]

(97) I.e. $v=z+v^6/z^3$ in easily iterable form. (98) 'itaq;' (consequently) is cancelled.

(99) In the text as reproduced in the *Commercium Epistolicum* (note (2)) at this point (on p. 14) one of its editors, almost certainly Newton himself, added the asseveration 'N.B. Hic describitur Methodus per Fluentes & earum Momenta. Hæc momenta a D. *Leibnitio* Differentiæ postmodum vocata sunt: Et inde nomen *Methodi Differentialis*' (Note this well. Here is described the method of operating by fluents and their moments. These moments were afterwards called differences by Mr *Leibniz*: and so came the name of *differential method*).

integer. As in this example $y^3 - xy^{\frac{1}{2}} + x^{\frac{4}{3}} = 0$ on setting $y^{\frac{1}{2}} = v$ and $x^{\frac{1}{3}} = z$ there will result $v^6 - z^3v + z^4 = 0$,(97) whose root is $v = z + z^3 \ldots$ or, on restoring the values, $y^{\frac{1}{2}} = x^{\frac{1}{3}} + x \ldots$ and, on squaring $y = x^{\frac{2}{3}} + 2x^{\frac{4}{3}} \ldots$.

Let these observations on investigating the areas of curves suffice. Indeed, since every problem on the length of curves, the quantity and surface of solids and the centre of gravity may ultimately be reduced to an inquiry into the quantity of a plane surface bounded by a curve line, there is no necessity to adjoin anything about them here. However, I shall speak very briefly of the method by which I there operate.

The ap[p]- cation o[f] aforesai[d] other problem[s] the kind

Let *ABD* be any curve and *AHKB* a rectangle whose side *AH* or *BK* is unity. And consider(99) that the straight line *DBK* describes the areas *ABD* and *AK* as it moves uniformly away from *AH*; that *BK*(1) is the moment by which *AK*(x) gradually increases and *BD*(y) that by which *ABD* does so; and that, when given continuously the moment of *BD*, you can by the foregoing rules investigate the area *ABD* described by it or compare it with *AK*(x) described with a unit moment.(101) Now, by the same means as the surface *ABD* is elicited by the foregoing rules from its continuously given moment, any other quantity will be elicited from its moment thus given. The matter will be clarified by example. Let *ADLE* be a circle whose arc length *AD* is to be discovered. On drawing the tangent *DHT*, completing the indefinitely small rectangle *HGBK* and setting $AE = 2AC = 1$,(102) there will then be *BK* or *GH* (the moment of the base *AB*) to *DH* (the moment of the arc *AD*)

Such as find the lengths curves.

$$= BT:DT = BD(\surd[x-x^2]):DC(\tfrac{1}{2}) = 1(BK):1/2\surd[x-x^2]\ (DH),$$

so that $1/2\surd[x-x^2]$ or $\surd[x-x^2]/2(x-x^2)$ is the moment of the arc *AD*. When reduced this becomes $\frac{1}{2}x^{-\frac{1}{2}} + \frac{1}{4}x^{\frac{1}{2}} + \frac{3}{16}x^{\frac{3}{2}} + \frac{5}{32}x^{\frac{5}{2}} + \frac{35}{256}x^{\frac{7}{2}} + \frac{63}{512}x^{\frac{9}{2}} \ldots$. Therefore by Rule 2 the length of the arc *AD*(103) is

$$x^{\frac{1}{2}} + \tfrac{1}{6}x^{\frac{3}{2}} + \tfrac{3}{40}x^{\frac{5}{2}} + \tfrac{5}{112}x^{\frac{7}{2}} + \tfrac{35}{1152}x^{\frac{9}{2}} + \tfrac{63}{2816}x^{\frac{11}{2}} \ldots,$$

or $x^{\frac{1}{2}}$ multiplied into $1 + \frac{1}{6}x + \frac{3}{40}x^2 \ldots$. Likewise, on setting $CB = x$ and the radius $CA = 1$, you will find the arc *LD* to be $x + \frac{1}{6}x^3 + \frac{3}{40}x^5 + \frac{5}{112}x^7 \ldots$.(104)

(100) 'sit' (Jones, 1711).

(101) Compare Problem 5 of the October 1666 tract (I, **2**, 7) and the 'Præparatio pro regula prima demonstranda' below. As before, of course, $AB\ (= (ABKH)/BK) = x$ and hence its 'momentum' (fluxional velocity) dx/dt is unity. Likewise the 'momentum' of the area $(ABD) = \int y.dx$ is $(d/dt)(ABD) = BD = y$. When the latter is known as a sequence of terms $\sum_i (a_i x^i)$ the former may be evaluated by Rule 2.

(102) With regard to the following argument a note at this point (p. 15) in the *Commercium* text (note (2)) asserts that it is an 'Exemplum calculi per Momenta Fluentium'.

(103) That is, $\sin^{-1}\surd x$ since $\widehat{AD} = AE\sin^{-1}(AD/AE)$.

(104) A corollary of the preceding expansion since $\widehat{LD} = CD\sin^{-1}(BC/CD) = \sin^{-1}x$, but we could use a binomial expansion to derive the series analogously as $\int_0^x (1-x^2)^{-\frac{1}{2}}.dx$.

Sed notandum est quod unitas ista quæ pro momento ponitur est superficies cùm de solidis, & linea cum de superficiebus, & punctū cum de lineis (ut in hoc exemplo) agitur.(105) Nec vereor loqui de unitate in punctis sive lineis infinitè parvis, siquidem proportiones ibi jam contemplantur Geometræ dum utuntur methodis Indivisibilium.

Ex his fiat conjectura de superficiebus & quantitatibus solidorum ac de centris gravitatum.(106) Verum si e contra ex areis vel longitudine &c curvæ alicujus datæ longitudo Basis AB desideratur, ex æquationibus per præcedentes regulas inventis extrahatur radix de x.

ctorum versum

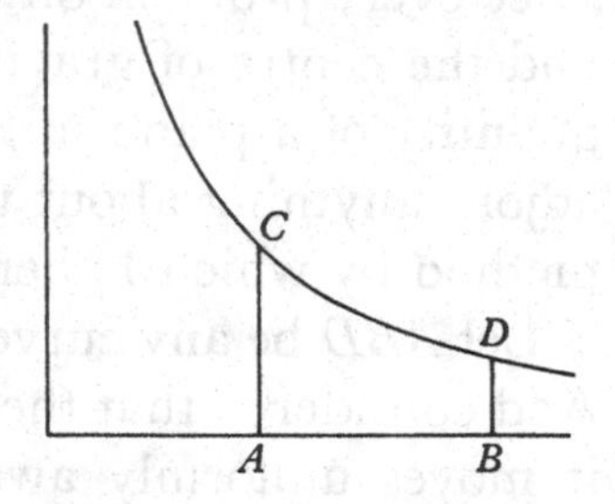

nventio ex area data.

Ut si ex area $ABDC$ Hyperbolæ $\left(\dfrac{1}{1+x}=y\right)$ datâ cupio basin AB cognoscere,(107) areâ ista z nominatâ, radicem hujus $z(ABCD)=x-\dfrac{x^2}{2}+\dfrac{x^3}{3}-\dfrac{x^4}{4}$ &c: extraho, neglectis illis terminis in quibus x est plurium dimensionum quam z in quotiente desideratur. Ut si vellem quod z ad quinqꝫ tantùm dimensiones in quotiente ascendat, negligo omnes $-\dfrac{x^6}{6}+\dfrac{x^7}{7}-\dfrac{x^8}{8}$ &c, & radicem hujus tantùm $\frac{1}{5}x^5-\frac{1}{4}x^4+\frac{1}{3}x^3-\frac{1}{2}x^2+x-z=0$ extraho.

		$(z+\frac{1}{2}z^2+\frac{1}{6}z^3+\frac{1}{24}z^4+\frac{1}{120}z^5$ [&c](108)
$z+p=x.)$	$+\frac{1}{5}x^5$	$+\frac{1}{5}z^5$ &c.
	$-\frac{1}{4}x^4$	$-\frac{1}{4}z^4-z^3p$ &c.
	$+\frac{1}{3}x^3$	$+\frac{1}{3}z^3+z^2p+zpp$ &c.
	$-\frac{1}{2}x^2$	$-\frac{1}{2}z^2-zp-\frac{1}{2}pp.$
	$+x$	$+z$ $+p.$
	$-z$	$-z$
$\frac{1}{2}z^2+q=p.)$	$+zp^2$	$+\frac{1}{4}z^5$ &c.
	$-\frac{1}{2}p^2$	$-\frac{1}{8}z^4-\frac{1}{2}z^2q$ &c.
	$-z^3p$	$-\frac{1}{2}z^5$ &c.
	$+z^2p$	$+\frac{1}{2}z^4$ $+z^2q.$
	$-zp$	$-\frac{1}{2}z^3$ $-zq.$
	$+p$	$+\frac{1}{2}z^2$ $+q.$
	$+\frac{1}{5}z^5$	$+\frac{1}{5}z^5$
	$-\frac{1}{4}z^4$	$-\frac{1}{4}z^4$
	$+\frac{1}{3}z^3$	$+\frac{1}{3}z^3$
	$-\frac{1}{2}z^2$	$-\frac{1}{2}z^2$
$1-z+\frac{1}{2}z^2)$ $\frac{1}{6}z^3-\frac{1}{8}z^4+\frac{1}{20}z^5$ $(\frac{1}{6}z^3+\frac{1}{24}z^4+\frac{1}{120}z^5$		

(105) Compare I, 1, 3, §2.

But it must be noted that that unity which is set for the moment is a surface when the question concerns solids, a line when it relates to surfaces and a point when (as in this example) it has to do with lines.[105] Nor am I afraid to talk of a unity in points or infinitely small lines inasmuch as geometers now consider proportions in these while using indivisible methods.

From these remarks may be inferred the procedure for surfaces, the quantities of solids and indeed for centres of gravities.[106] But if conversely it is desired to discover the length of the base AB from the areas or length, and so on, of any given curve, the root x is to be extracted from the equations found by the preceding rules. So if from the area $ABDC$ of the hyperbola $(1/[1+x]=y)$ given I desire to know the base AB, naming that area z I extract the root of this equation z (or $ABCD$) $=x-\frac{1}{2}x^2+\frac{1}{3}x^3-\frac{1}{4}x^4 \ldots$, where those terms are neglected in which x is of more dimensions than z is desired to be in the quotient. So should I want z to rise merely to five dimensions in the quotient, I neglect all of

The converse of aforesai[d]

Such as, finding [the] base wh[en] the area given.

$$-\tfrac{1}{6}x^6+\tfrac{1}{7}x^7-\tfrac{1}{8}x^8\ldots,$$

and extract the root merely of this, $\frac{1}{5}x^5-\frac{1}{4}x^4+\frac{1}{3}x^3-\frac{1}{2}x^2+x-z=0$.

$$[x=]\, z+\tfrac{1}{2}z^2+\tfrac{1}{6}z^3+\tfrac{1}{24}z^4+\tfrac{1}{120}z^5\,[\ldots]^{(108)}$$

$z+p=x.$	$+\frac{1}{5}x^5$	$+\frac{1}{5}z^5\ldots$
	$-\frac{1}{4}x^4$	$-\frac{1}{4}z^4-z^3p\ldots$
	$+\frac{1}{3}x^3$	$+\frac{1}{3}z^3+z^2p+zp^2\ldots$
	$-\frac{1}{2}x^2$	$-\frac{1}{2}z^2-zp-\frac{1}{2}p^2.$
	$+x$	$+z\ +p.$
	z	$-z.$
$\frac{1}{2}z^2+q=p.$	$+zp^2$	$+\frac{1}{4}z^5\ldots$
	$-\frac{1}{2}p^2$	$-\frac{1}{8}z^4-\frac{1}{2}z^2q\ldots$
	$-z^3p$	$-\frac{1}{2}z^5\ldots$
	$+z^2p$	$+\frac{1}{2}z^4+z^2q.$
	$-zp$	$-\frac{1}{2}z^3-zq.$
	$+p$	$+\frac{1}{2}z^2+q.$
	$+\frac{1}{5}z^5$	$+\frac{1}{5}z^5$
	$-\frac{1}{4}z^4$	$-\frac{1}{4}z^4$
	$+\frac{1}{3}z^3$	$+\frac{1}{3}z^3$
	$-\frac{1}{2}z^2$	$-\frac{1}{2}z^2$

$$1-z+\tfrac{1}{2}z^2)\ \tfrac{1}{6}z^3-\tfrac{1}{8}z^4+\tfrac{1}{20}z^5\ (\tfrac{1}{6}z^3+\tfrac{1}{24}z^4+\tfrac{1}{120}z^5$$

(106) Newton presumably intends 'gravitatis' (simple constant gravity).

(107) 'cupiam basim *AB* investigare' (Jones, 1711).

(108) $x=e^z-1=\sum_{1\leqslant i\leqslant 5}\left(\frac{z^i}{i!}\right)$ is of course the inverse, to $O(z^6)$, of $z=\log(1+x)$.

[margin, cut off: ...æc duo ...s adno- ...essent, ...tum in ...nentem ...nt cùm ...lutione ...ationis ...lis hæc ...ba (*Sin* ... *tantùm* ... *plures* ... *quotienti* ... *os* &c) ...bui.(109)]

Analysin ut vides exhibui propter adnotanda duo sequentia.

1. Quòd inter substituendum, istos terminos semper omitto quos nulli deinceps usui fore prævideam. Cujus rei regula esto, quòd post primum terminum ex qualibet quantitate sibi collaterali resultantem non addo plures terminos dextrorsum quàm istius primi termini index dimensionis ab indice dimensionis maximæ distat unitatibus. Ut in hoc exemplo ubi maxima dimensio est 5 omisi omnes terminos post z^5, post z^4 posui unicum, & duos tantùm post z^3. Cùm radix extrahenda (x) sit parium ubiꝗ, vel imparium dimensionum; Hæc esto regula; Quod post primum terminum ex qualibet quantitate sibi collaterali resultantem non addo plures terminos dextrorsum, quàm istius primi termini index dimensionis ab indice dimensionis maximæ binis unitatibus distat; vel ternis unitatibus, si indices dimensionū ipsius x unitatibus ubiꝗ ternis a se invicem distant. & sic de reliquis.

2. Cùm videam(110) p q vel r &c: in æquatione novissimè resultante esse unius tantùm dimensionis, ejus valorem, hoc est, reliquos terminos quotienti addendos, per divisionem quæro. Ut hic vides factū.(111)

[margin, cut off: ...ex data ...itudine curvæ.]

Si ex dato arcu αD sinus AB desideratur; æquationis $z=x+\frac{x^3}{6}+\frac{3x^5}{40}+\frac{5x^7}{112}$ &c supra inventæ (posito nempe $AB=x$, $\alpha D=z$ & $A\alpha=1$,) radix extracta erit

$$x=z-\tfrac{1}{6}z^3+\tfrac{1}{120}z^5-\tfrac{1}{5040}z^7+\tfrac{1}{362880}z^9 \text{ \&c.}$$

Et præterea si cosinum $A\beta$ ex isto arcu dato cupis, fac

$$A\beta(=\surd:1-xx:)=1-\frac{1}{2}z^2+\frac{z^4}{24}-\frac{z^6}{720}+\frac{z^8}{40320}-\frac{z^{10}}{3628800} \text{ \&c.}^{(112)}$$

[margin, cut off: ...De serie ...progres- ...sionum ...uanda.]

Hic obiter notetur, q^d 5 vel 6 terminis istarum radicum cognitis, eas plerumꝗ(113) ex analogia observata poteris ad arbitrium producere. Sic hanc $x=z+\frac{1}{2}z^2+\frac{1}{6}z^3+\frac{1}{24}z^4+\frac{1}{120}z^5$ &c produces dividendo ultimum terminum per

---

(109) This important marginal note, omitted by Jones from his 1711 text (and in consequence by all later editors of the 'De Analysi'), suggests forcibly that Newton composed his present manuscript somewhat hurriedly and without leaving himself time to rewrite it in a more finished form. For easy comprehension we have italicized his quotation of his own words earlier in the work: for the reference compare note (71) above.

(110) 'video' is cancelled and is also the reading in Jones' 1711 text.

(111) The terms $p, q, r, \ldots$ are of respective orders $z^2, z^3, z^4, \ldots$ in the present example. In general, when the given equation $f(x,z)=0$ has been reduced, by setting

$$x=\alpha+\beta z+\gamma z^2+\ldots+\lambda_i$$

(where λ_i is of the order of z^i), to the form

$$Az^i+Bz^{i+1}+\ldots+\lambda_i(A'+B'z+\ldots)+\lambda_i^2(A''+B''z+\ldots)+\ldots=0,$$

then x may be approximated to $O(z^{2i})$ by making $\lambda_i=(A+Bz+\ldots)z^i/(A'+B'z+\ldots)$ to that order. In the example $\lambda_3=q$ is evaluated as

$$(\tfrac{1}{6}-\tfrac{1}{8}z+\tfrac{1}{20}z^2+O(z^3))z^3/(1-z+\tfrac{1}{2}z^2+O(z^3)) = (\tfrac{1}{6}+\tfrac{1}{24}z+\tfrac{1}{120}z^2)z^3+O(z^6).$$

I have displayed the analysis as you see it so that I might add the two following remarks:

These t[…] observa[…] would h[…] been no[…] previou[…] they ha[…] to mind[…] the tim[…] wrote t[…] words r[…] to the r[…] lution o[…] algebra[…] equatio[…] *should I merely t[…] many terms. . .*

1. In the course of substituting I invariably omit those terms which I foresee will be of no use later. Let the rule for this be: that after the first term resulting from any quantity alongside it I do not add more terms on the right than the index of the dimension of that first term differs in units from the index of the greatest dimension. So in this example, in which the greatest dimension is 5, I have omitted all terms beyond z^5, setting a unique one after z^4 and two only after z^3. But when the root (x) to be extracted is everywhere of even or odd dimensions, let this be the rule: that after the first term resulting from any term alongside it I do not add more terms on the right than the index of the dimension of that first term differs in twos from the index of the greatest dimensions; or in threes, if the indexes of the dimensions of x differ everywhere in threes one from another. And so for the rest.

2. Since I see that $p, q, r, \ldots$ in the equation most recently resulting are of one dimension only, I seek its value, that is, the remaining terms to be added to the quotient, by division. As here you see done.(111)

If it is desired to find the sine AB from the arc αD given, of the equation $z = x+\frac{1}{6}x^3+\frac{3}{40}x^5+\frac{5}{112}x^7 \ldots$ found above (supposing namely that $AB = x$, $\widehat{\alpha D} = z$ and $A\alpha = 1$) I extract the root, which will be

or whe[…] curve's length [i]s given.

$$x = z-\tfrac{1}{6}z^3+\tfrac{1}{120}z^5-\tfrac{1}{5040}z^7+\tfrac{1}{362880}z^9 \ldots.$$

If, moreover, you want the cosine $A\beta$ of that given arc, make

$$A\beta(=\surd[1-x^2]) = 1-\tfrac{1}{2}z^2+\tfrac{1}{24}z^4-\tfrac{1}{720}z^6+\tfrac{1}{40320}z^8-\tfrac{1}{3628800}z^{10} \ldots.^{(112)}$$

Let it be noted here, by the way, that when you know 5 or 6 terms of those roots you will for the most part(113) be able to prolong them at will by observing analogies. Thus you may prolong this $x = z+\frac{1}{2}z^2+\frac{1}{6}z^3+\frac{1}{24}z^4+\frac{1}{120}z^5 \ldots$ by

On con[…]tinuing sequen[…] progres[…]

(112) These series for the sine and cosine, which here appear for the first time in a European manuscript, had, as we now know, already been displayed in a 1639 Malayālam compendium, the *Yuktibhāṣā*, which itself professes to be based on an early sixteenth-century Sanskrit original, the *Tantasaṅgraha*, perhaps the work of the little-known Hindu mathematician Nilakaṇṭha. (See A. R. A. Iyer's and R. Tampurān's modern Malayālam edition of the first part of the *Yuktibhāṣā* (Trichur, 1948): 204 ff.; and its review by C. T. Rajagopal and A. Venkotraman in *Mathematical Reviews* **12** (1951): 309–10. The latter manuscript is still unpublished.) The Hindu approach, which depends, for example, on repeated iteration of the identity

$$\sin x = \int[1-\int \sin x\, dx].dx,$$

is both wholly distinct from Newton's and was described in Europe only in the early nineteenth century (C. M. Whish, 'On the Hindu Quadrature of the Circle and the Infinite Series of the Proportion of the Circumference to the Diameter Exhibited in the Four Sastras...' *Transactions of the Royal Asiatic Society of Great Britain and Ireland*, **3** (1835): 509–23), so that there can be no question of any influence upon Newton in his independent rediscovery of the series.

(113) This Wallisian remark is somewhat over-optimistic. (Compare I, **1**, **3**, §3.)

hos ordine numeros 2. 3. 4. 5. 6. 7. &c,[114] et hanc $x=z-\frac{z^3}{6}+\frac{z^5}{120}-\frac{z^7}{5040}$ &c per hos $2\times 3.\ 4\times 5.\ 6\times 7.\ 8\times 9.\ 10\times 11.$ &c, & hanc $x=1-\frac{z^2}{2}+\frac{z^4}{24}-\frac{z^6}{720}$ &c per hos $1\times 2.\ 3\times 4.\ 5\times 6.\ 7\times 8.\ 9\times 10.$ &c. Et hanc $z=x+\frac{1}{6}x^3+\frac{3x^5}{40}+\frac{5x^7}{112}$ &c multiplicando per hos $\frac{1\times 1}{2\times 3}.\frac{3\times 3}{4\times 5}.\frac{5\times 5}{6\times 7}.\frac{7\times 7}{8\times 9}$ &c.[115] Et sic de reliquis.

plicatio ctorum curvas anicas.

Et hæc de curvis Geometricis dicta sufficiant. Quinetiam si curva Mechanica est[116] Methodum tamen nostram nequaquam respuit. Exemplo sit Trochoides, *ADFG* cujus vertex *A* & axis *AH*, & *AKH* rota qua describitur. Et quæratur superficies *ABD*. Jam posito $AB=x$, $BD=y$ ut supra, & $AH=1$; primò quæro longitudinem ipsius *BD*. Nempe ex natura Trochoidis [est] $KD=$ arcui *AK*, quare tota $BD=BK+$arc AK. Sed est $BK(=\surd:x-x^2)=x^{\frac{1}{2}}-\frac{1}{2}x^{\frac{3}{2}}-\frac{1}{8}x^{\frac{5}{2}}-\frac{1}{16}x^{\frac{7}{2}}$ &c, & (ex prædictis)[117] arcus $AK=x^{\frac{1}{2}}+\frac{1}{6}x^{\frac{3}{2}}+\frac{3}{40}x^{\frac{5}{2}}+\frac{5}{112}x^{\frac{7}{2}}$ &c. Ergo tota $BD=2x^{\frac{1}{2}}-\frac{1}{3}x^{\frac{3}{2}}-\frac{1}{20}x^{\frac{5}{2}}-\frac{1}{56}x^{\frac{7}{2}}$ &c. Et (per Reg 2) area $ABD=\frac{4}{3}x^{\frac{3}{2}}-\frac{2}{15}x^{\frac{5}{2}}-\frac{1}{70}x^{\frac{7}{2}}-\frac{1}{252}x^{\frac{9}{2}}$ &c.[118]

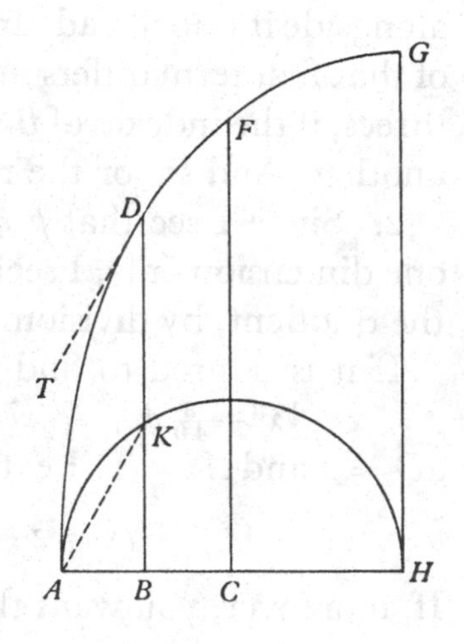

Vel brevius six: Cùm recta *AK* tangenti *TD* parallela sit[,] erit *AB* ad *BK* sicut momentum lineæ *AB*, momento lineæ *BD*, hoc est

$$x.\surd x-xx::1.\frac{1}{x}\surd:x-xx:=x^{-\frac{1}{2}}-\frac{1}{2}x^{\frac{1}{2}}-\frac{1}{8}x^{\frac{3}{2}}-\frac{1}{16}x^{\frac{5}{2}}-\frac{5}{128}x^{\frac{7}{2}}\ \&c.^{(119)}$$

Quare (per Reg 2) $BD=2x^{\frac{1}{2}}-\frac{1}{3}x^{\frac{3}{2}}-\frac{1}{20}x^{\frac{5}{2}}-\frac{1}{56}x^{\frac{7}{2}}-\frac{5}{576}x^{\frac{9}{2}}$ &c et superficies $ABD=\frac{4}{3}x^{\frac{3}{2}}-\frac{2}{15}x^{\frac{5}{2}}-\frac{1}{70}x^{\frac{7}{2}}-\frac{1}{252}x^{\frac{9}{2}}-\frac{5}{3168}x^{\frac{11}{2}}$ &c.

Non dissimili modo (posito *C* centro circuli & $CB=x$) obtinebis aream *CBDF* &c.[120]

Sit area *ABDV* Quadratricis *VDE* (cujus vertex est *V*, & *A* centrū circuli interioris *VK* cui aptatur) invenienda. Ducta qualibet *AKD* demitto perpendi-

(114) Or, as we would say, the ith term is $z^i/i!$. Similarly, the general terms in the two following expansions are $(-1)^i z^{2i+1}/(2i+1)!$ and $(-1)^i z^{2i}/(2i)!$ respectively.

(115) Here the general term is (compare note (104)) $(-1)^i\binom{-\frac{1}{2}}{i}\Big/(2i+1)$, that is, as Newton remarks, $1^2\times 3^2\times 5^2\times\ldots\times(2i-1)^2/(2i+1)!$.

(116) 'curva etiamsi Mechanica sit' (Jones, 1711).

(117) See note (103).

(118) Since $BD=BK+\widehat{AK}$, the Cartesian equation of the cycloid is

$$y=\surd[x-x^2]+\sin^{-1}\surd x,$$

dividing the last term by these numbers in order 2, 3, 4, 5, 6, 7, ...; and this $x = z-\frac{1}{6}z^3+\frac{1}{120}z^5-\frac{1}{5040}z^7 \ldots$ by these $2\times 3, 4\times 5, 6\times 7, 8\times 9, 10\times 11, \ldots$; and this $x = 1-\frac{1}{2}z^2+\frac{1}{24}z^4-\frac{1}{720}z^6 \ldots$ by these 1×2, 3×4, 5×6, 7×8, 9×10, ...; while this $z = x+\frac{1}{6}x^3+\frac{3}{40}x^5+\frac{5}{112}x^7 \ldots$ you may produce by multiplying by these $\frac{1\times 1}{2\times 3}, \frac{3\times 3}{4\times 5}, \frac{5\times 5}{6\times 7}, \frac{7\times 7}{8\times 9}, \ldots$.(115) And so for others.

The app
tion of t
precedir
mechan
curves.

Let these remarks suffice for geometrical curves. But, indeed, if the curve is mechanical it yet by no means spurns our method. Take, for example, the cycloid $ADFG$ whose vertex is A and axis AH while AKH is the 'wheel' by which it is generated. And let the surface ABD be sought. Setting now $AB = x$, $BD = y$ (as above) and $AH = 1$, I seek in the first instance the length of BD. Precisely, by the nature of a cycloid, KD is equal to the arc AK, and therefore the whole line $BD = BK+$ the arc AK. But

$$BK(=\sqrt{[x-x^2]}) = x^{\frac{1}{2}}-\tfrac{1}{2}x^{\frac{3}{2}}-\tfrac{1}{8}x^{\frac{5}{2}}-\tfrac{1}{16}x^{\frac{7}{2}}\ldots$$

and (from the preceding)(117) the arc $AK = x^{\frac{1}{2}}+\frac{1}{6}x^{\frac{3}{2}}+\frac{3}{40}x^{\frac{5}{2}}+\frac{5}{112}x^{\frac{7}{2}}\ldots$, so that in consequence the whole line $BD = 2x^{\frac{1}{2}}-\frac{1}{3}x^{\frac{3}{2}}-\frac{1}{20}x^{\frac{5}{2}}-\frac{1}{56}x^{\frac{7}{2}}\ldots$. And (by Rule 2) the area $ABD = \frac{4}{3}x^{\frac{3}{2}}-\frac{2}{15}x^{\frac{5}{2}}-\frac{1}{70}x^{\frac{7}{2}}-\frac{1}{252}x^{\frac{9}{2}}\ldots$.(118)

Or more briefly thus. Since the straight line AK is parallel to the tangent TD, AB will be to BK as the momentum of the line AB to the momentum of the line BD, that is,

$$x:\sqrt{[x-x^2]} = 1:x^{-1}\sqrt{[x-x^2]} \quad \text{or} \quad x^{-\frac{1}{2}}-\tfrac{1}{2}x^{\frac{1}{2}}-\tfrac{1}{8}x^{\frac{3}{2}}-\tfrac{1}{16}x^{\frac{5}{2}}-\tfrac{5}{128}x^{\frac{7}{2}}\ldots.^{(119)}$$

Hence (by Rule 2) $BD = 2x^{\frac{1}{2}}-\frac{1}{3}x^{\frac{3}{2}}-\frac{1}{20}x^{\frac{5}{2}}-\frac{1}{56}x^{\frac{7}{2}}-\frac{5}{576}x^{\frac{9}{2}}\ldots$ and the surface $ABD = \frac{4}{3}x^{\frac{3}{2}}-\frac{2}{15}x^{\frac{5}{2}}-\frac{1}{70}x^{\frac{7}{2}}-\frac{1}{252}x^{\frac{9}{2}}-\frac{5}{3168}x^{\frac{11}{2}}\ldots$.

In a not dissimilar way you will (on setting C as the circle's centre and $CB = x$) obtain the area $CBDF$, and so on.(120)

Suppose the area $ABDV$ of the quadratrix VDE (whose vertex is V with A the centre of the inner circle VK to which it is applied) is to be found. Drawing

---

and therefore we deduce at once

$$(ABD) = \int_0^x y.dx = \tfrac{1}{4}(1+2x)\sqrt{[x-x^2]}-\tfrac{1}{4}(1-4x)\sin^{-1}\sqrt{x} = AB\times BD-(ABK).$$

(119) The ratio of the 'momenta' of the co-ordinates $BD = y$ and $AB = x$ is, of course, the slope $dy/dx = \sqrt{[x^{-1}-1]}$ of the cycloid at the point D, while the area

$$(ABD) = \int_0^x \int_0^x \frac{dy}{dx}.dx\,dx.$$

(120) In this case the Cartesian defining equation of the cycloid is

$$(BD) \text{ or } y = \tfrac{1}{2}\sqrt{[1-4x^2]}+\tfrac{1}{2}\cos^{-1}2x,$$

so that the area $(CBDF)$ is then

$$\int_0^x y.dx = \tfrac{1}{4}(1-(1-x)\sqrt{[1-4x^2]})+\tfrac{1}{8}(2\pi x+(1-4x)\sin^{-1}2x).$$

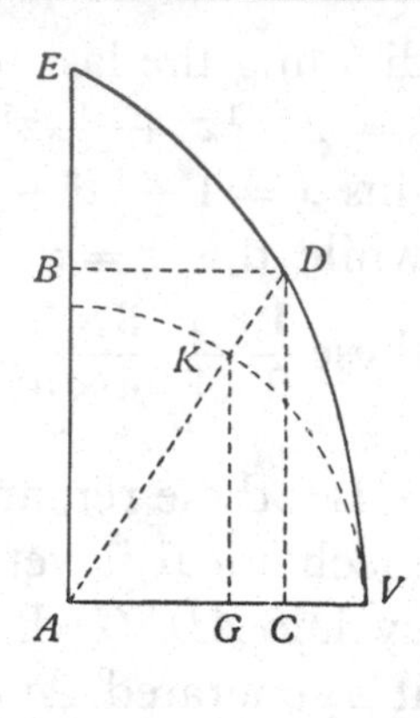

culares DB, DC, KG. Eritq; KG. AG:: $AB(x)$. $BD(y)$. sive $\frac{x \times AG}{KG} = y$. Verum ex natura Quadratricis est $BA(=DC) =$ arcui VK,[121] sive $VK = x$. Quare posito $AV = 1$[122] erit $GK = x - \frac{1}{6}x^3 + \frac{1}{120}x^5$ &c ex supra ostensis, &

$$GA = 1 - \tfrac{1}{2}x^2 + \tfrac{1}{24}x^4 - \tfrac{1}{720}x^6 \text{ \&c.}$$

Adeóq; $y\left(=\frac{x \times AG}{KG}\right) = \frac{1 - \frac{1}{2}x^2 + \frac{1}{24}x^4 - \frac{1}{720}x^6 \text{ [\&c]}}{1 - \frac{1}{6}xx + \frac{1}{120}x^4 - \frac{1}{5040}x^6 \text{ [\&c]}}$ sive, divisione facta, $y = 1 - \frac{1}{3}x^2 - \frac{1}{45}x^4 - \frac{2}{945}x^6$ &c & (per Reg 2) area $AVDB = x - \frac{1}{9}x^3 - \frac{1}{225}x^5 - \frac{2}{6615}x^7$ &c.[123]

Sic longitudo Quadratricis VD, licet calculo difficiliori, determinabilis est.[124] Nec quicquam hujusmodi scio ad quod hæc methodus idq; varijs modis, sese non extendit. Imo tangentes ad curvas Mechanicas (si quando id non alias fiat) hujus ope ducantur.[126] Et quicquid Vulgaris Analysis per æquationes ex finito terminorum numero constantes (quando id sit possibile) perficit, hæc per æquationes infinitas semper perficiat: Ut nil dubitaverim nomen Analysis etiam huic tribuere. Ratiocinia nempe in hâc non minùs certa sunt quàm in illâ, nec æquationes minùs exactæ;[127] licet omnes earum terminos nos homines &

nclusio, òd hæc ethodus alytica nsenda est.[125]

(121) A variant on the classical defining property of the quadratrix ($AB:AE = \widehat{VK}:\widehat{VKb}$, where the circle arc $\widehat{VK}$ meets AB in b) which enlists the derived result that

$$AV = \tfrac{1}{2}\pi . AE = \widehat{VKb}.$$

Evidently $AB = DC \propto \widehat{VK}$ and since, when DC and $\widehat{VK}$ pass (together) into zero their first increments are equal, the proportionality factor is unity. Puritanically, we might suggest that this elegant approach unnecessarily complicates the derivation of the curve's Cartesian equation $y = x \cot x$ (when, as here, $AE = \frac{1}{2}\pi$): this follows immediately in the form

$$x = \tan^{-1}(x/y)$$

from the defining proportion $AB:AE = \widehat{VK}:\widehat{VKb} = V\hat{A}K:\frac{1}{2}\pi$ and it is not directly relevant to introduce the equivalent of the limit consideration $\lim_{x\to 0}(x \cot x) = 1$.

(122) It follows that $AE = \frac{1}{2}\pi$, $GK = \sin x$ and $AK = \cos x$, the two latter of which Newton proceeds to expand as infinite series.

(123) Newton expands $y = x \cot x$ as an infinite series in powers of x and then evaluates

$$(AVDB) = \int_0^x y.dx \left[= x \log \sin x + \int_{\frac{1}{2}\pi}^{\frac{1}{2}\pi - x} \log \cos u.du\right]$$

similarly. When $x = \frac{1}{2}\pi$ we may deduce that

$$(AVE) = \int_{\frac{1}{2}\pi}^{0} \log \cos u.du = \tfrac{1}{2}\pi \log 2,$$

but there would seem no way of attaining the finite quadrature of the general sector $(AVDB)$ in terms of elementary functions. (Johann Bernoulli thought otherwise in the late 1730's, for in his critical 'Remarques sur le Livre intitulé *Analyse des infinimens petits, comprenant le Calcul integral dans toute son étendue*, &c. Par M. Stone...Imprimé à Paris en 1735' he took that author

AKD arbitrarily I let fall the perpendiculars *DB*, *DC* and *KG*. It will then be $KG:AG = AB(x):BD(y)$ or $x \times (AG/KG) = y$. But by the nature of the quadratrix $BA(=DC)$ is equal to the arc *VK*,(121) or $VK = x$. Hence, on setting $AV = 1$,(122) there will be $GK = x - \frac{1}{6}x^3 + \frac{1}{120}x^5 \ldots$ from what has been shown above, and again $GA = 1 - \frac{1}{2}x^2 + \frac{1}{24}x^4 - \frac{1}{720}x^6 \ldots$. Accordingly

$$y\left(= x \times \frac{AG}{KG}\right) \quad \text{is equal to} \quad \frac{1 - \frac{1}{2}x^2 + \frac{1}{24}x^4 - \frac{1}{720}x^6 [\ldots]}{1 - \frac{1}{6}x^2 + \frac{1}{120}x^4 - \frac{1}{5040}x^6 [\ldots]},$$

that is, on performing the division, $y = 1 - \frac{1}{3}x^2 - \frac{1}{45}x^4 - \frac{2}{945}x^6 \ldots$, and (by Rule 2) the area $AVDB = x - \frac{1}{9}x^3 - \frac{1}{225}x^5 - \frac{2}{6615}x^7 \ldots$.(123)

In this manner the length of the quadratrix' arc *VD* is determinable, though by a more difficult computation.(124) Nor do I know anything of this kind to which this method does not extend itself, and then in various ways. Indeed (if at any time it cannot be done otherwise) tangents may be drawn by its help to mechanical curves. And whatever common analysis performs by equations made up of a finite number of terms (whenever it may be possible), this method may always perform by infinite equations: in consequence, I have never hesitated to bestow on it also the name of analysis. To be sure, deductions in the latter are no less certain than in the other, nor its equations less exact,(127) even

My con
sion: th
method
be judg
analytic

to task for repeating Newton's present quadrature of the quadratrix as an infinite series: '...Avec tout cela il ne donne pas la quadrature par une expression finie, comme nous en pouvons donner une, quoique les Logarithmes y entrent' (*Johannis Bernoulli Opera Omnia*, 4 (Lausanne and Geneva, 1742): 177: 'Pag. 70. Exemple XX. *Quarrer un espace quelconque...de la Quadratrice &c*').) Newton communicated his quadratrix series in his *Epistola prior* of 13 June 1676 (*Correspondence of Isaac Newton*, **2** (1960): 28) and it was first published by John Wallis in his *Treatise of Algebra* (London, 1685: Chap. xcv, 341–7, especially 344) along with the other expansions given by Newton to Leibniz in that letter. In the manuscript the coefficient of the term in x^6 in the expansion of y was first entered as $\frac{11}{1890}$ and, correspondingly, that of x^7 in the value of $(AVDB)$ as $\frac{11}{13230}$. Since these reappear both in Collins' copy and the excerpts Leibniz made from it in 1676 (Appendix 1 below: especially note (36)), we conclude the correction was made by Newton at a fairly late date.

(124) In fact, $\widehat{DV} = \int_0^x \sqrt{[1 + (\cot x - x \operatorname{cosec}^2 x)^2]} . dx$

and we may evaluate the integral by extracting the root term by term in powers of x and then applying Rule 2 to each. Precisely, since

$$\cot x - x \operatorname{cosec}^2 x = -\tfrac{2}{3}x - \tfrac{4}{45}x^3 - \ldots,$$

it follows that $\widehat{DV} = x + \frac{2}{27}x^3 + \frac{14}{2025}x^5 + \frac{604}{893025}x^7 + \ldots$.

This series, too, was communicated in the *Epistola prior* and published by Wallis in 1685 (note (123)).

(125) This marginal annotation is not reproduced in Jones' text in 1711 and is omitted in all subsequent editions in consequence.

(126) Compare Problem 1 of the October 1666 tract (I, **2**, 7).

(127) Newton, of course, implicitly restricts his remark to those infinite series which converge to a finite limit.

rationis finitæ nec designare neque ita concipere possumus, ut quantitates inde desideratas exactè cognoscamus: Sicut radices surdæ finitarum æquationum nec numeris nec quavis arte Analytica ita possunt exhiberi ut alicujus quantitas a reliquis distinct[è][128] & exactè cognoscatur. [129]Geometricè quidem exhiberi possunt, quod hisce non conceditur: Imò et istis dimensionum duabus tribusve plurium, ante curvas in Geometriam nuper inductas, constructio nulla fuit habita.[130] Deniqȝ ad Analyticam merito pertinere censeatur cujus beneficio curvarum areæ & longitudines &c (id modò fiat)[131] exactè & Geometricè determinentur. Sed ista narrandi non est locus.

Respicienti, duo præ reliquis demonstranda occurrunt.

paratio regula prima demonstranda.

1. Quadratura[132] curvarum simplicium in Reg 1.[133] Sit itaqȝ curvæ alicujus $AD\delta$ Basis $AB=x$, perpendiculariter applicata $BD=y$ & area $ABD=z$ ut prius. Item sit $B\beta=o$[,] $BK=v$, et rectangulum $B\beta HK(ov)$ æquale spatio $B\beta\delta D$.[134] Est ergo $A\beta=x+o$ & $A\delta\beta=z+ov$. His præmissis, ex relatione inter x & z ad arbitrium assumptâ quæro y isto quem sequentem vides modo.

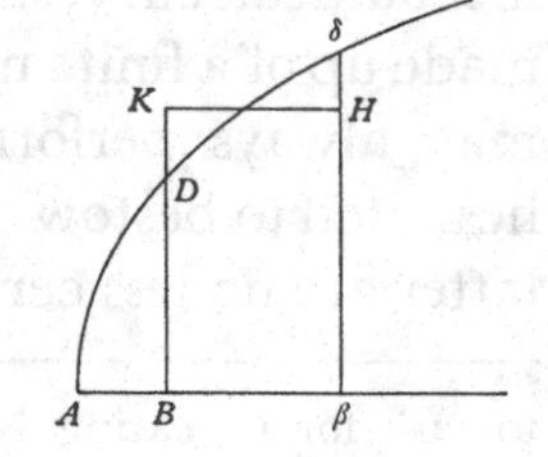

Pro lubitu sumatur $\frac{2}{3}x^{\frac{3}{2}}=z$ sive $\frac{4}{9}x^3=zz$. Tum $x+o(A\beta)$ pro x, & $z+ov(A\delta\beta)$ pro z substitutis prodibit

$$\tfrac{4}{9} \text{ in } x^3+3xxo+3xoo+o^3 = (\text{ex natura curvæ})\ zz+2zov+oovv.$$

Et sublatis $(\frac{4}{9}x^3$ & $zz)$ æqualibus, reliquisqȝ per o divisis, restat

$$\tfrac{4}{9} \text{ in } 3xx+3xo+oo=2zv+ovv.$$

Si jam supponamus $B\beta$ esse infinite parvam,[136] sive o esse nihil, erunt v & y

---

(128) The manuscript reading seems to be 'distincta' though the final letter may just possibly be a malformed 'e'. In his 1711 text Jones omitted the following '&' and transcribed 'distincta exacte', an illogical emendation which has been incorporated in all subsequent published versions.

(129) The following sentence is lacking in Jones' 1711 text and all subsequent published versions.

(130) Newton refers obliquely to Descartes' introduction of the trident into his *Geometrie* and his use of it there (in Book 3) to construct the general sextic by its meets with a defined circle. As we have seen (I, 3, §2: especially notes (13) ff.) Newton himself in May 1665 had preferred to introduce a Wallis cubic for the same purpose and had indicated how the construction of general 'geometrical' equations might be performed in a similar way.

(131) A note here on p. 18 of the *Commercium Epistolicum* text (note (2)) adds: 'N.B. Quadratura Curvarum per Æquationes infinitas, quæ nonnunquam terminantur & finitæ evadunt' (Note well. The quadrature of curves by infinite equations which sometimes terminate and come out finite). Newton in this comment (for it is surely his?) perhaps claims too much. We have been unable to find any documentary evidence that he had at this time any general

though we, mere men possessed only of finite intelligence, can neither designate all their terms nor so grasp them as to ascertain exactly the quantities we desire from them: just as surd roots of finite equations can be displayed neither numerically nor by any analytical artifice such that the quantity of some one of them may be distinctly and exactly known from the rest.(129) Geometrically, indeed, they may be displayed, but that is not here allowable: in fact, before certain curves were recently introduced into geometry, no construction at all was available for those of equations of more than two or three dimensions.(130) It should, finally, merit consideration as a relevant part of analysis since by its aid the areas and lengths of curves and so on (provided that may be done)(131) may be exactly and geometrically determined. But this is not the place to dwell on these matters.

As I look back, two points stand out above all others as needing proof.

Prepara[tion] for dem[on]strating first rul[e]

1. The quadrature of simple curves in Rule 1.(133) Let then any curve $AD\delta$ have base $AB = x$, perpendicular ordinate $BD = y$ and area $ABD = z$, as before. Likewise take $B\beta = o$, $BK = v$ and the rectangle $B\beta HK(ov)$ equal to the space $B\beta\delta D$.(134) It is, therefore, $A\beta = x+o$ and $A\delta\beta = z+ov$. With these premisses, from any arbitrarily assumed relationship between x and z I seek y in the way you see following.

Take at will(135) $\frac{2}{3}x^{\frac{3}{2}} = z$ or $\frac{4}{9}x^3 = z^2$. Then, when $x+o(A\beta)$ is substituted for x and $z+ov(A\delta\beta)$ for z, there arises (by the nature of the curve)

$$\tfrac{4}{9}(x^3+3x^2o+3xo^2+o^3) = z^2+2zov+o^2v^2.$$

On taking away equal quantities ($\frac{4}{9}x^3$ and z^2) and dividing the rest by o, there remains $\frac{4}{9}(3x^2+3xo+o^2) = 2zv+ov^2$. If we now suppose $B\beta$ to be infinitely small, that is, o to be zero, v and y will be equal and terms multiplied by o will

method for integrating algebraic functions comparable with that included by him in his *Epistola posterior* of 24 October 1676 destined for Leibniz: that, namely, which yields the indefinite integral of $dz^{\theta}(e+fz^{\eta})^{\lambda}$ as a series (usually unbounded) of algebraic terms and to which the present note makes an implicit reference.

(132) 'Demonstratio quadraturæ' (Jones, 1711).

(133) With regard to the following the *Commercium Epistolicum* text adds the pointer, 'Exemplum luculentum Calculi per momenta Fluentium' (A shining example of computation by the moments of fluents).

(134) An assumption that the arc $\widehat{AD\delta}$ is simply convex is here implicit. The idea of equating the curved segment ($B\beta\delta D$) to the 'median' rectangle ($B\beta HK$) is presumably taken over from Newton's 1665 researches in generalization of Heuraet's rectification procedure. (See I, 2, 5, §1: especially note (51); and compare the way in which the rectangles (πs), (βx), ... in the accompanying diagram are set equal to the corresponding curved segments bounded by the arc $\widehat{\psi d}$.)

(135) That is, arbitrarily.

(136) 'in infinitum diminui & evanescere' (Jones, 1711).

æquales & termini per o multiplicati evanescent, quare restabit $\frac{4}{9}\times 3xx = 2zv$, sive $\frac{2}{3}xx(=zy) = \frac{2}{3}x^{\frac{3}{2}}y$, sive $x^{\frac{1}{2}}\left(=\frac{x^2}{x^{\frac{3}{2}}}\right) = y$. Quare e contra si $x^{\frac{1}{2}} = y$ erit $\frac{2}{3}x^{\frac{3}{2}} = z$.(137)

Demonstratio.

Vel in genere(138) si $\frac{n}{m+n}\times ax^{\frac{m+n}{n}} = z$; sive, ponendo $\frac{na}{m+n} = c$ & $m+n = p$, si $cx^{\frac{p}{n}} = z$, sive $c^n x^p = z^n$; tum $x+o$ pro x & $z+ov$ (sive, quod perinde est,(139) $z+oy$) pro z substitutis prodit c^n in $x^p + pox^{p-1}$ &c $= z^n + noyz^{n-1}$ &c, reliquis nempe terminis qui tandem evanescerent omissis. Jam sublatis $c^n x^p$ & z^n æqualibus, reliquisq; per o divisis, restat $c^n px^{p-1} = nyz^{n-1}\left(=\frac{nyz^n}{z}\right) = \frac{nyc^n x^p}{cx^{\frac{p}{n}}}$. Sive, dividendo per $c^n x^p$, erit $px^{-1} = \frac{ny}{cx^{\frac{p}{n}}}$. sive $pcx^{\frac{p-n}{n}} = y$; vel restituendo $\frac{na}{m+n}$ pro c & $m+n$ pro p, hoc est m pro $p-n$ & na pro pc, fiet $ax^{\frac{m}{n}} = y$. Quare e contra si $ax^{\frac{m}{n}} = y$ erit $\frac{n}{m+n}ax^{\frac{m+n}{n}} = z$. Q.E.D.

[I]nventio [cu]rum(140) possunt [qu]adrari.

Hinc in transitu notetur modus quo curvæ tot quot placuerit, quarum areæ sunt cognitæ, possunt inveniri;(141) sumendo nempe quamlibet æquationem pro relatione inter aream z & basin x ut inde quæratur applicata y. Ut si supponas $\sqrt{:aa+xx:} = z$, ex calculo invenies $\frac{x}{\sqrt{aa+xx}} = y$. Et sic de reliquis.

[Demon]stratio [res]olutionis [æqua]tionum [aff]ectarū.

[2.] Alterum demonstrandum, est literalis æquationum affectarum resolutio. Nempe quòd quotiens, cum x sit satis parva, quo magis producitur eo magis veritati accedit, ut distantia sua (p, q, vel r &c)(142) ab exacto valore ipsius y, tandem evadat minor quavis data quantitate; et in infinitum producta sit ipsi y æqualis. Quod sic patebit

1: Quoniam ex ultimo termino æquationum quarum p, q, r &c sunt radices, ista quantitas in qua x est minimæ dimensionis (hoc est, plusquam dimidium istius ultimi termini, si supponis x satis parvam)(143) in qualibet operatione perpetuò tollitur; iste ultimus terminus (per 1. 10 Elem) tandem evadet minor quavis data quantitate; et prorsus evanescet si opus infinite continuatur.(144)

(137) Newton asserts that $\int x^{\frac{1}{2}}.dx = \frac{2}{3}x^{\frac{3}{2}}$ is an immediate corollary of his proof (from first principles) that $(d/dx)(\frac{2}{3}x^{\frac{3}{2}}) = x^{\frac{1}{2}}$. The careful geometrical justification of the inverse nature of the two procedures in I, 2, 5, §2 has now all but disappeared.

(138) Jones' 1711 text (which incorporates the marginal addition alongside into the general paragraph title) reads 'generaliter'.

(139) To $O(o^2)$ since $v = y$ to $O(o)$.

(140) A first continuation 'quarum areæ sunt cognitæ' (whose areas are known) is cancelled.

(141) The *Commercium Epistolicum* text (note (2)) adds the comment that 'Hac Propositione ex æquatione Fluentes involvente inveniuntur Fluxiones' (By this proposition from an equation involving fluent quantities the fluxions are found).

vanish and there will consequently remain $\frac{4}{9}\times 3x^2 = 2zv$ or $\frac{2}{3}x^2(=zy) = \frac{2}{3}x^{\frac{3}{2}}y$, that is, $x^{\frac{1}{2}}(=x^2/x^{\frac{3}{2}}) = y$. Conversely therefore if $x^{\frac{1}{2}} = y$, then will $\frac{2}{3}x^{\frac{3}{2}} = z$.(137)

Proof.

Or in general if $[n/(m+n)]ax^{(m+n)/n} = z$, that is, by setting $na/(m+n) = c$ and $m+n = p$, if $cx^{p/n} = z$ or $c^nx^p = z^n$, then when $x+o$ is substituted for x and $z+ov$ (or, what is its equivalent,(139) $z+oy$) for z there arises

$$c^n(x^p+pox^{p-1}\ldots) = z^n+noyz^{n-1}\ldots,$$

omitting the other terms, to be precise, which would ultimately vanish. Now, on taking away the equal terms c^nx^p and z^n and dividing the rest by o, there remains $c^npx^{p-1} = nyz^{n-1}(=nyz^n/z) = nyc^nx^p/cx^{p/n}$. That is, on dividing by c^nx^p, there will be $px^{-1} = ny/cx^{p/n}$ or $pcx^{(p-n)/n} = y$; in other words, by restoring $na/(m+n)$ for c and $m+n$ for p, that is, m for $p-n$ and na for pc, there will come $ax^{m/n} = y$. Conversely therefore if $ax^{m/n} = y$, then will $[n/(m+n)]\,ax^{(m+n)/n} = z$. As was to be proved.

The fin[d]ing of curve[s] which c[an be] squared

Here in passing may be noted a method by which as many curves as you please whose areas are known may be found:(141) namely, by assuming any equation at will for the relationship between the area z and from it in consequence seeking the ordinate y. So if you should suppose $\sqrt{[a^2+x^2]} = z$, by computation you will find $x/\sqrt{[a^2+x^2]} = y$. And similarly in other cases.

The de[mon]stration [of] the reso[lu]tion of affected equatio[ns]

[2.] The second point to be demonstrated is the algebraic resolution of affected equations: precisely, that, when x is small enough, the more the quotient is extended the more it approaches the truth, to the end that its difference ($p, q, r, \ldots$ successively) from the exact value of y finally shall come to be less than any given quantity you please and that the quotient extended to infinity shall be equal to y. This will be evident by the following argument:

1. Since from the last term of the equations whose roots are $p, q, r, \ldots$ that quantity in which x is of least dimension (that is, more than half that last term if you suppose x sufficiently small)(143) is forever removed in any of the operations, that last term (by *Elements*, x, 1) will at length come to be less than any given quantity, and will consequently vanish if the procedure is continued infinitely.

(142) 'defectus (...) quo distat' (Jones, 1711).

(143) We expect 'adéoq;' (and moreover) instead of 'hoc est', since the clause is an added assumption. Implicitly Newton demands that the given equation $f(x, y) = 0$ be a continuous function, so that its explicit expansion $y = \phi(x)$, where $\phi(0)$ is finite, will always be continuous as $x \to 0$ and hence convergent for at least arbitrarily small values of x. It then follows that, for small enough x, the term in which x is of its lowest power will be greater than half the total remaining portion of the expansion of y: if, that is, y is expansible as $\lim_{n\to\infty} \sum (a_i x^i)$ and the first m terms $\sum_{0\leqslant i\leqslant (m-1)} (a_i x^i)$ have been evaluated, then for small enough x Newton's assumption that $a_m x^m > \frac{1}{2} \lim_{n\to\infty} \sum_{m\leqslant i\leqslant n} (a_i x^i)$ will be true.

(144) Newton first wrote in a following sentence, 'Hoc est si radices æquationis resolvendæ gradatim augeantur per negativos vel diminuantur per affirmativos terminos quotienti con-

Nempe si $x=\frac{1}{2}$, erit x dimidium omnium $x+x^2+x^3+x^4$ &c & x^2 dimidiū omnium $x^2+x^3+x^4+x^5$ &c. Itaqȝ si $x \sqsupset$[(145)] $\frac{1}{2}$ erit x plusquā dimidium omnium $x+x^2+x^3$ &c: & x^2 plusquam dimidium omniū $x^2+x^3+x^4$ &c. Sic si $\frac{x}{b} \sqsupset \frac{1}{2}$ erit x plusquam dimidiū omnium $x+\frac{x^2}{b}+\frac{x^3}{bb}$ &c. Et sic de reliquis. Et numeros coefficientes quod attinet, illi plerumqȝ descrescunt perpetuò, vel si quando increscant, tantum opus est ut x aliquoties adhuc minor supponatur.[(146)]

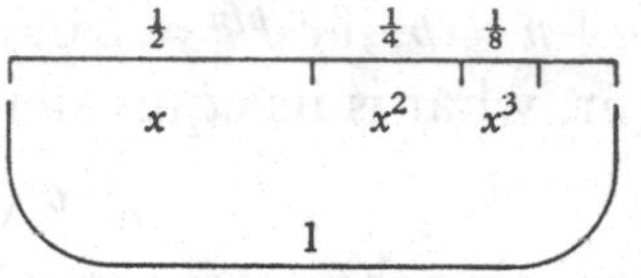

2. Si ultimus terminus alicujus æquationis continuò diminuatur donec tandem evanescat, una ex ejus radicibus etiam diminuetur donec cum ultimo termino simul evanescit.[(147)]

3. Quare quantitatum p, q, r &c unus valor continuo decrescit donec tandem, cùm opus in infinitum producitur, penitus evanescat.

4. Sed valores istarum $p\ q$ vel r &c unà cum quotiente eatenus extractâ adæquant radices æquationis propositæ. (Sic in resolutione æquationis $y^3+aay+axy-2a^3-x^3=0$. supra ostensâ, percipies

$$y=a+p=a-\frac{1}{4}x+q=a-\frac{1}{4}x+\frac{xx}{64a}+r\ \&c:)$$

Unde satis liquet propositum quod quotiens infinite producta est una ex valoribus de y.[(148)]

Idem patebit substituendo quotientem pro y in æquationem propositam.[(149)] Videbis enim terminos illos sese perpetuò destruere in quibus x est minimarum dimensionum.[(150)]

---

tinuo annexos, ejus ultimus terminus perpetuò decrescet, donec opere in infinitum continuato tandem evanescit' (That is, if the roots of the equation to be resolved are step by step increased by negative terms or diminished by positive ones continually appended to the quotient, its final term will perpetually decrease until, when the procedure has been continued into infinity, it at length vanishes). Evidently the added paragraphs 2–4 below are its more detailed revision. The first proposition of Euclid's tenth book is, of course, the 'Eudoxean' axiom that a procedure which at each stage takes away more than half the quantity remaining from a given quantity must at length take away its whole. We need not labour the point that this is a restrictive definition of the manifold to which it is applied.

(145) Barrow's symbol for lesser inequality, taken from the latter's *Notarum explicatio* which introduces his edition of Euclid's *Elements*. (See I, 1, Introduction: note (28).) It is merely, of course, an inversion of the equivalent symbol which William Oughtred listed in his *Notæ seu symbola quibus in sequentibus utor* which opens his own *Elementi Decimi Euclidis Declaratio* (I, 1, Introduction, Appendix 2: note (17)).

(146) Compare note (144). When after some mth stage all the coefficients a_i in the expansion $y = \lim_{n\to\infty} \sum_{0\leqslant i\leqslant n} (a_i x^i)$ are each less than unity, we may strictly apply Newton's preceding

Thus, if $x = \frac{1}{2}$, then will x be half of all of $x+x^2+x^3+x^4\ldots$ and x^2 half of all of $x^2+x^3+x^4+x^5\ldots$. And so if $x < \frac{1}{2}$, then x will be more than half of all of

$$x+x^2+x^3\ldots$$

and x^2 more than half of all of $x^2+x^3+x^4\ldots$. Similarly, if $x/b < \frac{1}{2}$, x will be more than half of all of $x+(x^2/b)+(x^3/b^2)\ldots$. And so for other cases. And, as for the numerical coefficients, for the most part they decrease continually, or if at any time they increase you need merely suppose x somewhat less still.(146)

2. If the last term of any equation be perpetually diminished till it at last vanishes, one of its roots will also be diminished till it vanishes along with that last term.

3. Therefore one value of the quantities $p, q, r, \ldots$ perpetually decreases until at last, when the procedure is extended to infinity, it entirely vanishes.

4. But the values of those quantities $p, q, r, \ldots$ together with the quotient so far extracted successively equal the roots of the equation proposed. (Thus in the resolution of the equation $y^3+a^2y+axy-2a^3-x^3 = 0$ shown above, you will perceive that $y = a+p = a-\frac{1}{4}x+q = a-\frac{1}{4}x+(x^2/64a)+r$, and so on.) Hence the premise that the quotient infinitely extended is one of the values of y is manifest enough.(148)

The same thing will be evident by substituting the quotient for y in the equation proposed.(149) For you will see that those terms continually destroy each other in which x is of least dimensions.(150)

---

remarks on the convergence of the geometric series $\lim_{n\to\infty} \sum_{m\leqslant i\leqslant n} (x^i)$, proved by him in Archimedean fashion from Euclid, x, 1 only for $x \leqslant \frac{1}{2}$: for then

$$y - \sum_{0\leqslant i\leqslant (m-1)} (a_i x^i) = \lim_{n\to\infty} \sum_{m\leqslant i\leqslant n} (a_i x^i) \leqslant \lim_{n\to\infty} \sum_{m\leqslant i\leqslant n} (x^i) = x^m/(1-x).$$

(See Props. 18–24 of Archimedes' *Quadrature of the Parabola*, and compare E. J. Dijksterhuis, *Archimedes* (Copenhagen, 1956): 133. It is a further comment on Newton's relative unfamiliarity with published contemporary mathematical research that he does not appear to know Grégoire de Saint-Vincent's *Opus Geometricum Quadraturæ Circuli et Sectionum Coni* (Antwerp, 1647) whose Book 2 'De Progressionibus Geometricis' (pp. 51–177) establishes the convergence of the geometric series for, in modern equivalent, $|x| < 1$.)

(147) More correctly, read 'evanescat' (Jones' emendation in 1711).

(148) Newton's syllogistic expansion of his preceding suppressed sentence (note (144)). In so far as these later paragraphs cloud the significance of convergence in the first they are perhaps an unwise insertion.

(149) So Jones' 1711 text but that of the *Commercium Epistolicum* (note (2)) reads 'in æquatione proposita'.

(150) Its writing suggests that this last paragraph was an afterthought. In our previous terminology, if $f(x, y) = 0$ determines $y = \phi(x)$, then evidently $f(x, \phi(x))$ is zero and hence all powers of x in it will vanish identically.

APPENDIX 1. LEIBNIZ' 'EXCERPTA' FROM NEWTON'S 'DE ANALYSI'

[October 1676][1]

From the original in the Niedersächsische Landesbibliothek, Hanover[2]

Excerpta ex tractatu Neutoni Mso.
De Analysi per æquationes numero terminorum infinitas

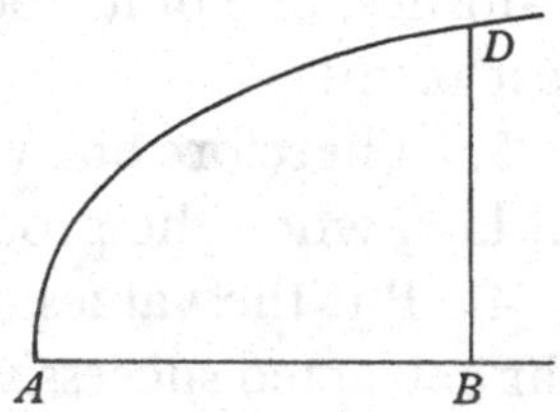

[1][3] $AB\sqcap^{(4)}x$. $BD\sqcap y$. a. b. c. quantitates datæ. $m.n.$ numeri integri. si $ax^{\frac{m}{n}}\sqcap y$, erit $\frac{na}{m+n}\sqcap(\frac{f}{y})^{(4)}$ areæ ABD. Ubi notandum hoc exemplum: Si $\frac{1}{x^2}\,(\sqcap x^{-2})\sqcap y$ id est si $a\sqcap 1$. $n\sqcap 1$. et $m\sqcap -2$ erit $\left(\frac{1}{-1}x^{\frac{-1}{1}}\sqcap\right)-x^{-1}\left(\sqcap\frac{-1}{x}\right)\sqcap \alpha BD$ infinitè versus α protensæ, quam calculus ponit negativam, quia jacet ex altera parte lineæ BD. Si $\frac{1}{x}\,(\sqcap x^{-1})\sqcap y$ erit $\frac{1}{0}x^{\frac{0}{1}}\sqcap\frac{1}{0}x^0\sqcap\frac{1}{0}x^1$ $\left(\textit{deberet dici } \frac{1}{0}\times 1^{(5)}\right)\sqcap\frac{1}{0}\sqcap$ infinitæ, quali est hyperbolæ area ex utraqꝫ parte.

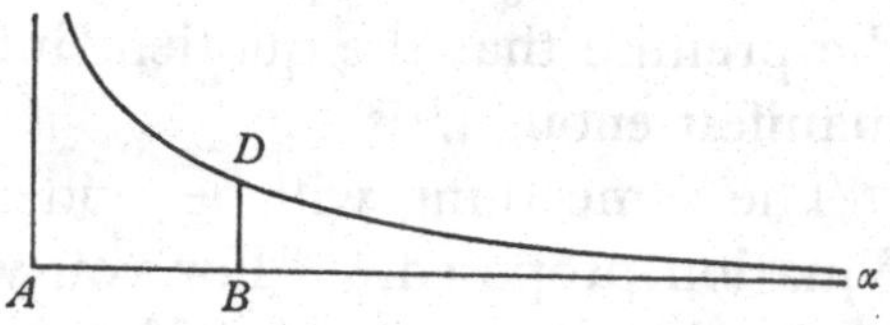

(1) Leibniz, then on his way back to Hanover after a four-year sojourn in Paris, made his annotations from Collins' copy of the 'De Analysi' (3: note (2) above) in the course of his second visit to London in mid-October 1676, probably between the 8th and 19th (O.S.) of that month. They were first described and partially printed by C. I. Gerhardt in his 'Leibniz in London' (*Sitzungsberichte der Königlich Preussischen Akademie der Wissenschaften zu Berlin*, Jahrgang 1891, x: 157–76, particularly 163–4 and 175–6) and have more recently been authoritatively discussed in relation to Leibniz' mathematical development during his Paris stay by J. E. Hofmann (*Die Entwicklungsgeschichte der Leibnizschen Mathematik während des Aufenthaltes in Paris (1672–1676) unter Mitbenutzung bisher unveröffentlichen Materials dargestellt* (Munich, 1949): 181–94, especially 182–3). (A not wholly accurate English rendering of Gerhardt's article with added critical notes was published by J. M. Child in *The Monist*, **22** (1917): 524–59 and is repeated in the latter's *The Early Mathematical Manuscripts of Leibniz* (Chicago and London, 1920): 159–95.) For a qualitative discussion of the importance of these 'excerpta' see the concluding portion of the introduction to the present Part 2.

(2) Leibniz–Handschriften **33**, VIII, 19: 1r–2r. The original is a folded folio sheet over whose first two and a half sides Leibniz has entered his annotations in his usual densely packed and not wholly legible scribble. On the lower half of f. 2r are, his partial 'Excerpta ex Epist. Neutoni 20 Aug. 1672 ad Neuton [sc. Collinsium]' (I, **3**, 3, §1: note (6)) which were presumably made on the same occasion. (See Gerhardt's 'Leibniz in London' (note (1)): 164, 176; and Hofmann's *Entwicklungsgeschichte* (note (1)): 183). The text we reproduce is Leibniz' except that, for easier comprehension, we have grouped his notes under numbered paragraph

[2](6) Si $\frac{1}{1+x^2}\sqcap y$. dividendo prodibit $y \sqcap 1-x^2+x^4-x^6$ &c. et

$$ABDC\sqcap\frac{x}{1}-\frac{x^3}{3}+\frac{x^5}{5} \text{ \&c.}$$

vel si terminus x^2(7) in divisione primus: fiet: $x^{-2}-x^{-4}+x^{-6}-x^{-8}$ pro valore

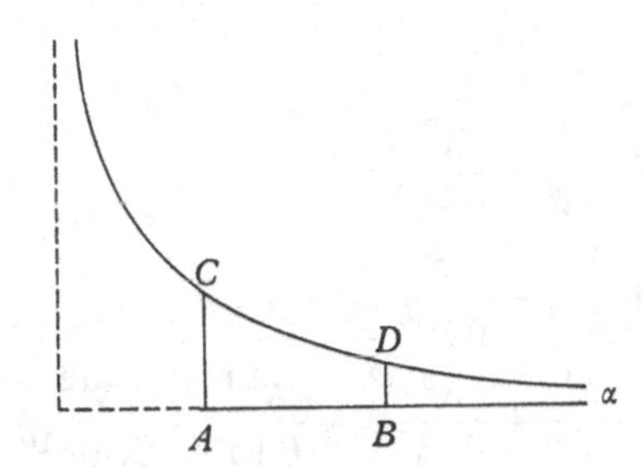

ipsius y. unde $BD\alpha\sqcap-\frac{x^{-1}}{1}+\frac{x^{-3}}{3}-\frac{x^{-5}}{5}$ &c. Priori modo procede cum x satis parva, posteriore cum satis magna.

[3](8) Extractionis radicum specimen $\sqrt{a^2+x^2}\sqcap y$. Radicem sic extraho.

---

headings and have set his bracketed comments on Newton's tract in italics: in addition, we have everywhere expanded the contraction '9' into 'con' and have in a few cases capitalized initial lower-case letters. We have not thought it necessary to repeat in facsimile Leibniz' hasty but substantially correct drawings of Collins' copied figures, but in their place make appropriate reproduction of the corresponding Newtonian originals.

(3) Omitting Newton's introductory paragraph, Leibniz begins his annotations with 'Regula I'.

(4) Here and universally below Leibniz employs the sign '$\sqcap$' for equality. The symbol '$\int \bar{y}$' (that is, 'S[umma]$\bar{y}$') is his notation for the operation of integrating the variable y indefinitely and with respect, it is understood, to the base variable x: later he will write it virtually in our modern form $\int y.dx$. (Compare F. Cajori, *A History of Mathematical Notations*, **2** (Chicago, 1929): 187, 201–3.)

(5) 'It ought to say $\frac{1}{0}\times 1$'. But Leibniz has not realized that the 'x' is Collins' malformed multiplication sign.

(6) Passing in silence over Rules 2 and 3 he notes the second example 'dividendo' which follows. This interests him, of course, because it affords a simple proof for the inverse tangent series which he had himself discovered three years before by applying the integral transform $z = y-x(dy/dx)$ to the circle $y^2 = x(2a-x)$ and then squaring the resulting curve

$$x(a^2+z^2) = 2az^2$$

(the 'versiera'). Compare Hofmann's *Entwicklungsgeschichte* (note (1)): 32–5.

(7) Understand Newton's verb 'ponatur'.

(8) The first of the 'Exempla Radicem extrahendo'. Note in the following scheme that in all but the last row of terms to be subtracted Leibniz has inverted the signs of Newton's original and then added them to the preceding row.

$$+a^2+x^2\left(a+\frac{x^2}{2a}-\frac{x^4}{8a^3}+\frac{x^6}{16a^5}-\frac{5x^8}{128a^7}\ \&c.\right.$$

$$\frac{-a^2}{0+x^2}$$

$$\frac{-x^2-\dfrac{x^4}{4a^2}}{0-\dfrac{x^4}{4a^2}}$$

$$\frac{+\dfrac{x^4}{4a^2}+\dfrac{x^6}{8a^4}-\dfrac{x^8}{64a^6}}{0+\dfrac{x^6}{8a^4}-\dfrac{x^8}{64a^6}}$$

$$\frac{+\dfrac{x^6}{8a^4}+\dfrac{x^8}{64a^6}{}^{(9)}-\dfrac{x^{10}}{64a^8}+\dfrac{x^{12}}{256a^{10}}}{0-\dfrac{5x^8}{64a^6}+\dfrac{x^{10}}{64a^8}-\dfrac{x^{12}}{256a^{10}}\ \&c.^{(10)}}$$

[4](11) $\dfrac{\sqrt{1+ax^2}}{\sqrt{1-bx^2}}\sqcap y$. cujus quadratura dat longitudinem curvæ Ellipticæ, extrahendo dabit: $\dfrac{1+\frac{1}{2}ax^2-\frac{a^2}{8}x^4+\frac{a^3}{16}x^6-\frac{5a^4}{128}x^8\ \&c.}{[1]^{(12)}-\frac{1}{2}bx^2-\frac{b^2}{8}x^4-\frac{b^3}{16}x^6-\frac{5b^4}{128}x^8\ \&c}$. Et dividendo sicut fit in fractionibus decimalibus fiet:

$$\begin{array}{llllll} 1 & +\frac{1}{2}bx^2 & +\frac{3}{8}b^2x^4 & +\frac{5}{16}b^3x^6 & +\frac{35}{128}b^4x^8 & \&c.^{(13)} \\ & \frac{1}{2}a & +\frac{1}{4}ab & +\frac{3}{16}ab^2 & +\frac{5}{32}ab^3 & \\ & & -\frac{1}{8}a^2 & -\frac{1}{16}a^2b & -\frac{3}{64}a^2b^2 & \\ & & & +\frac{1}{16}a^3 & +\frac{1}{32}a^3b & \\ & & & & -\frac{5}{128}a^4 & \end{array}$$

Brevior fuisset operatio, (*sed credo idem prodijsset*(14)) si ante reduxissemus denominatorem ad rationem multiplicando utcumq; terminum fractionis $\dfrac{\sqrt{1+ax^2}}{\sqrt{1-bx^2}}$ per $\sqrt{1-bx^2}$ ita fuisset opus tantum extractione ex numeratore et divisione per nominatorem finitum rationalem.(15)

---

(9) Read '$\dfrac{x^8}{16a^6}$'.

(10) In Newton's *epistola prior* of 13 June 1676 to him the series development of $\sqrt{[c^2+x^2]}$ had been presented as a particular case (Example 1) of the general binomial expansion and Leibniz was now doubtless intrigued to see it develop out of an algebraic root-extraction procedure. The following series expansions of $\int(a^2\pm x^2)^{\frac{1}{2}}.dx$ and $\int(x-x^2)^{\frac{1}{2}}.dx$ are omitted, presumably because Leibniz was familiar with their structure: Newton, of course, had also communicated in his *epistola prior* the slightly generalized expansion of $2\int(dx-x^2)^{\frac{1}{2}}.dx$. (See *Correspondence of Isaac Newton*, 2: 21, 30.)

[5][16] *De Resolutione æquationum affectarum.*

Modum quo ego utor in æquatione numerali exemplo primúm illustrabo. Sit $y^3-2y-5 \sqcap 0$ æquatio resolvenda, et sit 2 numerus qui minus quàm decimà sui parte differt à radice quæsita. Tum pono $2+p \sqcap y$. et substituo hunc valorem in æquationem et inde nova prodibit $p^3+6p^2+10p-1 \sqcap 0$, cujus radix p exquirenda est, ut quotienti addatur, nempe neglectis p^3+6p^2 ob parvitatem $10p-1 \sqcap 0$ sive $p \sqcap 0{,}1$ veritati prope est itaq; scribo 0,1. inquotiente, & suppono $0{,}1+q \sqcap p$. et hunc ejus valorem ut prius substituo, unde prodit $q^3+6{,}3q^2+11{,}23q+0{,}061 \sqcap 0$ et cum $11{,}23q+0{,}061$ veritati prope accedit sive ferè sit $q \sqcap -0{,}0054$, (dividendo nempe donec tot eliciantur figuræ quot locis primæ figuræ hujus et principalis quotientis exclusive distant) scribo, $-0{,}0054$ inferiori parte quotientis cum negativa sit, et supponens $0{,}0054$[17]$+r \sqcap q$ hunc ut prius substituo et operationem sic produco quousq; placuerit. Verum si ad bis tot figuras tantùm quot in quotiente jam reperiuntur, una demta operam continuare

$$\begin{array}{r} +2{,}10000000 \\ -0{,}00544853 \\ \hline 2{,}09455147 \end{array}$$

$2+p \sqcap y$)	y^3	$+8+12p+6p^2+p^3$
	$-2y$	$-4-2p$
	-5	-5
	Summa $\sqcap$	$-1+10p+6p^2+p^3$
$0{,}1 \mid q \sqcap p$)	p^3	$+0{,}001+0.03q+0[,]3q^2+q^3$
	$+6p^2$	$+0{,}06 \quad +1{,}2 \quad +6{,}0$
	$+10p$	$+1 \quad +10$
	-1	-1
	Summa $\sqcap$	$+0{,}061+11{,}23q+6{,}3q^2+q^3$
$-0{,}0054+r \sqcap q$)	$6{,}3q^2$	$+0{,}000183708-0{,}06804r+6{,}3r^2$
	$+11{,}23q$	$-0{,}060642 \quad +11{,}23$
	$+0{,}061$	$+0{,}061$
	Summa $\sqcap$	$+0{,}000541708+11{,}16196r+6{,}3r^2$
$-0{,}00004853$)[18]		

(11) The final example 'radicem extrahendo'.

(12) Leibniz' manuscript accidentally omits this term.

(13) The following 'quadratura' which yields the series expansion of the ellipse arc is left unnoted.

(14) 'but with the same result, I believe'. The remark is evidently just.

(15) Newton's concluding observation is disregarded.

(16) The following is a fairly complete transcript of Newton's 'Exempla per resolutionem Æquationum affectarum'. The 'Newtonian' resolution of numerical and algebraic equations was evidently wholly new to Leibniz at this time and his interest in it is patent.

(17) Evidently copying the present section somewhat automatically Leibniz has failed to notice that the original here reads '$-0{,}0054$'.

(18) Leibniz here first added and then cancelled a phrase from the accompanying text, 'prodit $6{,}3r^2+11{,}16196r+0{,}0000541708$ ferè'.

cupio, pro q substituo $-0{,}0054+r$ in hunc, $6{,}3q^2+11{,}23q+0{,}061$, (scilicet primo ejus termino q^3, propter exilitatem suam neglecto) et prodit

$$6{,}3r^2+11{,}16196r+0{,}000541708 \sqcap 0$$

ferè sive rejecto $6{,}3r^2$, $r \sqcap -\dfrac{0{,}000541708}{11{,}16196} \sqcap -0{,}00004853$ ferè quam scribo in negativa parte quotientis, deniq; negativam ab affirmativa subducens habeo 2,09455147 quotientem quæsitam. Aequationes plurium terminorum nihilo secius extrahuntur & operam sub fine ut hic factum fuit levabis, si primos ejus terminos gradatim omiseris. Præterea notandum est quod in hoc exemplo si dubitarem an $0{,}1 \sqcap p$, veritati satis accederet, pro $10p-1 \sqcap 0$. finxissem $6p^2+10p-1 \sqcap 0$ et ejus radicis primam figuram in quotiente scripsissem, et secundam tertiamve quotientis figuram sic explorare convenit, ubi in æquatione ista ultimò resultante quadratum coefficientis penultimi non sit decies major quam factus ex ultimo termino ducto in coefficientem termini antepenultimi. Imò laborem plerumq; minues, præsertim in æquationibus plurimarum dimensionum, si figuras omnes quotienti addendas dicto modo, (hoc est extrahendo minorem radicum ex tribus ultimis terminis æquationis novissimè resultantis) exquiras, isto enim modo figuras duplo plures qualibet vice quotienti lucraberis. Demonstratio ejus patet ex operandi modo unde et facilè in memoriam revocatur. Et semper relinquitur æquatio cujus radix una cum acquisita quotiente adæquat radicem æquationis datæ. Unde examinatio operis æquè poterit institui ac in reliqua arithmetica, auferendo nempe quotientem à radice primæ æquationis sicut analystis notum est, ut inde æquatio ultima vel termini ejus duo tresve ultimi producantur.(19)

His ita in numeris ostensis prim[o] literalis æquationum affectarum resolutio est explicanda. Sit æquatio literalis $y^3+a^2y-2a^3+axy-x^3 \sqcap 0$. Quæro primum valorem hujus æquationis ponendo x esse nullam, sive quasi esset tantùm $y^3+a^2y-2a^3 \sqcap 0$ et invenio esse $+a$. (*seu* $y \sqcap a$) Itaq; scribo a in quotiente, et ponendo $a+p \sqcap y$. substituo hunc valorem. Et novam produco æquationem. Ex qua assumo has partes $4a^2p+a^2x$, ubi p et x seorsim sunt minimarum dimensionum et eas nihilo ferè æquales pono sive $p \sqcap \dfrac{-x}{4}$ ferè, sive $p \sqcap \dfrac{-x}{4}+q$. Et scribens $-\dfrac{x}{4}$ in quotiente, substituo $-\dfrac{x}{4}+q$ pro p et terminos inde resultantes iterum in margine scribo, ut vides in annexo schemate, et inde assumo quanti-

(19) Leibniz passes over Newton's following account of his preferred method for reducing an equation by a given quantity. Presumably he was familiar with the procedure.

(20) 'ergo'.

(21) The accompanying tabular scheme which fills in the details of the extraction of this root is not noted.

tates $+4a^2q-\frac{1}{16}ax^2$ in quibus q et x seorsim sunt minimarum dimensionum et fingo $q \sqcap \frac{x^2}{64a}$ ferè sive $q \sqcap \frac{x^2}{64a}+r$, et annectens $\frac{x^2}{64a}$ quotienti substituo $\frac{x^2}{64a}+r$ pro q. et ita procedo quousꝗ placuerit, et fiet ∴[20]

$$a-\frac{x}{4}+\frac{x^2}{64a}+\frac{131x^3}{512a^2}+\frac{509x^4}{16384a^3} \text{ \&c.}^{(21)}$$

Sin duplo tantùm plures quotienti terminos uno demto jungendos adhuc vellem, primo termino q^3 æquationis novissimè resultantis misso, et ista etiam parte $-\frac{3}{4}xq^2$ secundi, ubi x est tot dimensionum quot in penultimo termino quotientis, in reliquos terminos $3aq^2+4a^2q$ &c. substituo $\frac{x^2}{64a}+r$ pro q, et ex ultimis duobus terminis $\left(\frac{15x^4}{4096a}-\frac{131x^3}{128}+\frac{9x^2r}{32}-\frac{1}{2}axr+4a^2r\right)$, æquationis inde resultantis facta divisione $4a^2-\frac{1}{2}ax+\frac{9}{32}x^2.\Big)+\frac{131}{128}x^3-\frac{15x^4}{4096[a]}$ elicio

$$+\frac{131x^3}{512a^2}+\frac{509x^4}{16384a^3}$$

quotienti annectendas. Servit hoc cum tanto magis vero ipsius y valori acceditur, quanto x est minor. Sed contra si velis ut valor inventus tanto magis accedat vero, quanto x major, tunc exemplum esto, $y^3+axy+x^2y-a^3-2x^3 \sqcap 0$. Resoluturus excerpo terminos $y^3+x^2y-2x^3$, in quibus x et y vel seorsim vel simul multiplicata sunt et plurimarum & æqualium ubiꝗ dimensionum. Et ex ijs quasi nihilo æqualibus radicem elicio, quam invenio esse x, et hanc in quotiente scribo vel quod eodem recidit, ex y^3+y-2 (unitate pro x substituta) radicem (1) elicio et eam per x multiplico et factum (x) in quotiente scribo, deinde pono $x+p \sqcap y$. et procedo ut in priori exemplo donec sit quotiens

$$x-\frac{a}{4}+\frac{a^2}{64x}+\frac{131a^3}{512x^2}+\frac{509a^4}{16384x^3} \text{ \&c.}^{(22)}$$

Hic modus particularis, *inquit*, nec applicabilis curvis in orbem ad modum ellipsium flexis, prior verò est generalis. De qua noto sequentia.

(22) Surprisingly, Newton's following evaluation of $\int y.dx$ as an infinite series with its unexplained term

$$\text{‘}\boxed{\frac{aa}{64x}}\text{’} \left(\text{an exact equivalent of Leibniz' ‘}\int\overline{\frac{a^2}{64x}}\text{’}\right)$$

is omitted. This is surely the clearest proof that at this period he remained totally unimpressed with the former's integration procedure and notation. The further omission of Newton's next paragraph in its entirety would appear to reveal on Leibniz' part, at this period at least, a lack of interest in the geometry of higher curves.

(1) Si quando accidit, quod valor ipsius y cum x nulla est, sit quantitas surda vel penitus ignota, licebit illam litera aliqua designare. Ut in exemplo

$$y^3+a^2y+axy-2a^3-x^3 \sqcap 0.$$

si radix hujus $y^3+a^2y-2a^3$, fuisset surda vel ignota, finxissem pro ea quamlibet (b) et scribens b in quotiente suppono $b+p \sqcap y$, et istum pro y substituo ut vides[(23)] unde nova p^3+3bp^2 &c. resultat rejectis terminis $b^3+a^2b-2a^3$, qui nihilo sunt æquales propterea quod b supponitur radix hujus $y^3+a^2y-2a^3 \sqcap 0$[,] deinde termini $3b^2p+a^2p+abx$ dant $\frac{-abx}{3b^2+a^2}$ quotienti apponendam et $\frac{-abx}{3b^2+a^2}+q$ substituendam pro [p] &c. $\left(b-\frac{3abx}{a^2+3b^2}\text{ \&c.}\right)$ Completo opere sumo numerum aliquem pro a et hanc $y^3+a^2y-2a^3 \sqcap 0$. resolvo, sicut de numerali æquatione ostensum supra et radicem ejus pro b substituo.

(2) Si dictus valor sit nihil, hoc est si in æquatione resolvenda nullus sit terminus nisi qui per x vel y sit multiplicatus, ut in hac $y^3+axy+x^3 \sqcap 0$. tum terminos $(-axy+x^3)$ seligo, in quibus x seorsim et y etiam seorsim si fiat, aliàs (*obscure*)[(24)] per x multiplicata sit minimarum dimensionum. Et illi dant $\frac{+x^2}{a}$ pro primo termino quotientis et $\frac{x^2}{a}+p$ pro y substituendam. In hac

$$y^3-a^2y+axy-x^3 \sqcap 0$$

licebit primum terminum quotientis vel ex a^2y[(25)]$-x^3$ vel ex y^3-a^2y elicere. (*hoc servit etiam pro inveniendis surdis.*)[(26)]

(3) Si valor iste sit imaginarius ut in hac

$$y^4+y^2-2y+6-x^2y^2-2x+x^2+x^4 \sqcap 0$$

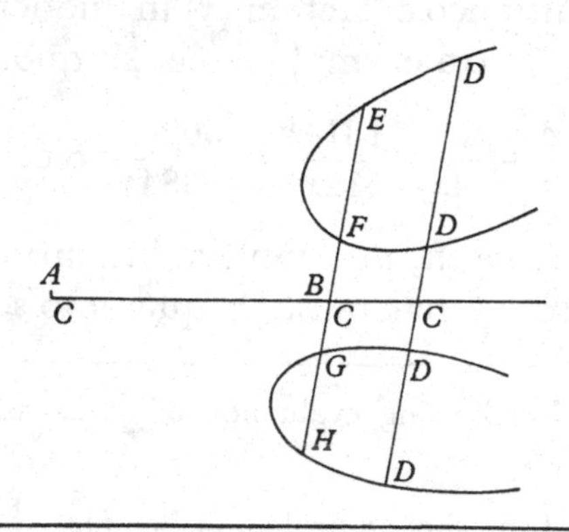

(23) A meaningless transcription since Leibniz does not reproduce Newton's accompanying scheme which lays out the computation in detail.

(24) That is, 'obscurè dictum' (obscurely pronounced).

(25) An automatic transcription: Leibniz does not notice that Newton's text should read '$-a^2y$'.

augeo vel minuo quantitatem x donec dictus valor evadat realis. Sic in annexo schemate[27] cum $AC(x)$ nulla est tum $CD(y)$ est imaginaria. Sin minuatur AC per datam AB, ut BC fiat x. tunc posito quod $BC(x)$ sit nulla $CD(y)$ erit valore quadruplici (CE, CF, CG & CH) realis, quarum radicum (CE, CF, CG, CH) utravis esto primus terminus quotientis, prout superficies $BEDC$, $BFDC$, $BGDC$, vel $BHDC$ desideratur. In alijs etiam casibus si quando hæsitas, te hoc modo extricabis. (*in varijs radicibus*)

Deniq; si index rationis de x vel y sit fractio reduco ad integrum, ut in hoc exemplo $y^3 - xy^{\frac{1}{2}} + x^{\frac{4}{3}} \sqcap 0$. posito $y^{\frac{1}{2}} \sqcap v$. et $x^{\frac{1}{3}} \sqcap z$ resultabit $v^6 - z^3 v + z^4 \sqcap 0$ (*male scriptum erat, & hic est*[28]) cujus radix est $v \sqcap z + z^3$ &c. (*nota: non manet eadem curva tali indeterminatarum mutatione*)[29] sive restituendo $y^{\frac{1}{2}} \sqcap x^{\frac{1}{3}} + x$ &c. et quadrando $y \sqcap x^{\frac{2}{3}} + 2x^{\frac{4}{3}}$ &c.

[6][30] Prædictorum conversum[,] ut inventio basis ex area data: Ex æquationibus infinitis per præcedentes regulas inventis extrahatur radix. Exempli

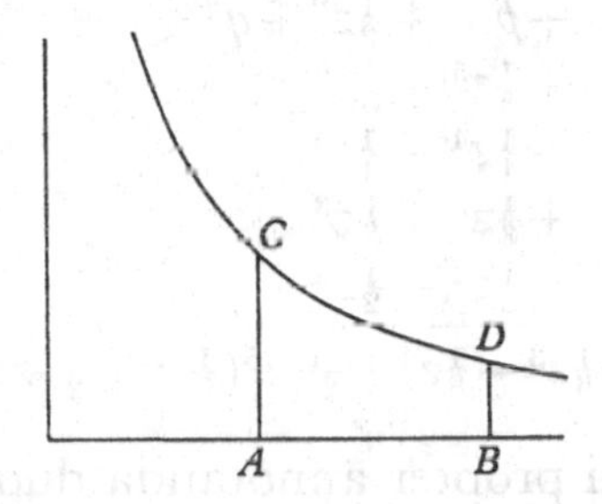

gratia ex area $ABDC$ hyperbolæ $\left(\frac{1}{1+x} \sqcap y\right)$ data cupio basin AB cognoscere. area istà z nominata, radicem hujus $z(ABCD) \sqcap \frac{x}{1} - \frac{x^2}{2} + \frac{x^3}{3} - \frac{x^4}{4}$ &c extraho, neglectis illis terminis in quibus x est plurium dimensionum quàm z in quotiente desideratur, ut si vellem quod z ad quinq; tantùm dimensiones in

(26) 'This serves also for finding surd quantities.'

(27) The accompanying diagram is (note (2)) our reproduction of Newton's original sketch. In his transcript, in fact, Leibniz makes $A\hat{C}D$ right following Collins' copy.

(28) '*correctum*' or some equivalent word is lacking. (Leibniz' remark refers, of course, to Collins' not wholly adequate copy.)

(29) 'Note: the curve does not remain the same under such a change of variables.' Presumably Leibniz intends the remark that the curve $y^3 - xy^{\frac{1}{2}} + x^{\frac{4}{3}} = 0$ is not identical with that $v^6 - z^3 v + 2^4 = 0$ defined with regard to Cartesian co-ordinate axes z and v: in other words, that $y = v^2$ and $x = z^3$ does not transform the Cartesian point (x, y) into a second Cartesian point (z, v).

(30) Ignoring Newton's brief fluxional account of indefinite integration which follows, Leibniz passes immediately to Newton's method of reversing the series $z = f(x)$ by 'extracting the root' $x = f^{-1}(z)$ of $f(x) - z = 0$.

quotiente assurgeret, negligo omnes $-\frac{x^6}{6}+\frac{x^7}{7}$ &c et radicem hujus tantùm $\frac{1}{5}x^5-\frac{1}{4}x^4+\frac{1}{3}x^3-\frac{1}{2}x^2+\frac{1}{1}x-z\sqcap 0$ extraho, quæ est $z+\frac{1}{2}z^2+\frac{1}{6}z^3+\frac{1}{24}z^4+\frac{1}{120}z^5$ &c.

$z+p\sqcap x)$	$+\frac{1}{5}x^5$	$+\frac{1}{5}x^5$ &c
	$-\frac{1}{4}x^4$	$-\frac{1}{4}z^4-z^3p$ &c
	$+\frac{1}{3}x^3$	$+\frac{1}{3}z^3+z^2p+zp^2$ &c
	$-\frac{1}{2}x^2$	$-\frac{1}{2}z^2-zp-\frac{1}{2}p^2$
	$+x$	$+z+p$
	$-z$	$-z$
$\frac{1}{2}z^2+q\sqcap p)$	$+zp^2$	$+\frac{1}{4}z^5$ &c
	$-\frac{1}{2}p^2$	$-\frac{1}{8}z^4-\frac{1}{2}z^2q$ &c
	$-z^3p$	$-\frac{1}{2}z^5$ &c
	$+z^2p$	$+\frac{1}{2}z^4+z^2q$
	$-zp$	$-\frac{1}{2}z^3-zq$
	$+p$	$+\frac{1}{2}z^2+q$
	$+\frac{1}{5}z^5$	$+\frac{1}{5}z^5$
	$-\frac{1}{4}z^4$	$-\frac{1}{4}z^4$
	$+\frac{1}{3}z^3$	$+\frac{1}{3}z^3$
	$-\frac{1}{2}z^2$	$-\frac{1}{2}z^2$

$$1-z+\tfrac{1}{2}z^2)\,\tfrac{1}{6}z^3-\tfrac{1}{8}z^4+\tfrac{1}{20}z^5\,(\tfrac{1}{6}z^3+\tfrac{1}{24}z^4+\tfrac{1}{120}z^5$$

Analysin ut vides exhibui propter annotanda duo.

(1) quod inter substituendum istos terminos semper omitto quos nulli deinceps usui fore prævideo. Cujus rei regula esto, quod post primum terminum ex qualibet quantitate sibi collaterali resultantem non addo plures dextrorsum, quam istius primi termini dimensio à dimensione maxima unitatibus distat, ut in hoc exemplo ubi maxima dimensio est 5, omisi omnes post z^5, post z^4 posui unicum, & duos tantùm post z^3. Cum radix extrahenda (x) sit parium ubiq; vel imparium dimensionum hæc esto regula, quod post primum terminum ex qualibet quantitate sibi collaterali resultantem non addo plures dextrorsum quam istius primi termini dimensio à dimensione maxima binis unitatibus distet (*error in descriptione*)(31), vel ternis unitatibus, si dimensiones ipsius x unitatibus ubiq; ternis à se invicem distant. & sic de reliquis.

(31) There is no such error, of course.

(32) Leibniz was especially interested in this series expansion for the cosine since, in unconscious rediscovery of the Hindu approach (3: note (112) above), he had a few months before begun to develop it by iterating the double integral $\cos z = 1-\iint \cos z.dz\,dz$ term by term (Hofmann, *Entwicklungsgeschichte* (note (1)): 156).

(33) A summary of Newton's 'De serie progressionum continuanda'.

(34) Leibniz does not see that Newton intends the successive fractions to be multiplied continually together.

(2) Cùm video *p. q.* vel *r* &c in æquatione novissimè resultante esse tantum unius dimensionis, ejus valorem, hoc est reliquos terminos quotienti addendos per divisionem quæro, ut hic vides factum.

Sic si ex dato arcu sinus quæratur, æquationis $z \sqcap x + \frac{x^3}{6} + \frac{3x^5}{40} + \frac{5x^7}{112}$ &c supra inventæ posito $AB \sqcap x$, $\alpha D \sqcap z$ et $A\alpha \sqcap 1$. radix extracta erit

$$z - \frac{z^3}{6} + \frac{z^5}{120} - \frac{1}{5040}z^7 + \frac{1}{362880}z^9 \text{ [\&c]}.$$

Sin $A\beta$ co-sinum ex dato arcu quæris, fac

$$A\beta (\sqcap \surd : 1 - x^2) \sqcap 1 - \frac{1}{2}z^2 + \frac{z^4}{24} - \frac{z^6}{720} + \frac{z^8}{40320} - \frac{z^{10}}{3628800} \text{ \&c.}^{(32)}$$

[7][33] Porro facile apparere solet serierum continuatio, ut

$$z \sqcap x + \frac{1}{6}x^3 + \frac{3x^5}{40} + \frac{5x^7}{112} \text{ \&c}$$

multiplicando per hos $\frac{1,1}{2,3} \cdot \frac{3,3}{4,5} \cdot \frac{5,5}{6,7} \cdot \frac{7,7}{8,9}$ &c. (*non consentiunt. debet esse error in describendo.*)[34]

[8] Eadem ad Curvas mechanicas facile applicantur, item tangentes curvarum mechanicarum hoc modo facilè ducuntur, si aliter id fieri commodè non possit.

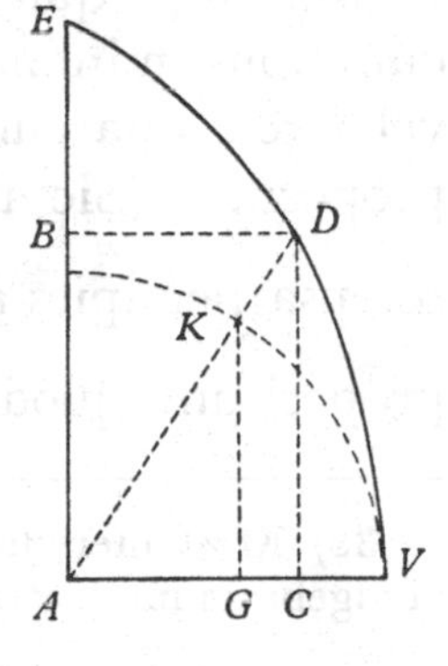

[35][Sit] *ABDV* area quadratricis *VDE*, cujus vertex *V* et *A* centrum circuli interioris *VK* cui aptatur, invenienda. Ducta qualibet *AKD* demitto perpendiculares *DB*, *DC*, *KG* eritq; $KG.AG :: AB(x).BD(y)$ sive $y \sqcap \frac{x, AG}{KG}$. Verum ex natura quadratricis $BA(\sqcap DC) \sqcap$ arcui *VK*, sive $VK \sqcap x$. Quare posito $VA \sqcap 1$. erit $GK \sqcap x - \frac{1}{6}x^3 + \frac{1}{120}x^5$ &c per priora, et $GA \sqcap 1 - \frac{1}{2}x^2 + \frac{1}{24}x^4 - \frac{1}{720}x^6$ &c adeoq;

$$y \left(\sqcap \frac{x, AG}{KG}\right) \sqcap \frac{1 - \frac{1}{2}x^2 + \frac{1}{24}x^4 - \frac{1}{720}x^6 \text{ \&c}}{1 - \frac{1}{6}x^2 + \frac{1}{120}x^4 - \frac{1}{5040}x^6 \text{ \&c}}$$

sive divisione facta $y \sqcap 1 - \frac{1}{3}x^2 - \frac{2}{45}x^4 - \frac{11}{1890}x^6$ [36] &c. unde area facile habetur.

[9][37] Putat (*rectè*) hanc methodū esse analyticam.

(35) Familiar with Newton's first example following (the cycloid) Leibniz passes it by.

(36) Compare 3: note (123) above.

(37) Leibniz' succinct summary of (and stated agreement with) Newton's conclusion 'quod hæc methodus Analytica censenda est'.

[10][38] Subjicienda demonstratio est, qua ostenditur resolutionem æquationum affectarum literalem esse probam: Nempe quod quotiens cum x satis parva est, quo magis producitur eo magis veritati accedit, ita ut distantia ejus p. q. vel r. ab exacto valore ipsius y, tandem evadat minor quâvis data quantitate, et in infinitum producta sit ipsi y æqualis, quod sic patebit.

(1) Quoniam ex ultimo termino æquationum quarum p. q. r. sunt radices ista quantitas in qua x est minimæ dimensionis (hoc est plusquam dimidium istius ultimi termini si supponis x satis parvam), in qualibet operatione perpetuò tollitur, iste ultimus terminus per 1. 10. Elem. tandem evadet minor quavis data quantitate, & prorsus evanescet si opus infinitè continuetur. Nempe si $x \sqcap \frac{1}{2}$ erit x dimidium omnium $x+x^2+x^3+x^4$ &c. et x^2 dimidium omnium

$$x^2+x^3+x^4+x^5 \quad \&c.$$

Itaqȝ si $x \sqsupset \frac{1}{2}$ erit x plusquam dimid. omnium $x+x^2+x^3$ &c. & x^2 plusquam dimidium omnium $x^2+x^3+x^4$ &c. Si sit $\frac{x}{b} \sqsupset \frac{1}{2}$ erit $x \sqsupset$[39] dimidium omnium $x+\frac{x^2}{b}+\frac{x^3}{b^2}$ &c. et sic de reliquis. & numeros coefficientes quod attinet, illi plerumqȝ decrescunt perpetuò vel si aliquando increscunt tantùm opus [est] ut x nonnihil minor assumatur.[40]

(2) Si ultimus terminus alicujus æquationis continuè diminuatur donec tandem evanescat, una ex ejus radicibus etiam diminuetur, donec cum ultimo termino simul evanescat.

(3) Quare quantitatum p. q. r. unus valor continuò decrescit, donec tandem cum opus in infinitum producitur penitus evanescat. Sed valores istarum p. q. vel r &c. una cum quotiente eatenus extracta adæquant radices æquationis propositæ. (Sic in resolutione æquationis $y^3+a^2y+axy-2a^3-x^3 \sqcap 0$ supra ostensa percipies $y \sqcap a+p \sqcap a-\frac{1}{4}x+q \sqcap a-\frac{1}{4}x+\frac{x^2}{64a}+r$ &c.) Unde satis liquet propositum, quod quotiens infinitè producta est una ex valoribus de y. Idem

(38) Most interestingly, Leibniz entirely ignores Newton's proof of the fundamental theorem of algebraic integration,

$$\int ax^p.dx = \frac{a}{p+1}x^{p+1} \quad (p = m/n).$$

(39) Correctly 'plusquam' (greater than) is rendered '$\sqsubset$' in Barrow's terminology (3: note (145) above). If Leibniz read Newton's '$\sqsupset$' thus in the preceding argument, it could not have made sense to him.

(40) A small square space is left blank on the left side of the manuscript, presumably intended for the insertion of the *De Analysi*'s final diagram.

(41) Leibniz ends his annotations from the 'De Analysi' at this point though his manuscript continues without any break to record his 'Excerpta ex Epist. Neutoni 20 Aug. 1672....' (note (2)).

patebit substituendo quotientem pro y in æquationem propositam, videbis enim terminos illos sese perpetuò destruere, in quibus x est minimarum dimensionum.[41]

APPENDIX 2. LEIBNIZ' REVIEW OF THE 'DE ANALYSI'

[February? 1712][1]

Extracted from the *Acta Eruditorum* for February 1712[2]

Analysis per quantitatum series, fluxiones ac differentias cum enumeratione linearum tertii Ordinis.

Londini, ex officina Pearsoniana, 1711. 4[°].

Plag. 16 & figg. æn.[3]

Guilielmus Jones, edita *Synopsi palmariorum Matheseos* (vid. Acta A. 1707 p. 178)[4] clarus, cum in scriniis D. Collinsii inter plurima a celebribus ejus temporis Mathematicis, præsertim Magnæ Britanniæ, ipsi communicata[5]

---

(1) The wider historical implications of this review are discussed in the Introduction to Part 2 above. On a narrow viewpoint this is a review, in the style of other similar recensions in Mencke and Pfautz's periodical, of William Jones' 1711 collection of Newtonian mathematical pieces under the title *Analysis/Per Quantitatum/Series, Fluxiones,/ac/Differentias:/cum Enumeratione Linearum/Tertii Ordinis./Londini:/Ex Officina Pearsoniana. Anno M. DCC. XI.* In breakdown this work contains: (a[1r] – c2r]) *Præfatio Editoris*; (1–21) *De Analysi Per Æquationes Numero Terminorum Infinitas*; (23–38) *Excerpta Ex Epistolis D. Newtoni Ad Methodum Fluxionum et Serierum Infinitarum Spectantibus*; (39–66) *Tractatus De Quadratura Curvarum*; (67–92) *Enumeratio Linearum Tertii Ordinis*; and (93–101) *Methodus Differentialis*. The review is unsigned and Ravier does not include it in his bibliography of Leibniz' published work, but we make our attribution on the strong internal evidence of the piece itself. In particular, in a portion we here omit, the review's author refers back to a previous critique (*Acta Eruditorum*, January 1705: 30–36) with the words 'De utroque diximus in Actis A. 1705 p. 30 & seqq.': this latter review, as O. Mencke's letter to Leibniz on 12 November 1704 makes clear, is indisputably Leibniz'. (Compare Émile Ravier, *Bibliographie des Œuvres de Leibniz* (Paris, 1937): 89–92, 105 note [243].)

(2) *Acta Eruditorum Anno MDCCXII publicata,...Lipsiæ, MDCCXII:* N°. II. (Mensis Februarii): 74–7. We have departed from early eighteenth-century convention in setting the piece titles quoted (and not the individuals named) in italics.

(3) That is, copper engravings. To be precise Jones' book is set in folio.

(4) *Acta Eruditorum* (April 1707): 178–81. The review was Wolff's, though as with others of this period Leibniz may have helped him with it.

(5) Compare Jones' *Præfatio*: 'Etiam secundus jam agitur annus ex quo scrinia D. Collinsii (qui, uti notum est, amplissimum cum sui sæculi Mathematicis commercium habuit) meas in manus inciderunt; & in illis plurima reperi à cunctis fere totius Europæ eruditis ipsi communicata' (a[1r]).

quædam reperiret, quæ a Viro summo Isaaco Newtono venerant; de his edendis cogitavit, non male profecto facturus, si integri commercii epistolici Collinsiani aut certe uberiorum excerptorum publicatione orbem eruditum sibi devinciret. Cum enim nec nos fugiat, Virum celeberrimum commercium litterarium habuisse eximium; nulli dubitamus, multa in literis ad ipsum scriptis præclara contineri, ad historiam Matheseos ipsumque hujus scientiæ incrementum profutura. Nec obstat, quod forte maxima pars eorum jam aliis occasionibus typis descripta prostet; norunt enim harum rerum intelligentes, quantum intersit nosse, quo tempore Viri præclari in meditationes suas inciderint.(6) Equidem Cl. Jonesius cum animadverteret, Newtoniana a Collinsio asservata ferme idem cum iis argumentum habere, quæ Vir illustris Newtonus jam ipse in lucem edidit, consilium suum mutavit;(7) Tractatum tamen de Curvarum quadratura cum luculenter ac concinne conscriptum & ad instruendos alios maxime accommodatum judicaret, eundem cum venia Autoris in lucem emisit(8) & alia nonnulla Analytica Newtoni inventa addidit, de quibus jam dicemus.

Tractatum istum de Quadraturis Curvarum inscripsit Newtonus *de Analysi per æquationes numero terminorum infinitas*(9) et in eo demonstrat,

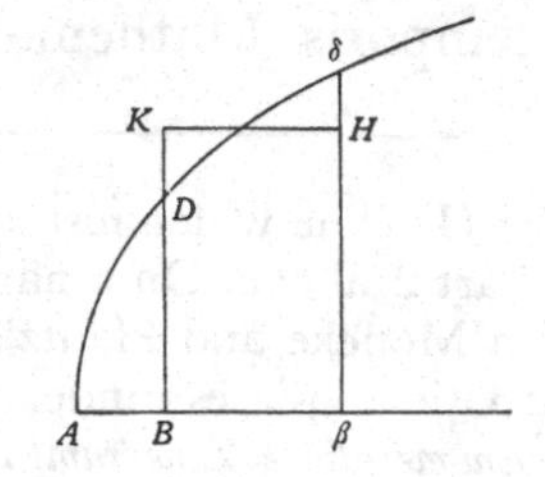

... si fuerit $ax^{m:n}=y$, aream esse $\frac{an}{m+n}x^{m+n,:n}$ sequentem in modum:(10) Sit $AB=x$, $BD=y$, area $ABD=z$, $\beta B=o$, $BK=v$ & rectangulum $B\beta HK$ æquale spatio $B\beta\delta D$. Pro lubitu sumatur $\frac{2}{3}x^{3:2}=z$, sive $\frac{4}{9}x^3=z^2$ & $x+o$ pro x, $z+ov$ pro z substitutis, prodibit

$$\tfrac{4}{9}\text{ in } x^3+3x^2o+3xo^2+o^3=z^2+2zov+o^2v^2,$$

& sublatis æqualibus $\frac{4}{9}x^3$ & z^2 reliquisque per o divisis, restat

$$\tfrac{4}{9}\text{ in } 3x^2+3xo+o^2=2zv+ov^2.$$

(6) This phrase recalls the opening of Leibniz' later autobiographical account of his discovery of algorithmic calculus: 'Utilissimum est cognoscere veras inventionum memorabilium origines, præsertim earum quæ non casu sed vi meditandi innotuere....' (C. I. Gerhardt, *Historia et Origo Calculi Differentialis a G. G. Leibnitio conscripta* (Hanover, 1846): 1).

(7) '...& inter ea non pauca, quæ à Viro Cl. D. Newtono scripta fuerunt; quæ cum tantæ molis essent, ut simul Tractatum breviusculum possent conficere, cœpi de iis edendis cogitare. Quum autem animadvertissem scripta ejus quæ jam in lucem prodierunt ferme idem cum hisce argumentum habere, haud operæ pretium me facturum, si typis mandarem, existimavi' (Jones' *Præfatio*: $a[1^r]$–$a[1^v]$).

(8) 'Unus tamen erat brevis de Curvarum Quadratura Tractatus adeo luculenter & concinne scriptus, atque ita accommodatus ad instituendos eos, qui nondum totam istam Methodum perspectam habeant, ut abstinere non potuerim, quominus Auctoris licentiam eundem edendi peterem. Quam Ille non solum summa cum humanitate concessit; sed & insuper veniam dedit reliqua ipsius colligendi, quæ ad idem argumentum spectabant' (Jones, *Præfatio*: $[a1^v]$).

Si jam supponatur, o esse nihil, erunt v & y æquales & termini per o multiplicati evanescent. Quare restabit $3x^2.\frac{4}{9}=2zv$, sive $x^{1:2}=y$. Ergo e contrario, si $x^{1:2}=y$, erit $\frac{2}{3}x^{3:2}=z$. Hanc methodum Newtonus applicat primum ad Curvas simplices, in quibus valor ipsius y unico termino constat, ut si fuerit $4\surd x=y$ vel $1:x^2=y$; dein ad compositas, in quibus e. gr. $x^2+x^{3:2}=y$, vel $x^2+x^{-2}=y$. Tandem valorem ipsius y in seriem resolvit vel per divisionem, exemplo Mercatoris, e. gr. si fuerit $a^2:(b+x)$, vel per extractionem radicis, ut si fuerit $\surd(a^2+x^2)=y$. Ubi simul exponit methodum extrahendi radices tam simplicium, quam affectarum æquationum a Wallisio in *Algebra* c. 94 f.m. 381(11) ex ipsius ad Oldenburgium literis jam repræsentatam. Eadem methodus quomodo ad longitudines Curvarum inveniendas, nec non ad curvas mechanicas quadrandas adhibeatur, aliquot exemplis docetur....

... ...(12)

Ceterum quod Cl. Editor methodum rationum primarum & ultimarum methodo quantitatum infinite parvarum præfert;(13) sciendum est, variari tantum in modo loquendi & pro rigorosa demonstratione utramque ad methodum Archimedeam revocari debere, ut error quovis dato minor osten-

(9) 'Hicce Tractatus, quem D. Collinsii manu exaratum comperi, inscriptus fuit *De Analysi per Æquationes Infinitas*' (Jones' *Præfatio*: [a1v]).

(10) See Newton's 'Præparatio pro regula prima demonstranda'. We have seen that Leibniz did not bother to note this passage in his 1676 'Excerpta' (Appendix 1: note (38) above) but now, with the outbreak of the priority dispute imminent, it becomes of crucial importance as a historical document.

(11) *Johannis Wallis S.T.D....De Algebra Tractatus; Historicus & Practicus. Anno 1685 Anglice editus: Nunc Auctus Latine. Cum variis Appendicibus....Operum Mathematicorum Volumen Alterum*, (Oxford, 1693): 381–3: Cap. XCIV, 'Nova Methodus extrahendi Radices tum Simplicium, tum Affectarum Æquationum' (reprinted from his 1685 *Treatise of Algebra*...: 338–40). Compare 3: note (2) above. (The review of this volume in the *Acta Eruditorum* (June, 1696): 249–60, partially quoted in the *Commercium Epistolicum* (3: note (2): 98–9), is Leibniz'.)

(12) In this omitted passage of some twenty lines Leibniz summarized the remaining contents of Jones' *Analysis*, noting the excerpts from Newton's *epistolæ prior et posterior* to Oldenburg and from his letter to Collins of 8 November 1676 (there printed for the first time), and then briefly characterized the following tracts *De Quadratura Curvarum*, *Enumeratio Linearum Tertii Ordinis* and *Methodus Differentialis*. With regard to the *Enumeratio*, a sophisticated and more comprehensive epitome of the cubic researches developed in **1**, 1 above, he remarked: 'Sed egregie de Geometris meritus fuisset Cl. Editor, si demonstrationem numeri linearum tertii ordinis, quam petenti non denegaturus erat Newtonus, una exhibuisset: immo adhuc bene mereri poterit, si per modum appendicis aut alia occasione eandem edat'.

(13) 'Hujus Geometriæ non minimam esse laudem duco, quod dum per limites Rationum Primarum & Ultimarum argumentatur, æque demonstrationibus Apodicticis ac illa Veterum munitur; utpote quæ haud innititur duriusculæ illi Hypothesi quantitatum Infinite parvarum vel Indivisibilium, quarum Evanescentia obstat quominus eas tanquam quantitates speculemur' (Jones' *Præfatio*: c[2r]). This refers more particularly, as we shall see in our seventh volume, to Newton's preface to his tract *De Quadratura Curvarum* but Leibniz, of course, extends its reference to include the *De Analysi*.

datur. Cumque in calculo præcedente adhibetur *o* & *ov*, quis non videt, revera adhiberi infinite parvas, nempe *o* pro *dx* & *ov* pro *dz*. Sane *o* jam Fermatius aliique[(14)] in talibus casibus adhibuere. Sed calculo illustris Leibnitii differentiali invento, non jam simpliciter nullæ, sed speciales quædam quantitates nullescentes adhibentur, nempe exprimentes, ex qua decrescente quantitate ad evanescentiam venerint. Ita *dx* vel *dz* est quantitas specialiter ad quantitatem *x* vel *z* relata, seu affectio quædam ipsius *x* vel ipsius *z*, nempe duarum *x* vel duarum *z* differentia, sed nullescens. Et ita non multiplicantur quantitates, quarum affectionibus ad curvas exprimendas est opus: atque adeo æquationes etiam curvarum transcendentium per solarum ordinatarum abscissarumque relationem habentur. Leibnitius noster ad exemplum tam Cavallerii, quam Robervallii, nunc unam, nunc alteram exprimendi rationem, hoc est, nunc infinite parvam, nunc motum seu continuum transitum, sive fluxum adhibuit, prout visum est commodius; & usum transitus hujus ultra Geometricam ad Physicam ipsam promovit, nova quadam consideratione inventa, quam *Legem continuitatis* vocat, per quam tanquam lapidem lydium[(15)] multa erronea in physicis redargui possunt. Proposuit eam ante multos annos in *Novellis Reip. Literariæ* Bælianis[(16)] & exemplis illustravit.[(17)]

---

(14) Leibniz is confused here, for Fermat himself never used the '*o*' notation for a variable's increment but invariably the majuscule vowels '*A*' or '*E*'. It would seem that the first to do so, about the late summer of 1638, was Jean de Beaugrand, who introduced it systematically into his 'De la Maniere de trouver les Tangentes des Lignes Courbes par l'Algèbre et des Imperfections de celle du S[eigneur] des C[artes]' (first printed by C. de Waard in his *Supplément* to the *Œuvres de Fermat* (Paris, 1922): 98–114, especially 102–13). Leibniz may well have seen this manuscript exposition of Fermat's tangent method during his stay in Paris between 1672 and 1676 and have confused the two in retrospect. As we have seen (I, **3**, Appendix 1: note (21)), Newton made wide use of the '*o*' notation from September 1664 while James Gregory was the first to employ it in a printed work, his *Geometriæ Pars Universalis* (Padua, 1668). Leibniz was well aware of Fermat's calculus anticipations for in his copy of the *Commercium Epistolicum* text (3: note (2)) of the *De Analysi* he entered against Newton's Rule 1 'Hoc jam notum Fermatio alijsq;' (Dietrich Mahnke, 'Zur Keimesgeschichte der Leibnizschen Differentialrechnung', *Sitzungsberichte der Gesellschaft zur Beförderung der gesamten Naturwissenschaften zu Marburg*, **67** (Berlin, 1932): 31–69, especially 66–7).

(15) That is, a touchstone (for testing the 'gold' of physical theories).

(16) 'Baylian', a reference to the first 'Auteur' of the *Nouvelles*, Pierre Bayle. The present editor was Jacques Bernard.

(17) See Leibniz' 'Remarque sur l'Article V. des *Nouvelles*...du mois de Février, 1706. Envoyée à l'Auteur de ces *Nouvelles*', *Nouvelles de la Republique des Lettres*, **27** (Amsterdam, November 1706): Article III: 521–8. In the February issue of the *Nouvelles* its editor, Bernard, had contributed a 'Relation abrégée' of Fontenelle's 'éloge funébre' of Jakob Bernoulli read out in public assembly at the Paris Académie des Sciences the previous 14 November 1705 (N.S.). In particular he had made the offensive summary remark that 'Mr. [Jacques] Bernoulli & un autre des ses fréres, encore plus jeune, qui avoit suivi ses traces, méditérent si profondément sur ces foibles rayons échapez à Mr. Leibnitz, qu'ayant résolu de lui enlever la gloire

APPENDIX 3. NEWTON'S COUNTER OBSERVATIONS ON LEIBNIZ' REVIEW

[1713][1]

From the unpublished originals in the University Library, Cambridge[2]

[1] In the *Acta Leipsica* of ye month of February 1712, an Extract is given of a collection of Tracts published the year before by Mr Jones & entituled *Analysis per Quantitatum series, fluxiones ac differentias cum enumeratione linearum tertij ordinis*. And whereas in the beginning of this Acc^t^ the author wishes that Mr Jones had given a fuller Acc^t^ of the Commercium epistolicum of Mr Collins[3] found

de l'invention, ils y réüssirent, & publiérent même avant Mr. Leibnitz le Systême, qu'il avoit trouvé il y avoit long-tems, & le firent, si juste, que Mr. Leibnitz lui-même...avoüa publiquement, que l'honneur en étoit autant du à Mess. Bernoulli qu' à lui' (Article V: 212). Leibniz' correction, tactfully submitted beforehand to the surviving brother Johann in a letter on 15 April, in briefly recounting the various stages by which he came to develop an algorithmic calculus and to apply it to physical problems claimed priority for himself in no uncertain terms. Though he did not then know it, when he penned such phrases as 'mon nouveau calcul' and 'mon droit d'Inventeur' he tempted providence and the combined wrath of Newton's English supporters. However, Bernard, Fontenelle and Johann Bernoulli each, explicitly or tacitly, allowed Leibniz his priority and the contretemps was settled amicably. Somewhat surprisingly, despite its dogmatic assertions and Leibniz' present reference to it, the article was not quoted during the fluxional priority dispute.

(1) For the historical background of these three draft criticisms of Leibniz' published review of Newton's *De Analysi* see the Introduction to the present Part 2. In the first piece the remark is made that 'of the Commercium epistolicum of Mr Collins found in amongst his papers [s]omething further has since been communicated by Mr Jones' and it was evidently composed a little while before the appearance of the 'Royal Society's' *Commercium Epistolicum D. Johannis Collins, et Aliorum de Analysi Promota* (3: note (2)) in the early spring of 1713. By the time Newton came to draft its extended version ([2] below) that collection of letters had 'been lately published'. The final version we reproduce was apparently composed somewhat later still and reads much like the many drafts (ULC. Add. 3968: passim) made between late 1713 and the end of 1714 for Newton's anonymous review of the *Commercium Epistolicum* in the *Philosophical Transactions* (**29**: No. 341 (for January and February 1713/14): 173–224: II. 'An Account of the Book entituled *Commercium Epistolicum*...')and for his equally anonymous appendix (pp. 97–123) to Joseph Raphson's *The History of Fluxions, Shewing in a Compendious Manner The First Rise of, and various Improvements made in that Incomparable Method* (London, 1715). How Newton came upon Leibniz' review we do not know. Certainly, he was never a voracious reader of periodicals and we might well suppose that someone, possibly John Keill, brought it to his notice. (The latter had already called Newton's attention to Leibniz' 'Schediasma de Resistentia Medii et Motu Projectorum Gravium in Medio Resistente', *Acta Eruditorum* (January, 1689): 38–47, in his letter of 3 April 1711 (ULC. Add. 3985.1).)

(2) Add. 3968.32: 463^r^–463^v^, 460^r^–461^v^ and 461^v^–462^v^. As before we have set book titles and quotations in italics for easier comprehension.

(3) It is an interesting conjecture that Newton may have taken this, the title of his *Commercium Epistolicum* (3: note (2)), from the equivalent phrase with which Leibniz opened his review (Appendix 2 above).

in amongst his papers[4] Something further has since been communicated by Mr Jones,[5] & if Mr Leibnitz who has a correspondence wth Mr Oldenburg & by [t]his means wth Mr Collins, would publish the Letters remaining in his custody relating to that correspondence or such extracts of them as may[6] conduce to complete what is wanting in the collection of Mr Jones,[7] he would equally oblige the world.

In the next place the Author of the Extract gives an acct of the method used in ye *Analysis* wth ye application thereof to the solving of Problemes & in ye end of the extract subjoins: *Cæterum quod Cl. Editor methodum rationū primarum & ultimarum methodo quantitatum infinite parvarum præfert: sciendū est, variari tantum in modo loquendi & pro rigorosa demonstratione utramꝗ ad methodum Archimedeam revocari deberi, ut error quovis dato minor ostendatur. Cumꝗ in calculo præcedente adhibetur o & ov, quis non videt revera adhiberi infinite parvas nempe o pro dx & ov pro dz.* By wch words I perceive that the Author of ye Abstract doth not yet understand ye Method of the first & last ratios. For in this method quantities are never considered as infinitely little nor are right lines ever put for arches neither are any lines or quantities put by approximation for any other lines or quantities to wch they are not exactly equal, but the whole operation is performed exactly in finite quantities by Euclides Geometry untill you come to an equation & then the equation is reduced by rejecting the terms wch destroy one another & dividing the residue by the finite quantity *o* & making this quantity *o* not to become infinitely little but totally to vanish. For Mr Newtons[8] words in explaining this method are: *Jam supponamus Bβ in infinitum diminui et evanescere, sive o esse nihil.* Had *Bβ* or *o* been considered as infinitely little, he would not have said *Jam supponamus Bβ in infinitum diminui*. Now by the vanishing of *o* there will remain an equation wch solves the Probleme. And this way of working being throughout as evident exact & demonstrative as any thing in Geometry is justly preferred by Mr Jones to ye method of infinitely little quantities:[9] wch proceeding frequently

(4) Newton has cancelled a first continuation, 'If ye Author pleases to signify what inventions of any of his friends who corresponded wth Mr Collins want an inquiry into their originalls, Mr Jones will be desired to search the scrinia of Mr Collins for that purpose'.

(5) Precisely, that portion of the Collins papers which were later deposited in the Royal Society and eventually bound up by Jones himself (compare **1**, 3, Appendix **2**: note (12)) to form the present *Commercium Epistolicum* volume (MS LXXXI).

(6) 'give light to the times in [wch]' is cancelled.

(7) A first continuation '& make a fuller discovery of the times when any inventions were [made]' is cancelled.

(8) As Leibniz, Newton in his counterblast preserves his anonymity.

(9) Newton first wrote 'of computing infinitely little quantities by approximations'.

(10) A first cancelled continuation reads: 'gives the name of moments to infinitely little quantities & represents them by the same characters, putting *o* for an infinitely little quantity'.

(11) 'consideration of' is cancelled.

by approximations is less Geometrical & more liable to errors, but yet may be usefull in some cases. And upon both these methods Mr Newton founded his method of fluxions as is manifest by this *Analysis* written in the year 1669, where he(10) sometimes considers quantities as increasing or decreasing by continual motion or fluxion, & gives the name of moments to their momentane[o]us increases or decreases; Fluxions or motions being finite quantities & the method of first & last ratios consisting in the consideration of nothing but finite quantities, & being exact & demonstrative & free from approximations: Mr Newton chose to call this sort of Analysis the method of fluxions rather than ye method of moments, or the method of Indivisibles or Infinitesimals. But yet he intended not thereby to exclude the working in(11) moments & infinitely little figures whenever it should be thought convenient[,] this way of working being expedite. And this he has sufficiently explained in the Introduction to his *Quadratura Curvarum.*

---

[2](12) *Observationes*

in synopsin Analyseos per quantitatum series fluxiones ac differentias cum enumeratione linearum tertij Ordinis.

The style & spirit of this Synopsis shew that it was writ by the Author of the synopsis of the Book of *Quadratures,*(13) & therefore deserved to [be] noted.

It begins thus: *Gulielmus Jones edita synopsi palmariorum Matheseos clarus, cum in scrinijs D. Collinsij inter plurima a celebribus...in meditationes suas inciderint.* The principal part of the *Commercium epistolicum Collinsianum*(14) relating to the subject of this *Analysis* has been lately published, together wth the dates of the Letters for discovering the times when things were invented.(15) And because the time when the *Analysis per æquationes numero terminorum infinitas* was written is not described in this synopsis, it may not be amiss to tell the Reader,(16) that it was communicated to Mr Collins in July 1669 as appears by three Letters of Dr Barrow & by several other Lette[r]s of Mr Collins & Mr James Gregory about things conteined therein & therefore the things conteined therein were invented before

---

(12) This more elaborately composed piece was evidently a draft for an improved version intended for publication, perhaps in the *Philosophical Transactions*, but presumably never completed. Ultimately, by way of the final draft which follows, Newton subsumed its content into his review of his *Commercium Epistolicum* (note (1)), especially, pp. 174–82.

(13) That is, Newton's *De Quadratura Curvarum*, reviewed by Leibniz in the *Acta Eruditorum* in 1705. See Appendix 2: note (1) above.

(14) This is Leibniz' phrase, of course (note (3) above).

(15) In the *Commercium Epistolicum D. Johannis Collins, et Aliorum de Analysi Promota* (3: note (2)).

(16) Compare note (12).

that time. It was the Compendium mentioned by Mr Newton in his letter of 24 Octob. 1676.(17)

The Author of the synopsis in the next place produces out of the *Analysis* a part of the computation by wch Mr Newton demonstrated the first of the three Rules upon wch the *Analysis* was founded. And by this Computation compared wth other things in the *Analysis* it appears that Mr Newton when he wrote that *Analysis* used the very same method of fluxions which he uses at present. He represents time by any quantity wch flows uniformly, the fluxion of time by an unit, a moment of time by the letter *o*, other flowing quantities & their fluxions by any other symbols, & their moments by their fluxions drawn into a moment of time, the name of moments being taken from the moments of time in wch they are generated & that of fluxions from the fluxion of time. If he is enquiring after truth or resolving a Probleme, he uses the letter *o* for an infinitely little moment & for the greater dispatch neglects to express it, putting the symbol of the fluxion alone both for the fluxion & for the moment but usually neglecting to express the coefficient *o* when it signifies the moment. If he is demonstrating any Proposition he always expresses the letter *o* & uses it for a finite moment or an indefinitely (not infinitely) little part of time. In the former case he uses any approximations wch he foresees will create no error in the conclusion. In the latter he proceeds in finite quantities exactly by vulgar Geometry without any approximations & when he has finished the computation he supposes that the finite moment *o* decreases in infinitum & vanishes. And by this method of fluxions he applies æquations both finite & infinite to the resolution of Problemes. This is his method at present, this was his method when he wrote his two Letters of 1676 & five years before when he wrote the Tract(18) mentioned in the latter of those two letters(19) & that this was his method in the year 1669 when he communicated his *Analysis* to Dr Barrow & by Dr Barrow to Mr Collins, appears by the *Analysis* itself.

This *Analysis* is founded on three Rules the two first of which are equipollent to the solution of this Problem, *Data æquatione fluentes duas quantitates involvente se invicem non multiplicantes fluxiones invenire; & contra*. The third Rule directs the resolution of finite equations [&] infinite ones when there is occasion. These Rules Mr Newton illustrates wth various examples & applies them to the quadrature of curves & then adds that all Problems concerning the length of

(17) 'Eo ipso...tempore quo [N. Mercatoris Logarithmotechnia] prodiit communicatum est ab D. Barrow...ad D. Collinsium Compendium quoddam methodi harum serierum, in quo significaveram areas et longitudines curvarum omnium et solidorum superficies et contenta, ex datis rectis et vice versa ex his datis rectas determinari posse, & methodum ibi indicatam illustraveram diversis seriebus' (*Correspondence of Isaac Newton*, **2** (1960): 114). Compare *Commercium Epistolicum* (3: note (2)): 70, note †.

(18) The 1671 tract on fluxions and infinite series: this we will reproduce in our next volume.

curves & the contents & surfaces of solids & centers of gravity may be reduced to Quadratures after the following manner.[20] *Sit ABD curva quævis...dato elicietur.* Here the fluents are represented by the Areas ABD & AK & their moments by the Ordinates $BD=y$ & $BK=1$, & the area AK is supposed to flow uniformly, or in proportion to time & an unit is put for its moment, the coefficient o wch makes the moments y & 1 infinitely little being neglected. And after he had set down an example by computing the length of the arch of a circle from its moment he subjoyns. *Sed notandum est quod unitas quæ pro momento ponitur est superficies cum de solidis & linea cum de superficiebus & punctum cum de lineis agitur. Nec vereor loqui de unitate in punctis, sive lineis infinite parvis, siquidem proportiones ibi jam contemplantur Geometræ dum utuntur methodis indivisibilium.* So then by a point Mr Newton understands here an infinitely short line, & by a line an infinitely narrow surface & when he calls these moments & represents them by an unit it is to be understood that this unit is multiplied by an infinitely small quantity o, or moment of time, to make it infinitely little. The moment 1 is $1\times o$ & the moment y is $y\times o$, but the coefficient o for shortning ye operation is not written down but understood. If o be not understood the lines 1 and y represent the fluxions of the areas BK & ABD, but if o be understood, those fluxions multiplied by o become the moments of BK & AD. For fluxions are finite quantities but moments here are infinitely little. Thus you see his Notation when he wrote this *Analysis* is of the same kind wth that wch he uses at present.[21]

So in demonstrating the first of his three Rules by this method, in the equation $c^n x^p = z^n$ he supposes x & z to increase & be augmented by the moments o & ov & to become $x+o$ & $z+ov$, & by those moments understands the rectangles $o\times 1$ & $o\times v$ conteined under the fluxions 1 & v & the moment o. Then in the said æquation writing $x+o$ for x & $z+ov$ for z there arises

$$c^n \text{ in } x^p + pox^{p-1} + \&c = z^n + novz^{n-1} + \&c.$$

Where the first termes $c^n x^p$ & z^n destroy one another & the next divided by o vizt $c^n px^{p-1}$ & nvz^{n-1} become equal & determin the proportion of the fluxions 1 & v or of their moments o & ov. And its here observable that the series

$$x^p + pox^{p-1} + \&c$$

(19) Newton to Oldenburg, 24 October 1676: 'Et ante annos quinqꝫ cum, suadentibus amicis, consilium cœperam edendi Tractatum de refractione Lucis et coloribus quem tunc in promptu habebam; cœpi de his seriebus iterum cogitare, & tractatum de iis etiam conscripsi ut utrumqꝫ simul ederem' (*Correspondence of Isaac Newton*, **2** (1960): 114).

(20) See Newton's 'Applicatio prædictorũ ad reliqua istiusmodi Problemata' (p. 232 above).

(21) A sly remark. Newton had indeed used the o-notation exactly as he is at pains here to describe it since late 1664 but the unwary reader is seduced into believing that his companion dot notation for fluxions likewise dates from this period. In the form in which Newton presented it to the world in his *De Quadratura* that latter notation was adapted only in late 1691 from the various multiple dot conventions which, as we have seen (I, 2, 4, §3, 5, §§2 and 5, and 7: passim), he had employed in his early fluxional researches in the mid-1660's.

is the same wth that in the beginning of Mr Newtons Letter of 13 Jun 1676,[22] & being produced becomes

$$x^p+pox^{p-1}+\frac{pp-p}{2}oox^{p-2}+\frac{p^3-3pp+2p}{6}o^3x^{p-3}+\&c.$$

And the like is to be understood of the series $z^n+[n]\,oz^{n-1}+\&c$ in like manner to be produced. But Mr Newton having shewn before that all the terms after the second would vanish with the moment o, neglected them. He saw therefore in those days that the second terms of these series gave the fluxions & moments of the dignities of any fluent quantity x or $x+o$; & by this property of these series he demonstrated the first of the three Rules upon wch he founded his *Analysis*. And when he understood this he could not be long without seeing the use of the third terms of these series & of the rest of the terms wch follow the third. For that he knew the use of the third terms in those days is evident from his letter of 10 Dec. 1672, where he saith that by his method he determined ye curvature of curves.[23]

After this Demonstration Mr Newton subjoyns the following conclusion from it. *Hinc in transitu notetur modus quo Curvæ quotcunꝙ, quarum Areæ sunt cognitæ possunt inveniri; sumendo nempe quamlibet æquationem pro relatione inter aream z & basem (vel abscissam) x ut inde quæratur (ordinatim) applicata. Ut si supp[o]nas* $\sqrt{aa+xx}=z$, *ex calculo invenies* $\dfrac{x}{\sqrt{aa+xx}}=y$. *Et sic de reliquis.* And this is the second Proposition of his book of *Quadratures*, & it is as much as to say that the method by wch Mr Newton had now demonstrated the first of his three Rules was general, &

(22) Newton refers to his exposition of the binomial series in his *epistola prior* to Oldenburg for Leibniz (Jones' *Analysis*: 23 = *Correspondence of Isaac Newton*, **2** (1960): 21): that is, in equivalent terms

$$[P^{m/n}(1+Q)^{m/n}=\,](P+PQ)^{m/n}=P^{m/n}+\frac{m}{n}P^{(m/n)-1}(PQ)+\frac{m}{n}\times\frac{m-n}{2n}P^{(m/n)-2}(PQ)^2+\ldots.$$

The present expansion is the particular case of this when $P=x$, $Q=ox^{-1}$, $m=p$ and $n=1$.

(23) *Commercium Epistolicum* (3: note (2)): 30 (*Correspondence of Isaac Newton*, **1** (1959): 247): 'Hoc est unum particulare, vel corollarium potius Methodi generalis, quæ extendit se, citra molestum ullum calculum, non modo ad ducendum Tangentes ad quasvis Curvas, sive Geometricas, sive Mechanicas, vel quomodocunꝙ rectas lineas aliasve Curvas respicientes; verum etiam ad resolvendum alia abstrusiora Problematum genera de Curvitatibus, Areis, Longitudinibus, centris Gravitatis Curvarum, &c' (This S^r is one particular, or rather a Corollary of a Generall Method wch extends it selfe without any troublesome calculation, not onely to the drawing tangents to all curve lines whether Geometrick or mechanick or how ever related to streight lines or to other curve lines but also to the resolving other abstruser kinds of Problems about the crookedness, areas, lengths, centers of gravity of curves &c). Newton had at this time just come under fire from Johann Bernoulli for not understanding the second derivative of a function in the latter's 'De Motu Corporum Gravium, Pendulorum, & Projectilium in medijs non resistentibus & resistentibus supposita Gravitate uniformi & non uniformi atque ad quodvis datum punctum tendente...' (*Acta Eruditorum* for February/March 1713: 77–95/115–32). The justice of Bernoulli's general attack (centred on Book 2, Proposition X of the

extended to the determination of this Problem *Data Æquatione fluentes duas quantitates involvente fluxiones invenire*. And that he extended it also to more then two fluents appears by his letter of 10 December 1672 where he saith that his method stuck not at surds.[(24)] For surds are in this method considered as fluents. So then the two first Propositions of the book of *Quadratures* were then known to Mr Newton.

In this Tract of *Analysis* Mr Newton writes also that his Method extends to such Curve lines as were then called Mechanical. And instances in the Quadratrix by shewing how to find the Ordinate & Area of this Curve & adding that its length may be found by the same method. And then he subjoyns *Nec quicquam hujusmodi scio ad quod hæc methodus, idq varijs modis sese non extendet. Imo tangentes ad Curvas mechanicas (siquando id non alias fiat) hujus ope ducuntur. Et quicquid vulgaris Analysis per æquationes ex finito terminorum numero constantes (quando id sit possibile) perfecit, hæc per æquationes infinitas semper perficit: Ut nil dubitaverim nomen Analysis etiam huic tribuere. Ratiocinia nempe in hac non minus certa sunt quam in illa, nec æquationes minus exactæ.—Deniq ad Analyticam merito pertinere censeatur, cujus beneficio Curvarum areæ & longitudines &c (id modo fiat) exacte et Geometrice determinentur. Sed ista narrandi non est locus.* These words (*id modo fiat*) have respect to a sort of series wch sometimes breake off & give the Quadrature in finite equations.[(25)] One of these series is set down by Mr Newton in his Letter of 24 Octob 1676 as the first of certain Theoremes for Quadratures wch he had formerly found by his method of fluxions. This is the 5th Proposition in his book of *Quadratures* & the first of those for squaring a given curve & the sixt is the second of the same kind, & these two depend on the 3d & 4th & those on the 1st & 2d. So that the first six Propositions of that book were known to him when he wrote his Letter of 24 Octob 1676, & even when he wrote his *Analysis*, as Mr Collins

first edition of Newton's *Principia*) we will examine in our final volume but we may note the essence of that part of it which relates to the present point (pp. 93–4, §33): '$\overline{a+o}^p$...more Newtoniano in seriem conjecta dabit... $a^p+\frac{p}{1}a^{p-1}o+\frac{p\times p-1}{1\times 2}a^{p-2}oo+\frac{p\times p-1\times p-2}{1\times 2\times 3}a^{p-3}o^3+\&c.$ cujus termini ex opinione Newtoni exprimerent successive differentiales omnium in eodem ordine graduum...; interim excepto primo & secundo termino reliquos omnes a veris differentialibus abludere, communis differentiandi regula docebit'. Newton was hard put to it to refute this unfounded interpretation for none of his researches in the early 1690's on the Taylor expansion (ULC. Add. 3960.10/11: passim, to be reproduced in our sixth volume) had been published. Here, perhaps for the first time, he attempted to suggest by the present reference that he had applied the second mean value theorem to the problem of curvature in or before 1672. There is no documentary evidence for this and it seems highly unlikely.

(24) *Commercium Epistolicum* (3: note (2)): 30 (*Correspondence of Isaac Newton*, 1: 247–8): 'Neq (quemadmodum *Huddenij* methodus de *Maximis* & *Minimis*) ad solas restringitur æquationes illas, quæ quantitatibus surdis sunt immunes' (Nor is it (as Huddens method de maximis et minimis...) limited to æquations wch are free from surd quantities).

(25) Compare 3: note (131) above.

in his letter to Mr Strode dated 26 July 1672 explains in these words: *By the same method may be obteined the Quadrature or Area of the figure accurately when it can be done, but always infinitely near.*(26)

Next after the *Analysis* Mr Jones has printed(27) Mr Newtons Letter of 13 June 1676, wherein the Rule for reducing binomials into infinite series is set down at large & explained by Examples. And all this was known to Mr Newton when he wrote his *Analysis*, (the two first terms of this Rule being there set down,) as was also the reduction of finite æquations into infinite series by extraction of roots out of affected æquations, & the Quadratures of Curves by those series, wch wth the approximation of Quadratures makes up the body of the Epistle.

... ...(28)

Now as he understood this method [of fluxions] in ye year 1671 when he wrote the said Tract upon it, so the *Analysis* shews that he understood it when he wrote that Tract & in the Introduction to his Tract of *Quadratures* he tells us that he found it gradually in the years of 1665 & 1666. The method of Series he found in ye year 1665 & the second terms of the series gave him the moments of the first terms.(29)

... ...(30)

---

[3](31) Fermat in his method de maximis et minimis & Gregory(32) in his method of Tangents & Newton [in his] method of the first & last ratios use the letter *o* to signify a quantity not infinitely but indefinitely small & Barrow in his method of tangents uses the letters *a* & *e* in the same manner.(33) For they all make the letters *o*, *a* & *e* to become infinitely little. So Mr Newton in his *Analysis* in demonstrating the first rule, so soon as he has done the calculation, uses these words *Si jam supponamus Bβ in infinitum diminui et evanescere, sive o esse nihil*. And so when he has finished the calculation in ye Demonstration of the first Proposition

---

(26) '...patet illam Methodum a dicto Newtono aliquot annis antea excogitatam & modo universali applicatam fuisse: ita ut ejus ope in quavis Figura Curvilinea proposita...Quadratura vel Area dictæ Figuræ, accurata si possibile sit, sin minus infinite vero propinqua ...obtineri queant' (*Commercium Epistolicum* (3: note (2)): 28–9). Newton quotes from the unpublished English original in Royal Society MS LXXXI, we presume.

(27) Jones' *Analysis*: 23–30.

(28) In this omitted portion Newton analysed the remaining excerpts printed by Jones (from his letters of 24 October 1676 to Oldenburg, of August 1692 to Wallis and of 8 November 1676 to Collins), pinpointing the ways in which they confirmed his asserted chronology of his development of fluxional calculus after 1669.

(29) Compare I, 2, 4, §3: passim.

(30) We omit Newton's concluding paragraphs which briefly discuss the content of the three remaining tracts (*De Quadratura Curvarum*, *Enumeratio* and *Methodus Differentialis*) printed by Jones in his 1711 *Analysis*.

in his Book of *Quadratures* he saith: *Minuatur quantitas o in infinitum*. By this means the whole calculation is done in finite figures by the Geometrical Propositions of Euclide, & so is demonstrative. And upon this Account Mr Jones commends the method in the Preface to his *Analytical Collection. Hujus Geometriæ Newtonianæ,* saith he, *non minimam laudem esse duco quod dum per limites Rationum primarum et ultimarum argumentatur, æque demonstrationibus Apodicticis ac illa Veterum munitur; utpote quæ haud innititur duriusculæ illi Hypothesi quantitatum infinitè parvarum, vel Indivisibilium, quarum evanescentia obstat quo minus eas tanquam quantitates speculemur.* But Mr L. in the Account wch he has given of the *Analytical Collection* of Mr Jones in y^e *Acta Eruditorum mensis Februarij 1712* pag 76 corrects Mr Jones in the following manner. *Ceterum quod Cl. Editor* (Jonesius) *methodum rationum primarum et ultimarum methodo quantitatum infinite parvarum præfert; sciendum est variari tantum in modo loquendi, et pro rigorosa demonstratione utramꝗ ad methodum Archimedeam revocari debere ut error quovis dato minor ostendatur. Cumꝗ in calculo præcedente* (i.e. in prædicta Demonstratione Regulæ primæ sub finem *Analyseos*) *adhibetur o et ov, quis non videt revera adhiberi infinite parvas, nempe o pro dx et ov pro dz. Sane o jam Fermatius alijꝗ in talibus casibus adhibuere. Sed calculo illustris Leibnitij differentiali invento* &c. But here Mr L. has misrepresented both the methods. For in the method of first & last ratios the error is not proved to be less then any given ratio but to be none at all. In this method there are no errors at all. The letter *o* represents a finite quantity indefinitely small till the calculation be finished & the whole calculation is performed in finite quantities by the Geometry of Euclide without any error. And then the quantity *o* decreases in infinitum & becomes nothing & leaves the ratio ultima of quantities without any error. In the other method, the calculation is not grounded upon Euclides Geometry. There is not one Proposition in Euclide concerning quantities or figures infinitely small. The calculation proceeds by approximations[,] putting the infinitely small arcs of curves & their chords, sines & tangents equall to one

(31) In this final counterblast Newton adopts a much more forthright tone, and names Leibniz outright (or at least 'Mr L.') as the author of the anonymous *Acta Eruditorum* review (Appendix 2 above). Very probably he has now received news of the propagation of Leibniz' anonymous *Charta volans* of 29 July 1713, which printed the essential passages of Johann Bernoulli's letter to him of the preceding 7 June with its heavy criticism of the 'Royal Society's' *Commercium Epistolicum*. At all events the gloves are now on. (An example of the original uncut doubled *Charta* is preserved in ULC. Add. 3968.34 along with its complete transcript in Newton's hand and several drafts of his 'Observations' upon it. On 482^r Newton states that he first received a copy of the 'flying paper in Latin' from Chamberlayne 'in autumn 1713'.)

(32) James Gregory, that is. Compare Appendix 2: note (14) above.

(33) Isaac Barrow, *Lectiones Geometricæ: In quibus* (*præsertim*) *Generalia Curvarum Linearum Symptomata declarantur* (appended to his *Lectiones* XVIII...; *In quibus Opticorum Phænomenωn Genuinæ Rationes Investigantur, ac exponuntur*) (London, 1670): Lectio x: 80–4. On his p. 80 Barrow remarked that he introduced this variant on Fermat's and Gregory's methods 'ex Amici [Newton!] consilio'.

another & frequently using other approximations w[ch] want a rigorous Demonstration that the error is less then any given error. Without such rig[or]ous Demonstrations the method is not geomet[r]ical, nor are the Propositions found thereby to be admitted into Geometry till they are rigorously demonstrated.(34) Nor is an Analytical Demonstration alone sufficient to make a Proposition Geometrical. It ought to be demonst[r]ated Synthetically & that in words at length to be read by people not skilled in Analysis.

But, saith M[r] L. *Cum in calculo præcedente adhibeatur o, et ov, quis non videt revera adhiberi infinite parvas, nempe o pro dx, et ov pro dz.* M[r] Newtons words above cited imply that they are considered as finite & only indefinitely small till the calculation is ended, & he supposes *o* to be diminished in infinitum & become nothing. And M[r] L. should have told his reader that the *Analysis* was sent to M[r] Collins in June 1669 & by consequence that the *dx* & *dz* of M[r] Leibnitz were put for the *o* & *ov* of M[r] Newton, & not on the contrary that y[e] *o* & *ov* of M[r] Newton were put for y[e] *dx* & *dz* of M[r] Leibnitz.

M[r] L. subjoyns: *Sane o jam Fermatius aliję in talibus casibus adhibuere. Sed calculo illustris Leibnitz differentiali invento* &c. W[ch] is as much as to say that the invention of a new method lay in the invention of new symbols. But M[r] Newton still uses the letter *o*, as may be seen in his book of *Quadratures*, And its for the honour of M[r] Fermat that it should still be used. And if M[r] Newtons method was the same when he wrote his *Analysis* that it is at present, the Question will be what M[r] Le[i]b[n]itz hath added to this method besides a new notation or what advantages have been brought to it by this notation.

M[r] L. saith: *dx vel dz est quantitas specialiter ad quantitatem x vel z relata, seu affectio quædam ipsius x vel ipsius z, nempe duarum x vel z differentia, sed nullescens. Et ita non multiplicantur quantitates quarum affectionibus ad Curvas exprimendas est opus: atę adeo æquationes etiam curvarum transcendentium per solarum ordinatarum abscissarumę relationem habeantur.* This is the great advantage of the differential notation. And so in the *Acta Eruditorum mensis Junii A. 1697*(35) pag 297 he saith *Malo autem dx et similia quam literas pro illis, quia istud dx est modificatio quædam ipsius x, et ita ope*

(34) Newton has cancelled a long following (and scarcely relevant) passage: 'Mr Newton found many of the Propositions in his *Principia Mathematica Philosophiæ* by this infinitesimal Analysis, but he did not propose them as Geometrical Propositions till he had rigorously demonstrated them. And in order to demonstrate them he spent a whole section in demonstrating Lemmas by the Method of the first & last Ratios. For Geometry is not to be laid aside or corrupted by an Arithmetical method. All Algebra is Arithmetic in species. And tho for the improvement of invention it be applied to magnitudes yet [it] is nothing more then an Arithmetical Analysis of Geometrical Problems & the Ancients admitted no Propositions into Geometry before they were demonstrated in words at length by the direct method w[ch] they called Synthesis or Composition. And by that means they have transmitted down to us an excellent Geometry'. Various existing drafts (in ULC. 3968.9 and in private possession) of a preface to an abortive revision of the second edition of his *Principia*, which Newton began to prepare at this time, develop this theme at some length.

ejus fit ut sola quando id fieri opus est litera x, cum suis scilicet potestatibus & differentialibus calculum ingrediatur et relationes transcendentes inter x et aliud exprimantur. Qua ratione etiam lineas transcendentes æquatione explicare licet. He acknowledges here that in the differential method he might have used letters for the differences a[s] Dr Barrow used *a* &c in his method of tangents: but he chose rather to use the symbol *dx* & the like for the sake of the advantages here mentioned. If he had used letters he must have defined their significations in every new Problem or at least in every Tract or Mathematical Book, but by using *dx* & *dy*, it suffices to define them once for all: & this is the great advantage for the sake of wch he chooses to use these symbols....(36) But certainly no man would call this Notation a new method of Analysis....

But let us compare the symbols of Mr Newton & Mr Leibnitz & see wch are the older & the better.

Mr Newton in his *Analysis* sometimes represents fluents by the areas of curves & their fluxions by y^e Ordinates, & moments by the Ordinates drawn into y^e letter o. So where the Ordinate is $\frac{aa}{64x}$ he puts $\boxed{\frac{aa}{64x}}$ for the area.(37) And so if the Ordinate be v or y the Area will be $\boxed{v}$ or $\boxed{y}$. And in this way of notation the moments will be $\frac{aao}{64x}$, vo, yo. Mr Leibnitz instead of the Notes $\boxed{\frac{aa}{64x}}$, $\boxed{v}$, $\boxed{y}$ uses the notes $\int\frac{aa}{64x}$, $\int v$, $\int y$. Mr Newtons are much the older being used by him in the year 1669.

When letters are put for fluents (as is commonly done) Mr Newton puts for the fluxions sometimes other letters, sometimes the same letters with a prick, sometimes the same letters in a different form or magnitude & still uses any of these notations without confining his method to any one of them. Mr Leibnitz has no proper symbols for fluxions:(38) these being finite quantities & the quantities being velocities of motion & [y^e] differences *dx*, *dy* &c being infinitely little ones....For moments Mr Newton puts the symbols of fluxions multiplied by the letter o w^ch (as was said) represents an infinitely little quantity answering to a moment of time: Mr Leibnitz puts ye symbols of the fluents wth the letter d before them.(39)

(35) Read '*1686*'. Newton's reference is to 'G. G. L. De Geometria Recondita et Analysi Indivisibilium atque infinitorum, Addenda his quæ dicta sunt in Actis a. 1684, Maji. p. 223; Octob, p. [4]64; Decemb. p. 586'. *Acta Eruditorum* (June 1686): 292–300, especially 297, lines 21–6.

(36) We omit some irrelevant Newtonian irony.

(37) Compare 3: note (78) above.

(38) Leibniz would, of course, represent the fluxion of x by dx/dt, in which t is an independent variable of time.

(39) The manuscript leaves off here somewhat abruptly.

an [illegible] *Geometria* [illegible] quantitatum [illegible] he acknowledges there that in the differential calculus [illegible] have used [illegible] for the differences [illegible] use the symbol [illegible] the like for the sake of [illegible] If he had [illegible] must have [illegible] or at least [illegible] of Mr [illegible] Cock, but [illegible] to define them [illegible] the sake of [illegible] these symbols. [illegible] But certainly [illegible] would call this [illegible] method of [illegible].

But let us compare the symbols of Mr Newton & Mr Leibnitz & see which are the [illegible] & the better.

Mr Newton in his [illegible] sometimes represents fluents by the areas of curves & their fluxions by the Ordinates or moments by the Ordinates drawn into [illegible]. So when the Ordinate is $\frac{aa}{64x}$ [illegible] or the area [illegible] Ordinate [illegible] Area will be [illegible]. And in this way of notation the moments will be $\frac{aa}{64x}$ [illegible] instead of the Notes $\left|\frac{aa}{64x}\right|$ [illegible] in the [illegible] being [illegible] by [illegible] in the [illegible].

When letters are put for fluents [illegible] Mr Newton puts for the fluxions [illegible] sometimes the same letters with a prick, sometimes the same letters in a different [illegible] & [illegible] of these notations without confining his method to any one of them. Mr Leibnitz has no [illegible] symbols for fluxions [illegible] the quantities [illegible] [illegible] velocities of motion [illegible] [illegible] being [illegible] [illegible] Mr Newton [illegible] [illegible] by the letter [illegible] quantity [illegible] to [illegible] Mr Leibnitz [illegible] symbol of the [illegible] the letter [illegible] before them.

[illegible]

PART 3

RESEARCHES IN ALGEBRA AND THE CONSTRUCTION OF EQUATIONS

(*c.* 1670)

INTRODUCTION

In this final part we reproduce two short works which together represent Newton's researches in the theory and construction of equations in the two or three years preceding 1671. The first of these, *In Algebram Gerardi Kinckhuysen Observationes*, gathers his emendations and additions to Nicolaus Mercator's manuscript Latin version of a newly published Dutch introduction to Cartesian algebra: before it, since Kinckhuysen's printed book is now exceedingly rare and even those with access to it may not find its language immediately comprehensible, we have set an annotated transcript of Mercator's Latin translation.(1) The second Newtonian text, untitled by its author but named by us 'Problems for construing æquations',(2) is a little known, densely packed tract on the geometrical construction of algebraic equations. As before, our detailed elucidation and criticism of these pieces will be found in the footnotes we have added at appropriate places, but we may here make some general observations on their historical background and content.

Gerard Kinckhuysen's *Algebra Ofte Stel-konst*, published in Low Dutch at his native town of Haerlem in the year of his death (1661), was composed by him as an algebraic companion volume to his exposition of the elements of Cartesian analytical geometry, *De Grondt der Meet-konst*, which had appeared the previous year. If we are to believe Kinckhuysen's own testimony in the preface to his *Algebra*, the latter work was not widely appreciated by his compatriots because they were ignorant of the algebraic foundations on which analytical geometry is based. Since no attempt had been made to publish an elementary algebraic work in the Dutch vernacular (since the days of Stevin and van Ceulen at least) it was his design in the *Algebra* to present recent advances succinctly in Cartesian notation and in a form intended to initiate the beginner into the higher mysteries of the art: 'hoc opus tyronibus tantùm conscriptum [est], quò viam insistant primulùm' (this work is written merely for the novice to take his first steps on the road). In the achievement of that aim, however, the work must have been no less a failure than its predecessor. As a popular expositor Kinckhuysen lacked the lightness of touch and confident grasp of his subject requisite to persuade

(1) After having disappeared from public knowledge in the middle 1670's this manuscript Latin version, interleaved by John Collins in a set of sheets of the original Low Dutch printing (see 1, §1, note (2) below), was finally traced in the summer of 1963 by Christoph J. Scriba: having apparently passed to John Wallis when the latter was composing his own *Algebra* (London, 1685), the interleaved Dutch and Latin version is now located in a bound miscellany of mathematical works and pamphlets which came from his library. Our debt in the following account to Scriba's 'Mercator's Kinckhuysen-Translation in the Bodleian Library at Oxford' (*British Journal for the History of Science*, **2** (1964): 45–58) will be evident.

(2) See 2, §2, note (1).

and delight the uninitiated, while his choice of content is seen in retrospect to be an unlikely blend of basic elements and long-winded technicality—a fundamental inconsistency which would have been obvious to Newton when he came to examine the work.

Though he had not succeeded in writing his intended essay in *haute vulgarisation* Kinckhuysen had nevertheless composed a useful if somewhat pedantic summary of recently published work in algebra. Substantially, the *Algebra Ofte Stel-konst* is a reordered version of the third book of Descartes' *Geometrie* filled out with additions from Frans van Schooten's *Commentarii*, Johann Hudde's *De Reductione Æquationum* and, on occasion, his own researches. To this is prefaced an introductory account of basic algebraic operations, founded upon Erasmius Bartholin's *Principia Matheseos Universalis* and Ludolf van Ceulen's *De Arithmetische en Geometrische fondamenten*, while from the latter work and elsewhere is appended a set of worked problems (all leading, in fact, to quadratic equations) to illustrate the application of the previous theory to both determinate and indeterminate cases. For their use in the solution of quadratics and cubics the extraction of the square and cube roots of surd quantities is explained at the close of the first part, while in the main section following Kinckhuysen discusses at length the effect of operating arithmetically on the roots of a general equation, the Cartesian sign-rule for determining bounds to the number of positive and negative ('true' and 'false') roots (though with an attempt, necessarily a failure, to add a counter-example), and the exact resolution of the reduced cubic and quartic equations, adding a sketch of Stevin's crude method for isolating the real roots of numerical equations. It is true that in the hundred or so pages of the *Algebra* a good deal of dross was also included, but we need not be surprised that the book caught John Collins' eye when some years later he was

(3) 'Gerard Kinckhuysen wrote an Algebra in low Dutch in 4to, printed at Haerlem 1661, which he intends but as introductory for understanding a treatise of conics he published there the year before, algebraically performed' (S. P. Rigaud, *Correspondence of Scientific Men of the Seventeenth Century*, **1** (Oxford, 1841): 118). About the same time Collins wrote to Wallis that it was 'an excellent introduction' (*ibid.* **2** (1841): 484).

(4) Confusing Kinckhuysen with Ian Stampioen, Collins asserted to Pell that the former 'was the first that put that tedious question in Des Cartes of three sticks, and in his Conics hath now published his own geometric and analytic solution thereof (Rigaud's *Correspondence* (note (3)), **1**: 118; compare **2**: 484).

(5) See Collins' letter to Pell, 9 April 1667: 'Being once in Mr. Thomson's shop, I met with a Cambridge scholar, who suggested that the small anonymous Jesuit's Euclid, printed by Mr. Martyn, [*Euclidis Elementa Geometrica, Novo Ordine ac Methodo fere, Demonstrata*, London, 1666] was now in good request amongst tutors and their scholars there, that they wanted the like for arithmetic and algebra.... I acquainted the gentleman that Kinckhuysen in low Dutch did write an introduction to Algebra in 1661, which had in it the doctrine of surd numbers, binomials and residuals and their roots, and more than either [Van Schooten's] Principia [Matheseos] or yours' (Rigaud's *Correspondence* (note (3)), **1**: 125–6).

searching for a text which, suitably pruned and translated, might serve the English reader as a primer of contemporary algebraic knowledge.

Collins became acquainted with Kinckhuysen's work some time before 28 August 1666, when he briefly described the *Algebra* in a letter to John Pell[(3)] with the added comment that 'The author seems very learned', though his judgement may well have been coloured by a false estimate of the Dutch author's originality.[(4)] When early the following year a 'Cambridge scholar' (Isaac Barrow?) conversed with him in a London bookshop on the need for an up-to-date university textbook in algebra, he was happy to recommend Kinckhuysen's 'Introduction' as a stop-gap[(5)] but was told in return that already 'one Mr. Jefferies did agitate with Mr. Martyn [the London bookseller] about it' and was indeed on the point of travelling to Holland to purchase a copy of the book for translation purposes. Ever generous, Collins immediately lent Jefferies his own copy. However, Martyn's rival stationer Moses Pitt (then about to publish Pell's augmented edition of Rahn's German *Algebra* in Thomas Brancker's English version and, no doubt, fearing for his profits should Kinckhuysen's book prove the better seller) immediately tried to procure a copy for himself and for a time there was some talk of having extracts from the Dutch *Algebra* appended to the Rahn edition or 'admitted' into Pell's introduction to it.[(6)] Whether or not Pell or Brancker refused to have the additions in their book we do not know, but the *Teutsche Algebra* finally appeared in 1668 without any insertions from Kinckhuysen. Nothing daunted, Collins pressed on with his design and arranged with Nicolaus Mercator for a complete Latin translation of the *Algebra Ofte Stel-konst* to be made. By the early summer of 1669 Mercator had finished his version, but in the meanwhile Collins' attention had been drawn to a newly published second Dutch work on algebra, Johan Jacob Ferguson's *Labyrinthus Algebræ*. For this work, too, Collins was filled with enthusiasm and, backed by the Royal Society's president, William Brouncker, he conceived the revised scheme of appending the most significant passages of Ferguson's book to an unabridged Latin edition of Kinckhuysen's. With

(6) Collins to Pell, 9 April 1667: 'I being acquainted with Mr. Jeffries procured him a sight of Kinckhuysen. The said Mr. Jefferies is upon going over into Holland with the ambassadors, and hath promised to procure it there and get it translated for Mr. Martyn, which Mr. Pitt is willing to prevent by having it himself, or by prevailing with yourself or Mr. Branker that some appendix about these matters may be further added to your book' (Rigaud's *Correspondence* (note (3)), **1**: 126). Compare Collins to Brancker, June [1667]: 'I would not that [Mr. Kersey's Algebra] should either come out sooner than the Doctor's [Pell's], or that his Introduction should have a better esteem. This moved me in my last...to write to you to incline the Doctor to admit the first seven sheets of the Introduction, enlarged out of Kinckhuysen...to come out as your translation, as soon as may be; the Doctor taking what time he pleaseth to supply the defect at the beginning, and to enlarge and complete the book' (*ibid.* **1**: 135–6).

Brouncker's continued support he had Ferguson's section on the resolution of cubics and quartics translated (presumably into Latin rather than English), wrote to John Wallis at Oxford informing him of the project,[7] and then announced his intentions publicly at the end of a glowing review of Ferguson's *Labyrinthus* which appeared in the *Philosophical Transactions* in July 1669:

...we think fit to intimate, that divers good Treatises of *Algebra* have been lately publish't in *Low Dutch*....*Gerhard Kinckhuysen* hath of late years publish'd several distinct *Quarto*-Books, *viz.* A *Tract* of *Analytical Conicks*: A *Collection* of *Geometrical Problems, Analytically* solv'd; as also such an acceptable *Introduction* to *Algebra*, that by the encouragement of some of the *R. Society* it hath been Translated into *Latine*, and fitted for the Press; to which will be annexed the *Methods* and *Examples* of *Ferguson* about the *Roots of Æquations*. And we have little reason to doubt, but that the just now mention'd *Introduction* will meet with such an acceptance, as shall quicken the Stationer to proceed in the Translating and Printing of the rest of the Books above-mentioned, or others of the like kind.[8]

We may be sure, too, that Collins would not have been slow to inform his friend and regular correspondent, Isaac Barrow, of his plans, if only to sound him out on the possible sale of the joint *Algebra* in Cambridge. Barrow, however, then on the point of resigning his Lucasian chair, was doubtless unwilling to be involved in Collins' plans for publication and when in the late autumn of 1669 he was requested to look Mercator's Latin version of Kinckhuysen's book over for the press he was seemingly content to turn the matter over to his newly appointed successor, Newton.

Collins first met the new Lucasian professor in London about the beginning of December 'somewhat late upon a Saturday night at his Inne... And

(7) Collins to Wallis, 17 June 1669: 'One John Jacob Ferguson hath lately, in 1667, at the Hague, published, in low Dutch, a book entitled *Labyrinthus Algebræ*, wherein he solves cubic and biquadratic equations by such new methods as render the roots in their proper species, when it may be done, to wit, in whole or mixed numbers, fractions, or surds, either simple, compound, or universal, and likewise improves the general method; thereby accomplishing as much as Hudden, in annexis Geometriæ Cartesianæ, seemed to promise about it. This part is translated by Mr. Old, and by me almost transcribed, which, annexed to Kinkhuysen's Introduction (with your help, advice, or assistance, which the Lord Brounker may possibly crave, the books being translated at his Lordship's desire,) will render the said Introduction very acceptable. I hope, ere long, to send you both to peruse' (Rigaud's *Correspondence* (note (3)), **2**: 515–16).

(8) 'Labyrinthus Algebræ, Auth. Joh. Jac. Ferguson. Printed at the Hague in 4°. 1667' (*Philosophical Transactions:* No. 49 [for 19 July 1669]: 996–9 especially 998–9). The review is not signed but, apart from a wealth of internal evidence the attribution to Collins is placed beyond doubt by the opening sentence: 'What we mention'd in *Numb*. 46. p. 931. sect. 8. about new methods, pretended by some to be found out for giving the Roots of all *Cubick* and *Biquadratick Æquations*, albeit those Roots are Fractions or Surds, Binomials or Residuals; We find since to be already accomplished by this Dutch Writer'. (See *Philosophical Transactions*: No. 46 [for 12 April 1669]: 929–34: 'An Account Concerning the Resolution of Equations in Numbers;

againe...the next day having invited him to Dinner', and we may suppose that the preparation of the *Algebra* for the press would have been touched upon briefly in the 'little discourse...about Mathematicks' they then had.[(9)] During the following winter of 1669/70 Newton began unhurriedly to annotate Kinckhuysen's book, and early in the New Year at the end of a first long letter to Collins reassured him of his leisurely progress.[(10)] The latter, no doubt eager to have the Dutch algebra 'improved' by an expert, would seem to have suggested to Newton in reply that he was free to augment the work with any additions of his own choosing and that he might like at the same time to revise Kinckhuysen's earlier work on analytical conics in a similar way. But Newton was unwilling to be drawn:

You seeme to apprehend as if I was about writing elaborate Notes upon Kinck-huyson: I understand from Mr Barrow y^t your desire was only to have ye booke reveived: that if any thing were defective or amisse it might be amended, & to that purpose about two Months since I reveived it & made some such observations upon it. But though the booke bee a good introduction I think it not worth the paines of a formall comment,

---

imparted by Mr. Iohn Collins'.) Since Ferguson's work is now rare and of some importance in the composition of Newton's Kinckhuysen observations we may quote that part of Collin's review relating to the resolution of cubic and quartic equations:

'3. [Ferguson] gives one General Rule (where others make more Cases of it) for finding the Roots of all *Cubick* Æquations, in which the Second term...is wanting, and then shows, how all other *Cubick* Æquations, wherein it is present, may be reduc'd thereunto, by taking it away. Moreover, when such Æquations...[are] not explicable but by a *quàm proximè*, [he finds the Roots] according to the general method of *Vieta*....

'4. When he comes to *Bi-quadratick Æquations*, he intimates, that all such Æquations may be reduced into two *Quadratick* Æquations, but not without the aid of a *Cubick* Æquation. And first, when the second Term or *Cubick Species* is not wanting, he shews how to find the said *Adjutant Cubick* Æquation, by placing the two highest terms of the Æquation on one side, and the rest of the Termes on the other, and then finds such Quantities, which, added to either side, render the same capable of a square Root; and this preparation being made, he thereby obtains the *Cubic* Æquation and the Root thereof, which serves...to divide the *Bi-quadratick* Æquation propos'd into two *Quadratick* Æquations, and so solved....And then, in regard it often happens, that Æquations are not otherwise explicable than by a *quàm proximè*, he proceeds according to the General method of *Vieta*, as in *Cubicks*....

'The whole Doctrine is illustrated with great variety of choice Examples, and the Author intending hereafter to treat more fully of *Algebra*, promiseth to extend his methods to Æquations of *higher* degrees, and to render the same more general....

'*Ferguson* about the Matter mention'd is more full than either the *Algebra* of *Frans van der Huips*, an *Octavo* Book in *Low Dutch*, 1654. or Kinckhuysen: neither do we find, that Ferguson ascribes the Invention of those Methods to himself.'

(9) See Collins' letter of 24 December 1670 to James Gregory, which recounts his first meeting with Newton (*James Gregory Tercentenary Memorial Volume* (London, 1939): 154 = *Correspondence of Isaac Newton*, **1** (Cambridge, 1961): 53).

(10) Newton to Collins, mid-January 1669/70: 'Your Kinck-Huysons *Algebra* I have made some notes upon. I suppose you are not much in hast of it, wch makes me doe y^t onely at my leisure' (*Correspondence of Isaac Newton*, **1**: 20).

There being nothing new or notable in it wch is not to bee found in other Authors of better esteeme.

You make mention of another book of the same Author translated badly into latten by a German Gunner; wch you would have mee correct.(11) I understand not Dutch & would not willingly doe the Author soe much wrong as to undertake to correct a translation where I understand not the originall.(12)

Collins must have been upset at this low estimate of the quality and originality of the *Algebra*. It is clear, however, that in his reply (now lost) he persuaded Newton to elaborate his first cautious annotations on Mercator's rendering of Kinckhuysen's text into a substantial scheme for 'correcting' and augmenting the work. In addition, he sent along Ferguson's *Labyrinthus Algebræ* (together, we presume, with Old's translation of its more important passages) so that Newton might incorporate portions of it at discretion.

Over the spring of 1670 Newton worked steadily away and his 'notes... intermixed wth the Authors discourse' took shape. At length on 11 July he sent them up to Collins in London along with the interleaved Dutch and Latin versions of the *Algebra*, adding:

I know not whither I have hit your meaning or noe but I have added & altered those things wch I thought convenient to bee added or altered, & I guess that was your desire I should doe. All & every part of what I have written I leave wholly to your choyse whither it shall bee printed together wth your translation or not. If you think fit to print any of it ye directions I have writ in english will shew you where it is to bee inserted. But if you have a mind not to change ye Author soe much, I would not have you recede from your intentions upon ye accompt of wt I have done. For I assure you I writ wt I send you not so much wth a design yt they should bee printed as yt your desires should bee satisfied to have me revise ye booke.(13)

With Collins' suggestion that he incorporate extracts from Ferguson in his commentary he was less than happy, however:

In a letter you hinted somthing to bee supplyed out of Ferguson's *Labyrinthus* about ye extraction of cubick roots; if you meant pure roots, I have done yt in as brief plaine & full a manner as I can. But if you meant affected roots, tis already done by Kinck-

(11) Thomas Brancker's Latin version of *De Grondt der Meet-konst*: compare Collins to Pell, 9 April 1667 (Rigaud's *Correspondence* (note (3)), **1**: 127).

(12) Newton to Collins, 6 February 1669/70 (*Correspondence of Isaac Newton*, **1**: 24).

(13) Newton to Collins, 11 July 1670 (*Correspondence*, **1**: 30). For the detailed contents of this first version of his 'observations' compare 1, §2: notes (1) and (2) below.

(14) *Correspondence*, **1**: 30–1.

(15) When two years afterwards John Wallis was told of this by Collins, he gave his unhesitant opinion that 'I have no acquaintance with [him] but...As to that [piece] of Kinkhuysen, I know not whether Mr. Newton were not better, and he might with as much ease, publish what he hath as a treatise of his own, rather than by way of notes on [it].' (Wallis to Collins, 25 January 1671/2 = Rigaud's *Correspondence* (note (3)), **2**,: 528–30 especially 529.)

Huyson pag 91 as well as by Ferguson. Indeed Ferguson seems to have done more in so much as to comprehend all cases of cubick equations w[th]in y[e] same rules; but that *more* is inartificiall because it supposes y[e] extraction of cubick roots out of imaginary binomiums, w[ch] how to doe hee hath not taught us.... Not but that it may bee done, & I know how to doe it, but I think it not worth y[e] inserting into Kinckhuyson, yet if you think it convenient (& indeed it may bee congruently enough inserted into him at pag 91) I will send you it done in my next letter.(14)

After this somewhat severe estimate of Ferguson's usefulness Newton concluded his covering letter in a typically self-effacing manner:

There remains but one thing more & thats about the Title page if you print these alterations wch I have made in the Author: For it may bee esteemed unhandsom & injurious to Kinckhuysen to father a booke wholly upon him w[ch] is soe much alter'd from what hee had made it.(15) But I think all will bee safe if after y[e] words *nunc e Belgico Latinè versa*, bee added *et ab alio Authore locupletata*.(16) or some other such note.

...As for y[e] coppys of Kinck-huysen you mentioned to me, I know tis usually not w[th]out some unwillingnesse that Math: books are printed. And I will not soe far discourage y[e] printing of it as to have any coppys reserved for mee. I had rather purchase your freinship then bookes. Yet if you please to send mee one coppy I shall acknowledg my selfe your debter for that....(17)

Collins was quick to acknowledge receipt of Newton's observations two days later, but felt he should justify his previous commendation of Ferguson:

I received yours with Kinckhuysens Introduction, and perceive you have taken great paines which god willing shall be inserted into y[e] Translation and printed with it. Hereby you have much obliged the young Students of Algebra and the Bookseller [Moses Pitt?] who was at y[e] commencem[t](18) on Munday and Tuesday but was so taken up and concerned you were so too that he did not see you.... I formerly intimated that I thought Kinck[huysen] had too slightly handled the doctrine of Surd Numbers, and that the same might be transcribed from others. I find you in the same opinion for in one of your Marginall Notes(19) you say thus, *The Author having slipped over the Addition Substraction Multiplication and Division of all but quadratick Surds* &c and he acknowledgeth as much by consequence himselfe referring the Reader to Wassenare *Onwissen Wisconstenaer*.(20) However he cheifely thereby intended y[e] roots of Binomialls which you have supplyed being unwilling the young Student should be referred to other bookes

(16) 'and enriched by another author'. The phrase was, of course, intended to be added to Mercator's Latin title-page of his version of Kinckhuysen's *Algebra*.

(17) Newton to Collins, 11 July 1670 (*Correspondence of Isaac Newton*, **1**: 31).

(18) The formal undergraduate disputations, in the schools or in Great St Mary's Church, which were the climax of the Cambridge academic year. By statute these were open to the public and usually well attended. (Compare W. T. Costello, *The Scholastic Curriculum at Early Seventeenth-Century Cambridge* (Harvard, 1958): 15 ff.) As a University professor Newton was, no doubt, usually required to be present.

(19) See 1, §2: note (7).

(20) *Algebra Ofte Stel-konst* (1, §1): 38. Compare note (4).

which are scarce for the Doctrine of Surds. I will a little further presume and therefore crave your judgem[t] of what you thinke necessary to be taken either out of Scheubelius, Van Ceulen, or Humes(21) which bookes I herewith send....

I was apt to beleive that Ferguson had done more then Kinckhuysen in these 3 particulars

1 In applying one generall rule to both kinds of Cubick Æquations, to wit as well those that are solved by meane Proportionalls as those that require Trisection.

2 In rendring the rootes of Cubick and Biquadratick Æquations properly that is to say in giving the rootes when they are explicated by fractions or Surds exactly, and not by a *quamproximè*.

3 In emprooving the generall Method.

But having failed in the first I conceive it opportune to shew at least wherein he hath failed, and if you please (which is by you offred and seemes desirable) supply his defect.

Lastly why you should desire to have your Name unmentioned I see not, but if it be your will and command so to have it, it shall be observed.(22)

Newton replied at once on the 16th that he was willing to augment his Kinckhuysen notes appropriately (and the section on the roots of binomial surds in particular) if the *Algebra* was not to go immediately to press:

I sometimes thought to have altered & enlarged Kinkhuysen his discourse upon surds but judging those examples I added would in some measure supply his defects I contented my selfe wth doing that onely. But since you would have it more fully done, if the booke goe not immediately into y[e] presse I desire you'le send it back wth those notes I have made (since you are resolved to print them also) & I will doe something more to it or if you please to send all but the first sheete or two, while y[t] is printing, Ile reveiw the rest & not only supply ye wants about surds but that about Æquations soluble by

(21) Johann Scheubel, *Algebræ Compendiosa Facilisque Descriptio* (Paris, 1551); Ludolf van Ceulen, *De Arithmetische en Geometrische fondementen* (Leyden, 1615) (see 1, §1: note (15)); James Hume, *Algebre de Viete* (Paris, 1636). Collins went on to note in his letter that 'There are two other Authors have excellently handled Surds, those are Frans van der Huyps in his Low Dutch *Algebra* in 1654 and ye Seiur d Taneur [Sieur Le Tenneur] in french in his Tract of irrationall quantities and Commentaries on 10 Euclid [*Traité des quantitez Incommensurables et le dixieme livre d'Euclide*] at Paris in 1640'. Evidently Collins sent along only the latter two of the three books he mentions for in his reply (*Correspondence*, **1**: 35) Newton thanked him 'for your two last bookes' alone.

(22) Collins to Newton, 13 July 1670 (*Correspondence*, **1**: 32–3). Compare his present defence of Ferguson's *Labyrinthus* with his review of the book the previous year (note (8) above).

(23) Kinckhuysen's 'Pars Tertia' (*Algebra* (1, §1): 96–108). For Newton's improvements and additions see pages 28–39 of his 'Observationes' (1, §2).

(24) 'Not to distinguish between known and unknown quantities and to apply a due process of reasoning'. Compare Oughtred's *Clavis Mathematicæ* (Oxford, $_3$1652): 50: Cap. XVI. 'De Æquatione & De quæstionibus per Æquationem solvendis'; and Descartes, [*Geometrie*, 1637: 300 =] *Geometria*, $_2$1659: 4: 'Deinde nullo inter lineas hasce cognitas & incognitas facto discrimine, evolvenda est Problematis difficultas, eo ordine, quo omnium naturalissimè pateat, quâ ratione dictæ lineæ à se invicem dependeant, donec inventâ fuerit via eandem quantitatem duobus modis exprimendi, id quod Æquatio vocatur.'

trisections, & somthing more I would say in the chapter *Quomodo quæstio alia ad æquationem redigatur*,(23) that being the most requisite & desirable doctrine to a Tyro & scarce touched upon by any writer unles in generall circumstances bidding them onely *Nota ab ignotis non discernere & adhibere debitum ratiocinium*.(24)

But on his previous low estimate of Ferguson's *Labyrinthus* he stood his ground:

As to Fergusons rendering the roots of [cubick] Æquations soluble by trisection his defect will appeare by example.(25)...In generall I see not wt hee hath done more then in Cardans rules....Nor doe I see wt hee hath done more then Descartes in his Solution of biquadratick equations: for both goe ye same way to worke in reducing them first to Cubick & then to quadratick æquations. Lastly I see not in what case his rules will render the roots of cubick or biquadratick Æquations *in proprio genere* where those of Cardan or Descartes will not.(26)

In immediate response to Newton's announcement that he was 'willing to take some more paines at present with Kinckhuysen' Collins three days later returned both the interleaved copy of the *Algebra* and Newton's notes upon it, adding reassuringly in his covering letter:

...doe not presse your selfe in time, your paines herein will be acceptable to some very eminent Grandees of the R. Societie who must be made acquainted therewith,... forasmuch as Algebra may receive a further Advancemt from your future endeavours and that you are more likely than any man I know, herein to oblige the Republick of Learning....

But to returne into the Way, when I had Fergusons Papers,(27) I only viewed his Examples and that cursorily. It seemes he soared but *Icari fine* to accomplish what Hudden promised page 503 *in annexis Geometriæ Cartesianæ*.(28) I scrupled his rootes of

(25) The example chosen by Newton is $x^3 = 6x+4$ (from page 12 of Ferguson's *Labyrinthus*): 'In order to solve this hee bidds extract ye cubick root of these binomiums $2+\sqrt{-4}$, & $2-\sqrt{-4}$. To doe this his rule pag 4 is:...put it in pure numbers &c:...but $[2+\sqrt{-4}]$ hath noe pure number answering to it. His rule therefore failes & the like difficulty is in his 3d example & in all other such cases. ...in this instance Cardans rule will give you

$$x=\sqrt{C:\overline{2+\sqrt{-4}}}+\sqrt{C:\overline{2-\sqrt{-4}}},$$

in wch ye only difficulty as before is to extract ye rootes of ye binomiums $2+\sqrt{-4}$ & $2-\sqrt{-4}$. Which roots indeed are $-1+\sqrt{-1}$ & $-1-\sqrt{-1}$, as he assignes them, but tells not how to extract them.' For Newton's improved explanation of this approach (in the 'impossible' case of the reduced cubic which has all its roots real) see pages 14 and 27–8 of his 'Observationes' (1, §2).

(26) Newton to Collins, 16 July 1670 (*Correspondence*, **1**: 34–5).

(27) Old's manuscript Latin version of portions of Ferguson's *Labyrinthus*, presumably (see note (7)).

(28) That is, of his 1657 tract *De Reductione Æquationum* (*Geometria*, ₂1659: 503): 'Diversas adhuc alias Regulas in paratu habeo, quas hîc simul adjungerem, si non aliquid in futurum reservare animus esset.' (See Karlheinz Haas, 'Die mathematischen Arbeiten von Johann Hudde (1628–1704), Bürgermeister von Amsterdam', *Centaurus*, **4** (1956): 235–84, especially §2: 238–49.) Hudde never published his promised rules.

negative quadratick quantities, and imagined that they expunged one another being affected with contrary Signes but conceited there might be more done in Cubicks then authors yet insist on.... Are not both wayes [Trisection and finding of 2 Meanes] the Solution of the Cubick æquation of the same kind[?][(29)]

Collins had stressed that there was no urgency in preparing the *Algebra* for Latin publication and Newton took his correspondent at his word. He was perhaps not wholly pleased to have his Kinckhuysen notes back again but, after the flurry of correspondence over the past week, doubtless content to settle down to their revision, working away without disturbance in his Cambridge study.[(30)] For two months he left Collins' letter unacknowledged and made no great haste to answer a second which arrived a few week's later.[(31)] At length, however, Newton made his reply:

The receipt of your last letter staying mee from sending back your Kinck-Huysens Introduction, I have hitherto deferred writing to you, waiting for Dr Barrows returne from London that I might consult his Library about what you propounded in your last letter but one to mee concerning ye solutions of Cubick æquations, before I sent you my thoughts upon it....I cannot...yet bee convinced that any one problem can be solved both those ways....[(32)]

Upon the receipt of your last letter I sometimes thought to have set upon writing a compleate introduction to Algebra, being cheifely moved to it by this that some things I had inserted into Kinck-Huysen were not so congruous as I could have wished to his manner of writing. Thus having composed somthing pretty largely about reducing problems to an æquation when I came to consider his examples (wch make ye 4th[(33)] part of his booke) I found most of them solved not by any generall Analyticall method but by particular & contingent inventions, wch though many times more concise then a generall method would allow, yet in my judgment are lesse propper to instruct a learner,

(29) Collins to Newton, 19 July 1670 (*Correspondence*, **1**: 36–7).

(30) Trinity College's book of exits and redits for the period shows that Newton remained in Cambridge during the whole of 1670 (J. Edleston, *Correspondence of Sir Isaac Newton and Professor Cotes* (London, 1850): lxxxv). He was for the most part, of course, pre-occupied with his own researches and in preparing his third set of Lucasian lectures on optics for delivery in the autumn.

(31) This, the 'last letter' referred to in the sequel, is lost. Presumably Collins repeated his assurance that there was no pressing need for Newton to complete his revision or to return the interleaved copy of Kinckhuysen so that the Latin version might be sent to the printer (Moses Pitt).

(32) The continuation of this passage is quoted in 1, §2: note (101), where its significance is discussed.

(33) Read '3rd' (since the examples in Kinckhuysen's 'Pars Tertia' occupy only pages 97–108 or less than one-eighth of the *Algebra*).

(34) Compare 1, §2: notes (110) and (111).

(35) John Pell's augmented edition (in Branker's Latin translation) of J. H. Rahn's *Teutsche Algebra* had appeared in London only two years before, while Oughtred's *Clavis*

as Acrostick's & such kind of artificiall Poetry though never soe excellent would bee but improper examples to instruct one y^{t} aimes at Ovidian Poetry.(34) But considering that by reason of several divertisements I should bee so long in doing it as to tire you[r] patience wth expectation, & also that there being severall Introductions to Algebra already published(35) I might thereby gain ye esteeme of one ambitious among ye croud to have my scribbles printed, I have chosen rather to let it passe wthout much altering what I sent you before. Yet because you seeme to bee most sollicitous about the doctrine of surds delivered in it, I desire y^{t} when your leisure will permit you to write you would intimate the particulars in wch you think it most defective. For at my reviewing ye papers, I judged it not so imperfect as I thought it had beene when I sent for them back againe & soe have hitherto added two or three examples(36) onely more then was done before.(37)

What Collins replied we do not know but evidently it was his wish, as Newton's, to be finished with the additions to Kinckhuysen and by Christmas 1670 Newton's revised version of his 'Observations', as we reproduce it here (1, §2 below), was complete. There remained, apparently, only the straightforward if laborious task of seeing the *Algebra* through the press. Meanwhile, in recognition of the painstaking care he had devoted to the preparation of his Kinckhuysen notes Newton was soon to be presented by its prospective publisher with a copy of Wallis' *Mechanica*.(38)

But the time when the 'Introduction to algebra' could profitably be published in Britain was already past: indeed, a new and voluminous (though not very penetrating) English *Algebra*, John Kersey's, was now advanced in its press-proof in London and, as Collins grew increasingly to realize, to stand any chance of capturing in the face of native competition the limited home market for technical works the *Algebra Ofte Stel-konst* would have to be sold not on the

Mathematicæ had come out in a fourth edition at Oxford in 1667. In addition, as many knew, Kersey's elaborate English *Algebra* was, after long years in press, on the point of being published. Newton may also have in mind Collins' recent review of Ferguson's *Labyrinthus Algebræ* (note (8)).

(36) Three, in fact (together with some addition to the sections on cubic surds and the resolution of cubic equations with three real roots). See §2, 1: note (133).

(37) Newton to Collins, 27 September 1670 (*Correspondence*, 1: 42–4). In his concluding paragraph Newton noted that 'I have sent back your Humes, Van Ceulen, Fergusons *Labyrinthus Algebræ* both parts of it [Dutch original and Old's Latin version?], & Kinck-Huysen on ye Con-sections. But his *Algebra* I presume to keepe by me till you have occasion for it'. Compare notes (11) and (21).

(38) See Collins to J. Gregory, 14 March 1671/2 (*Correspondence of Isaac Newton*, 1: 119); and compare Newton to Collins, 25 May 1672 (*ibid.*, 1: 161). Collins, we may note, had previously written to Gregory on 24 December 1670 that 'Kinckhuysens Introduction to Algebra with notes thereon and additions thereto made by the learned Mr Isaac Newton of Cambridge (at the request of Dr Barrow) is ready for the Presse' (*ibid.*, 1: 56)—perhaps a little hastily, for it was only on 20 July 1671 that he was told by Newton that 'The last winter I reviewed the Introduction & made some few additions to it' (*ibid.*, 1: 68).

name of its Dutch author but on that of the occupant of the Cambridge mathematical chair. As he told Newton on 5 July 1671:

The Bookseller Pitts is not desirous as yet to put the Introduction to Algebra to the Presse, and I conceive you have made so many usefull additions thereto, that when it comes to the Presse it may very well beare the Title of your Introduction, and thereby find the better entertainement, and more Speedy Sale.(39)

Newton had no outright objection to this change of author but was, however, somewhat concerned that what was apparently his first published work should be merely an augmented version of an algebraic primer in which he could see no outstanding merit—he naturally wished his first appearance in print to reflect the power and extent of his mathematical researches over the previous half dozen years. Could he, he wrote back, add to the *Algebra* the much amplified version(40) of his 'De Analysi' which he was in the process of composing and which might indeed show the world his true worth as a mathematician?

The last winter...partly upon Dr Barrows instigation I began to new methodiz ye discourse of infinite series, designing to illustrate it wth such problems as may (some of them perhaps) be more acceptable then ye invention it selfe of working by such series. But being suddainly diverted by some buisinesse in the Country,(41) I have not yet had leisure to returne to those thoughts, & I feare I shall not before winter. But since you informe me there needs no hast, I hope I may get into ye humour of completing them before ye impression of the introduction, because if I must helpe to fill up its title page, I had rather annex somthing wch I may call my owne & wch may bee acceptable to Artists as well as ye other to Tyros.(42)

The thought was still in his mind a year later, in May 1672, though already his first contacts with the time-wasting frustrations of contemporary publishing had embittered him:

Your kindnesse to me...in profering to promote the edition of my [optical] Lectures wch Dr Barrow told you of I reccon amongst the greatest, considering the multitude of buisinesse in wch you are involved. But I have now determined otherwise of them;

(39) Collins to Newton, 5 July 1671 (*Correspondence*, **1**: 66).

(40) The 1671 tract on fluxions and infinite series, usually identified (though not by Newton himself) by Jones' title of *Geometria Analytica*. We will reproduce it in our next volume from his unfinished, incompletely published manuscript original.

(41) Newton was away from Cambridge between 17 April and 11 May (Edleston (note (30)): lxxxv).

(42) Newton to Collins, 20 July 1671 (*Correspondence*, **1**: 68).

(43) As he went on to say, 'The Book here in Presse is Varenius his *Geography*, for wch I have described Schemes; & I suppose it will be finished about six weeks hence': however, only on 30 July was he able to tell Collins that 'Varenius is newly out of Presse' (*Correspondence*, **1**: 222). According to John Conduitt the book's title (*Bernhardus Varenius. Geographia Generalis*

finding already by that little use I have made of the Presse, that I shall not enjoy my former serene liberty till I have done with it; wch I hope will be so soon as I have made good what is already extant on my account....[43] The additions to Kinkhuysens *Algebra* I have long since augmented with what I intended, & particularly with a discourse concerning invention or the way of bringing Problems to an Æquation.[44] And those are at your command. If you have not determined any thing about them I may possibly hereafter review them & print them with the discourse concerning infinite series.[45]

As Collins reported to him some time in June, however, Moses Pitt wished to renounce his intention of printing the Kinckhuysen translation and was willing to sell his rights in it. Writing 'out of Northamptonshire' on 13 July Newton was deeply concerned:

I will inquire of some of our Booksellers [in Cambridge] whether they will purchase Mr Pitts his copy of Kinckhuysen & if not I will send it you. In the meane while I would know whether Mr Pitts thinks it will be more advantageous to print ye Author without alteration, or to insert those notes wch you formerly saw, y^t^ I may according send them wth ye Copy or detain them.[46]

And as soon as he arrived back in Cambridge a fortnight later, without waiting for Collins' response he added:

Yesterday I spoke wth a Bookseller here about ye translation of Kinkhuysen who upon my motion was willing to take it of Mr Pitts his hand at 3^llb^, but he has not yet seen the Book.[47]

In Collins' mind there was now no shadow of doubt that without Newton's annotations of Kinckhuysen's book it was 'not worth the printing especially at this time that Mr Kerseys booke is in ye Presse'. Moreover, he went on,

the enlarging Kinck[huysen] was according to your owne intimation the occasion of sending it back and both Mr Pitts and others are very desirous that you vouchsafe to affoard that paines you have taken and intended about that Introduction which I could rather wish might solely beare your Name as an Introduction necessary for the understanding of what you have else to publish.[48]

...*Summa cura quam plurimis in locis emendata, & xxxiii Schematibus novis, ære incisis, una cum Tabb. aliquot quæ desiderabantur aucta & illustrata, ab Isaaco Newtono*) is misleading, for 'Sr I. N. told Mr Jones all he did in the edition of Varenius before wch is put Curante Isaaco Newtono was to draw the schemes wch in the Elsever Edition were referred to, & were not there' (King's College, Cambridge, Keynes MS, 130.5).

(44) See 1, §2: note (110).

(45) Newton to Collins, 25 May 1672 (*Correspondence*, **1**: 161).

(46) Newton to Collins, 13 July 1672 (*Correspondence*, **1**: 215).

(47) Newton to Collins, 30 July 1672 (*Correspondence*, **1**: 222).

(48) Collins to Newton, 30 July 1672 (*Correspondence*, **1**: 223).

Two days later, on receiving Newton's second letter with its offer to negotiate the rights to Kinckhuysen's book, Collins was able to affirm that Pitt was

willing to take 3lib for his Interest...if the Bookseller will moreover give him 10 Coppies in quires when printed, and if you cannot procure so many I shall if he insi[st] on it compensate what falls short the rather in regard I am willing to shun the trouble at the presse which I promised to undertake, but chiefly because I thinke that Introduction proper to accomp[any] your doctrine of infinite Series.(49)

So it was agreed, and Moses Pitt received the lump sum of £4 for the copyright (apparently out of Newton's own pocket).(50) Whether either Collins or Newton yet knew it the payment was merely token. No one in London would now publish the 'Introduction', with or without Newton's improvements, and it soon became clear that the Cambridge booksellers were equally wary. Finally, when last we hear of it three years afterwards Newton had begun to incorporate his 'Observations' into his new series of Lucasian lectures on arithmetic and algebra, having abandoned any idea of adding them to the *Algebra*. As he told Collins in September 1676:

I have nothing in ye press, only Kinckhuysens *Algebra* I would have got printed here to satisfy ye expectation of some friends in London, but our Press cannot do it....(51) It is now in ye hands of a Bookseller here to get it printed: but if it doe come out I shall ad nothing to it.(52)

With that the project was laid finally to rest. Probably through Collins(53) Mercator's Latin version passed ultimately into the possession of John Wallis,

(49) Collins to Newton, 1 August 1672 (*Correspondence*, **1**: 226).

(50) When Collins drew up a list of Newton's papers on algebra for John Wallis about the end of 1677 he noted 'An Introductory part from Kinckhuysen out of low Dutch turned by Mercator into Latin, which he bought...' (*Correspondence*, **2**: 242). Newton was to have to remind Pitt, ever out for money, of the terms of the transaction before a year was out: 'It comes now into my mind yt when I sent Mr Pitts 4lib for Kinckhuysen he further urged a promise of some copies. When you have opportunity you will oblige me to remember him that his proposall was 4lib absolutely or 3lib wth some copies' (Newton to Collins, 9 April 1673, *ibid.*, **1**: 271).

(51) The Cambridge 'Press', as distinct from the unnamed 'Bookseller', is evidently one (or more) of the trio of printers who were at this time by the University (strictly, by 'the Chancellor or his vicegerent and three doctors') licensed to print and sell approved works under the University crest—perhaps John Hayes, the printer of Varenius' *Geography* (note (43)). H. W. Turnbull has given it as his opinion, presumably on the basis of the present letter, that 'Newton persisted slowly with the [Kinckhuysen] revision until it was finished, and then offered it in 1676 to the Cambridge University Press. It was refused and no further effort was made to find a publisher. This in itself may have mattered little, but for [the] circumstance [that] with the book the tract [on fluxions and infinite series] was also laid aside; the results of this lost opportunity were deplorable in no common degree' (*Correspondence of Isaac Newton*, **1**: 23, note (14)). With this judgement we must disagree on several counts. In his present

as we have seen, while Newton's annotations, once they had served their use as a source for his university lectures, were locked away in the mass of his unpublished scientific papers and passed into oblivion.

* * *

With the last piece (2, §2) we reproduce we must necessarily be brief, for we know little more of its historical background than that it was communicated by Newton to Collins in the summer of 1672, having been composed at an unspecified time previously for his own purposes. It is evidently in part a revision of his May 1665 paper 'Of the construction of Problems'(54) and also includes an application of the organic construction of curves, a method whose details he explored two or three years afterwards.(55) From an examination of its writing style we may, with some hesitancy, confirm that it was composed about 1670.

The subject of the geometrical resolution of algebraic equations may very well have come up in the 'little discourse we had about Mathematicks'(56) when Newton first met Collins in the autumn of 1669. The numerical solution of equations was, indeed, the latter's abiding interest and he had but recently published a lengthy discussion of the topic in the *Philosophical Transactions*.(57) It would appear that three years later Collins sought(58) to refresh his memory of what Newton had told him of his improvement of Descartes' construction of the general sextic by use of a *locus linearis*. Newton, however, was not immediately

remarks Newton makes it clear that the *Algebra* offered to the Press was Mercator's unadorned Latin version without any Newtonian additions or appendages of any kind, and its rejection could not have prejudiced the chances of his own (never completed) fluxional tract. The latter, indeed, was an augmented version of the 'De Analysi per æquationes numero terminorum infinitas' (2, 3 above), which, as the continuation of the present letter shows, Collins was then—not wholly with Newton's approval—endeavouring to have printed in London. Of course, any publisher willing to print the 'De Analysi' would presumably prefer to market its improved version, given the chance. The action, finally, of placing the *Algebra* in the hands of a local stationer 'to get it printed' even after the Press had rejected it is proof enough that Newton did not abandon all hope of having it published. But whatever the truth of this matter it should not be left unsaid that it is the direct descendants of the seventeenth-century 'Academiæ typographi', the modern Cambridge University Press syndics, who now after almost three centuries fulfil John Collins' wish, making it at last possible for Mercator's Latin version of Kinckhuysen's 'Introduction' together with its Newtonian additions to appear side by side in print.

(52) Newton to Collins, 5 September 1676 (*Correspondence*, **2**: 95).

(53) Compare note (1).

(54) I, **3**, **3**, §2.

(55) See 2, §2: notes (121) and (134); compare **1**, **3**, §2 above.

(56) See note (9).

(57) 'An Account Concerning the Resolution of Equations in Numbers' (note (8)).

(58) About the beginning of July 1672 in a letter now lost (summarily restored by H. W. Turnbull in *Correspondence of Isaac Newton*, **1**: 214–15).

willing to search out the details of his researches but in reply merely set down for his correspondent what he could remember of them offhand:

The way of resolving æquations of 5 or 6 dimensions, by a *locus linearis* was I beleive by the intersection of that & a Conick Section, something after the manner y^t^ Des-Cartes hath done it, but more conveniently in my opinion, because the same *locus linearis* once described will serve for ye resolving of all equations of those dimensions.(59) And as I remember the calculations to y^t^ intent are shorter & lesse intricate.(60)

This, of course, merely aroused Collins' curiosity the more and he tried (with the utmost tact) to pump Newton for further details:

I accept of and thanke you for the kind tender of communicating your Way of doing it,(61) and of resolving of Æquations of 5 or 6 Dimensions by ayd of a *Locus Linearis* (that varies not) intersected by a Conick Section to which the Calculations are not so tedious or abstruse, as those now in use. Of the like kind Dr Wallis for Cub[ick] Æ[quations] makes mention at the end of his *Op*[*era*] *Math* in 1657 ***pars prima*** [*Adversus Meibomium*] pag 46, but hath suited no Cal[culation]s thereto.(62) This I must confesse is a curiosity I am unacquainted with, but am loath to put you to the trouble of Writing of such Papers, which if sent I shall transcribe and returne with thankes.(63)

Newton evidently could not refuse so courteous a request and accordingly sent along the present compilation of his methods for resolving equations by geometrical and mechanical means, unfinished though it was. In his covering letter, however, he asked that Collins keep its contents strictly to himself:

I herewith send you a set of Problems for construing æquations wch that I might not forget them I heretofore set down rudely as you'le find. And therefore I think 'em not fit to be seen by any but your selfe, wch therefore you may return when you have perused them. How y^e^ afforesaid descriptions of y^e^ Conick sections(64) are to be applyed to these constructions I need not tell you.(65)

---

(59) Both in his May 1665 paper (I, **3**, 3, §2) and the present one (2, §2 below) Newton chooses as his *locus linearis* the cubical parabola $a^2x = y^3$, which may without loss of generality be taken as $x = y^3$: this is a parabola of determinate shape and may therefore, as he wrote in 1665, be traced 'uppon a plate' once for all instead of being drawn anew for each equation.

(60) Newton to Collins, 13 July 1672 (*Correspondence*, **1**: 215). The gain in simplicity when the constructing parabola is taken to be $x = y^3$ is usually balanced by an unnecessary complication in the defining equation of the conic which it intersects.

(61) A nomographic construction of equations by 'Gunters line' (*ibid.*, **1**: 215; compare I, **3**, 3, §1: note (6)) which Newton had promised to communicate. He had not, however, offered to describe his construction of the sextic by a *locus linearis*, as Collins somewhat slyly infers.

(62) See John Wallis, *Adversus Marci Meibomii De Proportionibus Dialogum, Tractatus Elencticus* (*Opera Mathematica*, **1** (Oxford, 1657)): *Dedicatio*: 43–6. On transposing his somewhat clumsily expressed argument into standard Cartesian form, we may say that Wallis there constructs the cubic equation $x^3 = mx+n$ geometrically as the meet of the cubical parabola

$$k^2(x+n/m) = y^3$$

Collins, it would seem, respected Newton's provisions: the original autograph of the 'Problems for construing æquations', no doubt speedily returned, has since Newton's day remained undisturbed in the corpus of his mathematical papers, while the transcript Collins then made[66] was circulated only long after his death when, in Newton's old age, it passed into William Jones' possession.

One of the privileged few to whom Jones showed the Collins copy, James Wilson, afterwards recorded his impressions of the paper in print and we may fittingly quote his excellent outline of its content:

About the year 1708, all the papers of Mr. John Collins, who had kept a correspondence with the most eminent geometers in Europe, fell into the possession of Mr. William Jones, then a teacher of the mathematicks in this city [London]....Amongst them was a small tract divided into two parts. In the first leaf was written 'Constructiones Geometricæ Æquationum per D. Isaac Newtonum. Ex Apographo D^ni^ Collins'.

The first part is a sort of Elements of geometry, where motion is introduced, in order to effect several problems after a different manner from what is common, and there is given constructions of solid problems (some of which are transcribed into his Algebra)[67] by placing a streight line of a given length between two other lines given in position, so as to pass through a given point, &c. and from the facility, wherewith this may be performed mechanically, by moving the given line between the others, till it arrives at the position required; he proposes to admit into geometry such motions, as an additional postulate to those of the elements.

The second part, which answered to the title, somebody[68] had prefixt to the whole, delivered in eight problems the construction of quadratic, cubic, biquadratic, and equations of higher powers by the circle, conic-sections, and cubical parabola. The 9th and last problem was this, 'Quomodo problemata solvenda sunt, ubi per intricatam

---

with the straight line $y = x\sqrt[3]{(k^2/m)}$ (drawn by him parallel to the parabola's tangent at $x = \sqrt{[\frac{1}{27}m] - n/m}$). Collins had already drawn attention to this passage (which clearly he did not fully comprehend) in his 1669 article ('An Account Concerning the Resolution of Equations in Numbers' (note (8)): 934: 'Dr. Wallis...hath...excellently resolved and constructed all Cubick *Æquations* at the end of the first Treatise of his *Opera Mathematica* by aide of a Cubick *Parabolaster*'). Isaac Barrow, we may note, in *Lectio XIII* of his recently published *Lectiones Geometricæ: In quibus (præsertim) Generalia Curvarum Linearum Symptomata Declarantur* (London, 1670) had suggested the rather inefficient construction of the general cubic

$$x^3+bx^2+c^2x = n$$

as the meet of the straight line $y = n$ with the 'hyperboliformis' $x^3+bx^2+c^2x = y$ (where we have substituted Cartesian variables x and y for his Fermatian ones a and e).

(63) Collins to Newton, 30 July 1672 (*Correspondence*, **1**: 223).

(64) This preceding passage (which sketches the organic description of conics from a straight line directrix) is reproduced in **1**, 3, Appendix 2 above.

(65) Newton to Collins, 20 August 1672 (**1**, 3, Appendix 2: 159 = *Correspondence*, **1**: 231).

(66) See 2, §2: notes (1) and (2) below.

(67) See 2, §2: note (1).

(68) William Jones, in fact.

terminorum complicationem non licet ad æquationes commode pervenire'; but the solution was wanting.[69] And indeed this copy was very imperfect; as I learn from some passages extracted from another manuscript.[70]

After Wilson only Samuel Horsley is known to have studied Newton's paper, either in its original or Collins' transcript (both of which disappeared into private possession for more than a century afterwards). Subsequent students of Newton's mathematical papers may well have been dissuaded from its detailed study by the Bishop's unfavourable comments as to its quality and originality[71] but future scholarly opinion, we feel sure, will set a higher value upon it.

---

(69) Compare 2, §2: note (144).

(70) *Mathematical Tracts of the late Benjamin Robins, Esq:...*, **2** (London, 1761): Appendix: 346–7. Wilson probably refers in his last sentence to an incomplete copy of Newton's covering letter of 20 August 1672 (note (65)) made by Collins when he returned the original (now ULC. Add. 3977.10) and its enclosure to Newton. This copy, now in private possession, was perhaps the version consulted by Leibniz when he made extracts from Newton's letter in October 1676 (compare **2**, 3, Appendix 1: note (2)).

(71) See 2, §2: note (1). Horsley's signed estimate of the tract, dated 20 October 1777, is entered on a slip of paper now preserved with Newton's manuscript in Cambridge University Library (Add. 3963.9).

(1) The historical background of Gerard Kinckhuysen's *Algebra Ofte Stel-konst* (Haerlem, 1661), submitted in Mercator's Latin version to Newton by Collins for comment in the early winter of 1669/70, is discussed in the introduction to the present Part 3 and in Christoph J. Scriba's 'Mercator's Kinckhuysen-Translation in the Bodleian Library at Oxford', *British Journal for the History of Science*, **2** (1964): 45–58. We have felt it necessary to reproduce this translation of Kinckhuysen's work without abridgement because Newton introduced revised versions of substantial passages from it into his Lucasian lectures on arithmetic and algebra from 1673 onwards (ULC. Dd. 9.68, later published by Whiston against Newton's wishes as the *Arithmetica Universalis: sive De Resolutione et Compositione Arithmetica Liber* (Cambridge, 1707)) and the nuances of his debt to Kinckhuysen would otherwise be lost.

(2) This version, written out in Mercator's own large, upright hand, is interleaved in a copy (Savile G. 20 (4)) of the Dutch edition of Kinckhuysen's book now found in a miscellany of mathematical books and pamphlets which once belonged to John Wallis. When the set was bound some time in the eighteenth century the sheets on which the translation was written, originally somewhat larger than the quarto ones of the Dutch printing, were slightly amputated with the occasional loss of a line of text at the foot of the page. As we would expect the end-papers show signs of active use, being soiled on their outside faces with ink stains and other dirt. The Dutch printed version is not quite complete, lacking the concluding sheet of 'Druck-fauten, die my in't naer sien ontmoet zijn' (with overleaf the final advertisement 'TE HAERLEM, Ghedrukt by *Isaac van Wesbuch*, Boeck-drucker in de korte Zijl-straet, in de groote Druckery. 1661'), though these are for the most part incorporated in the manuscript translation.

Mercator, it would appear, renders the Dutch text quite accurately but does not usually repeat the displayed mathematical passages in Kinckhuysen's work: these blanks we have silently filled in the version we reproduce. On occasion Newton himself has corrected slight omissions on Mercator's part (on page 13, for example, inserting 'Divide...per..., prodit...' for Kinckhuysen's 'Divide...door..., komt...'): elsewhere he has added asterisks to the

1

KINCKHUYSENS 'ALGEBRA' AND NEWTON'S 'OBSERVATIONS' UPON IT

[1670]

§1. MERCATOR'S LATIN VERSION OF KINCKHUYSEN'S 'ALGEBRA'.(1)

From the original in the Bodleian Library, Oxford(2)

ALGEBRA, / sive / Logistica Speciosa, / Tyronum usui / conscripta / à / Gerardo Kinckhuysen,(3) / et edita primùm / Haerlemi / Anno 1661. Nunc è Belgico Latinè versa.(4)

[3] ||Lectori.(5)

Superiori anno edidi libellum cui titulus est *Fundamentum Geometriæ*,(6) in quo utramꝗ ferè paginam facit *Logistica Speciosa*, cujus interim paucos in hac urbe(7)

Latin version to draw Collins' attention (and that of the future reader) to corresponding notes in his *Observationes* (§2 below), and on page 97 has appended a marginal amplification of Kinckhuysen's explanation of a problem. In our own text these are, asterisks apart, printed in italic between square brackets while appropriate mention is made in footnote. In addition we have ourselves numbered a few of Kinckhuysen's problems for purposes of identification. The break in Mercator's pages, which closely follows that of Kinckhuysen's text, we indicate at the division by the double vertical rule || and in the margin by ||[91] (where the bracketed number is that of Mercator's page).

(3) We omit the Cartesian diagram inserted here (from p. 92 below) purely to decorate the Dutch title page. The appearance of the Latin and Dutch titles may be seen in the reduced photocopy (Plate V) which faces page 54 of Scriba's 'Mercator's Kinckhuysen-Translation' (note (1)).

(4) Kinckhuysen's original Dutch title reads *Algebra/Ofte/Stel-konst,/Beschreven/Tot dienst van de Leerlinghen,/Door/Gerard Kinckhuysen./Tot Haerlem,/By Passchier van Wesbuch, Boekverkooper op de Marckt, in den beslaghen Bybel. Anno 1661.* Here (and correspondingly on his page 5 below) Mercator had some difficulty in finding an appropriate Latin equivalent for the Dutch 'Stel-konst' (symbol-art): his final *mot juste* 'Logistica Speciosa' replaces a cancelled first translation 'Ars Analytica'.

(5) Kinckhuysen's 'Tot den Leeser' (to the reader).

(6) *De Grondt der Meet-konst* [, *Ofte Een korte verklaringe der Keegel-sneeden, Met een By-vœghsel. Door Gerard Kinckhuysen*, Haerlem, 1660].

(7) That is, Haerlem.

deprehendi peritos esse, quod inde adeò fieri credas, quia parum extat de primis ejus principijs, prout in usu nunc sunt, Belgico idiomate vulgatum. Istiusmodi igitur tyronibus qui fortè hâc ope destituuntur excitandis et animandis suscepi laborem hoc opusculum in lucem edendi. Quod attinet scripta, quæ passim hac de re prodierunt in publicum, ea maximè probo, quæ *Renatus Descartes* prodidit in *Geometria* sua, nec non quæ ijsdem illustrandis edidit *Franciscus à Scho*[*o*]*ten*, ubi plures Authores peritiam suam ostenderunt. Horum ego vestigia, quò clarior et apertior foret expositio, quàm potui, pressim ||[4] sequutus, non erubui quibusdam in locis ipsissima Clarissimi ||Viri verba recitare, quemadmodum Lectori conferenti patebit. Nec dubito, quin alij hoc idem nostro idiomate multò rectiùs tradidissent, quos hac ratione excitatum iri confido, oblectante sese interim his nostris Tyrone, cui paucula hæc fidem facient, Logisticam hanc non modò ad abstrusiora Arithmeticæ sese extendere; sed clavem quoqꝫ esse Geometriæ, imò universæ Matheseos; neminemqꝫ adeò Mathematicum fore, quin Algebræ quodammodo peritus sit. Cúmqꝫ exercitia Mathematica ob certitudinem suam aptissima sint acuendo ingenio; non immeritò hæc ipsa, quæ potissimam et ingeniosissimam partem constituunt, inter optima exercitia censebuntur. Qui provectiores sunt studio, quàm nostra scriptio assequatur, cogitabunt, hoc opus tyronibus tantùm conscriptum, quò viam insistant primulùm, id quod meo judicio sufficit: nam nasutis(8) satisfacere non est instituti mei. Carpat hinc quivis quod commodum sibi duxerit, et gratias agamus Optimo Deo, facultatem nobis tribuenti hæc et similia comprehendendi.

[G. K.](9)

||[5] ||ALGEBRA / sive / Logistica Speciosa.(10)

Algebra est Logistica, vulgaris Arithmeticæ vestigia legens, et usum habet præcipuè in quæstionibus, quæ non facilè solvi posse videntur, nisi ipsum quæsitum, vel aliquod simile notum sit. Exempli gratiâ: Vendit quis equum 144 florenis et lucratur tantundem in quovis centenario quot florenis ipsi equus constiterat; quæritur quantum id sit, quod lucratur.(11)

Ubi videri poterat numerum florenorum, quo ipsi equus constitit, ad ineundum calculum oportere notum esse. In istiusmodi quæstionibus igitur, pro quantitate ignota, quam vellemus esse notam, ponitur symbolum quodvis, quod

(8) 'nosy' and so presumably those who have a delicate discriminatory sense, 'connoisseurs' (in translation of Kinckhuysen's 'Neus-wijsen').

(9) That is, 'G[erard] K[inckhuysen]'.

(10) Compare note (4) above.

(11) The first problem in Kinckhuysen's 'Pars Tertia' below.

(12) Newton adds a slight amplification to Mercator's version, perhaps because it does not render the full force of Kinckhuysen's '*plus*, ofte meer... *minus*, ofte min'.

inter operandum tractatur, quasi notum foret, redigiturꝗ quæstio, quâ commodissimè viâ potest, ad æquationem, et qua ||porrò ignotum redditur notum, quemadmodum deinceps patebit. [6]

Symbola pro rebus ignotis substituta, qualia sint, nihil interest, dummodò promptè nosci queant; *Descartes* ubiꝗ in sua *Geometria* adhibet ultimas literas alphabeti, nimirum x, y, vel z, quem et nos eâ in re sequimur.

Sæpe etiam fit, ut quæstio aliqua, quæ aliàs difficilior foret, solvi possit per Regulam, vel Theorema quoddam, id quod in Geometria frequentem usum habet. Ejusmodi Theoremata sese produnt, quando inter ratiocinandum pro numeris vel lineis notis etiam literæ vel symbola adhibentur, pro quibus rebus notis *Descartes* discriminis ergo sæpe ponit primas literas a, b vel c.

Usurpantur præterea brevitatis causâ signa quædam, veluti:

+ significat *plus* [, *hoc est subsequentem quantitatem addendam esse præcedenti*].
− significat *minus* [, *hoc est subsequentem quantitatem subducendam esse*].(12)

Iam veró, quo pacto hæ literæ signantes quantitates tam notas, quàm ignotas, calculo suscipiantur; et quomodo ad æquationem redigantur; et quomodo æquationes extricentur ac solvantur deniꝗ, id ipsum est circa quod occupatur *Algebra*. Ideoꝗ digerimus hoc opus in tres partes, quarum Prima aget de computo specierum, tam integrarū quàm fractarum et radicalium;(13) Secunda de Æquationibus; Tertia, eademꝗ postrema, de modo quæstionem aliquem redigendi ad Æquationem.(14)

||Pars Prima. / De computo Specierum.(15) [7]

In computandis speciebus vulgaris Arithmeticæ methodo procedimus, incipiendo à Speciebus integris indeꝗ progrediendo ad fractas, et sic deinceps. Quamobrem oportet, priusquā ad Algebram accedat quisquam, ut vulgarem Arithmeticam calleat, ad quam etiam refero extractionem radicū ex numeris vulgaribus. Nos interim rem ipsam aggredimur.

(13) That is, 'surdarum' (which Mercator has cancelled).

(14) These three divisions are developed by Kinckhuysen on his pp. 7–38, 39–95 and 96–108 respectively which follow.

(15) It is evident that Kinckhuysen borrows heavily in this part from Erasmius Bartholin's *Principia Matheseos Vniversalis, sev Introdvctio ad Geometriæ Methodvm Renati des Cartes* [= *Geometria*, **2** (1661): 1–47]. Less obvious is his debt to Ludolf van Ceulen's *De Arithmetische en Geometrische fondamenten... Met het ghebruyck van dien In veele verscheydene constiche questien, soo Geometrice door linien, als Arithmetice door irrationale ghetallen, oock door den regel Coss, ende de tafelen sinuum ghesolveert* (Leyden, 1615). Van Ceulen died before he could finish his Dutch work (*Fondamenten*: 271) and in translating it into Latin the same year (*Fundamenta Arithmetica et Geometrica cum eorundem usu In varij[s] problematis, Geometricis, partim solo linearum ductu, partim per numeros irrationales, & tabulas sinuum, & Algebram solutis*... (Leyden, 1615) Willebrord Snell revised, completed and partially abridged it.

Additio in integris. Si addendi sint duo vel plures numeri quibus idem signū est præfixum, atq; idem suffixum; colliguntur in unam summam numeri, collectisq; eadem signa junguntur. Sic si addenda veniant $3x$, $5x$, et $6x$; prodeunt $14x$. Rursus $-4x$ et $-7x$ addita faciunt $-11x$. Sin duæ quantitates addendæ sunt, quarū utraq; peculiare signum habet suffixum; conjunguntur planè, ut sunt. Sic $3x$ et $6xx$ addita faciunt $3x+6xx$ (ubi enim signum nullum præfigitur, subintelligitur $+$.) Sic a et b addita faciunt $a+b$. Et $-2x$ et $-5xx$ faciunt $-2x-5xx$. Sic $-ab$ et $-bb$ addita faciunt $-ab-bb$, et sic de cæteris.

||[8] ||Si addendæ sint duæ quantitates, quibus idem signum suffixum sit quidem, at alteri præfixum sit signum $+$, alteri vero $-$; subtrahitur minor ex majori, et præfigitur signum majoris. Sic addenda sint $+12x$ et $-9x$, fiunt $+3x$. Rursus addenda sint $-15x$ et $+10x$, fiunt $-5x$. Sed si utraq; peculiare signum habeat suffixum, conjunguntur ut sunt. Sic $+ax$ et $-by$ addita faciunt $+ax-by$. Item $-5xx$ et $+4x$ addita faciunt $-5xx+4x$, et sic in cæteris, quemadmodum latinè[16] videre est in subjectis exemplis:

$$
\begin{array}{r} 10x+8 \\ 8x+6 \\ \hline 18x+14. \end{array}
\quad
\begin{array}{r} 8y+7 \\ 6y-12 \\ \hline 14y-5. \end{array}
\quad
\begin{array}{r} a+b \\ a-b \\ \hline 2a \quad . \end{array}
\quad
\begin{array}{r} -6xx+11x \\ 5x^3-7x \\ \hline 5x^3-6xx+4x. \end{array}
$$

$$
\begin{array}{r} a+b \\ 2a-3b \\ 3a+4b \\ \hline 6a+2b. \end{array}
\quad
\begin{array}{r} yy+6y+9 \\ -6y-36 \\ -14 \\ \hline yy \; * \; -41. \end{array}
\quad
\begin{array}{r} aa+ab+ac \\ -ab-bb \\ -ac \\ \hline aa \quad -bb. \end{array}
$$

$$
\begin{array}{r} yy+ay+\frac{1}{4}aa \\ -ay-\frac{1}{2}aa \\ +\frac{1}{2}aa \\ -\frac{1}{2}bb \\ \hline yy*+\frac{1}{4}aa-\frac{1}{2}bb. \end{array}
\quad
\begin{array}{r} y^3-ayy+\frac{1}{3}aay-\frac{1}{27}a^3 \\ +ayy-\frac{2}{3}aay+\frac{1}{9}a^3 \\ -by+\frac{1}{3}ab \\ +c \\ \hline y^3*-\frac{1}{3}aay-by+\frac{2}{27}a^3+\frac{1}{3}ab+c. \end{array}
$$

||[9] ||*Subtractio in integris.* In Subtractione si occurrant duo numeri, quibus idem signum est præfixum et suffixum, sitq; numerus superior major, quàm inferior; subtrahitur inferior ex superiori, et reliquo eadem signa junguntur. Sic $+6x$ si auferanda sint ex $+8x$, fiunt $+2x$. Et $-3y$ ablata ex $-7y$ relinquunt $-4y$. Sin inferior numerus excedat superiorem; subtrahitur superior ex inferiori, et reliquo præfigitur signum contrarium. Sic $+10z$ si auferantur ex $+8z$ restant[17] $-2z$. Et $-10y$ ablata ex $-7y$ relinquunt $+3y$.

(16) This word is superfluous: Kinckhuysen wrote '...ghelijck breeder in de volghende voorbeelden te sie[n] is'.

(17) Newton has cancelled this and substituted 'relinquunt'.

(18) Newton clarifies an ellipsis on Mercator's part.

Si duo numeri dentur, quibus idem signum sit suffixū, at alteri præfigatur + alteri verò −; colliguntur in unam summam et præfigitur signum superioris. Sic $-2y$ si auferantur ex $+8y$, restant $+10y$; et $+4x$ ablata ex $-6x$ restat $-10x$.

Postremò, si duarū quantitatū utraq; peculiare signum habeat suffixum; conjunguntur præponendo superiorem, et subnectendo inferiorem cum signo contrario. Sic $3x$ si auferenda sint ex xx, scribe $xx-3x$. Subducturus $-ab$ ex $-bb$, scribe $-bb+ab$. Aufer $+bc$ ex $-ab$, restant $-ab-bc$. Item $-bx$ ablata ex $+cx$ relinquunt $cx+bx$. Atq; ita in cæteris, quemadmodum ulteriùs liquet ex subjectis exemplis.

Quod si proba[*tione*]m(18) instituas Subtractionis per Additionem, quemadmodum in vulgari Arithmetica usitatum est, tum singula [in sequentibus invertenda sunt.](19)

[10] ||

$$\begin{array}{ccccc}
14x+7 & 24ab-4b & 8xx-6x & 10y+6 & 7z^3-6zz \\
6x+4 & 20ab-3b & 10xx-8x & 12y-7 & 6z^3+7zz-9z \\
\hline
8x+3. & 4ab-b. & -2xx+2x. & -2y+13. & z^3-13zz+9z.
\end{array}$$

$$\begin{array}{ccccc}
a^3-2aa+6a & -2ab+cd-cc & ab+cd & zz & qq+zz \\
2a^3+aa\ -a & +ab-bc & de-ef & yy-2ay+aa & qq+yy-2aq \\
\hline
-a^3-3aa+7a. & -3ab+cd-cc+bc. & ab+cd-de+ef. & zz-yy+2ay-aa. & zz-yy+2aq.
\end{array}$$

Multiplicatio in integris. Cùm duæ quantitates multiplicandæ veniunt, ducuntur in se invicem numeri subjunctis signis utriq; numero adhærentibus. Sic ducturus $6x$ in $4x$, pone $24xx$, atq; hoc si denuò multiplicandum sit per $2x$, fiunt $48xxx$, vel brevitatis causâ $48x^3$. Et multiplicando x per y, fit xy, quasi dicas x ductū in y. Tum xy ductum in x, facit xxy. Et $3xx$ ductum in $2xx$ facit $6xxxx$, vel brevitatis causâ $6x^4$. Ducturus $3ab$ in ab, pone $3aabb$, et sic in alijs.

[11] ||Cùm signis + et − agendum [est], ut sequitur. Si + ducatur in +, vel − in −, productum erit +; sin + ducatur in −, vel − in +, productum erit −. Sic si a ducatur in b, fit $+ab$; et si $-a$ ducatur in $-b$, fit quoq; $+ab$. At si $+a$ ducatur in $-b$ vel $-a$ in $+b$, fit $-ab$. Operatio porrò peragitur quemadmodum in vulgari Arithmetica, incipiendo ab anterioribus vel posterioribus quemadmodum ex sequentibus exemplis liquet.

Si cui scrupulum moveat, cur + ductum in −, vel − in + producat −; id provectiores deprehendent ut sequitur: Ducendum sit $a-b$ in c; sit $a-b$ æquale d, productū erit cd. Est verò etiam a æquale $b+d$, quare ducto utroq; in c, fit ac æquale $bc+cd$; ergo $ac-bc$ æquale superiori producto $+cd$. Patet igitur quod $-b$ ductum in $+c$ producit $-bc$.

(19) The guillotine (note (2)) has taken away the close of the sentence and it is restored from Kinckhuysen's '...soo sal al dit ghestelde openbaer zijn'.

Sic quoqȝ, cur − ductum in −, producat +, deprehenditur ita: Ducendum sit $a-b$ in $-c$; sit $a-b$ æquale d, productum erit $-cd$. Est verò etiam a æquale $b+d$, quare ducto utroqȝ in $-c$, fit $-ac$ æquale $-bc-cd$; ergo $-ac+bc$ æquale est superior producto $-cd$. Unde liquet $-b$ ductum in $-c$ producere $+bc$.

||[12] ||

$$\begin{array}{r} 4x+6 \\ 5x \\ \hline 20xx+30x. \end{array} \qquad \begin{array}{r} 2ac-b \\ b \\ \hline 2abc-bb. \end{array} \qquad \begin{array}{l} 3y+4 \\ 4y+6 \\ \hline 12yy+16y \\ \qquad +18y+24 \\ \hline 12yy+34y+24. \end{array} \qquad \begin{array}{r} 6x+4 \\ 4x-3 \\ \hline -18x-12 \\ +24xx+16x \qquad \\ \hline +24xx-2x-12. \end{array} \qquad \begin{array}{l} x-c \\ x-d \\ \hline xx-cx \\ \qquad -dx+cd \\ \hline xx-cx-dx+cd. \end{array}$$

$$\begin{array}{l} \qquad y^3+ayy-by \\ \qquad\qquad y-a \\ \hline \qquad -ay^3-aayy+aby \\ +y^4+ay^3-byy \\ \hline +y^4 \quad * -aayy-byy+aby. \end{array} \qquad \begin{array}{l} aab-abb+b^3 \\ aa+ab \\ \hline a^4b-a^3bb+aab^3 \\ \qquad +a^3bb-aab^3+ab^4 \\ \hline a^4b \qquad\qquad\qquad +ab^4. \end{array}$$

Solemus etiam sæpe uti compendio quo ducturi $ab+bc-cc$ in xx, scribimus $\overline{ab+bc-cc}$ in xx, ubi lineola imminens speciebus $ab+bc-cc$ indicat, quousqȝ extendantur quantitates, quas in xx ductas volumus. Sic quoqȝ ducturi

$$ab+bc-cc \text{ in } xx-ax,$$

scribere solemus $\overline{ab+bc+cc}$ in $\overline{xx-ax}$. Atqȝ ita in alijs.

||[13] ||*Divisio in integris.* Divisio opponitur Multiplicationi, itidem ut in vulgari Arithmetica. Quare dividendo $24xx$ per $6x$ oritur $4x$. Et dividendo a^3 per a, oritur aa. Sic dividendo $aabcd$ per abd, oritur ac, et sic porrò.

Cum signis + et − agitur, ut sequitur: Dividendo + per +, vel − per −, oritur semper +; at dividendo − per + vel + per −, semper oritur −, planè ut in Multiplicatione dictum fuit.(20)

Peraguntur autem hæ Divisiones, non aliter quàm in vulgari Arithmetica solet, incipiendo ab anterioribus, quemadmodum sequentia exempla ostendunt.

Divide $20abcd$ per $5cd$, oritur $4ab$.

$$\begin{array}{r|l} 20abcd & 4ab \\ 5\ \ cd & \end{array}$$

[*Divide*] $\frac{1}{2}aab^3d$ [*per*] $\frac{3}{4}aab$, [*prodit*] $\frac{2}{3}bbd$.

$$\begin{array}{l|l} \frac{1}{2}aab^3d & \frac{2}{3}bbd \\ \frac{3}{4}aab & \end{array}$$

(20) For Newton's later comment at this point, see §2: note (5) below.

[*Divide*] $2xx-6x$ [*per*] x, [*prodit*] $2x-6$.

$$\begin{array}{r|l} 2xx-6x & 2x-6 \\ x \quad x & \end{array}$$

[14] || [*Divide*] $2xx+4x-30$ [*per*] $x+5$, [*prodit*] $2x-6$.

$$\begin{array}{r|l} -6x & \\ 2xx+4x-30 & 2x-6 \\ x+5 & \\ x+5 & \end{array}$$

Dic, quemadmodum in vulgari Divisione: quoties x continetur in $2xx$, Resp. $2x$; tum $2x$ ductum in $x+5$ facit $2xx+10x$ subducenda ex $2xx+4x$, restant $-6x$. Porrò, quoties x continetur in $-6x$, Resp. -6; tum -6 ductum in $x+5$ facit $-6x-30$, quibus subductis ex $-6x-30$, restat nihil.

[Divide] $a^3-3aab+3abb-b^3$ [per] $a-b$, [prodit] $aa-2ab+bb$.

$$\begin{array}{r|l} -2aab+\ abb & \\ a^3-3aab+3abb-b^3 & aa-2ab+bb \\ a-\ b \qquad\qquad & \\ a \qquad b \qquad & \\ a \qquad -b & \end{array}$$

Potest quoq; operatio harum Divisionum paulò aliter disponi, quemadmodum in proximè superiori exemplo, quod hîc resumimus.

[Divide] $a^3-3aab+3abb-b^3$ [per] $a-b$.

$$\begin{array}{l} a^3-aab \\ \hline 0-2aab \\ \quad -2aab+2abb \\ \hline \qquad 0+\ abb \\ \qquad\quad abb-b^3 \\ \hline \qquad\quad 0 \quad 0 \end{array} \qquad \begin{array}{c|c|c|c} a^3 & aa.\ -2aab & -2ab.\ +abb & +bb. \\ a & a & a & \end{array}$$

quotus[21] $\quad aa-2ab\ +bb$.

[15] || Hoc peragitur ut sequitur. Divide a^3 per a (primam partem Divisoris) oritur aa, quod pones in quoto, et per ipsum multiplica divisorem $a-b$, fit a^3-aab quod subduces ex a^3-3aab, restat $-2aab$. Hoc ipsum $-2aab$ divide denuò per a, oritur $-2ab$, quod pones in quoto et per ipsum multiplica divisorem $a-b$, fit $-2aab+2abb$, quo subducto ex $-2aab+3abb$, restat $+abb$. Hoc rursus divide per a, oritur $+bb$, quod pones in quoto et per ipsum multiplica divisorem $a-b$, fit $abb-b^3$, quo subducto ex $abb-b^3$, restat 0, quod arguit, nihil superesse.

(21) Mercator has cancelled the more exact rendering 'oritur' of Kinckhuysen's 'uytkomst'.

Nec abhorret ab usu(22) incipere divisionem a posterioribus, ut in exemplo sequenti.

$$\text{[Divide]}\ +y^6 - 8y^4 - 124yy - 64\ \text{[per]}\ yy-16.$$

$$\begin{array}{rrrr} +y^6 & +\ 8y^4 & +\ 4yy & -64 \\ \hline 0 & -16y^4 & -128yy & 0 \\ & -16y^4 & -128yy & \\ \hline & 0 & 0 & \end{array} \qquad \frac{-64}{-16}\Big|\,4. \quad \frac{-128yy}{-16}\Big|+8yy. \quad \frac{-16y^4}{-16}\Big|+y^4.$$

$$\text{quotus}\quad y^4+8yy+4.$$

Incipe ab ultimo termino -64, eumq; divide per -16, oritur $+4$, quod pones in quoto, et per ipsum multiplica divisorem $yy-16$, fit $+4yy-64$, quod subduces ex posterioribus, restat $-128yy$. Hoc ipsum $-128yy$ divide denuò per -16, oritur $+8yy$, quod pones in quoto, et per ipsum multiplica divisorem, fit $8y^4-128yy$, quod rursus subduces ex posterioribus, restabit $-16y^4$; quod rursus divides per -16, oritur y^4, hoc quoq; pone in quoto, idemq; multiplica ||[16] ||per divisorem, fit y^6-16y^4, id quod etiamnum subduces, restabitq; nihil; itaq; quotus absq; ullo residuo est $y^4+8yy+4$.

Descartes hoc ipsum paulò aliter docet,(23) cujus methodum accipe exemplo sequenti:

$$\text{[Divide]}\ +y^6 \begin{matrix}+aa\\-2cc\end{matrix} y^4 \begin{matrix}-a^4\\+c^4\end{matrix} yy \begin{matrix}-a^6\\-2a^4cc\end{matrix} \quad \text{[per]}\ yy-aa-cc.$$

$$\begin{array}{cccc} -y^6 & \begin{matrix}-2aa\\+cc\end{matrix} y^4 & \begin{matrix}-a^4\\-aacc\end{matrix} yy & -aac^4 \\ \hline 0 & -aa-cc & -aa-cc & -aa-cc \\ \hline \end{array}$$

$$\text{[oritur]}\quad +y^4 \begin{matrix}+2aa\\-cc\end{matrix} yy \begin{matrix}+a^4\\+aacc\end{matrix}.$$

Incipe, ut in proximè superiori, ab ultimo termino, eumq; divide per $-aa-cc$, oritur $+a^4+aacc$, quod pones in quoto, idemq; multiplica per yy, fit $\begin{matrix}+a^4\\+aacc\end{matrix}yy$, quod cum signis contrarijs addes penultimo termino (signa enim + vel − semper contraria scribenda sunt ijs, quæ ex multiplicatione prodeunt;) summam divide rursus per $-aa-cc$, oritur $\begin{matrix}+2aa\\-cc\end{matrix}yy$, quod pones in quoto, idemq;

(22) The cancelled phrase 'Fert quoq; usus' is a more literal translation of Kinckhuysen's 'Het heest mede gebruyck'.

(23) R. Descartes, *Geometrie* (Leyden, 1637): 381–3. Both the preceding and following examples are there divided by the latter method, and Kinckhuysen in the preceding paragraph has made only slight changes from Descartes' equivalent account on his pp. 381–2.

multiplica per yy, et sic $\begin{matrix}-2aa\\+cc\end{matrix}y^4$ statues sub proximo dividuo, summamq; $\begin{matrix}-aa\\-cc\end{matrix}y^4$ divide rursus per $-aa-cc$, oritur $+y^4$, quod pones in quoto; tum adde $-y^6$ ad $+y^6$, atq; ita nîl deest, nec abundat.

Cùm incidimus in Divisionem, quæ non exit rotundè absq; reliquo; tum id quod superest dividui supra lineam collocamus, et divisorem infra, unde oritur fractio. Veluti si bc dividendum sit per a, scribimus $\frac{bc}{a}$, et divisuri x^3-2xx per [17] $7x-12$, scribimus $\frac{x^3-2xx}{7x-12}$. ||Sic quoq; divisuri $20y^3+12yy+6y-8$ per $5y+3$, ponimus $\frac{20y^3+12yy+6y-8}{5y+3}$. Verùm hoc quoq; (si compendium subesse deprehendatur) dividi potest, faciendo ex residuo fractionem, oriturq; $4yy+\frac{6y-8}{5y+3}$.

De fractis. Ex quantitatibus quæ absq; residuo dividi nequeunt, oriuntur fracti, qui itidem ut in vulgari Arithmetica, hîc censentur in Numeratore et Denominatore, sic $\frac{ab}{a+c}$ est fractio cujus integrum sive unitas valet $\frac{a+c}{a+c}$, et ab vocatur Numerator et $a+c$ Denominator; calculus quoq; hîc instituitur ut in Arithmetica vulgari, quemadmodum ex sequentibus videre est.

De reductione fractorū ad minores terminos.

Quando Numerator et Denominator per eandem quantitatem dividi possunt absq; residuo, diminui potest fractio; veluti si detur $\frac{10xx}{8x}$, dividi potest tam quod est supra, quàm quod infra lineam per $2x$, quare substitui potest $\frac{5x}{4}$; et dato $\frac{ab+bc}{ac+cc}$, divide tam superiorem quàm inferiorem quantitatem per $a+c$, et emergit $\frac{b}{c}$; item dato $\frac{xx-cc}{xx+2cx+cc}$, divide superiorem et inferiorem per $x+c$, et habebis $\frac{x-c}{x+c}$, atq; ita in alijs.

[18] ||Additio fractorum.

Si fractiones addendæ communem Denominatorem habeant, colliguntur in unam summam Numeratores, cui subjungitur communis iste Denominator; Sic additurus $\frac{bx}{a+b}$ ad $\frac{ab}{a+b}$, scribe $\frac{bx+ab}{a+b}$; atq; ita in similibus. Sin diversum habeant Denominatorem, reducendi sunt ad minimum communem Denomi-

natorem, sicut fit in vulgari Arithmetica, quemadmodum exempla sequentia ostendunt.

Adde $\frac{3x}{x+3}$ ad $\frac{5x}{xy+3y}$, fit $\frac{3xy+5x}{xy+3y}$.

$$\begin{array}{lll} & & (xy+3y \\ \frac{3x}{x+3} & - - - & 3xy \\ \frac{5x}{xy+3y} & - - - & 5x \\ \hline & & \frac{3xy+5x}{xy+3y} \end{array}$$

Adde $\frac{ab}{ad+cd}$ et $\frac{cd}{ab+bc}$, fit $\frac{abb+cdd}{abd+bcd}$.

Adde $\frac{a}{b}$, $\frac{a-b}{a+b}$ et $\frac{aa+bc}{ab+bb}$, fit $\frac{2aa+2ab-bb+bc}{ab+bb}$.

Potest quoque additio duorum fractorum institui ut in exemplo sequenti.

||[19] ||

$$\begin{array}{c} 2ab+ac \\ \overline{ab+ac} \quad ab \\ \text{Adde } \frac{a}{b} \text{ ad } \frac{a}{b+c}, \text{ fit } \frac{2ab+ac}{bb+bc}. \\ \overline{bb+bc} \end{array}$$

Peragitur autem sic: multiplica fractos decussatim, habebis $ab+ac$ et ab, quæ addita exhibent Numeratorem quæsitum $2ab+ac$, multiplica deinde Denominatores, et prodit Denominator quæsitus $bb+bc$.

Subtractio fractorum.

Si fracti auferendi ab invicem communem habeant Denominatorem, subtrahitur numerator à numeratore, et quod restat statuitur supra lineam, et communis denominator infra; Veluti si $\frac{bx}{a+b}$ auferendū sit ex $\frac{ab}{a+b}$, scribes $\frac{ab-bx}{a+b}$. Sin diversos habeant denominatores, reducuntur ad communem denominatorem, prorsus ut in Additione factum, et exempla sequentia ostendunt.

Subtrahe $\frac{3}{8x}$ ex $\frac{9}{16x}$, restat $\frac{3}{16x}$.

Subtrahe $\frac{7y-8}{6}$ ex $\frac{6y-10}{6y-10}$, restat $\frac{-21yy+59y+2}{18y-30}$.

Subtrahe $\frac{a}{c}$ ex $\frac{ab+bc}{ab}$, restat $\frac{abc+bcc-aab}{abc}$.

Potest quocꝫ Subtractio institui, quemadmodum in exemplo sequenti.

[20] || $bb+bc$ — ac

$bb+bc$ ac

Ex $\frac{b}{c}$ aufer $\frac{a}{b+c}$, restat $\frac{bb+bc-ac}{bc+cc}$.

$bc+cc$

Hi fracti multiplicati decussatim, producunt $bb+bc$, et ac, quæ si auferantur ab invicem restat numerator quæsitus $bb+bc-ac$, et denominatores multiplicati exhibent denominatorem quæsitum $bc+cc$.

Circa hunc operandi modum observandum [est], quando fractorum loco objiciuntur integr[i], omnia reducenda esse ad fractos, veluti si dentur $ab+\frac{abc}{d}$, multiplicatur integrum, nimirum ab, per denominatorem fracti, videlicet d, et extat $\frac{abd+abc}{d}$.

Multiplicatio fractorum.

Hîc numerator multiplicatur per numeratorem, et denominator per denominatorem, et quod inde oritur est ipsum quæsitum; Veluti si $\frac{a}{b}$ multiplicatum sit per $\frac{c}{d}$, fit $\frac{ac}{bd}$, ita in exemplis sequentibus.

Multiplica $\frac{x+3}{x}$ per $\frac{6}{x}$, oritur $\frac{6x+18}{xx}$.

Multiplica $\frac{a-b}{b}$ per $\frac{a+b}{b}$, oritur $\frac{aa-bb}{bb}$.

Multiplica $\frac{aa+2ab+bb}{ab}$ per $\frac{bb}{a+b}$, oritur $\frac{ab+bb}{a}$.

[21] ||Hîc fracti primùm decussatim diminuuntur, nimirū $aa+2ab+bb$ et $a+b$ utrumcꝫ dividitur per $a+b$, et ab atcꝫ bb utrumcꝫ per b, id quod operationem reddit faciliorem.

Multiplica $\frac{a-b}{b+c}$ per $\frac{aa+bb}{a-b}$, oritur $\frac{aa+bb}{b+c}$.

Si integri admiscentur fractis, omnia reducenda sunt ad fractos; veluti si multiplicandum sit $a+\frac{bc}{a}$ per $b+\frac{ac}{d}$, multiplicantur integri, utercꝫ per denominatorem fracti sibi adhærentis, et emergit $\frac{aa+bc}{a}$ multiplicandum per $\frac{bd+ac}{d}$, unde producitur $\frac{aabd+a^3c+bbcd+abcc}{ad}$.

Multiplica $2y-20$ per $\frac{yy-6y}{10}$, oritur $\frac{y^3-16yy+60y}{5}$.

Divisio fractorum.

Dividuntur fracti, multiplicando ipsos decussatim, hoc est denominatorem hujus per numeratorem alterius, et vicissī; statuiturq; divisor ante dividendum; veluti si dividendum sit $\frac{a}{b}$ per $\frac{c}{d}$, ordinantur sic; $\frac{c}{d}$, $\frac{a}{b}$, tum si multiplicentur decussatim, prodit quæsitum $\frac{ad}{bc}$; ita in exemplis sequentibus.

Divide $\frac{y+3}{5}$ per $\frac{20x}{3}$, prodit $\frac{3y+9}{100x}$.

Divide $\frac{aa+bb}{c}$ per $\frac{bb+cc}{d}$, prodit $\frac{aad+bbd}{bbc+c^3}$.

||[22] Possunt quoq; brevitatis gratiâ, antequam multipli=||cetur decussatim, numeratores per communem mensuram diminui, quemadmodum et denominatores, veluti si dividendum sit $\frac{a^3-abb}{c-d}$ per $\frac{aa+2ab+bb}{c-d}$, numeratores diminui possunt per $a+b$, et denominatores per $c-d$, ita ut tandem prodeat quæsitum $\frac{aa-ab}{ab}$. Sic quoq; $\frac{yy-6}{y+20}$ si dividendum sit per $yy-6$, positâq; 1 pro denominatore integri, prodit quæsitum $\frac{1}{y+20}$. Ita in exemplo sequenti

Diviso $6x+12$ per $xx+\frac{8x+4}{3}$, prodit $\frac{18}{3x+2}$.(24)

Extractio radicum. Hîc notandum [est], quadrati aa radicem esse a, sicut quantitatis $aabb$ radix quadratica est ab, et ipsius a^4 radix aa; sic radix cubica ipsius a^3 est a, ipsius a^3b^3 est ab, et ipsius a^6 est aa; atq; ita de altioribus radicibus æstimandum.

Extractio radicum hîc peragitur eodem modo, quo in vulgari Arithmetica, usurpaturq; hîc eadem tabula geneseos.(25)

Quod si igitur radix quadratica extrahenda sit ex $aa+2ab+bb$, operatio instituenda, ut sequitur:

$$\begin{array}{ll} aa+2ab+bb \\ \hline a \qquad +b \\ \hline +2a \end{array} \quad \text{vel} \quad \begin{array}{lll} aa+ & 2ab+ & bb \\ aa & +2ab+ & bb \\ \hline 0 & 0 & 0 \\ \hline a & +b & \\ \hline & 2a & \end{array}$$

(24) $3x^2+8x+4 = (x+2)(3x+2)$.

(25) That is, the expansion of the binomial $(a+b)^n$, n integral. (Compare 1, 1, 2, §1: note (11).) At this point Newton has added a sentence in amplification in his *Observationes* (§2: note (6) below).

[23] ||Dic: radix quadratica ipsius aa est a, hoc statues loco quæsiti, et dic porrò a in a producit aa, quod si auferatur ex aa, restat 0, tum multiplica idem a per 2 (qui est numerus geneseos) prodit $2a$, tum dic, quoties continetur $2a$ in $2ab$, prodit $+b$, quod statues loco quæsiti, tum $+2a$ in $+b$ facit $2ab$, quo ablato ex $2ab$ restat 0; tum b in b, facit bb, quo ablato ex bb, restat iterum 0, ita ut nihil tandem restet, et quæsitum extet $a+b$.

Eodem modo radix quadratica si extrahatur ex $x^4-8x^3+28xx-48x+36$, prodit $xx-4x+6$.

Radix cubica extrahenda si sit ex $a^3+3aab+3abb+b^3$, operatio est hujusmodi:

$$\begin{array}{lll} \underline{a^3+3aab+3abb+b^3} & a & aa \\ \underline{a \qquad\qquad\qquad +b} & \underline{3} & \underline{3} \\ \quad 3aab+3abb+b^3 & 3a & 3aa \end{array}$$

$$\begin{array}{ccc} b^3 & bb & b \\ \hline b^3 & 3abb & 3aab \end{array}$$

Dic: Radix cubica ipsius a^3 est a, quod statues loco quæsiti, tum hoc ipsum a et ejus quadratum aa pone seorsim, et multiplica eadem per numeros geneseos ipsius cubi, nimirū 3 et 3, prodeunt $3a$ et $3aa$; tum dic, quoties continetur $3aa$ in $3aab$, prodit $+b$, quod statues loco quæsiti, et multiplica ipsum per $3aa$, et ejus quadratum per $3a$, deniq; ejus cubum per 1, et quod inde oritur subduc à residuo dati cubi, restatq; 0, et quæsitum absq; residuo extat $a+b$.

Eodem modo radix cubica si extrahatur ex

$$\begin{array}{l} a^3x^6+3aabx^5+3aacx^4+6abcx^3+3accxx+3bccx+c^3, \ \lfloor\text{oritur } axx+2bx+c.\rfloor \\ \qquad +3abbx^4 \quad +b^3x^3+3bbcxx \end{array}$$

[24] ||Atq; ita in altioribus radicibus. Fundamentum tabulæ geneseon patet ex sequentibus, nimirū si $a+b$ multiplicetur per $a+b$, prodit $aa+2ab+bb$, hoc rursus multiplicatum per $a+b$, producit $a^3+3aab+3abb+b^3$, hoc denuò in $a+b$, producit $a^4+4a^3b+6aabb+4ab^3+b^4$, atq; ita deinceps, tum demum numeri speciebus præfixi dictam tabulam exhibebunt.

Quod hîc de extractione radicum ex integris dictum est, idem de fractis intelligendum, ita radix quadratica ipsius $\frac{aabb}{cc}$ est $\frac{ab}{c}$, et radix quadratica ipsius $\frac{aa+2ab+bb}{bb}$ est $\frac{a+b}{b}$, sic radix cubica ipsius $\frac{a^3+3aab+3abb+b^3}{c^3}$ est $\frac{a+b}{c}$, atq; ita in cæteris.

Si quantitates occurrant, quarum radices absq; residuo extrahi nequeunt, his præfigitur nota $\surd$, quæ significat radicem quadraticam quantitatis, et si plures sint quantitates, lineola ijs superducitur, denotans quousq; extendantur quantitates, quarum radix innuitur, sic radix quadratica extrahenda si sit ex ab, scribe $\surd ab$, sin ex $aa+bb$, pone $\sqrt{aa+bb}$. Radix cubica extrahenda si sit, ponitur $\surd C.$ vel $\surd^3$,

sic radix cubica ipsius *aab* est $\surd C.\, aab$, et ipsius $abb+b^3$, est $\surd \overline{C.\, abb+b^3}$. Extracturi radicem quadrato-quadraticam, scribimus $\surd\surd$, vel $\surd^4$, atq; ita deinceps. Cæterùm numeri, quibus signum $\surd$ præfigitur, vocantur radicales.(26)

||[25] ||*De radicalibus.*(27) Hi radicales sæpius per alium radicalem diminui possunt, quando scilicet ex hoc altero radix extrahi potest; veluti si detur $\surd 50$, dividi hoc potest per $\surd 25$, hoc est per 5, quare scribere licebit $5\surd 2$, hoc est quinquies $\surd 2$. Dato $\surd\frac{16}{11}$, pro eo ponere licet $4\surd\frac{1}{11}$, et dato $\sqrt{80aabb}$ poni potest $4ab\surd 5$,[*](28) et pro $\surd C.\, 275$ scribere licet $5\surd C.\, 3$, et dato $\sqrt{\frac{aacc}{bb}+\frac{aac}{b}}$ ponere licet $\frac{a}{b}\sqrt{cc+bc}$, nam si reducantur ad communem denominatorem, prodit $\sqrt{\frac{aacc+aabc}{bb}}$, quod si dividas per $\sqrt{\frac{aa}{bb}}$, hoc est per $\frac{a}{b}$, prodit $\sqrt{cc+bc}$, atq; ita in similibus.

Si per nullum numerū quadratū dividi possit absq; residuo, tales dividere licet per numerū quadratū quemvis, statuendo loco orti fractionem; sic dato $\sqrt{14aabb}$, substitui potest (dividendo 14 per 4, 16, 25, et deinceps) $2ab\surd 3\frac{1}{2}$, vel $4ab\surd\frac{7}{8}$, vel $5ab\surd\frac{14}{25}$, et deinceps.

De radicalibus communicantibus.(29)

Si sint duo vel plures radicales, qui per eundem radicalem divisi, numeros exhibent, quorū radix quadratica extrahi potest; vocantur illi communicantes; Sic datis $\surd 75$ et $\surd 48$, diviso utroq; per $\surd 3$, prodeunt isthic $\surd 25$, hîc $\surd 16$; quare ||[26] hi $\surd 75$ et $\surd 48$, vel $5\surd 3$ et $4\surd 3$ sunt communicantes, suntq; ut ||5 ad 4, nam alter est quinquies $\surd 3$, et alter quater $\surd 3$.

Sic quoq; communicantes sunt $\sqrt{a^3+aab}$, et $\sqrt{abb+b^3}$, diviso utroq; per $\sqrt{a+b}$, prodeunt $\surd aa$ et $\surd bb$, hoc est *a* et *b*, quare scribere licet $a\sqrt{a+b}$ et $b\sqrt{a+b}$, et hæ quantitates se habent ut *a* ad *b*,[*](30) atq; ita deinceps.

Communicantes si ducantur in se invicem, semper producunt numerum, ex quo radix quadratica extrahi potest, sic $\surd 75$ et $\surd 48$ multiplicata faciunt $\surd 3600$, ex quo radix quadratica extrahi potest. Quod si igitur tres sint continuè proportionales, veluti $\surd a$, $\surd b$, et $\sqrt{\frac{bb}{a}}$, erunt extremi duo, nempe $\surd a$ et $\sqrt{\frac{bb}{a}}$

(26) In the first instance Mercator rendered Kinckhuysen's 'Wortel-ghetallen' here and below as 'surdi'.

(27) Compare van Ceulen's *Fondamenten* (note (15)): 45–68 [= *Fundamenta:* 1–31: 'Ludolphi à Ceulen Surdorum Arithmetica'].

(28) The asterisk which is inserted at this point seems to be Newton's but he has made no corresponding entry in his *Observationes* below. Perhaps he intended to add an algebraic example of a compound cube root $\surd C.\, A^3B/C$.

communicantes. Unde Regula invenitur, inveniendi communem divisorem non aliud faciendo, quàm dividendo utrumq; per alterutrum, veluti hoc loco per $\surd a$, et prodeunt $\surd 1$, et $\sqrt{\frac{bb}{aa}}$, unde sequitur pro $\surd a$ et $\sqrt{\frac{bb}{a}}$ substitui posse $1\surd a$, et $\frac{b}{a}\surd a$. Ita si communis divisor inveniendus sit hisce quantitatibus $\sqrt{8\surd 10\frac{1}{4}-24}$, et $\sqrt{27\surd 10\frac{1}{4}-83\frac{1}{2}}$, dividuntur ambæ per $\sqrt{8\surd 10\frac{1}{4}-24}$, prodeunt $\surd 1$ et $\sqrt{2\frac{5}{8}-\frac{1}{4}\surd 10\frac{1}{4}}$, vel extrahendo ex utraq; radicem, 1 et $\sqrt{\frac{41}{16}}-\frac{1}{4}$,(31) quare pro numeris quæsitis statuuntur $1-\sqrt{8\surd 10\frac{1}{4}-24}$, et $\sqrt{\frac{41}{16}}-\frac{1}{4}\sqrt{8\surd 10\frac{1}{4}-24}$, vel $2\sqrt{2\surd 10\frac{1}{4}-6}$, et $\surd 10\frac{1}{4}-\frac{1}{2}\sqrt{2\surd 10\frac{1}{4}-6}$, atq; ita in alijs.

[27] ||Additio radicalium.

Primùm considerandum [est], utrum sint communicantes quod si ita adduntur ut sequitur; datis $\surd 75$ et $\surd 48$, hoc est $5\surd 3$ et $4\surd 3$, adduntur 5 et 4, fiunt 9, et huic subjungitur $\surd 3$, et emergit quæsitum $9\surd 3$.

Ita si addenda sint $\sqrt{a^3+aab}$ et $\sqrt{abb+b^3}$, hoc est $a\sqrt{a+b}$, et $b\sqrt{a+b}$, adduntur a et b, et fit $a+b$, cui subjungitur ipsum $\sqrt{a+b}$, et emergit quæsitū

$$\overline{a+b}\sqrt{a+b},[*]^{(32)}$$

et sic in alijs.

Est et alius modus addendi communicantes, qui multùm usurpatur, estq; talis: Sint rursus addenda $\surd 75$ et $\surd 48$.

	75	75
	48	48
	123	3600
[Adde]	120	60
	$\surd 243$	2
		120

Adduntur 75 et 48, fiunt 123, tum multiplicantur invicem, prodeunt 3600, cujus radix quadratica est 60, et hujus duplum 120, quod additum ad 123, exhibet quæsitum $\surd 243$, æquale isti $9\surd 3$ suprà invento.(33)

(29) Here and below Mercator first rendered Kinckhuysen's 'Communicanten in de Wortel ghetallen' as 'commensurabiles surdi'. The terminology is that of van Ceulens' *Fondamenten* (note (15)): 45–68, especially 49: 'Korter Reghel, om alle Communicanten te addeeren'. In his Latin paraphrase (*Fundamenta* (note (151): 9) Snell speaks of the 'factum per communem symmetriæ mensuram'.

(30) The asterisk is Newton's (§2: note (7)).

(31) That is, by the following rule, $\surd[\frac{1}{2}(\frac{21}{8}+\surd\{(\frac{21}{8})^2-\frac{41}{64}\})]-\surd[\frac{1}{2}(\frac{21}{8}-\surd\{(\frac{21}{8})^2-\frac{41}{64}\})]$.

(32) Newton's asterisk (§2: note (8)).

(33) The procedure converts $\surd a^2c+\surd b^2c$ into $\surd[a^2c+b^2c+2\surd(a^2c)(b^2c)]$.

Sin occurrant radicales, qui isto modo addi nequeant, conjunguntur per signum +, veluti si addere oporteat $\surd 8$ et $\surd 12$, scribimus $\surd 8+\surd 12$, et addituri $\sqrt{a+b}$ ad $\sqrt{a-b}$, ponimus $\sqrt{a+b}+\sqrt{a-b}$, sic addito $a+b$ ad $\sqrt{aa+bb}$, fit $a+b+\sqrt{aa+bb}$, et sic in alijs.

||[28] ||Subtractio radicalium.

Subtractio additioni similis est in omnibus, nisi quod hîc subtrahitur, quando illic additur, quare eadem exempla hîc repetemus. Si sint communicantes, subtrahuntur, ut sequitur: Si $\surd 48$ oporteat subtrahere ex $\surd 75$, hoc est $4\surd 3$ ex $5\surd 3$, subtrahitur 4 ex 5, restat 1, cui subnectitur ipsum $\surd 3$, prodit quæsitum $1\surd 3$, hoc est $\surd 3$.

Ita quoq, si $\sqrt{abb+b^3}$ subtrahendum sit ex $\sqrt{a^3+aab}$, hoc est $b\sqrt{a+b}$ ex $a\sqrt{a+b}$, scribimus $\overline{a-b}\sqrt{a+b}$, quod est ipsum quæsitum;[*](34) et ita in alijs.

Possint quoq communicantes subtrahi, ut sequitur. Rursus subtrahendum sit $\surd 48$ ex $\surd 75$.

	75	75
	48	48
	123	3600
Subtrahe	120	60
restat	$\surd 3$	2
		120

Hoc peragitur ut in additione, nisi quod hîc 12[0] auferuntur ex 12[3], quæ illic addebantur, et restat $\surd 3$, quod est ipsum quæsitum, et sic in alijs.(35)

Sin radicales occurrant, qui isto modo subtrahi nequeant, tum res peragitur signo −, veluti si $\surd 8$ subtrahendum sit ex $\surd 12$, scribitur $\surd 12-\surd 8$; et subtracturi $\sqrt{a-b}$ ex $\sqrt{a+b}$, ponimus $\sqrt{a+b}-\sqrt{a-b}$; sic quoq subtracto $\sqrt{aa+bb}$ ex $a+b$, restat $a+b-\sqrt{aa+bb}$; atq ita in alijs.

||[29] ||Multiplicatio radicalium.

Si sint communicantes, operatio instituitur ut sequitur; Multiplicandum sit $\surd 75$ per $\surd 48$, hoc est $5\surd 3$ per $4\surd 3$, itaq multiplicatur 4 per 5, fiunt 20, atq hoc rursus per quadratum ipsius $\surd 3$, hoc est per 3, et prodit quæsitum 60.

Sic quoq multiplicaturi $\sqrt{a^3+aab}$ per $\sqrt{abb+b^3}$, hoc est $a\sqrt{a+b}$ per $b\sqrt{a+b}$, ducimus a in b, producitur ab, quod deinde rursus ductum in $a+b$ (nimirū quadratū ipsius $\sqrt{a+b}$) facit quæsitum $aab+abb$; atq ita in alijs.

(34) See Newton's accompanying observation (§2: note (8)).

Si non sint communicantes, semper multiplicantur radicales per radicales; veluti si multiplicandum sit $\sqrt{12}$ per $\sqrt{8}$, prodit $\sqrt{96}$, et multiplicaturi $\sqrt{a+b}$ per $\sqrt{a-b}$, ducimus $a+b$ in $a-b$, et emergit quæsitum $\sqrt{aa-bb}$.

Sed si alter sit radicalis, alter verò secus, oportet eos priùs ad eandem denominationem reducere, et quæsitum erit ejusdem denominationis cum istis. Veluti si multiplicandum sit $a+b$ per $\sqrt{a-b}$, reducendi sunt priùs ad eandem denominationem, nimirum, multiplicatur $a+b$ quadraticè, fit $\sqrt{aa+2ab+bb}$, tum ipsi radicales multiplicantur, et emergit quæsitum $\sqrt{a^3+aab-abb-b^3}$. Cæterùm in hujusmodi multiplicationibus sæpe brevitatis causâ scribimus potiùs $\overline{a+b}\sqrt{a-b}$.

[30] ||Sequuntur [hîc] alia exempla.

$$\begin{array}{rl}
\text{Mult.} & 6+\sqrt{20} \\
\text{per} & 6+\sqrt{20} \\ \hline
 & 36+6\sqrt{20} \\
 & \quad +6\sqrt{20}+20 \\ \hline
\text{prodit} & 56+12\sqrt{20}.
\end{array}
\qquad
\begin{array}{rl}
\text{Mult.} & \sqrt{8}-\sqrt{6} \\
\text{per} & \sqrt{12}-\sqrt{2} \\ \hline
 & \sqrt{96}-\sqrt{72} \\
 & \quad -\sqrt{16}+\sqrt{12} \\ \hline
\text{prodit} & \sqrt{96}-4-\sqrt{72}+\sqrt{12}.
\end{array}$$

$$\begin{array}{rl}
\text{Mult.} & \sqrt{ab}+\sqrt{aa-bb} \\
\text{per} & \sqrt{ab}-\sqrt{aa-bb} \\ \hline
 & ab+\sqrt{a^3b-ab^3} \\
 & -aa+bb-\sqrt{a^3b-ab^3} \\ \hline
\text{prodit} & ab-aa+bb
\end{array}$$

, verùm in hujusmodi casu multiplicandum est solùm $\sqrt{ab}$ per $\sqrt{ab}$, prodit ab; et $\sqrt{aa-bb}$ per $-\sqrt{aa-bb}$, prodit $-\overline{aa-bb}$, hoc est $-aa+bb$, nam $-$ in $\overline{aa-bb}$ producit $-aa+bb$.

Divisio radicalium.

Si sint communicantes, instituitur operatio ut sequitur; dividendum sit $\sqrt{75}$ per $\sqrt{48}$, hoc est $5\sqrt{3}$ per $4\sqrt{3}$; ergo dividitur 5 per 4, et prodit quæsitum $\frac{5}{4}$.

Sic divisuri $a\sqrt{a+b}$ per $b\sqrt{a+b}$, dividimus a per b, et prodit $\frac{a}{b}$.

[31] ||Eodem modo, diviso $5a\sqrt{3}$ per $\sqrt{3}$, emergit $5a$; atq; ita in alijs.

Si non sint communicantes, semper dividendi sunt radicales per radicales, veluti si dividere oporteat $\sqrt{96}$ per $\sqrt{8}$, prodit $\sqrt{12}$; et diviso $\sqrt{abbc-b^3c}$ per $\sqrt{ab-bb}$, prodit $\sqrt{bc}$.

At si alter sit radicalis, alter verò secùs, reducendi sunt ambo ad eandem denominationem, et quæsitum erit ejusdem denominationis.

(35) More generally, $\sqrt{a^2c}-\sqrt{b^2c} = \sqrt{[a^2c+b^2c-2\sqrt{(a^2c)(b^2c)}]}$.

Sic si dividendum sit $\sqrt{a^3+aab-abb-b^3}$ per $a+b$: multiplicatur $a+b$ in seipsum, et habetur ejus loco $\sqrt{aa+2ab+bb}$, tum institutâ divisione, prodit $\sqrt{a-b}$.

Si dividendum sit $a^3+abb+ab\sqrt{aa+bb}$ per $a\sqrt{aa+bb}$, divide primùm a^3+abb per $a\sqrt{aa+bb}$, prodit $\sqrt{aa+bb}$, tum $ab\sqrt{aa+bb}$ per $a\sqrt{aa+bb}$, prodit b, et quæsitum est $\sqrt{aa+bb}+b$.

Si dividere oporteat $\sqrt{20}+\sqrt{14}$ per $\sqrt{10}+\sqrt{6}$; ut divisor fiat numerus simplex sive vulgaris, multiplicatur utrumq; per $\sqrt{10}-\sqrt{6}$, prodit

$$\sqrt{200}+\sqrt{140}-\sqrt{120}-\sqrt{84},$$

dividendum per 4, hoc est per $\sqrt{16}$, et emergit quæsitum

$$\sqrt{12\tfrac{1}{2}}+\sqrt{8\tfrac{3}{4}}-\sqrt{7\tfrac{1}{2}}-\sqrt{5\tfrac{1}{4}}.$$

Aliàs scribi potest $\dfrac{\sqrt{20}+\sqrt{14}}{\sqrt{10}+\sqrt{6}}$.

Occurrunt quandoq; quantitates quæ absq; residuo non dividi quidem, attamen diminui possunt: Veluti si dividendum sit $x^3-6xx+14x-24$ per $8\sqrt{x-4}$, hîc $x^3-6xx+14x-24$, dividi potest per $x-4$, quare scribere licet $\dfrac{\overline{xx-2x+6} \text{ in } \overline{x-4}}{8\sqrt{x-4}}$, vel breviùs $\dfrac{xx-2x+6}{8}\sqrt{x-4}$; atq; ita in alijs, prout diversitas casuum requirit.

||[32] ||*Extrahere radicem quadraticam è binomijs.*(36) Extrahendæ radici quadraticæ è binomijs, inserviet hæc Regula:

Quadrata à nominibus auferantur ab invicem, et residui radix quadratica est differentia quadratorū nominum radicis quæsitæ.

Differentiam sic inventam adde nomini dato majori, prodit duplum quadratum nominis quæsiti majoris.

Eandem differentiam aufer à nomine dato majori, et restat duplum quadratum nominis quæsiti minoris.

---

(36) Compare van Ceulen's *Fondamenten* (note (15)): 59–60: 'Van de Extractij der quadraet Wortel uyt Binomische ghetallen' [= Snell's *Fundamenta*: 19–20: Cap. VII 'De analysi lateris quadrati in irrationalibus compositis'].

(37) Kinckhuysen does not make clear how this procedure is applied in the sequel, but its justification is immediate. If we set $\sqrt{[\alpha+\sqrt{\beta}]} = \sqrt{a}+\sqrt{b}$, then we may identify $\alpha = a+b$ and $\beta = 4ab$ so that $a-b = \sqrt{[\alpha^2-\beta]}$ and hence $a = \frac{1}{2}(\alpha\pm\sqrt{[\alpha^2-\beta]})$, $b = \frac{1}{2}(\alpha\mp\sqrt{[\alpha^2-\beta]})$. Compare van Ceulen's *Fondamenten* (note (15)): 59: 'Regel. Substraheert de quadraten der deelen van malcander, uyt Rest treckt den Wortel, dese addeert tot het grootste deel des

Hoc patet, ut sequitur; sit radix quadratica $\surd a+\surd b$, ejus quadratum erit $a+b+2\surd ab$, quorum nominum quadrata si auferantur ab invicem, restat $aa-2ab+bb$, cujus radix quadratica est $a-b$, quæ est differentia quadratorū nominum radicis.

Adde $a-b$ ad nomen datum majus $a+b$, prodit $2a$, quod est duplum quadrati nominis quæsiti majoris.

Et aufer eadem ab invicem, restat $2b$, quod est duplum quadrati nominis quæsiti minoris; ita ut radix quæsita sit $\surd a+\surd b$.(37)

Dato igitur $33+\surd 800$, extrahitur ejus radix quadratica, ut sequitur.

33]

$$\begin{array}{rrrl}
\|33+\surd 800 & & & \\
33 & & & \\
\hline
1089 & 33 & 33 & \\
800 & 17 & 17 & \\
\hline
289 & 50 & 16 & \\
\hline
17 & 25 & 8 & \text{prodit } 5+\surd 8.
\end{array}$$

Quadrata nominum ablata ab invicem relinquunt 289, cujus radix quadratica est 17, quæ addita nomini majori 33, facit 50, cujus [semis] 25 est quadratum nominis quæsiti majoris; et eadem 17 ablata ex 33, relinquunt 16, cujus semis 8 est quadratum nominis quæsiti minoris, atq; ita quæsitum est $5+\surd 8$.

Eodem modo radix quadratica extracta ex $26-\surd 80-\surd\overline{640-64\surd 80}$, invenitur $4-\surd\overline{10-\surd 80}$; ut patet ex operatione subjecta:

$$\begin{array}{lrrr}
 & 26-\surd 80-\surd\overline{640-64\surd 80} & & \\
 & 26-\surd 80 & & \\
\cline{2-2}
 & +676-26\surd 80 & 26-\surd 80 & 26-\surd 80 \\
 & +80-26\surd 80 & 6+\surd 80 & 6+\surd 80 \\
\cline{2-4}
 & 756-52\surd 80 & 32 & 20-2\surd 80 \\
 & 640-64\surd 80 & 16 & 10-\surd 80 \\
\cline{2-2}
\text{restat} & 116+12\surd 80 & & \\
\text{ejus radix} & 6+\surd 80 & \text{prodit} & 4-\surd\overline{10-\surd 80}.
\end{array}$$

Binomiums, den worteluyt de helft der somme is het eerste deel, de vorige helft substraheert wijders van het grootste deel, dan is den wortel, uyt de rest, het tweede deel des Binomium'. The rule is of great antiquity: as a geometrical proposition its cases are developed in detail by Euclid in his *Elements*, x, 90–6, while its algebraic equivalent was known to a wide group of sixteenth-century mathematicians, including Michael Stifel (*Arithmetica Integra*, Nuremberg, 1544) and Pierre de la Ramée (*Arithmeticæ Libri Tres* (Paris, 1555): compare Snell's *Fundamenta* (note (15)): 20).

||[34] ||Radix quadratica ex $mm+\frac{pxx}{m}+x\sqrt{4pm}$ est $m+x\sqrt{\frac{p}{m}}$, quod efficitur modo sequenti:

$$
\begin{array}{llll}
mm+\frac{pxx}{m}+x\sqrt{4pm} & & & \\
mm+\frac{pxx}{m} \quad x\sqrt{4pm} & & & \\
\hline
m^4+pmxx \quad 4pmxx & & & \\
+pmxx+\frac{ppx^4}{mm} & & mm+\frac{pxx}{m} & mm+\frac{pxx}{m} \\
\hline
m^4+2pmxx+\frac{ppx^4}{mm} & & mm-\frac{pxx}{m} & mm-\frac{pxx}{m} \\
4pmxx & & 2mm & +2\frac{pxx}{m} \\
\hline
\text{restat } m^4-2pmxx+\frac{ppx^4}{mm} & & mm & \frac{pxx}{m} \\
\text{ejus radix } mm-\frac{pxx}{m} & & \text{prodit } m+x\sqrt{\frac{p}{m}}. &
\end{array}
$$

Sic quoq; radix quadratica extracta ex $\overline{a+b}\sqrt{ab}+2ab$ invenitur $\sqrt{\sqrt{a^3b}}+\sqrt{\sqrt{ab^3}}$. Operatio subjicitur:

$$
\begin{array}{lll}
\overline{a+b}\sqrt{ab} \quad +2ab & & \\
\overline{a+b}\sqrt{ab} \quad 2ab & & \\
\hline
aa+2ab+bb \quad 4aabb & & \\
ab & \overline{a+b}\sqrt{ab} & \overline{a+b}\sqrt{ab} \\
\hline
\text{prodit } \quad a^3b+2aabb+ab^3 & \overline{a-b}\sqrt{ab} & \overline{a-b}\sqrt{ab} \\
\text{aufer } \quad 4aabb & 2a\sqrt{ab} & 2b\sqrt{ab} \\
\hline
\text{restat } \quad a^3b-2aabb+ab^3 & a\sqrt{ab} & b\sqrt{ab} \\
\text{ejus radix } \quad \overline{a-b}\sqrt{ab} & &
\end{array}
$$

prodit quæsitum $\sqrt{a\sqrt{ab}}+\sqrt{b\sqrt{ab}}$ vel $\sqrt{\sqrt{a^3b}}+\sqrt{\sqrt{ab^3}}$.

||[35] ||Sin occurrat binomium, cujus radix extrahi nequeat, veluti $\frac{1}{2}b+\sqrt{\frac{1}{4}bb-aa}$, ponimus $\sqrt{\frac{1}{2}b+\sqrt{\frac{1}{4}bb-aa}}$, atq; ita in alijs.

---

(38) The binomial is, of course, understood to be surd, and the generalization is easily justified: if $(\alpha+\sqrt{\beta})^n = A+B\sqrt{\beta}$, then $(\alpha-\sqrt{\beta})^n = A-B\sqrt{\beta}$, and hence

$$(\alpha^2-\beta)^n = A^2-(B\sqrt{\beta})^2.$$

(39) The rule is not well stated by Kinckhuysen nor is its validity here apparent. In its correct, generalized form we may state it in the following way: If $x+\sqrt{y} = \sqrt[n]{\{\alpha+\sqrt{\beta}\}}$, where x, y, α and β are integers (so that $x^2-y = \sqrt[n]{\{\alpha^2-\beta\}} = \gamma$, say) and if we determine the integer $2m$ such that $m > \sqrt[n]{\{\alpha+\sqrt{\beta}\}} > m-\frac{1}{2}$, then $x = [\frac{1}{2}(m+\gamma/m)]$. (When γ is not rational we may replace it by $\gamma^n = \alpha^2-\beta$ on multiplying throughout the first equation by $\gamma^{\frac{1}{2}(n-1)}$, for then $(x+\sqrt{y})\gamma^{(\frac{1}{2}n-1)} = \sqrt[n]{\{(\alpha+\sqrt{\beta})(\alpha^2-\beta)^{\frac{1}{2}(n-1)}\}}$ in which $\sqrt[n]{\{(\alpha^2-\beta)(\alpha^2-\beta)^{n-1}\}} = \alpha^2-\beta$.) The

Extrahere radicem cubicam è binomijs. Binomium quodcunque multiplicatum in seipsum cubicè, semper producit aliud Binomium, et differentia quadratorū à nominibus hujus posterioris, semper æqualis est cubo differentiæ quadratorū à nominibus prioris binomij, sive radicis.

Veluti si detur $a+\surd ab$, vel $a-\surd ab$, et multiplicetur in seipsum cubicè, prodit $a^3+3aab+\overline{3aa+ab}\surd ab$, vel $a^3+3aab-\overline{3aa+ab}\surd ab$, quadrata nominum ablata ab invicem relinquunt $a^6-3a^5b+3a^4bb-a^3b^3$, qui est cubus ipsius $aa-ab$, vel $-a^6+3a^5b-3a^4bb+a^3b^3$, qui est cubus ipsius $-aa+ab$, nimirū differentiæ quadratorum ab a et $\surd ab$, sive aa majus sit, sive contrà ab. Idem deprehenditur eodem modo se habere in radicibus altioribus.(38)

Liquet igitur, dato binomio, cujus radix cubica extrahi potest, si quadrata nominum radicis auferantur ab invicem, hoc residuum semper esse cubum differentiæ quadratorum à nominibus radicis.

Quare ex quovis binomio extractâ radice cubicâ prope verum, in numeris rationalibus innotescit summa (vel differentia) nominum radicis, et ex præcedentibus ||nota est differentia quadratorum à nominibus radicis, unde ipsa nomina radicis, utrumque seorsim, inveniri potest.

[36]

Nam duorum quorumvis numerorū, differentiâ quadratorum divisâ per summam ipsorū numerorū, id quod inde oritur (si addatur sumæ numerorū, prodit duplum numeri majoris) sin auferatur à summa amborum, restat duplum numeri minoris.(39)

Veluti si sint a et b, differentia quadratorum est $aa-bb$, quæ si dividatur per summam amborum, videlicet $a+b$, prodit $a-b$, cui si addatur $a+b$, prodit $2a$, duplum majoris; idemque $a-b$ si auferatur ab $a+b$, restat $2b$, duplum minoris.

Quod si igitur radix cubica extrahenda veniat ex $20+\surd 392$, primùm extrahitur radix cubica in numeris rationalibus modo sequenti; radix quadratica

rule, first published in Jacob van Waessenaer's *Aenmerckingen* in 1639 (see next note), is in fact Descartes'. The latter, indeed, communicated his 'Bewijs van onsen Regel' for the cubic case in his letter to Waessenaer on 1 February 1640 (*Œuvres de Descartes*, **3** (1899): 30–2) and his attempted demonstration there is essentially that printed as the concluding portion of Schooten's *Additamentum* to his *Geometria* ($_1$1649: 295–336, especially 329–36 = $_2$1659: 369–400, especially 394–400). In that 'Demonstratio' he correctly seeks to derive the inequality

$$0 \leqslant (\tfrac{1}{2}(m+\gamma/m)-x) \leqslant \tfrac{1}{2},$$

in which the general rule is patently contained, but is deceived into thinking it enough merely to justify the lower bound. The inequality is not, however, difficult to establish. If for some integer k we find $(k-1)^2+1 \leqslant y \leqslant k(k-1)$, then $k-1 < \surd y < k-\frac{1}{2}$ and $m = x+k-\frac{1}{2}$, so that $1/(4(x+k)-2) \leqslant (m+\gamma/m-2x) \leqslant (4k-7)/(4(x+k)-2)$: alternatively, if

$$k(k-1)+1 \leqslant y \leqslant k^2, \quad \text{then} \quad k-\tfrac{1}{2} < \surd y < k \quad \text{and} \quad m = x+k,$$

so that $0 \leqslant (m+\gamma/m-2x) \leqslant (k-1)/(x+k)$. In either case, accordingly,

$$0 \leqslant (m+\gamma/m-2x) < 1.$$

ex $\surd 392$ est paulò minor quàm 20, quæ addita ad 20, faciunt 40, cujus radix cubica est $3\frac{1}{2}$ paulò major verâ, à qua tamen haud differt semisse.(40)

$$\begin{array}{c} 20+\surd 392 \\ 20 \\ \hline 400 \\ 392 \\ \hline \surd C\, 8 \\ \hline 2. \end{array}$$

Prodit differentia quadratorum à nominibus radicis 2, datur igitur jam amborũ summa, nimirum $3\frac{1}{2}$, et differentia quadratorum ab ijsdem, videlicet 2; quare per præce=||dentia, si dividatur hæc differentia quadratorum 2, per amborum numerorum summam $3\frac{1}{2}$, emergit $\frac{4}{7}$, quo addito ad $3\frac{1}{2}$, prodit $4\frac{1}{14}$, duplum majoris, cujus semis est 2, nempe nomen rationale, à cujus quadrato si auferas differentiam quadratorum, nimirũ 2, restat 2 pro numero radicali; quare quæsitum extat $2+\surd 2$, quod si multiplicetur cubicè, restituitur $20+\surd 392$, unde liquet, radicem modò inventam, esse veram. ||[37]

Sin rationale nomen minus sit radicali, quærendum est duplum minoris, veluti si detur

$$\begin{array}{cc} 44+ & \surd 1944 \\ 44 & \surd 1944 \\ \hline 1936 & 1944 \\ & 1936 \\ \hline & 8 \end{array}$$

[cujus] radix cubica est, differentia quadratorum à nominibus radicis. Radix cubica ex $44+\surd 1944$ in numeris rationalibus incidit inter 4 et $4\frac{1}{2}$. Jam verò differentia quadratorum 2 si dividatur per summam amborũ $4\frac{1}{2}$, prodit $\frac{4}{9}$, quod ablatum à summa amborum $4\frac{1}{2}$ relinquit $4\frac{1}{18}$, duplum minoris, eritq; nomen rationale 2, cujus quadrato si addas differentiam

(40) This essential restriction is sprung at the reader without any justification, and it would seem that Kinckhuysen did not understand its necessity.

(41) The designedly harsh alliteration of Jacob van Waessenaer's *Den on-wissen Wis-Konstenaer [I. I. Stampioenius ontedekt. Door sijne ongegronde Weddinge ende mis-luckte Solutien van sijne eygene Questien. Midtsgaders eenen generalen Regel om de Cubic-wortelen ende alle andere to trecken uyt twee-namighe ghetallen: dewelcke voor desen niet bekent en is geweest....* (Leyden, 1640): 88 pp., 4°] is lost in Mercator's translation. We might perhaps render it as 'The unmathematical Mathematician [J. J. Stampioen revealed. By his imprudent Match and mis-begotten Solutions to his own Questions. Together with a general Rule for Cubic-roots and for treating all other binomial expressions: which has never before been known.... ']. Waessenaer's work was the last of a series of pamphlets which appeared during his altercation with Jan Stampioen de Jonge between 1638 and 1640. Stampioen, newly appointed mathematics professor at The Hague, had following the custom of his day attempted to reinforce his position by publishing in broadsheet several mathematical problems set as a challenge to the populace but intendedly

quadratorum, prodit 6 quadratum majoris, quare quæsitum invenitur $2+\sqrt{6}$, quod si rursus cubicè in seipsũ multiplicetur, restituitur $44+\sqrt{1944}$, unde sequitur, radicem sic inventam, esse veram. Atq; ita in alijs.

Nota; cùm constat numerum quæsitum esse integrum, licitum est quærere [38] fractum, qui paulò major sit quæsito; ||sed oportet differentiam minorem esse unitate, unde sequi[tur] majus integrum, contentum in fracto, esse numerũ quæsitum.

Porrò, fieri potest, ut in radice cubica alicujus binomij, existant fracti, quod ubi veremur, possumus radicem cubicam binomij dati accuratiùs paulò investigare in numeris rationalibus, appositis aliquot cyphris, veluti si detur $18+\sqrt{325}$, hujus radix cubica est $1\frac{1}{2}+\sqrt{3\frac{1}{4}}$.

Qui plura hîc desiderat, inspiciat libellum, cui titulus est, *Incertus Mathematicus*, &c. authore *Jacobo à Waessenaer.*(41)

---

[39] ## ||Pars Secunda. De Æquationibus.(42)

Quo pacto numeri et species computandæ sint, jam satis ostensum est; itaq; pergo nunc ad æquationes. Æquatio dicitur, quando eadem quantitas duobus modis exprimitur, vel cùm ignota quantitas reperitur æqualis quibusdam notis, vel mixtim notis atq; ignotis simul.

Hæ æquationes dicuntur tot dimensionum, quot sunt dimensiones in ignota quantitate, quæ plurimis dimensionibus constat; veluti cùm datur $z=b$, hoc est z æquale ipsi b (nam signum $=$ brevitatis causâ ponitur pro æquali)(43)

---

so difficult that only Stampioen himself would be able to resolve them. One of these broadsheets, the *Problema astronomicum et geometricum voorghestellt door Iohan Stampioen d'Ionge, Mathematicus* (to whose content we shall return in our fifth volume), was resolved by Waessenaer, a young Utrecht 'Landmeeter' (surveyor) and friend of Descartes, soon after it appeared in late 1638. His solution, by Cartesian analytical methods, was superior to Stampioen's trigonometrical approach (first published by him in his *Algebra ofte Nieuwe Stel-Regel*... (The Hague, 1639)) and more general than it: the latter, fearing for his reputation, tried in consequence to decry it in a series of pamphlets over the next year and a half. Descartes took Waessenaer's side in the dispute, but anonymously: jointly they swamped Stampioen with their *Aenmerckingen op den Nieuwen Stel-Regel*... (Leyden, 1639) and finally crushed him with the present work. (See Paul Tannery's full documentation in *Œuvres de Descartes*, **2**: 581–2, 611–13; and **3**: 30–2: compare also Charles Adam's *Vie & Œuvres de Descartes = Œuvres de Descartes, Supplément* (Paris, 1910): 272–7.)

(42) This main portion of the *Stel-konst*, which takes up this and its following 56 pages, is heavily indebted in structure, technical detail and nomenclature to the third book of Descartes' *Geometrie* (Leyden, 1637) (I, **1**, Introduction, Appendix 2: note (1)). In his translation, we may note, Mercator on the whole faithfully preserves this Cartesian atmosphere, but universally renders Kinckhuysen's (and Descartes') sign '$\propto$' for equality by the standard contemporary English equivalent '='.

(43) Compare the preceding note.

vocatur ea æquatio unius dimensionis; dato $zz = -az + b$, duarum dimensionum; dato $z^3 = +azz + bbz - c^3$, trium, atq; ita deinceps.[44]

Tractationem harum æquationum dividimus in quatuor capita: quorum primum aget de transformatione æquationum, non mutatis radicibus. Secundum, de natura æquationum, respectu radicum. Tertium, de transformatione æquationum, ubi radices simul mutantur. Quartum, quomodo æquatio aliqua solvenda sit.[45]

||[40] ||*Caput primum. De transformatione æquationum, non mutatis radicibus.*

Dum circa quæstionem aliquam ratiocinamur, sæpe incidimus in æquationem, quæ in simpliciorem vel commodiorem transformari potest, vel utrinq; æqualia addendo vel detrahendo, vel utrumq; æqualium per eundem numerum multiplicando vel dividendo, quibus medijs[46] æquatio transformatur quidem, at quantitates nihilominus manent æquales, quemadmodum contingit in bilance æquilibrata, ubi additis vel demtis utrinq; æqualibus, vel ponderibus utrinq; æquè multiplicatis vel divisis, ipsa nihilominus perstat in æquilibrio. Idem intelligendum de radicibus ejusdem speciei utrinq; extractis. Quæ cuncta fieri possunt, nihil immutato valore radicis, quemadmodum in sequentibus ostendemus.

Quomodo superflua æquationis eliminari possint, formaq; ejus mutari, addendo vel subtrahendo.

Veluti si detur $3x - 30 = 50$, et addatur utrinq; $+30$, prodit $3x = 80$. Dato ||[41] $x - 5 = 0$, si utrobiq; addantur 5, fit $x = 5$. ||Sic quoq; dato $dx - ad = bc - cx$, addito utrinq; $cx + ad$, prodit $cx + dx = bc + ad$; atq; ita in similibus.

Verùm dato $y + a = b$, ablato utrinq; $+a$, prodit $y = b - a$. Dato

$$xx + 6x + 25 = 3x + 75,$$

ablato utrinq; $3x + 75$, prodit $xx + 3x - 50 = 0$. Item dato $az - ab = -bz - cc$, et ablatâ æquatione utrinq; ex 0, prodit $-az + ab = +bz + cc$, atq; ita in similibus.

Quomodo fractiones alicujus æquationis mutentur in integra, et radicales in rationales, multiplicando.

Dato $x + \frac{3}{x} = 10$, multiplicantur omnia per x, prodit $xx + 3 = 10x$. Dato $\frac{ab + bb}{a + z} = c$, multiplicantur omnia per $a + z$, prodit $ab + bb = ac + cz$. Dato

---

(44) See Descartes' *Geometrie* (note (42)): 301.

(45) These four chapters occupy pp. 40–3, 43–9, 49–63 and 64–95 respectively of Kinckhuysen's tract.

(46) Mercator's rendering of 'door welcke dinghen' (by which means)!

$\frac{yy}{b}+\frac{by-dd}{c}=a$, multiplica omnia per bc, prodit $cyy+bby-bdd=abc$. Et dato $\frac{cx}{a}=\frac{xx+bb}{b+c}$, multiplica omnia per a et deinde omnia per $b+c$, hoc est, multiplica decussatim, nimirum a per $xx+ab$, prodit $axx+aab$, et cx per $b+c$, prodit $\overline{bc+cc}\,x$, eritq; $axx+aab=\overline{bc+cc}\,x$.

Sed oblato $\frac{cyy}{ab}=\frac{bbx+b^3}{aa}$, possunt denominatores diminui, si dividatur uterq; per a, et deinde multiplicando decussatim, producitur $b^3x+b^4=acyy$. Ita quoq; numeratores diminui possunt, veluti si detur $\frac{ax}{b}=\frac{\overline{b+c}\,x}{y}$; diviso utroq; numera-[42] tore per x, et ||multiplicando decussatim prodit $bb+bc=ay$. Porrò dato $x=\sqrt{aa+bb}$, multiplica utrumq; æqualium in seipsum, prodit $xx=aa+bb$. Sic dato $\sqrt{y}=\sqrt{\frac{1}{2}b+\sqrt{\frac{1}{4}bb+aa}}$, si libeat eliminare radicales, multiplica utrumq; æqualium in seipsum, prodit $y=\frac{1}{2}b+\sqrt{\frac{1}{4}bb+aa}$, tum ablato utrinq; $\frac{1}{2}b$, prodit $y-\frac{1}{2}b=\sqrt{\frac{1}{4}bb+aa}$. Tum rursus ducto utroq; æqualium in sese, prodit $yy-by+\frac{1}{4}bb=\frac{1}{4}bb+aa$, deniq; ablato utrinq; $\frac{1}{4}bb+aa$, restat $yy-by-aa=0$.

Eodem modo si detur $x=3+\sqrt{5}$, aufer utrinq; 3, prodit $x-3=\sqrt{5}$, et ducto utroq; æqualium in se, prodit $xx-6x+9=5$, tum ablato utrinq; 5, restat $xx-6x+4=0$. Nota, hoc modo apparet, quando talis æquatio datur, quâ ratione valor ipsius x recuperari possit, nam rationalis numer[u]s in binomio $3+\sqrt{5}$ est 3, nimirum semis ipsius 6[rij], et numerus radicalis est differentia inter quadratum ejusdem 3[rij] et 4[rium] datum. Atq; ita in alijs.

Quomodo æquatio diminuatur, dividendo, et extrahendo radices.

Veluti si detur $3xx=36x$, divide omnia per $3x$, prodit $x=12$. Dato $cz-bz=ab-bb$, divide omnia per $c-b$, prodit $z=\frac{ab-bb}{c-b}$. Item dato

$$aax-acx+acd-ccd=axx-cxx+a^3-aac,$$

divide utrumq; æqualium per $a-c$, prodit $ax+cd=xx+aa$. Atq; ita in [43] similibus. ||Rursus dato $xx=36$, extrahe utrinq; radicem quadraticam, prodit $x=6$. Dato $aa+2ab+bb=xx$, extrahe utrinq; radicem quadraticam, prodit $a+b=x$. Dato $yy=aa+bb$, extractâ utrinq; radice quadraticâ prodit $y=\sqrt{aa+bb}$. Dato $zz=aa+bc+2a\sqrt{bc}$, extrahe utrinq; radicē quadraticam, prodit $z=a+\sqrt{bc}$. Dato $x^4-10x^3+25x[x]=11+\sqrt{2}$, extractâ utrinq; radice quadraticâ, prodit $xx-5x=\sqrt{11+\sqrt{2}}$, atq; ita in alijs.

Idem intelligendum [est] de æquationibus, ex quibus radix cubica, vel etiam altior, extrahi potest.

Caput Secundum. De natura æquationum, respectu radicum.(47)

Cùm exhibetur æquatio aliqua, plura statim ex ipsa forma notari possunt, quæ ad solutionem non parùm conferant, sed conducible est, æquationem priùs transformari suprà memoratis modis, in ejusmodi aliquam, in qua omnes quantitates simul æquales sunt nihilo, siquidem naturam æquationum isto modo describemus.

||[44] ||Quot radices quævis æquatio habere possit.(48)

Concipiendum est, unamquamq; æquationem, cujus ignota quantitas aliquot dimensiones habet, productam esse ex multiplicatione aliarum quarundam æquationum, veluti si detur $xx-5x+6=0$, concipimus hanc æquationem prodijsse ex multiplicatione aliarum duarum, nimirum $x-2=0$, ductâ in $x-3=0$, hoc est, ex $x=2$, et $x=3$; adeò ut hæc æquatio $xx-5x+6=0$, habeat duas radices, hoc est, valor ipsius x est 2, atq; etiam 3. Sic $x^3-9xx+26x-24=0$, concipitur producta ex multiplicatione trium aliarū unius dimensionis, quemadmodum utiq; producta est ex multiplicatione ipsius $x-2=0$, et $x-3=0$, et $x-4=0$, ita ut in hac æquatione $x^3-9xx+26x-24=0$ (cujus quantitas ignota x tres dimensiones habet) reperiantur tres radices, nimirum 2, 3, et 4; unde infertur, quamvis æquationem tot radices habere posse, quot dimensiones habet quantitas ignota.

Quænam radices sint falsæ.(49)

Quæcunq; radix minor est nihilo, dicitur falsa, veluti si occurrat $x=-5$, hoc est $x+5=0$, hoc si multiplicetur per $x^3-9xx+26x-24=0$, prodit

$$x^4-4x^3-19xx+106x-120=0,$$

quæ æquatio jam quatuor habet radices, nimirum tres veras, 2, 3, et 4, et unam falsam -5.

||[45] ||Quæ sit summa omnium radicum alicujus operationis.(50)

Consideratis hoc pacto æquationibus, deprehenditur omnes radices collectas in unam summam, tantundem valere, quantum quantitas nota secundi termini; de signis $+$ et $-$ statuitur secundum regulam sequentem: Si primus terminus sit $+$, et secundus $-$; vel primus $-$, et secundus $+$; omnes radices valent tantundem, quantū $+$ nota quantitas secundi termini: sin primus et

(47) Compare Descartes' *Geometrie* (note (42)): 371ff.

(48) *Geometrie*: 372: 'Combien il peut y auoir de racines en chasq; Equatiō'.

(49) *Geometrie*: 372: 'Quelles sont les fausses racines'.

(50) This paragraph is based on Frans van Schooten's amplification (*Commentarii in Librum III*, *C* = *Geometria* ($_2$1659): 284–5) of Descartes' phrase 'Au moyen de quoy on diminue d'autant ses dimensions' (*Geometrie*: 372).

secundus terminus ambo sint +, vel ambo −; omnes radices valent, quantum − nota quantitas secundi termini. Id quod sic ostendi potest. Sint $x-a=0$, $x-b=0$, et $x+c=0$, quibus multiplicatis, prodit hæc æquatio

$$\begin{array}{l} +x^3-axx+abx+abc=0. \\ -b -ac \\ +c -bc \end{array}$$

Ubi apparet, notam quantitatem secundi termini constare ex summa trium radicum $+a$, $+b$, et $-c$, quanquam signa sunt contraria, quia primus terminus est +.

Hinc sequitur, in æquationibus duarum dimensionum, datâ unâ radice, dari et alteram, atq; in æquationibus trium dimensionum datis duabus radicibus, dari simul tertiam, atq; ita deinceps.

[46] ||Quodnam sit productum omnium radicum alicujus æquationis in se ductarum.

Ductis omnibus radicibus æquationis in invicem continuè, productum æquale est ultimo termino noto; nam in æquatione, cujus tres radices erant a, b, et c, ultimus terminus est abc.

Quomodo dimensiones æquationis diminui possint, quando una radix nota est.[51]

Ex omnibus hactenus dictis sequitur, datâ æquatione, cujus una radix nota sit, posse eam propter hoc unâ dimensione minui. Nam nihil amplius faciendum [est], quàm ut dividamus æquationem per ignotam quantitatem minutam radice notâ, si radix sit verâ; sin falsa sit, per ignotam quantitatem auctam radice notâ.

Quomodo investigari possit, num data quantitas sit una ex radicibus.[52]

Oblatâ æquatione, quæ per ignotam quantitatem +vel − quantitate datâ, dividi potest absq; residuo, sequitur istam quantitatem datam esse unam ex radicibus, Sin à divisione aliquid restet, argumento est, datam quantitatem nequaquam esse radicem; veluti hæc æquatio

$$x^4-4x^3-19xx+106x-120=0:$$

[47] ||Dividi quidem potest per $x-2$, per $x-3$, per $x-4$, et per $x+5$, at non per $x+$vel$-$ulla alia quantitate, quin aliquid supersit, id quod arguit, eam non posse habere alias radices præter hasce quatuor 2, 3, 4, et -5.

(51) *Geometrie*: 372: 'Coment on peut diminuer le nombre des dimensions d'vne Equation lorsqu'on connoist quelqu'vne des ses racines'.

(52) *Geometrie*: 372–3: 'Coment on peut examiner si quelque quantité donnée est la valeur d'vne racine'.

Quot veras radices quævis æquatio habere potest.(53)

Quoties in aliqua æquatione signum − sequitur signū +, vel signum + sequitur signum −, tot veras radices ipsa æquatio habere potest: et tot falsas, quoties duo signa +, vel duo signa − succenturiantur; veluti in hac æquatione $x^4-4x^3-19xx+106x-120=0$, ubi post $+x^4$ sequitur $-4x^3$, et post $-19xx$ habetur $+106x$, et post $106x$ extat -120, constat inde tres esse radices veras, et unicam falsam, quia semel tantùm duo signa similia se excipiunt, videlicet $-4x^3$ et $-19xx$.

Si qua æquatio objiciatur, in qua unus terminus desit, inseritur loco vacuo $+0$, si primus terminus sit +, aliàs −; veluti si detur $y^3*+py-q=0$, licebit scribere $y^3+0yy+py-q=0$, et deprehendetur una radix vera et duæ falsæ.

Verùm hæc omnia sic accipienda sunt, quando in æquationibus, ex quarum multiplicatione data æquatio producta est, nullus terminus defuit: nam si multiplicetur $y^3*+py-q=0$ per $y-b=0$, producitur

$$y^4-by^3+\begin{matrix}pyy\\ \end{matrix}-\begin{matrix}qy\\ -bp\end{matrix}+bq=0,$$

ubi non valet regula suprà tradita.(54)

||[48] ||Quo pacto veræ radices æquationis mutentur falsis, et vicissim.(55)

Perfacile est, in æquatione eâdem omnes falsas radices efficere veras, et contrà, nihil aliud agendo, quàm omnia signa + et − mutando, quæ reperiuntur loco secundo, quarto, sexto, deniq; quovis loco pari, non mutatis interim signis loco primo, tertio, quinto, omnive loco impari occurrentibus; sic pro

$$+x^4-4x^3-19xx+106x-120=0$$

scribitur

$$+x^4+4x^3-19xx-106x-120=0$$

et habetur æquatio, in qua una est radix vera, nimirum $+5$, et tres falsæ, nimirum -2, -3, et -4.

(53) *Geometrie*: 373: 'Combien il peut y auoir de vrayes racines en chasque Equatiō'. The 'peut y auoir' (Kinckhuysen's 'hebben kan') expresses, of course, an upper limit only to the number of real roots.

(54) As Newton duly noted in his *Observationes* (§2: note (87) below) Kinckhuysen's exception is invalid. From his remarks in the two preceding paragraphs (where, for example, Mercator's 'esse'—his 'zijn'—should more accurately be 'esse posse') it is clear that he has fallen into Roberval's mistake of supposing that the sign rule determines the number of real roots absolutely and not merely an upper bound (which may not be achieved) to that number. (See Carcavi to Descartes, 9 July 1649 in *Œuvres de Descartes*, **5** (1903): 369–74, especially 374: Observation 2. In his reply to Carcavi on 17 August 1649 (*Œuvres*, **5**: 391–401, especially 397) Descartes was firm about what he intended at this point in his *Geometrie*: '...ie n'ay pas dit, dans la page 373, ce que [M. Roberval] veut que i'aye dit, à sçauoir qu'il y a autant de vrayes racines que les signes + & − se trouuent de fois estre changez, ny n'ay eu aucune

Ita nullum est discrimen inter has duas æquationes:

$$z^3 = * + pz + q$$
$$z^3 = * + pz - q$$

nisi quód radices, quæ in hac sunt veræ, in alterâ falsæ sunt, ita ut hæ et similes eodem modo solvi possint.

Radices tam veras quàm falsas, posse vel reales, vel imaginarias.(56)

Radices tam veræ, quàm falsæ non semper reales sunt, sed quandoq; solùm [49] imaginariæ, hoc est, posse ‖semper in quavis æquatione tot radices imaginatione concipi, quot modò memoravimus, sed fieri quandoq;, ut nulla sit quantitas, quæ consentiat cum conceptis: sic, licèt imaginemur, hanc æquationem

$$x^3 - 6xx + 13x - 10 = 0,$$

habere tres radices, non habet tamen nisi unam realem, quæ est 2; at reliquæ duæ, quicquid moliamur, non nisi imaginariæ manent.

Caput Tertium. De transformatione æquationum, ubi radices simul mutantur.

Transformari possunt æquationes ita, ut radices, licèt ignorentur, datâ quantitate augeantur vel diminuantur. Possunt quoq; transformationis ope, radices ignotæ per datum numerum multiplicari vel dividi, quæ transformationes jam tradentur, et habent multiplicem usum.

Quomodo radices alicujus æquationis, licèt ignotæ sint, augeri vel diminui possint.(57)

Quando lubet radices alicujus æquationis quamvis ignotas, datâ quantitate augere vel minuere; tum loco signi, quo quantitatem ignotam notaveramus, [50] ‖aliud signum substituimus, quod tanto majus vel minus sit, quanto radices auctas vel minutas cupimus, atq; hoc posterius signū ubiq; usurpamus loco prioris.

Veluti si radices hujus æquationis $x^4 + 4x^3 - 19xx - 106x - 120 = 0$, libeat augere 3rio, ponendum est y loco ipsius x, et cogitandum, hanc quantitatem y

---

intention de le dire. I'ay dit seulement qu'il y en peut autant auoir; & i'ay monstré expressément, dans la page 380, quand c'est qu'il n'y en a pas tant, à sçauoir quand quelques-vnes de ces vrayes racines sont imaginaires'.)

(55) *Geometrie*: 373–4: 'Com̄ent on fait que les fausses racines d'vne Equation deuienēt vrayes, & les vrayes fausses'.

(56) *Geometrie*: 380: 'Que les racines, tant vrayes que fausses peuuent estre reelles ou imaginaires'. Compare note (54).

(57) *Geometrie*: 374–5: 'Com̄ent on peut augmenter ou diminuer les racines d'vne Equation sans les connoistre'.

3rio majorem esse quantitate x, adeò ut $y-3$ æquale sit ipsi x, et loco xx ponendum est quadratum ipsius $y-3$, nimirum $yy-6y-9$, et loco x^3 ponendus est cubus istius, nimirum $y^3-9yy+27y-27$, deniq; loco x^4 ponitur

$$y^4-12y^3+54yy-108y+81,$$

quemadmodum sequitur:

$$\begin{array}{lrl} \text{scribe} & y^4-12y^3+54yy-108y+81 & \text{loco ipsius } x^4. \\ & +\ 4y^3-36yy+108y-108 & \text{----} +4x^3. \\ & -19yy+114y-171 & \text{----} -19xx. \\ & -106y+318 & \text{----} -106x. \\ & -120 & \text{----} -120. \\ \hline & y^4-\ 8y^3-\ 1yy+\ \ 8y\ \ \ * & =0. \end{array}$$

vel $y^3-8yy-y+8=0$; radix vera, quæ antea erat 5, nunc est 8, quia ternario aucta est: hîc deprehenditur 3rium esse unam ex falsis radicibus, quòd per eum æquatio unâ dimensione minuatur.

Contrà si radices ejusdem æquationis 3rio minuendæ sint, statuendum est $y+3=x$, et $yy+6y+9=xx$, et sic deinceps, adeò ut loco ipsius

||[51]

$$\|x^4+4x^3-19xx-106x-120=0$$

$$\begin{array}{lr} \text{ponatur} & y^4+12y^3+54yy+108y+81 \\ & +\ 4y^3+36yy+108y+108 \\ & -19yy-114y-171 \\ & -106y-318 \\ & -120 \\ \hline & y^4+16y^3+171yy-\ \ \ 4y-420=0. \end{array}$$

Auctis radicibus veris, diminui falsas, et contrà.(58)

Consentaneum est, quando veræ radices certa quantitate augentur, falsas tantundem diminui, ideò quia sunt $-$, et si tam veræ quàm falsæ radices minuantur quantitate aliquâ, quæ ijs sit æquales, abire easdem in nimium, et si quantitas eas excedat, tum veras evadere falsas et vicissim; veluti hîc ubi vera radix erat 5, augebaturq; 3rio, oportet ut quælibet falsarū radicum 3rio minuantur, ita ut quæ fuerat -4, nunc sit -1, et quæ erat -3, nunc [sit] nihil, quæq; erat -2 facta [sit] vera, valetq; $+1$, nam $-2+3$ facit $+1$. Quare in hac æquatione $y^3-8yy-y+8=0$ non nisi tres sunt radices, quarum duæ sunt veræ, putà 1, et 8, et una est falsa, -1.

Hinc apparet, quomodo omnes radices alicujus æquationis veræ fieri possint, hoc solum agendo, ut æquationem augeamus quantitate quæ maximam falsam radicem excedat.

(58) *Geometrie*: 375–6: 'Qu'en augmentant les vrayes racines on diminue les fausses, & au contraire'.

[52] ||Quomodo secundus terminus æquationis tolli possit.(59)

Suprà numeratum est, in quavis æquatione, tot esse radices, quot ignota quantitas habet dimensiones, item summam omnium radicum æqualem esse notæ quantitati secundi termini, licèt cum signo contrario + vel −, si primus terminus sit +; unde sequitur, si libeat secundum terminū æquationis redigere ad nihilum, tum notam quantitatem secundi termini dividendam esse per numerum radicum, atq; eo quod oritur, radices minuendæ sunt, si primus terminus habeat signum +, et secundus −; vel eâdem quantitate augendæ sunt, si ambo habeant signum +, vel ambo signum −. Ita si tollendus sit secundus terminus præcedentis æquationis, quæ erat

$$y^4+16y^3+71yy-4y-420=0,$$

dividimus 16 per 4, quia y^4 habet quatuor dimensiones, prodit 4, ergo pono $z-4=y$, et scribo

$$\begin{array}{rrrrrcr} z^4 & -16z^3 & +\ 96zz & -256z & +256 & \text{pro} & +y^4 \\ & +16z^3 & +192zz & +768z & -1024 & \text{---} & +16y^3 \\ & & +\ 71zz & -568z & +1136 & \text{---} & +71yy \\ & & & -\ \ 4z & +\ \ 16 & \text{---} & -4y \\ & & & & -\ 420 & \text{---} & -420 \\ \hline z^4 & * & -\ 25zz & -\ 60z & -\ \ 36=0. & & \end{array}$$

[53] Vera radix, quæ erat 2, nunc est 6, siquidem aucta ||est 4rio, et falsæ, quæ erant −5, −6, et −7, nunc sunt tantum −1, −2, et −3.

Ita quoq;, si secundus terminus tollendus sit in hac æquatione

$$\begin{array}{l} x^4-2ax^3+2aaxx-2a^3x+a^4=0, \\ \qquad\qquad\qquad\ -cc \end{array}$$

quoniam $2a$ divisa per 4 exhibent $\frac{1}{2}a$, ponitur $z+\frac{1}{2}a=x$, et scribitur

$$\begin{array}{rrrrr} z^4 & +2az^3 & +\frac{3}{2}aazz & +\frac{1}{2}a^3z & +\frac{1}{16}a^4 \\ & -2az^3 & -3aazz & -\frac{1}{2}a^3z & -\frac{1}{4}a^4 \\ & & +2aazz & +2a^3z & +\frac{1}{2}a^4 \\ & & -\ \ cczz & -\ accz & -\frac{1}{4}aacc \\ & & & -2a^3z & -a^4 \\ & & & & +a^4 \\ \hline z^4 & * & +\frac{1}{2}aazz & -a^3z & +\frac{5}{16}a^4=0. \\ & & -\ cc & -acc & -\frac{1}{4}aacc \end{array}$$

Deinde invento valore ipsius z, addendum est $\frac{1}{2}a$, et prodit valor ipsius x.

---

(59) *Geometrie*: 376: 'Coment on peut oster le second terme d'vne Equation'.

Quomodo hac ratione æquationes duarum dimensionum solvantur.(60)

Oblatâ æquatione duarum dimensionum, potest ea solvi hâc ratione; veluti si detur $xx+px+q=0$, hæc æquatio habet duas radices, quæ simul faciunt $-p$, quare tollendi secundi termini gratiâ pone $y-\frac{1}{2}p=x$ et $yy-py+\frac{1}{4}pp=xx$, ut ||[54] sequitur:

$$\begin{array}{lrcr} \|\text{pone} & yy-py+\frac{1}{4}pp & \text{pro} & +xx \\ & +py-\frac{1}{2}pp & \text{---} & +px \\ & +\ q & \text{---} & +q \\ \hline \text{Prodit} & yy\ *\ -\frac{1}{4}pp+q=0. & & \end{array}$$

ergo $yy=+\frac{1}{4}pp-q$ et $y=\sqrt{\frac{1}{4}pp-q}$ et $x, =y-\frac{1}{2}p, =-\frac{1}{2}p+\sqrt{\frac{1}{4}pp-q}$; jam quia ambæ radices simul faciunt $-p$, oportet ut altera radix valeat $-\frac{1}{2}p-\sqrt{\frac{1}{4}pp-q}$. Dato igitur $xx+4x-96=0$, erit $p=4$, et $q=-96$, et prodit valor radicum quæsitarum $-2+\sqrt{4+96}$, hoc est, $-2+\sqrt{100}$ vel $+8$ [et] $-2-\sqrt{4+96}$, hoc est, $-2-\sqrt{100}$ vel -12. Unde sequitur

Regula pro æquationibus quadraticis

Datis $+xx.\ px.\ q.\ =0$, pone pro radicibus quæsitis $\frac{1}{2}p+\sqrt{\frac{1}{4}pp\cdot q}$.
$\frac{1}{2}p-\sqrt{\frac{1}{4}pp\cdot q}$.

Signa + et −, quæ non apposui, scribe ut sequitur:

Si habetur $-p$, statue in radicibus $+\frac{1}{2}p$, aliàs $-\frac{1}{2}p$.
Si habetur $-q$, statue in radicibus $+q$, aliàs $-q$.

Exempli gratiâ, detur $xx-38x+336=0$, ubi p valet 38, et q 336. Sunt igitur radices $+19+\sqrt{361-336}$. hoc est $+19+\sqrt{25}$. [vel] 24.
$+19-\sqrt{361-336}$. hoc est $+19-\sqrt{25}$. [vel] 14.

||[55] ||Nam $\frac{1}{2}p$ valet 19, atq; ejus quadratum, hoc est $\frac{1}{4}pp$, valet 361; atq; ita in omnibus.

Quomodo hac ratione omnes æquationes cubicæ reducantur ad tres casus.(61)

Cùm tollitur secundus terminus æquationis cubicæ, reducitur ipsa ad unum ex his tribus casibus

$$y^3=*-py+q.$$
$$y^3=*+py+q.$$
$$y^3=*+py-q.$$

(60) Compare Schooten's *Commentarii in Librum III*, *G* = *Geometria* ($_2$1659): 288–93, especially 290–2.

(61) See *Geometrie*: 397–8; and Schooten's *Commentarii in Librum III*, *G* = *Geometria* ($_2$1659): 292–3. Compare also the latter's *Appendix de Cvbicarvm Æqvationvm Resolutione* [Leyden, 1646 =] *Geometria* ($_2$1659): 345–68.

Veluti si detur $x^3+axx+bx+c=0$. Hæc æquatio potest habere tres radices, quæ addita faciunt $-a$, quare auctâ unaquæq; radicum $\frac{1}{3}a$, summa trium radicum evadit 0. Quare pone $y-\frac{1}{3}a=x$, et $yy-\frac{2}{3}ay+\frac{1}{9}aa=xx$, et

$$y^3-ayy+\frac{1}{3}aay-\frac{1}{27}a^3=x^3,$$

ut sequitur:

$$\begin{array}{llll} y^3 & -ayy & +\frac{1}{3}aay & -\frac{1}{27}a^3 \\ & +ayy & -\frac{2}{3}aay & +\frac{1}{9}a^3 \\ & & +\;by & -\frac{1}{3}ab \\ & & & +\;c \\ \hline y^3 & * & -\frac{1}{3}aay & +\frac{2}{27}a^3=0. \\ & & +\;b & -\frac{1}{3}ab \\ & & & +\;c \end{array}$$

[56] Dato igitur $x^3-4xx-15x+18=0$, erit $a\|=-4$, $b=-15$, et $c=18$, quare substitutis numeris loco specierum, prodit

$$\begin{array}{lll} y^3 * & -\frac{16}{3}y & -\frac{128}{27}=0 \\ & -15y & -20 \\ & & +18 \\ \hline y^3 * & 20\frac{1}{3}y & 6\frac{20}{27}=0, \text{ ct } y+\frac{4}{3}-x. \end{array}$$

Unde invenitur

REGULA, DE TOLLENDO SECUNDO TERMINO ÆQUATIONUM CUBICARUM

Dato $+x^3.\ axx.\ bx.\ c.\ =0$, substitue ejus loco $+y^3 * -\frac{1}{3}aay\ .\ \frac{2}{27}a^3=0.$

$$\begin{array}{cc} b & \frac{1}{3}ab \\ & c \end{array}$$

Signa + et −, quæ isthîc statuta non sunt, pone ut sequitur:

Si habetur $+a$, pone $+\frac{2}{27}a^3$, aliàs −.
Si habetur $+b$, pone $+b$, aliàs −.
Si habetur $+a+b$ vel $-a-b$ pone $-\frac{1}{3}ab$, sed si alterum sit +, alterum −, vel prius − et alterū +, pone $+\frac{1}{3}ab$.
Si habetur $+c$, pone $+c$, aliàs −.
Et si habetur $+a$, erit $x=y-\frac{1}{3}a$, aliàs $x=y+\frac{1}{3}a$.

Exempli gratiâ, si detur $\overset{a.}{x^3-1xx}\overset{b.}{-36x}\overset{c.}{-324}=0$, substituitur ejus loco

$$\begin{array}{lll} y^3 * & -\frac{1}{3}y & -\frac{2}{27} \\ & -36 & -12 \\ & & -324 \\ \hline y^3 * & -36\frac{1}{3}y & -336\frac{2}{27}=0, \text{ et } y+\frac{1}{3}=x. \end{array}$$

[57] ‖Ex hac æquatione cùm innotuerit valor ipsius y, addendus est triens, ut emergat valor ipsius x; et sic in alijs.

Quo pacto Regula inveniatur pro[62] tollendo secundo termino cujusvis æquationis biquadraticæ.

Usu venit quandoq;, ut æquationis biquadraticæ secundus terminus tollendus sit, cui rei Regula invenitur, si in casu quodam, ubi omnes termini signo + affecti sunt, secundus terminus tollatur, unde emergit hæc sequens regula.

Dato $+x^4.\ ax^3.\ bxx.\ cx.\ d. = 0.$ substituitur ejus loco

$$\begin{array}{llll} y^4 * -\frac{3}{8}aayy. & \frac{1}{8}a^3y & -\frac{3}{256}a^4 & =0. \\ \quad b & \frac{1}{2}ab & \frac{1}{16}aab & \\ & c & \frac{1}{4}ac & \\ & & d & \end{array}$$

Signa + et −, quæ isthic omissa sunt, pone ut sequitur:

Dato $-a$, pone $-\frac{1}{8}a^3$. [vel si] $+a$, pone $+\frac{1}{8}a^3$.

Dato $-a-b$, pone $-b-\frac{1}{2}ab-\frac{1}{16}aab$. [vel si] $-a+b$, pone $+b+\frac{1}{2}ab+\frac{1}{16}aab$. [vel si] $+a+b$, pone $+b-\frac{1}{2}ab+\frac{1}{16}aab$. [vel si] $+a-b$, pone $-b+\frac{1}{2}ab-\frac{1}{16}aab$.

||[58] ||Dato $-a-c$, pone $-c-\frac{1}{4}ac$. [vel si] $-a+c$, pone $+c+\frac{1}{4}ac$. [vel si] $+a-c$, pone $-c+\frac{1}{4}ac$. [vel si] $+a+c$, pone $+c-\frac{1}{4}ac$.

Dato $+d$, pone $+d$. [vel si] $-d$, pone $-d$.

Exempli gratiâ, si detur $\overset{a}{x^4+4x^3}\overset{b}{-3xx}\overset{c}{-8x}\overset{d}{+4}=0$, substituitur ejus loco

$$\begin{array}{rrrr} y^4 * & -6yy & +8y & -3 \\ & -3 & +6 & -3 \\ & & -8 & +8 \\ & & & +4 \\ \hline y^4 * & -9yy & +6y & +6=0, \text{ et } y-1=x. \end{array}$$

Atq; ita in omnibus alijs.

Quomodo penultimus terminus tolli possit.

Si cupias penultimum æquationis terminum redigere ad nihilum, primùm tollendus est secundus terminus, sicut monstravimus suprà, deinde procedimus modo sequenti: Detur $x^4 * -qxx+rx-s=0$. Pone ultimum terminum $=xz$, ||[59] eritq; $\frac{s}{x}=z$, et $\frac{s}{z}=x$, $\frac{ss}{zz}=xx$, et $||\frac{s^4}{z^4}=x^4$, tolle x in æquatione, et habebis

(62) This word, added to Mercator's translation by a caret, is in a minute character. Indeed, Newton in his quotation of this section heading (§2: note (93) below) missed it out entirely.

(63) *Geometrie*: 378–9: 'Comment on fait que toutes les places d'vne Equation soient remplies'.

$\frac{s^4}{z^4} * -\frac{qss}{zz}+\frac{rs}{z}-s=0$. Duc omnia in z^4, prodit $s^4 * -qsszz+rsz^3-sz^4=0$, subtrahe totam æquationem à 0, et divide omnia per s, et habebis

$$z^4-rz^3+qszz * -s^3=0,$$

invento valore ipsius z, datur valor ipsius x, nam $z=\frac{s}{x}$, atq; ita in alijs.

Quomodo faciendum ut omnes loci æquationis impleantur.(63)

Oblatâ æquatione, in qua aliqui termini desint, quod nollemus, oportet valorem radicum paulùm augeri, veluti si detur $x^5 * * * * -b=0$, cujus loco optemus habere æquationem sex dimensionum, in qua nullus terminorum desit, substituitur istius loco $x^6 * * * * -bx=0$, tum posito $y-a=x$, habebitur

$$\begin{array}{l} y^6-6ay^5+15aay^4-20a^3y^3+15a^4yy-6a^5y+a^6=0. \\ \qquad\qquad\qquad\qquad\qquad\qquad\qquad\quad -b \quad +ab \end{array}$$

Apparet quantulacunq; ponatur quantitas a, omnes locos æquationis impletum iri.

Quomodo radices multiplicari vel dividi possint, etiamsi ignotæ sint.(64)

Possunt etiam radices æquationis, licèt ignotæ, multiplicari vel dividi, per [60] quantitatem quamvis datam, ||veluti si detur $x^3-axx+bx-c=0$, libeatq; ignotos valores ipsius x multiplicare per f, ponitur $y=fx$, eritq; $\frac{y}{f}=x$, et $\frac{yy}{ff}=xx$, et $\frac{y^3}{f^3}=x^3$, et substituuntur hæ quantitates loco istarum, et habebitur $\frac{y^3}{f^3}-\frac{ayy}{ff}+\frac{by}{f}-c=0$. et quò fractiones tollantur, multiplicantur omnia per f^3, proditq; $y^3-fayy+ffby-f^3c=0$. Unde sequitur, quando radices alicujus æquationis per quantitatem datam multiplicandæ vel dividendæ sunt, oportere notam quantitatem secundi termini per eandem multiplicare vel dividere, et tertium terminum per ejus quadratum, quartum per ejus cubum, atq; ita porrò usq; ad ultimum.

Quomodo fractiones æquationis mutentur in integra, et radicales in rationales, et alia nonnulla quæ usui esse possunt.(65)

Hæc multiplicatio nonnunquam usum habere potest in mutandis fractis et radicalibus, in æquatione occurrentibus, in integros et rationales, veluti si detur

(64) *Geometrie*: 379: 'Commẽt on peut multiplier ou diuiser les racines sans les connoistre'.

(65) *Geometrie*: 379–80: 'Coñent on reduist les nombres rompus d'vne Equation a des entiers'.

$x^3-xx\surd 3+\frac{26}{27}x-\frac{8}{27\surd 3}=0$, cujus loco desideretur alia, cujus omnes termini sint rationales, ponendum est $y=x\surd 3$, eritq; $\frac{y}{\surd 3}=x$, $\frac{yy}{3}=xx$, et $\frac{y^3}{3\surd 3}=x^3$, quæ scribenda sunt loco priorum, et habebimus $\frac{y^3}{3\surd 3}-\frac{yy\surd 3}{3}+\frac{\frac{26}{27}y}{\surd 3}-\frac{8}{27\surd 3}=0$. Tum ||[61] ||multiplicatis omnibus per $3\surd 3$, exit $y^3-3yy+\frac{26}{9}y-\frac{8}{9}=0$, vel aliter, brevitatis causâ, duc notam quantitatem secundi termini in $\surd 3$, tertium terminum in ejusdem quadratum, nempe 3, et ultimum terminum in ejus cubum $3\surd 3$.

Quod si alia etiamnum æquatio desideretur, loco istius in qua notæ quantitates exprimantur numeris integris, ponendum est $z=3y$, hoc est $\frac{z}{3}=y$, et multiplicandum 3 per 3, $\frac{26}{9}$ per 9, et $\frac{8}{9}$ per 27, ut sequitur[:]

$$\begin{array}{l} y^3-3yy+\frac{26}{9}y-\frac{8}{9}=0 \\ 1 \quad 3 \quad 9 \quad 27 \\ \hline \end{array}$$

habebimus $z^3-9zz+26z-24=0$, cujus radices sunt 2, 3, et 4, unde scitur, quoniam hæ triplo majores sunt prioribus, radices prioris æquationis valere $\frac{2}{3}$, 1, et $\frac{4}{3}$, quæ si dividantur per $\surd 3$, emergunt radices primæ æquationis $\frac{2}{9}\surd 3$, $\frac{1}{3}\surd 3$, et $\frac{4}{9}\surd 3$.

Hæc multiplicatio radicum etiam usui esse potest, quando æquatio nullam habet radicem rationalem, quæ appropinquando quærenda sit, veluti si detur $x^3 * -3x+1=0$, primùm multiplicari possunt radices per 1000, quemadmodum hîc videre est.

$$\begin{array}{llll} x^3+ & 0xx & -3x & +1=0 \\ 1 & 1000 & 1000000 & 1000000000 \\ \hline \end{array}$$

prodit $y^3 \quad * -3000000x+1000000000=0$.

Valores hujus æquationis appropinquando quæsiti prodeunt ferè $+1532$, $+347$, et -1879. Quod si jam rursus quælibet radix (cùm per 1000 multiplicata sit) dividatur per 1000, prodeunt valores primæ æquationis $+1\frac{532}{1000}$, $+\frac{347}{1000}$, et $-1\frac{879}{1000}$, atq; ita in alijs.

||[62] ||Interdum dividendo radices, numeri æquationis diminui possunt, veluti si dentur $y^3 * -576y-25920=0$, dividanturq; ignotæ radices per 12, ut sequitur[:]

$$\begin{array}{llll} y^3+0yy & -576y & -25920=0 \\ 1 \quad 12 & 144 & 1728 \\ \hline \end{array}$$

$z^3 \quad * \quad -4z \quad -15=0$, in hac æquatione valor ipsius $z=3$, quare valor ipsius y [est] $=36$.

Quando æquatio aliqua solvenda venit per parabolam et circulum,[66] atq; ad

(66) Compare *Geometrie*: 390–5: 'Facon generale pour construire tous les problesmes solides, reduits a vne Equatiõ de trois ou quatre dimensions'.

omnes æquationes unicâ parabola uti lubet, et latus rectum ipsius parabolæ semper ponitur unitas; sæpe contingit, ut numeri nimis sint grandes, qui proinde hâc divisione radicum, quàm velis, exigui reddentur, neq; refert, etiamsi fractiones occurrant, tamen usui erit, quando radices æquationis præter propter novisse lubet.

Quomodo nota quantitas alicujus termini æqualis fieri possit cuivis alteri.[67]

Potest quoq; notarum quantitatum quævis et cujusvis termini æqualis fieri cuivis alteri datæ, veluti si detur $x^3 * -bbx+c^3=0$, et loco ipsius bb malimus habere $3aa$, tum multiplica primùm radices per z, eritq; $y=zx$, vel $\frac{y}{z}=x$, $\frac{yy}{zz}=xx$, [63] et $\frac{y^3}{z^3}=x^3$. Sublato x in æquatione, prodit $\|\frac{y^3}{z^3} * [-]\frac{bby}{z}+c^3=0$, quæ omnia ducta in z^3, producunt $y^3 * [-]bbzzy+c^3z^3=0$, [sed] desideratur hîc, ut $bbzz$ sit æquale $3aa$, erit igitur $zz=\frac{3aa}{bb}$, et $z=\sqrt{\frac{3aa}{bb}}$, itaq; y vel $zx-x\sqrt{\frac{3aa}{bb}}$, vel $y=\frac{ax}{b}\sqrt{3}$, porró in ultima æquatione sublato z, ut sequitur:

$$\begin{array}{llll} x^3+ & 0xx- & bbx+ & c^3=0 \\ 1. & \frac{a}{b}\sqrt{3}. & \frac{3aa}{bb}. & \frac{3a^3}{b^3}\sqrt{3} \\ \hline \text{prodit } y^3 & * & -3aay+ & \frac{3a^2c^3}{b^3}\sqrt{3}=0. \end{array}$$

Quomodo interdum ignotæ radices æquationis quadraticæ multiplicentur, quò fit, ut dimensiones diminuantur.

Accidit quandoq;, æquationes ita dari, ut ignotæ radices quadraticè multiplicari possint, veluti si detur $x^4-bxx-aa=0$, ponitur $xx=y$, et $x^4=yy$, atq; ita porrò, tam si in æquatione ubiq; x tollatur, extabit $yy-by-aa=0$, et valor ipsius y tum erit quadratum ipsius x.

Idem tenendum [est] quando valor radicum cubicè vel altiùs multiplicari potest.

Hinc sequitur, æquationes, in quibus hoc fieri potest, tot dimensionibus diminui posse, ideoq; $y^6-8y^4-124yy-64=0$, non nisi cubicam æquationem dicendam,[68] et $x^4-bxx-aa=0$, non nisi quadraticam, atq; ita porrò.

---

(67) *Geometrie*: 380: 'Comment on rend la quantité connuë de l'vn des termes d'vne Equation esgale a telle autre qu'on veut'.

(68) Compare Descartes' example in his *Geometrie*: 381.

||[64] ||*Caput Quartum. Quomodo æquatio solvenda sit.*

Quando æquatio aliqua solvenda venit, si fractiones habeat, vel radicales, hi primùm tollendi sunt, et tum proceditur ad solutiones sequentes. De quadraticis æquationibus hîc nihil dicetur, quæ jam ante(69) solutæ fuerunt tollendo secundum terminum.

De solutione æquationum cubicarum quando problema est planum.(70)

Cùm æquatio cubica solvenda venit, oportet ordinè tentare omnes quantitates, quæ ultimum terminum absq; fractione dividere possunt, utrum aliqua ex ijs juncta quantitati ignotæ per signum + vel −, possit dividere totam æquationem absq; residuo, quod si sic, tum nota quantitas hujus binomij est radix quæsita, et æquatione per hoc terminum divisâ, prodit æquatio duarum dimensionum, quæ, cùm volumus, circino et regulâ solvi potest, et problemata, ex quibus ejusmodi æquationes ortæ sunt, vocantur plana.

Exempli gratiâ, dato $y^6 - 8y^4 - 124yy - 64 = 0$, potest ultimus terminus 64, ||[65] absq; fractione dividi ||per 1, 2, 4, 8, 16, 32, et 64, adeoq; ordinè tentandum utrū hæc æquatio dividi possit per aliquod horum binomiorū $yy-1$ vel $yy+1$, $yy-2$ vel $yy+2$, $yy-4$ vel $yy+4$, et sic porrò; et deprehendetur, quòd absq; residuo dividi possit per $yy-16$, quod arguit valorem ipsius yy esse 16.(71)

Ita dato $\begin{array}{l} y^6 + aay^4 - a^4yy - a^6 \\ \quad\; -2cc \;\; +c^4 \quad -2a^4cc \\ \qquad\qquad\qquad\quad\; -aac^4 \end{array} = 0$, potest ultimus terminus absq; fractione dividi per a, aa, $aa+cc$, a^3+acc, et similes, sed sufficit duas ex his notâsse, nam cæteræ plures vel pauciores dimensiones producunt, quàm reperiantur in nota quantitate penultimi termini. Postremò si tentetur binomium $yy-aa-cc$, deprehendi divisionem fieri posse absq; residuo, quod arguit $aa+cc$ esse radicem quæsitam.(72) Quod si per nullam quantitatem divisio fieri possit, ista problemata, ex quibus tales æquationes oriuntur, vocantur solida, quòd per rectas lineas et circulos solvi nequeant.

Nota, quando in hujusmodi [æquatione], qualis erat hæc ultima, tentare lubet, per quas quantitates ultimus terminus dividi possit, primùm quantitates ultimi termini ordinè statuendi sunt, veluti hîc $-a^6-2a^4cc-aac^4$, atq; tum idem est, ac si æquatio dividenda foret, cujus ignota quantitas valet aa, et inquirendum venit per quas quantitates hic ultimus terminus aac^4 dividi possit, unde quæsitæ mox inveniuntur.

---

(69) On Kinckhuysen's pages 54 and 55 above.

(70) *Geometrie*: 380–3: 'La reduction des Equatiōs cubiques lorsque le problesme est plan'.

(71) The other two values of y^2 (namely, $-4 \pm \sqrt{12}$) are surd and cannot be isolated by factorization into rational 'binomials'.

[66] ||Cùm accidit, ut in æquationis speciosæ quolibet termino multæ sint species, et tentare libeat, utrum ea dividi possit, solet loco quarundam poni aliquis numerus, vel cyphra, nimirū, quando ultimus terminus per eum tolli nequit, nam pro quavis specie notarū quantitatum numerus quivis poni potest, quo æquatio brevior evadit, et si tum absq; residuo dividi nequeat, non opus est quærere ulteriùs; sic ultima hæc, si 1 ponatur utiq; pro a et c, habetur $y^6-y^4 * -4=0$, hoc dividi potest per $yy-2$. Verùm si eadem per nullam quantitatem absq; residuo dividi potuisset, nequicquam sumeretur opera, si quæreretur ulteriùs, quemadmodum experientia satis docebit.

De solutione æquationum quatuor dimensionum, quando problema est planum.(73)

Cùm occurrit æquatio, in qua ignota quantitas quatuor dimensionibus constat, oportet proximè superiori modo considerare, utrum binomium reperiri possit, quod totam æquationem absq; residuo dividat, quo binomio invento, quantitas ejus nota est radix quæsita. Hæc igitur æquatio istâ divisione mutata erit in aliam trium dimensionum, quam rursus eodem modo perscrutari oportet; sed si nullum tale binomium reperiatur, oportet secundum terminum [67] æquationis tollere, atq; eam deinde reducere ad duas ||æquationes quadraticas, mediante solutione unius æquationis cubicæ, ut sequitur:

Detur $+x^4 * +pxx+qx+r=0$, quam pono ortam esse ex multiplicatione duarū æquationum quadraticarum, nimirum ex

$$\begin{array}{l} xx+yx+z=0 \\ \underline{xx-yx+v=0} \\ x^4+yx^3+zxx \\ \quad -yx^3-yyxx-yzx \\ \qquad\quad \underline{+vxx+yvx+vz} \\ x^4 \; * \;\; +pxx \;+qx+r=0. \end{array}$$

Hinc igitur oriuntur tres æquationes[, nimirum]

$$+p=+z-yy+v. \text{ estq; } yy-z+p=v.$$

$$+q=-yz+yv. \text{ estq; } +z+\frac{q}{y}=v.$$

$$+r=vz.$$

(72) Kinckhuysen has already given the factorization

$$(y^2-a^2-c^2)\,(y^4+(2a^2-c^2)y^2+a^2(a^2+c^2))$$

of the equation's left-hand member as an example of algebraic division. (Compare note (23) above and *Geometrie*: 382.)

(73) *Geometrie*: 383–7: 'La reduction des Equations qui ont quatre dimēsions, lorsque le problesme est plan...'.

Hi duo valores ipsius v sunt igitur æquales, nimirum $yy-z+p$ æquale est $z+\frac{q}{y}$. adde utrobiq; $+z-\frac{q}{y}$. tum divide per 2, prodit $\frac{1}{2}yy+\frac{1}{2}p-\frac{q}{2y}=z$. Antea $yy-z+p$ æquale erat v, tolle z, prodit $\frac{1}{2}yy+\frac{1}{2}p+\frac{q}{2y}=v$. Deniq; r æquale est vz, tolle v et z ut sequitur[:]

$$\frac{1}{2}yy+\frac{1}{2}p-\frac{q}{2y}=z$$

$$\frac{1}{2}yy+\frac{1}{2}p+\frac{q}{2y}=v$$

$$\frac{1}{4}y^4+\frac{1}{4}pyy-\frac{1}{4}qy$$

$$+\frac{1}{4}pyy+\frac{1}{4}pp-\frac{1}{4}\frac{pq}{y}$$

$$+\frac{1}{4}qy+\frac{1}{4}\frac{pq}{y}-\frac{1}{4}\frac{qq}{yy}$$

||[68] prodit $r=\frac{1}{4}y^4+\frac{1}{2}pyy+\frac{1}{4}pp \quad * \quad -\frac{1}{4}\frac{qq}{yy}$, aufer utrinq; ||$r$, et duc omnia in $4yy$, et prodit hæc æquatio cubica $y^6+2py^4+\begin{matrix}ppyy\\-4r\end{matrix}-qq=0$. Duæ æquationes quadraticæ, ex quibus data æquatio ponitur orta, erant $xx+yx+z=0$ [et] $xx-yx+v=0$. Substituto $\frac{1}{2}yy+\frac{1}{2}p-\frac{q}{2y}$ pro z, et $\frac{1}{2}yy+\frac{1}{2}p+\frac{q}{2y}$ pro v, habetur

$$xx+yx+\frac{1}{2}yy+\frac{1}{2}p-\frac{q}{2y}=0$$

$$\text{[et]}\quad xx-yx+\frac{1}{2}yy+\frac{1}{2}p+\frac{q}{2y}=0.$$ (74)

Quòd si igitur ex inventa æquatione cubica eruatur valor ipsius y, eumq; ponamus loco ipsius y, et ejus quadratum pro yy, habetur quæsitum; verùm si valor ipsius y inveniri nequeat, problema, unde processit, solidum est.

Istâ igitur operatione invenitur hæc

Regula de distribuenda æquatione quatuor dimensionum in duas quadraticas, ope æquationis cubicæ.

Dato $+x^4 * \quad pxx.\ qx.\ r=0$ scribitur primò hæc æquatio cubica $+y^6\,[+]\,2py^4+\begin{matrix}ppyy\\ [-]4r\end{matrix}-qq=0$.

(74) Compare *Geometrie*: 383, 385. Descartes, however, does not there justify the form of his reduced cubic.

Quod ad signa $+$ et $-$, quæ non sunt apposita, ea scribes ut sequitur:

Dato $+p$ pone $+2p$, aliàs $-2p$.
Dato $+r$ pone $-4r$, aliàs $+4r$.

69] ||Exempli gratiâ,(75) dato $\overset{}{+x^4} * \overset{p}{-4xx} \overset{q}{-8x} \overset{r}{+35}=0$ scribitur

$$y^6-8y^4-124yy-64=0.$$

nam quantitas p valet 4, r valet 35, quare ponitur $\begin{matrix}+\ 16\\-140\end{matrix}yy$, hoc est $-124yy$, loco $\begin{matrix}+pp\\-4r\end{matrix}yy$, postremò q valet 8, ergo statuitur -64 pro $-qq$.

Similiter pro $x^4 * -9xx+6x+6=0$ scribitur $y^6-18y^4+57yy-36=0$. nam 18 duplum est ipsius 9, 57 est ejus quadratum minus 4 vicibus 6, et 36 est quadratum ipsius 6.

Sic quoqȝ(75) pro $\begin{matrix}+z^4 * +\frac{1}{2}aazz-a^3z+\frac{5}{16}a^4=0\\ \quad -cc \quad -acc \quad +\frac{1}{4}aacc\end{matrix}$

$$\text{scribitur } \begin{matrix}y^6+aay^4-a^4yy-a^6=0.\\ -2cc \quad +c^4 \quad -2a^4cc\\ \qquad\qquad -aac^4\end{matrix}$$

nam p est $+\frac{1}{2}aa-cc$, et pp est $\frac{1}{4}a^4-aacc+c^4$, et $4r$ est $-\frac{5}{4}a^4+aacc$, et tandem $-qq$ est $-a^6-2a^4cc-aac^4$.

Ex hac æquatione trium dimensionum deinde quærendus est valor istius yy, eo, quo ostendimus modo, et si nullum reperiatur binomium quod dividat æquationem absqȝ residuo, non ultra procedendum est, nam id arguit, problema unde oritur, solidum esse; sin reperiatur, ejus ope data æquatio dividitur in duas æquationes quadraticas, nimirum pro $+x^4 * pxx.\ qx.\ r.\ =0$ scribuntur aliæ hæ duæ:

$$+xx-yx+\tfrac{1}{2}yy\,[+]\tfrac{1}{2}p\cdot\frac{q}{2y}=0 \quad \text{et}$$

$$+xx+yx+\tfrac{1}{2}yy\,[+]\tfrac{1}{2}p\cdot\frac{q}{2y}=0.$$

Signa $+$ et $-$, quæ apposita non sunt, scribuntur ut sequitur:

[70] ||Dato $+p$, scribe in utraqȝ $+\frac{1}{2}p$, aliàs $-\frac{1}{2}p$.

Dato $+q$, scribe $+\frac{q}{2y}$ in ea, in qua reperitur $-yx$, et in altera, in qua habetur $+yx$, scribe $-\frac{q}{2y}$, aliàs contrarium.(76)

(75) *Geometrie*: 384.
(76) A straight translation of *Geometrie*: 384–5.

Exempli gratiâ,[77] quoniam deprehenditur yy valere 16, dum

$$y^6-34y^4+313yy-400=0,$$

ponitur pro $x^4 * -17xx-20x-6=0$, oportet pro hac æquatione

$$x^4 * -17xx-20x-6=0$$

scribere alias has duas $+xx-4x-3$, et $+xx+4x+2=0$, nam y est 4, $\frac{1}{2}yy$ est 8, p est 17, et q est 20; ita ut $+\frac{1}{2}yy-\frac{1}{2}p-\frac{q}{2y}$ valeat -3, et $+\frac{1}{2}yy-\frac{1}{2}p+\frac{q}{2y}$ valeat 2, et extractis radicibus ex hisce duabus æquationibus, invenitur una vera, nimirum $\surd 7+2$, et tres falsæ, nempe $\surd 7-2$, $2+\surd 2$, et $2-\surd 2$.

Sic quoqȝ dato $x^4 * -3xx+6x+6=0$, cùm radix ex $y^6-18y^4+57yy-36=0$, nimirum yy valeat 3, scribendum est $xx-x\surd 3-3+\surd 3[=0]$, et

$$xx+x\surd 3-3-\surd 3[=0].$$[78]

Sic dato[79] $z^4 * \begin{matrix}+\frac{1}{2}aazz \\ -cc\end{matrix} \begin{matrix}-a^3z \\ -acc\end{matrix} \begin{matrix}+\frac{5}{16}a^4 \\ +\frac{1}{2}aacc\end{matrix}=0$, cùm pro yy inveniatur $aa+cc$, scribendum est

$$zz-z\sqrt{aa+cc}+\tfrac{3}{4}aa-\tfrac{1}{2}a\sqrt{aa+cc}=0, \text{ et}$$
$$zz+z\sqrt{aa+cc}+\tfrac{3}{4}aa+\tfrac{1}{2}a\sqrt{aa+cc}=0,$$

nam y est $\sqrt{aa+cc}$, et $+\frac{1}{2}yy+\frac{1}{2}p$ est $\frac{1}{4}aa$, et $\frac{q}{2y}$ est $\frac{1}{2}a\sqrt{aa+cc}$, unde valor ipsius z invenitur

$$\tfrac{1}{2}\sqrt{aa+cc}+\sqrt{-\tfrac{1}{2}aa+\tfrac{1}{4}cc+\tfrac{1}{2}a\sqrt{aa+cc}}$$
$$\text{vel } \tfrac{1}{2}\sqrt{aa+cc}-\sqrt{-\tfrac{1}{2}aa+\tfrac{1}{4}cc+\tfrac{1}{2}a\sqrt{aa+cc}}.$$

||[71] Et cùm suprà posuerimus $z+\frac{1}{2}a=x$, invenitur ||quantitas x valere

$$+\tfrac{1}{2}a+\tfrac{1}{2}\sqrt{aa+cc}-\sqrt{\tfrac{1}{2}aa+\tfrac{1}{4}cc+\tfrac{1}{2}a\sqrt{aa+cc}}.$$

Possunt etiam alio modo æquationes biquadraticæ ope unius æquationis cubicæ dividi in duas æquationes quadraticas, non tollendo secundum terminum, ut sequitur:[80]

Detur $x^4+px^3+qxx+rx+s=0$, quam productam esse pono ex multiplicatione sequentium duarum æquationum quadraticarum:

$$\begin{array}{l} xx+vx+e=0 \\ xx+zx+f=0 \\ \hline x^4+vx^3+exx \\ \quad +zx^3+vzxx+ezx \\ \quad\quad +fxx+vfx+ef \\ \hline x^4+px^3+qxx+rx+s=0. \end{array}$$

(77) *Geometrie*: 385–6 rendered word for word.

(78) In consequence, $x=+\frac{1}{2}\surd 3\pm\surd[3\frac{3}{4}-\surd 3]$ or $-\frac{1}{2}\surd 3\pm\surd[3\frac{3}{4}+\surd 3]$.

Habemus igitur hîc quatuor has æquationes:

$$p=v+z, \text{ est}\&\ v=p-z. \quad q=e+vz+f. \quad r=ez+vf. \quad s=ef.$$

pone $e+f=y$, et $p-z$ pro v, habetur& $q=y+pz-zz$, et $z=\frac{1}{2}p+\sqrt{\frac{1}{4}pp-q+y}$ est& $v=\frac{1}{2}p-\sqrt{\frac{1}{4}pp-q+y}$. $e+f$ positum est $=y$, ergo pone $y-f$ pro e, et inventas

[72] quantitates pro z et v, et prodit ‖$r=\frac{1}{2}py+y\sqrt{\frac{1}{4}pp-q+y}-2f\sqrt{\frac{1}{4}pp-q+y}$ et $\frac{1}{2}y+\frac{\frac{1}{2}py-r}{2\sqrt{\frac{1}{4}pp-q+y}}=f$. Est& [ergo] $\frac{1}{2}y-\frac{\frac{1}{2}py-r}{2\sqrt{\frac{1}{4}pp-q+y}}=e$ [et] prodit

$$s=\frac{1}{4}yy-\frac{\frac{1}{4}ppyy-pry+rr}{pp-4q+4y}=ef$$

[vel] $spp-4qs+4sy+\frac{1}{4}ppyy-pry+rr=\frac{1}{4}ppyy-qyy+y^3$. Deniq&

$$\begin{aligned}+y^3-qyy-4sy&-spp=0.\\ +pr&+4qs\\ &-rr\end{aligned}$$

Duæ æquationes quadraticæ, ex quibus data æquatio ponitur orta esse, erant $\begin{matrix}xx+vx+e=0.\\ xx+zx+f=0.\end{matrix}$ Quòd si igitur pro v, z, e, et f, ponantur quantitates inventæ, habetur

$$xx+\frac{1}{2}px-x\sqrt{\frac{1}{4}pp-q+y}+\frac{1}{2}y-\frac{\frac{1}{2}py-r}{2\sqrt{\frac{1}{4}pp-q+y}}=0.$$

$$xx+\frac{1}{2}px+x\sqrt{\frac{1}{4}pp-q+y}+\frac{1}{2}y+\frac{\frac{1}{2}py-r}{2\sqrt{\frac{1}{4}pp-q+y}}=0.$$

Tum si numeri loco specierum in inventa æquatione cubica positi, ne& ipsa dividi possit per $y+$ vel $-$ aliquo numero, id arguit, ipsum $e+f$ non esse numerum rationalem integrum, vel æquationem esse solidam.

Sequitur exemplum, dato $x^4+4x^3-3xx-8x+4=0$. Hîc est $p=4$, $q=-3$, $r=-8$, et $s=4$. Invenitur& $y^3+3yy-16y-64=0$

$$\begin{aligned}-32y&-48\\ &-64\\ \hline y^3+3yy-48y&-176=0.\end{aligned}$$

hoc abs& residuo dividi

[73] potest per $y+4=0$, est& $y=-4$. ‖Prodit tandem $\begin{matrix}xx+\overline{2-\sqrt{3}}x-2=0.\\ xx+\overline{2+\sqrt{3}}x-2=0.\end{matrix}$ et sic in alijs.

(79) *Geometrie*: 386–7.

(80) This generalization of Descartes' quartic method is due to Johann Hudde, who expounded it in a 1657 tract *De Reductione Æquationum* communicated to Schooten on 1 April 1658 and published by the latter in his second Latin edition of the *Geometrie* (*Geometria* ($_2$1659): 406–506, especially, 494–6: 'XX. Regula').

Hac extractione acquirimus hanc regulā,
quando habetur $+x^4.\,px^3.\,qxx.\,rx.\,s.\ =0.$
scribitur $y^3.\,qyy.\,4sy.\,spp.\ =0.$

$$\begin{array}{ll} pr & 4qs \\ & -rr \end{array}$$

Signa + et −, quæ non apposita, statue ut sequitur:

Dato $+q$, pone $-q$, aliàs +.
Dato $+s$, pone $-4s$, et $-spp$, aliàs ambo +.
Dato $+p+r$ [vel] $-p-r$, pone $+pr$, sin alterum sit + et alterum −, vel hoc − et illud +, pone $-pr$.
Dato $+q$ [et] $+s$ [vel] $-q$ [et] $-s$, pone $+4qs$, sin diversa sint signa, pone $-4qs$.

Deinde scribitur $+xx.\,\frac{1}{2}px+x\sqrt{\frac{1}{4}pp.\,q.\,y.}\,\frac{1}{2}y+\dfrac{\frac{1}{2}yp.\,r}{2\sqrt{\frac{1}{4}pp.\,q.\,y}}$

et $+xx.\,\frac{1}{2}px-x\sqrt{\frac{1}{4}pp.\,q.\,y.}\,\frac{1}{2}y-\dfrac{\frac{1}{2}yp.\,r}{2\sqrt{\frac{1}{4}pp.\,q.\,y}}.$

Dato $+p$, pone $+\frac{1}{2}p$, aliàs −.
Dato $+q$, pone $-q$, aliàs +.
Dato $+r$, pone $-r$, aliàs +.
Si valor ipsius y sit radix vera, pone $+y$ et $+\frac{1}{2}y$, aliàs −.
Dato $+y$ [et] $+p$ [vel] $-y$ [et] $-p$, pone $+\frac{1}{2}yp$, sin diversa sint signa pone $-\frac{1}{2}yp$.

||[74] ||Quomodo æquatio solvi possit, quando conditiones quædam priorum æquationum, ex quarum multiplicatione data orta est, notæ sunt.

Suprà ostensum est satis, æquationem, quæ aliquot dimensionibus constat, ita considerandum, quasi ex multiplicatione duarum vel plurium orta sit. Quod si igitur conditiones quædam earum æquationum, quas multiplicatas esse supponimus, notæ sint, poterunt istæ conditiones nobis inservire ad solutionem, neqȝ aliud faciendum [est], quàm ut multiplicatione talium æquationum, quæ ejusmodi conditiones habent, producamus æquationem quandam; cujus deinde singuli termini æquales sunt singulis æquationis datæ, quemadmodum in sequentibus videre est.

(81) That is: given the sum of two roots of an equation whose coefficients are known, to determine each severally in terms of those coefficients and the given sum. The details of the following resolution appear to be Kinckhuysen's but the method is, of course, Cartesian: setting y and $b-y$ for the roots of given sum b, he supposes the third root n of his cubic arbitrarily, equates the coefficients to the appropriate homogeneous products of the roots and then eliminates n.

Quomodo æquatio solvatur, in qua valor duarum radicum additarum æqualis est quantitati datæ.(81)

Dato $x^3+pxx-qx+r=0$, ubi valor duarū radicum additarum æquipollet ipsi b, scire velimus, quanta sit utraq; radicum seorsim; supponimus æquationem ipsam ortam ex multiplicatione æquationū sequentium:

[75]

$$\begin{array}{l} \|x-y \quad =0 \\ \underline{x-b+y=0} \\ xx-bx+by-yy=0 \\ \underline{x+n \qquad\quad =0} \\ x^3-bxx+byx \\ \qquad\quad -yyx \; +byn \\ \quad +nxx-bnx \\ \qquad\qquad\quad \underline{-yyn} \\ x^3+pxx-qx+r=0. \end{array}$$

Hîc dantur tres æquationes, nimirum:

$$+p=-b+n, \quad \text{ergo} \quad p+b=n.$$

$$-q=+by-yy-bn, \quad \text{ergo} \quad \frac{by-yy+q}{b}=n.$$

$$+r=+byn-yyn, \quad \text{ergo} \quad \frac{r}{by-yy}=n.$$

$p+b$, igitur [est] $=\dfrac{by-yy+q}{b}$ et $yy-by+pb+bb-q=0$, prodit

$$y=\tfrac{1}{2}b+\sqrt{q-\tfrac{3}{4}bb-pb}. \quad \text{et} \quad b-y=\tfrac{1}{2}b-\sqrt{q-\tfrac{3}{4}bb-pb}.$$

Quomodo æquatio solvatur quando priorum æquationum ex quarum multiplicatione data æquatio orta est, unus terminus notus est.(82)

Data æquatione, veluti $x^4+px^3+qxx+rx+s=0$, quam suppono ortam esse ex multiplicatione duarum sequentium, quarum terminus h notus est.

[76]

$$\begin{array}{l} \|xx+kx+l=0 \\ \underline{xx+zx+h=0} \\ x^4+kx^3+lxx \\ \quad +zx^3+kzxx+lzx \\ \qquad\qquad +hxx+khx+hl \\ \end{array}$$

prodit quæsita $\quad x^4+px^3+qxx+rx+s=0.$

(82) Compare Hudde's *De Reductione Æquationum* (note (80)): 483–5: 'XVIII. Regula'.

Hîc emergunt quatuor æquationes, ut sequitur:

$$p=k+z, \text{ ergo } k=p-z.$$
$$q=l+kz+h.$$
$$r=lz+kh.$$
$$s=hl, \text{ ergo } l=\frac{s}{h}.$$

Tolle in tertia æquatione k et l, et prodit $[z]\frac{s}{h}+ph-zh=r$, et $z=\frac{rh-phh}{s-hh}$ vel $\frac{r-ph}{\frac{s}{h}-h}$, quo substituto pro z, dividi potest æquatio per $xx+\frac{r-ph}{\frac{s}{h}-h}x+h=0$.

Posito $l=h$, hoc est, $\sqrt{s}=l$ vel h, erit $q=2h+kz$, sublato k prodit $2h+pz-zz=q$, et $z=[\frac{1}{2}p]\pm$(83)$\sqrt{\frac{1}{4}pp+2h-q}$. Substituto hoc valore ipsius z pro z, dividi potest æquatio per $xx+z[x]+h=0$. Nota, divisâ quantitate s per h, prodit l. Sequitur exemplum: Dato $x^4-2x^3-45xx-2x+24=0$, si terminus h valeat -4, hîc est $p=-2$, $r=-2$, et $s=24$, tum æquatio hæc dividi potest per $xx+\frac{-2-8}{\frac{24}{-4}+4}x-4=0$, hoc est $xx+5x-4=0$, et sic in alijs.

||[77] ||Quomodo æquatio solvatur cujus duæ radices habent rationem datam.(84)

Dato $y^3-pyy-qy+r=0$, si duæ radices veræ se habeant ut 1 ad a, pono alteram radicum $y=e$, erit altera $y=ae$, deinde procedo ut sequitur:

$$\begin{array}{l} y-e\ =0 \\ \underline{y-ae=0} \\ yy\begin{smallmatrix}-e\\-ae\end{smallmatrix}y+aee=0 \\ \underline{y+z\qquad\qquad\quad} \\ y^3\begin{smallmatrix}-e\\-ae\end{smallmatrix}yy+aeey \\ \quad+zyy\begin{smallmatrix}-ez\\-aez\end{smallmatrix}y+aeez \\ \hline y^3-pyy-qy+r=0, \end{array}$$

et habentur tres hæ æquationes:

$$-p=-e-ae+z, \text{ ergo } e+ae-p=z.$$
$$-q=+aee-ez-aez, \text{ ergo } \frac{q+aee}{e+ae}=z.$$
$$+r=+aeez, \text{ ergo } \frac{r}{aee}=z.$$

(83) Mercator has here substituted Oughtred's symbol '±' (F. Cajori, *A History of Mathematical Notations*, **1** (Chicago, 1928): 245) for Kinckhuysen's Schootenian '♉', explained in an accompanying marginal note not reproduced by Mercator 'Merckt het teken ♉ beteeckent

Hîc valor ipsius z datur tribus modis, unde innumeræ æquationes produci possunt, v. gr. primus valor est æqualis secundo, primus æqualis tertio, secundus [78] ||æqualis tertio, duplum primi æquale secundo et tertio simul, et sic porrò.

Habemus $e+ae-p=\dfrac{q+aee}{e+ae}$ [, hoc est] $ee+aee+aaee-pe-pae-q=0$. Substituto igitur y pro e, habetur $yy+ayy+aayy-py-pay-q=0$. Talis æquatio statim obtineri potest ex data æquatione, si ducatur in progressionem, qualis hîc sequitur:

$$\begin{array}{llllll} & y^3 & -pyy & -qy & +r=0 \\ \text{Mult.} & 1+a+aa & 1+a & 1 & 0 \\ \hline \text{differ.} & & aa & a & 1 \\ \hline \text{prodit} & \multicolumn{4}{l}{y^3+ay^3+aay^3-pyy-apyy-qy\ * =0.} \\ \hline y & \multicolumn{4}{l}{yy+ayy+aayy-py-apy-q=0.} \end{array}$$

Progressio per quam hîc multiplicatum est, ejus est natura, ut differentia inter primum (incipiendo à dextrâ) et secundum sit 1, inter secundum et tertium a, inter tertium et quartum aa, et sic deinceps, ita ut soluturo tales æquationes nihil amplius faciendum sit, quàm ut multiplicet æquationem datam per progressionem quandam, cujus differentiæ à dextra versus sinistram crescant in ratione data duarum radicum, et sic obtinetur valor ipsius $y=e$.

Veluti, si detur $x^3-3xx-28x+60=0$; sintꝗ duæ radices in ratione 1 ad 3.

[79]

$$\begin{array}{lllll} & \|x^3-3xx-28x+60=0 & & \text{[vel]} & x^3-3xx-28x+60=0 \\ \text{Mult.} & 13.\quad 4.\quad 1.\quad 0. & & & 0.\quad -1.-1\tfrac{1}{3}.-1\tfrac{4}{9}. \\ \hline \text{differ.} & \quad 9.\quad 3.\quad 1. & & & \quad 1.\quad \tfrac{1}{3}.\quad \tfrac{1}{9}. \\ \hline & 13x^3-12xx-28x\ * =0 & & & [*]+3xx+37\tfrac{1}{3}x-86\tfrac{2}{3}=0 \\ \hline x & 13xx-12x-28=0 & & [\tfrac{1}{3}] & 9xx+112x-260=0. \end{array}$$

Ita habentur hîc duæ æquationes, multiplica priorem per 9, et alteram per 13, prodit $117xx-108x-252=117xx+1456x-3380$, et $x=2$.

Nota, licèt progressionem ordiamur hîc à 0, arbitrarium est tamen eandem ordiri à numero quovis et differentia quavis; saltim ponitur 0 ut æquatio unâ dimensione diminuâtur.

+ of −' (note that the sign ȣ means + or −). For Schooten's use of this apparent variant of the Greek ligature 'ȣ' (for ου), perhaps on the analogy of the Cartesian equality sign (note (42)), compare his *Commentarii in Librum III, D = Geometria*, [(₁1649): 242 =] (₂1659): 287; and *Exercitationum Mathematicarum Libri Quinque* (Leyden, 1657): 438. Compare I: 119.

(84) This extension of Descartes' double-root method (*Geometrie*: 347 ff.) is apparently Kinckhuysen's own discovery though, as we have seen (I, 3, 3, §4) Newton found it independently in May 1665 and made a similar use of 'Huddenian' multipliers in its application. The complexities of Kinckhuysen's approach are, however, unnecessary. If a function of $y, f(y)$, has the factors $y-e$ and $y-ae$, then $f(ay)$ has the factor $y-e$ also: hence if $y=e$, then $f(y)=f(ay)=0$ and the result follows by evaluating $[f(y)-f(ay)]/(1-a)=0$.

Sed in æquatione priùs inventa $ee+aee+aaee-pe-pae-q=0$, ponatur y loco ipsius ae, hoc est $\frac{y}{a}=e$, et $\frac{yy}{aa}=ee$, emergit $\frac{yy}{aa}+\frac{yy}{a}+yy-\frac{py}{a}-py-q=0$, quibus omnibus multiplicatis per aa prodit $yy+ayy+aayy-apy-aapy-aaq=0$. Talis æquatio[85] statim obtinetur si æquatio multiplicetur per sequentem progressionem:

$$\begin{array}{llccc} & y^3 & -pyy & -qy & +r=0 \\ \text{Mult.} & 1+a+aa & a+aa & aa & 0 \\ \hline \text{differ.} & 1 & a & aa & \\ \hline \text{prodit} & \multicolumn{4}{l}{y^3+ay^3+aay^3-apyy-aapyy-aaqy\ * =0} \\ y & & & & \\ \hline & \multicolumn{4}{l}{yy+ayy+aayy-apy-aapy-aaq=0.} \end{array}$$

||[80] ||Progressio est ejus naturæ, ut differentia inter primum terminum (incipiendo à sinistra) et secundum sit 1, et inter secundum et tertium a, inter tertium et quartum aa, et sic porrò.

Hac ratione obtinetur valor ipsius $y=ae$. Resumto nunc priori exemplo $x^3-3xx-28x+60=0$, et posito duas radices esse in ratione 1 ad 3,

$$\begin{array}{cccc} x^3-3xx & -28x & +60 & =0 \\ 0 \quad 1 & 4 & 13 & \\ \hline 1 & 3 & 9 & \\ \hline \multicolumn{4}{l}{[*]-3xx-112x+780=0} \end{array} \quad [\text{vel}] \quad \begin{array}{ccccl} x^3 & -3xx & -28x & +60=0 & \\ 1\frac{4}{9} & 1\frac{1}{3} & 1 & 0 & \\ \hline & \frac{1}{9} & \frac{1}{3} & 1 & \\ \hline \multicolumn{4}{l}{1\frac{4}{9}x^3-4xx-28x\ * =0} & \frac{9}{x} \\ \hline \multicolumn{4}{l}{13xx-36x-252=0.} & \end{array}$$

Hîc sunt duæ æquationes, multiplica priorem per 13, et alteram per -3, prodit $-39xx-1456x+10140=-39xx+108x+756$, et $x=6$.[86] Et sic in alijs.

Ex præcedentibus sequitur, positâ æquatione, quæ duas habeat radices æquales, hoc est in ratione 1 ad 1, futurū ut omnes differentiæ progressionis, per quam æquatio multiplicatur, sint æquales inter se, nam a tunc quoqꝫ valet 1; quare oblatâ ejusmodi æquatione, potest illa saltim multiplicari per Arithmeticam progressionem, ascendendo à sinistra dextrorsum, vel à dextra sinistrorsum, si libeat;[87] veluti si detur æquatio, quæ habeat duas radices æquales, nimirum $y^3-pyy-qy+r=0$, et quæratur valor istius y, proceditur modo sequenti:

(85) In the terminology of the previous note this is $[f(y)-f(a^{-1}y)]/(1-a^{-1})$.

(86) In fact, $x^3-3x^2-28x+60 = (x-2)(x-6)(x+5)$.

(87) This is Johann Hudde's rule for equations with a double root, but differently derived as the limit of the preceding section when $a\rightarrow 1$. (See Hudde's *De Reductione Æquationum* (note (80)): 433–9: 'X. Regula'; and his *Epistola de Maximis et Minimis* [to Schooten on 8 February 1658] = *Geometria* ($_2$1659): 507–16.) The theory of the algorithm and its extensions are discussed in the 'Historical Note' which introduces 1, **2**, 2.

[81] ||
$$\begin{array}{lcccccc} & y^3- & pyy- & qy+r & =0 \\ \text{Mult.} & 0 & 1 & 2 & 3 \\ \text{Mult.} & 3 & 2 & 1 & 0 \\ \hline \text{prodit} & * & -pyy & -2qy & +3r=0 \\ \text{et} & 3y^3 & -2pyy & -qy & * \; =0. \end{array}$$

Hæ duæ æquationes si multiplicantur prior per p et posterior per $-p$,(88) prodit

$$-3pyy-6qy+9r=-3pyy+2ppy+pq,$$

quare $9r-pq=2ppy+6qy$, et $\frac{9r-pq}{2pp+6q}=y$.

Sequuntur adhuc duo exempla ex Geometria *Cartesij*, de lineis tangentibus.(89) Prius est $yy+\frac{qr-2qv}{q-r}y+\frac{qvv-qss}{q-r}=0$, in quo valores ignoti habent duas radices æquales, quæritur valor ipsius v, cumq; in ultimo termino occurrat vv, ideò sub eo statuo 0, ut sequitur:

$$\begin{array}{llll} & yy+ & \frac{qr-2qv}{q-r}y+ & \frac{qvv-qss}{q-r}=0 \\ \text{Mult.} & 2. & 1. & 0. \\ \hline \text{prodit} & 2yy+ & \frac{qr-2qv}{q-r}y & * \; =0 \\ \hline \text{vel} & \multicolumn{3}{l}{2qy-2ry+qr=2qv \text{ et } y-\frac{ry}{q}+\tfrac{1}{2}r=v.} \end{array}$$

Alterum est æquatio sequens, et quæritur valor ipsius v; cùm igitur in quinto termino occurrat vv, ideò sub eo statuo 0, ut sequitur:

$$\begin{array}{lccccccc} & y^6 & -2by^5 & -2cdy^4 & +4bcdy^3 & -2bbcdyy & -2bccddy & +bbccdd=0 \\ & & & +bb & -2ddv & +ccdd & & \\ & & & +dd & & -ddss & & \\ & & & & & +ddvv & & \\ \text{Mult.} & & 4. & 3. & 2. & 1. & 0. & -1. \quad -2. \\ \hline \text{prodit} & & 4y^6 & -6by^5 & -4cdy^4 & +4bcdy^3 & * & +2bccddy-2bbccdd=0. \\ & & & +2bb & -2ddv & & & \\ & & & +2dd & & & & \end{array}$$

[82] ||

Divisis omnibus per $2ddy^3$, prodit $v=\frac{2y^3}{dd}-\frac{3byy}{dd}+\frac{bby}{dd}-\frac{2cy}{d}+y+\frac{2bc}{d}+\frac{bcc}{yy}-\frac{bbcc}{y^3}$.

(88) Mercator faithfully translates Kinckhuysen's text, which should read '$-p/y$'.

(89) The reference is to Descartes' *Geometrie*: Livre Second: 343/347 and 343–4/348–9 respectively, but the reduction by Huddenian multipliers is copied from the first two examples given by Hudde in illustration of the use of his method 'in inveniendis Tangentibus, determinandis Maximis & Minimis, & quibusvis extremis' in his *De Reductione Æquationum* (note (80)): 436–7.

Postremò sequitur adhuc exemplum: Hujus æquationis

$$y^4-py^3+qyy-ry+s=0,$$

valor ignotus duas habet radices æquales, quæritur valor ipsius y.

	y^4-	py^3	$+qyy$	$-ry$	$+s=0$
Mult.	0.	1.	2.	3.	4.
	4.	3.	2.	1.	0.
	−1.	0.	1.	2.	3.
	3.	2.	1.	0.	−1.
prodit	*	$-py^3$	$+2qyy$	$-3ry$	$+4s=0.$
	$4y^4$	$-3py^3$	$+2qyy$	$-ry$	$*=0.$
	$-y^4$	*	$+qyy$	$-2ry$	$+3s=0.$
	$3y^4$	$-2py^3$	$+qyy$	*	$-s=0.$

Primâ harum quatuor æquationum ductâ in 4,(90) et secundâ in p, si emergentes duæ æquationes addantur, prodit æquatio duarum dimensionum. Porrò tertia ducta in 3 si addatur quartæ, prodit æquatio trium dimensionum; tum quarta ducta in 3 si addatur tertiæ, prodit etiam æquatio trium dimensionum. Deinde ex duabus æquationibus trium dimensionum obtinetur æquatio duarum dimensionum, atq; ex hac unà cum alterâ istâ duarū dimensionū priùs inventâ, obtinetur valor ipsius y.

||[83] ||Nota, ex primis duabus æquationibus, obtinetur æquatio duarum dimensionum, quod hîc ad solutionem sufficit; verùm ulteriùs procedo tamen, ut ostendam quomodo agendum sit cum æquationibus altioribus.

Quomodo æquatio solvatur, quæ ex multiplicatione aliarum orta est, in quarum una terminus unus vel plures defuerunt.(91)

Sequentes duæ æquationes multiplicatæ in se invicē[:]

$$\begin{array}{llll} x^3+exx & +fx & +g=0 & \\ x+h & & =0 & \\ \hline x^4+ex^3 & +fxx & +gx & \\ \quad +hx^3 & +ehxx & +fhx & +gh \\ \hline \end{array}$$

producāt $x^4+px^3 \quad +qxx+rx+s=0$. Posito igitur $e=0$ erit

(90) Read '$4y$' here (and in Kinckhuysen's original).

(91) Compare Hudde's *De Reductione Æquationum* (note (80)): 439–58: 'XI. Regula, Quæ modum docet reducendi omnes æquationes, sive literales, sive numerales, quæ produci possunt ex multiplicatione duarum aliarum, in quarum alterutra unus pluresve termini deficiunt'.

(92) *De Reductione Æquationum*: 440: XI. Regula, 1ma Pars [Cas. 2].

$$p=h.$$
$$q=f.$$
$$r=g+fh, \text{ [vel] } r=g+pq, \text{ et } r-pq=g.$$
$$s=gh, \text{ [ergo est] } \frac{s}{h}=r-pq, \text{ et } h=\frac{s}{r-pq}.$$

Quòd si jam hæ quantitates inventæ substituantur pro h, dividi potest æquatio per $x+p=0$, et per $x+\frac{s}{r-pq}=0$.(92) Porrò posito $f=0$, et $p=0$, erit

$$p-e+h=0, \text{ et } e=-h.$$
$$q=eh, \text{ vel } q=-hh, \text{ vel } \surd-q=h.$$
$$r=g.$$
$$s=gh \text{ vel } s=rh, \text{ et } \frac{s}{r}=h.$$

84] ||Quòd si igitur hæ quantitates inventæ substituantur pro h, dividi potest hæc æquatio per $x+\surd-q=0$, et per $x+\frac{s}{r}=0$;(93) et sic in alijs.

Sequentes æquationes si multiplicentur in se invicem

$$x^3+exx+fx\ +g=0$$
$$xx+h\qquad\qquad =0$$
$$x^5+ex^4+fx^3+gxx$$
$$+hx^3+ehxx+fhx+gh$$

prodit $x^5+px^4+qx^3+rxx+sx+t=0$

estq; $p=e.$
$$q=f+h, \text{ vel } q-h=f.$$
$$r=g+eh, \text{ vel } r-ph=g.$$
$$s=fh, \text{ vel } s=qh-hh, \text{ vel } h=\tfrac{1}{2}q[\pm]\surd\tfrac{1}{4}qq-s.$$
$$t=gh, \text{ vel } t=rh-phh, \text{ vel } h=\frac{r}{2p}[\pm]\sqrt{\frac{rr}{4pp}-\frac{t}{p}}.^{(94)}$$

Quòd si igitur hæ quantitates inventæ substituantur pro h, dividi potest hæc æquatio per $xx+\tfrac{1}{2}q\pm\surd\tfrac{1}{4}qq-s=0$, et per $xx+\frac{r}{2p}\pm\sqrt{\frac{rr}{4pp}-\frac{t}{p}}=0$.(95)

(93) *Ibid.*: 1ma Pars [Cas. 5].

(94) These equations, displayed by Kinckhuysen's printer in the *Algebra*, are not copied in Mercator's translation but in reproducing them here we have seen fit to insert the latter's preferred sign '±' for the original's '♉'. (See note (83).)

(95) Compare Hudde's *De Reductione Æquationum* (note (80)): XI. Regula, 2da Pars [1, Cas. 6]: 443.

Porrò si in postrema multiplicatione ponatur $f=0$, habetur

$$p=e.$$
$$q=h.$$
$$r=g+eh, \text{ vel } r-ph=g.$$
$$t=gh, \text{ vel } t=rh-phh, \text{ vel } h=\frac{r}{2p}\pm\sqrt{\frac{rr}{4pp}-\frac{t}{p}}.$$

Quòd si hæ quantitates inventæ substituantur pro h, dividi potest hæc æquatio per $xx+q=0$, et $xx+\frac{r}{2p}\pm\sqrt{\frac{rr}{4pp}-\frac{t}{p}}=0$.(96) Et sic in alijs.

||[85] ||Sequentes duæ æquationes si ducantur in se invicem:

$$\begin{array}{l} x^3+exx+fx\ \ +g=0 \\ xx+hx\ +y\ \ \ \ \ \ =0 \\ \hline x^5+ex^4+fx^3\ +gxx \\ \quad +hx^4+ehx^3+fhxx+ghx \\ \quad\quad +yx^3\ +eyxx+fyx+gy \\ \hline \end{array}$$

prodit $x^5+px^4+qx^3+rxx+sx+t=0$. Posito jam $e=0$ erit

$$p=h.$$
$$q=f+y, \text{ ergo } q-y=f.$$
$$r=g+hf, \text{ ergo } r-pq+py=g.$$
$$s=gh+fy.$$
$$t=gy, \text{ ergo } t=ry-pqy+pyy, \text{ et } y=\tfrac{1}{2}q-\frac{r}{2p}[\pm]\sqrt{\tfrac{1}{4}qq-\frac{qr}{2p}+\frac{rr}{4p}+\frac{t}{p}}.$$

Quòd si igitur quantitates inventæ substituantur pro h et y, dividi potest hæc æquatio per $xx+px+\frac{1}{2}q-\frac{r}{2p}[\pm]\sqrt{\frac{1}{4}qq-\frac{qr}{2p}+\frac{rr}{4pp}+\frac{t}{p}}=0$.(97) Et posito $e=0$, $q=0$, et $s=0$, erit

$$p=h.$$
$$q=f+y=0, \text{ ergo } -y=f.$$
$$r=g+hf, \text{ ergo } r+py=g.$$
$$s=gh+fy=0, \text{ vel } pr+ppy=yy, \text{ ergo } y=\tfrac{1}{2}pp[\pm]\sqrt{\tfrac{1}{4}p^4+pr}.$$
$$t=gy, \text{ ergo } t=ry+pyy, \text{ et } y=-\frac{r}{2p}[\pm]\sqrt{\frac{rr}{4pp}+\frac{t}{p}}.$$

Quòd si igitur quantitates inventæ substituantur pro h, et y, dividi potest hæc æquatio per $xx+px+\frac{1}{2}pp[\pm]\sqrt{\frac{1}{4}p^4+pr}=0$, et $xx+px-\frac{r}{2p}[\pm]\sqrt{\frac{rr}{4pp}+\frac{t}{p}}=0$,(98) et sic in alijs.

(96) *Ibid.*: 2da Pars [1, Cas. 7]: 443. Note that $f=0$ implies $s=0$.
(97) *Ibid.*: 3a Pars [Cas. 1]: 447.

[86] ||Sequentes duæ æquationes si multiplicentur invicem

$$\begin{array}{l} x^4+ex^3+fxx+gx+h=0 \\ xx+yx+z=0 \\ \hline x^6+ex^5+fx^4+gx^3+hxx \\ \quad +yx^5+eyx^4+fyx^3+gyxx+hyx \\ \qquad +zx^4+ezx^3+fzxx+gzx+hz \\ \hline \end{array}$$

prodit $x^6+px^5+qx^4+rx^3+sxx+tx+v=0$. Et posito $e=0$ erit

$p=y.$

$q=f+z$, vel $q-z=f.$

$r=g+fy$, vel $r-pq+pz=g.$

$s=h+gy+fz$, vel $s=\frac{v}{z}+rp-ppq+ppz+qz-zz.$

$t=hy+gz$, ergo $t=\frac{vp}{z}+rz-pqz+pzz$, vel $\frac{t}{p}=\frac{v}{z}+\frac{r}{p}z-qz+zz.$

$v=hz$, vel $\frac{v}{z}=h.$

Quantitates inventas s et $\frac{t}{p}$ adde invicem, prodit $s+\frac{t}{p}=\frac{2v}{z}+rp-ppq+ppz+\frac{r}{p}z$, hâc æquatione ductâ in z et divisâ per $\frac{r}{p}+pp$, prodit $zz-\dfrac{ppqz+rpz-\frac{t}{p}z-sz}{\frac{r}{p}+pp}+2v=0.$[99] Quòd si igitur valores inventi substituantur pro y et z, dividi potest æquatio per $xx+yx+z=0$.

[87] ||Postremò, sequentes æquationes si multiplicentur invicem

$$\begin{array}{l} xx+ex+f=0 \\ xx+gx+h=0 \\ \hline x^4+ex^3+fxx \\ \quad +gx^3+egxx+fgx \\ \qquad +hxx+ehx+fh \\ \hline x^4+kx^3+lxx+mx+n=0 \\ xx+yx+z=0 \\ \hline x^6+kx^5+lx^4+mx^3+nxx \\ \quad +yx^5+kyx^4+lyx^3+myxx+nyx \\ \qquad +zx^4+kzx^3+lzxx+mzx+nz \\ \hline \end{array}$$

prodit $x^6+px^5+qx^4+rx^3+sxx+tx+v=0$. Tum posito $e=0$, et $l=0$, erit in prima multiplicatione:

(98) *Ibid.*: 3ª Pars [Cas. 6]: 447.

(99) *Ibid.*: 4ª Pars, A: 450.

$k=g.$
$l=f+h=0$, et $h=-f.$
$m=fg.$
$n=fh.$

et in secunda multiplicatione

$p=k+y.$
$q=ky+z.$
$r=m+kz.$
$s=n+my.$
$t=ny+mz.$
$v=nz.$

Quòd si igitur pro k statuatur g, pro m statuatur fg, pro n statuatur fh, habetur

||[88] ||$p=g+y$, vel $p-y=g.$
$q=gy+z$, vel $q-py+yy=z.$
$r=fg+gz$, vel $\frac{r}{p-y}-q+py-yy=f$, ergo $fg=r-pq+ppy-2pyy+y^3.$
$s=fh+fgy$, vel $s-fgy=fh.$
$t=fhy+fgz$, vel $\frac{t-fgz}{y}=fh.$
$v=fhz$, vel $\frac{v}{z}=fh.$

Hîc sunt tres æquationes, quæ sunt æquales ipsi fh, quare $s-fgy=\frac{t-fgz}{y}$ et $sy-fgyy+fgz-t=0$, multiplicato jam valore ipsius fg per valorem ipsius $-yy+z$, positoq; producto pro $-fgyy+fgz$, tum divisis omnibus per $-p$, prodit

$$\begin{array}{llll} y^4-\frac{q}{p}y^3 & +3qyy & -2pqy & +\frac{t}{p}=0.^{(100)} \\ -2p & +pp & -\frac{qq}{p} & -\frac{q}{r} \\ & & +r & +qq \\ & & -\frac{s}{p} & \end{array}$$

Quòd si igitur valor ipsius y ponatur pro ipso y, et valor ipsius z pro ipso z, dividi potest æquatio per $xx+yx+z=0$.

---

(100) *Ibid.*: 4ª Pars, C: 450.

(101) This section is apparently of Kinckhuysen's own invention but, patently, his method is modelled on the Cartesian techniques developed by Hudde in his *De Reductione Æquationum* (note (80)). On the whole it would seem to take longer to substitute in Kinckhuysen's derived

Quomodo absq; divisione nosci queat, quando æquationis alicujus radix una nota est, quænam æquatio cæteris radicibus congruat.(101)

Veluti, si detur $x^4+px^3+qxx+rx+s=0$, quæ dividi possit per $x+b$, semper [89] inveniri ||potest quæsitum absq; divisione, modo sequenti:

$$\begin{array}{l} x^3+kxx+lx\quad +m=0 \\ x\;+b \qquad\qquad\quad =0 \\ \hline x^4+kx^3+lxx\;+mx \\ \quad +bx^3+bkxx+blx+bm \\ \hline x^4+px^3+qxx\;+rx\;+s=0. \end{array}$$

Estq; $p=k+b$, et $k=p-b$.
$q=l+bk$, sublato k prodit $l=q-pb+bb$.
$r=m+bl$, sublato l prodit $m=r-qb+pbb-b^3$.
$s=bm$, prodit $m=\frac{s}{b}$.

Quòd si igitur in hac æquatione $x^3+kxx+lx+m=0$, tollantur k, l, et m, prodit

$$\begin{array}{llll} x^3 & +pxx & +qx & +r=0. \\ & -b & -pb & -qb \\ & & +bb & +pbb \\ & & & -b^3 \end{array}$$

Vel quia valor ipsius m duobus modis habetur

$$\begin{array}{lll} x^3 & +pxx & +qx+\frac{s}{b}=0. \\ & -b & -pb \\ & & +bb \end{array}$$

Sequitur exemplum, $x^4-10x^3+27xx-6x-24=0$, dividi potest per $x-4=0$, hîc est $p=-10$, $q=27$, $s=-24$, et $b=-4$, itaq; prodit

$$\begin{array}{llll} x^3 & -10xx & +27x & \frac{-24}{-4}=0, \\ & +4 & -40 & \\ & & +16 & \end{array}$$

[90] hoc est ||$x^3-6xx+3x+6=0$, et tantundem prodibit, diviso

$$x^4-10x^3+27xx-6x-24=0,$$

per $x-4=0$.

Porrò cùm p est$=0$, prodit $\begin{array}{l} x^3-bxx+qx+\frac{s}{b}=0. \\ \qquad\qquad\quad +bb \end{array}$

Quando q est$=0$, prodit $\begin{array}{l} x^3+pxx-pbx+\frac{s}{b}=0. \\ \qquad\;\; -b \quad +bb \end{array}$

---

equation than to divide through by the linear factor straightforwardly. In the following example note that $s/b = r-qb+pb^2-b^3$, that is, $b^4-pb^3+qb^2-rb+s = 0$, since $x+b$ is a factor of $x^4+px^3+qx^2+rx+s = 0$.

Quando s est $=0$, etiam m est $=0$, tunc habetur $\begin{array}{lll} xx+px & +q & =0, \\ \quad\ -b & -pb & \\ & +bb & \end{array}$ vel $\begin{array}{l} xx+px+\frac{r}{b}=0. \\ \quad\ -b \end{array}$

Quando r et s est $=0$, etiam l et m est $=0$, et habetur $x+p=0$, vel $\begin{array}{l} x+\frac{q}{b}=0, \\ \ \ -b \end{array}$ et sic in alijs.

Quomodo æquationis cubicæ radix inveniri possit, extrahendo radicem cubicam ex binomio.(102)

Cùm æquationes cubicas solvere volumus extrahendo radicem cubicam ex binomio, oportet priùs tollere secundum æquationis terminum, si adsit, et emerget aliquis ex tribus hisce casibus:

$$x^3 = * - px + q.$$
$$x^3 = * + px + q.$$
$$x^3 = * + px - q.$$

Talis æquatio ut constituatur multiplicando, ponitur $x=y+z$, quo utrinq; ||[91] cubicè multiplicato extat $x^3=y^3+3zyy+3zzy+z^3$, tum posito x loco ||ipsius $y+z$, emergit $x^3=3zyx+y^3+z^3$, quæ æquatio si comparetur cum secundo casu, habebuntur duæ hæ æquationes:

$$+p=3zy, \text{ ergo } \frac{+p}{3z}=y.$$
$$+q=y^3+z^3, \text{ ergo } y^3=q-z^3.$$

Pone $\frac{+p}{3z}$ loco ipsius y, habebis $\frac{+p^3}{27z^3}=q-z^3$, vel $+p^3=27qz^3-27z^6$, et $z^6-qz^3+\frac{1}{27}p^3=0$. Et $v=z^3=\frac{1}{2}q+\sqrt{\frac{1}{4}qq-\frac{1}{27}p^3}$, ergo $q-z^3=y^3=\frac{1}{2}q-\sqrt{\frac{1}{4}qq-\frac{1}{27}p^3}$. Ita prodit $z=\sqrt{C.\ \frac{1}{2}q+\sqrt{\frac{1}{4}qq-\frac{1}{27}p^3}}$, et $y=\sqrt{C.\ \frac{1}{2}q-\sqrt{\frac{1}{4}qq-\frac{1}{27}p^3}}$. Ergo $z+y$ vel $x=\sqrt{C.\ \frac{1}{2}q+\sqrt{\frac{1}{4}qq-\frac{1}{27}p^3}}+\sqrt{C.\ \frac{1}{2}q-\sqrt{\frac{1}{4}qq-\frac{1}{27}p^3}}$, pro secundo casu. Jam si ponatur $-p$ pro $+p$, extat pro primo casu:

$$x=\sqrt{C.\ \tfrac{1}{2}q+\sqrt{\tfrac{1}{4}qq+\tfrac{1}{27}p^3}}+\sqrt{C.\ \tfrac{1}{2}q-\sqrt{\tfrac{1}{4}qq+\tfrac{1}{27}p^3}}.$$

---

(102) Compare Descartes' *Geometrie* (note (23)): 397–400; and Johann Hudde's *De Reductione Æquationum* (note (80)): 499–503.

(103) The rule is, of course, 'Cardan's' (Girolamo Cardano, *Artis Magnæ sive de Regulis Algebraicis Liber Unus* (Nuremberg, 1545): Cap XI) but Kinckhuysen takes his proof more directly from Hudde. Compare also I, 1, 3, §3.5. As Newton was to point out in his *Observationes* (§2: note (102) below) Cardan's rule is not, in fact, restricted as Kinckhuysen implies in the sequel to the 'reducible' cubics (with a unique real root) for which $\frac{1}{4}q^2 \geqslant \frac{1}{27}p^3$ and there is no theoretical necessity to resort to Descartes' 'trisection' method in dealing with the complementary 'irreducible' cubic (all of whose roots are real) when $\frac{1}{4}q^2 < \frac{1}{27}p^3$. Note, too, that

Et si $-q$ ponatur pro $+q$, extat pro tertio casu:

$$x=\sqrt{C.-\tfrac{1}{2}q+\sqrt{\tfrac{1}{4}qq-\tfrac{1}{27}p^3}}+\sqrt{C.-\tfrac{1}{2}q-\sqrt{\tfrac{1}{4}qq-\tfrac{1}{27}p^3}}.^{(103)}$$

Exemplum pro primo casu, dato $x^3=*-3x+36$, erit $x=\sqrt{C.18+\sqrt{325}}+\sqrt{C.18-\sqrt{325}}$, nam $\frac{1}{2}q$ valet 18, $\frac{1}{4}qq$ valet 324, et $\frac{1}{27}p^3$, hoc est cubus ipsius $\frac{1}{3}p$, valet 1. Extractâ radice cubicâ ex utroq; binomio extat $x=1\frac{1}{2}+\sqrt{3\frac{1}{4}}+1\frac{1}{2}-\sqrt{3\frac{1}{4}}$, hoc est $x=3$.

Hîc notandum, in hoc primo casu non posse nisi unã veram radicem esse, et duas falsas non nisi imaginarias esse, de quo casu nonnihil indicatũ [erat] fol. 47.

[92] ||Exemplum pro secundo casu, dato $x^3=*+6x+40$. Erit

$$x=\sqrt{C.20+\sqrt{392}}+\sqrt{C.20-\sqrt{392}}.$$

Nam $\frac{1}{2}q$ valet 20, $\frac{1}{4}qq$ valet 400, et $\frac{1}{27}p^3$, hoc est cubus ipsius $\frac{1}{3}p$ valet 8. Extractâ radice cubica ex utroq; horum binomiorum, extat $x=2+\sqrt{2}+2-\sqrt{2}$, hoc est $x=4$.

Tertius casus idem est cum secundo, nisi quod radices, quæ in secundo casu veræ sunt, in tertio sunt falsæ. Hîc notandum, in secundo et tertio casu ipsum $\frac{1}{4}qq$ semper oportere majorem esse $\frac{1}{27}p^3$. Verùm si contingat $\frac{1}{27}p^3$ majorem esse quàm $\frac{1}{4}qq$, solutio fieri potest per anguli trisectionem, quemadmodum hîc sequitur in adjecto circulo,(104) ubi ponatur radius $NO=\sqrt{\frac{1}{3}p}$, et chorda $NP=\frac{3q}{p}$, tum arcûs NTP et NVP dividuntur uterq; in tres partes æquales, erit summa chordarum TP et PV vera radix in secundo casu, hoc est WP, posito TV æquali [93] WV; et TP et PV erunt duæ falsæ. ||At in tertio casu chordæ TP et PV erunt radices veræ, et WP erit falsa.(105) Hoc pacto valor radicis præter propter inveniri potest per tabulam sinuum.

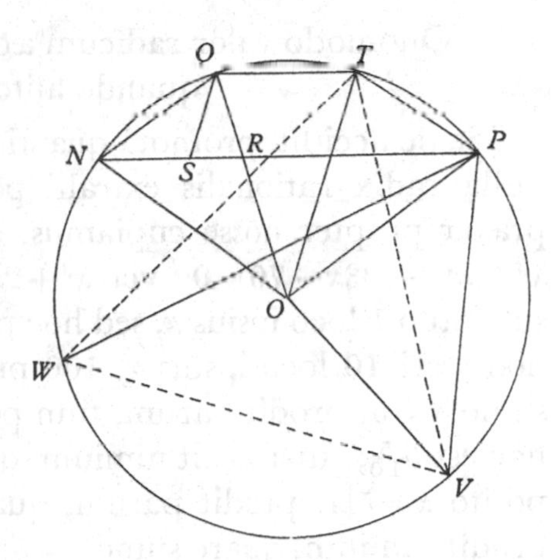

Kinckhuysen does not in his two examples concern himself with the complex pairs of roots $\frac{1}{2}\omega(1+\sqrt{13})+\frac{1}{2}\omega^{-1}(1-\sqrt{13}) = -\frac{1}{2}(3\pm\sqrt{-39})$ and $\omega(2+\sqrt{2})+\omega^{-1}(2-\sqrt{2}) = -2\pm\sqrt{-6}$ respectively.

(104) Apart from the triangle TVW, added by Kinckhuysen in broken line, the figure is Descartes' (*Geometrie*: 396).

(105) If we set $P\hat{O}T = \vartheta$, $W\hat{P}O = \frac{1}{2}\phi$ (or $\vartheta+\phi = \frac{1}{3}\pi$) and $r = \sqrt{[\frac{1}{3}p]}$, we may easily show that $\frac{1}{3}N\hat{P}O$ (or $\frac{1}{2}\pi-\frac{3}{2}\vartheta$) $= W\hat{P}O$: also, on defining

$$S(\vartheta) = 2r\sin\tfrac{1}{2}\vartheta \quad\text{and}\quad CS(\phi) = 2r\cos\tfrac{1}{2}\phi,$$

that

$$NP = 3q/p = S(3\vartheta) = CS(3\phi)$$

with

$$PT = S(\vartheta) = CS(\phi+\tfrac{2}{3}\pi), \quad -PW = S(\vartheta+\tfrac{2}{3}\pi) = CS(\phi)$$

Si quis dubitet, utrum WP æquale sit $TP+PV$, id explorari potest modo sequenti;(106) quoniam TP est $\frac{1}{3}$ arcûs NTP, et $PV=\frac{1}{3}$ arcûs NVP, ergo $TV=\frac{1}{3}$ integræ peripheriæ, idéoq; TVW triangulum est æquilaterū, cujus latus pono $=z$, $TP=x$, et $PV=y$, duo rectāgula TP, WV et WT, PV simul æqualia sunt rectangulo TV, WP, quare $xz+yz$ si dividatur per $TV=z$, prodit $WP=x+y$, hoc est $=TP+PV$.

Cur NO ponatur $=\sqrt{\frac{1}{3}p}$, et $NP=\frac{3q}{p}$, sic patebit. Sit $NO=m$, $NP=n$, et $NQ=x$, tum dic ut $NO=m$ ad $NQ=x$, ita $NQ=x$ ad $QR=\frac{xx}{m}$, et sic $[QR=]\frac{xx}{m}$ ad $\frac{x^3}{mm}$ pro SR. Porrò $SR+NP$(107) æquale est ter NQ, hoc est $\frac{x^3}{mm}+n=3x$, ductis omnibus in mm, prodit $x^3=3mmx-mmn$, quare dato $x^3=+px-q$, duæ sunt æquationes

$+p=+3mm$, quare $\frac{1}{3}p=mm$, et $\sqrt{\frac{1}{3}p}=m$.

$-q=-mmn$, prodit $\frac{q}{mm}=n$, pone $\frac{1}{3}p$ loco ipsius mm, prodit $\frac{3q}{p}=n$.

Quomodo valor radicum æquationis investigatur appropinquando, quando aliter inveniri nequeunt.(108)

||[94] ||Sæpe accidit, promotâ quæstione ad æquationem, ut tū ||ex ista æquatione nulla radix rationalis extrahi possit, tum si valorem radicum nihilominus præter propter nôsse cupiamus, quæritur is appropinquando, veluti si detur $x^3+2xx-23x-70=0$, vel $x^3+2xx=23x+70$. Primùm pono $x=1$, hoc est substituo 1 loco ipsius x, sed hoc parum esse deprehenditur, quare pono $x=10$, hoc [est] 10 loco ipsius x, 100 pro xx, et 1000 pro x^3, prodit nimium, quare sumo $x=5$, prodit parum, tum pono $x=6$, prodit nimium, tum pono $x=5\lfloor 5$, hoc est $5\frac{5}{10}$, rursus exit nimium, quare pono $x=5\lfloor 2$, adhuc erit nimium, quare posito $x=5\lfloor 1$, prodit parum, quare rursus pono medium, nimirum $x=5\lfloor 15$, prodit nimium, quare sumo $x=5\lfloor 13$, prodit parum, et $x=5\lfloor 14$, prodit nimium quare pono $x=5\lfloor 135$, prodit nimium, et $x=5\lfloor 134$, prodit parum, ita ut valor ipsius x major sit quàm $5\frac{134}{1000}$, minor autem quàm $5\frac{135}{1000}$; sic continuando, tam prope ad verum accedere licet, quantum calculū prosequi lubet; et sic in alijs.

---

and $PV = S(\vartheta+\frac{4}{3}\pi) = CS(\phi-\frac{2}{3}\pi)$. In these terms Kinckhuysen's second case reduces to $(x/r)-3(x/r)^3 = CS(3\phi)/r$ and the third to $3(x/r)-(x/r)^3 = S(3\vartheta)/r$: in the former, consequently, $x_i = CS(\phi+\frac{2}{3}i\pi)$, $i = -1, 0, 1$ and in the latter $x_i = S(\vartheta+\frac{2}{3}i\pi)$, $i = 0, 1, 2$, while in both the sum of the roots are zero. Compare Frans van Schooten's *Commentarii in Librum, III, X = Geometria* ($_2$1659): 330–43; and also 1, 3, 2, §2.4: Theorem 1.

(106) The method, which depends on an application of Ptolemy's theorem on the products of diagonals and opposite sides of a quadrilateral inscribed in a circle (*Almagest*, Book 1), is evidently Kinckhuysen's.

(107) That is, $NR+SP$ (or $2NR$) $= 3NR$.

Quo pacto sciri possit, quibus limitibus quantitas verarum radicum contineatur.(109)

Dato $x^3 * -qx+r=0$, vel $x^3+r=qx$, erit qx majus quàm r, diviso utroq; per q, prodit x majus quàm $\frac{r}{q}$; est quoq; qx majus quàm x^3, et q majus quàm xx, ergo $\surd q$ majus quàm x. Quare invenimus x majus esse quàm $\frac{r}{q}$, et minus quàm $\surd q$.

Dato $x^3 * +qx-r=0$, vel $x^3+qx=r$, erit r majus quàm qx, et $\frac{r}{q}$ majus quàm x; est quoq; r majus quàm x^3, et $\surd C.r$ majus quàm x, ergo $x\surd C.rr$ majus quàm x^3, quare $x\surd C.rr+qx$ majus est quàm r, et x majus quàm $\frac{r}{\surd C.rr+q}$. Invenimus [95] ||igitur x majus quàm $\frac{r}{\surd C.rr+q}$, et minus quàm $\frac{r}{q}$.

Dato $x^3-pxx+qx-r=0$, hoc est $x^3-pxx=r-qx$. Si x magis sit quàm p, erit $\frac{r}{q}$ majus quàm x, sed si p majus sit quàm x, erit x majus quàm $\frac{r}{q}$, quare invenimus valorem ipsius x consistere inter p et $\frac{r}{q}$.

Dato $x^3-pxx-qx+r=0$, hoc est $x^3+r=pxx+qx$, erit $pxx+qx$ majus quàm r, et x majus quàm $-\frac{q}{2p}+\sqrt{\frac{qq}{4pp}+\frac{r}{p}}$; porrò $pxx+qx$ majus est quàm x^3, hoc est $px+q$ majus quàm xx, quare $\frac{1}{2}p+\sqrt{\frac{1}{4}pp+q}$ majus est quàm x; invenimus igitur x majus esse quàm $-\frac{q}{2p}+\sqrt{\frac{qq}{4pp}+\frac{r}{p}}$ et minus quàm $[\frac{1}{2}p]+\sqrt{\frac{1}{4}pp+q}$.

Idem aliter. $x^3+r=pxx+qx$, ergo $px+q$ majus quàm xx, quòd si igitur x majus sit quàm $\surd q$, erit $px+x\surd q$ majus quàm xx, et $p+\surd q$ majus quàm x, et si $\surd q$ majus sit quàm x, erit $p+\surd q$ multò major quàm x. Porrò $pxx+qx$ majus est quàm r, et cùm $p+\surd q$ majus sit quàm x, erit $ppx+px\surd q$ majus quàm pxx, quare $ppx+px\surd q+qx$ majus quàm r, et x majus quàm $\frac{r}{pp+p\surd q+q}$; invenimus igitur x majus quàm $\frac{r}{pp+p\surd q+q}$ et minus quàm $p+\surd q$.

---

(108) This cumbrous but effective approach is Stevin's: see his *Appendice algebraique* ((Leyden, 1594): all copies lost) to his *Arithmetique...Contenant les computations des nombres Arithmetiques ou vulgaires* ((Leyden, ₂1625): Prob. LXXVII, Corollarium): and compare D. J. Struik's summary in *The Principal Works of Simon Stevin*, **2** (Amsterdam, 1958): 475–6. In his Dutch original Kinckhuysen, we may note, employed the decimal point in its modern English usage, but Mercator has here rendered it by Oughtred's somewhat clumsier '⌊' sign.

(109) Compare Erasmius Bartholin's *De Æquationum Natura, Constitutione, & Limitibus Opuscula Duo. Incepta à Florimondo de Beaune...: Absoluta verò, & post mortem ejus edita* (Amsterdam, 1661) = *Geometria*, **2** (1661): 49–152, especially 122ff.

Dato $x^4 * -qxx-rx-s=0$, erit $x^4-qxx=rx+s$, quare xx majus quàm q, x majus quàm $\sqrt{q}$ et $x^3\sqrt{q}$ majus quàm qxx; porrò x^4-rx est $=qxx+s$, ergo x^3 majus quàm r, x majus quàm $\sqrt{C.r}$ et $x^3\sqrt{C.r}$ majus quàm rx, deinde x^4-s est $=qxx+rx$, ergo x^4 majus quàm s, x majus quàm $\sqrt{}\sqrt{s}$, et $x^3\sqrt{}\sqrt{s}$ majus quàm s. Etiam $x^4=qxx+rx+s$, ergo $x^3\sqrt{q}+x^3\sqrt{C.r}+x^3\sqrt{}\sqrt{s}$ majus quàm x^4, et

$$\sqrt{q}+\sqrt{C.r}+\sqrt{}\sqrt{s}$$

majus quàm x; invenimus igitur x majus quàm $\sqrt{q}$, vel $\sqrt{C.r}$, vel $\sqrt{}\sqrt{s}$, at minus quàm $\sqrt{q}+\sqrt{C.r}+\sqrt{}\sqrt{s}$, et sic in alijs.

Atq; hæc de æquationibus dicta sufficiant, nam mihi propositum est tantùm tyronem in viam introducere.

||[96]

||Pars Tertia. Quomodo quæstio aliqua ad æquationem Redigatur.

(110)Restat nunc ut ostendamus, quomodo quæstio aliqua ad æquationem redigatur, ubi commodissima via occurrit, si hîc solutionem apponam quorundam problematum. Interim notandum [est], in qualibet quæstione, tot æquationes inveniendas esse, quot positæ fuerint quantitates ignotæ, quæ deinde ad unicam reducitur talem, in qua non nisi unius generis quantitates ignotæ restent. Quòd si igitur accidat ut tot æquationes inveniri haud possint, licèt agatur quicquid potest, argumentum est pauciora esse data quàm oportet, atq; tum pro qualibet istarum quantitatum ignotarum, quibus æquatio inveniri nequit, prout natura quæstionis postulat, talem numerum ponere licet qualem lubet; unde sequitur ejusmodi quæstiones multas habere solutiones. Quæ res ex sequentibus hisce problematis, alijsq; quæ quilibet sibi ipsi proponere potest, satis intelligitur. Delegi in hanc finem quæstiones, quæ æquationes quadraticas non excedunt, quarum aliquæ ex quæstionibus artificiosis *Ludolphi à Ceulen* sumtæ sunt.(111)

||[97] ||[1] *Vendit quis equum 144 florenis,*(112) *et lucratur pro quovis centenario totidem, quot ipsi florenis equus constitit. Quæritur de lucro.*

quum illi constitisse x florenis, et lucrum erit . Quare cùm pro quovis centenario lucretur x oc est cùm pretium quo constitit equus sit ad t 100 ad x sive cum sit $x. 144-x::100.x$, onendo æqualitatem inter factos extremorū et diorum) $xx=14400-100x$. et $x=80$. quare 64$(=144-x)$ est lucrum quæsitum.(113)]

Pone equum illi constitisse x florenis, quare ut 100 ad $100+x$, ita x ad 144. Nota quatuor proportionalium rectangulum sub medijs æquale est rectangulo sub extremis, quare $100x+xx=14400$, vel $xx+100x-14400=0$, et $x=80$, pretium scilicet quo equus constitit, quare lucrū quæsitum est 64 flor.

---

(110) Newton in 1670 rightly judged this introductory paragraph to be inadequate and wrote in its stead an extended preface (§2: note (111) below) to this 'most noble & difficult part of the [algebraick] art'.

[2] *Quinam sunt numeri quorum summa primi et secundi ducta in tertium facit 264, at secundi et tertij summa ducta in primum facit 102, deniq; summa primi et tertij [ducta] in secundum producit 234.*

Pro numeris quæsitis pone x, y, et z, eritq;

$$\begin{array}{r} xz+yz=264 \\ xy+xz=102 \\ yx+yz=234 \\ \hline 2xz+2yz+2xy=600 \\ \hline xz\ +yz\ +xy=300 \\ xz\ +yz\ \ \ \ \ \ \ =264 \\ \hline xy=\ 36\ [\text{et}]\ y=\dfrac{36}{x}. \end{array} \qquad \begin{array}{r} xy+xz=102 \\ xy\ \ \ \ \ \ \ =\ 36 \\ \hline xz=\ 66\ [\text{et}]\ z=\dfrac{66}{x}. \end{array}$$

Im prima æquatione pone $\frac{36}{x}$ loco ipsius y, et $\frac{66}{x}$ loco ipsius z, extabit $66+\frac{2376}{xx}=264$, quare $198xx=2376$, et $xx=12$, proditq; $x=\sqrt{12}$, $y=\frac{36}{\sqrt{12}}$, et $z=\frac{66}{\sqrt{12}}$, vel $[x=]2\sqrt{3}$, $[y=]6\sqrt{3}$, et $[z=]11\sqrt{3}$.

[98] || [3](114) *Sunt tres continuè proportionales, quorum summa est 20, et quadratorū summa est 140; quæritur quinam sint.*

Pro summa duorum extremorum pone x, erit medius $20-x$, tum posito y pro primo, erit tertius $x-y$. Atqui rectangulum sub extremis æquale est quadrato medij, prodit $xy-yy=400-40x+xx$, vel $yy-xy+400-40x+xx=0$, quare $y=\frac{1}{2}x+\sqrt{-400+40x-\frac{3}{4}xx}$, pro primo numero, et

$$x-y=\tfrac{1}{2}x-\sqrt{-400+40x-\tfrac{3}{4}xx},$$

pro tertio; summa quadratorū à tribus hisce numeris, nimirū $40x-400$, æqualis est 140,(115) prodit $40x=540$, et $x=13\frac{1}{2}$, quòd si igitur in tribus inventis numeris ponatur ubiq; $13\frac{1}{2}$ loco ipsius x, prodit medius $6\frac{1}{2}$, et duo extremi $6\frac{3}{4}+\sqrt{3\frac{5}{16}}$, et $6\frac{3}{4}-\sqrt{3\frac{5}{16}}$.

(111) 'uyt de konstighe Vraghen van *Ludolf van Ceulen*' (Kinckhuysen). The reference is to Ceulen's *De Arithmetische en Geometrische fondamenten* (note (15)).

(112) 'florenæ' (florins) is Mercator's anglicization of Kinckhuysen's 'gulden' (guilders).

(113) Kinckhuysen's resolution of this first problem is, indeed, somewhat elliptically expressed and not easy to follow, but it is not clear why Newton singled it out for extended revision in the margin. Perhaps he intended to do likewise for others below, and then changed his mind.

(114) Newton will take this as the first of his explanatory examples in his reworking of Kinckhuysen's 'Pars Tertia'. See §2: note (115) below.

(115) More simply, $40x-400 = y^2-xy+x^2 = y^2+y(x-y)+(x-y)^2 = 140$.

[4] *Quatuor continuè proportionalium, duo medij simul constituunt 12, et duo extremi 20; quæritur quinā sint.*

Pro primo statue $10-y$, erit quartus $10+y$, tum pro secundo statue $6-x$, erit tertius $6+x$. Rectangulū sub extremis æquale est rectangulo sub medijs, quare $100-yy=36-xx$, prodit $xx=yy-64$, et $x=\sqrt{yy-64}$, hoc igitur substituo pro x, quatuor numeri sunt $10-y$, $6-\sqrt{yy-64}$, $6+\sqrt{yy-64}$, et $10+y$. Rursus rectangulum sub primo et tertio æquale est quadrato medij, hoc est:

$$60-6y\begin{smallmatrix}+10\\-y\end{smallmatrix}\sqrt{yy-64}=yy-28-12\sqrt{yy-64}.$$

Item rectangulum sub secundo et quarto æquale est quadrato tertij, hoc est:

||[99] ||$$60+6y\begin{smallmatrix}-10\\-y\end{smallmatrix}\sqrt{yy-64}=yy-28+12\sqrt{yy-64}.$$

Hæ duæ æquationes si auferantur ab invicem, restat

$$24\sqrt{yy-64}=12y-20\sqrt{yy-64},$$

hoc est $44\sqrt{yy-64}=12y$, quibus divisis per 4, si orti multiplicentur quadraticè, prodit $121yy-7744=9yy$ et $yy=69\frac{1}{7}$, quare $y=\sqrt{69\frac{1}{7}}$, vel $y=22\sqrt{\frac{1}{7}}$, quod si ubiq; in quatuor numeris inventis substituatur pro y, prodeunt quatuor numeri quæsiti, $10-22\sqrt{\frac{1}{7}}$, $6-6\sqrt{\frac{1}{7}}$, $6+6\sqrt{\frac{1}{7}}$, et $10+22\sqrt{\frac{1}{7}}$.

[5] *Quinam sunt numeri quorum primus et tertius simul constituunt duplum secundi; et primus ductus in secundum facit 20; deniq; rectangulū sub tertio et primo unà cum quadrato tertij facit 80.*

Pro tribus numeris statue x, $\frac{20}{x}$, et y, erit $x+y=\frac{40}{x}$, et $y=\frac{40}{x}-x$. Rursus est $xy+yy=80$, et $y=\sqrt{\frac{1}{4}xx+80}-\frac{1}{2}x$. Hi duo valores ipsius y sunt invicē æquales, nimirum $\frac{40}{x}-x=\sqrt{\frac{1}{4}xx+80}-\frac{1}{2}x$, ergo $\frac{80-xx}{2x}=\sqrt{\frac{1}{4}xx+80}$, quibus utrobiq; multiplicatis quadraticè prodit $\frac{6400-160xx+x^4}{4xx}=\frac{1}{4}xx+80$, quare $xx=13\frac{1}{3}$, et $x=\sqrt{13\frac{1}{3}}$, $\frac{20}{x}=\sqrt{30}$, et $y=\frac{40}{x}-x=\sqrt{53\frac{1}{3}}$. Quare tres numeri quæsiti sunt $2\sqrt{\frac{10}{3}}$, $3\sqrt{\frac{10}{3}}$, et $4\sqrt{\frac{10}{3}}$.

[6] *Quinq; numeri sunt in progressione Arithmetica, quorum quatuor primorum summa ducta in quintū facit 6000, et summa quatuor postremorum ducta in primum facit 4488. Quæritur quinā sint.*

||[100] ||Pro primo numero pone x, et pro differentia numerorum y, sunt igitur x, $x+y$, $x+2y$, $x+3y$, et $x+4y$. Summa quatuor primorū, nimirū $4x+6y$, si ducatur in $x+4y$, prodit $4xx+22xy+24yy=6000$. Summa quatuor postremorū,

nimirum $4x+10y$, si ducatur in x, prodit $4xx+10xy=4488$. Hæ duæ æquationes ablatæ ab invicem relinquunt $12xy+24yy=1512$. Quare $x=\frac{126-2yy}{y}$, quo substituto in prima æquatione pro x, et ejus quadrato pro xx, prodit

$$\frac{63504+756yy-4y^4}{yy}=6000.$$

Quare $y^4+1311yy-15876=0$, et $yy=12$,(116) $x=\frac{126-2yy}{y}$ valebit $\surd 867$. Est igitur $x=17\surd 3$, et $y=2\surd 3$, prodeunt tandem numeri quæsiti $17\surd 3$, $19\surd 3$, $21\surd 3$, $23\surd 3$, et $25\surd 3$.

Hæc quæstio etiam sic tractari potest: Duæ æquationes erant

$$4xx+22yx+24yy=6000, \text{ ergo } x=\sqrt{\tfrac{25}{16}yy+1500}-\tfrac{11}{4}y.$$
$$\text{et } 4xx+10yx \qquad =4488, \text{ ergo } x=\sqrt{\tfrac{25}{16}yy+1122}-\tfrac{5}{4}y.$$

Hi duo valores ipsius x æquales sunt invicem; quare

$$\sqrt{\tfrac{25}{16}yy+1500}-\tfrac{11}{4}y=\sqrt{\tfrac{25}{16}yy+1122}-\tfrac{5}{4}y,$$

adde utrobiq; $\frac{5}{4}y$ et multiplica deinde utrūq; æqualium quadraticè, prodit $\frac{61}{16}yy+1500-3y\sqrt{\frac{25}{16}yy+1500}=\frac{25}{16}yy+1122$, quare

$$\tfrac{9}{4}yy+378=3y\sqrt{\tfrac{25}{16}yy+1500},$$

multiplicato rursus utroq; æqualium quadraticè, ablatoq; uno æqualium ex altero, restat $9y^4+11799yy-142884=0$, vel $y^4+1311yy-15876=0$, et ut ante $y=\surd 12$, quare $x=\sqrt{\frac{25}{16}[\times]12+1500}-\frac{11}{4}\surd 12$, vel $x=22\frac{1}{2}\surd 3-5\frac{1}{2}\surd 3$, hoc est $x=17\surd 3$.

[101] ||[7] *Est progressio Arithmetica aliquot terminorum; si sum̄a omnium præter primum ducatur in primū prodit 70. Sin summa excepto secundo ducatur in secundum, fit 198. Deniq; si summa omnium excepto tertio ducatur in tertium, fit 310. Quot sunt termini progressionis, et qui?*

Pro primo numero pone x et pro differentia numerorum y, et pro numero terminorum z. Erit ultimus numerus $x+zy-y$. Etenim numerus terminorum minus unitate si ducatur in differentiam terminorum, prodit differentia primi et ultimi. Adde primum ultimo, prodit $2x+zy-y$, quo ducto in semissem numeri terminorum, hoc est in $\frac{1}{2}z$, prodit sum̄a omnium numerorum

$$xz+\tfrac{1}{2}yzz-\tfrac{1}{2}yz,$$

et ut vitemus molem magnorū numerorum, pro isto substituo v. Ab hac summa

(116) $(27\surd 10\frac{1}{4}-83\frac{1}{2})(27\surd 10\frac{1}{4}+83\frac{1}{2}) = \frac{1}{4}(27^2\times 41-167^2) = 500.$

ablato primo, et reliquo ducto in primū prodit $vx-xx=70$. A summa numerorum ablato secundo, nimirum $x+y$, et reliquo ducto in $x+y$, prodit

$$vx-xx+vy-2xy-yy=198.$$

A summa numerorum ablato tertio, nimirum $x+2y$, et reliquo ducto in $x+2y$, prodit $vx-xx+2vy-4xy-4yy=310$. Subtrahe primam æquationem à secunda, restat $vy-2xy-yy=128$. Tum subtrahe secundam æquationem à tertia, restat $vy-2xy-3yy=112$. Hæ duæ æquationes inventæ si auferantur ab invicem, ||[102] restat $2yy=16$, ergo $y=\surd 8$. ||Suprà inventum est $vy-2xy-yy=128$, substitue $\surd 8$ pro y, erit $v=\frac{136}{\surd 8}+2x$. Rursus $vx-xx=70$, sublato v prodit $\frac{136}{\surd 8}x+xx=70$, quare $x=\surd 2$. Ergo v valet $\frac{136}{\surd 8}+2\surd 2=xz-\frac{1}{2}yzz-\frac{1}{2}yz$, tolle x et y, prodit $36\surd 2=z\surd 2+zz\surd 2-z\surd 2$, hoc est $36\surd 2=zz\surd 2$, quo diviso per $\surd 2$, prodit $36=zz$, et $6=z$, pro numero terminorum; sunt igitur numeri quæsiti $\surd 2$, $3\surd 2$, $5\surd 2$, $7\surd 2$, $9\surd 2$, et $11\surd 2$.

[8] *Quærantur quatuor numeri, quorum differentia secundi et quarti sit $\frac{3}{8}$ differentiæ primi et tertij; item differentia secundi et tertij sit $\frac{2}{9}$ differentiæ primi et quarti. Tum si ducatur primus in quartum, productum sit 20, at secundi in tertium productum sit 16.*

Pro numeris quæsitis pone x, y, $x-z$, et $y-\frac{3}{8}z$, differentia inter secundum et tertium est $y-x+z$, et differentia inter primum et quartum est $x-y[+\frac{3}{8}]z$, prior ducatur in 9, et posterior in 2, prodit $9y-9x+9z=2x-2y+\frac{3}{4}z$, adde utrobiꝗ $-2x+2y-\frac{3}{4}z$, prodit $11y-11x+8\frac{1}{4}z=0$, et $z=\frac{4}{3}x-\frac{4}{3}y$, quo substituto pro z, sunt quatuor numeri x, y, $\frac{4}{3}y-\frac{1}{3}x$, et $1\frac{1}{2}y-\frac{1}{2}x$. Quare $1\frac{1}{2}xy-\frac{1}{2}xx=20$. et $\frac{4}{3}yy-\frac{1}{3}xy=16$. vel $yy-\frac{1}{4}xy-12=0$. Est igitur $y=\frac{1}{8}x+\sqrt{\frac{1}{64}xx+12}$, hoc substitue pro y in priori harum duarū æquationum, prodit $\frac{3}{16}xx+1\frac{1}{2}x\sqrt{\frac{1}{64}xx+12}-\frac{1}{2}xx=20$. ||[103] Ergo $24x\sqrt{\frac{1}{64}xx+12}=320+5xx$. Multiplica utrinꝗ quadraticè, ||et subtrahe æqualia ab invicem, prodit $16x^4-2712xx+102400=0$, quo diviso per 16, prodit $x^4-232xx+6400=0$. Quare $xx=200$ vel 32, et $x=10\surd 2$ vel $4\surd 2$; y est $=\frac{1}{8}x+\sqrt{\frac{1}{64}xx+12}$, sublato x prodit $y=4\surd 2$ vel $3\surd 2$; sunt igitur numeri quatuor quæsiti $10\surd 2$, $4\surd 2$, $2\surd 2$, et $\surd 2$, vel $4\surd 2$, $3\surd 2$, $2\frac{2}{3}\surd 2$, et $2\frac{1}{2}\surd 2$.

[9] *Est progressio Arithmetica quatuor terminorum, quorū summa primi et secundi ducta in tertium facit 20, at secundi et tertij summa in quartū* [*ducta*] *facit 40; Quæritur quinam sint numeri.*

Pro quatuor numeris quæsitis pone x, $x+y$, $x+2y$, et $x+3y$, summa primi et secundi $2x+y$ ducta in tertium $x+2y$, producit $2xx+5xy+2yy=20$. Quare, $yy+2\frac{1}{2}xy+xx-10=0$, est prima æquatio. Deinde summa secundi et tertij $2x+3y$ ducta in quartum $x+3y$ producit $2xx+9xy+9yy=40$, quare

$$yy+xy+\frac{2}{9}xx-4\frac{4}{9}=0,$$

quæ æquatio si auferatur à priori, restat $1\frac{1}{2}xy+\frac{7}{9}xx-5\frac{5}{9}=0$, ergo $y=\frac{100}{27x}-\frac{14x}{27}$, vel $\frac{100-14xx}{27x}$, quo substituto pro y in prima æquatione, erit

$$\frac{10000}{729xx}-\frac{3340}{729}-\frac{20xx}{729}=0,$$

vel $10000-3340xx-20x^4=0$, et $500-167xx-x^4=0$, quare $xx=\sqrt{7472\frac{1}{4}}-83\frac{1}{2}$, et $x=\sqrt{27\sqrt{10\frac{1}{4}}-83\frac{1}{2}}$. Porrò $y=\frac{100-14xx}{27x}$, tolle x, prodit $y=\frac{47-14\sqrt{10\frac{1}{4}}}{\sqrt{27\sqrt{10\frac{1}{4}}-83\frac{1}{2}}}$, vel divisâ fractione $\sqrt{8\sqrt{10\frac{1}{4}}-24}$. Hæc divisio peragitur modo sequenti: primùm multiplico $47-14\sqrt{10\frac{1}{4}}$ quadraticè, prodit $4218-1316\sqrt{10\frac{1}{4}}$, quo diviso per $27\sqrt{10\frac{1}{4}}-83\frac{1}{2}$, nimirū multiplicando priùs per $27\sqrt{10\frac{1}{4}}+83\frac{1}{2}$, extat

$$4000\sqrt{10\tfrac{1}{4}}-12000$$

[104] per 500 [dividendum],(116) prodit $8\sqrt{10\frac{1}{4}}-24$, unde extractâ radice ||quadraticâ, prodit $\sqrt{8\sqrt{10\frac{1}{4}}-24}$ pro valore ipsius y. Valor ipsius x, et valor ipsius y sunt communicantes, quoniam multiplicatis eorum quadratis, ex producto radix extrahi potest. Quære igitur communem illorum divisorem, prodit

$$x=\sqrt{10\tfrac{1}{4}}-\tfrac{1}{2}\sqrt{2\sqrt{10\tfrac{1}{4}}-6}, \quad \text{et} \quad y=2\sqrt{2\sqrt{10\tfrac{1}{4}}-6};$$

sunt igitur numeri quæsiti $\sqrt{10\frac{1}{4}}-\frac{1}{2}\sqrt{2\sqrt{10\frac{1}{4}}-6}$, $\sqrt{10\frac{1}{4}}+1\frac{1}{2}\sqrt{2\sqrt{10\frac{1}{4}}-6}$,

$$\sqrt{10\tfrac{1}{4}}+3\tfrac{1}{2}\sqrt{2\sqrt{10\tfrac{1}{4}}-6}, \quad \text{et} \quad \sqrt{10\tfrac{1}{4}}+5\tfrac{1}{2}\sqrt{2\sqrt{10\tfrac{1}{4}}-6}.$$

Vel $\sqrt{27\sqrt{10\frac{1}{4}}-83\frac{1}{2}}$, $\sqrt{7\sqrt{10\frac{1}{4}}-13\frac{1}{2}}$, $\sqrt{3\sqrt{10\frac{1}{4}}+8\frac{1}{2}}$, et $\sqrt{15\sqrt{10\frac{1}{4}}-17\frac{1}{2}}$.

[10] *Sunt quatuor numeri continuè proportionales, quorum summa est 10, et quadratorū summa est 80. Quæritur quinam sint.*

Pone summam duorū mediorum $=y$, et summam extremorum $10-y$, rectangulū sub medijs æquale est rectangulo sub extremis, quod ponitur $=r$, deinde pro primo numero pone z, et pro secundo x. Rectangulū sub medijs est $yx-xx=r$, quare $x=\frac{1}{2}y-\sqrt{\frac{1}{4}yy-r}$, et $=\frac{1}{2}y+\sqrt{\frac{1}{4}yy-r}$ pro duobus medijs. Rectangulum sub extremis est $\genfrac{}{}{0pt}{}{10}{-y}z-zz=r$, quare $z=5-\frac{1}{2}y-\sqrt{25-5y+\frac{1}{4}yy-r}$, et $=5-\frac{1}{2}y+\sqrt{25-5y+\frac{1}{4}yy-r}$, pro duobus extremis. Summa quadratorum horum numerorum est $100-20y+2yy-4r$, quæ est $=80$, ergo $r=5-5y+\frac{1}{2}yy$, quo substituto pro r, prodeunt quatuor numeri, ut sequitur:

$$\text{Medij}\begin{cases}\frac{1}{2}y-\sqrt{-\frac{1}{4}yy+5y-5}.\\ \frac{1}{2}y+\sqrt{-\frac{1}{4}yy+5y-5}.\end{cases}$$

[105] ||

$$\text{Extremi}\begin{cases}5-\frac{1}{2}y-\sqrt{-\frac{1}{4}yy+20}.\\ 5-\frac{1}{2}y+\sqrt{-\frac{1}{4}yy+20}.\end{cases}$$

(116) See note on p. 357.

Restat inquirendus valor ipsius y, vitandorum magnorū numerorū gratiâ, substitue $\begin{cases}\frac{1}{2}y-\surd b,\\ \frac{1}{2}y+\surd b,\end{cases}$ et $\begin{cases}5-\frac{1}{2}y-\surd c.\\ 5-\frac{1}{2}y+\surd c.\end{cases}$ Rectangulum sub secundo et quarto æquale est quadrato tertij, prodit $2\frac{1}{2}y-\frac{1}{4}yy+\frac{1}{2}y\surd c-5+\frac{1}{2}y\surd b-\surd\overline{bc}=\frac{1}{4}yy+b+y\surd b$. Item rectangulum sub primo et tertio æquale est quadrato secundi, prodit $2\frac{1}{2}y-\frac{1}{4}yy-\frac{1}{2}y\surd c+5-\frac{1}{2}y\surd b-\surd\overline{bc}=\frac{1}{4}yy+b-y\surd b$. Has duas æquationes aufer ab invicem, restat $y\surd c-10+y\surd b=2y\surd b$, vel $y\surd c=\overline{10+y\surd b}$. Pone rursus $\surd\overline{-\frac{1}{4}yy+20}$ loco ipsius $\surd c$, et $\surd\overline{-\frac{1}{4}yy+5y-5}$ loco ipsius $\surd b$, prodit

$$y\surd\overline{-\tfrac{1}{4}yy+20}=\overline{10+y}\surd\overline{-\tfrac{1}{4}yy+5y-5},$$

vel $-\frac{1}{4}y^4+20yy=-\frac{1}{4}y^4*+70yy+400y-500$. Prodit [ergo] $yy+8y-10=0$, et $y=\surd 26-4$. Hoc jam pone in quatuor numeris suprà inventis, loco ipsius y, habebitur pro

$$\text{Medijs}\begin{cases}\surd 6\frac{1}{2}-2-\surd\overline{-35\frac{1}{2}+7\surd 26}.\\ \surd 6\frac{1}{2}-2+\surd\overline{-35\frac{1}{2}+7\surd 26}.\end{cases}$$

$$\text{Extremis}\begin{cases}7-\surd 6\frac{1}{2}-\surd\overline{9\frac{1}{2}+2\surd 26}.\\ 7-\surd 6\frac{1}{2}+\surd\overline{9\frac{1}{2}+2\surd 26}.\end{cases}$$

Quod attinet quæstiones, in quibus tot æquationes non reperiuntur, quot ponuntur quantitates ignotæ, eæ [sunt] similes tribus hisce sequentibus.

|[106] ||[11] *Emit quis 3 poma, 4 pyra, 5 citria 6 solidis; et eodem pretio 2 poma, 5 pyra, et 7 citria 7 solidis et $10\frac{2}{3}$ denarijs.*(117) *Quæritur quanti veneant singula.*

Pone pro pretio cujusq; pomi x den. cujusq; pyri y den. cujusq; citrij z den. Habemus igitur hîc tres quantitates ignotas, et non nisi duas æquationes, nimirum: $3x+4y+5z=72$ [et] $2x+5y+7z=94\frac{2}{3}$. In priori æquatione est $x=24-\frac{4}{3}y-\frac{5}{3}z$, quo substituto pro x in secunda æquatione, prodit

$$48+\tfrac{7}{3}y+\tfrac{11}{3}z=94\tfrac{2}{3},$$

ergo $y=20-\frac{11}{7}z$. Pro x invenimus $24-\frac{4}{3}y-\frac{5}{3}z$, pone $20-\frac{11}{7}z$ pro y, prodit $x=\frac{3}{7}z-\frac{8}{3}$. Pretium quæsitum igitur cujusq; pomi est $\frac{3}{7}z-\frac{8}{3}$, cujusq; pyri $20-\frac{11}{7}z$, et cujusq; citrij z. Pro valore z ponere licet, quod libet, dummodo $\frac{3}{7}z$ excedant $\frac{8}{3}$, et $\frac{11}{7}z$ subsistant infra 20; vel ita ut z plus sit quàm $6[\frac{2}{9}]$, minus autem quàm $12\frac{8}{11}$.

Quod si igitur pro z ponatur 8, constat quodq; pomū $\frac{16}{21}$ den. quodq; pyrum $7\frac{3}{7}$ den. et quodq; citrium 8 den.

(117) Mercator renders Kinckhuysen's money symbols '*ß*' and '*g*' by their Latin equivalents 'solidi' (shillings) and 'denarij' (pence): in the next question correspondingly, 'libræ' (pounds) renders the Dutchman's '*ℒ*'.

(118) 100 'solidi' = 5 'libræ'.

[12] *Triginta homines, partim viri, partim fœminæ, liberiꝗ et servi consumserunt 5 libras, pendente quolibet viro 5 solidos, quavis fœmina tres, quovis liberorum duos, et quovis servorum unum. Quot fuerunt singuli?*

Pone x viros, y fœminas, z liberos, et $30-x-y-z$ servos.

07]

$$\begin{array}{cccc} \|x. & y. & z. & 30\ -x\ -y\ -z \\ 5 & 3 & z & 1 \\ \hline 5x & 3y & 2z & 30\ -x\ -y\ -z \\ & & & +5x+3y+2z \\ \hline & & \text{prodit} & 30+4x+2y+\ z=100.^{(118)} \end{array}$$

Est igitur $z=70-4x-2y$; hîc desunt duæ æquationes, ita ut hîc solutio sit x viri, y fœminæ, $70-4x-2y$ liberi, et $-40+3x+y$ servi. Posito igitur $x=10$, et $y=12$, extant 10 viri, 12 fœminæ, 6 liberi, et duo servi. Sed in ponendis numeris observandum [est] ut 70 excedant $4x+2y$, et ut 40 subsistant infra $3x+y$.

[13] *Tres mulierculæ*(119) *vendiderunt citria mala duobus diebus, prima 10, secunda 30, tertia 50; primo die quælibet certum numerum, et singula mala eodem pretio; altero die quæꝗ quod reliquum habebat, et rursus eodem singula pretio. Venditis omnibus deprehenduntur tantundem æris accepisse quælibet. Quæritur quot mala quæꝗ vendiderit quovis die, et quanto pretio.*

Pone pretium cujusꝗ citrij primo die v solidos, et secundo die w solidos: porrò pone quamꝗ vendidisse primo die, primam x, secundam y, tertiam z citria, ergo altero die vendiderunt, prima $10-x$, secunda $30-y$, et tertia $50-z$ citria.

108]

$$\begin{array}{ccc} \|x. & y. & z \\ v & v & v \\ \hline \end{array} \qquad\qquad \begin{array}{ccc} 10-x. & 30-y. & 50-z. \\ w & w & w \\ \hline \end{array}$$

prodit $vx\ vy\ vz$, quod acceperunt primo die [et] $10w-wx\ 30w-wy\ 50w-wz$, quod acceperunt secundo die. Quare quæꝗ accepit duobus istis diebus ut sequitur: $10w-wx+vx$. $30w-wy+vy$. $50w-wz+vz$. Hîc habentur duæ æquationes, nimirum: $10w-wx+vx=30w-wy+vy$. et $10w-wx+vx=50w-wz+vz$. Per primam invenitur $x+\frac{20w}{w-v}=y$. Per secundam invenitur $x+\frac{20w}{w-v}=z$. Ita ut hîc desint tres æquationes, et pro numero malorū quem quæꝗ primo die vendidit, nihil aliud poni possit quàm x, $x+\frac{20w}{w-v}$, et $x+\frac{40w}{w-v}$. Possunt igitur pro his quantitatibus ignotis poni numeri pro arbitrio, observando ut $x+\frac{40w}{w-v}$ minus sit quàm 50. Quod si igitur ponatur $v=\frac{1}{2}$ sol. $w=5\frac{1}{2}$ sol. et $x=4$; vendiderit quæꝗ primo die, prima 4, secunda 26, et tertia 48 citria, unumquodꝗ

(119) Kinckhuysen's 'Appel-wijven'.

$\frac{1}{2}$ solido, et altero die, prima 6, secunda 4, et tertia 2 citria, unumquodqꝫ $5\frac{1}{2}$ solidis.

||[109] ||MONITUM.(120)

Sub finem libri cui titulum feci *Fundamentum Geometriæ*,(121) mentio fit de generibus curvarum; cùm verò inaudiverim reperiri quosdam, quibus iste modus enumerandi genera displiceat, oportunum duxi hoc loco pauca de eo memorare.

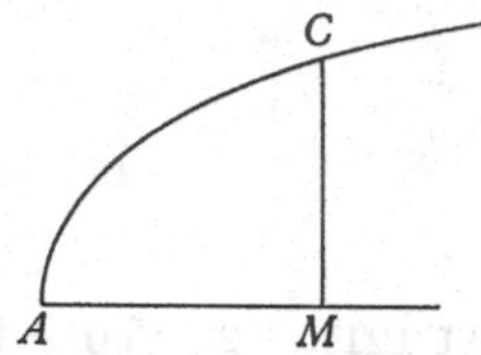

Posui fol: 53. cùm latus rectum statuitur $=r$, $AM=y$, atqꝫ tum CM deprehenditur valere $\sqrt{ry}$ (hoc est, posito $CM=x$, ut sit $xx=ry$), talis curva est linea primi generis. Eodem modo posito $x^3=ry$, lineā istam voco secundi generis; et quando x^4 æqualis est ry, eam voco curvam tertij generis, et sic deinceps. Quo innuere volui me numerare genera ut sequitur: omnes curvas ubi æquatio est duarum dimensionum colloco sub primo genere; ubi trium sub secundo; ubi quatuor, sub tertio, et sic porrò. Hæc enumeratio mihi videtur commodissima, et consentit cum operatione *Cartesij*, nimirum, ubi curvæ uno motu unum genus addit, quando ego pono, parabolas esse, quemadmodum videre [est] in Geometria ejus Gallica fol: 322, ubi respiciens [figuram] illic positam(122) scribit; quando linea CNK recta est, tum curva CE linea est primi generis; quando linea CNK est curva primi generis, tum linea CE est secundi generis; quando CNK est curva secundi generis, tum linea CE est curva tertii generis, et sic deinceps.(123) Per hanc additionem quævis curvæ, si sint parabolæ, in quovis genere

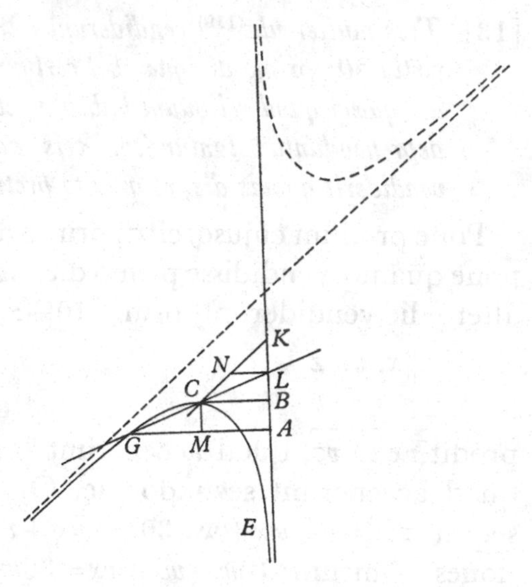

||[110] unâ dimensione auge[n]tur, quemadmodum in dictâ conclu=||sione ostenditur fol: 63, ubi in curva tertij generis pro CM emergit $\frac{-y^4+dy^3+fny-bdn}{ny}$.(124) Eodem modo genera æquationum enumerari poterant pro numero radicum,

(120) Kinckhuysen's appended 'Na reden' relates strictly to the passages of his previously published *De Grondt der Meet-konst* (note (6)) which discuss the classification of algebraic curves into 'genera', but we reproduce it here (in Mercator's version) since it was presumably intended to be printed with the Latin edition of the *Algebra* and because its subject was of continuing interest to Newton. Indeed, Kinckhuysen's preferred classification by algebraic degree in the *Meet-konst* (rather than by paired degrees, as in Descartes' *Geometrie*) is that on which Newton later settled when in 1695 he prepared the final verison of his cubic enumera-

quas habere possunt, verùm si respiciamus ad solutionem æquationum, possunt inter solvendum quæ tribus et quatuor dimensionibus constant, accipi pro uno genere, et quæ 5 vel 6 dimensionibus constant, itidem pro uno, et sic porrò. Nam in eadem conclusione dixi fol: 55 et 56, si ducantur duæ curvæ primi generis, quæ sese secent in aliquibus locis, eas inservire posse problematis ad tres vel quatuor dimensiones ascendentibus, et quoties curvæ additur genus unum, toties binis dimensionibus augetur æquatio (putà, quando altera curva manet primi generis).

Credo ego curvas trium et quatuor dimensionū accipi pro uno genere, quoniam deprehenditur, si loco parabolæ sumatur ellipsis, et aliquid pro DE, tum curvam ACN esse lineam 4 dimensionum,(126) sin pro DE ponatur nihil, tum ACN esse tantùm lineam trium dimensionum.(127) Isto modo curvæ duarum et trium dimensionum possint quoqꝫ pro uno genere accipi, nam si ponatur curva CDF parabola, et

D F E C G A M H N

Vide Figurar in conc Fundar Geome fol: 62.

tion for the press. To Kinckhuysen's unillustrated text we have added appropriate Cartesian figures.

(121) Kinckhuysen's 'de Grondt der Meet-konst'. (See notes (6) and (120).)

(122) To our accompanying reproduction of Descartes' figure (*Geometrie* (Leyden, 1637): 320–1), which illustrates the case when CK is a straight line and hence the curve (C) is a hyperbola, we have added the latter's second asymptote and conjugate branch.

(123) In general, Kinckhuysen intends by a curve of $(p-1)$th genus the general parabola of pth degree $n \times KB = (CB)^p$. In his *Geometrie* Descartes constructed his derived curve (C) as the meet of the given parabola CNK translated uniformly along AK with a straight line rotating round the fixed pole G and passing through a point L (in AK) fixed in the plane of the moving parabola. If, with Descartes and Kinckhuysen, we make $AB = MC = x$, $BC = AM = y$ the defining co-ordinates of the general point C with respect to the origin A and suppose $AG = d$, $KL = b$, it follows that $(y-d)(yp-bn)+nxy = 0$ is the defining equation of the derived curve (C) and consequently that it is of pth genus.

(124) The third term in the numerator should read '$+bny$'. This is, of course, the particular case of note (123) when $p = 3$.

(125) The figure we reproduce alongside, equivalent to Kinckhuysen's, is a simplified version of that on *Geometrie*: 404. (This illustrates the particular case when CD is an Apollonian parabola and the derived curve (C) is consequently a Cartesian trident: we have added its conjugate branch through F in line with the figure on *Geometrie*: 336.)

(126) Much as in note (123), if we take $HG = MC = x$, $GC = HM = y$, $HA = d$, $DE = b$ and if the defining relation between $GC = y$ and $DG = b+xy/(d-y)$ be the general ellipse (of main axis HGD) $CG^2 = r \times KB - (q/r)KB^2$, then the equation of the derived curve (C) is $qy^2(d-y)^2 = r(b(d-y)+xy)((q-b)(d-y)-xy)$, of fourth degree.

(127) When DE (or b) is zero, the curve (C) degenerates into the asymptote HGD $(y = 0)$ and the cubic $qy(d-y)^2 = rx(q(d-y)-xy)$.

sumatur aliquid pro DE, tum curva ACN est linea trium dimensionum.[128] sed si pro DE ponatur nihil, tum ACN est curva linea duarū dimensionum, nimirum, rursus parabola similis CDF.[129] Tantùm de enumeratione generū, quibus verò mea enumeratio non placet, ij numerare possunt genera cum excellentissimo *Cartesio*, à quo non discrepo, nisi in denominatione, quâ genus cujusqꝫ curvæ noscitur.

FINIS.[130]

§2. NEWTON'S 'OBSERVATIONS' ON KINCKHUYSEN'S 'ALGEBRA'

[1669–70][1]

From the original in the University Library, Cambridge[2]

||[1] ||IN ALGEBRAM GERARDI KINCKHUYSEN OBSERVATIONES.

[3]Substractio in integris.

Substractio peragitur addendo quantitatem substrahendam cum signo contrario.[4] Sic ad auferendum $+6x$ de $+8x$, addo $-6x$, et prodit $+2x$. Ad auferendum $-6x$ de $+8x$, addo $+6x$, et prodit $+14x$. Ad auferendum $3z-5a$ de $4z+b$, addo $-3z+5a$, et prodit $z+b+5a$. Atqꝫ ita in cæteris, quemadmodum ulteriùs liquet ex subjectis exemplis.

---

(128) In the terminology of notes (123) and (126), when $n \times KB = CB^2$ is the defining relation of the parabola, then the constructed curve (C) is the Cartesian trident

$$(y-d)(y^2-nb)+nxy = 0$$

(*Geometrie*: 336ff., 345–6, 502ff.).

(129) When DE (or b) is zero, the trident degenerates into the asymptote HGD $(y = 0)$ and the parabola $nx = y(d-y)$. All parabolas are, of course, similar but the constructed one (whose main axis is the perpendicular bisector of AH) lies parallel to the constructing parabola CDF.

(130) As we have mentioned (note (2)) Mercator has incorporated Kinckhuysen's following 'Druck-fauten' into his Latin text.

(1) The historical background to these notes is surveyed in our introduction to the present Part 3. In brief, a set of sheets of the 1661 printing of Kinckhuysen's *Stel-konst* interleaved with Mercator's handwritten Latin version (§1 above) had, by way of Isaac Barrow, been sent by John Collins to Newton for review and comment in the late autumn of 1669. The latter, insufficiently impressed by the work's originality to 'think it...worth the paines of a formall comment' (Newton to Collins, 6 February 1669/70 = *Correspondence of Isaac Newton*, **1** (1959): 24), at first approached the task lightly but over the spring of 1670 was encouraged by Collins to draft the first version of the present *Observationes*. Kinckhuysen's interleaved *Algebra* and his own 'notes wch I have intermixed wth the Authors discourse' Newton at length sent to Collins on 11 July, complementing the remarks in his accompanying letter (*Correspondence*, **1**: 30–1)

Translation

Observations on the Algebra of Gerard Kinckhuysen

[3]Subtraction in integers

Subtraction is performed by adding the quantity to be subtracted with opposite sign.[4] So to take away $+6x$ from $+8x$, I add $-6x$ and there comes out $+2x$. To take away $-6x$ from $+8x$, I add $+6x$ and there comes out $+14x$. To take away $3z-5a$ from $4z+b$, I add $-3z+5a$ and there comes out $z+b+5a$. And so of the rest, as will further appear from the examples subjoined.

---

with no less lengthy comments inserted in the margins of the *Observationes* itself. In his reply two days later Collins took up Newton's main criticism of the *Algebra* (that Kinckhuysen's discussion of binomial surds was badly deficient) and Newton responded on the 16th that if Collins would send his notes back he would 'not only supply ye wants about surds but that about Æquations soluble by trisections, & somthing more I would say in the chapter *Quomodo quæstio aliqua ad æquationem redigatur*...' (*Correspondence*, **1**: 34). In consequence, little more than a week after Newton had sent them, the interleaved *Algebra* and his observations thereon were remanded by Collins for correction and augmentation, and the latter seem never again to have left his possession. Between late July and 27 September 1670, when he again wrote to Collins (*Correspondence*, **1**: 42–4), Newton worked intermittently at revising his Kinckhuysen notes, but, apprehensive that his insertions 'were not so congruous as I could have wished to his manner of writing', he decided finally to defer the bulk of his new improvements for a 'compleate introduction to Algebra' he was himself planning (and which he later delivered as Lucasian lectures between 1673 and 1683), amending only the sections on binomial surds and the geometrical resolution of cubics and adding 'two or three' further geometrical examples at the end.

(2) Add. 3959.1: 2^r–21^r, described by Horsley (on a paper slip now loose in Add. 4005) as: 'From N° 4 of selected papers/Sir Isaac Newtons Additions to Kinckhuysens Algebra. This seems to have been written before the Arithmetica Universalis was composed. It contains nothing but what we have in a more finished form in that treatise & is therefore quite unfit to be published. S. Horsley. Octr 20th 1777'. (The manuscript now lacks one leaf and an inserted slip and these have been appropriately restored from the revised versions which Newton later incorporated in his Lucasian lectures on arithmetic and algebra (ULC. Dd. 9.68).) More recently, C. J. Scriba has briefly detailed its contents in his 'Mercator's Kinckhuysen-translation in the Bodleian Library at Oxford', *British Journal for the History of Science*, **2** (1964): 45–58, especially 48–9. Apart from the modern library foliation Newton himself has separately paginated the two versions of his *Observationes* (note (1)), the first having 30 pp. (2^r–6^v/10^r–19^v) and the latter (whose division we indicate) 39 pp. The accompanying marginal remarks to Collins have not been reproduced in the text or translation, but are set at appropriate places in the footnotes.

(3) See Mercator's pp. 9–10 in §1. In the margin alongside Newton remarks to Collins: 'Tis noe great matter whither you retain y^e^ Authors chapter of substraction or substitute this in its stead. This I have writ that you might have yor choyse'.

(4) Compare the simple forthrightness of Newton's explanation, at once an axiom and an algorithm, with the stilted clumsiness of Kinckhuysen's subtraction rule.

$$\begin{array}{ccccccc} -2z & 3a & -az & -bb & 14x+7 & 10y+6 & ab+cd \\ z & b & -2az & -ac & 6x+4 & 12y-7 & de-ef \\ \hline -3z & 3a-b & az & ac-bb & 8x+3 & -2y+13 & ab+cd-de+ef \end{array}$$

$$\begin{array}{cccc} 7z^3-6zz & -2ab+cd-cc & zz & qq+zz \\ 6z^3+7zz-9z & +ab \quad -cc & yy-2ay+aa & qq+yy-2ay \\ \hline z^3-13zz+9z & -3ab+cd & zz-yy+2ay-aa & zz-yy+2ay \end{array}$$

$$\begin{array}{c} \frac{1}{2}aa-\frac{3}{5}ab+\frac{2}{3}bb+\frac{1}{6}\frac{bb}{a} \\ \frac{1}{4}aa+\frac{2}{5}ab+\frac{1}{2}bb+\ \frac{bb}{a} \\ \hline \frac{1}{4}aa-\ ab+\frac{1}{6}bb-\frac{5}{6}\frac{bb}{a} \end{array}$$

The chapter of Division from these words *planè ut in multiplicatione dictum fuit*, (pag 1[3] lin 8) may be thus continued.

Sic $20abcd$ divisum per $5cd$ dat $4ab$. Et a^3b divisum per $-a$ dat $-aab$. Et $-6ab$ divisum per $2b$ dat $-3a$. Et $-a^4bbx$ divisum per $-abx$ dat a^3b. Porro $2xx-6x$ divisum per x dat $2x-6$. Et $\frac{1}{2}a^3z^3-\frac{2}{3}a^4bz$ divisum per $\frac{3}{4}az$ dat $\frac{2}{3}aazz-\frac{8}{9}a^3b$.

(5)Cùm divisor constat ex pluribus terminis signis $+$ & $-$ connexis, quantitas dividenda primò disponi debet secundum ordinem eundem. Quo facto, divisio peragitur ut in vulgari Arithmeticâ. Ut si $4x+2xx-30$ dividendum sit per $x+5$: Vel possunt disponi in hunc ordinem ‖ $2xx+4x-30$ & $x+5$, et divisio fiet ut sequitur, $2x-6$ prodeunte. ‖[2]

$$\begin{array}{l} \quad -6x \\ 2xx+4x-30|2x-6. \\ \quad x+5 \\ \qquad x+5 \end{array}$$

Dic quemadmodum in vulgari Arithmeti[c]â, quoties x contineatur in $2xx$. Resp: $2x$; Tum $2x$ in $x+5$ facit $2xx+10x$, quo de $2xx+4x$ subducto restat $-6x$. Porrò quoties x continetur in $-6x$? Resp: -6: Tum -6 in $x+5$ facit $-6x-30$, quo de $-6x-30$ subducto restat nihil.

Vel etiam possunt disponi in ordine contrario $-30+4x+2xx$, & $5+x$: divisione ut sequitur peractâ et $-6+2x$ prodeunte.

$$\begin{array}{l} \qquad +10x \\ -30+4x+2xx|-6+2x. \\ \quad 5\ +x \\ \qquad\quad 5\ +x \end{array}$$

Sic ad dividendum $y^6-2ccy^4+c^4yy-a^6-2a^4cc-aac^4+aay^4-a^4yy$ per

$$yy-aa-cc.$$

Vel dispono quantitatem dividendam et divisorem secundum dimensiones

$$\begin{array}{rrrrrrr}
-2z & 3a & -az & -bb & 14x+7 & 10y+6 & ab+cd \\
z & b & -2az & -ac & 6x+4 & 12y-7 & de-ef \\
\hline
-3z & 3a-b & az & ac-bb & 8x+3 & -2y+13 & ab+cd-de+ef
\end{array}$$

$$\begin{array}{rrrr}
7z^3-6z^2 & -2ab+cd-c^2 & z^2 & q^2+z^2 \\
6z^3+7z^2-9z & +ab \qquad -c^2 & y^2-2ay+a^2 & q^2+y^2-2ay \\
\hline
z^3-13z^2-9z & -3ab+cd & z^2-y^2+2ay-a^2 & z^2-y^2+2ay
\end{array}$$

$$\begin{array}{l}
\tfrac{1}{2}a^2-\tfrac{3}{5}ab+\tfrac{2}{3}b^2+\tfrac{1}{6}(b^2/a) \\
\tfrac{1}{4}a^2+\tfrac{2}{5}ab+\tfrac{1}{2}b^2+(b^2/a) \\
\hline
\tfrac{1}{4}a^2-ab+\tfrac{1}{6}b^2-\tfrac{5}{6}(b^2/a)
\end{array}$$

[The chapter of Division from these words *planè ut in multiplicatione dictum fuit*, (pag 13 lin 8) may be thus continued.]

Thus $20abcd$ divided by $5cd$ gives $4ab$. And a^3b divided by $-a$ gives $-a^2b$. And $-6ab$ divided by $2b$ gives $-3a$. And $-a^4b^2x$ divided by $-abx$ gives a^3b. Moreover, $2x^2-6x$ divided by x gives $2x-6$. And $\tfrac{1}{2}a^3z^3-\tfrac{2}{3}a^4bz$ divided by $\tfrac{3}{4}az$ gives $\tfrac{2}{3}a^2z^2-\tfrac{8}{9}a^3b$.

(5) When the divisor consists of several terms connected by the signs + and −, the quantity to be divided should first be arranged in the same order. Once that is done, the division is performed as in common arithmetic. Thus if $4x+2x^2-30$ is to be divided by $x+5$, they can either be arranged in this order $2x^2+4x-30$ and $x+5$ and the division will be as follows, producing $2x-6$:

$$\begin{array}{l}
\quad -6x \\
2x^2+4x-30 \,|\, 2x-6 \\
\;\; x\;+5 \\
\qquad\quad x+5
\end{array}$$

Say, as in common arithmetic: how many x's are there in $2x^2$? Reply: $2x$. Then $2x$ times $x+5$ makes $2x^2+10x$, and when this is taken from $2x^2+4x$ there remains $-6x$. Further, how many x's are there in $-6x$? Reply: -6. Then -6 times $x+5$ makes $-6x-30$, and when this is taken from $-6x-30$ there remains nothing.

Or they can also be arranged in the contrary order $-30+4x+2x^2$ and $5+x$, when the division is performed as follows, producing $-6+2x$.

$$\begin{array}{l}
\qquad +10x \\
-30+4x+2x^2 \,|\, -6+2x \\
\quad 5+\;x \\
\qquad\quad 5+\;x
\end{array}$$

Thus to divide $y^6-2c^2y^4+c^4y^2-a^6-2a^4c^2-a^2c^4+a^2y^4-a^4y^2$ by $y^2-a^2-c^2$, I either arrange the dividend and divisor according to the dimensions of y, and

---

(5) Newton has added the marginal comment for Collins: 'I judge these notes of ordering the termes & supplying those that are wanting, to bee very convenient for shunning confusion in Division'. Compare §1: note (2).

ipsius y, et fiunt $y^6 \begin{smallmatrix}+aa\\-2cc\end{smallmatrix} y^4 \begin{smallmatrix}-a^4\\+c^4\end{smallmatrix} yy \begin{smallmatrix}-a^6\\-2a^4cc\\-aac^4\end{smallmatrix}$, et $yy \begin{smallmatrix}-aa\\-cc\end{smallmatrix}$. Vel secundum dimensiones ipsius a, et fiunt $-a^6 \begin{smallmatrix}-yy\\-2cc\end{smallmatrix} a^4 \begin{smallmatrix}+y^4\\-c^4\end{smallmatrix} aa \begin{smallmatrix}+y^6\\-2y^4cc\\+yyc^4\end{smallmatrix}$, et $-aa \begin{smallmatrix}+yy\\-cc\end{smallmatrix}$. Vel etiam secundum dimensiones ipsius c, aut earundem literarum in ordine contrario. Quo facto, divisionem ut priùs instituo.

Quantitatibus ut dictum fuit in ordinem redactis, si termini aliqui desint, hoc est, si dimensiones literâe ad quam ordinatio fit, non in eâdem ubiꝗ progressione Arithmetica sed per saltum alicubi procedant: locis vacuis substituantur cyphræ. Quemadmodum videre est in sequentibus.

Divide $xx-aa$ per $x+a$, prodit $x-a$.

$$\begin{array}{l} \quad -ax \\ xx+0x-aa\,|\,x-a. \\ x+\;a \\ \quad\; x+a \end{array}$$

||[3] ||Divide y^4-a^4 per $yy-aa$, prodit $yy+aa$.

$$\begin{array}{l} \quad +aayy \\ y^4+0yy-a^4\,|\,yy+aa. \\ yy-aa \\ \qquad yy-aa \end{array}$$

Divide y^4-a^4 per $y-a$, prodit $y^3+ayy+aay+a^3$.

$$\begin{array}{l} \quad +ay^3+aayy+a^3y \\ y^4+0y^3\;+0yy\;+0y-a^4\,|\,y^3+ayy+aay+a^3. \\ y-a \\ \quad y \quad -a \\ \qquad\quad y \quad -a \\ \qquad\qquad y-a \end{array}$$

Divide $y^5+aby^3-a^3yy-2aabby+a^4b$ per $y^3+2aby-a^3$, prodit $yy-ab$.

$$\begin{array}{l} \quad +0 \quad -aby^3+0 \\ y^5+0y^4+aby^3-a^3yy-2aabby+a^4b\,|\,yy-ab. \\ y^3+0yy+2aby-a^3 \\ \qquad\qquad y^3+0yy+2aby \quad -a^3 \end{array}$$

Potest quoꝗ operatio harum divisionum paulò aliter disponi, quemadmodum dividendo $a^3-3aab+3abb-b^3$ per $a-b$, prodibit $aa-2ab+bb$, ad hunc modum.

they become $y^6+(a^2-2c^2)\,y^4-(a^4-c^4)\,y^2-(a^6+2a^4c^2+a^2c^4)$ and $y^2-(a^2+c^2)$; or according to the dimensions of a, and they become

$$-a^6-(y^2+2c^2)\,a^4+(y^4-c^4)\,a^2+(y^6-2y^4c^2+y^2c^4)$$

and $-a^2+(y^2-c^2)$; or indeed according to the dimensions of c, or of these same letters in reverse order. When that is done, I lay out the division as before.

If, when the quantities have been reduced to order in the way mentioned, some terms are lacking, if in other words the dimensions of the letter according to which the ordering is made do not everywhere progress in the same arithmetical progression but at times make a leap, then zeros are to be substituted for the vacant places. Exactly as is to be seen in the following:

Divide x^2-a^2 by $x+a$, there comes out $x-a$.

$$\begin{array}{l} \quad -ax \\ x^2+0x-a^2|x-a. \\ x\ +\ a \\ \qquad x+a. \end{array}$$

Divide y^4-a^4 by y^2-a^2, there comes out y^2+a^2.

$$\begin{array}{l} \quad +a^2y^2 \\ y^4+\ 0y^2-a^4|y^2+a^2. \\ y^2-\quad a^2 \\ \qquad y^2-a^2. \end{array}$$

Divide y^4-a^4 by $y-a$, there comes out $y^3+ay^2+a^2y+a^3$.

$$\begin{array}{l} \quad +ay^3+a^2y^2+a^3y \\ y^4+0y^3+0y^2\ +\ 0y-a^4|y^3+ay^2+a^2y+a^3. \\ y\ -\ a \\ \qquad y\ -a \\ \qquad\qquad y-a \\ \qquad\qquad\quad y-a. \end{array}$$

Divide $y^5+aby^3-a^3y^2-2a^2b^2y+a^4b$ by $y^3+2aby-a^3$, there comes out y^2-ab.

$$\begin{array}{l} \quad +0\quad -aby^3+0 \\ y^5+0y^4+aby^3-a^3y^2-2a^2b^2y+a^4b|y^2-ab. \\ y^3+0y^2+2aby\ \ -a^3 \\ \qquad\qquad y^3+0y^2+2aby-a^3. \end{array}$$

The working of these divisions may also be arranged a little differently: so, in dividing $a^3-3a^2b+3ab^2-b^3$ by $a-b$, there will come out $a^2-2ab+b^2$ in this manner:

$$\begin{array}{l} a-b)a^3-3aab+3abb-b^3(aa-2ab+bb. \\ \quad a^3-\ aab \\ \quad 0-2aab \\ \quad\quad -2aab+2abb \\ \quad\quad 0\ +abb \\ \quad\quad\quad abb-b^3 \\ \quad\quad\quad 0\quad 0 \end{array}$$

Divide a^3 per a (primam partem divisoris) oritur aa. /

From hence (pag 15 lin 1) the words of the Author may bee continued to the end of the chapter: Only (pag 15 lin 12) write thus.

Nec abhorret ab usu incipere divisionem a posterioribus: licèt tantundem egeris si inverso terminorum ordine incipias a prioribus. Sic in exemplo sequenti &c:

||[4] ||pag 22.(6) After the words *usurpaturq; hic eadem tabula geneseon* may bee added:

Et observandum est quod termini quantitatis cujus radix extrahenda est disponi debent secundum dimensiones alicujus literæ.

Quod si igitur radix quadratica extrahenda sit ex $aa+2ab+bb$, terminis secundum dimensiones ipsius a dispositis, operatio instituenda est ut sequitur. &c

---

Pag 26 lin 6.(7) After the words *ut a ad b*. write,

Et sic $\surd C:\overline{2a^3b+a^4}$ & $\surd C:\overline{ab^3+2b^4}$ sunt communicantes et possunt reduci ad $a\surd C:\overline{a+2b}$ et $b\surd C:\overline{a+2b}$. Atq; ita de cæteris.

Pag 27 lin 7.(8) After y^e words *et emergit quæsitum* $\overline{a+b}\surd\overline{a+b}$ write, Atq; ita $\surd C:\overline{a^4+2a^3b}$ & $\surd C:\overline{ab^3+2b^4}$ dant $\overline{a+b}\surd C:\overline{a+2b}$. Et sic in alijs.

Pag 28 lin 9.(8) After the words *quod est ipsum quæsitum* write,

Non secus ad subtrahendum $\surd C:\overline{ab^3+2b^4}$ ex $\surd C:\overline{a^4+2a^3b}$ scribimus

$$\overline{a-b}\surd C:\overline{a+2b}.$$

Et sic in alijs.

Pag 30. At y^e end of the chapter of Multiplication subjoyn

---

(6) Compare §1: note (25).

(7) Compare §1: note (30). In the margin alongside Newton has added the comment: 'The Author having slipped over the Addition substraction multiplication & divisiō of all but quadratick surds: I thought good to make these additions'. In his letter to Newton of 13 July 1670 thanking him for communicating the first version of the present *Observationes* Collins intimated: 'I formerly...thought Kinck[huysen] had too slightly handled the doctrine of

$$
\begin{array}{l}
a-b)a^3-3a^2b+3ab^2-b^3(a^2-2ab+b^2. \\
\quad a^3-\ a^2b \\
\quad 0-2a^2b \\
\quad\quad -2a^2b+2ab^2 \\
\quad\quad 0\ +ab^2 \\
\quad\quad\quad ab^2-b^3 \\
\quad\quad\quad 0\quad 0
\end{array}
$$

Divide a^3 by a (the first part of the divisor), there arises a^2....

[From hence (pag 15 lin 1) the words of the Author may bee continued to the end of the chapter: Only (pag 15 lin 12) write thus.]

Nor will it be unavailing to begin the division from the latter terms, though you will do precisely what you would should you invert the order and begin from the first ones. Thus in the following example....

[pag 22.(6) After the words *usurpaturq; hic eadem tabula geneseon* may bee added:]

And it should be observed that the terms of the quantity whose root is to be extracted ought to be arranged according to the dimensions of some letter.

If, then, the square root is to be extracted out of $a^2+2ab+b^2$, whose terms are arranged according to the dimensions of a, the working is to be laid out as follows....

[Pag 26 lin 6.(7) After the words *ut a ad b*. write,]

And so $\sqrt[3]{(2a^3b+a^4)}$ and $\sqrt[3]{(ab^3+2b^4)}$ are communicant and may be reduced to $a\sqrt[3]{(a+2b)}$ and $b\sqrt[3]{(a+2b)}$. And so of the rest.

[Pag 27 lin 7.(8) After ye words *et emergit quæsitum* $\overline{a+b}\sqrt{a+b}$ write,] And thus $\sqrt[3]{(a^4+2a^3b)}$ and $\sqrt[3]{(ab^3+2b^4)}$ give $(a+b)\sqrt[3]{(a+2b)}$. And so in other cases.

[Pag 28 lin 9. After the words *quod est ipsum quæsitum* write,] No differently, to subtract $\sqrt[3]{(ab^3+2b^4)}$ from $\sqrt[3]{(a^4+2a^3b)}$ we write $(a-b)\sqrt[3]{(a+2b)}$. And so in other cases.

[Pag 30. At ye end of the chapter of Multiplication subjoyn]

Surd Numbers, and that the same might be transcribed from others. I find you in the same opinion for in one of your Marginall Notes you say thus, *The Author having slipped over the Addition Substraction Multiplication and Division of all but quadratick Surds &c.* and he acknowledgeth as much by consequence himselfe referring the Reader to Wassenare Onwissen Wisconstenaer. However he chiefely thereby intended ye roots of Binomialls which you have supplyed being unwilling the young Student should be referred to other bookes which are scarce for the Doctrine of Surds' (*Correspondence of Isaac Newton*, 1 (1959): 32–3; and compare §1: note (41)).

(8) Compare §1: notes (32) and (34). This and the following additions are, of course, complements to the preceding.

Hæc de multiplicatione quadraticarum: Quod si radicales aliarum potestatum sint multiplicandæ, si sint ejusdem denominationis multiplicantur sub eodem signo radicali. Sic $\surd C\!:a$ mult: per $\surd C\!:b$ dat $\surd C\!:ab$. Et $\surd 5\!:aab^3$: per $\surd 5\!:a^3bb$ dat $\surd 5\!:a^5b^5$ sive ab. Et $\surd 6\!:ax^5$ per $\surd 6\!:abbx^3$ dat $\surd 6\!:aabbx^8$ sive $x\surd 3\!:abx$.[9]

Sin diversæ denominationis existant, reducendæ sunt ad eandem. Sic $\surd a$ mult: per $\surd 3\!:b$ sive $\surd 6\!:a^3$ per $\surd 6\!:bb$ dat $\surd 6\!:a^3bb$. Sic $a+b$ mult: per $\surd 3\!:abb$, sive $\surd 3\!:\overline{a^3+3aab+3abb+b^3}$: per $\surd 3\!:abb$ dat $\surd 3\!:\overline{a^4bb+3a^3b^3+3aab^4+ab^5}$. Vel scribi potest $\overline{a+b}\surd 3\!:abb$.

Pag 31. At the end of the chapter of Division subjoyn

[10]Hæc de radicalibus quadraticis. Quod si divisio peragenda sit in radicalibus aliarum potestatum: consule exempla sequentia. $\surd 3\!:aab^4$ divisū per $\surd 3\!:aab$ dat ‖[5] $\surd 3\!:b^3$ sive b. ‖$\surd 3\!:a$ per $\surd 3\!:b$ dat $\surd 3\!:\frac{a}{b}$. $\surd 3\!:aab$ per $\surd ab$ sive $\surd 6\!:a^4bb$ per $\surd 6\!:a^3b^3$ dat $\surd 6\!:\frac{a}{b}$. $\surd ab$ per $\surd 3\!:a+b$ sive $\surd 6\!:a^3b^3$ per $\surd 6\!:aa+2ab+bb$ dat $\surd 6\!:\dfrac{a^3b^3}{aa+2ab+bb}$. Et sic porro.

Quandoq; accidat quod Additio Substractio Multiplicatio vel Divisio peragendæ sint in radicalibus fractis. Sic adde $\frac{a-b}{c}\sqrt{\frac{bc}{a}}$ ad $\frac{a+b}{c}\sqrt{\frac{bc}{a}}$ et fit $\frac{2a}{c}\sqrt{\frac{bc}{a}}$: vel substrahe et fit $\frac{2b}{c}\sqrt{\frac{bc}{a}}$. Sic multiplica $\frac{a}{b}\sqrt{\frac{ac+bc}{a-b}}$ per $\frac{a}{b}\sqrt{\frac{ac-bc}{b}}$ et fit $\frac{aa}{bb}\sqrt{\frac{acc+bcc}{b}}$: vel divide et fit $\sqrt{\dfrac{ab+bb}{aa-2ab+bb}}$.

Extrahendi radicem quadraticam e Binomijs Regula.[11]

Differentiæ quadratorum a nominibus[12] radix addatur majori nomini et abinde subducatur: eruntq; semissium summæ illius et residui radices cum signo + vel − connexæ radix quæsita.

---

(9) Note Newton's silent change from the notation '$\surd C$:' to '$\surd 3$:'.

(10) Newton's following pages 4–15 are intended to replace Kinckhuysen's pages 32–38. As we shall see he had some difficulty in composing them, for to their first version sent to Collins in July 1670 he later added the four pages 11–14 and then further revised these additions on the inserted sheet 7r/7v.

(11) In an accompanying marginal remark to Collins Newton explains why he has replaced Kinckhuysen's section on extracting the square roots of 'affected' surds with the one which follows: 'I should not have intermedled wth this chapter but that you intimated it (in a letter) together with ye following chapter of extracting cubick roots of Binomialls to bee somthing

5

$\sqrt{3}: a$ per $\sqrt{3}: b$ dat $\sqrt{3}: \frac{a}{b}$. $\sqrt{3}: aab$ per $\sqrt{ab}$ sive $\sqrt{6}: a^4bb$ per $\sqrt{6}: abb$ dat $\sqrt{6}: \frac{a}{b}$. $\sqrt{ab}$ per $\sqrt{3}: a+b$ sive $\sqrt{6}: ab^3$ per $\sqrt{6}: aa+2ab+bb$ dat $\sqrt{6}: \frac{ab^3}{aa+2ab+bb}$. [illegible]

Quando accidat quod additio substractio multiplicatio vel divisio peragenda sit in radicalibus fractis. Sic adde $\frac{a-b}{c}\sqrt{\frac{bc}{a}}$ ad $\frac{a+b}{c}\sqrt{\frac{bc}{a}}$ et fit $\frac{2a}{c}\sqrt{\frac{bc}{a}}$: vel substrahe et fit $\frac{2b}{c}\sqrt{\frac{bc}{a}}$. Sic multiplica $\frac{a}{b}\sqrt{\frac{ac+bc}{a-b}}$ per $\frac{a}{b}\sqrt{\frac{ac-bc}{b}}$ et fit $\frac{aa}{bb}\sqrt{\frac{acc+bcc}{b}}$ vel divide et fit $\sqrt{\frac{ab+bb}{aa-2ab+bb}}$.

Extrahendi radicem quadraticam e Binomiis

Regula.

Differentia quadratorum a nominibus radix addatur majori nomini et abinde subducatur: summaeque reliquique semissium illius et residui radices cum signo + vel − connexa [illegible] radix quaesita.

Quemadmodum si radix extrahenda sit ex $33+\sqrt{800}$;

$33+\sqrt{800}$	33
33	17
1089	50. 16.
800	25. 8.
289	radix $5+\sqrt{8}$.
17	

Quadrata nominum sunt 1089 & 800, eorum differentia 289, cujus radix 17 addita majori nomini 33 dat 50, et ablata dat 16 quarum semisses sunt 25 & 8 earumque radices $5, \sqrt{8}$ componunt $5+\sqrt{8}$ radicem quaesitam.

Sic ex $5-\sqrt{24}$ radix est $\sqrt{3}-\sqrt{2}$.

Et ex $4+\sqrt{20}$ radix $\sqrt{:\sqrt{5}+1}+\sqrt{:\sqrt{5}-1}$.

Et ex $\sqrt{30}-\sqrt{24}$ radix $\sqrt{:1\frac{1}{2}+\sqrt{1\frac{1}{2}}}-\sqrt{:\sqrt{7\frac{1}{2}}-\sqrt{1\frac{1}{2}}}$.

Quinetiam ex $26-\sqrt{80}-\sqrt{640}-6+\sqrt{80}$ radix invenitur $4-\sqrt{:10-\sqrt{80}}$. quod patet ex operatione subjecta, accedendo $20-\sqrt{80}$ subeat vices unius nominis.

Quadrata nominum	$756-52\sqrt{80}$.	Majoris nominis $20-\sqrt{80}$. et
	$640-0+\sqrt{32}$.	inventa radicis $6+\sqrt{80}$
Quadratorum differentia	$116+12\sqrt{80}$.	summa 32. et $20-2\sqrt{80}$
cujus radix sunt in exterior extracta	$6+\sqrt{80}$.	Harum dimidia 16. $10-\sqrt{80}$. ex quibus radices dant $4-\sqrt{10-\sqrt{80}}$ radicem quaesitam.

I should not have intermedled with this chapter but that you intimated it in a letter [illegible] together wth the chapter of extracting the roots of Binomials [illegible] be something defective. this therefore I have somewhat altered & augmented wth examples partly taken out of Ferguson. You may substitute Ferguson's chapter of extracting the roots of Binomials in stead of the [illegible] but I think its better to substitute Schooten's writings of that subject [illegible] of his

Plate III. 'Observations' on Kinckhuysen's extraction of the square root of an algebraic binomial (3, 1, §2).

So much for the multiplication of quadratics. Should radicals of higher powers have to be multiplied, if they are of the same denomination they are to be multiplied beneath the same radical sign. So $\sqrt[3]{a}$ multiplied by $\sqrt[3]{b}$ gives $\sqrt[3]{ab}$. And $\sqrt[5]{a^2b^3}$ by $\sqrt[5]{a^3b^2}$ gives $\sqrt[5]{a^5b^5}$ or ab. And $\sqrt[6]{ax^5}$ by $\sqrt[6]{ab^2x^3}$ gives $\sqrt[6]{a^2b^2x^8}$ or $x\sqrt[3]{abx}$.

But if they are of different denominations, they must be reduced to the same one. So $\sqrt{a}$ multiplied by $\sqrt[3]{b}$ or $\sqrt[6]{a^3}$ by $\sqrt[6]{b^2}$ gives $\sqrt[6]{a^3b^2}$. So $a+b$ multiplied by $\sqrt[3]{ab^2}$, or $\sqrt[3]{(a^3+3a^2b+3ab^2+b^3)}$ by $\sqrt[3]{(ab^2)}$ gives $\sqrt[3]{(a^4b^2+3a^3b^3+3a^2b^4+ab^5)}$. Or there can be written $(a+b)\sqrt[3]{ab^2}$.

[Pag 31. At the end of the chapter of Division subjoyn.]

(10)So much for quadratic radicals. But if division is to be performed in radicals of other powers, consider the following examples. $\sqrt[3]{a^2b^4}$ divided by $\sqrt[3]{a^2b}$ gives $\sqrt[3]{b^3}$ or b. $\sqrt[3]{a}$ by $\sqrt[3]{b}$ gives $\sqrt[3]{(a/b)}$. $\sqrt[3]{a^2b}$ by $\sqrt{ab}$ or $\sqrt[6]{a^4b^2}$ by $\sqrt[6]{a^3b^3}$ gives $\sqrt[6]{(a/b)}$. $\sqrt{ab}$ by $\sqrt[3]{(a+b)}$ or $\sqrt[6]{a^3b^3}$ by $\sqrt[6]{(a^2+2ab+b^2)}$ gives $\sqrt[6]{\dfrac{a^3b^3}{a^2+2ab+b^2}}$. And so forth.

It may happen on occasion that addition, subtraction, multiplication or division are to be performed on fractional radicals. So add $\dfrac{a-b}{c}\sqrt{\dfrac{bc}{a}}$ to $\dfrac{a+b}{c}\sqrt{\dfrac{bc}{a}}$ and there comes $\dfrac{2a}{c}\sqrt{\dfrac{bc}{a}}$; or subtract them and there comes $\dfrac{2b}{c}\sqrt{\dfrac{bc}{a}}$. So multiply $\dfrac{a}{b}\sqrt{\dfrac{ac+bc}{a-b}}$ by $\dfrac{a}{b}\sqrt{\dfrac{ac-bc}{b}}$ and there comes $\dfrac{a^2}{b^2}\sqrt{\dfrac{ac^2+bc^2}{b}}$; or divide and there comes $\sqrt{\dfrac{ab+b^2}{a^2-2ab+b^2}}$.

Rule for extracting the square root of binomials(11)

Let the root of the difference of the squares of the terms(12) be added to the greater term and subtracted from it: then will the roots of the halves of that sum and remainder joined with the sign + or − be the required root.

---

defective. This therefore, I have somwt altered & augmented wth examples partly taken out of Ferguson'. The letter from Collins to which Newton refers (mentioned also in his covering letter of 11 July 1670, in a passage we reproduce in note (17) below) is apparently now lost and we do not know its precise content, but it is difficult to see how his objection to the adequacy of Kinckhuysen's rule could be well founded. Of course, Kinckhuysen's account of it is clumsy and overlong but it remains an accurate, efficient algorithm and, indeed, one handed down over many centuries (§1: note (37)). His examples, moreover, adequately illustrate the rule's application. In his present remoulding Newton successfully shortens and simplifies Kinckhuysen's exposition but essentially he can do little more than repeat the latter's examples together with additional ones taken over from Johann Jacob Ferguson's *Labyrinthus Algebræ... verhandelende de Ontbindinge der z, æ en zz Vergelijkingen* (The Hague, 1667). It would appear

Quemadmodum si radix extrahenda sit ex $33+\surd 800$;(13)

$$\begin{array}{l} 33+\surd 800 \\ 33 \\ \hline 1089 \\ 800 \\ \hline 289 \\ \hline 17. \end{array} \qquad \begin{array}{rr} & 33 \\ & 17 \\ \hline 50. & 16. \\ \hline 25. & 8. \\ \hline \end{array}$$

Quadrata nominum sunt 1089 & 800, eorum differentia 289, cujus radix 17 addita majori nomini 33 dat 50, et ablata dat 16, quorum semisses sunt 25 & 8 eorumq; radices 5, & $\surd 8$ componentes $5+\surd 8$ radicem quæsitam.

prodit $5+\surd 8$.

Sic ex $5-\surd 24$ radix est $\surd 3-\surd 2$.

Et ex $4+\surd 20$ radix $\surd:\overline{\surd 5+1}:+\surd:\overline{\surd 5-1}:$.

Et ex $\surd 30-\surd 24$ radix $\surd:\overline{\surd 7\tfrac{1}{2}+\surd 1\tfrac{1}{2}}:-\surd:\overline{\surd 7\tfrac{1}{2}-\surd 1\tfrac{1}{2}}:$.

Quinetiam ex $26-\surd 80-\sqrt{640-64\surd 80}$(14) radix invenietur $4-\surd:10-\surd 80$ quod pateat ex operatione subjectâ faciendo ut $26-\surd 80$ subeat vices unius nominis.

Quadrata nominum	$\begin{cases} 756-52\surd 80. \\ 640-64\surd 80. \end{cases}$	Majoris nominis inventæ radicis	$26-\surd 80$, et $6+\surd 80$
Quadratorū differentia	$116+12\surd 80.$	Summa 32.	diff: $20-2\surd 80.$
Ejusq; radix sicut in prioribus extracta	$6+\surd 80.$	Harum dimidia 16. Ex quibus radices dant 4.	$10-\surd 80.$ $\surd:10-\surd 80$ radicem quæsitam.

‖[6] ‖Porrò si radix extrahenda sit ex $mm+\frac{pxx}{m}+x\surd 4pm$,(15) prodit $m+x\sqrt{\frac{p}{m}}$.

$$\text{Quadrata nominum}\begin{cases} m^4+2pmxx+\dfrac{ppx^4}{mm}. & mm+\dfrac{pxx}{m}. \\[2ex] 4pmxx. & mm-\dfrac{pxx}{m}. \end{cases}$$

$$\begin{array}{lcc} \text{Restat} & m^4-2pmxx+\dfrac{ppx^4}{mm}. & 2mm.\ \ 2\dfrac{pxx}{m}. \\[2ex] \text{Ejus radix} & mm-\dfrac{pxx}{m}. & mm.\ \ \dfrac{pxx}{m}. \\[2ex] & \text{prodit} & m+x\sqrt{\dfrac{p}{m}}. \end{array}$$

---

that Collins sent along for the purpose not only the Dutch original of Ferguson's work but the Latin version 'translated by Mr Old, and by me almost transcribed' also. (See Collins' letter to Wallis on 17 June 1669 = S. P. Rigaud, *Correspondence of Scientific Men of the Seventeenth Century*, 2 (1841): 515. In his letter to Collins on 27 September 1670 (note (1)) Newton reported that he had 'sent back...Fergusons Labyrinthus Algebræ both parts of it'.)

(12) 'utriusq; nominis' (of each term) is cancelled.

As for instance if the root is to be extracted out of $33+\surd 800$;(13)

$33+\surd 800$	33	The squares of the terms are 1089 and 800, their difference 289, whose root 17 added to the larger term 33 gives 50 and taken from it gives 16: the halves of these are 25 and 8, and their roots 5 and $\surd 8$ together make $5+\surd 8$, the required root.
33	17	
1089	50. 16.	
800	25. 8.	
289		
17.		

There comes $5+\surd 8$.

So the root of $5-\surd 24$ is $\surd 3-\surd 2$.

And the root of $4+\surd 20$ is $\surd(\surd 5+1)+\surd(\surd 5-1)$.

And the root of $\surd 30-\surd 24$ is $\surd(\surd 7\tfrac{1}{2}+\surd 1\tfrac{1}{2})-\surd(\surd 7\tfrac{1}{2}-\surd 1\tfrac{1}{2})$.

Indeed, the root of $26-\surd 80-\surd(640-64\surd 80)$ will be found to be

$$4-\surd(10-\surd 80),$$

as may appear from the working below, by making $26-\surd 80$ take the turn of a single term.

The squares of the terms	$756-52\surd 80.$ $640-64\surd 80.$	Of the greater term of the root found	$26-\surd 80$ and $6+\surd 80$
The difference of these squares	$116+12\surd 80.$	The sum is 32.	difference $20-2\surd 80.$
		The halves of these 16.	$10-\surd 80.$
Its root extracted as in the preceding	$6+\surd 80.$	Whose roots 4.	$\surd(10-\surd 80),$
		give the required root.	

Again, if the root is to be extracted from $m^2+\frac{px^2}{m}+x\surd 4pm$,(15) there comes out $m+x\surd(p/m)$.

The squares of the terms	$m^4+2pmx^2+\frac{p^2x^4}{m^2}.$	$m^2+\frac{px^2}{m}.$
	$4pmx^2.$	$m^2-\frac{px^2}{m}.$
There remains	$m^4-2pmx^2+\frac{p^2x^4}{m^2}.$	$2m^2.\ 2\frac{px^2}{m}.$
Its root	$m^2-\frac{px^2}{m}.$	$m^2.\ \frac{px^2}{m}.$

There comes $m+x\sqrt{\frac{p}{m}}$.

(13) Kinckhuysen's first example (on his pp. 32–3). Newton has cancelled the continuation 'fac ut sequitur' (do as follows).

(14) Kinckhuysen's second example (on his p. 33).

(15) Kinckhuysen's third example (on his p. 34).

Sic quoq radix ex $\overline{a+b}\sqrt{ab}+2ab$(16) invenitur $\sqrt{4:a^3b}+\sqrt{4:ab^3}$.

$$\begin{array}{ccc} a^3b+2aabb+ab^3. & & \overline{a+b}\sqrt{ab}. \\ 4aabb. & & \overline{a-b}\sqrt{ab}. \\ \hline a^3b-2aabb+ab^3. & & 2a\sqrt{ab}. \quad 2b\sqrt{ab}. \\ \hline \overline{a-b}\sqrt{ab}. & & a\sqrt{ab}. \quad b\sqrt{ab}. \\ & \text{prodit} & \sqrt{a\sqrt{ab}}+\sqrt{b\sqrt{ab}}. \\ & \text{vel} & \sqrt{4:a^3b}+\sqrt{4:ab^3}. \end{array}$$

Haud secus radix ex $\sqrt{bc^3}-2a\sqrt{:\sqrt{bc^3}-aa:}$ invenietur $a-\sqrt{:\sqrt{bc^3}-aa:}$

$$\begin{array}{lll} & bc^3. & \sqrt{bc^3}. \\ & 4aa\sqrt{bc^3}-4a^4. & \sqrt{bc^3}-2aa. \\ \hline \text{Restat} & bc^3+4a^4-4aa\sqrt{bc^3}. & 2\sqrt{bc^3}-2aa. \quad 2aa. \\ \hline \text{radix ejus} & \sqrt{bc^3}-2aa, \text{ sicut} & \sqrt{bc^3}-aa. \quad aa. \end{array}$$

in prioribus extracta. prodit $a-\sqrt{:\sqrt{bc^3}-aa}$. vel $\sqrt{:\sqrt{bc^3}-aa:}-a$.

Quod si occurrat binomium cujus radix extrahi nequeat veluti

$$\tfrac{1}{2}b+\sqrt{\tfrac{1}{4}bb-aa},$$

ponimus $\sqrt{\tfrac{1}{2}b+\sqrt{\tfrac{1}{4}bb-aa}}$. Atq ita in alijs.

||[7] ||(17)Extrahendi radicem cubicem e Binomijs Regula.

Multiplicetur Binomium per 1000, et ejus pars surda in numeris puris posita addatur alteri parti et abinde subducatur; hujusq tum summæ tum residui

---

(16) Kinckhuysen's fourth example (on his p. 34).

(17) Newton's present replacement (pp. 7–15 following) for Kinckhuysen's section on extracting the cube root of a binomial was the fruit of much hard thought on his part. In a letter now lost (compare note (11)) Collins had 'intimated...y^e^...chapter of extracting cubick roots of Binomialls to bee somthing defective' and in its stead 'hinted somthing to bee supplyed out of Ferguson's *Labyrinthus*' (*Correspondence of Isaac Newton*, **1**: 30). As a cancelled marginal reply to Collins on pp. 5–6 of his *Observationes* makes clear it had been Newton's first reaction that the best way of amending Kinckhuysen on this point was not to insert extracts from Ferguson in his account but to replace it entirely with Schooten's version of the Descartes–Waessenaer rule (§1: note (39)), omitting only its inadequate 'Demonstratio': 'You may substitute Fergusons chapter of extracting cubick roots of Binomialls in stead of the Authors but I think its better to substitute w^t^ Van Scooten writes of that subject at ye end of his commentary of Descartes Geometry. Leaving out only the Demonstration at the latter end of it. For hee not only comprehends ye extraction of all other sorts of roots besids cubick, but doth it as plainly & fully as I think it can well bee done. Unlesse perhaps there should bee added two or three more examples at ye end of his discourse set in some such forme as Fergusons examples are'. However, in the first version (pp. 7–10/15) of the following pages, which he communicated to Collins on 11 July 1670, Newton decided to begin (p. 7) with Ferguson's rule and then (pp. 9/15) add his own algorithm for extracting the cube root of a rational

So also the root of $(a+b)\sqrt{ab}+2ab$[16] is found to be $\sqrt[4]{a^3b}+\sqrt[4]{ab^3}$.

$$\begin{array}{ll} a^3b+2a^2b^2+ab^3. & (a+b)\sqrt{ab}. \\ 4a^2b^2. & (a-b)\sqrt{ab}. \\ \hline a^3b-2a^2b^2+ab^3. & 2a\sqrt{ab}.\quad 2b\sqrt{ab}. \\ \hline (a-b)\sqrt{ab}. & a\sqrt{ab}.\quad b\sqrt{ab}. \end{array}$$

There comes $\sqrt{(a\sqrt{ab})}+\sqrt{(b\sqrt{ab})}$
or $\sqrt[4]{a^3b}+\sqrt[4]{ab^3}$.

No differently, the root of $\sqrt{bc^3}-2a\sqrt{(\sqrt{bc^3}-a^2)}$ will be found to be

$$a-\sqrt{(\sqrt{bc^3}-a^2)}.$$

$$\begin{array}{lll} & bc^3 & \sqrt{bc^3} \\ & 4a^2\sqrt{bc^3}-4a^4 & \sqrt{bc^3}-2a^2 \\ \text{There remains} & bc^3+4a^4-4a^2\sqrt{bc^3}. & 2\sqrt{bc^3}-2a^2.\quad 2a^2. \end{array}$$

Its root $\sqrt{bc^3}-2a^2$ extracted as before. $\sqrt{bc^3}-a^2.$ $a^2.$

There comes $a-\sqrt{(\sqrt{bc^3}-a^2)}$, or $\sqrt{(\sqrt{bc^3}-a^3)}-a$.

But should there present itself a binomial whose root cannot be extracted, such as $\frac{1}{2}b+\sqrt{(\frac{1}{4}b^2-a^2)}$, we set it $\sqrt{(\frac{1}{2}b+\sqrt{[\frac{1}{4}b^2-a^2]})}$. And so in other cases.

---

(17)Rule for extracting the cube root of binomials

Let the binomial be multiplied by 1000 and its surd part, expressed as a pure number, added to and taken away from the other part; then let the cube roots

binomial. (See note (29) below.) As he admitted, neither his own nor Ferguson's rules were intrinsically superior to 'Waessenaer's' but the Cartesian one is not easy to present simply and it may be also that Newton at this time did not appreciate its theoretical accuracy. (We will see that, when he came to deliver his Lucasian lectures on algebra, he in fact preferred his generalization of the Descartes–Waessenaer rule to both his own present one and Ferguson's.) The chief defect of all these rules was that as they were presented they could not deal with the case of a complex binomial, just the type needed to apply Cardan's cubic resolution in its 'irreducible' case (§1: note (103)). As Newton observed to Collins in his covering letters on 11 and 16 July, 'Ferguson seems to have done more [then Kinck-Huyson] in so much as to comprehend all cases of cubick equations w^th^in y^e^ same rules [Cardan's] but that more is inartificiall because it supposes y^e^ extraction of cubick roots out of imaginary binomiums, wch how to doe hee hath not taught us, his rule taught in pag 4 not extending to it. Thus his second example, ...$x^3 = 6x+4$, in pag 12. In order to solve this hee bids extract y^e^ cubick root out of these binomiums $2+\sqrt{-4}$, & $2-\sqrt{-4}$. To doe this his rule pag 4 is: *Multiply y^e^ binomium by 1000, put it in pure numbers &c*: Now $2+\sqrt{-4}$ in 1000 makes $2000+\sqrt{-4000000}$, but to put this in pure numbers is impossible for $\sqrt{-4000000}$ is an impossible quantity & hath noe pure number answering to it. His rule therefore failes & the like difficulty is in his 3^d^ example & in all other such cases...y^e^ cubick roote of $2+\sqrt{-4}$...extracted...indeed is $-1+\sqrt{-1}$, but I would know by w^t^ direct method hee teacheth to find it. Not but that it may bee done, & I know how to doe it, but I think it not worth y^e^ inserting into Kinckhuysen, yet if you think

radices cubicæ seorsim extrahantur in numeris integris quàm proximis, et (signis + ac − probè observatis) addantur. Summæ per 10 divisæ et in proximis numeris integris positæ dimidium erit pars pura radicis. Cujus quadrato subducto a parte purâ binomij propositi per eandem inventam partem radicis priùs divisâ, residui triens erit quadratum alterius partis. Quæ duæ partes cum signis + & − similiter connectendæ sunt atq; partes binomij propositi.[18]

Exemplum 1. Radix cubica ex $26+\surd 675$ est $2+\surd 3$.

Operatio.

Multiplica $26+\surd 675$ per 1000
prodit $26000+\surd 675000000$. hoc pone in numeris puris
fit 26000. 25981.
Aggreg: 51981. Resid 19. Ex utroq; extrahe radicem cubicam,
prodit 37, et 2.
Summa 39. hujus decima pars in proximis integris posita est 4.
Ejus dimidium 2, pars pura radicis: per quam 26 pars pura binomij
divisa fit 13. a qua quadrato ipsius 2 partis puræ radicis subducto
manet 9. cujus trientis radix quadratica
est $\surd 3$. pars altera radicis. Adeoq; $2+\surd 3$ radix desiderata;[19]
Non secus radix ex $26-\surd 675$ erit $2-\surd 3$.

[20]Cæterùm Binomium propositum debet fractionibus carere, et habere alterum nomen et alterius nominis quadratum, ut et radicem cubicam differentiæ quadratorum a Nominibus ejus, rationalia. Et si non sit ejusmodi, reducendum est ad talem formam multiplicando per denominatores fractionum, per partem surdam alterutrius nominis, et per[21] differentiam quadratorum a nominibus. Deinde radix per præcedentem regulam extracta dividenda est per radicem cubicam numerorum per quos Binomium primò propositum multiplicatum fuerit; et habes radicem desideratam.

---

it convenient (& indeed, it may bee congruently enough inserted into him at pag 91) I will send you it done in my next letter' (*Correspondence of Isaac Newton*, **1**: 30, 35, 30–1). Newton there refers to the extension of his cubic root algorithm to include the complex binomial, which, together with a revised set of examples for the real case, he incorporated in the new pages (11–14) added by him in the autumn of 1670 to the first version of his *Observationes* (see note (42) below).

(18) The rule is Ferguson's (*Labyrinthus Algebræ* (note (11)): 4). In proof, if

$$\sqrt[3]{[\alpha+\surd\beta]} = a+\surd b,$$

then (when β is non-square) $\sqrt[3]{[\alpha-\surd\beta]} = a-\surd b$ and so $a = \frac{1}{2}(\sqrt[3]{[\alpha+\surd\beta]}+\sqrt[3]{[\alpha-\surd\beta]})$ with $\alpha = a^3+3ab$ or $b = \frac{1}{3}(\alpha/a-a^2)$. The multiplication by 1000 and subsequent division by $\sqrt[3]{1000} = 10$ is a mere computational convenience, but the point is implicitly made in the following examples that if α and β are each integral, then so are $2a$ and $4b$ and consequently

of this sum and difference be each extracted in turn to the nearest whole number and added together (with the signs + and − properly observed). Half of the sum divided by 10 and expressed to the nearest integer will be the pure part of the root. And when its square is taken from the pure part of the propounded binomial previously divided by the same part of the root just now found, the third of the remainder will be the square of the second part. These two parts are to be connected with the signs + and − in the same way as the parts of the propounded binomial.(18)

Example 1. The cube root of $26+\surd 675$ is $2+\surd 3$.

Working

Multiply $26+\surd 675$ by 1000,
there comes $26{,}000+\surd 675{,}000{,}000$. Express this in pure numbers,
it becomes 26,000. 25,981.
Aggregate 51,981. Residue 19. Extract the cube root of each,
there comes 37, and 2.
Sum 39. Its tenth part expressed to the nearest integer is 4.
Half of this is 2, the pure part of the root: the pure part (26) of the binomial divided by it is 13. And when the square of the pure part (2) of the root is
taken away there remains 9, the square root of whose third part is $\surd 3$, the second part of the root. Hence $2+\surd 3$ is the desired root.(19)

Likewise, the root of $26-\surd 675$ will be $2-\surd 3$.

(20)But the propounded binomial must lack fractions and have one term, and the square of the second, as also the cube root of the difference of the squares of its terms, rational. And if it is not of this sort it must be reduced to such a form by multiplying through by the denominators of the fractions, by the surd part of either term, and by(21) the difference of the squares of the terms. Then the root extracted by the preceding rule must be divided by the cube root of the numbers by which the binomial first propounded was multiplied; and you have the desired root.

---

in such a case the numerical approximation of $\sqrt[3]{[\alpha \pm \surd \beta]}$ need be taken only to the nearest $\frac{1}{2}$ unit.

(19) Here $a = \frac{1}{2}([2+\surd 3]+[2-\surd 3]) = \frac{1}{2}(3{\cdot}73_{+}+0{\cdot}27_{-})$ with $b = \frac{1}{3}(26/2-2^2)$.

(20) The following paragraph, originally set before Example 1, has been here inserted in accordance with Newton's marginal note: 'These words *Cæterum Binomium &c* I thinke may be more propperly set immediately after the first of yᵉ foure following Examples.'

(21) 'numerum in parte surdâ radicis cubicæ' (the number in the surd part of the cube root) is cancelled.

||[8] ||Exemplum 2. Radix cubica ex $\sqrt{\frac{5}{4}}-1$ est $\sqrt{6:\frac{125}{256}}-\sqrt{3:\frac{1}{16}}$.

Operatio.

Duc $\sqrt{\frac{5}{4}}-1$ in 2 ut fractio tollatur
prodit $\sqrt{5}-2$. hoc in 1000
facit $\sqrt{5000000}-2000$. quod in numeris puris
est 2236.-2000.
Aggreg 236. Resid -4236. eorum cubicæ
radices 6. -16.
Summa -10. ejus decimæ partis
dimidium $-\frac{1}{2}$. pars pura in radice. Per quā -2 pars
pura binomij divisa dat 4. unde $\frac{1}{4}$ quadrato ipsius $-\frac{1}{2}$ subducto
restat $3\frac{3}{4}$. cujus trientis radix quadratica
est $\sqrt{1\frac{1}{4}}$ pars altera radicis. Adeoq ex
$\sqrt{5}-2$ radix cubica est $\sqrt{1\frac{1}{4}}-\frac{1}{2}$. Et hæc per $\sqrt{3}$: 2 divisa
dat $\sqrt{6:\frac{125}{256}}-\sqrt{3:\frac{1}{16}}$ radicem quæsitam.

Simili modo radices cubicæ ex $2+\sqrt{5}$ et ex $1+\sqrt{\frac{5}{4}}$ sunt $\frac{1}{2}+\sqrt{1\frac{1}{4}}$ et

$\sqrt{3:\frac{1}{16}}+\sqrt{6:\frac{125}{256}}$.(22)

Exemplum 3. Radix cubica ex $\sqrt{242}+\sqrt{243}$ est $\sqrt{2}+\sqrt{3}$.

Operatio.

Ex $\sqrt{242}+\sqrt{243}$ extrahendo quicquid rationale est
fit $11\sqrt{2}+9\sqrt{3}$. binomium igitur multiplico per $\sqrt{2}$ vel per $\sqrt{3}$ ut alterum e nominibus evadat rationale. Quemadmod[ū] multiplicando per $\sqrt{2}$
fit $22+\sqrt{486}$. Hujus partibus quadratis
fiunt 484. 486. Quorum differentia
est 2. rursus itaq multiplico binomium per 2
prodit $44+\sqrt{1944}$. Hoc in 1000
fac[i]t $44000+\sqrt{1944000000}$.
In numeris puris 44000. 44091.
Aggreg 88091. Resid -91.
Radices cubicæ 44. -4.
eorum Summa 40.
decimæ partis dimidium 2, pars pura in radice.
Divide 44 per 2, fit 22.
substrahe 4 restat 18.
trientis ejus radix quadrat: est $\sqrt{6}$. pars altera in radice.

||[9] ||Adeoq ex $44+\sqrt{1944}$ radix cubica est $2+\sqrt{6}$. Hanc itáq dividendo

(22) In summary, $\sqrt[3]{[\frac{1}{2}(\pm 2+\sqrt{5})]} = \sqrt[3]{\frac{1}{2}}\times\frac{1}{2}(\pm 1+\sqrt{5})$, where $\sqrt[3]{[\pm 2+\sqrt{5}]}$ is evaluated by the general rule.

Example 2. The cube root of $\sqrt{\frac{5}{4}}-1$ is $\sqrt[6]{\frac{125}{256}}-\sqrt[3]{\frac{1}{16}}$.

Working

Multiply $\sqrt{\frac{5}{4}}-1$ into 2 to remove the fraction,
there comes $\sqrt{5}-2$. 1000 times this
makes $\sqrt{5{,}000{,}000}-2000$, which in pure numbers
is 2236. -2000.
Aggregate 236. Residue -4236. Their cube
roots 6. -16.
Sum -10, and of its tenth part
one half is $-\frac{1}{2}$, the pure part in the root. The pure part (-2)
of the binomial divided by it gives 4. When $\frac{1}{4}$ (the square of $-\frac{1}{2}$) is subtracted
from it there remains $3\frac{3}{4}$, the square root of whose third part
is $\sqrt{1\frac{1}{4}}$, the second part of the root. Hence the cube
root of $\sqrt{5}-2$ is $\sqrt{1\frac{1}{4}}-\frac{1}{2}$. And this divided by $\sqrt[3]{2}$ gives

$\sqrt[6]{\frac{125}{256}}-\sqrt[3]{\frac{1}{16}}$, the required root.

In a similar fashion the cube roots of $2+\sqrt{5}$ and $1+\sqrt{\frac{5}{4}}$ are $\frac{1}{2}+\sqrt{1\frac{1}{4}}$ and $\sqrt[3]{\frac{1}{16}}+\sqrt[6]{\frac{125}{256}}$.(22)

Example 3. The cube root of $\sqrt{242} \mid \sqrt{243}$ is $\sqrt{2} \mid \sqrt{3}$.

Working

Out of $\sqrt{242}+\sqrt{243}$ by extracting all that is rational
there comes $11\sqrt{2}+9\sqrt{3}$. I therefore multiply the binomial by $\sqrt{2}$ or $\sqrt{3}$ so
that one or other of the terms comes to be rational. As for instance by multiplying
by $\sqrt{2}$ there comes $22+\sqrt{486}$. The parts of this when squared
become 484. 486. Their difference
is 2. Again, in consequence, I multiply the binomial by 2,
there comes $44+\sqrt{1944}$. 1000 times this
makes $44{,}000+\sqrt{1{,}944{,}000{,}000}$.
In pure numbers 44,000. 44,091.
Aggregate 88,091. Residue -91.
Cube roots 44. -4.
their sum 40.
half its tenth part 2, the pure part in the root.
Divide 44 by 2, there comes 22.
subtract 4, there remains 18.
The square root of its third part is $\sqrt{6}$, the second part in the root.

Hence the cube root of $44+\sqrt{1944}$ is $2+\sqrt{6}$. And consequently, by dividing

per $\sqrt{3:2}$ in $\sqrt{6:2}$ hoc est per $\sqrt{2}$ prodit $\sqrt{2}+\sqrt{3}$ radix cubica ex $\sqrt{242}+\sqrt{243}$.(23)

Siquando partes binomij habeant communes divisores, juvat plerumq; binomium sub initio operis per eos dividere. Sic in exemplo sequenti divido $\sqrt{96}+\sqrt{120}$ per $\sqrt{24}$.

Exempl. 4. $\sqrt{\text{cub}}$(24) ex $\sqrt{19\frac{1}{5}}+\sqrt{24}$ est $\sqrt{6:\frac{3}{40}}+\sqrt{6:9\frac{3}{8}}$.

Operatio.

$$\frac{\sqrt{19\frac{1}{5}}+\sqrt{24}.}{\sqrt{5}.}$$

$$\frac{\sqrt{5}}{\sqrt{24}}.\quad \sqrt{24})\overline{\sqrt{96}+\sqrt{120}}(2+\sqrt{5}.$$

$$\begin{array}{rr} & 1000 \\ \hline 2000+\sqrt{5000000}. & \\ 2000. & 2236. \\ 4236. & -236. \\ 16. & -6. \\ \hline . & 10. \\ & 1. \\ & \tfrac{1}{2})2(4. \\ & \tfrac{1}{4}. \\ \hline & 3\tfrac{3}{4}. \\ & \sqrt{1\tfrac{1}{4}}. \end{array}$$

Est ergo $\frac{1}{2}+\sqrt{1\frac{1}{4}}$ radix cubica ex $2+\sqrt{5}$. Quæ deniq; divisa per $\sqrt{3}:\frac{\sqrt{5}}{\sqrt{24}}$, hoc est per $\sqrt{6:\frac{5}{24}}$, dat $\sqrt{6:\frac{3}{40}}+\sqrt{6:9\frac{3}{8}}$, radicem quæsitam.(25)

Hâc methodo(26) binomiorum radices cubicæ solent extrahi. Sed alia concinnior occurrit et universalior, quæ nec egeat extractione radicis cubicæ in numeris proximis, nec tantâ præparatione binomij quâ numeri solent excrescere, et ad literales æquationes æque ac numerales se extendit. Quare cum nondum quod scio in usu sit, operæ pretium facturus videor si hic describam.

(23) Newton multiplies the binomial $11\sqrt{2}+9\sqrt{3}$ so that the difference of the squares of its terms is a perfect cube while its first term becomes an integer: since

$$(11\sqrt{2})^2-(9\sqrt{3})^2 = (-1)^3,$$

the required multiplier is $\sqrt{8} = 2\sqrt{2}$. The general rule then finds $\sqrt[3]{[44+18\sqrt{3}]} = 2+\sqrt{6}$.

(24) Read 'Radix cubica'.

(25) The multiplying factor $\sqrt{\frac{5}{24}}$ converts the binomial $\sqrt{19\frac{1}{5}}+\sqrt{24}$ into $2+\sqrt{5}$, whose first term is integral and the difference of the squares of whose terms is a perfect cube $(-1)^3$. By Example 2, $\sqrt[3]{[2+\sqrt{5}]} = \frac{1}{2}(1+\sqrt{5})$.

(26) Newton has rightly cancelled the following clause 'cujus inventio Jacobo a Waess-[enaer] a Schooten debetur' and the referent marginal note 'Vide Additament[um] ad Commentar^s in Geom: Cartesij' (for whose discovery Schooten is indebted to Jacob van Waessenaer: See his *Additamentum* to Descartes' Geometry). In his *Additamentum* (*Geometria*

this by $\sqrt[3]{2}$ times $\sqrt[6]{2}$, that is, by $\sqrt{2}$, there comes out $\sqrt{2}+\sqrt{3}$ the cube root of $\sqrt{242}+\sqrt{243}$.(23)

If at any time the parts of the binomial have common divisors, it is helpful for the most part to divide the binomial by them at the beginning of the work. So in the following example I divide $\sqrt{96}+\sqrt{120}$ by $\sqrt{24}$.

Example 4. The cube root of $\sqrt{19\frac{1}{5}}+\sqrt{24}$ is $\sqrt[6]{\frac{3}{40}}+\sqrt[6]{9\frac{3}{8}}$.

Working

$$\begin{array}{lll} & & \underline{\sqrt{19\frac{1}{5}}+\sqrt{24}.} \\ \dfrac{\sqrt{5}}{\sqrt{24}}. & & \sqrt{5}. \\ & \sqrt{24}) & \overline{\sqrt{96}+\sqrt{120}}(2+\sqrt{5}. \\ & & \qquad\qquad \underline{1000} \\ & & \qquad\qquad 2000+\sqrt{5,000,000}. \\ & & \qquad\qquad 2000. \quad 2236. \\ & & \qquad\qquad 4236. \quad -236. \\ & & \qquad\qquad \underline{\ \ 16. \quad\ \ -6.} \\ & & \qquad\qquad\quad 10. \\ & & \qquad\qquad\quad 1. \\ & & \qquad\qquad\quad \frac{1}{2})2(4. \\ & & \qquad\qquad\qquad \underline{\frac{1}{4}.} \\ & & \qquad\qquad\qquad 3\frac{3}{4}. \\ & & \qquad\qquad\qquad \sqrt{1\frac{1}{4}}. \end{array}$$

Therefore $\frac{1}{2}+\sqrt{1\frac{1}{4}}$ is the cube root of $2+\sqrt{5}$. This, finally, divided by $\sqrt[3]{\dfrac{\sqrt{5}}{\sqrt{24}}}$, that is, by $\sqrt[6]{\frac{5}{24}}$, gives $\sqrt[6]{\frac{3}{40}}+\sqrt[6]{9\frac{3}{8}}$, the root required.(25)

It is by this method(26) that the cube roots of binomials are usually extracted. But another approach suggests itself, neater and more universal in that it requires no extraction of a cube root to the nearest integer nor so great a preparation of the binomial (with its consequent increase in numerical complexity), while extending equally to literal as well as to numerical equations. Consequently, since as far as I know it is not in use, I believe it will be worthwhile if I describe it here.

($_2$1659): 369–400) Schooten did in fact state (p. 369) that the 'solutio artificiosissima' of the three staves problem came from Waessenaer's *Den onwissen Wis-konstenaer I. I. Stampioenius*, but the demonstration of the following 'regula generalis...extrahendi radices quaslibet ex quibuscunque Binomiis, radicem binomiam habentibus', which likewise had appeared in Waessenaer's book, he presented (p. 389) 'qualis à me inventa est'. As we have seen (§1: note (41)) Descartes was the author of both resolutions, but perhaps most curious of all is Newton's present confusion of Ferguson's and 'Waessenaer's' cube root algorithms, which are essentially distinct. That confusion appears convincing evidence that Newton at this time did not appreciate the generality or the nicety of Descartes' approach. (Compare note (17) above.)

||[10] ||Præparatio.

Binomium imprimis multiplicari debet per denominatores fractionum, siquas habeat, ac dividi per maximum communem divisorem nominum. Dein quærenda est differentia quadratorum nominum; et ex ejus radice cubicâ, ut et e nominibus extrahendum quicquid est rationale. Perspicuitatis gratiâ sint $A\surd B$ et $C\surd D$ nomina sic præparata, et $E\surd 3:F$ radix cubica differentiæ quadratorum nominum.

Regula.

Alterutrū nomen $A\surd B$ pro sequenti opere adhibe, illud si placet cujus factoris puri ad numerum in factore surdo (sive A ad B) minor est ratio, vel cujus factor purus pauciores divisores admittit.(27) Si $A\surd B$ sit minus nomen, e numeris integris quorum cubi non sunt majores quàm $\frac{F\times A}{4B}$ quære dividentes A sine residuo, quotientes asserva et unumquemq; divisorem (si plures existant) sigillatim duc in $\surd B$ et factum dic M, ac $\sqrt{MM+EF}$ dic N.

Sin $A\surd B$ sit majus nomen; e numeris quorum cubi non sunt majores quàm $\frac{F\times A}{B}$ nec minores quàm $\frac{F\times A}{4B}$ quære dividentes A sine residuo, quotientes asserva, et unumquemq; divisorem (si sint plures) sigillatim duc in $\surd B$ et factum dic M ac $\sqrt{MM-EF}$ dic N.

(28)Dein statue binomium $[M.N]$ signis $+$ et $-$ similiter atq; binomium $[A\surd B.C\surd D.]$ affectum ac divisum per $\surd 3:F$, esse radicem ejusdem

$$[A\surd B.C\surd D.],$$

si modò $MM+3NN$ æquetur præfato quotienti in F ducto; sin secus, rejice illos M et N. Et si nullæ inveniantur M et N istis conditionibus præditæ id arguit radicem non posse extrahi.(29)

---

(27) Newton's first version of the remainder of this paragraph reads: 'Si minus nomen adhibuisti, tunc e numeris quorum cubi non sunt majores quàm $\frac{F\times A}{4B}$ quære quinam dividunt A sine residuo, quotientes asserva et singulos (si plures ejusmodi reperias) sigillatim duc in $\surd B$ et factum dicito M. Dein quære $\sqrt{MM+EF}$ ac dicito N, dummodò præfatus quotiens in F ductus æquetur $MM+3NN$: aliàs rejice' (If you employed the lesser term, then from the numbers whose cubes are not greater than $FA/4B$ seek those which divide A without remainder, retaining the quotients, and (if you find several of the kind) multiply each separately into $\surd B$ and call the product M. Then seek $\surd[M^2+EF]$ and call it N, so long as the corresponding above-mentioned quotient multiplied into F be equal to M^2+3N^2: otherwise reject [the divisor]). Similarly, his first version of the following paragraph reads: 'Sin majus nomen adhibuisti; e numeris quorum cubi non sunt majores quàm $\frac{F\times A}{B}$ nec minores quàm $\frac{F\times A}{4B}$

Preparation

In the first instance the binomial should be multiplied by the denominators of its fractions, if it has any, and divided by the greatest common divisor of its terms. Then must be sought the difference of the squares of the terms, and from its cube root, as also that of the terms, whatever is rational must be extracted. For clarity's sake let $A\surd B$ and $C\surd D$ be the terms thus prepared, and $E\sqrt[3]{F}$ the cube root of the difference of the squares of the terms.

Rule

Employ one or other term $A\surd B$ for the following operation, that one if it suits whose pure factor bears to the number in the surd factor (that is, A to B) a lesser ratio, or whose pure factor admits of fewer divisors.(27) If $A\surd B$ be the lesser term, of the integers whose cubes are not greater than $FA/4B$ seek those which divide A without remainder, retain the quotients and multiply each divisor (if several are forthcoming) separately into $\surd B$, calling the product M and $\surd(M^2+EF)$ calling N. But if $A\surd B$ be the greater term, of the numbers whose cubes are not greater than FA/B nor less than $FA/4B$ seek those which divide A without remainder, retain the quotients and multiply each divisor (if there are several) separately into $\surd B$, calling the product M and $\surd(M^2-EF)$ calling N.

(28)Then fix on the binomial $M \pm N$, affected with the signs + and − in the same way as $A\surd B \pm C\surd D$ and divided by $\sqrt[3]{F}$, as the root of that same $A\surd B \pm C\surd D$, provided that M^2+3N^2 be equal to the above-mentioned quotient multiplied into F: but if not, reject those quantities M and N. And if no M and N are to be found satisfying those conditions, that shows the root cannot be extracted.(29)

---

quære quinam dividunt A sine residuo, quotientes asserva, et singulos (si sint ejusmodi plures) sigillatim duc in $\surd B$ et factum dicito M. Dein quære $\sqrt{MM-EF}$ et N dicito, si modò præfatus quotiens in F ductus æquatur $MM+3NN$: alioquin rejice.'

(28) A first version of this paragraph reads: 'Deniq; si tales divisores aut tale N non obtinuisti, conclude radicem non posse extrahi. Sin obtinuisti N (cujusmodi plura non obtinebis) id innuit binomium $[M.N]$ signis + et − similiter atq; binomium $[A\surd B.C\surd D]$ affectum ac divisum per $\surd 3: F$, esse radicem ejusdem $[A\surd B.C\surd D]$. Quod tamen majoris certitudinis causâ tentabis cubando ipsum' (Finally, if you have not obtained such divisors or such an N, conclude that the root is not extractable. But if you have obtained N (and you will not obtain several of the kind) that argues that the binomial $M \pm N$, affected with the signs + and − in the same way as the binomial $A\surd B \pm C\surd D$, is on division by $\sqrt[3]{F}$ the root of that very same $A\surd B \pm C\surd D$. For the sake of greater certainty you will test this by cubing it). Newton's remark about the uniqueness of the (real) cube root of a (real) binomial is, of course, exact.

(29) The 'preparation' of the binomial $\sqrt[3]{[A\surd B \pm C\surd D]}$ ensures that on multiplying through by $\sqrt[3]{F}$ the resulting binomial has the difference of the squares of its component terms, $(FA\surd B)^2-(FC\surd D)^2$, an exact cube ($\pm E^3F^3$). Hence, on setting

$$M+N = \sqrt[3]{F} \times \sqrt[3]{[A\surd B+C\surd D]},$$

[30]Res patebit exemplis quæ literis juxta positis explicui.[31]

||[11] ||Exempl 1. Ex $\surd 242 + \surd 243$ $\surd$cub est $\surd 2 + \surd 3$.

Præparatio.	Operis prosecutio.
$\surd 242 + \surd 243$,	sive $11\surd 2 + 9\surd 3$.
Quadr: 242 . 243.	Limes $\frac{11}{8}$ nempe $\frac{FA}{4B}$.
Diff: 1.	Divisor 1)11(11 Quotiens.
Rad cub diff 1.	$M = 1\surd 2$ seu $\surd 2$. $M^q = 2$.
Quare $E = 1$, & $F = 1$.	Adde $EF = 1$. Fit $N^q = 3$.

Ubi cùm videam $M^q + 3N^q$ seu $2 + 9$ æquari $11E$ seu 11, concludo radicem quæsitam esse $\surd 2 + \surd 3$ utpote $\frac{M+N}{\surd 3 : F}$.[32]

Hic adhibui minus nomen $11\surd 2$ pro divisoribus eliciendis, sin majus $9\surd 3$ adhibuissem, hæc fuisset operationis forma.

$9\surd 3 + 11\surd 2$.

Limites 3 & $\frac{3}{4}$ nempe $\frac{FA}{B}$ et $\frac{FA}{4B}$.

Divisor 1)9(9 Quotiens.

$M = 1\surd 3$ seu $\surd 3$. $M^q = 3$.

Subduc $EF = 1$. Restat $N^q = 2$. Ubi cum sit $M^q + 3N^q = 9E$ seu 9, concludo radicem quæsitam esse $\surd 3 + \surd 2$ utpote $\frac{M+N}{\surd 3 : F}$ ut supra.

---

since B and D are non-square, therefore $M - N = \sqrt[3]{F} \times \sqrt[3]{[A\surd B - C\surd D]}$ and so

$$M^2 - N^2 = \sqrt[3]{[(FA\surd B)^2 - (FC\surd D)^2]} = \pm EF.$$

Further, since $(M \pm N)^3 = F(A\surd B \pm C\surd D)$, we may identify corresponding parts of the equation by setting $M(M^2 + 3N^2) = FA\surd B$ and $N(3M^2 + N^2) = FC\surd D$, that is, on taking $M = m\surd B$, $m(M^2 + 3N^2) = FA$. If now $A\surd B < C\surd D$ (and so $M < N$) then

$$4M^3 < M(M^2 + 3N^2) = FA\surd B \quad \text{and} \quad m^3 < FA/4B$$

(or, as Newton says, m is 'bounded' by $\sqrt[3]{[FA/4B]}$): we have then to find some suitable m which satisfies this inequality and, on setting $(A\surd B)^2 - (C\surd D)^2 = -E^3F$, that is

$$M^2 - N^2 = -EF,$$

also satisfies $m(M^2 + 3(M^2 + EF)) = FA$. If, however, $A\surd B > C\surd D$ (and so $M > N$), then

$$4M^3 > M(M^2 + 3N^2) = FA\surd B > M^3 \quad \text{and} \quad FA/4B < m^3 < FA/B$$

(or, as Newton says, m is 'bounded' by $\sqrt[3]{[FA/B]}$ and $\sqrt[3]{[FA/4B]}$): accordingly, on setting $(A\surd B)^2 - (C\surd D)^2 = +E^3F$, that is, $M^2 - N^2 = +EF$, we must then seek an m which lies within those limits and satisfies $m(M^2 + 3(M^2 - EF)) = FA$.

[30]The process will be clear from examples, which I have set forth with algebraic equivalents placed alongside.[31]

Example 1. The cube root of $\sqrt{242}+\sqrt{243}$ is $\sqrt{2}+\sqrt{3}$.

Preparation.	The operation carried out.
$\sqrt{242}+\sqrt{243}$, or	$11\sqrt{2}+9\sqrt{3}$.
Squares 242 . 243	Bound $\frac{11}{8}$, that is, $FA/4B$.
Difference 1.	Divisor 1)11(11 Quotient.
Its cube root 1.	$M=1\sqrt{2}$ or $\sqrt{2}$. $M^2=2$.
Hence $E=1$ and $F=1$.	Add $EF=1$. There comes $N^2=3$.

Since there I see that M^2+3N^2 or $2+9$ equals $11E$ or 11, I conclude that the root required is $\sqrt{2}+\sqrt{3}$, namely $(M+N)/\sqrt[3]{F}$.[32]

Here I have made use of the lesser term $11\sqrt{2}$ for eliciting the divisors, but if I had used the greater $9\sqrt{3}$, this would have been the mode of operation:

$9\sqrt{3}+11\sqrt{2}$.
Bounds 3 and $\frac{3}{4}$, that is, FA/B and $FA/4B$.
Divisor 1)9(9 Quotient.
$M=1\sqrt{3}$ or $\sqrt{3}$. $M^2=3$.

Subtract $EF=1$. There remains $N^2=2$. Since there M^2+3N^2 is equal to $9E$ or 9, I conclude that the root sought is $\sqrt{3}+\sqrt{2}$, namely $(M+N)/\sqrt[3]{F}$, as before.

(30) In the version of his *Observationes* which (compare note (11)) Newton sent to Collins on 11 July 1670, the text at this point continued without break into the (subsequently cancelled) first paragraph at the head of p. 15 below. When some time between late July and the following September Newton inserted the following four pages (numbered 11–14) he added this sentence in introduction to his examples on p. 11.

(31) We reproduce the amended Examples 1 and 2 which Newton wrote out on the interleaved sheet $7^v/7^r$ (with a preliminary draft on 7^r). The layout of his original p. 11, which is somewhat more elaborate, is given in Appendix 1.

(32) Here $A=11$, $B=2$, $C=9$, $D=3$ and $-E^3F=242-243=-1$, and Newton supposes $E=F=1$. In the inverted scheme which follows, $A=9$, $B=3$, $C=11$, $D=2$ and $E^3F=243-242$ with $E=F=1$ as before.

Exempl 2. Ex $\sqrt{1896075}+\sqrt{1792252}$ rad cub est $\sqrt{75}+\sqrt{28}$.

Præparatio. Operis continuatio.

$\sqrt{1896075}+\sqrt{1792252}$, seu $795\sqrt{3}+506\sqrt{7}$.

[Qu]adr: 1896075. 1792252.	[Limites] 265 & $66\frac{1}{4}$.
Diff: 103823.	[Divisor] 5)795(159 [Quotiens].
Rad cub 47.	$[M=]5\sqrt{3}$. $[M^q=]75$.
Ergo $E=47$, & $F=1$.	$[-EF=]-47$. $[N^q=]28$. Prodit $\sqrt{75}+\sqrt{28}$.

Probatio. $75+3\times 28=159\times 1$.(33)

Nota. Numeri 795 plures esse divisores nempe 1, 3, 5, 15, 53, 159, 265, & 795 sed me omnes præter 5 neglexisse quia ejus solummodo cubus 125 intercessit limitibus 265 et $66\frac{1}{4}$.

Si alterum nomen ad eliciendos divisores adhibuissem hæc fuisset operis forma.

$506\sqrt{7}+795\sqrt{3}$.

[Limes] $18\frac{1}{4}$.

1)506(506.	[vel]	2)506(253.
$1\sqrt{7}$. 7.		$2\sqrt{7}$. 28.
+47. 54.		+47. 75.
Prod[it] $\sqrt{7}+\sqrt{54}$.		Prod[it] $\sqrt{28}+\sqrt{75}$.
Prob[atio]. $7+3\times 54=506\times 1$ falsè.		Prob[atio]. $28+3\times 75=2$[53 × 1] ve[rè].

Cum hic duo sint divisores numeri 506 nempe 1 et 2 quorum cubi 1 et 8 non excedunt limitem $18\frac{1}{4}$, opus primo per divisorem 1 tento sed frustra. Quare tento deinde per alterum divisore[m] et res succedit.

‖[12] ‖Cæterum ne inventio numerorum(34) E et F moram aliquam injiciat; nota quod F non potest esse alius numerus quàm 1, 2, vel 4.(35) Unde si neq; differentia quadratorum nominum, neq; dimidium ejus, neq; pars quarta sit cubus numerus; Binomij radix non potest extrahi. Sin eorum trium aliquis sit cubus numerus, pone radicem ejus pro E, et 1, 2, vel 4 respectivè pro F. Sic habitâ quadratorum differentiâ 4394, cùm non sit cubus numerus divido per 2, & prodit 2197 cubus numerus cujus radix est 13, pono ergo 13 pro E et 2 pro F.

Est insuper notandum quòd cum F existit unitas, radix aliquando potest esse fractio per 2 denominata.(36) Quod ubi suspicaris, vice divisorum adhibe tales

(33) Similarly, $A = 795$, $B = 3$, $C = 506$, $D = 7$ and $E^3F = 47^3$ and Newton supposes $E = 47$, $F = 1$. In the inverted scheme the values of A, B and C, D are interchanged.

(34) 'terminorum' (terms) is cancelled.

(35) If $F(A\sqrt{B}+C\sqrt{D}) = (\sqrt{\alpha}+\sqrt{\beta})^3$, where α and β are coprime, then F must be the

Example 2. The cube root of $\sqrt{1,896,075}+\sqrt{1,792,252}$ is $\sqrt{75}+\sqrt{28}$.

Preparation.	The work carried out.
$\sqrt{1,896,075}+\sqrt{1,792,252}$, or $795\sqrt{3}+506\sqrt{7}$.	
[Sq]uares 1,896,075 . 1,792,252.	[Bounds] 265 and $66\frac{1}{4}$.
Difference 103,823.	[Divisor] 5)795(159 [Quotient]
Cube root 47.	$[M=]\,5\sqrt{3}$. $[M^2=]\,75$.
Therefore $E=47$ and $F=1$.	$[-EF=]-47$. $[N^2=]\,28$. There comes $\sqrt{75}+\sqrt{28}$.
	Test. $75+3\times 28=159\times 1$.[33]

Note. The number 795 has several divisors, namely 1, 3, 5, 15, 53, 159, 265 and 795, but I have neglected all but 5 since its cube 125 alone lies within the bounds 265 and $66\frac{1}{4}$.

If I had employed the other term to elicit divisors, this would have been the way of working:

$506\sqrt{7}+795\sqrt{3}$.

[Bound] $18\frac{1}{4}$.

1)506(506.	[or]	2)506(253.
$1\sqrt{7}$. 7.		$2\sqrt{7}$. 28.
$+47$. 54.		$+47$. 75.
There comes $\sqrt{7}+\sqrt{54}$.		There comes $\sqrt{28}+\sqrt{75}$.
Test. $7+3\times 54=506\times 1$, falsely.		Test. $28+3\times 75=253\times 1$, correctly.

Since here the number 506 has two divisors, namely 1 and 2, whose cube 1 and 8 do not exceed the bound $18\frac{1}{4}$, I attempt in the first instance the operation with divisor 1 but without success. I therefore next attempt the procedure with the second divisor and it works.

But lest finding the numbers[34] E and F cause some delay, note that F can be no number other than 1, 2 or 4.[35] Hence, if neither the difference of the squares of the terms, nor its half, nor its fourth part is an exact cube, the binomial root cannot be extracted. But if some one of those three be a cube, take its root for E and 1, 2 or 4 correspondingly for F. Thus, supposing the difference of the squares to be 4394, since it is not an exact cube I divide by 2 and there comes out 2197, a cube whose root is 13, and I therefore take 13 for E and 2 for F.

It should be noted in addition that, when F turns out to be unity, the root may on occasion be a fraction of denominator 2.[36] When you suspect as much,

common factor of $3\alpha+\beta$ and $\alpha+3\beta$, and so of $4(\alpha+\beta)$ and $2(\alpha-\beta)$; that is, it can only be 1, 2 or 4. A factor 2 may, however, be present in one of B or D.

(36) In the terminology of the previous note, when $F=1$, then $\alpha=a^2B$ and $\beta=b^2D$, where a and b are non-surd: hence $a(a^2B+3b^2D)=A$, $b(3a^2B+b^2D)=C$, where of course

fractos numeros a 2 denominatos quorum numeratores dividunt A sine residuo. Sic proposito Binomio $95\sqrt{5}-144$: quadrata partium sunt 45125 & 20736. Differentiæ 24389 radix cubica 29. Quare est $E=29$ & $F=1$. Jam majus nomen $95\sqrt{5}$ dat 19 & $4[\frac{3}{4}]$ pro limitibus. His intercedit cubus ex $\frac{5}{2}$ cujus numerator 5 potest dividere 95. Pone itaq; $\frac{5}{2}\sqrt{5}$ pro M et $\sqrt{MM-EF}$ sive $\sqrt{\frac{125}{4}-29}$ dat $\frac{3}{2}$ pro N. Est ergo $\frac{5}{2}\sqrt{5}-\frac{3}{2}$ radix quæsita: quemadmodum ex eo pateat quod sit $\frac{5}{2})95(38=\frac{125}{4}+\frac{27}{4}$.(37)

Extractio cubicarum radicum e Binomijs literalibus non differt ab opere in numeris nisi quòd divisores hic non ex magnitudine sed dimensionibus limitantur quæ debent esse $\frac{1}{3}$ dimensionum quantitatis $\frac{A}{B}$. Unde oportet $\frac{A}{B}$ vel nullas vel tres vel sex dimensiones habere &c, aut radicem non posse extrahi. Idem concludi potest si radix cubica differentiæ quadratorū nominum non sit rationalis; nam F hic semper debet esse unitas.(38)

‖[13] ‖Exemplum.

Ex $\overline{3a^4b-8ab^4\sqrt{ab}+a^4\sqrt{a^4-3ab^3}}\ \sqrt{\text{cub}}$: est $b\sqrt{ab}+\sqrt{a^4-3ab^3}$.

Præparatio		Operatio.
	$\overline{3a^4b-8ab^4\sqrt{ab}+a^4\sqrt{a^4-3ab^3}}$.	
Divide per $a\sqrt{a}$, fit	$\overline{3a^3b-8b^4\sqrt{b}+a^3\sqrt{a^3-3b^3}}$.	$\overline{3a^3b-8b^4\sqrt{b}}\ [+a^3\sqrt{a^3-3b^3}]$.
Quadrata,	$9a^6b^3-48a^3b^6+64b^9$. $a^9-3a^6b^3$.	$b)3a^3b-8b^4(3a^3-8b^3$.
Diff:	$-a^9+12a^6b^3-48a^3b^6+64b^9$.	$b\sqrt{b}$. b^3.
Rad: cub: Diff:	$-a^3+4b^3$.	$-a^3+4b^3$. a^3-3b^3.

Prodit $b\sqrt{b}+\sqrt{a^3-3b^3}$.

Probatio. b^3 & $3a^3-9b^3=3a^3-8b^3$ Quotienti.(39)

Jam quia diviseram binomium per $a\sqrt{a}$, inventam radicem multiplico per $\sqrt{3}:a\sqrt{a}$ id est per $\sqrt{a}$ et prodit $b\sqrt{ab}+\sqrt{a^4-3ab^3}$ radix quæsita.

Supra adhibui majus nomen quale finxi fore $\overline{3a^3b-8b^4}\sqrt{b}$, siquidem ab ejus quadrato priùs subducebam quadratum alterius ad inveniendam differentiam. At operatio secundū alterū nomen sic potuit institui

A, B, C and D are integers. These conditions may evidently be met when $2a$ and $2b$ are each odd integers.

(37) That is, $A/m = M^2+3N^2$, where $A = 95$, $M = m\sqrt{5} = \frac{5}{2}\sqrt{5}$ and $N = -\frac{3}{2}$.

(38) For in the terminology of note (35), but where now A, B, C, D, α and β are general algebraic expressions, if by suitable 'preparation' $F(A\sqrt{B}+C\sqrt{D}) = (\sqrt{\alpha}+\sqrt{\beta})^3$ with the pairs A, C and B, D each without a common factor, then, since $\alpha-\beta = \sqrt[3]{[F^2(A^2B-C^2D)]}$, must A^2B-C^2D be the exact cube $(\alpha-\beta)^3$ on setting $F = 1$.

(39) Here Newton takes $A = 3a^3b-8b^4$, $B = b$, $C = a^3$ and $D = a^3-3b^3$, so that $E = \sqrt[3]{[A^2B-C^2D]} = -a^3+4b^3$ with $F = 1$ (note (38)). On trial by the divisor $m = b$ it follows

instead of simple divisors employ such fractions of denominator 2 as have their numerators divisors of A without remainder. So, should the binomial $95\sqrt{5}-144$ be proposed, the squares of its parts are 45,125 and 20,736, and the cube root of the difference 24,389 is 29, so that $E = 29$ and $F = 1$. Now the greater term $95\sqrt{5}$ gives 19 and $4\frac{3}{4}$ for the bounds. Between these lies the cube of $\frac{5}{2}$, whose numerator 5 can divide 95. In consequence, set $\frac{5}{2}\sqrt{5}$ for M and $\sqrt{(M^2-EF)}$ or $\sqrt{(\frac{125}{4}-29)}$ gives $\frac{3}{2}$ for N. Therefore $\frac{5}{2}\sqrt{5}-\frac{3}{2}$ is the required root: as may be evidenced from there being $\frac{5}{2})95(38=\frac{125}{4}+\frac{27}{4}$.[37]

The extraction of the cube roots of algebraic binomials does not differ from the operation in numerical cases except that here the divisors are limited not by their magnitude but by their dimensions, which ought to be $\frac{1}{3}$ of the dimensions of the quantity A/B. In consequence, A/B must have either zero or three or six dimensions, and so on, or alternatively the root cannot be extracted. The same conclusion may be drawn if the cube root of the difference of the squares of the terms is not rational, for F must here always be unity.[38]

Example

Of $(3a^4b-8ab^4)\sqrt{ab}+a^4\sqrt{(a^4-3ab^3)}$ the cube root is $b\sqrt{ab}+\sqrt{(a^4-3ab^3)}$.

Preparation | Operation

$(3a^4b-8ab^4)\sqrt{ab}+a^4\sqrt{(a^4-3ab^3)}$.

Divide by $a\sqrt{a}$, there comes

$(3a^3b-8b^4)\sqrt{b}+a^3\sqrt{(a^3-3b^3)}$. $\quad (3a^3b-8b^4)\sqrt{b}[+a^3\sqrt{(a^3-3b^3)}]$.

Squares, $9a^6b^3-48a^3b^6+64b^9$. $a^9-3a^6b^3$. $\quad b)3a^3b-8b^4(3a^3-8b^3$.

Difference $\quad -a^9+12a^6b^3-48a^3b^6+64b^9$. $\quad b\sqrt{b}$. $\quad b^3$.

Cube root of the difference $\quad -a^3+4b^3$. $\quad -a^3+4b^3$. $\quad a^3-3b^3$.

There comes $b\sqrt{b}+\sqrt{(a^3-3b^3)}$.

Test. b^3 plus $3a^3-9b^3 = 3a^3-8b^3$, the Quotient.[39]

Now since I divided the binomial by $a\sqrt{a}$, I multiply the root found by $\sqrt[3]{(a\sqrt{a})}$, that is, by $\sqrt{a}$, and there comes out $b\sqrt{ab}+\sqrt{(a^4-3ab^3)}$, the root required.

I employed the greater term above, which I there supposed to be

$$(3a^3b-8b^4)\sqrt{b}$$

inasmuch as from its square I previously took away the square of the second term to find the difference. But the operation could thus have been arranged according to the second term:

---

that $M = m\sqrt{b} = b\sqrt{b}$ and $N^2 = M^2-EF = a^3-3b^3$: correctly so, since on checking $A/m = 3a^3-8b^3 = M^2+3N^2$.

$$a^3\sqrt{a^3-3b^3}\,[+\overline{3a^3b-8b^4}\sqrt{b}].$$
$$1)a^3(a^3.$$
$$\sqrt{a^3-3b^3}.\ a^3-3b^3.$$
$$-a^3+4b^3.\ b^3.$$

Prodit $\sqrt{a^3-3b^3}+\sqrt{b^3}$, ut ante.

[Probatio.] $a^3-3b^3\ \&\ +3b^3=a^3$ Quotienti.[40]

Ad divisorum dimensiones quod attinet; in priori casu ex eo quod triens dimensionum quantitatis $\dfrac{3a^3b-8b^4}{b}$ fuerit, novi divisores unius dimensionis (qualis fuit b) proposito tantùm inservire posse. Sed in posteriori casu ex eo quod $\dfrac{a^3}{a^3-3b^3}$ fuerit nullius dimensionis, novi divisores nullarum dimensionum id est tantùm numerales adhibendos esse, cujusmodi solam unitatem reperi.

[41]Quicquid de extractione radicis cubicæ Binomiorū explicui facilè ad ||[14] quadrato-cubicas et alias magis compositas idq; totidem ||fere verbis applicatur, et exercitationis gratiâ cuiq; pro re natâ inveniendum linquo.[42] Quinetiam radices e Binomijs impossibilibus hâc ratione possunt extrahi: quod cùm aliqui fortè ad penum Analyticū pertinere judicent, siquidem ejus beneficio æquationes omnes cubicæ per Cardani regulas evadunt reducibiles; non gravabor illustrare.[43] Atq; equidem cùm utráq; pars binomij existit impossibilis radix extrahitur ut ante, signis + et − probè observatis. At si pars altera tantùm sit impossibilis, divisores nominum non observant præscriptos limites, et proinde rem aliter varietatis gratiâ sic aggredior.

Binomio ut ante præparato, sit $A\sqrt{-B}$ nomen impossibile, altero existente rationali. Duc hanc quadratorum progressionem 1, 4, 9, 16, 25 &c in B, et singulos factos qui subducti a $4E$[44] relinquunt quadratos numeros elige, cujuslibet et quadrati relicti radicem seorsim extrahe, divide per $2\sqrt{3}$: F, conjunge cum debitis signis + et −, ac tenta cubando num sit radix quæsita.[45]

(40) With the previous values of A and C, B and D interchanged the trial divisor $m = 1$ (and so $M^2 = a^3-3b^3$, $N^2 = M^2+EF = b^3$) satisfies $M^2+3N^2 = A/m = a^3$. We should add that Newton, lacking adequate space in his manuscript, presented this computation in a slightly different lay-out, and we have reshaped it slightly to make its pattern identical with that of the preceding numerical root extractions.

(41) This replaces the cancelled paragraph on Newton's p. 15 (note (51) below).

(42) Descartes' cube root algorithm (§1: note (39)) is, of course, immediately generalizable to the extraction of numerical kth roots, as Descartes himself well knew. The Newtonian algorithm in its generalization depends on identifying

$$(m\sqrt{B}\pm n\sqrt{D})^k = F(A\sqrt{B}\pm C\sqrt{D})$$

and rapidly becomes unwieldy as k increases.

$a^3\sqrt{(a^3-3b^3)}\,[+(3a^3b-8b^4)\sqrt{b}].$
$1)a^3(a^3.$
$\sqrt{(a^3-3b^3)}.\quad a^3-3b^3.$
$-a^3+4b^3.\qquad b^3.$
There comes $\sqrt{(a^3-3b^3)}+\sqrt{b^3}$, as before.
[Test.] a^3-3b^3 plus $+3b^3=a^3$, the Quotient.(40)

With regard to the dimensions of the divisors, in the former case I knew from its being one third of the dimension of the quantity $(3a^3b-8b^4)/b$ that the divisors of unit dimension (such as b was) could alone satisfy the purpose. But in the latter case I knew from $a^3/(a^3-3b^3)$ being of zero dimension that divisors of zero dimension, that is, merely numerical ones, were to be employed, and of that sort I found only unity.

(41)What I have set forth on the extraction of cube roots of binomials may easily be applied to fifth ones and to others more compounded, and that in almost as many words: I leave the matter as an exercise for anyone to investigate as circumstances allow. Indeed, the roots of complex binomials may be extracted by this method, and, since some may perhaps judge it pertinent to the store of analysis inasmuch as by its means all cubic equations come out reducible by Cardan's rules, I will not be reluctant to elucidate it.(43) To be sure, when each portion of the binomial proves to be impossible, the root is extracted as before, with proper observance of the signs + and −. But if one part or other only is impossible, the divisors of the terms do not observe the prescribed limits and, in consequence, for variety's sake I approach the problem differently in this manner.

When the binomial has been prepared as before, let $A\sqrt{-B}$ be the impossible term with the other rational. Multiply this square progression 1, 4, 9, 16, 25, ... into B, and select the products which taken singly from $4E$(44) leave an exact square: out of each of these and the square which is left separately extract the root, divide by $2\sqrt[3]{F}$, combine each pair appropriately with the signs + and −, and then try by cubing whether the result is the root required.(45)

(43) See note (17). Newton will make the application to the irreducible case of Cardan's rule on his inserted pp. 27/28 below.

(44) Read '$4EF$', in general. But Newton patently interests himself only in the restricted type of complex surd $\sqrt[3]{[\frac{1}{2}q\pm\sqrt{(\frac{1}{4}q^2-\frac{1}{27}p^3)}]}$ which arises in the Cardan solution of the reduced cubic $x^3=px+q$, and for this (as in his following examples)

$$\sqrt[3]{[(\tfrac{1}{2}q)^2-\{\pm\sqrt{(\tfrac{1}{4}q^2-\tfrac{1}{27}p^3)}\}^2]}=\tfrac{1}{3}p,$$

so that $F=1$. The cubic application, likewise, is evidently the reason for his restricting his attention to binomials whose real root is rational.

(45) Much as in note (29) above, we may set $M+N=\sqrt[3]{F}\times\sqrt[3]{[A\sqrt{-B}+C\sqrt{D}]}$ with $(A\sqrt{-B})^2-(C\sqrt{D})^2=-E^3F$ (or $M^2-N^2=-EF$) and $M=m\sqrt{-B}$. Now when A, B, C

De signis autem nota quod partes puræ : impuræ in utroq; binomio (proposito scilicet et invento) habebunt eadem si triplum : triens ablati facti sit minor quadrato relicto.(46)

Exempl: 1. Ex $-2+\surd-121\ \surd$cub est $-2+\surd-1$.

Præparatio.	Progr:	factus in -1.	resid[uū] a 20.	Amborum radices	Binomia de quibus tentandum sit.
$-2+\surd-121$, sive $-2+11\surd-1$.					
Quadrata 4. -121.	1.	-1.	19.	*	
Dif: 125.	4.	-4.	16.	$\surd-4$. 4.	$+\surd-1-2$.
$\surd C$: diff 5, ejus	9.	-9.	11.	*	
quadruplū 20.	16.	-16.	4.	$\surd-16$. 2.	$-\surd-4+1$.

Cùm 19 et 11 non sint quadrati numeri rejicio et e cæteris -4. 16. & -16. 4, obtineo $-2+\surd-1$ & $1-\surd-4$, quorum $-2+\surd-1$ cubando facit $-2+\surd-121$.(47)

Exemplum 2. Ex $10-\surd-243$, radix [cubica] est triplex [viz:] $-\frac{1}{2}+\frac{3}{2}\surd-3$, $-2-\surd-3$, & $\frac{5}{2}-\frac{1}{2}\surd-3$.

Præparatio.		Operatio.			
$10-\surd-243$, sive $10-9\surd-3$.	1	-3.	25.	$\surd-3$. 5.	$-\frac{1}{2}\surd-3+\frac{5}{2}$.
100. -243.	4	-12.	16.	$\surd-12$. 4.	$-\surd-3-2$.
343.	9	-27.	1.	$\surd-27$. 1.	$+\frac{3}{2}\surd-3-\frac{1}{2}$.
7. 28.					

Cubando Binomia tria sic inventa, video omnia producere $10-\surd-243$, adeoq; concludo esse radices ejus.(48)

Ad eundem ferè modum radix extrahi potest cum utraq; ꝑs binomij sit surda;(49) sed in his nimius fuisse videor.

and D (which Newton supposes to be unity) are integers, so are $2m$ and $4N^2$ (note (36)): hence, provided that it is rationally evaluable,

$$\sqrt[3]{[A\surd-B+C\surd D]} = \frac{2m\surd-B+\surd(4EF-(2m)^2B)}{2\sqrt[3]{F}},$$

where $2m$ is a positive integer which makes $4EF-(2m)^2B$ an exact square.

(46) Here the 'quadratum relictum' is $4EF-(2m)^2B = 4N^2$, while the 'factus ablatus' is $(2m)^2B = -4M^2$. Newton's observations then follow as a corollary of the identifications (note (45)) $M(M^2+3N^2) = FA\surd-B$ and $N(3M^2+N^2) = FC\surd D$.

(47) In fact, $1-\surd-4 = \sqrt[3]{[-11+2\surd-1]}$, for which E is also 5. The other two values of $\sqrt[3]{[-2+11\surd-1]}$ are $(-2+\surd-1)\times-\frac{1}{2}(1\pm\surd-3)$, that is, $\frac{1}{2}(2\pm\surd3-(1\pm2\surd3)\surd-1)$. Neither of these can be directly evaluated by Newton's present method, which finds solutions

$$a\surd-1+b,$$

Concerning the signs, however, note that the pure (impure) parts in either binomial (the one propounded, namely, and the one found) will have the same ones if three times (one third of) the product taken away be less than the square that is left.(46)

Example 1. The cube root of $-2+\sqrt{-121}$ is $-2+\sqrt{-1}$.

Preparation $-2+\sqrt{-121}$, or $-2+11\sqrt{-1}$		The product times -1	The residue from 20	The roots of both	The binomials on which trial is to be made
Squares 4, -121	1	-1	19	*	
Difference 125	4	-4	16	$\sqrt{-4}$, 4	$+\sqrt{-1}$, -2
Cube root of the difference 5,	9	-9	11	*	
and its quadruple 20	16	-16	4	$\sqrt{-16}$, 2	$-\sqrt{-4}$, $+1$

Since 19 and 11 are not squares I reject them and from the others -4, 16, and -16, 4, I obtain $-2+\sqrt{-1}$ and $1-\sqrt{-4}$: of these $-2+\sqrt{-1}$ on cubing makes $-2+\sqrt{-121}$.(47)

Examplc 2. Thc [cubc] root of $10-\sqrt{-243}$ is triple, [viz.] $-\frac{1}{2}+\frac{3}{2}\sqrt{-3}$, $-2-\sqrt{-3}$ and $\frac{5}{2}-\frac{1}{2}\sqrt{-3}$.

Preparation				Operation		
$10-\sqrt{-243}$, or $10-9\sqrt{-3}$	1	-3	25	$\sqrt{-3}$	5	$-\frac{1}{2}\sqrt{-3}+\frac{5}{2}$
100, -243	4	-12	16	$\sqrt{-12}$	4	$-\sqrt{-3}-2$
343	9	-27	1	$\sqrt{-27}$	1	$+\frac{3}{2}\sqrt{-3}-\frac{1}{2}$
7, 28						

By cubing the three binomials found in this way, I see they all produce

$$10-\sqrt{-243}$$

and consequently conclude that they are its roots.(48)

In almost the same manner the root may be extracted when either part of the binomial is surd;(49) but I seem to have been overlong in these matters.

---

where a and b are rational. The example is Bombelli's (see *Correspondence of Isaac Newton*, **1**: 36, note (4)).

(48) The three roots, in fact, are $-(2+\sqrt{-3})\omega$, where $\omega^3 = 1$. Newton does not evidently realize his good fortune in choosing a root of form $\alpha\sqrt{-3}+\beta$: namely, that

$$(\alpha\sqrt{-3}+\beta)\times-\tfrac{1}{2}(1\pm\sqrt{-3})$$

remains of the form $a\sqrt{-3}+b$, with a and b rational, which his method can isolate.

(49) See note (45).

|[15]|| Deniqꝫ quæ de radicibus cubicis fuerunt ostensa facilè applicantur ad radices surde-solidas(50) aliasqꝫ extrahendas.(51)

Pars Secunda. De Æquationibus.

(52)De computo generali(53) jam satis: Ut hoc ad aliquem scopum dirigam et ostendam usum, ad æquationes jam tractandas pergā. Soluturus igitur aliquod problema ad hanc metam semper collimet ut habeatur æquatio quâ mediante quæsita quantitas innotescat. Est autem æquatio quantitatum congeries quarum una pars æquatur alteri(54) vel quæ omnes æquantur nihilo. Ut $x+a=b$. vel $x+a-b=0$. Hoc est $x+a$ æquale b. vel $x+a-b$ æquale nihilo. Nam nota $=$ designat æqualitatem quantitatum quibus interponitur.

Aequationes duobus præcipuè modis considerandæ veniunt, vel ut ultimæ conclusiones ad quas in problematibus solvendis deventum est, vel ut media quorum ope finales æquationes acquirendæ sunt. Prioris generis æquatio ex unica tantùm incognitâ quantitate cognitis involuta(55) conflatur modò problema sit definitum et aliquid certi quærendum innuat. Sed eæ posterioris generis involvunt plures quantitates incognitas, quæ ideò debent inter se comparari & ita connecti ut ex omnibus(56) una tandem emergat æquatio nova cui inest unica tantum quam quærimus incognita quantitas admista cognitis. Quæ quantitas ut exinde facilius eliciatur: æquatio ista varijs plerumqꝫ modis transformanda est donec evadat simplicissima quæ potest, atqꝫ etiam similis alicui ex sequentibus earum gradibus.

$$
\begin{array}{lcl}
x=p & \text{vel} & x+p=0 \\
xx=px+q & & xx+px+q=0 \\
x^3=pxx+qx+r & & x^3+pxx+qx+r=0 \\
x^4=px^3+qxx+rx+s & & x^4+px^3+qxx+rx+s=0 \\
\quad \&\text{c.} & & \quad \&\text{c.}
\end{array}
$$

Ubi x designat quantitatem quæsitam et p, q, r, s quantitates ex quibus determinatis et cognitis etiam x determinatur et per methodos posthac docendas

---

(50) That is 'supersolidas' (of dimension 5). Compare Paul Tannery's note on the word in Descartes' *Œuvres*, **2** (Paris, 1898): 579.

(51) See notes (30) and (41). This short paragraph was cancelled some time in the late summer of 1670 when Newton inserted the previous four pages (pp. 11–14) into his first version of the *Observationes*.

(52) The following two and a half pages (pp. 15–17) of Newton's text are intended to replace Kinckhuysen's brief introduction to his Part 2 on p. 39. An accompanying marginal note to Collins clarifies the point: 'Judging it of maine importance for a new beginner to have a right notion of the designe & use of Æquations I thought it not amisse to substitute this larger introduction to them instead of the Authors'.

‖ What, finally, has been shown concerning cube roots may be easily applied to the extraction of sursolid[50] and other roots.[51]

Part Two. Equations

[52]Enough of computation in general: in order that I may direct this to some end and show its use I will now proceed to treat of equations. He therefore who is prepared to resolve some problem will always have this goal in view, that an equation may be had by whose means the quantity sought may become known. An equation is a set of quantities one portion of which equals the other[54] or which are all together equal to nothing. For instance, $x+a=b$ or $x+a-b=0$: that is, $x+a$ equal to b or $x+a-b$ equal to nothing. For the mark $=$ designates the equality of the quantities between which it is placed.

Equations come to be considered in two especial ways: either as ultimate conclusions arrived at in resolving problems, or as media by whose aid the final equations are to be acquired. An equation of the first sort is a conflation of but a single unknown quantity intermingled[55] with known ones, provided the problem be defined and convey that something certain is to be sought for. But those of the latter sort involve several unknown quantities, and for that reason must be compared one with another and so conjoined that from them all[56] shall at length emerge a single new equation, in which there is a unique unknown quantity for us to seek, intermixed with known ones. To elicit that quantity the more easily thereafter, that equation must usually be transformed in various ways until it comes to be the simplest possible, and similar indeed to some one of their following grades:

$$\begin{array}{lll} x=p, & \text{or} & x+p=0,\\ x^2=px+q, & & x^2+px+q=0,\\ x^3=px^2+qx+r, & & x^3+px^2+qx+r=0,\\ x^4=px^3+qx^2+rx+s, & & x^4+px^3+qx^2+rx+s=0,\\ \dots\ \dots\ \dots & & \dots\ \dots\ \dots \end{array}$$

Here x designates the quantity sought and p, q, r, s quantities from which, when they are determined and known, x also is determined and may be investigated

---

(53) This is written by Newton over the equivalent phrase 'computationibus in genere', cancelling it.

(54) A first version of this opening, perhaps that of the 'Observations' sent to Collins on 11 July 1670, reads 'Aequatio autem dicuntur quantitates quibus intercedit æqualitas' (An equation is the name of quantities between which there is equality).

(55) 'admista' (mixed) is cancelled.

(56) The continuation 'sic collatis' (thus brought together) is cancelled.

investigari poterit idq; vel calculo Arithmetico vel per geometriam ductu linearum.

||[16] De his autem notandum est quod termini ||secundum dimensiones incognitæ quantitatis ordinantur. Et primus terminus dicitur in quo x est plurimarum dimensionum vel potestatum (ut x. xx. x^3. x^4 &c). Secundus terminus est in quo dimensiones ipsius x sint una vice minores (ut p. px. px^2. px^3 &c) et sic præterea. Et quod signa terminorum attinet licèt hic affirmativa posuerī tamen possunt esse negativa; imò et unus vel plures ex intermedijs terminis aliquando deesse. Sic $x^3 * -bbx+b^3[=0]$ vel $x^3=bbx-b^3$ est æquatio tertij gradus et

$$z^4 \begin{smallmatrix}+a\\-b\end{smallmatrix} z^3 * [*] \begin{smallmatrix}+ab^3\\-b^4\end{smallmatrix} = 0$$

æquatio quarti. Nam gradus æquationum æstimantur ex maximâ dimensione quantitatis incognitæ, nullo respectu ad quantitates cognitas habito, nec ad intermedios terminos. Attamen ex defectu intermediorum terminorum æquatio plerumq; fit multò simplicior et nonnunquam ad gradum inferiorem quodammodo deprimitur: Sic enim $x^4=qxx+s$, æqu[atio] secundi gradus censenda est,(57) siquidem in duas æquationes secundi gradus resolvi potest. Nam supposito $xx=y$, et y pro xx scripto, prodibit $yy=qy+s$ æquatio secundi gradus; cujus ope cum y inventa est, æquatio $xx=y$ secundi etiam gradus dabit x.(58)

Quæsita quantitas eò difficiliùs elicitur ex æquationibus prout sunt altioris gradûs. Etenim in ijs primi gradûs nihil opus est nisi ut quantitas p noscatur: sed in ijs secundi gradûs quærenda est quantitas cujus quadratum æquale est ad q una cum facto sui ipsius in p ducti: et sic præterea. Hinc problemata fiunt alia alijs longè difficiliora. Ea sunt facillima quæ solvi possunt per æquationes primi gradûs sive unius dimensionis: et ea uno gradu difficiliora quorum solutio vendicat æquationem secundi gradûs: et sic deinceps. Et hinc Veteres distinxerunt problemata in plana solida et linearia. Plana quæ possunt deduci ad æquationes unius duarumve dimensionum; quippe quæ solvuntur lineis planis rectâ et circulari. Solida quæ ad æquationes 3 vel 4 dimensionum solummodò ||[17] deducantur; quippe quæ ad eorum solutionem figuras solidas ||(Parabolam, Hyperbolam vel Ellipsin) requirunt. Cætera dixerunt linearia propterea quod absq; lineis adhuc magis compositis non possunt solvi.(59)

Patet itáq; quàm necessarium sit ut Artifex non solùm addiscat methodos

(57) Newton first wrote '...esse quodammodo habeatur' (may be considered after a fashion).

(58) 'Et sic de cæteris' (And so of the rest) is cancelled.

(59) This recalls Descartes, [*Geometrie*: 315 =] *Geometria*, ($_2$1659): 17: 'Veteres optimè considerârunt, quòd Geometriæ Problematum alia sint Plana; alia Solida; alia denique Linearia; hoc est, quod quædam eorum construi possint, ducendo tantùm rectas lineas & circulos; cum alia construi nequeant, nisi ad minimum adhibeatur Conica aliqua sectio; ac reliqua denique, quin ad constructionem eorum assumatur alia quædam linea magis composita'. Descartes himself, we may conjecture, drew on the account of Greek geometrical

by methods to be taught subsequently, and then either by an arithmetical calculus or geometrically by the drawing of lines.

In this regard, however, it should be noted that the terms are ranged according to the dimensions of the unknown quantity. That term is said to be first in which x is of most dimensions or powers (as x, x^2, x^3, x^4 and so on). The second term is that in which the dimensions of x are one degree less (as p, px, px^2, px^3 and so on) and similarly for the rest. With respect to the signs of the terms, though I have here set them positive they may notwithstanding be negative; and indeed one or more of the intermediate terms may on occasion be lacking. Thus

$$x^3 * -b^2x+b^3[=0] \quad \text{or} \quad x^3 = b^2x-b^3$$

is an equation of third degree and $z^4+(a-b)\,z^3*[*]+ab^3-b^4=0$ one of fourth. For the degrees of equations are estimated from the greatest dimension of the unknown quantity, without regard to the known quantities or intermediate terms. Nevertheless, through the absence of intermediate terms an equation usually becomes much simpler and is not infrequently depressed some way or other to a lower degree. Thus, for instance, $x^4 = qx^2+s$ is to be conceived[(57)] of second degree inasmuch as it may be reduced to two equations of second degree. For if x^2 be supposed equal to y and y be written for x^2, there will arise $y^2 = qy+s$, an equation of second degree; and by its aid, when y has been found, the equation $x^2=y$ also of second degree will give x.[(58)]

The quantity sought is the more difficult to elicit from equations the higher their degree. So in those of first degree we need know only the quantity p; but in those of second degree we have to find the quantity whose square is equal to q together with the product of it multiplied into p: and so forth. On that account some problems come to be far more difficult than others. The easiest are those which are resolvable by equations of first degree or of a single dimension; and those are one degree more difficult whose solution demands an equation of second degree: and so on. In consequence of this the Ancients distinguished problems as plane, solid and linear. Plane problems are those reducible to equations of one or two dimensions, seeing that these are resolvable by plane curves (the straight line and circle); solid ones those which may only be reduced to equations of 3 or 4 dimensions seeing that their solutions require solid figures (the parabola, hyperbola or ellipse). The others they called linear because they cannot be solved except by lines of still greater complexity.[(59)]

It is consequently manifest how necessary it is for the practitioner of this art

method with which Pappus introduced the seventh book of his *Collectiones Mathematicæ* (ed. F. Commandino, Pesaro, 1588): from that edition, indeed, on pp. 304–6 of his *Geometrie* he quotes the Greek 3/4 line locus problem (to which his attention had been drawn by Jacob Gool in 1631). We have found nothing to suggest that Newton himself at this period was familiar with Pappus' work.

quibus Problemata ad conclusiones deducantur: sed etiam ut peritus sit in reducendis æquationibus, præsertim finalibus ad formas quas natura rei sinit simplicissimas. Ideóqȝ tractationem hanc de Aequationibus in sequentia quinqȝ(60) capita distinguo. Primum aget de transformationibus æquationis solitariæ; secundum de duabus pluribusve æquationibus in unam transformandis; tertium de natura Æquationum respectu radicum, ac de regulis generalibus quibus tentandum est an possunt deprimi ad gradus simpliciores; quartum de mutandis radicibus æquationis per assumptionem æquationis novæ ut emergat altera nova formæ desideratæ; et quintum de solutionibus æquationum. Ubi simul agetur de modis particularibus quibus deprimi possunt ad gradus simpliciores.

---

pag 40.(61) Caput primum.

De transformationibus æquationis solitariæ, non mutatis radicibus

In this chapter I see noe reason to change ye words of the Author.

Caput secundum(62)

De duabus pluribúsve æquationibus in unam transformandis, ut incognitæ quantitates exterminentur.(63)

Cum in alicujus Problematis solutionem(64) plures habeantur æquationes statum quæstionis comprehendentes quarum unicuiqȝ plures etiam incognitæ quantitates involvuntur: æquationes istæ (duæ per vices si modò sint plures duabus) sunt ita connectendæ ut una ex incognitis quantitatibus per singulas operationes tollatur et emergat æquatio nova. Sic habitis ||æquationibus $2x=y+5$ et $x=y+2$, demendo æquales ex æqualibus prodibit $x=3$. Et sciendum est quod per quamlibet æquationem una quantitas incognita potest tolli, atqȝ adeò cùm tot sint æquationes quot quantitates incognitæ, omnes possunt ad unam deniqȝ reduci in quâ unica manebit quantitas incognita. Sin quantitates incognitæ sint unâ plures quàm æquationes habentur, tum in æquatione ultimò resultante duæ manebunt quantitates incognitæ,(65) et si sint duabus plures quàm æquationes habentur, tum in æquatione ultimò resultante manebunt tres, & sic præterea.

||[18]

(60) Into Kinckhuysen's four chapters Newton will insert a wholly new 'Caput secundum' on reducing equations (see his pp. 17–22 below).

(61) That is, pp. 40–3 of Kinckhuysen's 'Algebra'.

(62) This is intended to be inserted on Kinckhuysen's p. 43 to fill a gap (as Newton sees it) in his presentation. An accompanying marginal note to Collins adds: 'Though this bee omitted by all that have writ introductions to this Art, yet I judge it very propper & necessary

not only to learn methods by which problems may be brought to their conclusions but also to be proficient in reducing equations, especially final ones, to the simplest forms which the nature of the matter allows. For that reason I divide this discussion of equations into the following five(60) chapters. The first deals with the transformation of a solitary equation; the second with transforming two or more equations into one; the third with the nature of equations in regard to their roots and with the general rules for testing whether they may be depressed to simpler degree; the fourth with changing the roots of an equation by assuming a new equation so that there may emerge a second new one of a desired form; and the fifth with the solution of equations. There we deal collectively with particular methods by which an equation may be depressed to a simpler degree.

Page 40.(61) Chapter one

The transformation of a solitary equation without altering its roots.

[In this chapter I see noe reason to change ye words of the Author.]

Chapter two(62)

The transforming of two or more equations into one so that unknown quantities may be eliminated.(63)

When for resolving some problem there are had several equations embracing the conditions of the question, and each of these, again, involve several unknown quantities, those equations are to be so conjoined (two at a time should there be more than two) that one of the unknowns is removed at each operation and there emerges a new equation. Thus if there be had the equations $2x = y+5$ and $x = y+2$, on taking equals from equals there arises $x = 3$. And it should be understood that a single unknown quantity may be removed by any equation, so that when there are as many equations as unknowns they may all finally be reduced to one in which there remains a unique unknown. But if the unknown quantities be one more in number than there are equations, then in the equation which finally results there will remain two unknowns,(65) and if there are two more than there are equations, then in the equation which finally results there will remain three unknowns, and so on.

to make an introduction compleate'. Newton's third, fourth and fifth chapters following, of course, revise Kinckhuysen's second, third and fourth.

(63) Here and several times below Newton has replaced the appropriate tense of 'elimino' by that of the equivalent verb 'extermino'.

(64) Newton first began equivalently with '...in ordine ad aliquod Problema solvendum'.

(65) The following clause is a late addition in the margin.

Possunt etiam(66) duæ vel plures quantitates incognitæ per duas tantùm æquationes fortasse tolli. Ut si sit $ax-by=ab-az$, et $bx+by=bb+az$: Tum æqualibus simul additis prodibit $ax+bx=ab+bb$, exterminatis utrisꝗ y et z.(67) Sed ejusmodi casus vel arguunt vitium aliquod in statu quæstionis latere, vel calculum erroneum esse aut non satis artificiosum. Modus autem quo una quantitas incognita per singulas æquationes tollatur ex sequentibus patebit.

Exterminatio quantitatis incognitæ per æqualitatem valorum ejus.

Cùm quantitas tollenda est unius tantùm dimensionis in utrâꝗ æquatione: valor ejus in utrâꝗ per regulas in præcedenti capite traditas quærendus est, et alter valor statuendus æqualis alteri.

Sic positis $a+x=b+y$, et $2x+y=3b$; ut eliminetur y æquatio prima dabit $a+x-b=y$, et secunda dabit $3b-2x=y$. Est ergo $a+x-b=3b-2x$. Sive ordinando $x=\frac{4b-a}{3}$.

Atꝗ ita $2x=y$, et $5+x=y$ dant $2x=5+x$ sive $x=5$.

Et $ax-2by=ab$ et $xy=bb$ dant $\frac{ax-ab}{2b}(=y)=\frac{bb}{x}$ sive ordinando

$$xx-bx[-2]\frac{b^3}{a}=0.$$

Item $\frac{bbx-aby}{a}=ab+xy$ et $bx+\frac{ayy}{c}=[2]aa$ tollendo x dant

$$\frac{aby+aab}{bb-ay}(=x)=\frac{2[a]ac-ayy}{bc}.$$

Sive in ordinem redigendo $y^3-\frac{bb}{a}yy\frac{-2aac-bbc}{a}y+bbc=0$.

||[19] ||Et sic $x+y-z=0$ et $ay=xz$, tollendo z dant $x+y(=z)=\frac{ay}{x}$. Sive $xx+xy=ay$.

Hoc idem quoꝗ perficitur subducendo alterum valorem quantitatis incognitæ ab altero et ponendo residuum æquale nihilo. Sic in exemplorū primo tolle $3b-2x$ ab $a+x-b$ et manebit $a+3x-4b=0$, sive $x=\frac{4b-a}{3}$.

Eliminatio(68) quantitatis incognitæ substituendo pro se valorem ejus.

Cùm in alterâ saltem æquatione tollenda quantitas unius tantùm dimensionis existat: valor ejus in ea quærendus est et pro se substituendus in æquationem alteram.

Sic propositis $xyy=b^3$ et $xx+yy=by-ax$: ut eliminetur x, prima dabit

(66) 'Nonnunquam accidat quòd' (It may sometimes happen that) is cancelled.

(67) A sentence, here deleted, read 'In primis æquationibus tres sunt incognitæ quantitates

It may also be that[66] two or more unknowns chance to be removed merely by two equations. So if $ax-by=ab-az$, and $bx+by = bb+az$, then on adding equals together there arises $ax+bx = ab+b^2$, from which both y and z have been eliminated.[67]. But cases of this kind either betray some concealed blemish in the statement of a question or argue that the calculation is erroneous or insufficiently subtle. The way, however, in which one unknown quantity may be removed throughout separate equations will be evident from the following.

Elimination of an unknown quantity by equating its values.

When the quantity to be removed is of one dimension alone in either equation, its value in each is to be sought by the rules laid down in the preceding chapter and one value set equal to the other.

Thus, on setting $a+x = b+y$ and $2x+y = 3b$, to eliminate y the first equation will give $a+x-b = y$ and the second $3b-2x = y$. There is consequently $a+x-b = 3b-2x$, or, on ordering, $x = \frac{1}{3}(4b-a)$.

And similarly $2x = y$ and $5+x = y$ give $2x = 5+x$, or $x = 5$.

And $ax-2by = ab$ and $xy = b^2$ give $\dfrac{ax-ab}{2b}(=y) = \dfrac{b^2}{x}$, or on ordering $x^2-bx[-2]b^3/a = 0$.

Likewise $\dfrac{b^2x-aby}{a} = ab+xy$ and $bx+\dfrac{ay^2}{c} = [2]\,a^2$ on removing x give $\dfrac{aby+a^2b}{b^2-ay}(=x) = \dfrac{2a^2c-ay^2}{bc}$: or by reducing it to order

$$y^3-\frac{b^2}{a}y^2-\frac{2a^2c+b^2c}{a}y+b^2c = 0.$$

And thus $x+y-z = 0$ and $ay = xz$ give, on removing z, $x+y(=z) = ay/x$, or $x^2+xy = ay$.

This same operation is also performed by subtracting one value of the unknown from the other and setting the residue equal to zero. Thus in the first of the examples take $3b-2x$ from $a+x-b$ and there will remain $a+3x-4b = 0$, or $x = \frac{1}{3}(4b-a)$.

Elimination of an unknown quantity by substituting its value for it.

When in one at least of the equations the quantity to be removed is of one dimension only, its value is to be sought in that one and substituted for it in the other equation.

So, when there are proposed $xy^2 = b^3$ and $x^2+y^2 = by-ax$, to eliminate x the

$x\ y$ & z, in ultimâ autem resultante manet unica x' (In the original equations there are three unknowns, x, y, and z, but in that finally resulting there remains only x).

(68) Newton has presumably forgotten to replace this by 'Exterminatio' in agreement with his other revised half titles (note (63)).

$\frac{b^3}{yy}=x$: quare in secundum substitue $\frac{b^3}{yy}$ pro x et prodit $\frac{b^6}{y^4}+yy=by-\frac{ab^3}{yy}$ sive ordinando $y^6-by^5+ab^3yy+b^6=0$.

Propositis $ayy+aay=z^3$ et $yz-ay=az$, ut y tollatur secunda dabit $y=\frac{az}{z-a}$. Quare substituo $\frac{az}{z-a}$ in primam, proditꝗ $\frac{a^3zz}{zz-2az+aa}+\frac{a^3z}{z-a}=z^3$ et reducendo $z^4-2az^3+aazz-2a^3z+a^4=0$.

Pari modo propositis $\frac{xy}{c}=z$, et $cy+zx=cc$, ad z tollendum pro eo substituo $\frac{xy}{c}$ in æquationem secundam et prodit $cy+\frac{xxy}{c}=cc$.

Cæterùm qui in hujusmodi computationibus exercitatus fuerit, sæpenumero contractiores modos percipiet quibus incognita quantitas exterminari possit. Sic habitis $ax=\frac{bbx-b^3}{z}$ et $x=\frac{az}{x-b}$, si æqualia multiplicentur æqualibus prodibunt æqualia $axx=abb$, sive $x=b$.(69) Sed casus ejusmodi particulares studiosis proprio marte(70) investigandos linquo.

||[20]

||Exterminatio quantitatis incognitæ cum plurium in utráꝗ Æquatione dimensionum existit.

In utrâꝗ Æquatione valor maximæ potestatis quantitatis exterminandæ quærendus est. Deinde si potestates istæ non sint eædem, æquatio potestatis minoris multiplicanda est per quantitatem eliminandam aut per ejus quadratum aut cubum &c donec evadat ejusdem potestatis cum æquatione altera. Tum valores illarum potestatum ponantur æquales et æquatio nova prodibit ubi maxima potestas sive dimensio quantitatis eliminandæ diminuitur. Et hanc operationem iterando quantitas eliminanda tandem auferetur.(71)

Quemadmodum si $xx+5x=3yy$ et $2xy-3xx=4$, ut x tollatur prima dabit $xx=3yy-5x$, et secunda dabit $xx=\frac{2xy-4}{3}$. Pono itaꝗ $3yy-5x=\frac{2xy-4}{3}$ ubi x ad unicam dimensionem reducitur, et proinde tolli potest per ea quæ paulò ante(72) ostendi. Scilicet æquationem novissimam ordinando, prodit

$$9yy-15x=2xy-4,$$

sive $x=\frac{9yy+4}{2y+15}$: hunc itaꝗ valorem pro x in aliquam ex æquationibus primò propositis (velut in $xx+5x=3yy$) substituo et oritur

$$\frac{81y^4+72yy+16}{4yy+60y+225}+\frac{45yy+20}{2y+15}=3yy.$$

(69) More precisely, $x=\pm b$, the positive root occurring only when $az=0$.

first will give $b^3/y^2 = x$: hence substitute b^3/y^2 for x in the second and there arises $(b^6/y^4)+y^2 = by-(ab^3/y^2)$ or, on ordering, $y^6-by^5+ab^3y^2+b^6 = 0$.

When $ay^2+a^2y = z^3$ and $yz-ay = az$ are proposed, to eliminate y the second will give $y = az/(z-a)$. I therefore substitute $az/(z-a)$ in the first and there comes out

$$\frac{a^3z^2}{z^2-2az+a^2}+\frac{a^3z}{z-a} = z^3$$

and after reduction $z^4-2az^3+a^2z^2-2a^3z+a^4 = 0$.

Equally, when $(xy/c) = z$ and $cy+zx = c^2$ are proposed, to remove z I substitute xy/c for it in the second equation and there comes $cy+(x^2y/c) = c^2$.

For the rest, he who is practised in this kind of computation will frequently perceive more contracted methods by which an unknown may be eliminated. Thus, when there is had $ax = (b^2x-b^3)/z$ and $x = az/(x-b)$, if equals be multiplied by equals there will arise the equals $ax^2 = ab^2$, or $x = b$.(69) But particular cases of this kind I leave for the student to investigate for himself.(70)

Elimination of an unknown quantity when it is of several dimensions in either equation.

In either equation the value of the greatest power of the quantity to be eliminated is to be sought. Next, if those powers are not the same, the equation of lower power must be multiplied by the quantity to be eliminated (or by its square or cube, and so on) until it comes to be of the same power as the other equation. Then the values of those powers are to be set equal and a new equation will arise in which the greatest power or dimension of the quantity to be eliminated is diminished. And by repeating this operation the quantity to be eliminated will at length be taken away.(71)

For instance, if $x^2+5x = 3y^2$ and $2xy-3x^2 = 4$, to remove x the first will give $x^2 = 3y^2-5x$ and the second $x^2 = \frac{1}{3}(2xy-4)$. I consequently set

$$3y^2-5x = \tfrac{1}{3}(2xy-4):$$

in this x is reduced to a single dimension and therefore may be removed by what I showed just above.(72) Precisely, on ordering this most recent equation there comes $9y^2-15x = 2xy-4$, or $x = (9y^2+4)/(2y+15)$: and so I substitute this value for x in one of the equations first proposed (say, in $x^2+5x = 3y^2$) and there arises

$$\frac{81y^4+72y^2+16}{4y^2+60y+225}+\frac{45y^2+20}{2y+15} = 3y^2.$$

(70) The phrase 'cùm occasio lata est' (when the occasion arises) is here cancelled.

(71) This scheme of reduction, we may note, is essentially that expounded by Euler in his 1748 paper on the topic (**2**, 1, §2: note (15)).

(72) In the previous section, 'Eliminatio quantitatis incognitæ substituendo pro se valorem ejus'.

Quæ ut in ordinem redigatur multiplico per $4yy+60y+225$ et prodit

$$81y^4+72yy+16+90y^3+40y+675yy+300=12y^4+180y^3+675yy.$$

Sive $69y^4-90y^3+72yy+40y+316=0$.(73)

Porrò si $y^3=xyy+3x$ et $yy=xx-xy-3$: ut y tollatur multiplico posteriorem æquationem per y et fit $y^3=xxy-xyy-3y$ totidem dimensionum quot prior: Jam ponendo valores de y^3 sibimet æquales habeo $xyy+3x=xxy-xyy-3y$, ubi y deprimitur ad duas dimensiones. Per hanc itaq; et simpliciorem ex æquationibus primò propositis $yy=xx-xy-3$ quantitas y prorsus tolletur ut in exemplo priori factum fuit.(74)

(75)Sunt et alij modi quibus hæc eadem perficiuntur idq; sæpenumero contractiùs. Quemadmodum ex $yy=\frac{2xxy}{a}+xx$, & $yy=2xy+\frac{x^4}{aa}$, ut y deleatur, ex utrâq; radicem quadraticam (ut postea docebitur) extraho et prodeunt $y=\frac{xx}{a}+\sqrt{\frac{x^4}{aa}+xx}$ et $y=x+\sqrt{\frac{x^4}{aa}+xx}$. Jam hos ipsius y valores ponendo æquales habeo $\frac{xx}{a}+\sqrt{\frac{x^4}{aa}+xx}=x+\sqrt{\frac{x^4}{aa}+xx}$. et auferendo $\sqrt{\frac{x^4}{aa}+xx}$ restat $\frac{xx}{a}=x$, vel $xx=ax$(76) et $x=a$.

Porro(77) ut ex æqu: $x+y+\frac{yy}{x}=20$, et $xx+yy+\frac{y^4}{xx}=140$ tollatur x; aufero y de partibus æquationis primæ et restat $x+\frac{yy}{x}=20-y$, cujus partes quadrando, fit $xx+2yy+\frac{y^4}{xx}=400-40y+yy$, et tollendo yy utrinq;, restat

$$xx+yy+\frac{y^4}{xx}=400-40y:$$

Quare cum $400-40y$ & 140 ijsdem quantitatibus æquentur, erit

$$400-40y=140.$$

Quæ reducta fit $y=6\frac{1}{2}$.

(73) That is, $-\begin{vmatrix} 1 & 5 & -3y^2 & 0 \\ 3 & -2y & 4 & 0 \\ 0 & 1 & 5 & -3y^2 \\ 0 & 3 & -2y & 4 \end{vmatrix} = -(-2y-15)(20-6y^3)+(4+9y^2)^2 = 0.$

(74) In fact, $-\begin{vmatrix} 1 & -x & 0 & -3x & 0 \\ 1 & x & -x^2+3 & 0 & 0 \\ 0 & 1 & -x & 0 & -3x \\ 0 & 1 & x & -x^2+3 & 0 \\ 0 & 0 & 1 & x & -x^2+3 \end{vmatrix} = x^6+18x^4-45x^2+27 = 0.$

(75) The three following paragraphs are a marginal afterthought, inserted, it would appear, when Newton elaborated the first of the worked examples (Kinckhuysen's third algebraic

To reduce this to order I multiply by $4y^2+60y+225$ and there comes

$$81y^4+72y^2+16+90y^3+40y+675y^2+300 = 12y^4+180y^3+675y^2,$$

or $69y^4-90y^3+72y^2+40y+316 = 0.$(73)

Moreover, if $y^3 = xy^3+3x$ and $y^2 = x^2-xy-3$, to remove y I multiply the latter equation by y and it becomes $y^3 = x^2y-xy^2-3y$, of as many dimensions as the former: now, by setting the values of y^3 equal one to the other I have $xy^2+3x = x^2y-xy^2-3y$, in which y is lowered to two dimensions. Through this, accordingly, and the simpler of the equations first proposed, $y^2 = x^2-xy-3$, the quantity y will be entirely removed, as was done in the former example.(74)

(75)There are other ways, too, for achieving this end and that frequently in a more contracted form. For instance, to delete y from

$$y^2 = \frac{2x^2y}{a}+x^2 \quad \text{and} \quad y^2 = 2xy+\frac{x^4}{a^2},$$

out of each I extract the quadratic root (as will afterwards be explained) and there comes

$$y = \frac{x^2}{a}+\sqrt{\left(\frac{x^4}{a^2}+x^2\right)} \quad \text{and} \quad y = x+\sqrt{\left(\frac{x^4}{a^2}+x^2\right)}.$$

Now on setting these values of y equal I have

$$\frac{x^2}{a}+\sqrt{\left(\frac{x^4}{a^2}+x^2\right)} = x+\sqrt{\left(\frac{x^4}{a^2}+x^2\right)}$$

and, on taking $\sqrt{([x^4/a^2]+x^2)}$ away, there remains $x^2/a = x$, or $x^2 = ax$(76) and $x = a$.

Again,(77) to remove x from the equations

$$x+y+(y^2/x) = 20 \quad \text{and} \quad x^2+y^2+(y^4/x^2) = 140$$

I take away y from the parts of the first equation and there remains

$$x+(y^2/x) = 20-y:$$

on squaring its parts, this becomes $x^2+2y^2+(y^4/x^2) = 400-40y+y^2$ and, on taking y^2 away on either side, there remains $x^2+y^2+(y^4/x^2) = 400-40y$. Hence, since $400-40y$ and 140 are equal to the same quantities, there will be

$$400-40y = 140:$$

this on reduction becomes $y = 6\frac{1}{2}$.

problem on p. 98 of his *Algebra* (§1)) with which he introduced his revision of Kinckhuysen's 'Pars Tertia' on his pp. 29–30 below.

(76) This is not, however, the full reduction: indeed, elimination of y between the two quadratics in the usual way yields $(x^2-ax)(3x^2+2ax+3a^2) = 0$, with the second factor arising (as Newton fails to mention) from the equally valid equality between $x^2/a+\sqrt{[x^4/a^2+x^2]}$ and $x-\sqrt{[x^4/a^2+x^2]}$. Note the anticipation here of Kinckhuysen's resolution of the general quadratic in the next chapter.

(77) See Newton's p. 30 below and note (75). A trivial solution $x=y=0$ exists.

Et sic opus in plerisq alijs æquationibus contrahetur.

||[21] ||Cæterùm cùm quantitas eliminanda multarum dimensionum existit, ad eam ex æquationibus auferendam calculus maximè laboriosus nonnunquam requiritur: sed labor tunc plurimùm minuetur per exempla sequentia tanquam regulas adhibita.

Reg 1. Ex $axx+bx+c=0$, et $fxx+gx+h=0$,

exterminato x prodibit

$$\overline{ah-bg-2cf}\times ah:+\overline{bh-cg}\times bf:+\overline{agg+cff}\times c=0.^{(78)}$$

Reg 2. Ex $ax^3+bxx+cx+d=0$, et $fxx+gx+h=0$,

exterminato x prodibit

$$\overline{ah-bg-2cf}\times ahh:+\overline{bh-cg-2df}\times bfh:+\overline{ch-dg}\times\overline{agg+cff}$$
$$+\overline{3agh+bgg+dff}\times df=0.^{(79)}$$

Verbi gratiâ ut ex æquationibus $xx+5x-3yy=0$, et $3xx-2xy+4=0$ exterminetur x: in regulam primam substituo $1, 5, -3yy; 3, -2y, \& 4$. Et signis
pro $a, b, c; f, g, \& h$.
$+$ & $-$ probè observatis oritur

$$\overline{4+10y+18yy}\times 4:+\overline{20-6y^3}\times 15:+\overline{4yy-27yy}\times -3yy=0.$$

Sive $16+40y+72yy+300-90y^3+69y^4=0.^{(80)}$

Simili ratione ut y deleatur ex æquationibus $y^3-xyy-3x=0$, et

$$yy+xy-xx+3=0:$$

substituo $1, -x, 0, -3x; 1, x, -xx+3, \& y$, in Reg: secundam proditq
pro $a, b, c, d; f, g, h, \& x$.

$$\overline{3-xx+xx}\times\overline{9-6xx+x^4}:\overline{-3x+x^3+6x}\times\overline{-3x+x^3}:+3xx\times xx:$$
$$+\overline{9x-3x^3-x^3-3x}\times -3x=0.$$

Tum delendo superflua et multiplicando fit

$$27-18xx+3x^4,\quad -9xx+x^6,\quad +3x^4,\quad -18x-6x^4=0.^{(81)}$$

et ordinando $x^6-27xx-18x+27=0$.

Ad eundem modum artifex alias sibi regulas conficiat pro æquationibus altiorum potestatum reducendis. Nam earum usus in æquationibus maximè compositis præsertim elucescit.

(78) That is, $-\begin{vmatrix} a & b & c & 0 \\ f & g & h & 0 \\ 0 & a & b & c \\ 0 & f & g & h \end{vmatrix} = -(ag-bf)(bh-cg)+(ah-cf)^2 = 0.$

And thus the procedure will be contracted in most other equations.

When, however, the quantity to be eliminated proves to be of many dimensions, to take it out of the equations requires not infrequently an extremely laborious computation: but in those cases the labour will be for the most part diminished by employing the following examples as guiding rules.

RULE 1. When from $ax^2+bx+c=0$ and $fx^2+gx+h=0$

x is eliminated, there comes

$$(ah-bg-2cf)\,ah+(bh-cg)\,bf+(ag^2+cf^2)\,c=0.^{(78)}$$

RULE 2. When from $ax^3+bx^2+cx+d=0$ and $fx^2+gx+h=0$

x is eliminated, there comes

$$(ah-bg-2cf)\,ah^2+(bh-cg-2df)\,bfh+(ch-dg)\,(ag^2+cf^2)$$
$$+(3agh+bg^2+df^2)\,df=0.^{(79)}$$

For example, to eliminate x from the equations $x^2+5x-3y^2=0$ and $3x^2-2xy+4=0$, in the first rule for $a, b, c; f, g$ and h I substitute $1, 5, -3y^2; 3, -2y$ and 4. And with proper observance of the signs $+$ and $-$ there arises $(4+10y+18y^2)\,4+(20-6y^3)\,15+(4y^2-27y^2)\,(-3y^2)=0$ or

$$16+40y+72y^2+300-90y^3+69y^4=0.^{(80)}$$

By a similar reasoning, to delete y from the equations $y^3-xy^2-3x=0$ and $y^2+xy-x^2+3=0$ for $a, b, c, d; f, g, h,$ and x I substitute $1, -x, 0, -3x; 1, x, -x^2+3$, and y in the second rule and there comes

$$(3-x^2+x^2)\,(9-6x^2+x^4)+(-3x+x^3+6x)\,(-3x+x^3)$$
$$+3x^2\times x^2+(9x-3x^3-x^3-3x)\,(-3x)=0.$$

On deleting superfluities and multiplying out, it then becomes $27-18x^2+3x^4, -9x^2+x^6, +3x^4, -18x-6x^4=0$(81) and, on reordering, $x^6-27x^2-18x+27=0$.

In much the same way the practitioner of this art will make up other rules for himself to reduce equations of higher powers. For their use in dealing with the most highly complex equations is particularly conspicuous.

---

(79) In modern determinantal form, $-\begin{vmatrix} a & b & c & d & 0 \\ f & g & h & 0 & 0 \\ 0 & a & b & c & d \\ 0 & f & g & h & 0 \\ 0 & 0 & f & g & h \end{vmatrix}=0.$

(80) It is curious that Newton has not bothered to collect the constant terms as '316'. Compare note (73).

(81) The last two terms on the left side should be '$-18xx+12x^4$'. In compensation the following should read '$x^6+18x^4-45xx+27=0$' (compare note (74)).

Hactenus de unicâ incognitâ quantitate ex duabus datis æquationibus tollendâ:

||[22] Quinetiam eodem recidit cùm plures ||dantur æquationes ad plures incognitas quantitates tollendas. Quemadmodum ex datis $ax=yz$, $x+y=z$, et $5x=y+3z$ si quantitas y investiganda sit, imprimis tollo x substituendo valorem ejus $\frac{yz}{a}$ (per æquationem primam inventum) in æquationem secundam ac tertiam. Quo pacto obtinebuntur $\frac{yz}{a}+y=z$, & $\frac{5yz}{a}=y+3z$. Ex quibus deinde tollo z per æqualitatem valorum ejus. Nam earum prior ordinando dat $z=\frac{ay}{a-y}$ & posterior dat $z=\frac{ay}{5y-3a}$. Quare est $\frac{ay}{a-y}=\frac{ay}{5y-3a}$ et multiplicando in crucem fit $5y-3a=a-y$(82) et ordinando $y=\frac{2a}{3}$.

De modo tollendi quantitates surdas ex Æquationibus.

Huc debet referri quantitatum surdarum exterminatio fingendo eas literis quibuslibet æquales: de quâ re consule exempla sequentia.

Sit $a\sqrt{ax}-x\sqrt{bx}=ab$. Scribendo itaq; y pro $\sqrt{ax}$ et z pro $\sqrt{bx}$ prodit $ay-xz=ab$, ex qua y et z tollenda sunt ope fictitiarum æquationum $y=\sqrt{ax}$ et $z=\sqrt{bx}$ sive $yy=ax$ et $zz=bx$. Per æquationem $ay-xz=ab$ invenio $y=\frac{ab+xz}{a}$ quem valorem substituo in æquationem $yy=ax$, et prodit $\frac{aabb+2abxz+xxzz}{aa}=ax$, sive (scribendo bx pro zz) $\frac{aabb+2abxz+bx^3}{aa}=ax$, et ordinando $z=\frac{a^3x-bx^3-aabb}{2abx}$(83).

Hunc itaq; valorem scribo pro z in æquationē $zz=bx$. Et emergit

$$\frac{a^6xx-2a^3bx^4-2a^5bbx+bbx^6-2aab^3x^3+a^4b^4}{4aabbxx}=bx,$$

ubi nulla reperitur surda quantitas.

Pari ratione in $\sqrt{ay}=3a+2\sqrt{C:\overline{ayy}}$ scribendo t pro $\sqrt{ay}$ et v pro $\sqrt{C:ayy}$ habebuntur æquationes $t=3a+2v$, $tt=ay$, & $v^3=ayy$; ex quibus tollendo t et v resultabit æquatio(84) libera ab omni asymmetriâ.

||[23] ||Caput tertium.

De naturâ æquationum respectu radicum.(85)

Postquam de computationibus quibus ad æquationes perveniatur, eædemq; redigantur in ordinem diximus; convenit ut quædā de formis earum proferantur, quorum notitia ad eas solvendas plurimùm conducunt.

So far we have been concerned with removing a single unknown quantity out of two given equations: but the procedure is much the same when several equations are given for the removal of several unknowns. For instance, if the quantity y is to be determined from the given trio $ax = yz$, $x+y = z$ and $5x = y+3z$, in the first place I remove x by substituting its value yz/a (found by the first equation) into the second and third equations. In this manner there will be obtained $(yz/a)+y = z$ and $(5yz/a) = y+3z$. Out of these I next remove z by equating its values: for, on reordering, the former gives $z = ay/(a-y)$ and the latter $z = ay/(5y-3a)$. Hence is $ay/(a-y) = ay/(5y-3a)$, and on multiplying crosswise it becomes $5y-3a = a-y$[82] and, on reordering, $y = 2a/3$.

The way of removing surd quantities from equations.

To this should be related the elimination of surd quantities by imagining them equal to arbitrary letters: on this point, consider the following examples.

Let there be $a\sqrt{ax}-x\sqrt{bx} = ab$. In consequence, upon writing y for $\sqrt{ax}$ and z for $\sqrt{bx}$ there comes $ay-xz = ab$, out of which y and z have to be removed with the help of the fictitious equations $y = \sqrt{ax}$ and $z = \sqrt{bx}$, that is, $y^2 = ax$ and $z^2 = bx$. From the equation $ay-xz = ab$ I find $y = (ab+xz)/a$: this value I substitute in the equation $y^2 = ax$ and there comes

$$(a^2b^2+2abxz+x^2z^2)/a^2 = ax,$$

that is (on writing bx for z^2) $(a^2b^2+2abxz+bx^3)/a^2 = ax$, and, by ordering it, $z = (a^3x-bx^3-a^2b^2)/2abx$.[83] I consequently write this value for z in the equation $z^2 = bx$ and there emerges

$$(a^6x^2-2a^3bx^4-2a^5b^2x+b^2x^6-2a^2b^3x^3+a^4b^4)/4a^2b^2x^2 = bx,$$

in which no surd quantity is found.

By a similar reasoning, on writing in $\sqrt{ay} = 3a+2\sqrt[3]{[ay^2]}$ t for $\sqrt{ay}$ and v for $\sqrt[3]{ay^2}$ there will be had the equations $t = 3a+2v$, $t^2 = ay$ and $v^3 = ay^2$; and on removing t and v out of them there will result an equation[84] free of all asymmetry.

Chapter three

The nature of equations in regard to their roots.[85]

After we have spoken of the computations by which equations are attained and reduced to order, it is convenient to say something of their forms: for knowledge of these things contributes very greatly to their solution.

---

(82) Or, of course, $ay = 0$.

(83) But more simply here, $a^2b^2 = (a\sqrt{ax}-x\sqrt{bx})^2 = a^3x-2ax^2\sqrt{ab}+bx^3$ with $zx = a\sqrt{ax}-ab$.

(84) $ay(y+27a)^2 = (8y^2+9ay+81a^2)^2$.

(85) Newton's replacement for the opening passages of Kinckhuysen's 'Caput secundum' (*Algebra* (§1): 43–4).

Quot radices quævis Æquatio habere possit.

Hence (pag 44) continue y^e words of the Author to these words *præter hasce quatuor 2, 3, 4 et −5* pag 47, where subjoyne w^t follows.

Idem expeditiùs tentari possit substituendo quantitatem datam in locum radicis. Ut in exemplo præcedente, quolibet ex numeris 2, 3, 4 vel −5 pro x in æquationem scriptis, termini se destruent; quod arguit eos esse radices ejus.

Radices tam veras quàm falsas nonnunquam imaginarias esse sive impossibiles.(86)

Radices tam veræ quàm falsæ non semper reales sunt, sed quandoꝗ solùm imaginariæ, hoc est, posse semper in quavis æquatione tot radices imaginatione concipi quot modò memoravimus, sed fieri quandoꝗ ut nulla sit quantitas quæ consentiat cum conceptis. Sic licèt imaginemur hanc æquationem

$$x^3-6xx+13x-10=0$$

habere tres radices, non habet tamen nisi unicam realem quæ est 2. Et reliquæ duæ quicquid moliamur, non nisi imaginariæ manent.

Quot veras radices quævis Æquatio habere possit.(87)

Quoties in aliquâ Æquatione signum − sequitur signum + vel signum + sequitur signum −, tot veras radices ipsa æquatio habere potest: et tot falsas, quoties duo signa + vel duo signa − succenturiantur. Veluti in hâc Æquatione $x^4-4x^3-19xx+106x-120=0$, ubi post $+x^4$ sequitur $-4x^3$ et post $-19xx$ habetur $+106x$ et post $+106x$ extat -120 constat inde tres esse radices veras; ‖[24] et unicam falsam, quia semel tantùm duo signa similia se excipiunt, ‖videlicet $-4x^3$ et $-19xx$.

Verùm hæc regula de æquationibus intelligenda est ubi nullæ radices sint imaginariæ. Quippe radices istæ propriè loquendo nec veræ sunt nec falsæ, sed indifferentèr pro utris'ꝗ sumi possunt. Sic in $x^3-pxx+3ppx-q^3=0$ signa monstrant tres veras radices: Unde si multiplicetur per $x+2p=0$, Æquatio prodiens $x^4+px^3+ppxx\begin{smallmatrix}+6p^3\\-q^3\end{smallmatrix}x-2pq^3=0$ haberet tres veras et unam falsam radicem: tamen signa ejus ostendunt tres falsas et unicam esse veram radicem. Concludendum est itáꝗ quòd æquationis primò propositæ duæ sunt imaginariæ radices, quæ nec veræ nec falsæ speciem utramꝗ per vices induunt.

(86) The final paragraph of Kinckhuysen's 'Caput secundum' (*Algebra* (§1): 48–9) here restated by Newton in a more logical position but otherwise unchanged. See note (87).

(87) Newton's revised version of Kinckhuysen's corresponding section (*Algebra* (§1): 47), changed only to delete the latter's invalid counter-example to Descartes' sign rule. As he noted for Collins' benefit in the margin, 'The Authors exception from ye preceding rule in these words *Verùm hæc omnia sic accipienda sunt, quando in &c* is invalid; as may appeare by the Æqua-

How many roots any equation may have.

[Hence (pag 44) continue ye words of the Author to these words *præter hasce quatuor 2, 3, 4 et −5* pag 47, where subjoyne w^t follows.]

The same may more speedily be tried by substituting a given quantity in place of the root. Thus in the preceding example should any of the numbers 2, 3, 4 or −5 be written for x in the equation the terms will destroy one another: a proof that they are its roots.

Both true and false roots may sometimes be imaginary or impossible.(86)

Both true and false roots are not always real but on occasion merely imaginary, that is, there may in any equation be conceived to be in imagination as many roots as we mentioned just now, but it may happen on occasion that there is no quantity which accords with those concepts. So though we may imagine this equation $x^3-6x^2+13x-10=0$ to have three roots, it has nevertheless only a single real one (2, that is). As for the other two roots, however hard we try, they remain but imaginary.

How many true roots any equation may have.(87)

As many times in some equation the sign − follows the sign + or the sign + follows the sign −, that many true roots may be possessed by the equation: and there may be as many false ones as occurrences together of two signs + or two signs −. So in this equation $x^4-4x^3-19x^2+106x-120=0$, where after $+x^4$ follows $-4x^3$, after $-19x^2$ there is found $+106x$ and after $+106x$ there comes -120, it is determined in consequence that there are three true roots; and also a single false one, since two like signs come together but once, namely, with $-4x^3$ and $-19x^2$.

But this rule should be understood to hold for equations, none of whose roots are imaginary: for those roots, properly speaking, are neither true nor false but may be assumed indifferently to be one or the other. Thus, in

$$x^3-px^2+3p^2x-q^3=0$$

the signs indicate three true roots: hence if it be multiplied by $x+2p=0$ the ensuing equation $x^4+px^3+p^2x^2+(6p^3-q^3)x-2pq^3=0$ should have three true roots and one false one: none the less its signs indicate three false ones and but a single real root. The conclusion must consequently be drawn that the equation originally proposed has two imaginary roots, which though neither true nor false take on the appearance of each in turn.

---

tion $x^4+px^3+ppxx\begin{matrix}+6p^3\\-q^3\end{matrix}x-2pq^3=0$, which has but one fals root & is produced by y^e multiplication of $x+2p$ [$=0$] & $x^3-pxx+3ppx-q^3$ [$=0$]. in wch noe termes are wanting. The exception that I have made præsupposing y^e knowledg of imaginary roots, hath made me change the order of these sections'.

Siqua Æquatio objiciatur in qua unus terminus desit inseritur loco vacuo vel $+0$ vel -0 pro arbitratu tuo. Veluti si detur $y^3-py+q=0$; licebit scribere $y^3+0yy-py+q=0$, vel $y^3-0yy-py+q=0$, et in utroꝗ casu signa monstrant duas veras radices et unicam falsam.[88] Quod si habeatur $y^3+py-q=0$, hæc $y^3+0yy+py-q=0$ indicat unicam veram radicem, et hæc $y^3-0yy+py-q=0$ indicat tres veras; concludo itáꝗ duas esse imaginarias.[89]

Quo pacto veræ radices mutentur falsis et vicissim.

Perfacile est &c.

Read y^e^ words of the Author to y^e^ end of this section. pag 48.

Quot imaginarias radices Æquatio quævis habere possit.[90]

||[25] ||Numerus impossibilium radicum semper est par: si una sit impossibilis necesse est ut sint etiam duæ impossibiles; si tres, ut sint etiam quatuor; et sic præterea. Hinc Æquationes imparium dimensionum unam ad minus radicem semper habent possibilem. At in ijs parium dimensionum radices omnes possunt esse imaginariæ.

Porrò cùm aliquis terminus Æquationis inter duo signa similia desit vel cùm duo termini simul desint, id arguit duas minimùm radices esse imaginarias. Et cùm tres termini desint inter duo signa similia, vel cùm quatuor simul desint, id arguit quatuor minimùm radices esse imaginarias: Et sic deinceps.

Veluti in $x^3+px-q=0$, cùm terminus inter duos x^3 et $+px$ ejusdem signi desit, concludo duas esse radices imaginarias.

Item in $x^5-3x^4-2x+5=0$, et in $x^5-3x^4+5=0$ video duas ad minus radices esse imaginarias; sed in $x^5-3x^4-5=0$ video quatuor imaginarias esse.

Deniꝗ in $x^5+4x^3-6xx-5=0$, concludo quatuor radices esse imaginarias propter duos terminos inter signa similia (secundum inter $+$ & $+$, ac penultimum inter $-$ & $-$) deficientes.

Caput quartum.

De transformatione Æquationum, ubi radices simul mutantur.[91]

In this chapter I see nothing needs bee altered unless you think fit to insert the following section next after[92] the section intitled *Quo pacto regula inveniatur*[93] *tollendo secundo termino cujusvis Æquationis biquadraticæ*. pag 57.

---

(88) All three roots will, of course, be real only when $\frac{1}{27}p^3 > \frac{1}{4}q^2$.

(89) Compare Kinckhuysen's *Algebra* (§1): 91.

(90) This section, wholly new with Newton, is intended by him for insertion at the end of Kinckhuysen's 'Caput secundum' (*Algebra* (§1): 49).

Should an equation be presented in which one of the terms is missing, $+0$ or -0 is inserted in the vacant place as you will. So if there be given $y^3-py+q=0$, it is permissible to write $y^3+0y^2-py+q=0$ or $y^3-0y^2-py+q=0$: in either case the signs indicate two true roots and a single false one.(88) But if there be had $y^3+py-q=0$, this form $y^3+0y^2+py-q=0$ reveals a single true root, while this $y^3-0y^2+py-q=0$ reveals three true ones: I conclude in consequence that two are imaginary.(89)

In what manner true roots may be changed into false ones, and the contrary.

[*Perfacile est &c.*

Read y^e words of the Author to y^e end of this section. pag 48.]

How many imaginary roots any equation may have.(90)

The number of impossible roots is always even: if one be impossible, there must necessarily also be two impossible; if three, there must also be four; and so forth. Hence equations of odd dimensions have always one root at least which is possible. But in those of even dimensions all the roots may be imaginary.

Further, when some term of an equation is lacking between two like signs or when two adjacent terms are lacking, that shows that at least two roots are imaginary. And when three terms are lacking between two like signs or when four adjacent ones are lacking, that shows there are at least four imaginary roots: and so on.

As, for instance, in $x^3+px-q=0$, since the term between the two x^3 and $+px$ of the same sign are lacking, I conclude that there are two imaginary roots.

Likewise, in $x^5-3x^4-2x+5=0$ and in $x^5-3x^4+5=0$ I see that two roots at least are imaginary; but in $x^5-3x^4-5=0$ I see there are four imaginary ones.

In $x^5+4x^3-6x^2-5=0$, finally, I conclude that four roots are imaginary because two terms are lacking between like signs (the second one between $+$ and $+$, and the next to the last between $-$ and $-$).

Chapter four

The transformation of equations by a simultaneous changing of the roots.(91)

[In this chapter I see nothing needs bee altered unless you think fit to insert the following section next after(92) the section intitled *Quo pacto regula inveniatur*(93) *tollendo secundo termino cujusvis Æquationis biquadraticæ*. pag 57.]

---

(91) That is (compare note (62)) Kinckhuysen's 'Caput Tertium' (*Algebra* (§1): 49–63).

(92) Precisely, on Kinckhuysen's p. 58 immediately before the following section 'Quomodo penultimus terminus tolli possit'.

(93) See §1: note (62).

Quomodo tertius terminus tolli possit.

Proponatur Æquatio $x^4-3x^3-6xx+2x-5=0$, cujus tertius terminus $-6xx$ tollendus est. Fingo itaq; $x=y+e$ et in æquatione proposita scribens $y+e$ pro x prodit

$$\begin{array}{llll} y^4+4ey^3+6ee\,yy+4e^3y+e^4 & =0, \\ \quad -3y^3-9e\quad -9ee\;+3e^3 & \\ \qquad\quad -6\quad -12e\;-6ee & \\ \qquad\qquad\quad +2\;\;+2e & \\ \qquad\qquad\qquad\;\; -5 & \end{array}$$

cujus terminum $6ee-9e-6$ pono ||[26] æqualem nihilo, et ||æquationem illam solvendo[94] habeo $e=2$ vel $e=-\frac{1}{2}$. Quamobrem in æquatione primò propositâ scribendo $y+2$ vel $y-\frac{1}{2}$ pro x, orietur æquatio $y^4+5y^3-26y-33=0$. vel $y^4-y^3+5\frac{1}{4}y-7\frac{1}{16}=0$ ubi tertius terminus deest.

Pro æquationibus autem biquadraticis in genere hoc esto regula quòd ex $x^4+px^3+qxx+rx+s=0$ tertius terminus tolletur substituendo

$$y-\tfrac{1}{4}p\pm\sqrt{\frac{pp}{16}+\frac{q}{6}}$$

pro x modò signa $+$ et $-$ probè observentur.[95] Verbi gratiâ ad tollendum tertium terminum Æquationis $x^4-x^3+3xx-x+2=0$, scribo -1 pro p et $+3$ pro q et exit $y+\frac{1}{4}\pm\sqrt{\frac{1}{16}+\frac{1}{2}}$ hoc est $y+1$ vel $y-\frac{1}{2}$ substituendum pro x.

Non secus in æquationibus magis compositis procedendū est. Quinetiam ex dictis facilè patebit modus tollendi quartum quintum aliumve quemvis terminum ex æquationibus quibuslibet: labore tamen eo difficiliori quo terminus tollendus magis abest a primo termino.[96] Quemadmodum enim secundus terminus tollitur ope simplicis æquationis et tertius ope quadraticæ; sic ad quartum eadem methodo tollendum requiritur solutio æquationis cubicæ, ad quintum tollendum requiritur solutio æquationis biquadrati[c]æ, et ita deinceps.

After y^e following section intitled *Quomodo penultimus terminus tolli possit*: pag 58 subscribe these words.[97]

Ad eundem modum terminus antepenultimus potest tolli si tertius terminus primò tollatur: & sic de reliquis.

---

(94) $6e^2-9e-6 = 3(e-2)(2e+1) = 0$.

(95) On substituting $x = y+e$, the coefficient of y^2 in the resulting quartic is $6e^2+3pe+q$ and Newton's condition is that this be zero. By a slip in sign he solves $6e^2+3pe-q=0$.

(96) Not quite true, perhaps. If it is required that we eliminate the mth term in the equation

$$\sum_{0\leqslant i\leqslant n}(a_i x^{n-i}) = 0$$

then we must substitute $x = y+e$, where

$$\sum_{0\leqslant j\leqslant (m-1)}\left[\binom{n-j}{m-j-1}a_j e^{m-j-1}\right] = 0.$$

How the third term may be removed.

Suppose the equation $x^4-3x^3-6x^2+2x-5=0$ to be propounded, and that its third term $-6x^2$ is to be removed. I suppose, consequently, that $x=y+e$ and, on writing $y+e$ for x in the equation proposed there comes

$$y^4+(4e-3)\,y^3+(6e^2-9e-6)\,y^2$$
$$+(4e^3-9e^2-12e+2)\,y+e^4-3e^3-6e^2+2e-5=0:$$

its term $6e^2-9e-6$ I set equal to zero and, by resolving that equation,(94) I have $e=2$ or $e=-\frac{1}{2}$. By writing, on that account, $y+2$ or $y-\frac{1}{2}$ for x in the equation first proposed, there will arise the equations $y^4+5y^3-26y-33=0$ or

$$y^4-y^3+5\tfrac{1}{4}y-7\tfrac{1}{16}=0,$$

in which the third term is missing.

For quartic equations in general, however, let this be the rule: that the third term will be removed from $x^4+px^3+qx^2+rx+s=0$ by substituting

$$y-\tfrac{1}{4}p\pm\surd(\tfrac{1}{16}p^2-\tfrac{1}{6}q)$$

for x, provided the signs + and − be properly observed.(95) For example, to remove the third term of the equation $x^4-x^3-3x^2-x+2=0$ I write -1 for p and -3 for q and there comes out $y+\frac{1}{4}\pm\surd(\frac{1}{16}+\frac{1}{2})$, that is, $y+1$ or $y-\frac{1}{2}$, to be substituted for x.

The procedure is no different in more complex equations. Indeed, from what has been said the way of removing the fourth, fifth or any term you like from any equations whatsoever is easily attainable: the labour, however, is the more(96) difficult the farther the term to be removed is from the first term. For, just as the second term is removed with the aid of a simple equation and the third with the aid of a quadratic, so the removal of the fourth by the same method requires the resolution of a cubic equation, the removal of the fifth requires the resolution of a quartic, and so on.

[After ye following section intitled *Quomodo penultimus terminus tolli possit*: pag 58 subscribe these words.(97)]

In the same way the next but one last term may be removed if the third term be first removed: and so of the rest.

The complexity of the operation will depend not merely on the size of m but on that of $n-m$ and of the binomial coefficients $\binom{n-j}{m-j-1}$.

(97) This is patently in line with the preceding insertion.

Caput quintum.

De solutione Æquationum.(98)

This chapter may bee continued to y^e^ end almost as the Author hath done it unlesse you think it convenient to put y^e^ particular solutions, begining at pag 74,(99) in manner of an Appendix, wth some such præface as this.

Appendix.

Præter generales regulas quas de solutione æquationum tradidi, alij etiam particulares modi ex aliqua earum conditione præcognitâ occurrunt. Quorum quidem notitia licèt non omninò necessaria sit, tamen cùm Artificem nonnunquam juvet, et (postquam in præcedentibus sat fuerit versatus) ingenium ejus quoad naturam æquationum penitiùs intelligendam excolat, placuit eos more Appendicis subnectere.

Supra ostensum est satis, æquationem &c.

Where continue the Authors words unto pag 92.(100)

||[27] ||pag 92 lin 5. after *hoc est* $x=4$. may be inserted what follows.(101)

(98) Kinckhuysen's 'Caput Quartum. Quomodo æquatio solvenda sit' (*Algebra* (§1): 64–95).

(99) And terminating, by implication at least, on p. 95 of Kinckhuysen's tract. Perhaps, however, Newton intended to exclude Kinckhuysen's discussion on his pp. 90–5 of Cardan's solution of the reduced cubic and its geometrical Cartesian complement (over whose revision he himself will take a deal of care in the two following pages) and also of Schooten's iterative method for isolating the real roots of numerical equations. Neither seems particularized in Newton's present sense, one which he amplified in an accompanying marginal note to Collins: 'I think its best to set these particular solutions as an appendix, not onely because they apperteine not to the generall doctrin of Algebra (the designe of this booke) but also because thereby the learner may have a caveat not to trouble & perplex himselve wth them till hee is Master of the generall doctrine tought before.' We should note that among the 'particular solutions' he now preferred to relegate to an appendix was an independent discovery of his own (see §1: note (84)).

(100) In the first version of his *Observationes* (which lacked the following pp. 27–8) Newton wrote '95', thus bridging the gap to p. 29 below.

(101) This relates, of course, to Kinckhuysen's exposition of Cardan's solution of the reduced cubic (see note (102) following) on pp. 90–2 of his *Algebra*, and Newton's present revision was made at Collins' direct request. The latter had been dissatisfied from the first with Kinckhuysen's Huddenian account and it would appear that in his lost letter to Newton in the spring of 1670 he suggested that its content be improved by inserting appropriate excerpts from Ferguson's *Labyrinthus Algebræ* (note (11)): as he wrote to Wallis on 17 June 1669, 'Ferguson...solves cubic and biquadratic equations by such new methods as render the roots in their proper species, when it may be done, to wit, in whole or mixed numbers, fractions, or surds, either simple, compound, or universal, and likewise improves the general method; thereby accomplishing as much as Hudden, *in annexis Geometriæ Cartesianæ*, seemed to promise about it' (Rigaud's *Correspondence* (note (11)), **2**: 515). When Newton sent along the first version of his *Observationes* on 11 July 1670 he did not at all agree: 'In a letter you hinted somthing to bee supplyed out of Ferguson's *Labyrinthus* about ye extraction of cubick roots;...

Chapter five

The solution of equations.[98]

[This chapter may bee continued to ye end almost as the Author hath done it unlesse you think it convenient to put ye particular solutions, begining at pag 74,[99] in manner of an Appendix, wth some such præface as this.]

Appendix.

Apart from the general rules which I have delivered for resolving equations, there are also other particular methods to be met with which are dependent on some element or other of their situation being previously made known. Though, indeed, a knowledge of these is not at all necessary, nevertheless since it may not infrequently be of help to the practitioner of the art and (after he has come to be sufficiently skilled in the preceding matters) may improve his intellectual grasp in so far as the deeper understanding of the nature of equations is concerned, it has seemed appropriate to subjoin these methods in manner of an appendix.

[*Supra ostensum est satis, æquationem &c.*

Where continue the Authors words unto pag 92.[100]

pag 92 lin 5. after *hoc est* $x = 4$. may be inserted what follows.][101]

---

if you meant affected roots, tis already done by Kinck-Huyson pag 91 as well as by Ferguson. Indeed Ferguson seems to have done more in so much as to comprehend all cases of cubick equations w^th^in y^e^ same rules; but that *more* is inartificiall because it supposes ye extraction of cubick roots out of imaginary binomiums, wch how to doe hee hath not taught us.... Not but that it may bee done, & I know how to doe it, but I think it not worth ye inserting into Kinckhuyson...' (*Correspondence of Isaac Newton*, **1**: 30–1; compare note (17)). This seems somewhat unfair to Ferguson, who had indeed gone beyond Kinckhuysen (and to be sure Hudde's printed account) in showing how by suitable manipulation of conjugate complex quantities a rational real root, when it existed, could sometimes be had from Cardan's cubic formula even in the 'irreducible' case. Collins seems to have thought as much but phrased his reply on 13 July somewhat unfortunately: 'I was apt to believe that Ferguson had done more then Kinckhuysen in these 3 particulars 1 In applying one generall rule to both kinds of Cubick Æquations, to wit as well those that are solved by meane Proportionalls as those that require Trisection...' (*Correspondence*, **1**: 33). Again, six days later, he wrote that 'when I had Fergusons Papers, I only viewed his Examples and that cursorily, it seemes he soared but *Icari fine* to accomplish what Hudden promised page 503 *in annexis Geometriæ Cartesianæ*. I scrupled his rootes of negative quadratick quantities, and imagined that they expunged one another being affected with contrary Signes but conceited there might be more done in Cubicks then authors yet insist on.... Are not both wayes [Trisection, and finding of 2 Meanes] the Solution of the Cubick æquation of the same kind[?]' (*ibid.*, **1**: 37). In his reply to Collins on 16 July Newton refused to see any saving graces in Ferguson's work, for 'I see not w^t^ hee hath done more then in Cardans rules.... Nor doe I see w^t^ hee hath done more then Descartes in his Solution of biquadratick equations: for both goe ye same way to worke in reducing them first to Cubick & then to quadratick æquations. Lastly I see not in what case his rules will render the roots of cubick or biquadratick Æquations *in proprio genere* where those of Cardan or Descartes will not' (*ibid.*, **1**: 35). Over Collins' suggestion that

Tertius casus idem est cum secundo nisi quòd radices quæ in secundo casu veræ sunt in tertio sunt falsæ. Sed in utroq casu notandum est quod si $\frac{1}{4}qq$ non sit major quàm $\frac{1}{27}p^3$, radix per has regulas designata est reverà impossibilis: et tamen per regulam extrahendi radices ex impossibilibus binomijs supra traditam reales radices exinde possunt obtineri.(102)

Quemadmodum dato $x^3=15x-4$, erit

$$x=\sqrt{C:\overline{-2+\sqrt{-121}}}+\sqrt{C:\overline{-2-\sqrt{-121}}},$$

nam $\frac{1}{2}q$ valet 2 et $\frac{1}{4}qq$ valet 4 et $\frac{1}{27}p^3$ valet 125. Extractâ radice ex utroq binomio(103) extat $x=-2+\sqrt{-1}-2-\sqrt{-1}$. hoc est $x=-4$. Inventâ hâc falsâ radice, ejus ope veras radices invenies dividendo propositam æquationem per $x+4$; prodibit enim $xx-19x+1=0$, et inde $x=\dfrac{9\pm\sqrt{77}}{2}$.(104)

Ad eundem modum dato $x^3=21x+20$, erit

$$x=\sqrt{C:\overline{10+\sqrt{-243}}}+\sqrt{C:\overline{10-\sqrt{-243}}},$$

et extractis radicibus ex utroq binomio,(105) prodibunt $x=\dfrac{5+\sqrt{-3}}{2}+\dfrac{5-\sqrt{-3}}{2}$, $x=-2+\sqrt{-3}-2-\sqrt{-3}$, & $x=\dfrac{-1-[3]\sqrt{-3}}{2}\dfrac{-1+[3]\sqrt{-3}}{2}$; hoc est $x=5$, $x=-4$, et $x=-1$.(106)

Cæterùm cùm binomiorum radices non possunt extrahi id arguit nullas e radicibus æquationis propositæ rationales esse, et proinde solutionem aliunde petendam. Quemadmodum si geometricè designare velimus; regulæ jam explicatæ (quarum inventionem Cardanus cuidam Scipioni Ferreo tribuit)(107)

---

a cubic resolvable by 'trisection' could be reduced to finding 'two meane proportionalls' he was suitably scathing in a second letter on 27 September, 'wch if it could, it would bee noe hard matter to take away both ye two middle termes of any cubick æquation. Which whoever performes I shall esteem as a great Apollo & admire as much as if hee had squared ye circle, because I judg both impossible. And my reason is this that æquations to what termes soever they are reduced their reall roots never becom imaginary nor their imaginary roots reall (though indeed their true roots may become false & false ones true). But could a cubick æquation wch hath 3 reall roots (and consequently is solvible by trisection) have its two middle termes taken away (& consequently become soluble by 2 meanes) two of its reall roots must bee transformed into imaginary ones, for all simple cubick æquations can have but one root reall & two imaginary' (*ibid.*, **1**: 43). Collins, of course, had misunderstood Ferguson on this point, and it was to clarify the relative status of Cardan's rule for the cubic *vis à vis* the Cartesian reduction to angular trisection that Newton, some time in the early autumn of 1670, composed the augmented version of Kinckhuysen's p. 92 which follows.

(102) In these cases, Cardan's rule for the reduced cubic $x^3 = px\pm q$ yields

$$x = \sqrt[3]{[a+\sqrt{-b}]}+\sqrt[3]{[a-\sqrt{-b}]},$$

where $a = \pm\frac{1}{2}q$ and $b = \frac{1}{27}p^3-\frac{1}{4}q^2$. If $\alpha+\sqrt{-\beta}$ is one value of $\sqrt[3]{[a+\sqrt{-b}]}$, then its two other values are $(\alpha+\sqrt{-\beta})\times-\frac{1}{2}(1\pm\sqrt{-3}) = -\frac{1}{2}(\alpha\pm\sqrt{3\beta}+[\pm\alpha\sqrt{3}+\sqrt{\beta}]\sqrt{-1})$ and therefore x has the three values 2α and $-\alpha\pm\sqrt{3\beta}$, all real for $b\geqslant 0$. Alternatively, $\frac{1}{27}p^3-\frac{1}{4}q^2\geqslant 0$ is

The third case is the same as the second except that the roots which are true in the second are false in the third. But in either case it should be noted that, if $\frac{1}{4}q^2$ be not greater than $\frac{1}{27}p^3$, the root specified by these rules is in fact impossible; and yet by the rule for extracting the roots out of impossible binomials delivered above real roots may be obtained from them.(102)

For instance, given that $x^3 = 15x - 4$, then will there be

$$x = \sqrt[3]{(-2+\sqrt{-121})} + \sqrt[3]{(-2-\sqrt{-121})},$$

for the value of q is 2, that of $\frac{1}{4}q^2$ is 4 and that of $\frac{1}{27}p^3$ is 125. And when the root of each binomial is extracted, there proves to be $x = -2+\sqrt{-1}-2-\sqrt{-1}$, that is, $x = -4$. Having found this false root, you may by its aid find the true roots on dividing the propounded equation by $x+4$, for there will come $x^2 - 19x + 1 = 0$, and thence $x = \frac{1}{2}(9 \pm \sqrt{77})$.(104)

In the same way, given $x^3 = 21x + 20$, there will be

$$x = \sqrt[3]{(10+\sqrt{-243})} + \sqrt[3]{(10-\sqrt{-243})},$$

and on extracting the roots of each binomial(105) there will come

$$x = \tfrac{1}{2}(5+\sqrt{-3}) + \tfrac{1}{2}(5-\sqrt{-3}), \quad x = -2+\sqrt{-3}-2-\sqrt{-3}$$

and $x = \frac{1}{2}(-1-[3]\sqrt{-3}) + \frac{1}{2}(-1+[3]\sqrt{-3})$; that is, $x = 5$, $x = -4$ and $x = -1$.(106)

For the rest, when the roots of the binomial cannot be extracted, that shows that none of the roots of the propounded equation are rational and consequently the solution has to be sought by other means. Should we wish, for instance, to determine it geometrically, the rules already explained (whose invention Cardan attributed to a certain Scipione del Ferro)(107) show that that may be

---

the condition (necessary and sufficient) which permits a Cartesian solution 'per trisectionem anguli': for in that application, in the terminology of §1: note (105), it is necessary that NP (or $3q/p$) be not greater than the diameter $2r = 2\sqrt{\frac{1}{3}p}$.

(103) See Newton's first example on his p. 14 above.

(104) This is the solution of the quadratic '$xx - 9x + 1 = 0$', but on correctly dividing the cubic by $x+4$ the quadratic factor is '$xx - 4x + 1 = 0$, et inde $x = 2 \pm \sqrt{3}$'! In proof (note (47)), where $\omega^3 = 1$, then

$$\sqrt[3]{[-2+11\sqrt{-1}]} = (-2+\sqrt{-1})\omega = -2+\sqrt{-1}, \text{ or } \tfrac{1}{2}(2 \pm \sqrt{3} - [1 \pm 2\sqrt{3}]\sqrt{-1}).$$

It has evidently escaped Newton's attention that the sum of his three roots, as they are given, is not zero though his cubic lacks a second term.

(105) See Newton's second example on his p. 14 above.

(106) In fact (note (48)), where $\omega^3 = 1$, $\sqrt[3]{[10 \pm \sqrt{-243}]} = (-2 \pm \sqrt{-3})\omega$, that is,

$$-2 \pm \sqrt{-3},\ \tfrac{1}{2}(2 \pm 3 + [\mp 1 + 2]\sqrt{-3}) \text{ or } \tfrac{1}{2}(2 \mp 3 - [\pm 1 + 2]\sqrt{-3}).$$

(107) Compare Descartes' [*Geometrie*: 398 =] *Geometria* ($_2$1659): 93: 'regula, cujus inventionem Cardanus cuidam, Scipioni Ferreo, tribuit'. The reference is to G. Cardano's *Artis Magnæ sive de Regulis Algebraicis Liber Unus* (Nuremberg, 1545): Cap. XI: 29^r: 'Scipio Ferreus Bononiensis iam annis abhinc triginta fermè capitulum hoc inuenit, tradidit uero Anthonio

monstrant id fieri posse per prima duorum mediorum proportionalium inter unitatem et ista binomia aut per latera cuborum quorum solidum contentum per ista binomia designantur: dummodò $\frac{1}{27}p^3$ non sit major quàm $\frac{1}{4}qq$. Et ‖[28] ejusmodi media ‖proportionalia, aut latera cuborum expeditè per tabulam Logarithmorum in proximis numeris adipiscuntur. Sin $\frac{1}{27}p^3$ sit major quam $\frac{1}{4}qq$, solutio potest fieri per trisectionem anguli, sicut in adjecto circulo(108) pateat ubi ponitur radius $NO = \sqrt{\frac{1}{3}p}$ et chorda $NP = \frac{3q}{p}$ ac arcus NTP et NVP dividuntur uterꝗ in tres partes æquales. Erit enim summa chordarum TP et PV vera radix in secundo casu, [&c]

Hence(109) continue y^e^ Authors words to y^e^ end of pag 95.

‖[29]

‖Pars Tertia. Quomodo quæstio aliqua ad æquationem redigatur(110)

(111)Postquam Tyro aliquamdiu exercitatus fuerit in speciebus computandis, in æquationibus secundum dimensiones incognitæ quantitatis aut literæ cujuslibet ordinandis, et in transformatione duarum aut plurium æquationum tot incognitas quantitates involventium in unam æquationem continentem unicam quam libuerit quantitatem incognitam: haud incommodum judico ut ingenij vires in problematîs facilioribus ad æquationem redigendis tentet(112) etsi fortè resolutiones earum nondum attigerit. Quinimò cùm in hoc mediocritèr versatus fuerit, et percipiat se artem nonnihil callere quà ex statu quæstionis tot eliciantur æquationes quot sufficiunt ad omnes ejus conditiones penitùs implendas, easꝗ

Mariæ Florido Veneto, qui cũ in certamen cũ Nicolao Tartalea Brixellense aliquando uenisset, occasionem dedit, ut Nicolaus inueneri & ipse, qui cum nobis rogantibus tradidisset, suppressa demonstratione, freti hoc auxilio, demonstrationem quæsiuimus, eamque in modos, quod difficillimum fuit, redactam sic subiecimus'.

(108) Understand the Cartesian figure on Kinckhuysen's p. 92 (§1: note (104)).

(109) From the last line of Kinckhuysen's p. 92.

(110) Newton's observations on Kinckhuysen's final Part 3 (*Algebra* (§1): 96–108) were clearly composed at two separate times. The first version (Newton's pp. 29–34) is evidently that which he communicated to Collins on 11 July 1670 in the first draft of his *Observationes*. However, after Collins in his reply two days later had shown himself eager that Newton further revise 'Kinckhuysen his discourse upon surds', the latter wrote back on 16 July that 'if the booke goe not immediately into ye presse I desire you'le send it back wth those notes I have made...Ile...not only supply ye wants about surds but that about Æquations soluble by trisections, & somthing more I would say in the chapter *Quomodò quæstio aliqua ad æquationem redigatur.* that being the most requisite & desirable doctrine to a Tyro & scarce touched upon by any writer unless in generall circumstances bidding them onely *Nota ab ignotis non discernere & adhibere debitum ratiocinium*' (*Correspondence of Isaac Newton*, **1**: 35). The *Observationes* were duly returned to him on the 19th but Newton subsequently decided to do little more than revise his previous additions to Kinckhuysen's 'Pars Tertia' (including perhaps an extension on the missing insert to Newton's p. 33) and append 'two or three' further geometrical worked

effected by taking the first of two mean proportionals between unity and those binomials or the sides of cubes whose solid contents are designated by those binomials, provided only that $\frac{1}{27}p^3$ be not greater than $\frac{1}{4}q^2$. Mean proportionals of this sort or the sides of cubes are speedily obtained in approximate numerical terms by means of a table of logarithms. But if $\frac{1}{27}p^3$ be greater than $\frac{1}{4}q^2$, the solution may be had through the trisection of an angle, as may appear in the adjoining circle[108] where the radius *NO* is set equal to $\sqrt{\frac{1}{3}p}$ and the chord *NP* to $3q/p$, while the arcs *NTP* and *NVP* are each divided into three equal parts. [*Erit enim summa chordarum TP et PV vera radix in secundo casu* &c.

Hence[109] continue y^e^ Authors words to y^e^ end of pag 95.]

Part Three. How any question may be reduced to an equation[110]

[111]After the novice has exercised himself some little while in algebraic computation, in ordering equations according to the dimensions of the unknown or of any arbitrary letter, and in the transformation of two or more equations involving the same number of unknowns into one equation containing any single one of those unknowns, I judge it not unfitting that he test his intellectual powers in reducing easier problems to an equation, even though perhaps he may not yet have attained their resolution. Indeed, when he is moderately well versed in this subject and conceives he has some degree of skill in the art of eliciting from the circumstances of a question as many equations as suffice to implement fully all its conditions and knows how to reduce all those equations (should there be

---

examples on pp. 36–9. As he told Collins on the following 27 September, 'I sometimes thought to have set upon writing a compleate introduction to Algebra, being cheifely moved to it by this that some things I had inserted into Kinck-Huysen were not so congruous as I could have wished to his manner of writing. Thus having composed somthing pretty largely about reducing problems to an æquation when I came to consider his examples (wch make ye 4th [sic] part of his booke) I found most of them solved not by any generall Analyticall method but by particular & contingent inventions, wch though many times more concise then a generall method would allow, yet in my judgment are lesse propper to instruct a learner, as Acrostick's & such kind of artificiall Poetry though never soe excellent would bee but impropper examples to instruct one y^t^ aimes at Ovidian Poetry. But considering that by reason of severall divertisements I should bee so long in doing it as to tire you[r] patience wth expectation, & also...there being severall Introductions to Algebra already published...I have chosen rather to let it passe w^th^out much altering what I sent you before...& soe have hitherto added two or three examples onely more...' (*ibid.*, **1**: 43–4).

(111) Newton has justified his rejection of Kinckhuysen's own introduction (on his p. 96) in an accompanying marginal aside to Collins: 'I have substituted this preface to ye following problems instead of the Authors, as being more congruous to w^t^ I have added to ye booke before & more profitable in respect of w^t^ follows, illustrating the most noble & difficult part of the art, namely the reductiō of problems to æquations.' Compare the previous note.

(112) The equivalent 'periclitetur' is cancelled.

omnes æquationes (modò plures sint) ad unam finalem quæstioni satisfacientem redigere noverit: tunc majori cum fructu et voluptate contemplabitur naturam et proprietates æquationum et earum resolutiones Algebraicas, Geometricas et Arithmeticas addiscet.

Proposito autem aliquo Problemate,(113) Artificis ingenium in eo præsertim requiritur ut omnes ejus conditiones totidem æquationibus designet. Ad quod faciendum perpendet imprimis an propositiones sive sententiæ quibus enunciatur sint omnes aptæ quæ terminis Algebraicis designari possint, haud secus quàm conceptus nostri characteribus Græcis vel latinis. Quod si contingat (uti solet in quæstionibus quæ ad numeros vel abstractas quantitates spectant) tunc nomina quantitatibus ignotis atq; etiam notis si opus fuerit imponat, & sensum quæstionis algebraicè designet; ejusq; status ad algebraicos terminos sic translatus(114) tot dabit æquationes quot ei solvendæ sufficiunt.

Quemadmodum si quærantur tres numeri continuè proportionales quorum ||[30] summa sit 20 & quadratorum summa || 140.(115) Posito x y et z pro tribus numeris quæsitis, Quæstio e latinis literis in Algebraicas vertetur ut sequitur.

Quæstio latinè enunciata.	Eadem algebraicè.
Quæruntur tres numeri, his conditionibus.	$x.y.z.$
Ut sint continuè proportionales.	$x.y::y.z.$ sive $xz=yy$.
Ut omnium summa sit 20.	$x+y+z=20$.
Et ut quadratorum summa sit 140.	$xx+yy+zz=140$.

Atq; ita quæstio deducitur ad æquationes $xz=yy$, $x+y+z=20$, &

$$xx+yy+zz=140,$$

quarum ope x y et z sunt investigandæ.

Cæterùm notandum est solutiones quæstionum eo magis expeditas et artificiosas evadere quo pauciores incognitæ quantitates sub initio ponuntur. Sic in hac quæstione posito x pro primo numero et y pro secundo erit $\frac{yy}{x}$ tertius continuè proportionalis; quem proinde ponens pro tertio numero, quæstionem ad æquationes sic reduco.

Quæstio latinè enunciata.	Eadem algebraicè.
Quæruntur tres numeri continuè proportionales,	$x.\ y.\ \frac{yy}{x}$.
Quorum summa sit 20.	$x+y+\frac{yy}{x}=20$.
Et quadratorum summa 140.	$xx+yy+\frac{y^4}{xx}=140$.

several) to a final one which satisfies the question, then will he with greater profit and enjoyment contemplate the nature and properties of equations and learn their algebraic, geometrical and arithmetical resolutions.

But when some problem has been proposed, the practitioner's(113) skill is particularly demanded when it comes to designating all its conditions by an equal number of equations. To do this, let him in the first place examine whether the propositions or phrases by which it is enunciated are all fit to be denoted in algebraic terms in the same way that we express our concepts in greek or latin characters. Should this happen (as usually is the case in questions which relate to numbers or abstract quantities) let him then set names on the unknowns and also on known quantities if need be, denoting the sense of the question algebraically, and its circumstances thus translated into algebraic terms will yield as many equations as suffice to resolve it.

For instance, if there are required three continuously proportional numbers whose sum is 20 and the sum of their squares 140,(115) on setting x, y and z for the three numbers sought the question will be changed from its verbal form into a corresponding algebraic expression as follows:

The question expressed verbally	The algebraic expression of the same
There are required three numbers subject to these conditions:	x, y, z.
they must be continuously proportional,	$x:y = y:z$ or $xz = y^2$.
their sum total must be 20,	$x+y+z = 20$.
and the sum of their squares 140.	$x^2+y^2+z^2 = 140$.

And so the question is brought down to the equations $xz = y^2$, $x+y+z = 20$ and $x^2+y^2+z^2 = 140$, by whose aid x, y and z are to be investigated.

But it should be noted that the solutions to questions come out the more speedily and skilfully the fewer the unknown quantities supposed at the beginning. So in this question, on setting x for the first number and y for the second, y^2/x will be the third continued proportional; and, in consequence, on setting this for the third number I reduce the question to equations in the following way:

The question expressed verbally	The algebraic expression of the same
There are sought three numbers in continued proportion,	$x, y, y^2/x$.
whose sum is 20	$x+y+y^2/x = 20$.
and the sum of their squares 140.	$x^2+y^2+y^4/x^2 = 140$.

(113) The hyperbole 'summum' (highest) is cancelled.
(114) Newton first wrote the more usual equivalent 'redactus'.
(115) Kinckhuysen's third worked example (on his p. 98): see §1: note (114).

de pag harum adver- sionum Habes itaq; æquationes $x+y+\frac{yy}{x}=20.$ & $xx+yy+\frac{y^4}{xx}=140.$ quarum ope x et y determinandi restant.

Aliud exemplum accipe. Mercator quidam nummos ejus triente quotannis adauget, demptis 100lib quas annuatim impendet in familiam; et post tres annos fit duplo ditior. Quæruntur num̄i.

Ad isthoc autem solvendum sciendum est quod plures latent propositiones quæ omnes sic eruuntur et enunciantur.

latinè.	algebraicè.
Mercator habet nummos quosdam	$x.$
Ex quibus anno 1mo expend[i]t 100lib	$x-100.$
Et reliquum adauget triente.	$x-100+\frac{x-100}{3}$, sive $\frac{4x-400}{3}$.
[31/32] ‖(117)[Annoq; secundo expendit 100lib	$\frac{4x-400}{3}-100$, sive $\frac{4x-700}{3}$.
Et reliquum adauget triente.	$\frac{4x-700}{3}+\frac{4x-700}{9}$, sive $\frac{16x-2800}{9}$.
Et sic anno tertio expendit 100lib	$\frac{16x-2800}{9}-100$, sive $\frac{16x-3700}{9}$.
Et reliquo trientem similiter lucratus est.	$\frac{16x-3700}{9}+\frac{16x-3700}{27}$, sive $\frac{64x-14800}{27}$.
Fitq; duplò ditior quàm sub initio.	$\frac{64x-14800}{27}=2x.$

Quæstio itaq; ad æquationem $\frac{64x-14800}{27}=2x$ redigitur cujus reductione eruendus est x. Nempe duc in 27 & fit $64x-14800=54x$. subduc $54x$ & restat $10x-14800=0$ seu $10x=14800$, et dividendo per 10 fit $x=1480$. Quare 1480lib sunt nummi sub initio, ut et lucrum.(118)

Vides itaq; quod ad solutiones quæstionum quæ circa numeros vel abstractas quantitatum relationes solummodò versantur, nihil aliud ferè requiritur quàm ut e sermone Latino vel alio quovis in quo Problema proponitur, translatio fiat in sermonem (si ita loquar) Algebraicum, hoc est in characteres qui apti sunt ut nostros de quantitatum relationibus conceptus designent. Nonnunquam verò

(116) Compare notes (75) and (77) above.

(117) The following leaf (Newton's pp. 31–2) is now lacking in Newton's manuscript, but we restore its main outline from the lightly corrected version which he later incorporated in his Lucasian lectures on algebra (ULC. Dd. 9.68: 40–1 = *Arithmetica Universalis* (Cambridge, 1707): 78–9).

(118) Since, of course, over the three-year period the merchant doubled his capital.

You have consequently the equations

$$x+y+y^2/x = 20 \quad \text{and} \quad x^2+y^2+y^4/x^2 = 140,$$

while x and y remain to be determined by their aid.

See pa 20(116) c these o vations

Take another example. A certain merchant increases his capital by a third each year, less £100 which he spends annually on domestic purposes, and after three years he becomes twice as rich. What was his capital?

To resolve this, it must be realized that several propositions are concealed in the enunciation, and these when made explicit may be expressed thus:

In verbal terms	Algebraically
A merchant has a certain sum of money,	x.
a £100 of which he spends in the first year	$x-100$.
and the remainder he increases by a third.	$x-100+\dfrac{x-100}{3}$ or $\dfrac{4x-400}{3}$.
(117)In the second year he spends £100	$\dfrac{4x-400}{3}-100$ or $\dfrac{4x-700}{3}$.
and the remainder he increases by a third.	$\dfrac{4x-700}{3}+\dfrac{4x-700}{9}$ or $\dfrac{16x-2800}{9}$.
And so in the third year he spends £100	$\dfrac{16x-2800}{9}-100$ or $\dfrac{16x-3700}{9}$.
and on the rest makes a profit of one third,	$\dfrac{16x-3700}{9}+\dfrac{16x-3700}{27}$ or $\dfrac{64x-14{,}800}{27}$.
so becoming twice as rich as when he began.	$\dfrac{64x-14{,}800}{27}=2x$.

The question is consequently reduced to the equation $\dfrac{64x-14{,}800}{27}=2x$ and from its solution x is to be evaluated. Precisely, multiply through by 27 and there comes $64x-14{,}800 = 54x$: subtract $54x$ and there remains

$$10x-14{,}800 = 0 \quad \text{or} \quad 10x = 14{,}800,$$

and on dividing by 10 there comes $x = 1480$. Hence the initial capital, and also the profit,(118) is £1480.

You see, consequently, that to resolve questions which are concerned only with numbers or the abstract relationships of quantities scarcely anything is required except the translation from latin or any other language in which the problem is proposed into that of algebra (if I may use the phrase), that is, into characters suitable to denote our concepts of quantitative relationships. But sometimes, indeed, it may happen that the language in which the circum-

potest accidere quòd sermo quocum status quæstionis exprimitur ineptus videatur qui in Algebraicum possit verti; sed parvis mutationibus adhibitis & ad sensum potiùs quam verborum sonos attendendo versio reddetur facilis. Sic enim quælibet apud Gentes loquendi formæ propria habent Idiomata: quæ ubi obvenerint, translatio ex unis in alias non verbo tenus instituenda est sed ex sensu determinanda. Cæterum ut hujusmodi problemata hac methodo ad æquationes redigendi familiaritatem convincam & illustrem, & cùm Artes exemplis facilius quàm præceptis addiscantur, placuit sequentium problematum solutiones adjungere.

Hence continue y^e Authors words on pag 97.][119]

||[33] ||After the Authors Algebraick problems[120] you may subjoyne these geometrick ones.

1. Trianguli rectanguli *ABC* perimetro et areâ datis, invenire hypotenusam *BC*.[121]

Esto perimeter $=a$, area $=bb$, $BC=x$, et $AC=y$. Eritq; $AB=\sqrt{xx-yy}$. Unde rursus perimeter $(BC+AC+AB)$ est $x+y+\sqrt{xx-yy}$, & area $(\frac{1}{2}AC\times AB)$ est $\frac{1}{2}y\sqrt{xx-yy}$. Adeóq; $x+y+\sqrt{xx-yy}=a$ et $\frac{1}{2}y\sqrt{xx-yy}=bb$.

Harum æquationum posterior dat $\sqrt{xx-yy}=\frac{2bb}{y}$, quare scribo $\frac{2bb}{y}$ pro $\sqrt{xx-yy}$ in æquatione priori ut asymmetria tollatur, & prodit $x+y+\frac{2bb}{y}=a$, sive multiplicando per y et ordinando $yy=ay-xy-2bb$. Porrò ex partibus æquationis prioris aufero $x+y$ & restat $\sqrt{xx-yy}=a-x-y$, cujus partes quadrando ut asymmetria rursus tollatur prodit

$$xx-yy=aa-2ax-2ay+xx+2xy+yy,$$

quæ in ordinem redacta et per 2 divisa fit $yy=ay-xy+ax-\frac{1}{2}aa$. Deniq; ponendo æqualitatem inter duos valores ipsius yy habeo $ay-xy-2bb=ay-xy+ax-\frac{1}{2}aa$, quæ reducta fit $\frac{1}{2}a-\frac{2bb}{a}=x$.

---

(119) Our restoration thus far would fill, we judge, a little more than a page of Newton's manuscript. On his p. 32 it is probable that Newton added some comments in revision of the 'particular' problems on Kinckhuysen's following pp. 97–108. The latter's third problem (on p. 98), already discussed by Newton in detail (see notes (75), (77) and (115) above), would now presumably be omitted. Otherwise, the only clue to his possible further suggested revisions here is the layout and technical content of the seventeen arithmetical problems which he inserted at the corresponding point in his Lucasian lectures on algebra (ULC. Dd. 9.68:

stances of the question are expressed seems unsuitable to be rendered in equivalent algebraic terms: however, by making a few changes and paying heed to the sense rather than to the sound of words the transliteration will be easily made. Thus, for instance, every human form of speech has its own peculiar idioms, and when these are met with, translation from one into another is not to be undertaken literally but to be determined from the sense. However, so that I may impress an intimate understanding of this method of reducing problems of this sort to equations and illustrate it, and because craft skills are more easily learnt by example than by precept, it seems appropriate to adjoin the solutions of the following problems.

[Hence continue y^e Authors words on pag 97.](119)

[After the Authors Algebraick problems(120) you may subjoyne these geometrick ones.]

1. Given the perimeter and area of the right-angled triangle ABC, to find the hypotenuse BC.(121)

Let the perimeter $= a$, the area $= b^2$, $BC = x$ and $AC = y$: then will

$$AB = \surd(x^2 - y^2).$$

Hence, again, the perimeter $(BC+AC+AB)$ is $x+y+\surd(x^2-y^2)$ and the area $(\frac{1}{2}AC\times AB)$ is $\frac{1}{2}y\surd(x^2-y^2)$, so that $x+y+\surd(x^2-y^2) = a$ and $\frac{1}{2}y\surd(x^2-y^2) = b^2$. The latter of these equations gives $\surd(x^2-y^2) = 2b^2/y$. I therefore write $2b^2/y$ for $\surd(x^2-y^2)$ in the former equation to remove asymmetry and there comes $x+y+(2b^2/y) = a$, that is, on multiplying by y and ordering, $y^2 = ay-xy-2b^2$. Again, from the parts of the former equation I take away $x+y$ and there remains $\surd(x^2-y^2) = a-x-y$: on squaring its parts to remove asymmetry once more there comes $x^2-y^2 = a^2-2ax-2ay+x^2+2xy+y^2$, and this reduced to order and divided through by 2 becomes $y^2 = ay-xy+ax-\frac{1}{2}a^2$. Finally, on supposing equality between the two values of y^2 I have

$$ay-xy-2b^2 = ay-xy+ax-\tfrac{1}{2}a^2$$

and this upon reduction becomes $\frac{1}{2}a-2b^2/a = x$.

41–52 = *Arithmetica Universalis* (Cambridge, 1707): 80–96) and to these we refer the reader so that he may make up his own mind.

(120) That is, in continuation of Kinckhuysen's p. 108.

(121) This and the three following problems are in the spirit of Sectio x ('De modo inveniendi triangula, quorum singula latera, segmenta basis & perpendicularis exprimantur per numeros rationales absolutos') of Frans van Schooten's *Exercitationum Mathematicarum Libri Quinque* (Leyden, 1657): *V. Sectiones Miscellaneæ Triginta*: 426–30. Newton, as we have seen (I, 1, 1, §3) made extensive notes on the section in late 1664.

Idem aliter. Esto $\frac{1}{2}$ perimeter $=a$, area $=bb$, et $BC=x$. eritq; $AC+AB=2a-x$. Jam cùm sit $xx(BC^q)=AC^q+BC^q$, et $4bb=2AC\times AB$: erit

$$xx+4bb=AC^q+AB^q+2AC\times AB$$
$$=\text{quadrato ex }(AC+AB)=\text{quadrato ex }2a-x=4aa-4ax+xx.$$

Hoc est $xx+4bb=4aa-4ax+xx$; quæ reducta fit $a-\dfrac{bb}{a}=x$.

2. Cujuslibet Trianguli ABC datis lateribus AB, AC, et basi BC quam perpendiculum AD ab angulo verticali demissum secat in D: invenire segmenta BD ac DC.

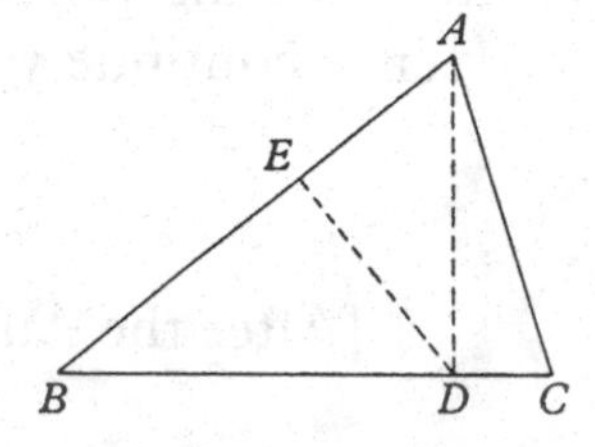

Sit $AB=a$, $AC=b$. $BC=c$ et $BD=x$ eritq; $DC=c-x$. Jam cùm $AB^q-BD^q(aa-xx)=AD^q$; et

$$AC^q-DC^q(bb-cc+2cx-xx)=AD^q:$$

Erit $aa-xx=bb-cc+2cx-xx$, quæ in ordinem redacta fit $\dfrac{aa-bb+cc}{2c}=x$.(122)

(123)Cæterùm ut pateat omnes omnium problematum difficultates per solam linearum proportionalitatem sine adminiculo prop 47 primi Elem:(124) licet non absq; circuitu enodari posse: placuit sequentem hujus solutionem ex abundanti subjungere. A puncto D in latus AB demitte DE normalem, et stantibus jam positis linearum nominibus, erit $AB.BD::BD.BE.$ et $BA-BE\left(a-\dfrac{xx}{a}\right)=EA$.

$$a \quad x \quad x \quad \frac{xx}{a}$$

Nec non $EA.AD::AD.AB$. [adeoq; $EA\times AB(aa-xx)=AD^q$. Et sic ratiocinando circa triangulum ACD invenietur iterum $AD^q=bb-cc+2cx-xx$. Quare $aa-xx=bb-cc+2cx-xx$; unde obtinebitur ut ante $x=\dfrac{aa-bb+cc}{2c}$.

3. Trianguli cujuscunq; ABC, datis area perimetro & uno angulorum A, cætera determinare.

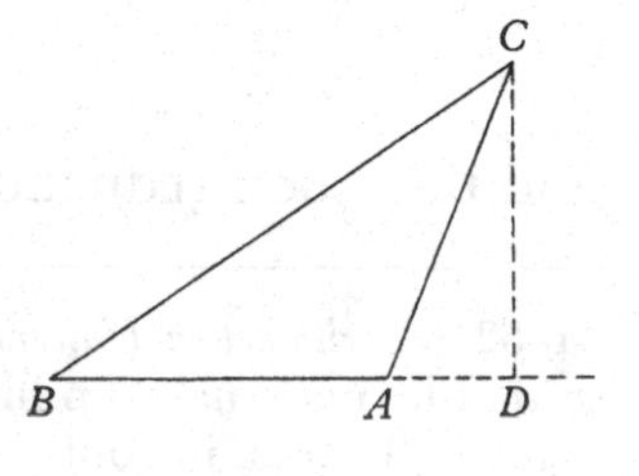

Esto perimeter $=a$ & area $=bb$, & ab ignotorum angulorum alterutro C ad latus oppositum AB demitte perpendiculum CD et propter angulum A datum erit AC and CD in data ratione, puta d ad e. Dic ergo $AC=x$ & erit $CD=\dfrac{ex}{d}$, per quam divide duplam aream et prodibit $\dfrac{2bbd}{ex}=AB$. Adde AD (nempe $\sqrt{AC^q-CD^q}$,

(122) An immediate deduction from Euclid 2, 9/10 or, equivalently, the cosine rule

$$AC^2 = AB^2+BC^2-2AB\times BC\times\cos\hat{B}.$$

The same result another way. Let the semi-perimeter $= a$, area $= b^2$ and $BC = x$: then will $AC+AB = 2a-x$. Now since there is $x^2(BC^2) = AC^2+BC^2$ and $4b^2 = 2AC\times AB$, there will be

$$x^2+4b^2 = AC^2+AB^2+2AC\times AB$$
$$= \text{the square of } (AC+AB) = \text{the square of } 2a-x = 4a^2-4ax+x^2.$$

That is, $x^2+4b^2 = 4a^2-4ax+x^2$ and this on reduction becomes $a-b^2/a = x$.

2. In any triangle ABC given the sides AB, AC and the base BC, which the perpendicular AD dropped from the vertex intersects in D, to find the segments BD and DC.

Let $AB = a$, $AC = b$, $BC = c$ and $BD = x$, and there will be $DC = c-x$. Now since

$$AB^2-BD^2(a^2-x^2) = AD^2 \quad \text{and} \quad AC^2-DC^2(b^2-c^2+2cx-x^2) = AD^2,$$

there will be $a^2-x^2 = b^2-c^2+2cx-x^2$, and this reduced to order becomes $(a^2-b^2+c^2)/2c = x$.(122)

(123)However, to make it evident that every difficulty of every problem may, though not without some deviousness, be unravelled by the proportionality of lines alone without the prop of *Elements* I, 47,(124) it seems appropriate to subjoin an extra solution of this problem. From the point D onto the side AB drop the normal DE and, on retaining the denomination of lines previously supposed, there will be $AB(a):BD(x) = BD(x):BE$ or x^2/a and $BA-BE(a-x^2/a) = EA$. In addition, $EA:AD = AD:AB$ and consequently $EA\times AB(a^2-x^2) = AD^2$. And by reasoning in this way about the triangle ACD it will again be found that $AD^2 = b^2-c^2+2cx-x^2$. Therefore $a^2-x^2 = b^2-c^2+2cx-x^2$, and from this will be obtained as before $x = (a^2-b^2+c^2)/2c$.

3. In any triangle ABC given the area, the perimeter and one of the angles, A, to determine its remaining elements.

Let the perimeter $= a$ and area $= b^2$, and from one or other of the unknown angles, C, to the opposite side AB drop the perpendicular CD, and because of the angle A given AC will be to CD in a given ratio, say, as d to e. Call, then, $AC = x$ and there will be $CD = ex/d$. By this divide twice the area and there will come $2b^2d/ex = AB$. Add AD (that is, $\surd[AC^2-CD^2]$ or $x\surd[d^2-e^2]/d$) and there will

(123) The handwriting of the manuscript suggests strongly that this *idem aliter* was added at the time (early autumn 1670) when Newton augmented the first version of his 'geometrick problems' with those on his pp. 36–9. The first portion of it is squashed in at the foot of p. 30 but its latter half and a following third problem were added (with a referent double asterisk) on an inserted sheet which is now lost. These we have restored from their revised versions which Newton later incorporated in his Lucasian lectures on algebra (ULC. Dd. 9.68: 68–9).

(124) In a Euclidean metric the postulate of similarity is strictly equivalent to the assumption of Pythagoras' theorem.

sive $\frac{x}{d}\sqrt{dd-ee}$) et emerget $BD=\frac{2bbd}{ex}+\frac{x}{d}\sqrt{dd-ee}$: cujus quadrato adde CD^q et orietur $\frac{4b^4dd}{eexx}+xx+\frac{4bb}{e}\sqrt{dd-ee}=BC^q$. Adhæc a perimetro aufer AC & AB, et restabit $a-x-\frac{2bbd}{ex}=BC$, cujus quadratum

$$aa-2ax+xx-\frac{4abbd}{ex}+\frac{4bbd}{e}+\frac{4b^4dd}{eexx}$$

pone æquale quadrato priùs invento & neglectis æquipollentibus erit

$$\frac{4bb}{e}\sqrt{dd-ee}=aa-2ax-\frac{4abbd}{ex}+\frac{4bbd}{e}.$$

Et hæc, assumendo $4af$ pro datis terminis $aa+\frac{4bbd}{e}-\frac{4bb}{e}\sqrt{dd-ee}$ et reducendo, evadit $xx=2fx-\frac{2bbd}{e}$, sive $AC=x=f\pm\sqrt{ff-\frac{2bbd}{e}}$.

Eadem æquatio prodijsset quærendo crus AB, nam crura AB et AC similiter se habent ad omnes conditiones problematis. Quare si AC ponatur

$$f-\sqrt{ff-\frac{2bbd}{e}}$$

erit $AB=f+\sqrt{ff-\frac{2bbd}{e}}$, & vicissim: atq; horum summa $2f$ subducta de perimetro relinquit tertium latus $BC=a-2f$.]

||[34] ||4. Datâ rectâ terminatâ BC a cujus extremitatibus duæ rectæ BA, CA ducuntur in datis angulis ABC, ACB: inveni[r]e AD altitudinem concursûs A supra datam BC.

Sit $BC=a$, et $AD=y$; et cùm angulus ABD detur tabula Sinuum vel Tangentium dabit rationem inter lineas AD et BD quam pono ut d ad e. est ergo $d.e::AD(y).BD$: Quare $BD=\frac{ey}{d}$. Similiter propter datum angulum ACD dabitur ratio inter AD et DC quam pone ut d ad f et erit $DC=\frac{fy}{d}$. At $[C]D+DB=BC$: hoc est $\frac{ey}{d}+\frac{fy}{d}=a$. Quæ reducta fit $y=\frac{ad}{e+f}$.(125)

5. Trapezium $ABCD$ cujus anguli et latera dantur datâ zonâ cingere inter lineas EF, FG, GH, HI et latera trapezij AB, BC, CD, DA pari ubiq; intervalla ab invicem dissitas comprehensâ.(126)

Sit istud intervallum, sive zonæ latitudo $=x$, et ejus quantitas $=aa$. Tum a punctis A, B, C, D ad lineas EF, FG &c demissis perpendicularibus AK, BL,

emerge $BD = 2b^2d/ex + x/d\surd(d^2-e^2)$: to its square add CD^2 and there will arise $4b^4d^2/e^2x^2 + x^2 + 4b^2\surd(d^2-e^2)/e = BC^2$. Again, from the perimeter take away AC and AB and there will remain $a-x-2b^2d/ex = BC$: the square of this,

$$a^2-2ax+x^2-4ab^2d/ex+4b^2d/e+4b^4d^2/e^2x^2,$$

set equal to the square previously found and, when equals are neglected, there will be $4b^2\surd(d^2-e^2)/e = a^2-2ax-4ab^2d/ex+4b^2d/e$. This, on assuming $4af$ for the given terms $a^2+4b^2d/e-4b^2\surd(d^2-e^2)/e$ and reducing, will come out as $x^2 = 2fx-2b^2d/e$, or $AC = x = f \pm \surd(f^2-2b^2d/e)$.

The same equation would have been produced by seeking the leg AB, for the legs AB and AC have a similar relationship to each of the conditions of the problem. Hence if AC is set as $f-\surd(f^2-2b^2d/e)$, then will

$$AB = f+\surd(f^2-2b^2d/e),$$

and conversely so; and their sum $2f$ subtracted from the perimeter will leave the third side $BC = a-2f$.

4. Given the straight line segment BC, from whose extremities the two lines BA, CA are drawn at the given angles $A\hat{B}C$, $A\hat{C}B$, to find AD, the height of their meet A above the given line BC.

Let $BC = a$ and $AD = y$; and, since the angle $A\hat{B}D$ is given, a table of sines or tangents will give the ratio between the lines AD and BD, which I suppose to be as d to e. There is then $d:e = AD(y):BD$ and therefore $BD = ey/d$. Similarly, from the angle $A\hat{C}D$ being given there will be given the ratio between AD and DC: this I set as d to f and there will be $DC = fy/d$. But $CD+DB = BC$, that is, $ey/d+fy/d = a$. After reduction this becomes $y = ad/(e+f)$.(125)

5. To encompass the quadrilateral $ABCD$, whose angles and sides are given, by a given belt comprised within the lines EF, FG, GH, HI and the quadrilateral's sides AB, BC, CD, DA, which are separated everywhere one from the other by the same distance.(126)

Let that distance or the width of the belt $= x$ and the latter's quantity $= a^2$. Then, when the perpendiculars AK, BL, BM, CN, CO, DP, DQ, AI are let fall

(125) In more modern terms $AD = BC/(\cot \hat{B} + \cot \hat{C})$.

(126) Newton's generalization of Schooten's 'fish-pond' problem (*Exercitationum Mathematicarum Libri Quinque* (Leyden, 1657): Liber 1, XLV: 96–8), viz: 'Trapezii *ABCD*...cognita sunt omnia latera.... Jam, si in eo piscina fodienda sit *EFGH*, datæ magnitudinis..., ita ut cingatur ambulacro ubique ejusdem latitudinis, quæritur, quanta sit futura ejus latitudo...?'. There, of course, it was required to set the 'trapezium' *EFGH* within the quadrilateral *ABCD*.

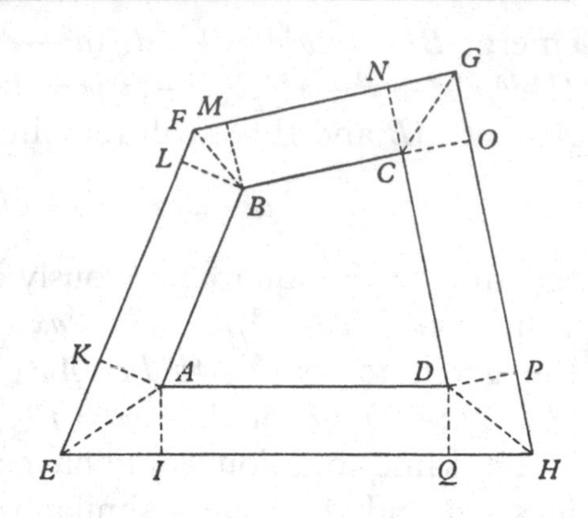

BM, CN, CO, DP, DQ, AI, zona dividetur in quatuor Trapezia IK, LM, NO, PQ, et in quatuor parallelogramma AL, BN, CP, DI, latitudinis x, et ejusdem longitudinis cum lateribus dati trapezij. Sit ergo summa laterum $(AB+BC+CD+DA)=b$, et erit summa parallelogrammorum $=bx$.

Porrò ductis AE, BF, CG, DH; cùm sit $AI=AK$, erit ang: $AEI=$ ang: $AEK=\frac{1}{2}IEK$ sive $=\frac{1}{2}DAB$. Datur ergo ang: AEI et proinde ratio ipsius AI ad IE, quam pone ut d ad e, et erit $IE=\frac{ex}{d}$. duc in $\frac{1}{2}AI$ sive $\frac{1}{2}x$ et fiet area trianguli $AEI=\frac{exx}{2d}$. Sed propter æquales angulos et latera, triangula AEI et AEK sunt æqualia, adeóq; trapezium $IK(=2$ triang: $AEI)=\frac{exx}{d}$. Simili modo ponendo $BL.LF::d.f$; $CN.NG::d.g$; et $DP.PH::d.h$ (nam illæ etiam rationes dantur ex datis angulis B, C, et D), habebitur trapez[ium] $LM=\frac{fxx}{d}$, $NO=\frac{gxx}{d}$, et $PQ=\frac{hxx}{d}$. Quamobrem $\frac{exx}{d}+\frac{fxx}{d}+\frac{gxx}{d}+\frac{hxx}{d}$ sive $\frac{pxx}{d}$ (scribendo p pro $e+f+g+h$) erit ||[35] æquale trapezijs ||quatuor $IK+LM+NO+PQ$. Et proinde $\frac{pxx}{d}+bx$ æquabitur toti zonæ aa. Quam æquationem $\frac{pxx}{d}+bx=aa$ reduco multiplicando per $\frac{d}{p}$ ut fiat $xx+\frac{bd}{p}x=\frac{aad}{p}$, & extrahendo radicem ut habeatur $x=\frac{-bd+\sqrt{bbdd+4aapd}}{2p}$. Zonæ latitudine sic inventâ, facile est ipsam describere.

Cæterùm notandum est quòd æquationis jam inventæ $xx+\frac{bd}{p}x=\frac{aad}{p}$, duæ sunt radices, una affirmativa quæ est latitudo datæ zonæ circumscribendæ (prout in hac quæstione requiritur;) altera verò negativa quæ est latitudo inscribendæ.(127)

6. Angulum rectum EAF datâ rectâ EF subtendere quæ transibit per datum punctum C a lineis rectum angulum comprehendentibus æquidistans.(128)

Quadratum $ABCD$ compleatur et linea EF bisecetur in G. Tum dic CB vel CD esse a, EG vel FG esse b, et CG esse x: eritq; $CE=x-b$, & $CF=x+b$. Dein cùm $CF^q-BC^q=BF^q$, erit $BF=\sqrt{xx+2bx+bb-aa}$. Deniq; propter similia triangula CDE, FBC, est $CE.CD::CF.BF$. sive

$$x-b.a::x+b.\sqrt{xx+2bx+bb-aa}.$$

from the points A, B, C, D to the lines $EF, FG, \ldots$, the belt will be divided into the four quadrilaterals IK, LM, NO, PQ and into the four parallelograms AL, BN, CP, DI of width x and of the same length as the sides of the given quadrilateral. Let then the sum of the sides $(AB+BC+CD+DA) = b$ and the sum of the parallelograms will be $= bx$.

Further, when AE, BF, CG, DH are drawn, since there is $AI = AK$ then will the angle $A\hat{E}I =$ angle $A\hat{E}K = \frac{1}{2}I\hat{E}K$, that is, $= \frac{1}{2}D\hat{A}B$. There is therefore given the angle $A\hat{E}I$ and consequently the ratio of AI to IE. When this is set as d to e, there will be $IE = ex/d$: multiply by $\frac{1}{2}AI$ or $\frac{1}{2}x$ and the area of the triangle AEI will become equal to $ex^2/2d$. But since their angles and sides are equal the triangles AEI and AEK are congruent, and consequently the quadrilateral $IK(= 2\triangle AEI) = ex^2/d$. In a similar manner, by setting

$$BL:LF = d:f, \quad CN:NG = d:g \quad \text{and} \quad DP:PH = d:h$$

(for those ratios also are given by reason of the given angles $\hat{B}$, $\hat{C}$ and $\hat{D}$) it will be found that the quadrilaterals $LM = fx^2/d$, $NO = gx^2/d$ and $PQ = hx^2/d$. In consequence of which $ex^2/d+fx^2/d+gx^2/d+hx^2/d$ or px^2/d (on writing p for $e+f+g+h$) will be equal to the four quadrilaterals $IK+LM+NO+PQ$. Hence $px^2/d+bx$ will equal the whole belt a^2. This equation $px^2/d+bx = a^2$ I reduce by multiplying through by d/p to make it $x^2+(bd/p)\,x = a^2d/p$ and by extracting the root have $x = [-bd+\sqrt{(b^2d^2+4a^2pd)}]/2p$. After the width of the belt has been found in this way it is easy to describe it.

But it should be noted that the equation now found $x^2+(bd/p)\,x = a^2d/p$ has two roots, a positive one which is the width of the given zone to be circumscribed (as the present question requires) and a second, negative one which is the width of that to be inscribed.(127)

6. To subtend the right angle $E\hat{A}F$ by the given straight line EF which shall pass through the point C equidistant from the lines comprising the right angle.(128)

Let the square $ABCD$ be completed and the line EF bisected in G. Then call CB or CD a, EG or FG b and CG x: then will $CE = x-b$ and $CF = x+b$. Next, since $CF^2-BC^2 = BF^2$, there will be $BF = \sqrt{(x^2+2bx+b^2-a^2)}$. Finally, on account of the similar triangles CDE, FBC there is $CE:CD = CF:BF$ or

$$x-b:a = x+b:\sqrt{(x^2+2bx+b^2-a^2)}.$$

(127) As in Schooten's problem (note (126)).

(128) This Apollonian problem was taken by Newton in May 1665 from Descartes (*Geometrie*: 387–8 = *Geometria* ($_2$1659): 82–4) and incorporated by him in his paper 'Concerning Equations when the ratio of their rootes is considered' (I, 3, 3, §4: notes (8) and (9)). The present 'Idem aliter', indeed, essentially repeats the Cartesian reduction, while the opening paragraph represents Newton's improvement in which $b \to 2b$ and $x \to x+b$.

Unde $ax+ab=\overline{x-b}\sqrt{xx+2bx+bb-aa}$. Cujus æquationis utrâq; parte quadratâ et in ordinem redactâ, prodit

$$x^4 = \begin{matrix} 2aa \\ +2bb \end{matrix} xx \begin{matrix} +2aabb \\ -b^4. \end{matrix}$$

Et extractâ radice sicut fit in æquationibus quadraticis, prodit

$$xx = aa+bb+\sqrt{a^4+4aabb};$$

adeóq; $x=\sqrt{aa+bb+a\sqrt{aa+4bb}}$.(129) CG sic inventa dat CE vel CF quæ determinando puncta E vel F problemati satisfaciunt.

Idem aliter.(130) Sit $CE=x$, $CD=a$, et $EF=b$. et erit $CF=x+b$ et

$$BF=\sqrt{xx+2bx+bb-aa}.$$

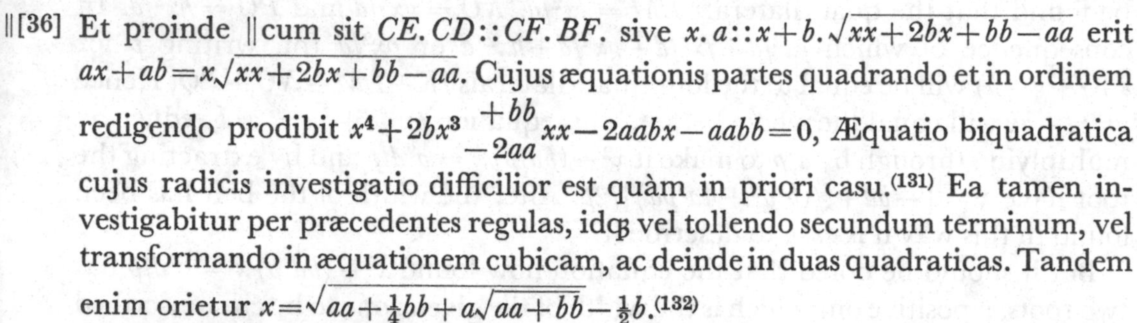

‖[36] Et proinde ‖cum sit $CE.CD::CF.BF$. sive $x.a::x+b.\sqrt{xx+2bx+bb-aa}$ erit $ax+ab=x\sqrt{xx+2bx+bb-aa}$. Cujus æquationis partes quadrando et in ordinem redigendo prodibit $x^4+2bx^3 \begin{matrix} +bb \\ -2aa \end{matrix} xx-2aabx-aabb=0$, Æquatio biquadratica cujus radicis investigatio difficilior est quàm in priori casu.(131) Ea tamen investigabitur per præcedentes regulas, idq; vel tollendo secundum terminum, vel transformando in æquationem cubicam, ac deinde in duas quadraticas. Tandem enim orietur $x=\sqrt{aa+\frac{1}{4}bb+a\sqrt{aa+bb}}-\frac{1}{2}b$.(132)

(133)Ex his occasionem nactus sum tradendi regulam qua terminos ad ineundum calculum maximè accommodatos prima fronte possis utplurimùm eligere. Scilicet cùm duorum terminorum talis accidit affinitas sive relationis ad cæteros quæstionis terminos adæquata similitudo,(134) ut oporteret æquationes per omnia similes ex utrovis adhibito produci, aut ambos si simul adhiberentur, easdem in æquatione finali dimensiones et eandem omninò formam (signis forte + et − exceptis) habituros esse; Id quòd facilè prospicitur: Tunc neutrum adhibere convenit sed eorum vice tertium quemvis eligere qui similem utriq; relationem gerit, puta $\frac{1}{2}$ summā, vel $\frac{1}{2}$ differentiam, vel medium proportionale

(129) More precisely '$x=\pm\sqrt{aa+bb\pm a\sqrt{aa+4bb}}$'. When $b > a\sqrt{2}$, there will be four real solutions to the problem, but only two when $b < a\sqrt{2}$, compare Newton's diagram in I, 3, 3, §4. The limit case $b=a\sqrt{2}$ yields three solutions ($x=\pm a\sqrt{6}, 0$).

(130) Newton has cancelled the weaker conjunction 'Sic proinde'.

(131) This is an overstatement. The quartic factorizes immediately as

$$(x^2+bx-a^2+a\sqrt{[a^2+b^2]})\,(x^2+bx-a^2-a\sqrt{[a^2+b^2]}) = 0,$$

while increasing its root by $\frac{1}{2}b$ transforms it into the simplified equation (much as in the pre-

Hence $ax+ab = (x-b)\surd(x^2+2bx+b^2-a^2)$. After either side of this equation has been squared and it has been reduced to order, there will come

$$x^4 = 2(a^2+b^2)\,x^2+2a^2b^2-b^4,$$

and when the root is extracted as is done in quadratic equations there comes $x^2 = a^2+b^2+\surd(a^4+4a^2b^2)$ and consequently $x = \surd[a^2+b^2+a\surd(a^2+4b^2)]$.(129) When CG has been found in this way it gives CE or CF, and these by determining the points E or F satisfy the problem.

The same done otherwise. Let $CE = x$, $CD = a$ and $EF = b$: and there will be $CF = x+b$ and $BF = \surd(x^2+2bx+b^2-a^2)$. In consequence, since there is $CE:CD = CF:BF$ or $x:a = x+b:\surd(x^2+2bx+b^2-a^2)$, there will be

$$ax+ab = x\surd(x^2+2bx+b^2-a^2).$$

By squaring the parts of this equation and reducing them to order there will come out $x^4+2bx^3+(b^2-2a^2)\,x^2-2a^2bx-a^2b^2 = 0$, a quartic equation the investigation of whose root is more difficult than in the former case.(131) It may, however, be investigated by the preceding rules, either, that is, by removing the second term or by transforming it into a cubic and then into two quadratics: for there will at length arise either way $x = \surd[a^2+\frac{1}{4}b^2+a\surd(a^2+b^2)]-\frac{1}{2}b$.(132)

(133)In this regard I take the opportunity to deliver a rule by which you may for the most part choose at first sight the terms most suitable to start the calculation from. Precisely, when there chances to be such an affinity or identity of relationship between one or other of two terms and the other terms of the question that, should either be employed, the equations produced would of necessity be wholly alike or, if they were employed together, both would have the same dimensions and entirely the same form (except perhaps for the signs + and −) in the final equation—which is easy to foresee—then it is convenient to employ neither but to choose in their stead some third term which bears a similar relationship to each, their semi-sum, say, or their semi-difference or maybe their

---

ceding paragraph) $x^4-(2a^2+\frac{1}{2}b^2)x^2-(\frac{1}{4}a^2b^2-\frac{1}{16}b^4) = 0$. Either way, the reduction to a quadratic is directly attained.

(132) More precisely, '$x = \pm\sqrt{aa+\frac{1}{4}bb\pm a\sqrt{aa+bb}}-\frac{1}{2}b$', which yields two or four real solutions according as b is less or greater than $2a\surd 2$ (much as in note (129)).

(133) The final portion of Newton's *Observationes* which follows was composed in the early autumn of 1670 in augmentation (see note (110)) of the appended 'geometrick problems' which he sent to Collins on 11 July. A few slight corrections to the manuscript, which we judge by their handwriting to have been made *c.* 1677 when Newton incorporated the present text in his Lucasian lectures on algebra (ULC. Dd. 9.68), are here silently omitted.

(134) Newton has cancelled a first continuation, 'ut ambobus relationem ad cæteros quæstionis terminos planè similem habentibus, nulla potest excogitari ratio cur unus pro alia debe[a]t eligi' (that, when both plainly bear a similar relationship to the remaining terms of the question, no reason can be conceived why one should be chosen in preference to another).

forsan, aut quamvis aliam quantitatem utriꝗ indifferenter et sine compare relatam. Sic in præcedenti problemate cum vidi lineam *EF* pariter ad utrámꝗ *AB* et *AD* referri (ut patet si ducas itidem in angulo *BAH*,) atꝗ adeò nullâ ratione suaderi posse cur *ED* potiùs quàm *BF* vel *AE* potiùs quàm *AF* vel *CE* potiùs quàm *CF* pro quærendâ quantitate adhiberetur: vice punctorum *C* et *F* unde hæc ambiguitas proficiscitur sumpsi intermedium *G* quod parem relationem ad utrámꝗ linearum *AB* et *AD* observat. Deinde ab hoc *G* non demisi perpendiculum ad *AF* pro quærendâ quantitate, quia potui eâdem ratione demisisse ad *AD*. Et eapropter in neutrum *CB* vel *CD* demisi, sed institui *CG* quærendum ||[37] esse ||quod nullum admittit compar. Et sic æquationem obtinui sine terminis imparibus.(135)

Potui etiam (considerando punctum *G* jacere in periferiâ circuli centro *A* radio *EG* descripti) demisisse *GK* perpendiculum in diagonalem *AC* et quæsivisse *AK* vel *CK*, quæ indifferenter etiam respectant lineas *AB* et *AD*: Atꝗ ita in æquationem quadraticam $yy = \frac{1}{2}ey + \frac{1}{2}bb$ incidissem, posito $AK = y$. $AC = e$. et $EG = b$.(136) Unde cùm habeam *AK*, erigo perpendiculum *KG* præfato circulo occurrens in *G*, per quod *CF* transibit.

Ad hunc modum in prob: 2 præcedente cum duo latera trianguli similiter respectant segmenta basis et e contra; si ipsorum vice $\frac{1}{2}$ summas et $\frac{1}{2}$ differentias adhibuissem conclusio evasisset simplicior. Nempe sit laterum $\frac{1}{2}$ summa a, $\frac{1}{2}$ diff: b; segmentorum basis $\frac{1}{2}$ summa c, et $\frac{1}{2}$ diff: x: et emerget $\frac{ab}{c} = x$. Et ob hanc rationem potius adhibui differentiam laterum in Prob 3, quàm eorum alterutrum quærendum esse [supposui]. Et sic in prob 1 si latera trianguli desiderata fuissent, promptiùs invenissem quærendo summā ac differentiam eorum. Sed observationis hujus utilitas e sequenti problemate magis elucescet.

7. Rectam *DC* datæ longitudinis in datam curvam *DAC* sic inscribere, ut per punctum *G* positione datum transeat.

Sit *AF* recta quævis ad quam curva simplicissimè refertur et a punctis *D*, *G*, et *C* ad hanc demitte normales *DH*, *GE*, et *CB*. Jam ad determinandam posi-

(135) In analytical terms, Newton's argument is that where the solution of some problem is reduced to that of the quadratic equation $x^2+ax+b = 0$ (or $x = -\frac{1}{2}a \pm \frac{1}{2}\sqrt{[a^2-4b]}$) it is a useful simplification to increase the variable x by the semi-sum $-\frac{1}{2}a$, for then the equation reduces to the form $x^2 = \frac{1}{4}a^2-b$ 'sine terminis imparibus'. In so far as he attempts to extend the argument to the general equation of degree n he is perhaps over-optimistic. Any n-degree algebraic equation may indeed be expressed as the product of real quadratic factors

$$x^2+a_i x+b_i = 0, \quad i = 1, 2, \ldots, [\tfrac{1}{2}n]$$

(with an additional linear factor also if n be odd), but the reduction of any one of these, say $x^2+a_k x+b_k = 0$, by the transform $x \to x-\frac{1}{2}a_k x$ will not usually simplify any of the others and in affording that hope his present example is perhaps not wisely chosen.

mean proportional or any other quantity related indifferently to each and without a fellow. Thus, in the preceding problem when I saw that the line EF was equally referred to both AB and AD (as is evident should you draw it in like manner in the angle $B\hat{A}H$) and consequently that no reason could be urged why ED should be employed for the quantity to be sought rather than BF, or AE rather than AF, or CE rather than CF, instead of the points C and F which are the source of this ambiguity I took their intermediate point G, which preserves an equal relationship to each of the lines AB and AD. Next, I did not drop a perpendicular from this point G to AF for the quantity to be sought since I might have dropped it for the same reason to AD. For that reason, likewise, I dropped it on neither CB or CD, but fixed on CG, which admits no fellow, as the quantity to be sought. And in this way I obtained an equation lacking odd terms.(135)

I might also (by considering the point G to lie in the circumference of a circle described on centre A and with radius EG) have dropped GK perpendicular to the diagonal AC and then sought AK or CK, which also have an indifferent regard to the lines AB and AD: and in this way I should have struck upon the quadratic equation $y^2 = \frac{1}{2}ey + \frac{1}{2}b^2$ on setting $AK = y$, $AC = e$ and $EG = b$.(136) When in consequence I have AK, I erect the perpendicular KG meeting the aforesaid circle in G, and through this point CF will pass.

In much the same way, since in Problem 2 preceding the two sides of the triangle have a similar regard to the segments of the base, and conversely so, if in their stead I had employed their semi-sums and semi-differences, the conclusion would have come out in simpler form. Precisely, let the semi-sum of the sides be a and their semi-difference b, the semi-sum of the base segments c and their semi-difference x, and there will emerge $ab/c = x$. And for this reason I would have done better to employ the difference of the sides in Problem 3 than suppose that one or other of them was to be sought. And thus in Problem 1 if the sides of the triangles had been desired, I would have found them more readily by seeking their sum and difference. But the usefulness of this observation will be more transparent from the following problem.

7. To inscribe the straight line DC of given length within the given curve DAC such that it passes through the point G given in position.

Let AF be the straight line to which the curve is most simply referred, and to this from the points D, G and C drop the normals DH, GE and CB. Now to determine the position of the line DC, the finding of the points D or C may perhaps

(136) In this case, on comparing with the terminology of Newton's first resolution above we find $AC^2 = 2CD^2$ or $e^2 = 2a^2$, and $CG^2 = CA^2+AG^2-2CA\times AK$ or $x^2 = e^2+b^2-2ey$ and the quartic $x^4 = (2a^2+2b^2)x^2+b^2(2a^2-b^2)$ reduces to the present quadratic on making appropriate substitution.

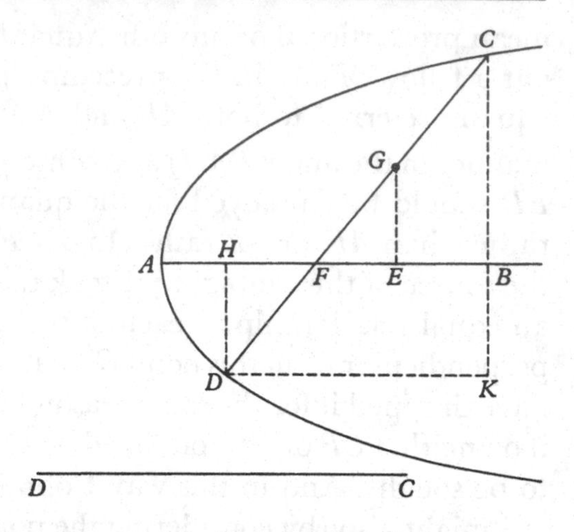

tionem rectæ DC, puncti D aut C inventio requiri(137) forte videatur: sed cum hæc sint paria, et ad invicem tam affinia ut ad alterutrum determinandum operatio similis evasura sit; sive quæram CG, CB, aut AB, sive comparia DG, DH, aut AH: eapropter de tertio aliquo puncto prospicio quod utrumq; D et C similiter respectet, et unà determinet. Et hujusmodi video esse punctum F.

Jam sit $AE=a$. $EG=b$. $DC=c$. $EF=z$. et præterea cùm relatio inter AB et BC habetur in æquatione quam suppono pro curvâ determinandâ datam esse, sit $AB=x$, et $BC=y$, et erit $FB=x-a[+]z$.

‖[38] Et propter $GE.EF::CB.FB$, erit iterum ‖$FB=\frac{yz}{b}$. Ergo $x-a+z=\frac{yz}{b}$.

His ita præparatis tolle x per æquationem quæ curvam designat. Quemadmodum si curva sit Parabola per æquationē $rx=yy$ designata: scribendo $\frac{yy}{r}$ pro x, orietur $\frac{yy}{r}-a+z=\frac{yz}{b}$. Et extractâ radice $y=\frac{rz}{2b}\pm\sqrt{\frac{rrzz}{4bb}+ar-rz}$. Unde patet $\sqrt{\frac{rrzz}{bb}+4ar-4rz}$ esse differentiam gemini valoris y, id est linearum $+BC$ et $-DH$; adeóq; (demisso DK in CB normali) valere CK. Est autem $FG.CE::DC.CK$. hoc est $\sqrt{bb+zz}.b::c.\sqrt{\frac{rrzz}{bb}+4ar-4rz}$. ducendóq; quadrata extremorum et mediorum in invicem, et facta ordinando, orietur

$$z^4=\frac{4bbrz^3\begin{matrix}-4abbr\\-bbrr\end{matrix}zz+4b^4rz\begin{matrix}-4ab^4r\\+b^4cc\end{matrix}}{rr}$$ æquatio quatuor tantum dimensionum, quæ octo dimensionū fuisset si quæsivissem CG vel CB aut AB.(138)

Ad hunc modum in Ellipsi et Hyperbola hoc idem perficitur, sed in curvis magis compositis ubi radix y ut supra non potest extrahi, convenit ut æquationis istius termini per progressionem multiplicentur qualem habes descriptam pag 77:(139) proportione radicum sive BC et HD hic existente y ad $y-\frac{bc}{\sqrt{bb+zz}}$, et inde progressione $0.\ y.\ 2y-\frac{bc}{\sqrt{bb+zz}}$. &c. Et sic in alijs hujusmodi problematibus procedi potest.(140) Sed hæc in transitu.

(137) 'necessaria' (necessary) is cancelled.

seem requisite:[137] but since these are mates and so related to each other that to determine either the operation would prove to be similar whether I should seek CG, CB or AB or their mates DG, DH or AH, I look out in consequence for some third point which shall bear a similar respect to both D and C and determine them together as one. And I see that F is a point of this kind.

Now let $AE = a$, $EG = b$, $DC = c$, $EF = z$ and furthermore, since the relationship between AB and BC is contained in the equation which I suppose given for determining the curve, let $AB = x$ and $BC = y$. There will then be

$$FB = x-a+z$$

and again, since $GE:EF = CB:FB$, $FB = yz/b$: hence $x-a+z = yz/b$.

After this preparation remove x by the equation which designates the curve. For instance, if the curve be a parabola designated by the equation $rx = y^2$, on writing y^2/r for x there will arise $y^2/r-a+z = yz/b$ and, when the root is extracted, $y = rz/2b \pm \surd(r^2z^2/4b^2+ar-rz)$. It is hence evident that

$$\surd(r^2z^2/b^2+4ar-4rz)$$

is the difference of the twin value of y, that is, of the lines $+BC$ and $-DH$, and so (on letting fall the normal DK onto CB) that it is the value of CK. But there is $FG:CE = DC:CK$, that is, $\surd(b^2+z^2):b = c:\surd(r^2z^2/b^2+4ar-4rz)$, and by multiplying the squares of the means and extremes into each other and ordering the product, there will arise $z^4 = (4b^2rz^3-(4ab^2r+b^2r^2)\,z^2+4b^4rz-4ab^4r+b^4c^2)/r^2$, an equation of only four dimensions, which would have been of eight dimensions if I had sought CG or CB or AB.[138]

In much this way this same procedure is enacted in the ellipse and hyperbola, but in more compound curves where the root y cannot be extracted as above, it is convenient to multiply the terms of this equation by a progression of the kind which you have had described on page 77:[139] here the proportion of the roots, namely, of BC and HD, is as y to $y-bc/\surd(b^2+z^2)$ and therefore the progression comes out $0.\ y.\ 2y-bc/\surd(b^2+z^2).\ \ldots$. And a similar procedure may be applied in other problems of this kind.[140] But this in passing.

(138) For, since $x-a = y^2/r-a = (z/b)\,(y-b)$ or $z = b(y^2-ar)/r(y-b)$, the quartic in z will be converted into an octic in y, and each of the 8 roots of that octic will determine uniquely correspondent values of x by $x = y^2/r$.

(139) That is, of Kinckhuysen's preceding tract (*Algebra* (§1): 77–80).

(140) If we suppose the given curve to be defined by $f(x, y) = 0$ and we seek its meets C, D with an arbitrary line $x-a = (z/b)\,(y-b)$ through $G(a, b)$, determined by its meet $F(FE = z)$ with AB, such that $CD = c$, Newton's technique for resolving the present problem is to equate $CK = (GE/GF) \times CD = bc/\surd[b^2+z^2]$ with the difference of a pair of roots of $f(a+(z/b)\,(y-b), y) = 0$. The method works well when $f(x, y) = 0$ is a conic for then the resulting equation in y is of the form $Ay^2+By+C = 0$ and we derive immediately

$$y_1-y_2 = \surd[(y_1+y_2)^2-4y_1y_2]$$

De electione verò terminorum ad ineundum calculum venit insuper notandum, quòd cùm duæ lineæ se indifferenter offerunt quarum altera tantùm (cæteris paribus) sit positione data, conclusio solet esse simplicior cùm illa adhibetur præ alterâ positione non determinatâ. Quemadmodum in hoc problemate punctum F æquè determinari potuisset quærendo lineam GF atq; lineam EF vel AF, nisi quod GF sit incertæ positionis. Adhibui itaq; EF et prodijt æquatio 4 tantùm dimensionum, quæ ad octo dimensiones sine terminis imparibus ascendisset si quæsivissem GF.(141) Quod si quæsivissem CB in æquationem totidem dimensionum cum terminis imparibus incidissem longè omnium perplexissimam, qualis nempe proditura est si valor z per æquationem $\frac{yy}{r}-a+z=\frac{yz}{b}$ inventum $\left(\text{id est } \frac{byy-bar}{ry-rb}\right)$ vice ejus in hac

$$z^4 = \frac{4bbrz^3 \begin{matrix} -4abbr \\ -bbrr \end{matrix} zz \ \&c}{rr}$$

||[39] scribatur. Quæ hic de lineis EF, GF, et CB animadversa sunt ||ad lineas etiam AK, CG et CE in Prob 6 possunt applicari.

Deniq; de hoc problemate notari potest quod puncto G extra curvam sito, si ponitur DC et inde CK, sive $\sqrt{\frac{rrzz}{bb}+4ar-4rz}=0$, habebo æquationem $zz=\frac{4bbz}{r}-\frac{4bba}{r}$, sive $z=\frac{2bb}{r}\pm\frac{2b}{r}\sqrt{ab-ar}$ pro determinando puncto F per quod GF in isto casu duci debet. Atqui DC evanescente portio curvæ CAD simul evanescit et proinde GF producta tunc non resecabit portionem a curva sed ipsam tantùm tanget. Patet itaq; modus quo recta tangens curvam possit a dato puncto duci. Sed minus expatiari videor præsertim cùm praxis ducendi tangentes ab alijs methodis fœliciùs depromitur.(142)

since $y_1+y_2 = -B/A$ and $y_1y_2 = C/A$. (In Newton's instance of the parabola $rx = y^2$, for example, $r[a+(z/b)(y-b)] = y^2$, so that $A = 1$, $B = -rz/b$ and $C = r(z-a)$.) For curves of higher degree finding the difference of two roots of the resultant equation in y is no longer simple: indeed, the problem is no longer uniquely defined since a straight line through G will meet an algebraic curve of degree n in up to n real points and we may make

$$\binom{n}{2} = \tfrac{1}{2}n(n-1)$$

choices of pairs of these intersections. (Compare the difficulties Newton ran up against in attempting to find the 'girth' of a curve in I, 2, 3, §1.) The alternative approach which Newton sketches in sequel is more widely applicable but somewhat cumbrous: since the ratio of the roots $BC = y_1$ and $-HD = y_2$ is

$$BC:BC-CK \text{ (or } CD\times[GE/GF]) = y:(y-bc/\sqrt{[b^2+z^2]}),$$

a pair of roots of $f(a+(z/b)(y-b), y) = 0$ must bear this ratio one to the other, and Newton applies his (and Kinckhuysen's) method for deriving a condition that this be so. Elimination

But, concerning the choice of terms to begin the calculation from, it occurs to me to note that when two lines proffer themselves indifferently, one of which alone (other things being equal) is given in position, the conclusion is usually simpler when that one is employed in preference to the other not determined in position. In this problem, for instance, the point F might have been determined by seeking the line GF equally as well as the line EF or AF, except that GF is of uncertain position. I therefore employed EF and there arose an equation of only 4 dimensions, which would have risen to eight dimensions without odd terms if I had sought GF.(141) But if I had sought CB, I would have fallen on an equation of the same number of dimensions with odd terms, by far the most intricate of all—of the sort, precisely, which would be forthcoming if the value of z found by the equation $y^2/r-a+z=yz/b$ (that is, $(by^2-bar)/[r(y-b)]$) were to be written in its place in this equation $z^4=[4b^2rz^3-(4ab^2r+b^2r^2)\,z^2\ldots]/r^2$. The remarks made here concerning the lines EF, GF and CB may also be applied to the lines AK, CG and CE in Problem 6.

It may finally be noted concerning this problem that when the point G is situated outside the curve, if there is set DC and consequently CK or

$$\surd(r^2z^2/b^2+4ar-4rz)=0,$$

I will have the equation $z^2-(4b^2z/r)-(4b^2a/r)$, or

$$z=(2b^2/r)\pm(2b/r)\surd(ab-ar),$$

by which to determine the point F through which GF ought in that case to be drawn. However, as DC vanishes the curve portion CAD vanishes with it and in consequence GF produced will not cut off a section of the curve but will merely be tangent to it. A way of drawing a tangent to a curve from a given point is therefore manifest. But I think I may be less expansive on this point especially when the practice of drawing tangents is more happily derived by other methods.(142)

---

of y between the two equations (by 'Reg 1' on Newton's p. 21 above) yields finally a resultant in z alone whose roots are values of FE which satisfy the problem. (In Newton's parabolic example we may derive the resultant quartic in z by multiplying the terms of

$$y^2-(rz/b)y+r(z-a)=0$$

by 0, y and $2y-bc/\surd[b^2+z^2]$ respectively and then eliminating y.)

(141) Since $GF=\surd[b^2+z^2]$.

(142) In general, when, in the terminology of note (140), $CD=c$ vanishes the successive coefficients of powers of y in $f(a+(z/b)(y-b),y)=0$ are multiplied by the Huddenian factors 0, y, $2y$, $3y$ and so on (Kinckhuysen's *Algebra* (§1): 78–80), so that the resulting equation is equivalent to its derivative with respect to y, that is, to $(z/b)f_x+f_y=0$, whence

$$b/z=-f_x/f_y=dy/dx$$

and the line CG is indeed tangent at C. Compare **2**, 1, §2: Problem 3 above.

8. Datum angulum per datum numerum multiplicare vel dividere.

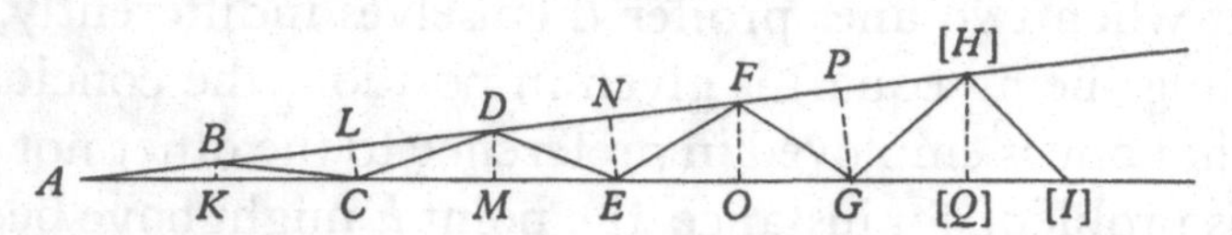

In angulo quovis FAG inscribe lineas AB, BC, CD, DE &c ejusdem cujusvis longitudinis, et erunt triangula ABC, BCD, CDE, DEF &c Isoscelia: Adeóq; per 32[,] 1 Elem erit ang CBD=ang $A+ACB=2$ ang A et

ang DCE=ang $A+ADC=3$ ang A

et ang EDF=ang $A+AED=4$ ang A et ang $FEG=5$ [ang] A et sic deinceps. Positis jam AB, BC, CD &c radijs ejusdem circuli, perpendicula BK in AC, CL in BD, DM in CE &c erunt sinus istorum angulorum et AK, BL, CM, DN &c sinus complementorum ad rectum. Vel posito AB diametro, illæ AK, BL, CM &c erunt chordæ. Sit ergo $AB=2r$, et $AK=x$, dein sic operabere.

$$\begin{matrix} AB.AK::AC.AL. \\ 2r.x \;::\; 2x.\dfrac{xx}{r}. \end{matrix} \quad \text{Et} \left.\begin{matrix} AL-AB \\ \dfrac{xx}{r}-2r \end{matrix}\right\} = BL, \text{ Duplicatio.}$$

$$\begin{matrix} AB.AK::AD(2AL-AB).AM. \\ 2r.x \;::\; \dfrac{2xx}{r}-2r.\dfrac{x^3}{rr}-x. \end{matrix} \quad \text{Et} \left.\begin{matrix} AM-AC \\ \dfrac{x^3}{rr}-3x \end{matrix}\right\} = CM, \text{ Triplicatio.}$$

$$\begin{matrix} AB.AK::AE(2AM-AC).AN. \\ 2r.x \;::\; \dfrac{2x^3}{rr}-4x.\dfrac{x^4}{r^3}-\dfrac{2xx}{r}. \end{matrix} \quad \text{Et} \left.\begin{matrix} AN-AD \\ \dfrac{x^4}{r^3}-\dfrac{4xx}{r}+2r \end{matrix}\right\} = DN, \text{ Quadruplicatio.}$$

$$\begin{matrix} AB.AK::AF(2AN-AD).AO. \\ 2r.x \;::\; \dfrac{2x^4}{r^3}-\dfrac{6xx}{r}+2r.\dfrac{x^5}{r^4}-\dfrac{3x^3}{rr}+x. \end{matrix} \quad \text{Et} \left.\begin{matrix} AO-AE \\ \dfrac{x^5}{r^4}-\dfrac{5x^3}{rr}+5x \end{matrix}\right\} = EO, \text{ Quintuplicatio.}$$

Et sic præterea. Quod si velis angulum in aliquot partes dividere; pone q pro BL, CM, DN &c [et] habebis $xx-2rr=qr$ ad bisectionem, $x^3-3rrx=qrr$ ad trisectionē, $x^4-4rrxx+2r^4=qr^3$ ad quadrisectionem &c.(143)

(143) Newton here borrows from his earlier researches into angular sections (I, **3**, 2, §2.1–6, especially 1 and 6). If in his present scheme we set $A = B_0$, $B = B_1$, $C = B_2$, $D = B_3$, ... and $K = K_1$, $L = K_2$, $M = K_3$, $N = K_4$, ..., then, where $\hat{A} = \frac{1}{2}\theta$ we may set

$$B_{n-1}K_n = 2r\cos\tfrac{1}{2}n\theta \equiv CS(n\theta),$$

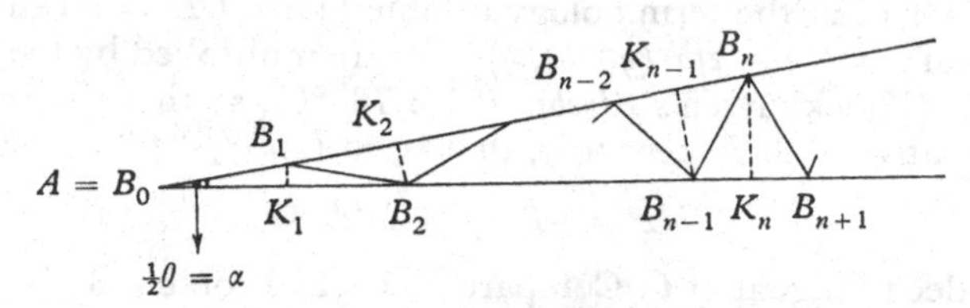

8. To multiply or divide a given angle by a given number.

In any angle $F\hat{A}G$ inscribe the lines AB, BC, CD, DE, ... of the same arbitrary length, and the triangles ABC, BCD, CDE, DEF, ... will then be isosceles. Consequently, by *Elements*, I, 32 there will be $C\hat{B}D = \hat{A}+A\hat{C}B = 2\hat{A}$,

$$D\hat{C}E = \hat{A}+A\hat{D}C = 3\hat{A},\quad E\hat{D}F = \hat{A}+A\hat{E}D = 4\hat{A},\quad F\hat{E}G = 5\hat{A},$$

and so on. If now AB, BC, CD, ... are supposed to be the radii of the same circle, the perpendiculars BK to AC, CL to BD, DM to CE, ... will be the sines of those angles and AK, BL, CM, DN, ... their cosines. Or, on setting AB as the diameter, those lines AK, BL, CM, ... will be their chords. Let then $AB = 2r$ and $AK = x$, and then proceed thus:

$$\left.\begin{array}{l} :AK = AC:AL \\ :x \quad = 2x:x^2/r \end{array}\right. \text{And} \left.\begin{array}{l} AL-AB \\ x^2/r-2r \end{array}\right\} = BL.\ \text{Duplication.}$$

$$\left.\begin{array}{l} :AK = AD(2AL-AB):AM \\ :x \quad = 2x^2/r-2r \quad :x^3/r^2-x \end{array}\right. \text{And} \left.\begin{array}{l} AM-AC \\ x^3/r^2-3x \end{array}\right\} = CM.\ \text{Triplication.}$$

$$\left.\begin{array}{l} :AK = AE(2AM-AC):AN \\ :x \quad = 2x^3/r^2-4x \quad :x^4/r^3-2x^2/r \end{array}\right. \text{And} \left.\begin{array}{l} AN-AD \\ x^4/r^3-4x^2/r+2r \end{array}\right\} = DN.\ \text{Quadruplication.}$$

$$\left.\begin{array}{l} :AK = AF(2AN-AD)\ :AO. \\ :x \quad = 2x^4/r^3-6x^2/r+2r:x^5/r^4-3x^3/r^2+x. \end{array}\right. \text{And} \left.\begin{array}{l} AO-AE \\ x^5/r^4-5x^3/r^2+5x \end{array}\right\} = EO.\ \text{Quintuplication.}$$

And so forth. But should you wish to divide the angle into some number of parts, set q for BL, CM, DN, and so on, and you will have $x^2-2r^2 = qr$ in the case of bisection, $x^3-3r^2x = qr^2$ in that of trisection, $x^4-4r^2x^2+2r^4 = qr^3$ in that of quadrisection, and so on.(143)

---

say, and then write the trigonometrical equivalent of Newton's argument as follows: since

$$B_0B_n:B_0K_n = B_0B_1:B_0K_1 = 2r:x$$

where $B_0B_n = 2B_0K_{n-1}-B_0B_{n-2}$ (or $B_{n-2}K_{n-1} = K_{n-1}B_n$), then

$$B_{n-1}K_n = (x/r)B_0B_n-B_0B_{n-1}$$
$$= 2(x/2r)B_{n-2}K_{n-1}-B_{n-3}K_{n-2},$$

that is, $2r\cos\frac{1}{2}n\theta = 2\cos\frac{1}{2}\theta\times 2r\cos\frac{1}{2}(n-1)\theta-2r\cos\frac{1}{2}(n-2)\theta$,

or $$CS(n\theta) = 2(CS(\theta)/CS(0))\times CS((n-1)\theta)-CS((n-2)\theta).$$

On the whole, this is a surprising choice on Newton's part of an illustrative example for Kinckhuysen's text, but perhaps he intended to show how the Cartesian trisection cubic (§1: note (105)) fits into a general scheme of angular sections.

APPENDIX

NEWTON'S EXTRACTION OF THE CUBE ROOTS OF BINOMIAL SURDS: CANCELLED EXAMPLES FROM PAGE 11 OF HIS KINCKHUYSEN 'OBSERVATIONS'.(1)

Exempl 1. Ex $\surd 242+\surd 243$ $\surd$cub est $\surd 2+\surd 3$.

Præparatio.	Operis prosecutio.			Explicatio ejus.	
$\surd 242+\surd 243$, sive	$11\surd 2+9\surd 3$,			$A\surd B+C\surd D$.	
Quadr: 242. 243.	Limes $\frac{11}{8}$. nempe			$\frac{FA}{4B}$.	
Diff: 1.	Divisor 1.)11(11 × 1 =			$*)[A]\ (*\times F=$	
$\surd$cub: Diff $1=E\surd 3:F$.	$1\surd 2$.	2.	2	$M.\ MM.$	MM
Est ergo $E=1$, F=1.	Adde	1.	+	$+EF.$	+
Et $EF=1$.	fit	3.	9.	$NN.$	$3NN.$

Quare prodit $\surd 2+\surd 3$, utpote $\dfrac{M+N}{\surd 3:F}$.

Hic adhibui minus nomen $11\surd 2$ pro divisoribus eliciendis; sin majus $9\surd 3$ adhibuissem, hæc fuisset operationis forma.

$9\surd 3+11\surd 2$.			$A\surd B+C\surd D$.	
Limites, 3 & $\frac{3}{4}$. nempe			$\frac{FA}{B}$ & $\frac{FA}{4B}$.	
Divisor 1.)[9](9 × 1 =			$*)[A]\ (*\times F=$	
$1\surd 3$.	3.	3	$M.\ MM.$	MM
Subduc	1.	+	$-EF.$	+
fit	2.	6.	$NN.$	$3NN.$

Prodit itaq; $\surd 3+\surd 2$ utpote $\surd 3:F)M+N$.

(1) See §2: notes (29) and (31). We reproduce this variant lay-out of Newton's procedure for extracting the cube root of a surd binomial because of its intrinsic interest and to illustrate yet again the thoroughness with which he approached the task of emending Kinckhuysen's *Algebra*.

Exemplū 2. Ex $3\frac{1}{27}-10\surd\frac{1}{3}\surd$cub: est $\frac{1}{3}-\surd 3$.

Præparatio.	Operatio.	
$3\frac{1}{27}-10\surd\frac{1}{3}$.		
Duc in 27, fit $82-90\surd 3$.		
Divide ꝑ 2, fit $41-45\surd 3$.	$-45\surd 3+41$.	
Quadrata, 1681. 6075.	[Limites] 30. $7\frac{1}{2}$	
Diff. 4394.	[Divisor] 3)45(15	$15\times 2=30$.
$\surd$cub: diff. $13\surd 3:2$.	$3\surd 3$. 27.	27
Est ergo $E=13$, $F=2$,	26.	$+$ $=30$.
Et $EF=26$.	1.	3.

Prodit $\dfrac{-3\surd 3+1}{\surd 3:2}$ radix cubica ex $-45\surd 3+41$.

Jam quia duxeram Binomium in 27 ac diviseram per 2, radicem inventam e contra multiplico per $\surd 3:2$ dividoq̧ per $\surd 3:27$, sive per 3, et oritur $\frac{1}{3}-\surd 3$ radix quæsita.

Potuit etiam operatio a minori nomine 41 sic institui.

$41-45\surd 3$, ubi B valet 1.

[Limes] $10\frac{1}{4}$.	
[Divisor] 1)41(41	$41\times 2=82$.
1. 1.	1
26.	$+$ $=82$.
27.	81.

Prodit $\dfrac{1-\surd 27}{\surd C:2}$, ut ante.

2

RESEARCHES IN THE GEOMETRICAL CONSTRUCTION OF EQUATIONS

[*c.* 1670?]

§1. MISCELLANEOUS PRELIMINARY CALCULATIONS.

From an original worksheet in private possession(1)

[1] $$\frac{axx+bx+c}{-dx-e}=z=\frac{fxx+gxy+hyy+kx+ly+m}{nx+oy+p}.$$

$$\begin{array}{llllll} na\,x^3+nb\,xx & +nc\,x+pc & +ao\,xxy & +bo\,xy & +co\,y & +dhxyy\ [=0]. \\ +df \quad\ +pa & +bp & +em+dg & +eg & +el & +ehyy \\ \quad\ +dk & +dm & & +dl & & \\ \quad\ +ef & +ek & & & & \end{array}$$

[Ubi] $e=0=dg+ao=dl+bo(=a=g=l=b)$ et non sit d, h, o, c, $p=0$. erit $\begin{array}{l} dfx^3+dkxx+nc\,x+pc+ocy+dhxyy=0. \\ \qquad\qquad\qquad +dm \end{array}$ et $\frac{c}{-dx}=z=\frac{fxx+hyy+kx+m}{p+oy+nx}$. [adeoq;] $hyy=oyz+pz+nxz-fxx-kx-m$. $y=\frac{oz}{2h}+\sqrt{\frac{oozz}{4hh}+\frac{pz+nxz-fxx-kx-m}{h}}$. Vel sic. $f=m=$nihilo$=k=\frac{nc}{+d}+m$. $[pc+ocy+dhxyy=0.]$(2)

---

(1) These calculations are written on the lower half of the loose waste-sheet from which the miscellany in **2**, 1, §1 above was taken, and we may suppose that they were made some time in the late 1660's. We reproduce them here since their common theme appears to be elimination between simultaneous equations in several variables in preparation for the geometrical construction of their eliminant.

(2) For some reason not apparent Newton eliminates z between

$$ax^2+bx+c+z(dx+e) = 0$$

and $$fx^2+gxy+hy^2+kx+ly+m = z(nx+oy+p),$$

evaluating the cubic eliminant in x and y. He then considers particular cases in which sets of the original coefficients are equated to zero. The technique of elimination, we may note, is that of Part 2, Chapter 2 of his Kinckhuysen 'Observationes' (**1**, §2 above). We may perhaps suppose that z is ultimately to be replaced by y and that Newton will seek the meets of the two conics which result.

[2] $^{(3)}y^3+bxyy+cxxy+dx^3+eyy+fxy+gxx+hy+kx+l=0.$

$$x=h.\ i.\ k.\ l.\ m.\ n.\ o.$$

$$y=p.\ q.\ r.\ s.\ t.\ v.\ w.^{(4)}$$

$$y=x^3.\ x^9+bx^7+ex^6+cx^5+fx^4+dx^3+gxx+kx+l[=0].$$
$$+h$$

$$yy+bxy+cxx+ey+fx+g[=0].\quad y=xx.$$

$$x^4+bx^3+cxx+fx+g[=0].$$
$$+e$$

Hæc sunt ad inveniendam Curvā Primi gradus[5] trium dimensionum unde Problemata novem dimensionum solvantur.[6]

---

(3) For simplicity Newton has cancelled an initial coefficient 'a'.

(4) The suggestion is that the cubic is to be made to pass through the seven pairs of points indicated. Two further determining conditions on the cubic remain, of course, to be defined.

(5) That is, in the terminology of **1**, **1**, §3, adiametral (and so most general).

(6) 'These are to find a first-grade curve of three dimensions by which problems of nine dimensions may be resolved.' In inverse manner to Problem 8 of §2.2 following Newton determines the intersections of a general (first grade) cubic with a cubic parabola by eliminating the variable y between their Cartesian defining equations, finding for their 9th-degree eliminant an equation in x which lacks its second term. He then considers the parallel case ($h=k=l=0$) in which the intersections of a general conic and the parabola $y=x^2$ are fixed by their general quartic eliminant.

§2. 'PROBLEMS FOR CONSTRUING ÆQUATIONS.'

[1670?][1]

From the original[2] in the University Library, Cambridge

[1][3] *Definitiones.*

1. Extensum vel extensio in data ratione movetur[4] quando motus ejus ab alijs extensis mutuo contactu determinatur.

2. Positio in qua movens sistendum est tunc dari dicitur ubi occurrit aliud extensum[5] quod mutuo contactu potest obsistere.

Cæterum punctum contactûs in alterutro contingentium aut ambobus fortè debet esse certum et cognitum aliter motus pro ignoto et mechanico æstimandus

(1) This is yet one more Newtonian mathematical paper of whose historical origins we know next to nothing. It is, beyond doubt, the 'set of Problems for construing æquations' sent by Newton to Collins on 20 August 1672 (*Correspondence of Isaac Newton*, **1** (1959): 231), for the copy of it which Collins made at that time (note (2) below) is identical with the present autograph manuscript. These problems were, however, not gathered together for Collins' especial benefit but had been composed at an unspecified earlier time ('heretofore') by Newton for his own private purposes and 'set down rudely...that I might not forget them'. The composition date (*c.* 1670) is conjectured somewhat tentatively from an analysis of Newton's handwriting and of the level of maturity of the paper's technical content. In confirmation, it will be evident that the later sections of the piece incorporate the substance of his May 1665 paper 'Of the construction of Problems' (I, **3**, 3, §2). Though Collins' copy was known to William Jones (into whose possession it later passed and who upon it wrote the subtitle, 'A Copy of a piece of S^r. Is. Newton in J. Collins's handwriting') and through him to James Wilson (*Mathematical Tracts of the late Benjamin Robins*... **2** (London, 1761): 346), the paper has never been published. Newton, however, subsequently repeated extracts from it with little change in his Lucasian lectures on algebra (ULC. Dd. 9.68: 211–51: 'Æquationum constructio linearis' = *Arithmetica Universalis* (Cambridge, $_1$1707): 279–326) and afterwards in 1695 in the final version of his cubic enumeration (ULC. Add. 3961.1: 49^v/50^r: 'Constructio æquationum per descriptionem curvarum' = *Enumeratio Linearum Tertii Ordinis* (*Opticks* (London, $_1$1704)): 161–2).

(2) Add. 3963.9: 70^r–106^v, a small booklet of folio pages folded into fours: only the 34 pages 70^r–86^v (numbered 1–34 by Newton himself at the top right-hand corner) are written upon, apart from two short Latin notes on 69^v in the hand of Samuel Horsley. Enclosed with the manuscript is Horsley's signed judgement of 20 October 1777 upon it: 'From N. 4 of selected papers./Æquationum constructiones./This tract is unfinished. The greater part of it it comprizd in Sr I. Ns appendix to the Arithmetica Universalis. There is indeed a construction of biquadratics & some directions concerning linear Problems, not containd in that appendix. The former might be not improperly annexed to that appendix, & the latter to the Enumeration of the lines of the third order. It may serve to elucidate the seventh section of that work. But all the rest of this tract is quite unfit for publication.' During Newton's lifetime the work passed out of Newton's possession twice at most, and then only briefly. On 20 August 1672 he lent it to John Collins in response to the latter's request for information concerning the geometrical resolution of the general sextic by the aid of a Wallisian parabola (*locus linearis*), but in his accompanying letter requested that it should not be shown to anyone else and that it be returned when Collins was finished with it. The latter, as we now know, made

Translation

[1][3] *Definitions*

1. An extended object or extension moves[4] in given respect when its motion is determined by other extended objects in mutual contact.

2. The position in which a moving object shall be brought to a standstill is then said to be given when it meets a second extended object which can stop it by mutual contact.

But the point of contact in one or other of the touching objects, or perchance in both, must be certain and known, otherwise the motion must be rated as

---

an accurate copy of the tract (the first portion, to the end of Problem 20, in a hurried English version and the remainder in straightforward Latin transcription) but we may be sure he quickly returned the original to Newton's keeping. Whether or not he showed his copy to anyone else during his lifetime we do not know but he spoke of it, in general terms at least, to Wallis in 1677. (See *The Correspondence of Isaac Newton*, **2** (1960): **243**.) When, a quarter century after Collins' death, his papers passed into the hands of William Jones the first six pages of the copy had gone astray but it would appear that Jones, then preparing a selection of Newton's mathematical pieces for publication, was allowed by Newton to make good the deficiency from the original: at least, the Collins copy (now in private possession) is now completed by six pages in Jones' hand. We may note, finally, that certain light cancellations made by Jones in that copy suggest that he was intending to publish an abridged version of Newton's work, presumably in the abortive sequel to his 1711 Newtonian collection, *Analysis per Quantitatum Series, Fluxiones ac Differentias*.

(3) In this first of the two parts into which Newton's text divides itself he attempts to set up an axiomatically defined plane geometry which may afford a rigorous foundation for the more technical constructions which conclude the present division and form the whole of the following section [2]. His model is evidently the second book of Frans van Schooten's *Exercitationes Mathematicæ*, which concerns itself with straight line and circle constructions (*Exercitationum Mathematicarum Libri Quinque* (Leyden, 1657): 123–49, 150–9: 'De Constructione Problematum Simplicium Geometricorum, seu, quæ solvi possunt, ducendo tantùm rectas lineas', and 'De Constructione Planorum Problematum Geometricorum, seu, quæ ductis rectis lineis, descriptisque circulorum circumferentiis, solvi possunt': compare Schooten's *Præfatio ad Lectorem*: 119–22), though of course Newton does not here restrict himself to plane and solid problems but takes the whole field of elementary algebraic geometry into his survey. Where, however, Schooten had a static conception of his constructional theory and was thereby led to make separate postulates for each type of locus he introduced (three for the straight line, one for the circle), Newton's viewpoint, closely related to his notion of fluxional increase in general, is essentially dynamical and allows him to admit any cuve at will (the circle in Problem 3 below, for example) as a mere defined locus, though he too restricts his discussion to the straight line and circle for the most part—or apparently so, at least, for his 'neusis' arguments frequently demand for their exact construction a general conic or a conchoid.

(4) Newton first wrote 'moveri dicitur' (is said to move). The phrase must have been cancelled by him after Collins returned his paper in 1672 for the latter's copy (note (2)) translates the original wording. Someone, Samuel Horsley we presume (note (2)), has added the clarification 'lege datâ moveri dicitur' on the blank facing page.

(5) The clarifying phrase 'positione datum' (given in position) has here been inserted in the same hand as that of the previous note.

est. Et hinc contactus rectarum et curvarum infinita obliquitate concurrentiū[(6)] excluduntur.

3. Extensa describi dicuntur quæ per cursus moventium vel positiones quiescentium designantur.

Postulata.

1. Postuletur ut quodlibet extensum quacunq; data ratione movere concedatur.

2. Et movens in quacunq; datâ positione sistere.

3. Item extensa describere quæ per cursus moventium vel positiones quiescentium designantur.

Prob 1.

A quovis puncto ad quodvis punctum rectam lineam ducere et continuò producere.[(7)]

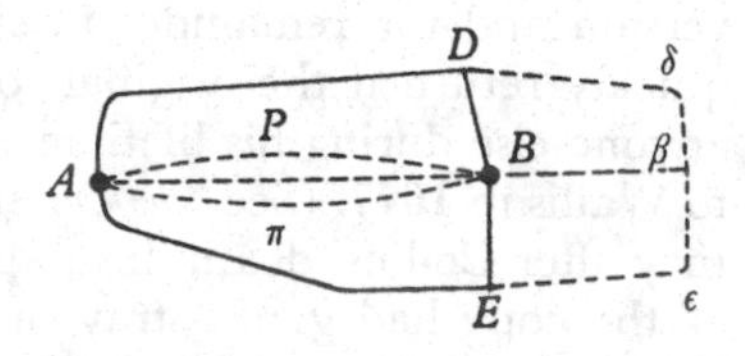

Sint A et B data puncta, et $ADBE$ spatium vel solidum quodvis interjectum utcunq; terminatum. Hoc solidū circa ista duo puncta tanquam polos convolvatur (pet 1) et lineam AB circa quam volvitur sive quæ transit per ejus omnia quiescentia puncta describe per pet 3. Dico factum. Si negas sit alia quævis APB recta circa præfatam AB volvens, sitq; $A\pi B$ eadem recta in alia positione: Et rectæ APB $A\pi B$ comprehendent spatium contra ax [14][(8)] primi Elem.

Producitur verò si spatium longiùs extensum puta ad $\delta\epsilon$ circa ipsā jam descriptam iterum convolvas.

Prob 2.

Planum per data tria puncta A, B, C transiturum ducere et continuò producere.

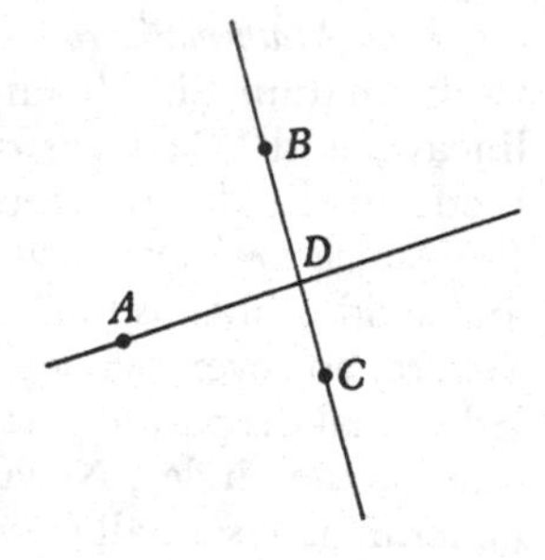

Per duo quælibet punctorum B et C duc rectam BC (prob 1)[,] dein circa tertium punctorum A tanquam centrum gyretur alia similiter ducta recta AD ita ut lineam BC continuò tangat (pet 1) et hæc describet planum per puncta A B et C transiturum.

Nam cum rectæ BC duo puncta B et C sunt in plano describendo (Hyp) ea tota cum eodem plano coincidet ([1, XI, *Elem.*]) et proinde concursus rectarū AD et BC semper erit in isto plano. At recta AD iterum in A cum

(6) The continuation 'et non decussantium' (and not crossing) is cancelled. Newton's present rejection of the angle of contact (or 'horn' angle) as being not well defined and so 'mechanical', that is, non-geometrical, is completely Cartesian. Compare *Geometrie* (Leyden, 1637): 319, where Descartes defines geometrical curves as those 'qui tombent sous quelque mesure précise & exacte'.

(7) This condenses Schooten's three 'petitiones' for the construction of 'simple' problems (*Exercitationes Mathematicæ* (note (3)): 123). Newton's present attempt to construct the straight

unknown and mechanical. And for this reason the contacts of straight and curved lines meeting[6] at an indeterminate inclination are excluded.

3. Extended objects are said to be described in so far as they are designated by the paths of moving things or the positions of ones at rest.

Postulates

1. That any extended object be allowed to move in any given respect.
2. And stop a moving object in any given position.
3. And, likewise, describe extended objects which are designated by the paths of moving things or the positions of ones at rest.

Problem 1

To draw a straight line from any point to any other, and produce it indefinitely.[7]

Let A and B be given points and $ADBE$ any intervening space or solid arbitrarily bounded. Let this solid be spun round those two points as poles (postulate 1) and, by postulate 3, describe the line AB round which it revolves or which passes through all its points at rest. I say it is done. If you deny that, let APB be any other straight line revolving round the aforesaid AB, and $A\pi B$ the same straight line in another position: the straight lines APB and $A\pi B$ will then encompass a space, contrary to *Elements*, I, axiom [14].[8]

It is produced, indeed, should you once more rotate a longer space, extended say to $\delta\epsilon$, around the straight line just now described.

Problem 2

To draw a plane which shall pass through three given points A, B, C and produce it indefinitely.

Through any two of the points, B and C, draw the straight line BC (problem 1), then around the third point A as centre let there circle a second straight line AD, similarly drawn, so as continuously to touch the line BC (postulate 1): this will describe the plane which is to pass through the points A, B and C.

For, since the two points B and C of the straight line BC are in the plane to be described (by hypothesis), that line will wholly coincide with this same plane ([*Elements*, XI, 1]) and consequently the intersection of the straight lines AD and BC will always be in that plane. But the straight line AD meets that plane a

line as an axis of rotation rather than, with Euclid and Schooten, accept it as a conventional axiom seems inefficient and arbitrary. Just as we might well deny the physical necessity of a straight line defined in Archimedean style as the minimal path between two points, we might with equal justice fault the physical validity of Newton's present construction.

(8) In Barrow's 1655 edition of the *Elements* (I, 1, Introduction: note (28)), 'Duæ rectæ lineæ spatium non comprehendunt': 'Two right-lines do not contain a space' in Haselden's English version (*Euclide's Elements: The whole Fifteen Books compendiously Demonstrated...by Isaac Barrow* (London, 1732): 7). The 'axiom' is, of course, a post-Euclidean addendum.

isto plano convenit (Hyp). Quare illa etiam *AD* tota cum eodem desiderato plano conveniet ([1, XI, *Elem.*]) atq; adeo per circumlationem describet ipsum [(]pet 3[)].(9)

Quæ sequuntur omnia in dato plano construenda esse intelligo.(10)

Prob 3.

Dato centro et intervallo circulum(11) *describere.*(12)

A centro ad datum intervallum duc rectam (prob 1) et circa centrum in dato plano convolve (pet 1) ejusq; mobilis extremitas describet ejusmodi circulum (def [15, 16] 1mi Elem.)(13)

Prob 4.

Cuilibet lineæ vel figuræ datæ similem et æqualem in data positione constituere.

A quolibet assignato puncto loci in quo linea vel figura constituenda est ad correspondens punctum datæ lineæ vel figuræ duc rectam et in hâc moveatur punctum istud lineæ donec attingat correspondens punctū loci[...](14)

Prob 4.

Ad datum punctum α datā rectā lineā AB in data positione ponere.(15)

Rectam *AB* (si opus est productam) converte circa quodvis ejus punctum donec datum punctum α attingat. Dein promoveatur secundum longitudinem ejus donec alterutra ejus extremitas *A* cum eodem α conveniat. Deniq; convolvatur circa istud α usq; dum assignatam positionem acquirat per pet 1 & 2 ac def 1 & 2.

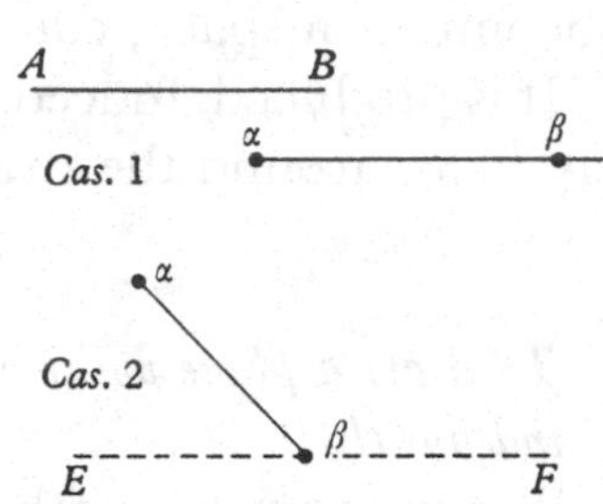

Schol. Hujus prob: duo sunt casus[:] primò ut recta data collocetur in alia datâ rectâ ad datum ejus punctum, secundò ut inter datum punctum et quamcunq; descriptam lineam (quando id possibile est) collocetur. Et his duobus fundamentis(16) innituntur solutiones omnium planorum problematum.

(9) In its detail this construction (and proof) seems original with Newton, though its structure is of course implicit in the Euclidean definition of a plane in Book 1 of the *Elements*. Schooten in his *Exercitationes Mathematicæ*, Liber II, accepts the existence of a 'campus' as an axiomatic truth, presumably from Euclid.

(10) This sentence is a late addition, inserted after the enunciation of the following problem was written down: the latter (see note (11)) was originally explicitly restricted by this condition.

(11) The phrase 'in dato plano' (in a given plane) is cancelled: see note (10).

(12) Compare Schooten's *Exercitationes* (note (3)): 150: 'Postulatum....ut è dato puncto ceu centro in dato intervallo circulum describere liceat'. In Newton's dynamic theory Schooten's postulate becomes a mere locus definition.

second time in A (by hypothesis). Hence that line AD also will wholly coincide with the same desired plane ([*Elements*, XI, 1]) and so will describe it by rotation (postulate 3).(9)

All that follows I understand to be constructed in a given plane.(10)

Problem 3

With given centre and radius(11) *to describe a circle.*(12)

From the centre draw a straight line to the length of the given radius (problem 1) and rotate it round that centre in the given plane (by postulate 1): its mobile end-point will describe such a circle (by *Elements*, I, definitions [15 and 16]).(13)

Problem 4

To place in given position a line or figure congruent to a given one.

From any arbitrarily assigned point of the location in which the line or figure is to be placed to the corresponding point of the given line or figure draw a straight line and in this let that point be moved till it reaches the corresponding point in the location....(14)

Problem 4

On the given point α to set the given straight line AB in given position.(15)

Rotate the straight line AB (produced if need be) round any point of itself till it reaches the given point α. Then let it be advanced lengthways until one or other of its extremities A coincides with this same point α. Finally, let it be rotated round α until it gains its assigned position (by postulates 1 and 2 and definitions 1 and 2).

Scholium. This problem has two cases: first, that a given straight line be positioned in another given straight line at a given point in it; secondly, that it be positioned between a given point and some described line (when that is possible). On these two foundations(16) are based the resolutions of all plane problems.

(13) 'per pet 2' (by postulate 2) is cancelled.

(14) Newton has presumably cancelled this version of the problem (and an unimportant first, minor redraft which we here omit) because his phrasing of it was not clear: in the revised version which follows he classifies his argument by reference to a figure.

(15) Compare Schooten's *Exercitationes* (note (3)): 148: 'V. Propositio. Ad datum punctum A in data recta indefinita AD rectam lineam in campo collocare AE, datæ rectæ terminatæ BC æqualem'.

(16) That is, Problems 3 and 4 preceding, which do indeed offer (together with the postulates of congruence and parallelism contained in the Euclidean metric understood) a sufficient foundation for the plane geometry of the straight line and circle.

Prob 5.

Inter duas datas lineas DG et EG datam rectam lineam AB in datâ positione collocare.

Data recta AB transferatur ut in priori prob: donec interjaceat duas datas DG et EG uti videre est ad $\alpha\beta$. Dein ita moveatur ut easdem lineas cum extremitatibus ejus perpetuò tangat (pet 1) idq̧ donec datam positionem acquirat [(]pet 2[)].

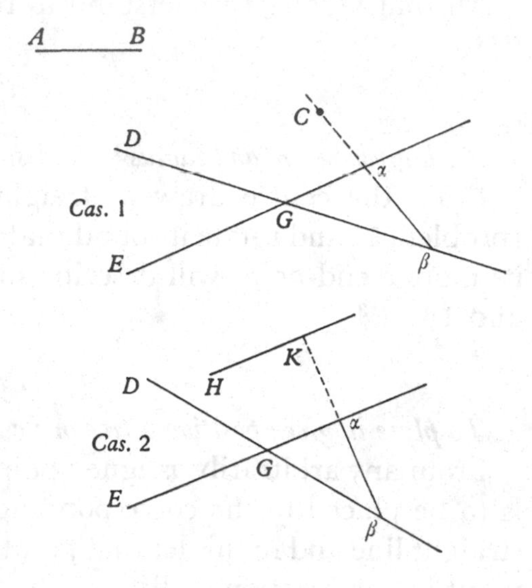

Hujus autem problematis duo sunt casus prout moventis rectæ positio vel occursu ejus cum dato puncto C determinatur vel occursu puncti alicujus K in eâ dati cū datâ quâcunq̧ lineâ HK. Et his duobus fundamentis innituntur constructiones solidorum et fortè magis compositorum problematum quatenus determinando positionem unicæ moventis rectæ possunt solvi.(17)

Prob 6.

Quamcunq̧ datam figuram in data positione collocare.

Constructio ejus ex præcedentibus manifestum est. Complectitur verò solutiones omnes problematum quæ fieri possunt absq̧ motuum complicatione.(18)

Prob 7.

A dato puncto A datæ rectæ lineæ BC ducere parallelam rectam lineam AD.(19)

Duc quamvis AB alicubi occurrentem BC puta in B & produc ad E ut sit $BE=AB$ [(]prob 4[)]; per E duc aliam quamvis EC secantem rectam BC in C [&] produc ad D ut sit $CD=EC$ [(]prob 4[)]. Et AD ducta erit parallela ad BC.

Prob 8.

Ad punctum quodvis A in data recta BC perpendiculum AD erigere.(20)

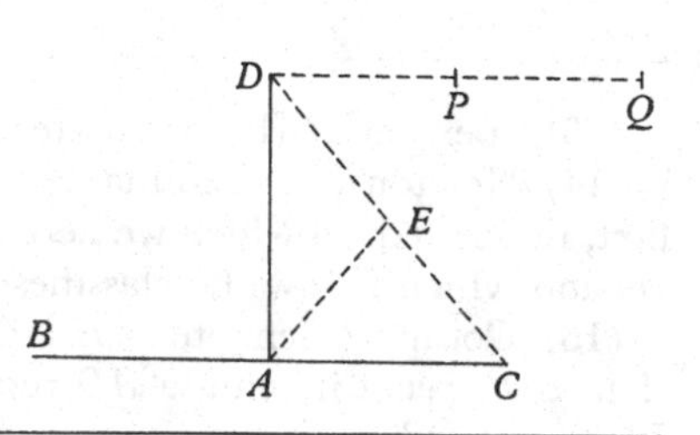

Ab A duc quamvis AE et ei æqualem inscribe EC[,] produc ad D ut sit $ED=EC$ ac iunge AD.

(17) The first case is the 'conchoidal' one, of which Newton will make continual use in his later problems: in general, since the positions of $\alpha\beta$ which satisfy the requirement are constructible by the meet of a conchoid of pole C and 'regula' $\alpha\beta$ with one or other of the given

Problem 5

Between the two given lines DG and EG to set the given straight line AB in given position.

Let the given line *AB* be transposed as in the previous problem until it lies between the two given ones *DG* and *EG*, as may be seen at $\alpha\beta$. Then let it be so moved as perpetually to touch these same lines with its extremities (by postulate 1) and that till it gain the given position (by postulate 2).

Of this problem also there are two cases according as the position of the moving right line is determined by its meet with the given point *C* or by the meet of some point *K* given in it with some given line *HK*. On these two foundations are based the constructions of solid problems and, it may be, of more compounded ones so far as they can be resolved by determining the position of a single moving line.(17)

Problem 6

To set any given figure in given position.

Its construction is manifest from the preceding. It embraces, indeed, all resolutions of problems which can be effected without compounding motions.(18)

Problem 7

From a given point A to draw the straight line AD parallel to the given line BC.(19)

Draw any line *AB* meeting *BC* in some arbitrary point, say in *B*, and produce it to *E* so that $BE = AB$ (problem 4); through *E* draw any other line *EC* intersecting the straight line *BC* in *C* and produce it to *D* so that $CD = EC$ (problem 4). When *AD* is drawn it will be parallel to *BC*.

Problem 8

At any point A in the given straight line BC to erect the perpendicular AD.(20)

From *A* draw any line *AE* and equal to it inscribe *EC*; then produce it to *D* so that $ED = EC$ and join *AD*.

---

straight lines *DG*, *EG*, the problem will be of fourth degree algebraically and so resolvable by the meets of two 'solid' loci (or conics) which may be taken to be a circle and a parabola. The second case is 'plane' for the locus of *L* fixed in $\alpha\beta$ as α moves in *EG*, β in *DG* will be an ellipse (I, 1, 1, §1.3*g*) and this will meet *HK* in two points: furthermore, the meets of *HK* with the ellipse may be constructed as the intersections of a circle (suitably defined) with *HK*.

(18) Those, that is, which are constructible by rotation and translation of a single given line-segment.

(19) Compare Schooten's *Exercitationes* (note (3)): 131: 'Problema III. Per datum punctum *C* rectam lineam ducere, datæ rectæ *AB* parallelam'. As Schooten, Newton in his construction makes the postulate of parallelism dependent on the equivalent one of similarity of figures.

(20) Compare Schooten's *Exercitationes* (note (3)): 137–9: 'Problema IV. Super data recta linea indefinita perpendicularem constituere'. With Schooten Newton constructs the right angle $D\hat{A}C$ as that in the semicircle of diameter *DC* and centre *E*.

Prob 9.

A dato puncto D ad datam rectam BC demittere perpendiculum DA.(21)

Duc quamvis DP et produc ad Q ut sit $PQ=DP$. Inscribe $DC=DQ$. Cape $DE=DP$ et inscribe EA ejusdem longitudinis. Erit DA perpendiculū.

Prob 10.

Datis lateribus triangulum conficere.(22)

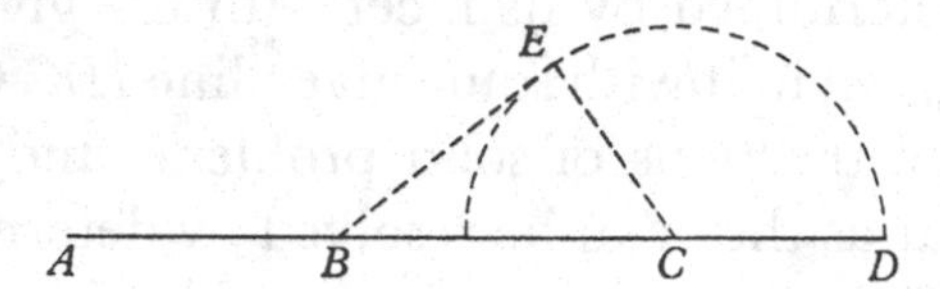

Coroll. Hinc angulus dato æqualis et in data positione constitui potest.(23)

Prob 11.

Invenire summam quadratorum.(24)

Prob 12.

Invenire differentiam quadratorum.(24)

Prob 13.

Invenire quartum proportionalem.(25)

Coroll 1. Invenire tertium proportionalem.(26)

2. Lineam rectam in data ratione secare.(27)

3. Figuram datæ figuræ similem super data basi constituere.(28)

4. Parallelogrammū datæ magnitudinis super data basi et cum datis angulis conficere.(29) Idem de parallelopipidis.

(21) Compare Schooten's *Exercitationes* (note (3)): 139: 'Problema V. Datâ rectâ lineâ indefinitâ AB, à puncto in ea vel extra ipsam dato C, rectam lineam ducere CF, quæ ad datum AB sit perpendicularis'. Newton's construction, however, in effect inverts that of the preceding problem, while Schooten makes cumbrous use of parallels.

(22) Compare Schooten's *Exercitationes* (note (3)): 156–7: 'III. Problema. Ex tribus datis rectis lineis terminatis A, B, & C, quarum duæ simul, quomodocunque sumptæ, tertiâ sunt majores, triangulum constituere.' Newton, omitting the necessary condition for real solution that any two of the given sides must be greater than the third, leaves his figure to explain itself. Given, of course, the three sides, AB, BC and CD he constructs the triangle sought on the segment BC, finding its third vertex E as the meet of the circles on centre B with radius AB and on centre C with radius CD. There will be two solutions, mirror-images one of the other with regard to BC.

(23) If, for instance, the angle $\hat{B}$ is to be constructed, set off BE and BC of arbitrary lengths on its sides: the length EC will then be determined, and we need only to construct (by Problem 6) the triangle congruent to BEC in given position.

Problem 9

From a given point D to let fall the perpendicular DA to the given straight line BC.[21]

Draw any line DP and produce it to Q so that $PQ = DP$. Inscribe $DC = DQ$, take $DE = DP$ and inscribe EA of the same length. Then will DA be the perpendicular.

Problem 10

To make up a triangle, given its sides.[22]

Corollary. Hence may be constructed an angle which is equal to a given one and in given position.[23]

Problem 11

To find the sum of squares.[24]

Problem 12

To find the difference of squares.[24]

Problem 13

To find a fourth proportional.[25]

Corollaries. 1. To find a third proportional.[26]

2. To cut a given straight line in given ratio.[27]

3. To construct a figure equal to a given figure upon a given base.[28]

4. To make a parallelogram of given magnitude upon a given base and with given angles.[29] The same for parallelepipeds.

---

(24) After each of these enunciated problems Newton has left blank spaces for future entry of their constructions. Evidently, their resolution will easily be effected by Pythagoras' theorem (*Elements*, I, 47). Compare Schooten's *Exercitationes* (note (3)): 152–3: 'II. Problema. Trianguli rectanguli datâ Hypotenusâ, & uno circa rectum angulum latere: invenire latus alterum circa rectum.'

(25) Following this also Newton has left a space blank for the intended entry of the problem's construction. Presumably this would depend on Problem 7, for if in its figure we set the three given quantities equal to EA, EB and AC, and then draw AD parallel to BC, ED will be the fourth proportional to them.

(26) The particular case of Problem 13 in which the middle terms of the proportion are equal.

(27) In the figure of Problem 7 take DE to be the given line and $AB:BE$ the given proportion.

(28) Since corresponding lengths in similar figures are proportional, the construction will be performed by suitable application of the preceding corollary.

(29) On the given base and with the given angles construct a parallelogram of arbitrary height and find its 'magnitude' (or area): this will be in proportion to the given area as the height of the parallelogram constructed to the height of that desired.

Prob 14.

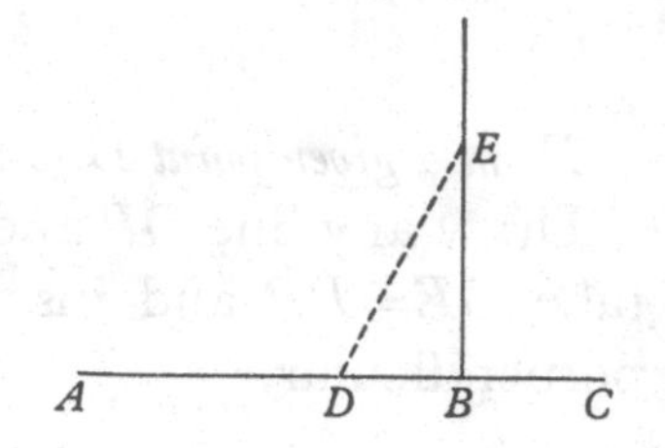

Invenire medium proportionale inter AB et BC.

Erige BE perp[endiculum] ad AC, biseca AC in D, inscribe $DE=AD$ et erit EB medium proport[ionale].(30)

Idem aliter modis infinitis. fac circulum quemvis per A et C transiturum sitꝗ ejus centrum F, biseca BF in D. Inscribe $ED=DB$, erunt $BA.\ BE.\ BC$ ∺(31)

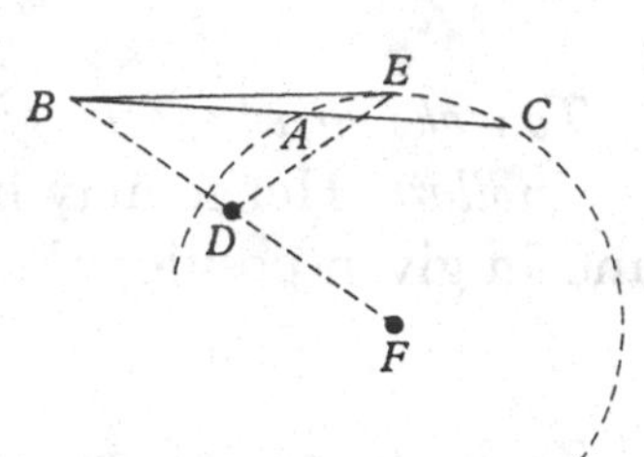

Coroll 1. Hinc rectangula et aliæ omnes rectilineæ figuræ in quadrata aut in alias figuras cujuscunꝗ datæ formæ convertuntur.(32)

Prob 15.

Invenire duo media proportionalia inter AK et AB.

Biseca AK in I. Sit AB ad angulos rectos. Cape $AH=BI$ et $AC=BH$. Punctis B et C inventis cape $AD=DE$ cujusvis longitudinis et centris D et E intervallis BD et CE describe duos circulos inter quos inscribe $GF=AI$ ita ut transeat per punctum A et erit AF primum e medijs proportionalibus. Scilicet $AB:AF:*:A[K]$ ∺(33)

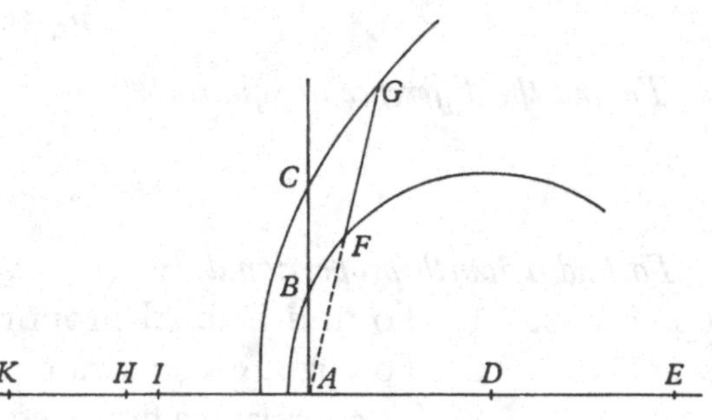

Idem aliter. Super DX fac isoscele triang: DPX cum latere $PX=\frac{1}{2}KX$, produc PD ad A ut sit $DA=DP$, junge AX et produc versus E. Inscribe $EY=PD$ quæ transeat per P (prob 5) et erūt $DX.\ PE.\ XY.\ XK$ ∺(34)

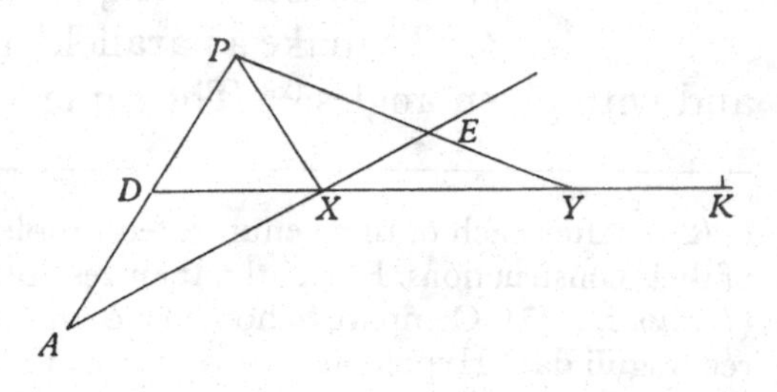

(30) Since E is on the circle of diameter AC, the angle $A\hat{E}C$ is right and hence the right triangles ABE, EBC are similar.

(31) By *Elements*, II, 36 (or through the proportionality of the triangles BAE, BEC). Note that the point B is no longer between A and C. A more exact generalization of the preceding would be to take an arbitrary circle through the points A and C, and then construct BE as the half-chord through B perpendicular to the diameter through that point.

(32) Any rectilinear figure is easily reduced to a triangle, and so a rectangle, of equal area and then, by this problem, that rectangle may be reduced to an equivalent square: a given figure is then converted into one of given shape but of equal area by application of the rule that the areas of similar figures are as the squares of corresponding line-lengths.

Problem 14

To find a mean proportional between AB and BC.

Erect BE perpendicular to AC, bisect AC in D and inscribe $DE = AD$: then will EB be the mean proportional.(30)

The same done otherwise in an infinity of ways. Make any circle pass through A and C and let its centre be F; bisect BF in D, and inscribe $ED = DB$: then will BA, BE and BC be continued proportionals.(31).

Corollary 1. Hence may rectangles and all other rectilinear figures be turned into squares or other figures of any given shape.(32)

Problem 15

To find two mean proportionals between AK and AB.

Bisect AK in I and let AB be at right angles to it. Take $AH = BI$ and $AC = BH$. When you have found the points B and C take $AD = DE$ of any length you wish and with centres D and E, radii BD and CE describe two circles; between these inscribe $GF = AI$ so as to pass through the point A. Then will AF be the first of the mean proportionals: that is, AB, AF, * and AK will be in continued proportion.(33)

The same another way. On DX construct the isosceles triangle DPX of side $PX = \frac{1}{2}KX$, produce PD to A so that $DA = DP$ and join AX, producing it in the direction of E. Inscribe $EY = PD$ so that it passes through P (by problem 5) and DX, PE, XY and XK will be in continued proportion.(34)

---

(33) We may easily restore Newton's analysis of his construction in the following way. If we suppose that $FG = a$ is to be so placed between circles of radii $DB = DF = b$ and $EC = EG = c$ that it passes through A in their common diameter, then on setting $AD = d$, $AE = e$ and $AF = x$ we deduce that $2\cos\hat{A} = (x^2+d^2-b^2)/dx = [(x+a)^2+e^2-c^2]/e(x+a)$ and consequently $x^3+\alpha x^2+\beta x+\gamma = 0$, in which $\alpha = a(e-2d)/(e-d)$,

$$\beta = [e(d^2-b^2)-d(a^2+e^2-c^2)]/(e-d) \quad \text{and} \quad \gamma = ae(d^2-b^2)/(e-d).$$

On identifying this with $x^3-m^2n = 0$ we find $\alpha = \beta = 0$ and $\gamma = -m^2n$, or $e = 2d$,

$$2a(d^2-b^2) = -m^2n \quad \text{and} \quad 2(d^2-b^2) = a^2+e^2-c^2.$$

For simplicity Newton takes $AB^2 = b^2-d^2 = m^2$ (or $AB = m$) and consequently

$$FG = a = \tfrac{1}{2}n$$

with $AC^2 = c^2-e^2 = \frac{1}{4}n^2+2m^2$: these two latter quantities he then constructs from $AK = n$.

(34) On setting, generally, $PX = a$, $PD = b$, $XD = c$, $DA = YE = d$ and supposing $PE = x$, $XY = y$ to be the two means between the quantities m and n (or $xy = mn$ and $x^3 = m^2n$), we find, since $DX \times YE \times PA = XY \times EP \times DA$, that $xy = c(b+d)$: also since $2\cos\hat{D} = (b^2+c^2-a^2)/bc = [b^2+(c+y)^2-(d+x)^2]/b(c+y)$, it follows that

$$x^4+2dx^3+(d^2-a^2)x^2+(b+d)(b^2-c^2-a^2)x-c^2(b+d)^2 = 0,$$

that is, $(x+b+d)(x^3-(b-d)x^2+(b^2-a^2)x-c^2(b+d)) = 0$. On equating the cubic factor with $x^3-m^2n = 0$ we deduce that $a = b = d$ and $c^2(b+d) = m^2n$ with $xy = mn = 2cd$, so that $c = m$ and $d = \frac{1}{2}n$. The construction is Nicomedes' (T. L. Heath, *Greek Mathematics*, **1**, 1921: 260).

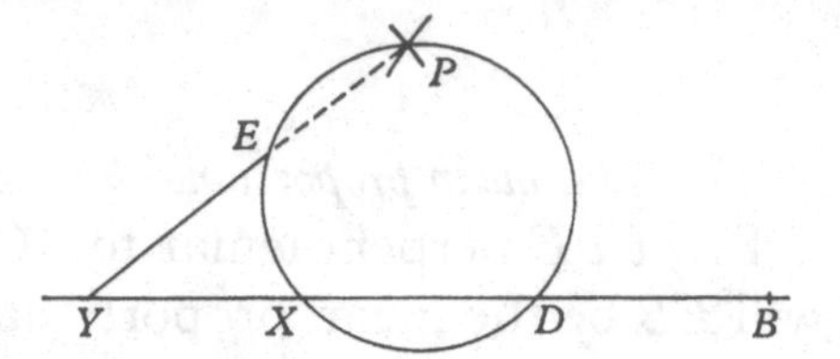

Idem aliter.[35] Isoscelis PBX biseca latus XB in D. Per puncta D P et X describe circulum inter quem et XD inscribe $EY = XD$ quæ transeat per P et erunt $\frac{1}{2}XP.\ XY.\ ED.\ XB \div\!\div$[36]

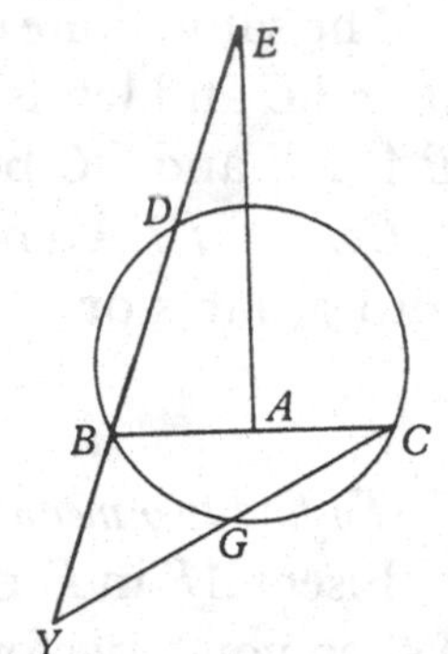

Vel sic. Cape $AB = AC = m$. Erige $AE \perp$,[37] inscribe $BE = n$, biseca in D. Per B D et C fac circ[ulum] ac inscribe $GY = BD$ quæ transeat per C, et erit $m.\ BY.\ GD.\ n \div\!\div$[38]

Idem aliter.[39] Cape $AD = m$. $AE = n$. et per A E D describe circulum. Item fac $DK = AD$. $KI = \frac{1}{8}AE = IH$. Age AH et IET secantes in E. Deniq; inter lineas KH et IE inscribe $PL = EH$ ita ut $PF = EA$ terminetur in circulum et ab F demitte $FG \perp$.[37] Et erit $AE.\ AG.\ FG.\ AD \div\!\div$[40]

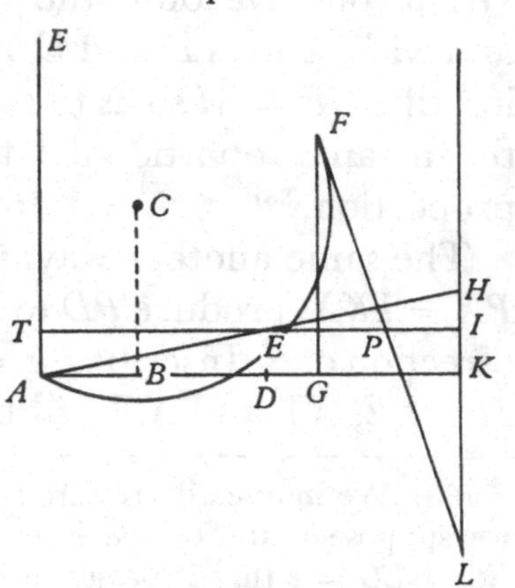

Vel sic generaliùs enunciatur infinitis modis. Cape $AB = \frac{1}{2}m$. $BC = \frac{1}{2}n$. Centro C radio CA describe circulum AF. Porro sumpto AD ad arbitrium fac $AB.\ AD :: AD.\ AK$ et $AK.\ AB :: BC.\ IK$. Duc parallelam $IE = KD$ et per E duc AH. Deniq; inter lineas EI et HK inscribe $LP = HE$ ita ut[41] posito $PF = EA$ ejus punctum F cadat in circulum, ac demitte $FG \perp$,[37] eritq; $m.\ FG.\ AG.\ n \div\!\div$[42]

(35) Newton has cancelled a first continuation 'Biseca XK in C et erige CP perp[endiculum]. Cape $DB = DX$ [&] in angulo BCP inscribe $BP = BX$' (Bisect XK in C and erect the perpendicular CP. Take $DP = DX$ and in the angle BCP inscribe $BP = BX$). In a first version of the accompanying figure, correspondingly, the normal PC to XD was drawn and K marked in the latter such that $XC = CK$.

(36) Evidently, if we suppose that BP meets the circle again in C, then $BC = \frac{1}{2}BP = YE$. Hence, on setting generally $XD = a$, $DB = b$, $BC = YE = c$, $XP = d$ and supposing $XY = x$, $ED = y$, we have by elementary circle properties $PB = b(a+b)/c$, $YP = x(a+x)/c$ and also $YX : XP = YE : ED$ or $xy = cd$. Further, since

$$2PX \cos P\hat{X}Y = (YX^2 + PX^2 - YP^2)/YX$$
$$= -(XD^2 + PX^2 - DP^2)/XD,$$

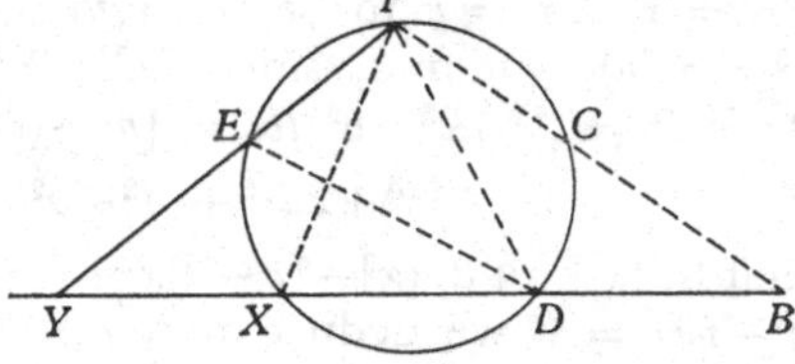

there comes $(x^2 + d^2 - YP^2)/x + (a^2 + d^2 - DP^2)/a = 0$: in a similar way there follows

$$(a^2 + DP^2 - d^2)/a + (b^2 + DP^2 - PB^2)/b = 0.$$

The same another way. Bisect the side XB of the isosceles triangle PBX in D. Through the points D, P and X describe a circle and between this and XD inscribe $EY = XD$ so that it passes through P. Then will $\frac{1}{2}XP$, XY, ED and XB be in continued proportion.(36)

Or this way. Take $AB = AC = m$, erect the perpendicular AE, inscribe $BE = n$ and bisect it in D. Through B, D and C construct a circle and inscribe $GY = BD$ so as to pass through C. Then will m, BY, GD and n be in continued proportion.(38)

The same but another way.(39) Take $AD = m$, $AE = n$ and through A, E, D describe a circle. Moreover, make $DK = AD$, $KI = \frac{1}{8}AE = IH$ and draw AH, and IET intersecting in E. Finally, between the lines KH and IE inscribe $PL = EH$ so that $PF = EA$ terminates on the circle and from F drop the perpendicular FG. Then will AE, AG, FG and AD be continued proportionals.(40)

Or, more generally, it may thus be expressed in an infinity of ways. Take $AB = \frac{1}{2}m$, $BC = \frac{1}{2}n$ and with centre C, radius CA describe the circle AF. Furthermore, assuming AD arbitrarily make $AB:AD = AD:AK$ and $AK:AB = BC:IK$. Draw IE parallel and equal to KD and through E draw AH. Finally, between the lines EI and HK inscribe $LP = HE$ so that,(41) on setting $PF = EA$, its end-point F shall fall on the circle, and then drop the perpendicular FG. Then will m, FG, AG and n be continued proportionals.(42)

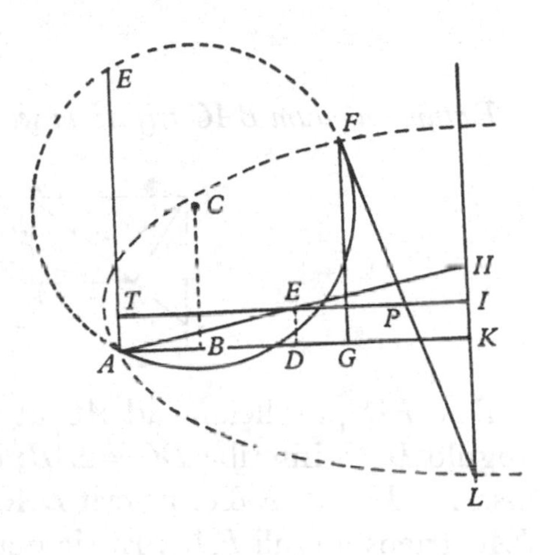

Finally, on eliminating PB, YP and DP there comes

$$x^4 + 2ax^3 + (a^2 - c^2)x^2 + [(a+b)(b^2 - c^2) - c^2d^2/(a+b)]x - c^2d^2 = 0,$$

that is, $(x+a+b)(x^3 + (a-b)x^2 + (b^2-c^2)x - c^2d^2/(a+b)) = 0$. If now x and y are the two means between m and n, then $xy = mn$ and $x^3 - m^2n = 0$. By the former $cd = mn$, and on identifying the latter with the preceding cubic factor we deduce that $a = b = c$ and $\frac{1}{2}cd^2 = m^2n$, so that $c = \frac{1}{2}n$ and $d = 2m$.

(37) Read 'perpendiculum'.

(38) A simple variant on the previous construction. Clearly $EC = EB = 2GY$.

(39) 'Adhuc aliter' (Still otherwise) is cancelled.

(40) This is the particular case of the following when D is the second meet of AB with the circle AF, that is, when, in the terminology of note (42), $k = m$ and the ellipse has for its defining equation $x^2 + 4y^2 - 4mx - ny = 0$. Note the two points E in Newton's figure.

(41) A first continuation '*PF* circulo intercept[a] sit æqual[is] *EA*' (*PF* cut off by the circle be equal to *EA*) is cancelled.

(42) If $AG = x$ and $GF = y$ are the two means between m and n, then $y^2 = mx$ and $x^2 = ny$. Geometrically, these are parabolas which intersect in two real points, the origin A and a second point F whose co-ordinates are the means required. Through these intersections,

Nota quod hæc constructio perinde fit ac si Ellipsis describeretur secans circulum in F.(43)

Prob 16.

Datum angulum BAC bisecare idq; continuò.(44)

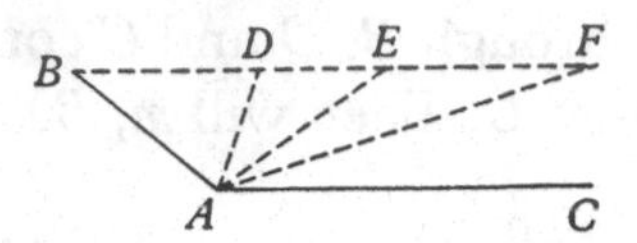

Duc quamvis BD parallelam alterutro lateri AC & secantem alterum latus AB in B. In eadem cape $BD=BA$, $DE=DA$, $EF=EA$ &c et AD bisecabit angulum BAC, AE bisecabit angulum DAC, AF bisecabit angulū EAC &c. Est enim ang $BAD=BDA=DAC$ [&c].

Prob 17.

Datum angulum BAC trifariam secare idq; continuò.

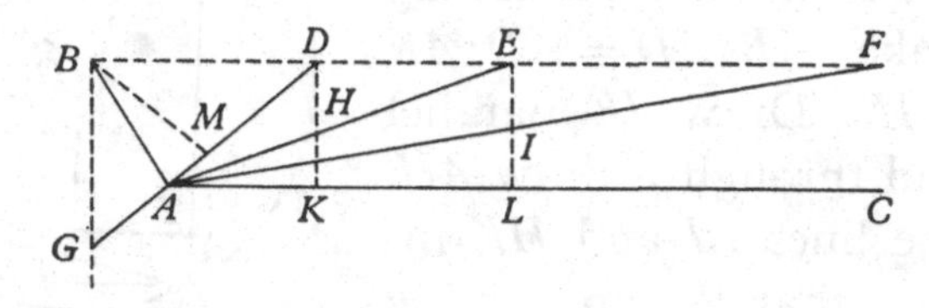

Duc BD parallelam ad AC ut ante et ad B erige BG perpendiculum, inq; angulo DBG inscribe $DG=2AB$; erige DK et inscribe $HE=2AD$; erige EL et inscribe $IF=2AE$ &c. Et erit DAC triens anguli BAC, EAC triens anguli DAC, FAC triens anguli EAC. Et sic deinceps.

Nam biseca GD in M et erit $BM=MD=AB$. Ergo ang DAC sive

$$ADB=DBM=\tfrac{1}{2}BMA=\tfrac{1}{2}BAM.$$

hoc est $DAC=\tfrac{1}{3}BAC$.(45) [&c]

Schol. Hinc si centro B radio BA describatur circulus semidiameter GM inter perpendiculum BG et periferiam inscripta determinabit punctum M. Sed hoc Theorema usui minus accommodatur. Sunt et alij quàm plurimi modi trisecandi angulum quemadmodum e sequentibus patebit, ubi etiam quinq; sectio(46) et septusectio docebitur.(47) Interea verò cum anguli recti divisio in partes æquales sit problema simplicius, lubet hic sigillatim ostendere.

---

moreover, may be drawn the circle $x^2+y^2-mx-ny = 0$ of centre $C(\frac{1}{2}m, \frac{1}{2}n)$ and radius $CA = \sqrt{[\frac{1}{4}m^2+\frac{1}{4}n^2]}$, and the family of ellipses $x^2-ny+4(k^2/m^2)(y^2-mx) = 0$ of centre $I(2k^2/m, m^2n/8k^2)$ and semi-axes $\sqrt{[(2k^2/m)^2+(mn/4k)^2]}$ and $\sqrt{[(m^2n/8k^2)^2+k^2]}$. The former is easily described on centre C through the origin A. The latter Newton constructs by an elliptic trammel FPL guided by the 'runners' HIL and TPI, where, on taking $AD = k$, $AK = 2k^2/m$,

Note that this construction is made just as though an ellipse were described intersecting the circle in F.(43)

Problem 16

To bisect a given angle $B\hat{A}C$, and that continually.(44)

Draw any line BD parallel to one of the sides, AC, and meeting the second, AB, in B. In it take $BD = BA$, $DE = DA$, $EF = EA$, and so on. Then AD will bisect the angle $B\hat{A}C$, AE will bisect the angle $D\hat{A}C$, AF will bisect the angle $E\hat{A}C$, and so on.

For the angles $B\hat{A}D$, $B\hat{D}A$ and $D\hat{A}C$ are equal, and so on.

Problem 17

To trisect a given angle $B\hat{A}C$, and that continually.

Draw BD parallel to AC as before, then at B erect the perpendicular BG and in the angle $D\hat{B}G$ inscribe $DG = 2AB$; erect DK and inscribe $HE = 2AD$; erect EL and inscribe $IF = 2AE$; and so on. Then will $D\hat{A}C$ be a third of the angle $B\hat{A}C$, $E\hat{A}C$ a third of the angle $D\hat{A}C$, $F\hat{A}C$ a third of the angle $E\hat{A}C$. And so forth.

For bisect GD in M and there will be $BM = MD = AB$. Therefore the angle $D\hat{A}C$ or $A\hat{D}B = D\hat{B}M = \frac{1}{2}B\hat{M}A = \frac{1}{2}B\hat{A}M$: that is, $D\hat{A}C = \frac{1}{3}B\hat{A}C$.(45) And so on.

Scholium. Hence if with centre B and radius BA be drawn a circle, the semidiameter GM inscribed between the perpendicular BG and the periphery will determine the point M. But this theorem is less suitable in practice. There are also innumerable other ways of trisecting an angle, as will be evident in the sequel, where too quinquesection and septisection will be taught.(47) But meanwhile, since division of a right angle into equal parts is a simpler problem, it is agreeable to expose it here separately.

$IK = m^2n/8k^2$ and $KH = mn/4k$, the trammel segments are $FL = AH = \sqrt{[AK^2+KH^2]}$ and $FP = AE = \sqrt{[AD^2+IK^2]}$.

(43) This is, of course, the ellipse (F) constructed by the trammel (note (42)).

(44) Compare Schooten's *Exercitationes* (note (3)): 124–8: 'Problema I. Datum angulum rectilineum *BAC* bifariam secare'. As we have seen (I, **1**, 1, §2) Newton's draft of this present problem exists, appropriately enough, in his library copy (Trinity College, Cambridge. NQ. 16. 184) of Schooten's work.

(45) And similarly, of course, for the other angles $E\hat{A}C$, $F\hat{A}C$, ... by bisecting HE, IF, ... and so on in succession. Compare Newton's undergraduate notes on Viète's *Supplementum Geometriæ* (I, **1**, 2, §2.1.)

(46) Read 'quinquesectio'.

(47) The completed portion of Newton's tract below does not discuss the general problem of angular section. Since, however, the division of an angle into a given odd number requires the resolution of an algebraic equation of corresponding degree (compare the final problem of Newton's Kinckhuysen observations (1, §2 above)), the general trisection of an angle, being resolvable by a cubic, would be 'solid' (constructible, that is, by the intersections of conics) in Newton's present terminology, but division into five or seven equal parts would be 'linear'.

Prob 18.

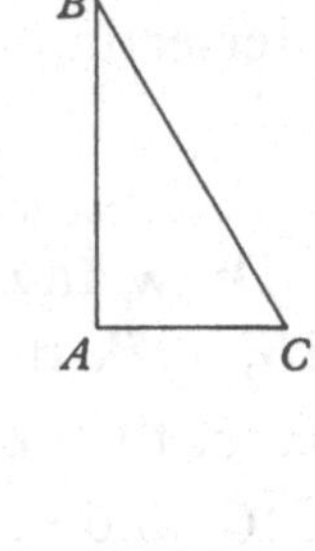

Anguli recti BAC trientem invenire.(48)
Inscribe $BC=2AC$.(49) Et erit CBA triens.

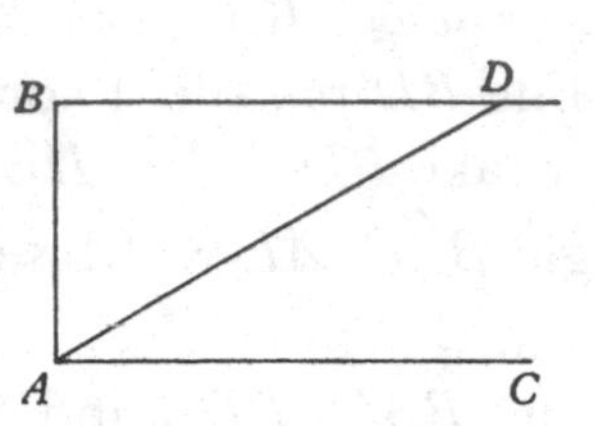

Vel sic. Agatur parallela BD et inscribe $AD=2AB$. [Erit DAC triens.](50)

Prob 19.

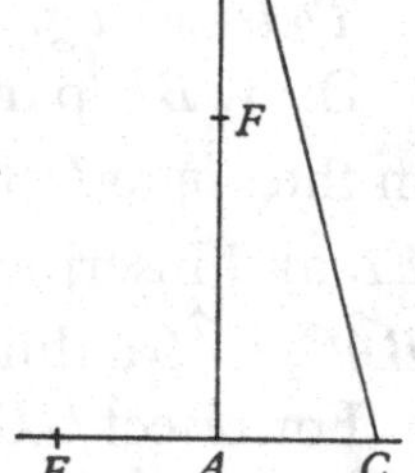

Anguli recti quintam partem invenire.(51)
Cape quamlibet AE et ejus duplum AF. Cape $EC=EF$, et a puncto C inscribe $CB=2AF$ eritq CBA quinta pars recti BAC.

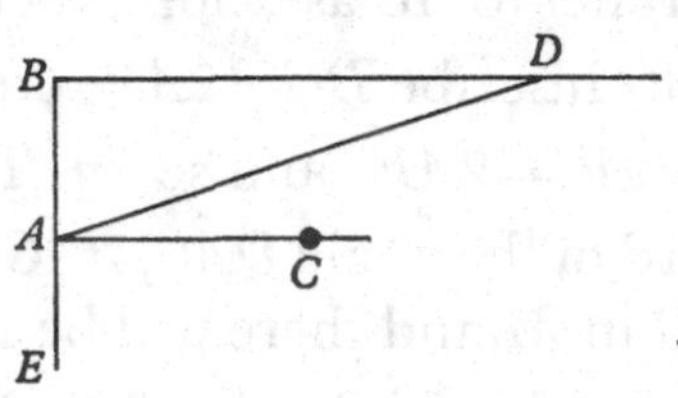

Vel sic. Cape quamlibet AE et ejus duplum AC, et $EB=EC$. Duc $DB \| AC$ et inscribe $AD=2AC$.(52)

Prob 20.

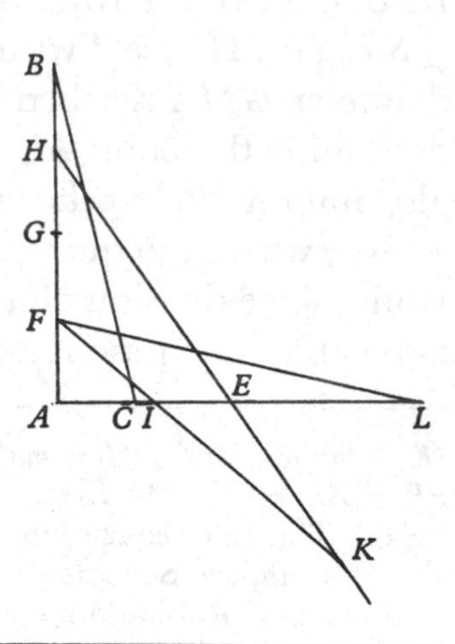

Anguli recti septusectio.
Cape $AF=FG=GH=HB$ cujusvis longitudinis et $AE=AG$. Junge HE et produc versus K. In angulo AEK inscribe $IK=AH$ quæ transeat per F. Fac $AC=IE$. Junge BC, et erit CBA septima pars recti BAC.(53)

---

(48) The equivalent enunciation 'Angulum rectum *BAC* trifariam secare' is cancelled.

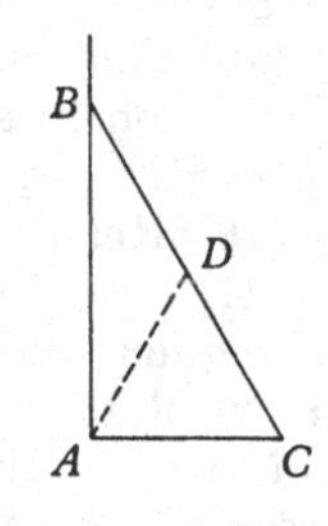

(49) Newton has cancelled a first continuation 'Cape $CD=AC$. junge AD' (Take $CD = AC$ and join AD) and deleted the line AD in his figure correspondingly. Of course, DA will be the trisector of the right angle $B\hat{A}C$.

(50) In analytical terms, if $BC = r$ and $A\hat{B}C = \frac{1}{6}\pi = \alpha$, then on setting $x = 2r\cos\alpha$ and $y = 2r\cos 2\alpha = (x^2-2r^2)/r$ it follows that

$$\cos 3\alpha = 0 = x^3-3r^2x = x(y-r), \quad \text{or} \quad y = r,$$

and so $AC = r\cos 2\alpha = \frac{1}{2}r$.

(51) A first version of this enunciation read equivalently 'Anguli recti quinquesectio'.

Problem 18

To find the third part of the right angle $B\hat{A}C$.

Inscribe $BC = 2AC$.(49) Then will $C\hat{B}A$ be its third part.

Or thus. Draw the parallel BD and inscribe $AD = 2AB$. [Then will $D\hat{A}C$ be its third part.](50)

Problem 19

To find the fifth part of a right angle.

Take any line AE and its double AF. Take $EC = EF$ and from the point C inscribe $CB = 2AF$: $C\hat{B}A$ will be the fifth part of the right angle $B\hat{A}C$.

Or thus. Take any line AE and its double AC, and also $EB = EC$. Draw BD parallel to AC and inscribe $AD = 2AC$.(52)

Problem 20

Division of the right angle into seven equal parts.

Take $AF = FG = GH = HB$ of any length and $AE = AG$. Join HE and produce it in the direction of K, and in the angle $A\hat{E}K$ inscribe $IK = AH$ so that it passes through F. Make $AC = IE$ and join BC. Then will $C\hat{B}A$ be the seventh part of the right angle $B\hat{A}C$.(53)

---

(52) A trivial variant of the preceding. Newton's construction is an immediate corollary of *Elements*, IV, 10, but its analytical derivation is straightforward. If $BC = r$ and $A\hat{B}C = \frac{1}{10}\pi = \alpha$, then on setting $x = 2r\cos\alpha$ and $y = 2r\cos 2\alpha = (x^2-2r^2)/r$ there follows

$$\cos 5\alpha = 0 = x^5-5r^2x^3+5r^4x = r^2x(y^2-ry-r^2), \quad \text{or} \quad y^2-ry-r^2 = 0,$$

and therefore $y = \frac{1}{2}(1+\sqrt{5})r$: hence $AC = r\sin\alpha = \frac{1}{2}r\sqrt{[2-2\cos 2\alpha]} = \frac{1}{4}(\sqrt{5}-1)r$.

(53) Newton has made a mistake in his construction, for BC (and not BA) should equal $4FA$. Much as in note (34) we may set $FE = a$, $FA = b$, $AE = c$, $AH = KI = d$ and suppose $FK = x$, $EI = y$: it then follows that $xy = c(b+d)$ and $x^3-(b-d)x^2+(b^2-a^2)x-c^2(b+d) = 0$, or $y^3-[(b^2-a^2)/c]y^2+(b^2-d^2)y-c(b+d)^2 = 0$. If now $A\hat{B}C = \frac{1}{14}\pi = \alpha$ and we set

$$x = 2r\cos\alpha, \quad y = 2r\cos 2\alpha = (x^2-2r^2)/r,$$

then $\cos 7\alpha = x^7-7r^2x^5+14r^4x^3-7r^6x = r^3x(y^3-ry^2-2r^2y+r^3) = 0$. When therefore the cubic factor is identified with the preceding cubic in y we find $b^2-a^2 = cr$, $b^2-d^2 = -2r^2$ and $-c(b+d)^2 = r^3$: that is, on setting $GA = EA = -c = r$, we have $b+d = r$ and so $b-d = -2r$ (or $b = -\frac{1}{2}r$, $d = \frac{3}{2}r$), and further $FE^2 = a^2 = b^2+c^2 = FA^2+AE^2$, or $F\hat{A}E$ is right. (Negative signs attached to line-lengths denote, of course, that they are to be taken in the opposite sense from their equivalents in Problem 15.) We should note that there are three possible positions of IK in Newton's figure correspondingly as $IE = y$ is a root of $y^3-ry^2-2r^2y+r^3 = 0$, that is, as $IE = 2r\cos 2\alpha$ satisfies the equality $\cos 7\alpha = 0$: precisely, $IE = AC$ may take on any of the three values

$$2r\cos(\pi/7) = 2r\sin(5\pi/14), \quad 2r\cos(3\pi/7) = 2r\sin(\pi/14), \quad 2r\cos(5\pi/7) = -2r\sin(3\pi/14).$$

On taking, correctly, $BC = 4FA = 2r$, the second does indeed construct $A\hat{B}C = \pi/14$, one seventh of a right angle.

Vel cape $FL = FK$ et erit FLA pars septima.(54)

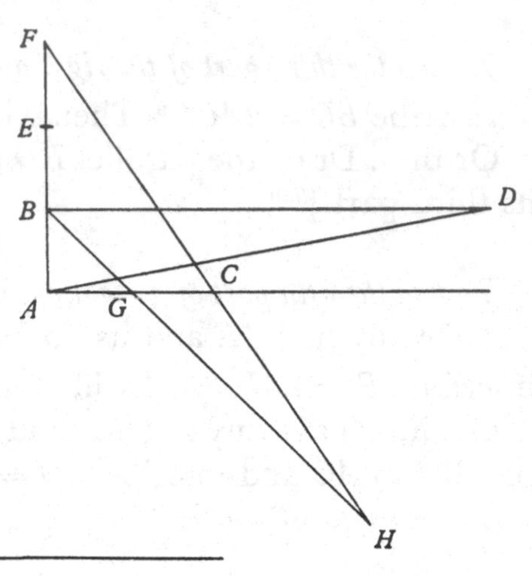

Vel sic. Cape $AB = BE = EF$, $AC = AE$, junge FC et produc [&] inscribe $GH = AF$ quæ transeat per B. Duc $BD \parallel AC$ & $= BH$ ac junge AD. [Erit BDA pars septima.](55)

[2] Sequuntur generalia problemata infinitis modis soluta.

Prob 1.

Æquationem quadraticam $xx = px + q$ solvere.(56)

Duc quamvis rectam lineam KC et in ea cape KA cujusvis longitudinis quam dic n.(57) Dein cape $KB = \frac{q}{n}$ versus A si habetur $+q$, aliter ad contrarias partes. Biseca AB in C et centro C radio AC fac circulum AX cui inscribe $AX = p$, et ut[r]inq; produc donec in Y et Z secet alium circulum quem centro C radio CK describes, et erunt AY et AZ radices desideratæ.(58)

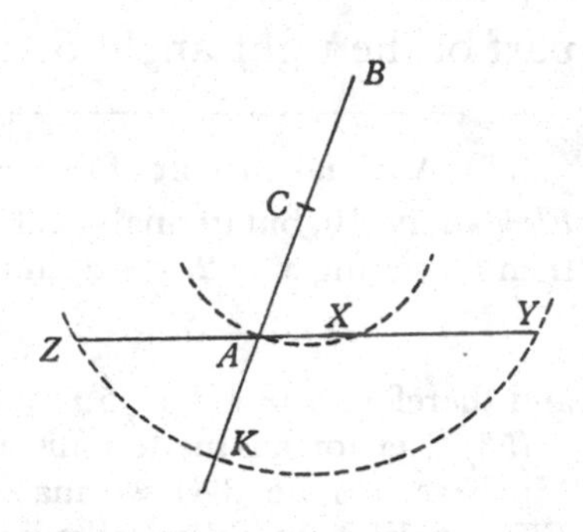

Et nota 1 quòd affirmativæ radices cadunt versus X si habetur $+p$, et negativæ ad contrarias ptes. Et contra si habetur $-p$.

2 Quòd convenit ut aliquis e divisoribus termini q(59) pro n adhibeatur, & quòd ubi habetur $+q$ differentia inter n et $\frac{q}{n}$ non debeat esse minor quàm p.

(54) This (correct) alternative construction was added when Newton cancelled the following paragraph. In the terminology of the previous note, since $|xy| = r^2$, when

$$IE = y = 2r \sin(\pi/14), \quad \text{then} \quad FK = FL = x = \tfrac{1}{2}r \operatorname{cosec}(\pi/14).$$

Alternatively, since $AH \times FK \times IE = FH \times IK \times AE$ and $AH = IK$, therefore

$$FK \text{ (or } FL) : \tfrac{1}{2}AE \text{ (or } FA) = 2FH \text{ (or } BC) : IE \text{ (or } AC).$$

(55) This trivially variant addendum is presumably cancelled because Newton noticed that AD (and not BD) should be equal to BH: compare notes (53) and (54).

Or take $FL = FK$ and $F\hat{L}A$ will be the seventh part.(54)

Or thus. Take $AB = BE = EF$, $AC = AE$, then join FC and produce it, inscribing $GH = AF$ so as to pass through B. Draw BD parallel to AC and equal to BH and join AD. [Then will $B\hat{D}A$ be the seventh part.](55)

---

[2] There follow general problems resolved in an infinity of ways.

Problem 1

To resolve the quadratic equation $x^2 = px + q$.(56)

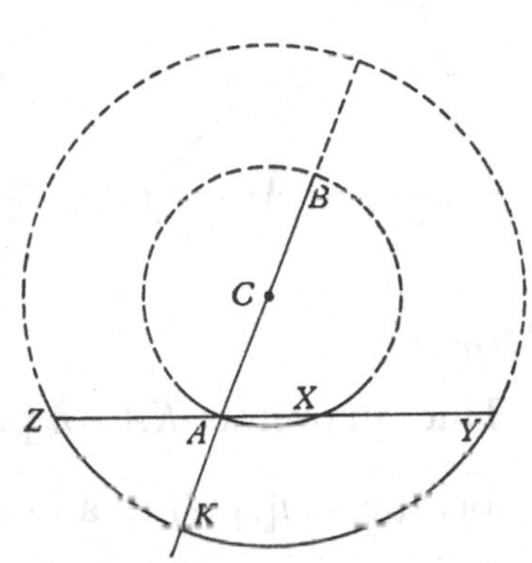

Draw any straight line KC and in it take KA of any length, say n.(57) Then take $KB = q/n$ towards A if $+q$ is supposed, otherwise in the contrary direction. Bisect AB in C and with centre C, radius AC construct the circle AX: in this inscribe $AX = p$, producing it either way till it meets in Y and Z a second circle which you are to describe with centre C and radius CK. Then will AY and AZ be the desired roots.(58)

Note also: 1. Positive roots fall towards X if p is supposed positive, and negative ones in the contrary direction. The contrary is true if p is supposed negative.

2. It is convenient to take some one of the divisors of the term q(59) for n, and when q is supposed positive the difference between n and q/n should not be less than p.

---

(56) Compare the opening section ('The resolution of plaine problems by ye Circle') of Newton's 1665 paper 'Of the construction of Problems' (I, 3, 3, §2). The present construction is, however, general in that the element KA may be chosen arbitrarily.

(57) A following sentence 'Convenit autem ut n sit aliquis e divisoribus termini q' (It is, however, convenient for n to be one of the divisors of the term q) has been cancelled, but its equivalent is added as 'Nota 2' below.

(58) As Newton shows in his 'Demonstratio' below, the construction is effected by factorizing each side of $x^2 - px = q$ as $x(x-p) = n(q/n)$ and then defining the quantities x, p, n and q/n in terms of a suitable geometrical model: precisely, the components x, $x-p$ and n, q/n are made chord-intercepts in a circle ZKY of diameter $n+q/n$. When the point A is fixed in the radius KC by setting $KA = n$, the problem is reduced to finding a chord through A whose intercepts AY and $ZA = XY$ differ by the quantity $AX = p$. Evidently X will be on a concentric circle AX through A. The familiar explicit solution of the quadratic follows immediately, for

$$x - \tfrac{1}{2}p = \tfrac{1}{2}XY = \sqrt{[CY^2 - CX^2 + \tfrac{1}{4}AX^2]} = \sqrt{[\tfrac{1}{4}(n+q/n)^2 - \tfrac{1}{4}(n-q/n)^2 + \tfrac{1}{4}p^2]}.$$

(59) A following phrase 'vel denominator ejus si sit fractio' (or its denominator if it be a fraction) is cancelled.

Demonstrat°. Est $AY \times AZ = AK \times \overline{AB + AK}$. Hoc est $x \times \overline{x - p} = n \times \frac{q}{n}$ sive $xx = px + q$.

Prob 2.

Æquationem Cubicam $x^3 = qx + r$ solvere.

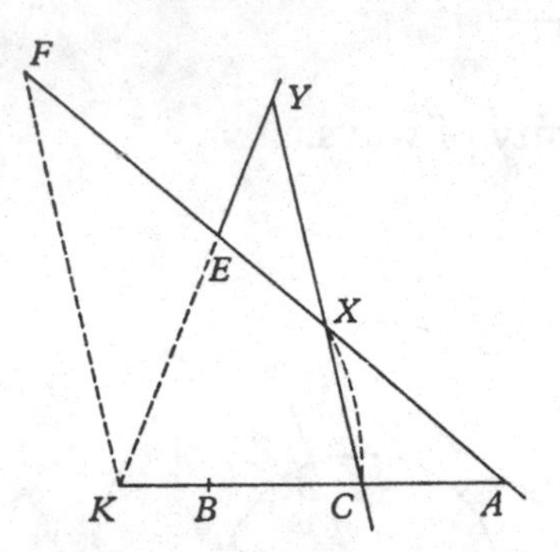

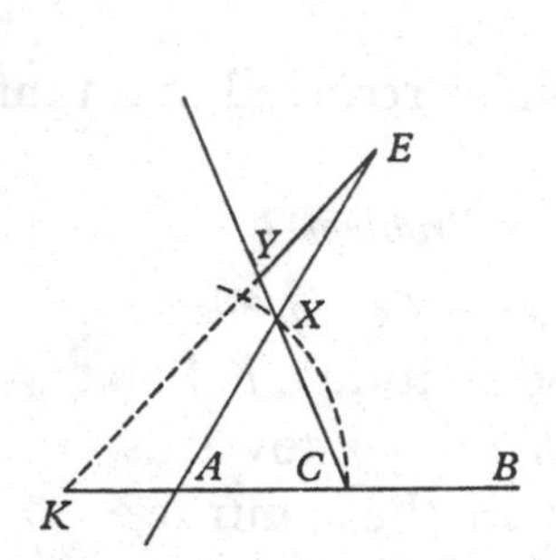

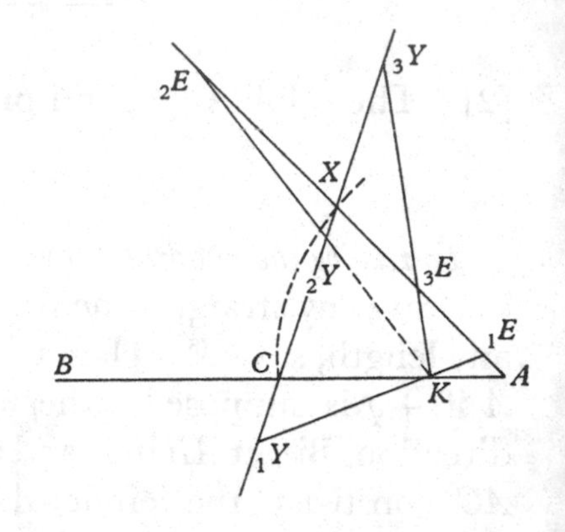

Mod^s 1.

Duc quamlibet $KA = n$ pro arbitrio, et $KB = \frac{q}{n}$ ad easdem partes cum KA si habeatur $-q$[,] aliter ad contrarias. Biseca BA in C et centro K radio KC fac circulum CX cui inscribe $CX = \frac{r}{nn}$, et produc(60) utrinq;. Item junge AX et produc(60) utrinq;. Deniq;(61) inter has lineas CX & AX inscribe EY ejusdem longitudinis cum CA ita ut transeat per K, et XY erit radix æquationis. Radices autem affirmativæ sunt quæ cadunt versus C si habeatur $-r$ et contra si habeatur $+r$.(62)

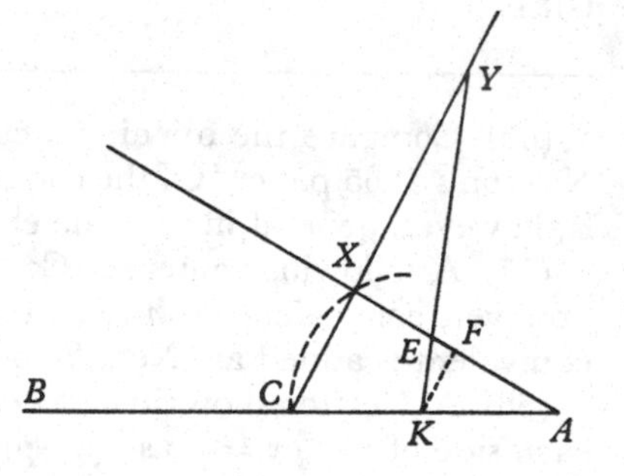

Demonstratio. Age KF parallelam ad CX et erit $AC.AK :: CX.KF$. et $YX.YE(AC) :: KF.KE$. Quare ex æquo perturbatè est $YX.AK :: CX.KE$. Adeóq; $KE = \frac{AK \times CX}{YX}$. Et præterea per compositionem est

$$YX.AK :: (CX + YX =)\, CY.KE + AK.$$

Adhoc per 12[,]2 Elem est

$$YK^q - CK^q = CY^q - CY \times CX = CY \times YX.^{(63)}$$

Hoc est $CY.YK - CK :: YK + CK.YX$. Sed est

$$YK - CK = YK - YE + CA - CK = KE + AK$$

et $YK + CK = YK - YE + CA + CK = BK + KE$. Adeoq; est

$$CY.AK + KE[::]\, BK + KE.YX$$

Demonstration. There is $AY \times AZ = AK(AB+AK)$, that is, $x(x-p) = n \times q/n$ or $x^2 = px+q$.

Problem 2

To resolve the cubic equation $x^3 = qx+r$.

Method 1

Draw any line $KA = n$ at will and $KB = q/n$ in the same direction as KA if there be supposed $-q$, otherwise in the contrary. Bisect BA in C and with centre K, radius KC construct the circle CX: in this inscribe $CX = r/n^2$ and produce it[60] either way. Likewise join AX and produce it[60] either way. Finally,[61] between these lines CX and AX inscribe EY of the same length as CA so as to pass through K, and XY will be a root of the equation. Positive roots are those which fall towards C if r is supposed negative, but the contrary is the case if r is supposed affirmative.[62]

Demonstration. Draw KF parallel to CX and there will be $AC:AK = CX:KF$ and $YX:YE$ (or AC) $= KF:KE$. Hence, on combining equals in contrary order, there is $YX:AK = CX:KE$ and so $KE = (AK \times CX)/YX$. Further, by compounding the ratio, $YX:AK = (CX+YX$ or) $CY:KE+AK$. Moreover, by *Elements*, II, 12, it is $YK^2-CK^2 = CY^2-CY \times CX = CY \times YX$,[63] that is, $CY:YK-CK = YK+CK:YX$. But $YK-CK = YK-YE+CA-CK = KE+AK$ and $YK+CK = YK-YE+CA+CK = BK+KE$, so that

$$CY:AK+KE = BK+KE:YX,$$

(60) The less general phrases, 'versus Y' (in the direction of Y) and 'versus E' (in the direction of E) respectively, are cancelled.

(61) The tautologous 'in angulo' (in the angle) is cancelled.

(62) Much as in notes (34) and (53) above, on setting $KX = a$, $KC = b$, $CX = c$, $CA = YE = d$ and supposing $XY = x$ we may find

$$x^3-[(b^2-a^2)/c]x^2+(b^2-d^2)x-c(b+d)^2 = 0.$$

When this is identified with $x^3-qx-r = 0$, there results $a = b$ (or $KX = KC$), $b^2-d^2 = -q$ and $c(b+d)^2 = r$: Newton then takes $KA = b+d$ of arbitrary length n, so that $CX = c = r/n^2$ and $KB = b-d = -q/n$. In the 'Demonstratio' which follows he gives a synthetic composition of this analysis in classical style.

(63) CX is twice the projection of CK normally on CY.

et connectendo similes rationes, $YK.AK::BK+KE.YX$. Quare cum fuerit $\frac{AK\times CX}{YX}=KE$. erit $YX^q=AK\times BK+\frac{AK^q\times CX}{YX}$, sive

$$YX^{cub}=AK\times BK\times YX+AK^q\times CX.$$

Hoc est restituendo præscriptos valores $x^3=qx+r$. Q.E.D.

(64)*Coroll:* Si $KB=0$. sive KA bisectur in C erit $AK.YX.KE.CX\div\div$(65)

Aut sic meliùs. *Lem 1.* $YK.AK::CX.KE$.

Lem 2. Est $YX.AK::CY.AK+KE$.

Lem 3. Est $BK+KE.YX::YX.AK$.

Demonstr. In primo Lem[mate] erat $YX.AK::CX.KE$. sive $\frac{AK\times CX}{YX}=KE$. quem valorem scribe pro KE in Lem 3 et orietur $BK+\frac{AK\times CX}{YX}.YX::YX.AK$. Hoc est $YX^{cub}=AK\times BK\times YX+AK^q\times CX$.

Mod^s 2. Æquationem cubicam $x^3=qx+r$ aliter construere.

Constructio.

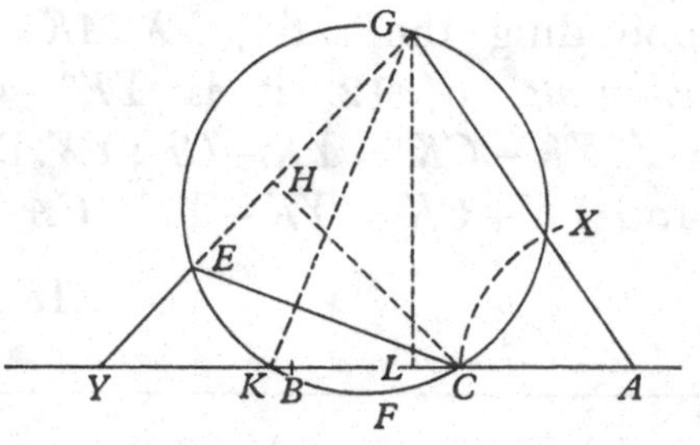

Duc quamlibet KA ad arbitrium quam dic n et $KB=\frac{q}{n}$ ad easdem partes cum KA si habeatur $+q$, aliter ad diversas. Biseca BA in C et centro A radio AC fac circulum CX cui inscribe $CX=\frac{r}{n^2}$. Et per puncta K C et X describe circulum. Junge AX et produc donec iterum secet circulum KCX in G. Deniq; inter hunc circulum et rectam KC utrinq; productam inscribe EY ejusdem longitudinis cum AC ita ut transeat per punctum G: Et EC ducta erit radix æquationis. Radices autem affirmativæ cadunt in majori segmento KGC si habeatur $+r$ et in minori KFC si habeatur $-r$.(66)

Ad Demonstrationem.

Lem 1. Est $CE.KA::CE+CX.AY$. Nam ducta KG est

$$AC.AK::CX.KG \quad \text{et} \quad CE.EY(AC)::KG.KY.$$

Quare ex æquo pertubatè est

$CE.KA::CX.KY$ (et per compositionem)$::CX+CE.KA+KY(AY)$.

---

(64) The following is a late insertion added in afterthought.

(65) See [1], Problem 15 (first construction) and compare note (34) above. When

$$KB=-q/n=0 \text{ (or } q=0),$$

the equation resolved becomes $x^3=r=KA^2\times CX$.

and on conjoining similar ratios $YK:AK = BK+KE:YX$. Hence, since KE was equal to $(AK\times CX)/YX$, there will be $YX^2 = AK\times BK+(AK^2\times CX)/YX$ or $YX^3 = AK\times BK\times YX+AK^2\times CX$. That is, on restoring allotted values, $x^3 = qx+r$. Which was to be demonstrated.

(64)*Corollary*. If KB is zero or KA is bisected in C, then will AK, YX, KE and CX be in continued proportion.(65)

Or better thus. *Lemma 1.* $YK:AK = CX:KE$.

Lemma 2. $YK:AK = CY:AK+KE$.

Lemma 3. $BK+KE:YX = YX:AK$.

Demonstration. In the first Lemma there was $YK:AK = CX:KE$ or

$$KE = (AK\times CX)/YX.$$

Write this value for KE in Lemma 3 and there will come out

$$BK+(AK\times CX)/YX:YX = YX:AK,$$

that is, $YX^3 = AK\times BK\times YX+AK^2\times CX$.

Method 2. To construct the cubic equation $x^3 = qx+r$ another way.

Construction.

Draw any line KA at will, say n, and $KB = q/n$ in the same direction as KA if q is supposed positive, otherwise in the opposite one. Bisect BA in C and with centre A, radius AC construct the circle CX and in it inscribe $CX = r/n^2$. Then through the points K, C and X describe a circle and, joining AX, produce it till it intersects the circle KCX again in G. Finally, between this circle and the straight line KC produced either way inscribe EY of the same length as AC so as to pass through the point G: when EC is drawn it will be a root of the equation. Positive roots fall within the greater segment KGC if r is supposed positive, but in the lesser KFC if r is supposed negative.(66)

For the demonstration.

Lemma 1. There is $CE:KA = CE+CX:AY$. For on drawing KG it is

$$AC:AK = CX:KG \quad \text{and} \quad CE:EY \text{ (or } AC) = KG:KY.$$

Hence on combining equals in contrary order

$CE:KA = CX:KY =$ (by compounding the ratio) $CX+CE:KA+KY$ (or AY).

---

(66) Much as before (see note (36)), on setting $KC = a$, $CA = b$, $AX = YE = c$, $KG = d$ and supposing $EC = x$ it follows that $x^3-[(a+b)(b^2-c^2)/cd]x^2-(b^2-a^2)x-(a+b)cd = 0$, and when this is identified with $x^3-qx-r = 0$ there comes $b = c$ (or $AC = AX$), $b^2-a^2 = q$ and $(a+b)cd = r$. Now, since the triangles ACX, AKG are isosceles with common angle $\hat{A}$ they are similar, and so $CX = CA\times KG/KA = dc/(a+b)$: hence on setting $KA = a+c = n$, arbitrarily, there is in consequence $KB = c-a = q/n$ and $CX = r/(a+b)^2 = r/n^2$. The composition of this analysis is set out synthetically in the 'Demonstratio' which follows.

Lem 2. Demisso CH ad YG perpendiculo est $2EY\times EH=CE\times CX$. Nam demisso etiam GL perpendiculo ad AY, propter sim[ilia] tri[angula] KGL, ECH erit $KG.KL::EC.EH$. Item per 13 2^{di} Elem est

$$KG^q-AG^q+AK^q(=KG^q)=2KA\times AL.$$

Adeoq 2KA. KG (sive 2CA. CX id est::2EY. CX)::$KG.KL$. Quare

$$2EY.CX::EC.EH.$$

ade[ó]q $2EY\times EH=EC\times CX$.

Lem 3. Est $BY.CE::CE.KA$. Nam per 12[,]2 Elem est

$$CY^q-EY^q=CE^q+2EY\times EH=(\text{per Lem 2})\ CE^q+CE\times CX.$$

Adeoq $CE+CX.CY+EY::CY-EY.CE$. Est autem $EY=CA=CB$ et inde $CY+EY=AY$ et $CY-EY=BY$. Atq ita

$$CE+CX.AY::BY.CE::(\text{per lem 1})\ CE.KA.$$

Demonstratio. Per Lem 1 est $CE.KA::CX.KY$. Adeoq $\dfrac{KA\times CX}{CE}=KY$ et $BK+\dfrac{KA\times CX}{CE}=BY$. Quare per Lem 3 est $BA+\dfrac{KA\times CX}{CE}.CE::CE.KA$. et multiplicando extrema et media &c[67] est

$$CE^{\text{cub}}=BK\times KA\times CE+KA^q\times CX.$$

Et restitutis valoribus supra assignatis $x^3=qx+r$.

Coroll. Si $BK=0$ sive $KC=CA$ erit $KA.CE::CE.KY::KY.CX$.[68]

Modus 3. Idem adhuc aliter.[69]

Prob 3. Æquationem tertio termino carentem $x^3=pxx+r$ infinitis modis construere.

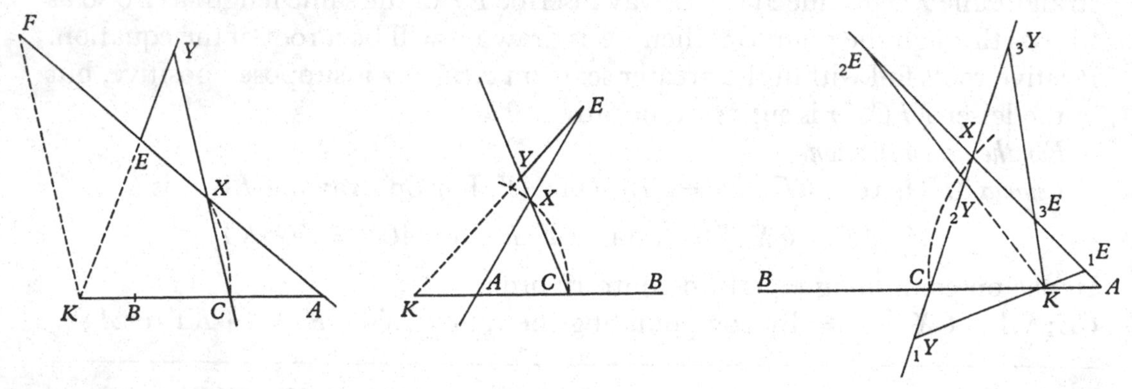

Assumpto quolibet n duc $KA=\dfrac{r}{nn}$ & $KB=p$ ad easdem partes si habeant eadem signa, aliter ad diversas: biseca BA in C, et centro K radio KC fac circulū, cui inscribe $CX=n$, et produc[60] utrinq. Item junge AX et produc[60] utrinq.

(67) Precisely, on multiplying through by CE.

Lemma 2. When there is let fall CH perpendicular to YG, then

$$2EY \times EH = CE \times CX.$$

For, on letting fall also GL perpendicular to AY, on account of the similar triangles KGL, ECH there will be $KG:KL = EC:EH$. Further, by *Elements*, II, 13, there is $KG^2 - AG^2 + AK^2 (= KG^2) = 2KA \times AL$, so that

$$2KA:KG \text{ (or } 2CA:CX\text{, that is, } 2EY:CX) = KG:KL.$$

Hence $2EY:CX = EC:EH$, and so $2EY \times EH = EC \times CX$.

Lemma 3. There is $BY:CE = CE:KA$. For by *Elements*, II, 12, it is

$$CY^2 - EY^2 = CE^2 + 2EY \times EH = \text{(by Lemma 2) } CE^2 + CE \times CX,$$

so that $CE + CX:CY + EY = CY - EY:CE$. However, $EY = CA = CB$ and thence $CY + EY = AY$ and $CY - EY = BY$. Accordingly,

$$CE + CX:AY = BY:CE = \text{(by Lemma 1) } CE:KA.$$

Demonstration. By Lemma 1 there is $CE:KA = CX:KY$, so that

$$KY = (KA \times CX)/CE \quad \text{and} \quad BY = BK + (KA \times CX)/CE.$$

Hence by Lemma 3 $BA + (KA \times CX)/CE:CE = CE:KA$ and, by multiplying extremes and middles and so on,[(67)] it is $CE^3 = BK \times KA \times CE + KA^2 \times CX$. And, when the values assigned above are restored, $x^3 = qx + r$.

Corollary. If BK is zero, or $KC = CA$, then $KA:CE = CE:KY = KY:CX$.[(68)]

Method 3. The same yet another way.[(69)]

Problem 3. To construct the equation $x^3 = px^2 + r$ lacking its third term in an infinity of ways.

Assuming any n you please draw $KA = r/n^2$ and $KB = p$ in the same direction should they have the same signs, otherwise in opposite ones: bisect BA in C and with centre K, radius KC construct a circle, and in it inscribe $CX = n$, producing it[(60)] either way. Likewise join AX and produce it[(60)] either way. Finally, between

(68) Compare [1], Problem 15, Idem aliter and see note (36) above. Evidently, when $BK = -q/n = 0$ (or $q = 0$) then $x^3 = r = KA^2 \times CX$.

(69) Newton has left the remainder of his page and the whole of the following one blank for the intended later insertion of his third construction. We cannot know for sure what this was but we may suppose it was a further application of one or other of the two preceding 'conchoidal' models. In the first, for instance, if (note (62)) we now suppose $KE = x$ we may deduce that $x^3 - (b-d)x^2 + (b^2 - a^2)x - c^2(b+d) = 0$; but in the second, on taking $KY = x$ there comes $x^3 + (a-b)x^2 + (b^2 - c^2)x - c^2d^2/(a+b) = 0$. If we identify these cubics with $x^3 - qx - r = 0$, in the former case there comes $b = d$, $b^2 - a^2 = -q$ and $c^2(b+d) = r$, or, on setting $c = n$, $b = d = \frac{1}{2}r/n^2$ and $a = \sqrt{[\frac{1}{4}r^2/n^4 + q]}$; while in the latter there follows $a = b$, $b^2 - c^2 = -q$ and $c^2d^2/(a+b) = r$, or, on setting $a = b = \frac{1}{2}r/n^2$, $c = \sqrt{[b^2 + q]}$ and $d = r/nc$. Neither, of course, is a particularly convenient approach.

[(60) See note (60), page 471.]

Deniq; inter has lineas CX et AX inscribe EY ejusdem longitudinis cum CA ita ut transeat per K; et KE erit radix æquationis. Radices autem affirmativæ sunt ubi Y cadit a parte $[K]$ versus C si modò habeatur $-r$[;] sin habeatur $+r$, affirmativæ sunt ubi Y cadit adversus C.(70)

Schemata(71) vero et Lemmata ad sequentem demonstrationem de priori propositione mutuò sumantur. [Modo 1mo.]

Demōstrat^o. Per Lem 3 erat $KB+KE.YX::YX.AK$ aut (sumpto KB ad easdem partes cum KA) $KE-KB.YX::YX.AK::$(Lem 1) $CX.KE$. Adeoq; $\overline{KE-KB}\times KE.YX\times KE(AK\times CX)::CX.KE$. Quare est

$$KE^{cub}-KB\times KE^q=AK\times CX^q.$$

Et restitutis valoribus supra assignatis orietur $x^3=pxx+r$.

Idem aliter.

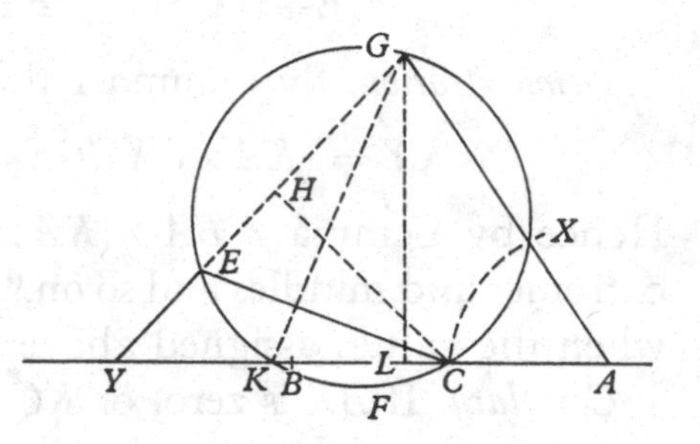

Assumpto quolibet n duc $KA=\frac{r}{nn}$ et $KB=p$ ad diversas(72) partes si habent eadem signa, aliter ad easdem. Biseca BA in C et centro $[A]$ radio $[A]C$(73) fac circulum CX cui inscribe $CX=n$. Junge AX et produc ad G ut sit $AG=AK$ et per puncta $K\,C\,X\,G$ describe circulum. Deniq; inter hunc circulum et rectam KC utrinq; productam inscribe EY ejusdem longitudinis cum AC ita ut transeat per punctum G: Et KY erit radix æquationis. Radices autem affirmativæ cadunt versus A si sit $-r$ et adversus A si sit $+r$.(74)

Schemata(75) itidem et Lemmata ad sequentem demonstrationem e priori prop[ositione] mutuò sumantur. Modo 2do.

Demonstrat^o. Per Lem 3 est $BY(KY-KB).CE::CE.KA::$(Lem 1) $CX.KY$. Adeoq; $\overline{KY-KB}\times KY.CE\times KY(KA\times CX)::CX.KY$ et inde

$$KY^{cub}-KB\times KY^q=KA\times CX^q.$$(76)

(70) In the terminology of notes (62) and (69), when we set $KE=x$ there follows

$$x^3-(b-d)x^2+(b^2-a^2)x-c^2(b+d)=0,$$

and on identifying this with $x^3-px^2-r=0$ there comes $a=b$ (or $KC=KX$), $KB=b-d=p$ and $c^2(b+d)=r$. Newton supposes c (or CX) $=n$ arbitrarily, so that $KA=b+d=r/n^2$.

(71) For convenience, we reproduce these figures a second time.

(72) The cancelled word 'easdem' (the same) betrays some initial doubt in Newton's mind.

(73) The text reads 'centro K radio KC', an evident slip of the pen.

(74) In the terminology of notes (66) and (69), it follows on setting $KY=x$ that

$$x^3+(a+b)x^2+(b^2-c^2)x-c^2d^2/(a+b)=0,$$

and, when this is identified with $x^3-px^2-r=0$, that

$$b=c \text{ (or } CA=CX \text{ and so } CX=CA\times KG/KA=cd/[a+c]),$$

these lines CX and AX inscribe EY of the same length as CA so as to pass through K; and KE will be a root of the equation. Positive roots are those in which Y falls on the side of K towards C provided that r is supposed negative; but if it be supposed affirmative, then positive ones are those in which Y falls on the side away from C.[70]

The illustrations,[71] however, and lemmas for the following demonstration are to be borrowed from the previous proposition[, method 1].

Demonstration. By Lemma 3 it was $KB+KE:YX = YX:AK$ or, when KB is taken in the same direction as KA,

$$KE-KB:YX = YX:AK = \text{(by Lemma 1) } CX:KE,$$

so that $(KE-KB)\,KE:YX\times KE$ (or $AK\times CX$) $= CX:KE$. Hence it is

$$KE^3-KB\times KE^2 = AK\times CX^2.$$

And when the values assigned above are restored there comes $x^3 = px^2+r$.

The same another way

Having assumed n arbitrarily, draw $KA = r/n^2$ and $KB = p$ in opposite[72] directions if they have the same signs, otherwise the same way. Bisect BA in C and with centre A, radius AC[73] construct the circle CX and in it inscribe $CX = n$. Join AX, producing it to G so that $AC = AK$, and through the points K, C, X and G describe a circle. Finally, between this circle and the straight line KC produced either way inscribe EY of the same length as AC so as to pass through the point G: KY will then be a root of the equation. Positive roots, however, fall towards A if r be negative and away from A if it be affirmative.[74]

The illustrations[75] and lemmas for the following demonstration are likewise to be borrowed from the previous proposition, method 2.

Demonstration. By Lemma 3 it is

$$BY \text{ (or } KY-KB):CE = CE:KA = \text{(by Lemma 1) } CX:KY,$$

so that $(KY-KB)\,KY:CE\times KY$ (or $KA\times CX$) $= CX:KY$ and thence

$$KY^3-KB\times KY^2 = KA\times CX^2.\text{[76]}$$

$KB = a-b = -p$ and $c^2d^2/(a+b)$, that is, $c^2d^2/(a+c)$, $= r$. In his construction Newton takes $CX = cd/(a+c) = n$ arbitrarily, so that $KA = a+b = r/n^2$.

(75) We repeat the preceding figure concerned for ease of reference.

(76) The following page is left blank, presumably because Newton thought he might later insert an 'Idem adhuc aliter'. Two further variants of the preceding constructions are indeed possible: namely, by identifying $x^3-px^2-r = 0$ with

$$x^3-[(b^2-a^2)/c]x^2+(b^2-d^2)x-c(b+d)^2 = 0$$

(which results when we suppose $KY = x$ in the former) or with

$$x^3-[(a+b)(b^2-c^2)/cd]x^2-(b^2-a^2)x-cd(a+b) = 0$$

(which is the consequence of setting $EC = x$ in the latter).

Prob 4. Æquationem quamcunq cubicam $x^3=pxx+qx+r$, sive aliquo sive nullo termino carentem solvere, idq modis dupliciter infinitis.

A puncto quovis B cape duas quascunq rectas BC, BE ad eandem plagam sitq BD medium proportionale, et BC dic n. Cape etiam $BA=\frac{q}{n}$, idq versus C si sit $-q$[,] aliter adversus. Et ad punctum A erige perpendiculum AI, inq eo cape $AF=FG=p$, $FI=\frac{r}{nn}$, et fac

$$BE.BC::FI.FH;$$

FI verò et FH capienda sunt versus G si p et r habeant eadem signa, aliter versus A. Dein parallelogramma $IACK$, & $HAEL$ compleantur centróq K intervallo KG describatur circulus. Item in linea HL capiatur HR ad utramvis partem puncti H ita ut sit $BE.BD::HL.HR$, et agatur GR secans EL in S.

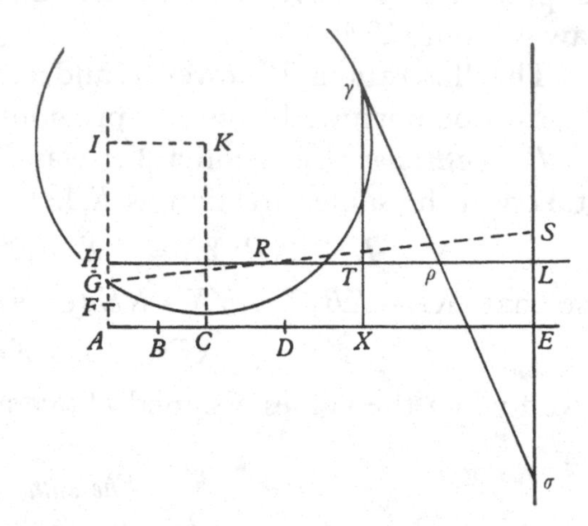

Deniq hæc linea GRS moveatur, puncto ejus R super HL et puncto S super EL incedente, donec tertium ejus punctum G circulo oc[c]urrat quemadmodum videre est in positione $\gamma\rho\sigma$: ac dimidium perpendiculi γX a puncto occursûs γ in rectam AE demissi erit radix æquationis. Potest autem punctum G circulo in tot punctis occurrere quot sunt possibiles radices, et radices affirmativæ sunt quæ cadunt ad eas partes ipsius AE versus quas punctum I sumitur a puncto F si sit $+r$, et contra si $-r$.(77)

Demonstratio.

Conclusio, 1. Notum est quod cum recta $\sigma\rho\gamma$ pro more assignato movetur ejus punctum γ describet Ellipsin cujus centrum est L et axes in rectis LE et LH

(77) Newton has well disguised his analysis of this construction but preliminary investigation will reveal that it must have been equivalent to the following. To the given cubic $x^3-px^2-qx-r=0$ add the fourth root $x-p=0$ and then suppose the resulting quartic $x^4-2px^3+(p^2-q)x^2+(pq-r)x+pr=0$ to be the eliminant of

$$x^2 = ax+by \quad\text{and}\quad y^2 = cy+dx+e,$$

or $x^4-2ax^3+(a^2-bc)x^2+b(ac-bd)x-b^2e=0$. On identifying coefficients we find $a=p$, $bc=q$, $b^2d=r$ and $b^2e=-pr$, or, if we choose $b=n$ arbitrarily, $c=q/n$, $d=r/n^2$ and $e=-pr/n^2$. The expressions which result,

$$\alpha \equiv x^2-px-ny=0 \quad\text{and}\quad \beta \equiv y^2-(q/n)y-(r/n^2)x+pr/n^2=0,$$

are in a normal Cartesian co-ordinate system the defining equations of two parabolas which intersect in points such that $x=p$ or $x^3=px^2+qx+r$, the cubic whose roots are to be constructed. Presumably to avoid numerical fractions in his subsequent computation Newton

Problem 4. To resolve any cubic equation $x^3 = px^2+qx+r$, lacking some one of its terms or none at all, and that in a doubly infinite manner

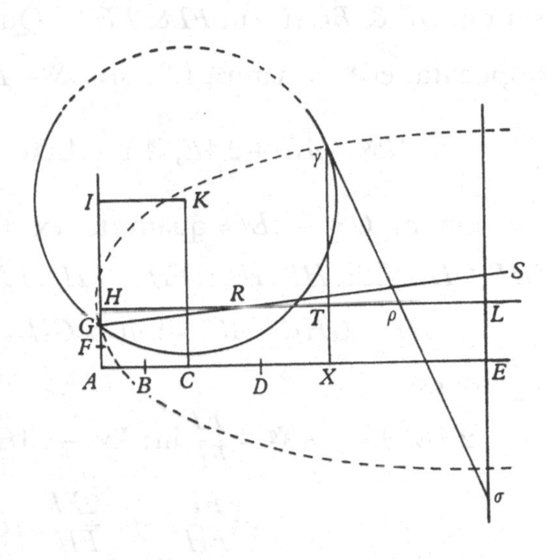

From any point B take any two straight lines BC, BE going the same way and let BD be their mean proportional. Call BC n and take also $BA = q/n$, towards C if q be negative, away from it otherwise. Further, at the point A erect the perpendicular AI and in it take $AF = FG = p$, $FI = r/n^2$ and make $BE:BC = FI:FH$; FI and FH, indeed, should be taken towards G if p and r have the same signs, otherwise towards A. Then let the parallelograms $IACK$ and $HAEL$ be completed and a circle be described on centre K with radius KG. Moreover, in the line HL let HR be taken on whichever side of the point H you wish such that $BE:BD = HL:HR$ and let there be drawn GR intersecting EL in S.

Finally, let this line GRS be moved with its point R travelling on HL and point S on EL till its third point G meet the circle, as may be seen in the position $\gamma\rho\sigma$: then half the perpendicular γX let fall from the meeting point γ onto the line AE will be a root of the equation. It is possible, however, for the point G to meet the circle in as many points as there are possible roots, while positive roots are those which fall that way from AE in which the point I is taken from the point F should r be positive, but the contrary if r is negative.(77)

Demonstration

Conclusion 1. It is known that when the straight line $\sigma\rho\gamma$ moves after its assigned fashion its point γ will describe an ellipse of centre L and with its axes

transforms to the doubled co-ordinates $X = 2x$ and $Y = 2y$, and then considers the family of ellipses (k free) $\alpha+(n^2/k^2)\beta \equiv X^2+(n^2/k^2)Y^2-2(p+r/k^2)X-2(k+q/k)(n/k)Y+4pr/k^2 = 0$ which are, of course, through the parabolic meets $\alpha = \beta = 0$ (and, in particular, through $(2p, 0)$). These have their centres at $(p+r/k^2, (k+q/k)k/n)$ and semi-axes g and $(k/n)g$, where $g^2 = (-p+r/k^2)^2+(k+q/k)^2$. In his figure Newton chooses $BC = n$ and $BD = k$ arbitrarily (his 'modi dupliciter infiniti') and sets $AF = FG = p$, $AB = q/n$, $FI = r/n^2$ and so $BE = k^2/n$, $FH = r/k^2$. In consequence, the circle on centre K (where $AI = p+r/n^2$, $IK = n+q/n$) through G (where $AG = 2p$) is the member of the ellipse family for which $n = k$, while G and γ (where $X\gamma = X$, $AX = Y$) are on the general ellipse of semi-axes $GR = g$, $GS = (k/n)g$ of centre L (where $AH = p+r/k^2$, $HL = AE = q/n+k^2/n$). As a final cloak thrown over his working, the latter ellipse is constructed by a trammel construction (note (78)).

positi [sunt], quorum qui in LE jacet æquatur $2\gamma\rho$ sive $2GR$ et alter in LH æquatur $2\gamma\sigma$ sive $2GS$.(78) Et horum ratio est HR ad HL sive BD ad $[B]E$, ratio nempe linearum sub initio assumptarum. Unde latus rectum et transversum sunt ut BE & BC sive ut FI & FH.(79) Quamobrem cum $T\gamma$ sit ad HL ordinatim applicata, erit ex naturâ Ellipsis $GS^q - LT^q = \frac{FI}{FH} \times T\gamma^q$. sive

$$GS^q - AE^q + 2AE, AX - AX^q = \frac{FI}{FH} \text{ in } X\gamma^q - 2AH, X\gamma + AH^q:$$

Est autem $GS^q - AE^q =$ quadrato ex $GH + LS$, hoc est $= \frac{FI}{FH} \times GH^q$, (nam $GH.GH+LS::HR.HL::FH.\surd FH \times FI$.) Item est

$$AH^q - GH^q (= AG^q + 2AG \times GH = 2AG \times \tfrac{1}{2}AG + GH) = 2AG \times FH.$$

Quare est

$$2AE, AX - AX^q = \frac{FI}{FH} \text{in}: X\gamma^q - 2AH, X\gamma(+AH^q - GH^{q\,(80)}) + 2AG, FH:$$
$$= \frac{FI}{FH} X\gamma^q - \frac{2FI}{FH} AH, X\gamma + 2AG, FI.$$

Conclus: 2. Porro ex natura circuli est $K\gamma^q - CX^q =$ quadrato ex $\gamma X - AI$. Sed est $K\gamma^q (GI^q + AC^q) - CX^q (-AC^q + 2AC, AX - AX^q) = GI^q + 2AC, AX - AX^q$, adeoq $= \gamma X^q - 2AI, \gamma X + AI^q$. Insuper est

$$AI^q - GI^q = AG^q + 2AG \times GI = 2AG \times :\tfrac{1}{2}AG + GI: = 2AG \times FI.$$

Quare etiam $2AC \times AX - AX^q = \gamma X^q - 2AI \times \gamma X + 2AG \times FI$.

Conclus: 3. Jam hi termini subducantur a prioribus per Ellipsin inventis et restabunt $2CE, AX = \frac{HI}{FH} X\gamma^q - \frac{2FI}{FH} AH, X\gamma + 2AI \times X\gamma$. Est autem $CE = \frac{HI, BC}{FH}$, ac $\frac{2FI, AH}{FH} - 2AI \left(= \frac{2FI, AH}{FH} - 2AH - 2HI = \frac{2HI \times AH}{FH} - 2HI \right) = \frac{2HI \times AF}{FH}$. Quare est $\frac{2HI}{FH} BC, AX = \frac{HI}{FH} X\gamma^q - \frac{2HI}{FH} AF, X\gamma$. Sive $2BC, AX = X\gamma^q - 2AF, X\gamma$. Atq adeò $AX.X\gamma - 2AF(X\gamma - AG)::X\gamma.2BC$.

Conclus: 4. A terminis ultimò inventis subducantur termini in conclusione 2dâ per circulum inventi et restabunt $-2AB, AX + AX^q = 2FI, X\gamma - 2AG, FI$. Unde est $AX.X\gamma - AG::2FI.AX - 2AB$ (adeóq)$::X\gamma.2BC$.(81)

(78) Compare note (77) and I, 1, 1, §1.3*f*.

(79) The ratio of the latus rectum $2GR^2/GS$ and latus transversum $2GS$ is

$$GR^2:GS^2 = n^2:k^2 = n:k^2/n \text{ (or } BC:BE) = r/k^2:r/n^2 \text{ (or } FH:FI).$$

(80) Read 'sive' for clarity: Newton means $AH^2 - GH^2 = 2AG \times FH$.

(81) By Conclusion 3.

placed along the lines LE and LH: of these that which lies along LE is equal to $2\gamma\rho$ or $2GR$ while the second along LH is equal to $2\gamma\sigma$ or $2GS$.(78) Their ratio, moreover, is HR to HL or BD to $[B]E$, the ratio, namely, of the lines assumed towards the beginning. Hence the *latus rectum* and major axis are as BE and BC or as FI and FH.(79) Therefore, since $T\gamma$ is ordinately applied to HL, by the property of the ellipse there will be $GS^2-LT^2=(FI/FH)\times T\gamma^2$, or

$$GS^2-AE^2+2AE\times AX-AX^2=\frac{FI}{FH}(X\gamma^2-2AH\times X\gamma+AH^2).$$

However, $GS^2-AE^2=$ the square of $(GH+LS)$, that is, $=FI/FH\times GH^2$ (for $GH:GH+LS=HR:HL=FH:\surd(FH\times FI)$). Further,

$$AH^2-GH^2=(AG^2+2AG\times GH=2AG(\tfrac{1}{2}AG+GH)=)\;2AG\times FH.$$

Hence $2AE\times AX-AX^2=\dfrac{FI}{FH}(X\gamma^2-2AH\times X\gamma+[AH^2-GH^2=]\;2AG\times FH)$

$$=\frac{FI}{FH}X\gamma^2-\frac{2FI}{FH}AH\times X\gamma+2AG\times FI.$$

Conclusion 2. Furthermore, by the property of the circle it is

$$K\gamma^2-CX^2=\text{the square of }(\gamma X-AI).$$

But

$K\gamma^2$ (or GI^2+AC^2) $-CX^2$ (or $-AC^2+2AC\times AX-AX^2$)

$$=GI^2+2AC\times AX-AX^2,$$

so that this is equal to $\gamma X^2-2AI\times\gamma X+AI^2$. In addition,

$$AI^2-GI^2=AG^2+2AG\times GI=2AG(\tfrac{1}{2}AG+GI)=2AG\times FI.$$

Hence also $2AC\times AX-AX^2=\gamma X^2-2AI\times\gamma X+2AG\times FI$.

Conclusion 3. Now let these terms be taken away from the former ones found by the ellipse and there will remain

$$2CE\times AX=\frac{HI}{FH}X\gamma^2-\frac{2FI}{FH}AH\times X\gamma+2AI\times X\gamma.$$

However, $CE=HI\times BC/FH$ and

$$\frac{2FI\times AH}{FH}-2AI=\left(\frac{2FI\times AH}{FH}-2AH-2HI=\frac{2HI\times AH}{FH}-2HI=\right)\frac{2HI\times AF}{FH}.$$

Hence

$$\frac{2HI}{FH}BC\times AX=\frac{HI}{FH}X\gamma^2-\frac{2HI}{FH}AF\times X\gamma,\quad\text{or}\quad 2BC\times AX=X\gamma^2-2AF\times X\gamma,$$

and so $AX:X\gamma-2AF$ (or $X\gamma-AG$) $=X\gamma:2BC$.

Conclusion 4. From the terms most recently found let there be taken away the terms in the second conclusion found by the circle and there will remain $-2AB\times AX+AX^2=2FI\times X\gamma-2AG\times FI$. Hence

$$AX:X\gamma-AG=2FI:AX-2AB=\text{(in consequence)}.\;X\gamma:2BC.^{(81)}$$

Conclus: 5. Hinc est

$$2BC\times 2FI\,.\,2BC\times AX\ (\text{sive } X\gamma^q-2AF, X\gamma)-2BC\times 2AB::X\gamma\,.\,2BC$$

et multiplicando extrema et media orietur

$$X\gamma^{\text{cub}}-2AF, X\gamma^q-4BC, AB, X\gamma=8BC^q, FI.$$

Et restitutis supra assignatis valoribus,(82) fit

$$x^3-pxx-qx=r \text{ sive } x^3=pxx+qx+r.^{(83)}$$

Coroll. Hinc si AF et AB ponantur nulla, per conclusionem 4 patet esse $2FI\,.\,AX::AX\,.\,X\gamma::X\gamma\,.\,2BC$. Unde constat inventio duorũ mediè proportionalium inter data quælibet FI et BC, idqȝ modis infinitis.(84)

Cæterùm hæc regula videtur omnium tam generalium(85) perfectissima quæ excogitari potest, sive spectes genus curvæ quam motus rectæ $\gamma\rho\sigma$ innuit esse describendam, sive simplicitatem constructionis et varietatem. Nam duæ sunt quantitates BC ac BE ad arbitrium assumendæ quarum altera BC potest esse unitas vel quilibet divisor terminorũ q et r aut alia quævis quantitas quæ constructionem abbreviabit[,] deinde altera BE in quavis ratione ad priorem BC sumi potest prout velis determinare speciem Ellipsis per motum puncti γ describendæ, aut efficere ut circulum minùs obliquè secet. Convenit autem (ad devitandum incommodum inveniendi medium proportionale BD et quarta proportionalia FH et HR per ductum linearum) ut BE ad BC sumatur in ratione quadrati numeri ad quadratum numerum.

Verùm ut hujus regulæ perfectio adhuc magis elucescat, notandum est quòd non modò competit Ellipsi sed cæteris etiam Conicis sectionibus: ut ut sola Ellipsis per motum unicæ tantùm rectæ ferè ad instar circuli(86) describitur, et inde huic rei præ cæteris accommodatur.

Quemadmodum si velis Hyperbolam adhibere cujus latus rectum ad latus transversum sit in quâlibet ratione puta d ad e: Cape BC ad arbitrium, et ad contrarias partes puncti B cape BE quæ sit ad BC sicut e ad d; nam species curvæ determinatur ex assumptione lineæ BC.(87) Dein puncta A, F, G, I, H, K, L

(82) $X\gamma = 2x$, $AF = p$, $BC = n$, $AB = q/n$, $FI = r/n^2$.

(83) It is instructive to interpret the steps of Newton's composition of his argument, here laid out in classical synthetic style, with the analytical equivalents suggested in note (77). It seems likely that he built up the main stages of that composition by a prior analytical argument, and then transformed them into their geometrical forms, filling in the gaps in his resulting synthetic argument in approved classical style.

(84) In the terminology of note (77) when $p = q = 0$, then $\alpha \equiv x^2-ny = 0$ and

$$\beta \equiv y^2-(r/n^2)x = 0,$$

so that, on setting $2x = X = X\gamma$, $2y = Y = AX$, $BC = n$ and $FI = r^2/n$, $\frac{1}{4}X\gamma^2 = \frac{1}{2}BC\times AX$ and $\frac{1}{4}AX^2 = \frac{1}{2}FI\times X\gamma$.

(85) It will be obvious from notes (34), (36), (62), (65) and (69) that the 'conchoidal' constructions of Problems 2 and 3 above may be applied to the resolution of the general cubic $x^3+px^2+qx+r = 0$.

Conclusion 5. Hence there is

$$2BC \times 2FI : 2BC \times AX \text{ (or } X\gamma^2 - 2AF \times X\gamma) - 2BC \times 2AB = X\gamma : 2BC,$$

and on multiplying extremes and middles there will arise

$$X\gamma^3 - 2AF \times X\gamma^2 - 4BC \times AB \times X\gamma = 8BC^2 \times FI.$$

And this, when the above assigned values(82) are restored, comes to be

$$x^3 - px^2 - qx = r \quad \text{or} \quad x^3 = px^2 + qx + r.^{(83)}$$

Corollary. Hence should AF and AB be set as zero, by conclusion 4 it is evident that $2FI : AX = AX : X\gamma = X\gamma : 2BC$. By this is settled the finding of two mean proportionals between any two given lines FI and BC, and that in an infinity of ways.(84)

For the rest, of all rules as general this(85) appears the most perfect that can be contrived whether you regard the order of the curve which the motion of the straight line $\gamma\rho\sigma$ implies to be described or the simplicity and variety of its construction. For there are two quantities BC and BE to be assumed at will: one of these, BC, can be unity or any divisor of the terms q and r or any other quantity you desire which will shorten the construction, and the other, BE, may then be assumed in any ratio whatever to the former, BC, according as you wish to determine the species of ellipse to be described by the motion of the point γ or make it intersect the circle at a less narrow angle. It is indeed convenient (in order to avoid the annoyance of having to find the mean proportional BD and the fourth proportionals FH and HR by drawing curves) to take BE to BC in the ratio of a square number to a square.

To be sure, to make the perfection of this rule yet more resplendent, it should be noted that it applies not merely to the ellipse but to the other conic sections also: however, the ellipse alone is described by the motion of but a single straight line very nearly as a circle is,(86) and for that reason is better suited in this case than those others.

If, for instance, you wish to employ a hyperbola whose latus rectum is in any proportion you please to its main axis, say, as d to e: take BC arbitrarily and on the opposite side of the point B take BE such that it is to BC as e to d (for the species of the curve is determined by the assumption of the line BC).(87) Then let the points A, F, G, I, H, K, L and R be determined as before, excepting only

(86) Indeed, in the case of a circle the trammel construction of the ellipse reduces to the revolution of a radius round the centre.

(87) In the terminology of note (77) the family of conics $\alpha \pm (n^2/k^2)\beta = 0$ is determined in species by the nature of the line-segment $BC = n$: precisely, $\alpha + (n^2/k^2)\beta = 0$ is (note (77)) a family of ellipses when neither n nor k are zero or infinity, and likewise $\alpha - (n^2/k^2)\beta = 0$ is (note (88)) a corresponding family of hyperbolas, while the parabolas $\alpha = 0$ and $\beta = 0$ are the limit-case when one of n or k is zero or infinity (compare note (93) below).

& R determinentur ut ante, excepto tantùm quòd FH debet sumi ad partes ipsius F contra I,[88] et quòd HR non in linea HL sed in linea AI a puncto H hinc inde capi debet, et vice rectæ GRS duæ aliæ rectæ a puncto L ad ista duo puncta R hinc inde duci pro Asymptotis Hyperbolæ.[89] Cum istis itaq; asymptotis LR describe Hyperbolam per punctum G transientem, ut et circulum centro K intervallo KG: et dimidia perpendiculorum ab eorum intersectionibus ad rectam AE demissorum erunt radices æquationis propositæ. Et hæc demonstrantur ut priùs si signa + & − probè mutentur.[90]

Sin velis Parabolam adhibere, punctum E in infinitum abibit, adeoq; nullibi capiendum est[,] et punctum H cum F coincidet.[91] Eritq; parabola circa axem HL cum latere recto BC[92] describenda quæ transeat per puncta G et A, abeatq; in infinitum ad partes versus quas BC sumitur a puncto B.[93]

Præterea cùm species curvæ ad arbitrium determinari possit,[94] ut patet; hinc problema nobilissimum absolvitur, construendi nempe quamcunq; cubicam æquationem per datam quamcunq; sectionem Conicam. Id quod sic explico.

Prob 5. Æquationem quamcunq; cubicam $x^3 = pxx.\ qx.\ r$, ope datæ cujuscunq; sectionis Conicæ modis infinitis construere.

Constructio.

A puncto quovis B cape duas quascunq; rectas BC, BE ad easdem plagas si data curva sit Ellipsis, vel ad contrarias si sit Hyperbola. Sit autem BC ad BE ut datæ sectionis conicæ latus rectum ad latus transversum. Et BC nominatâ n,

(88) For FH must now be set equal to $-r/k^2$.

(89) The ratio $HR:HL = BD:BE = n:k$ now, of course, determines the slope of the asymptotes and not the proportion of the semi-axes.

(90) Indeed, if in note (77) every occurrence of k^2 (actual or implicit) is replaced by $-k^2$,

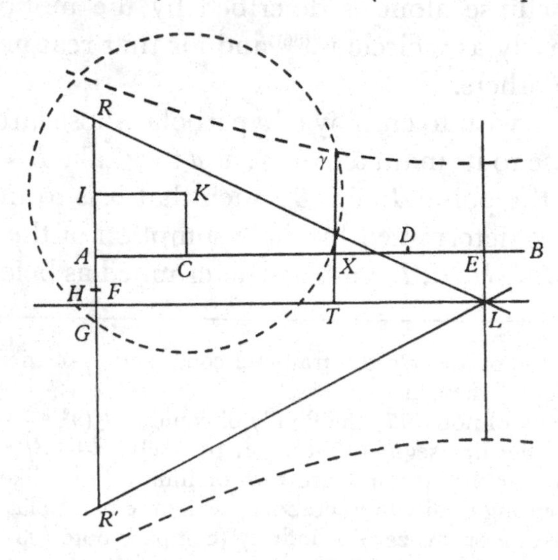

that FH should be taken on the opposite side of F to I(88) and that HR should be set not in the line HL but in the line AI on both sides of the point H, while in place of the line GRS two other straight lines must be drawn on either side from the point L to those two points R to serve as the hyperbola's asymptotes. Accordingly, with those asymptotes LR describe a hyperbola passing through the point G, and also a circle of centre K, radius KG: then will the halves of the perpendiculars let fall from their points of intersection to the line AE be the roots of the propounded equation. The proof of these assertions is as before providing the signs $+$ and $-$ are appropriately changed.(90)

But should you wish to employ a parabola, the point E will pass away to infinity, and so is not to be taken anywhere, while the points H and F will coincide.(91) The parabola must then be described about the axis HL with latus rectum BC(92) so as to go through the points G and A and pass away to infinity in the direction in which BC is taken from the point B.(93)

Furthermore, since the species of the curve may be determined at will,(94) as is evident, a most noble problem is in consequence solved: precisely, that of constructing any cubic equation by any given conic section. That I explain in this manner.

Problem 5. To construct any cubic equation $x^3 = \pm px^2 + qx \pm r$ with the help of any given conic section in an infinity of ways.

Construction.

From any point B you wish take any two straight lines BC, BE going the same way if the given curve be an ellipse, or the opposite if it be a hyperbola. And let BC be to BE as the latus rectum of the given conic section to its main axis. Then,

---

the analytical argument there displayed for the family of ellipses $\alpha + (n^2/k^2)\beta = 0$ will carry over with appropriate modification to the family of hyperbolas (γ) defined with respect to Cartesian co-ordinates $AX = Y$, $X\gamma = X$ by $\alpha - (n^2/k^2)\beta = 0$, where $\alpha \equiv X^2 - 2pX - 2nY$ and $\beta \equiv Y^2 - 2(q/n)Y - 2(r/n^2)X + 4pr/n^2$. Specifically, the hyperbolas have their centres at $L(p - r/k^2, -k^2/n + q/n)$ and semi-axes equal to $\sqrt{[(p + r/k^2)^2 + (-k + q/k)^2]}$, and each will pass through $G(2p, 0)$. (Newton's text has no accompanying figure: our accompanying sketch illustrates the case when p is taken negative.) The circle $G\gamma$ of centre K, defined by $\alpha + \beta = 0$, remains unchanged, of course.

(91) That is, on taking $k^2 = \infty$.

(92) Read '$2BC$'. Perhaps Newton is confusing his parabola with the parabolic eliminant $x^2 - px - ny = 0$ (note (77)).

(93) $k = \infty$ is (note (87)) one condition for the general conic $\alpha \pm (n^2/k^2)\beta = 0$ to be a parabola ($\alpha \equiv X^2 - 2pX - 2nY = 0$, in fact, passing through the origin A and $G(2p, 0)$).

(94) Precisely, by varying the ratio of n to k.

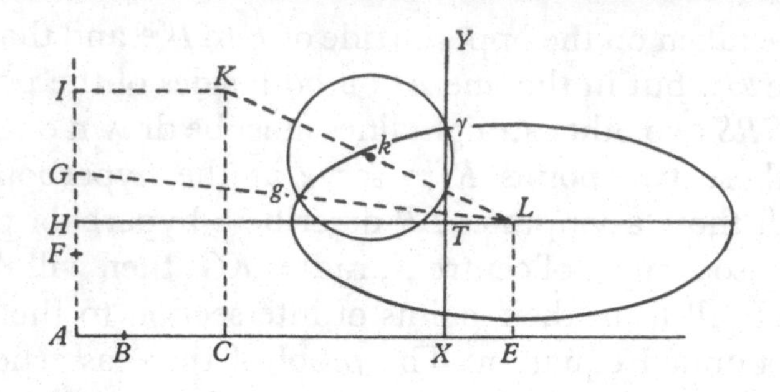

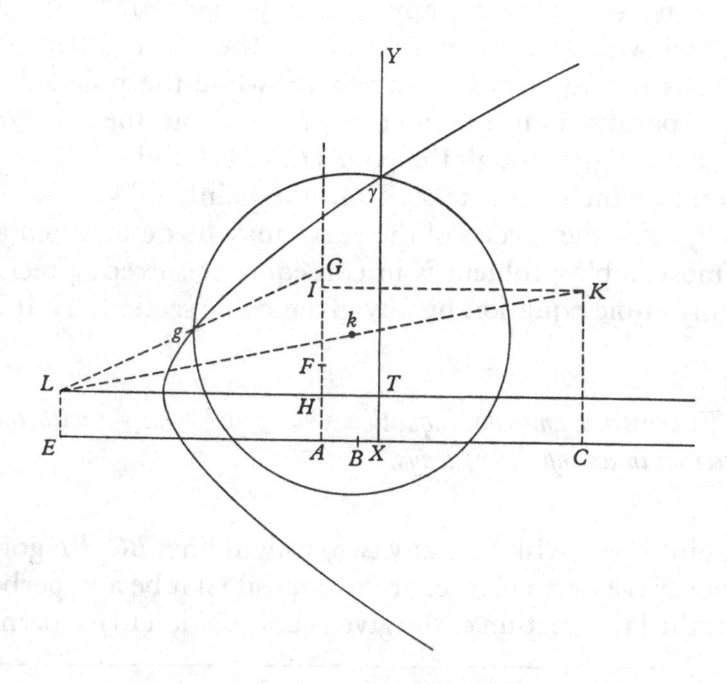

cape $BA = \frac{q}{n}$ idq; versus C si sit $-q$[,] aliter adversus. Et ad punctum A erige perpendiculum AI inq; eo cape $AF = FG = p$; Item $FI = \frac{r}{nn}$ versus G cape si p et r habeant eadem signa, aliter versus A. Dein fac $BE.BC::FI.FH$ et cape FH versus I si curva sit Ellipsis aut adversus I si sit Hyperbola. Porrò compleantur parallelogramma $IACK$ et $HAEL$, et hæ omnes jam descriptæ lineæ transferantur ad datam curvam, aut quod perinde est curva superponatur ita ut axis ejus sive transversa diameter cum rectâ LH conveniat et centrum cum puncto L.

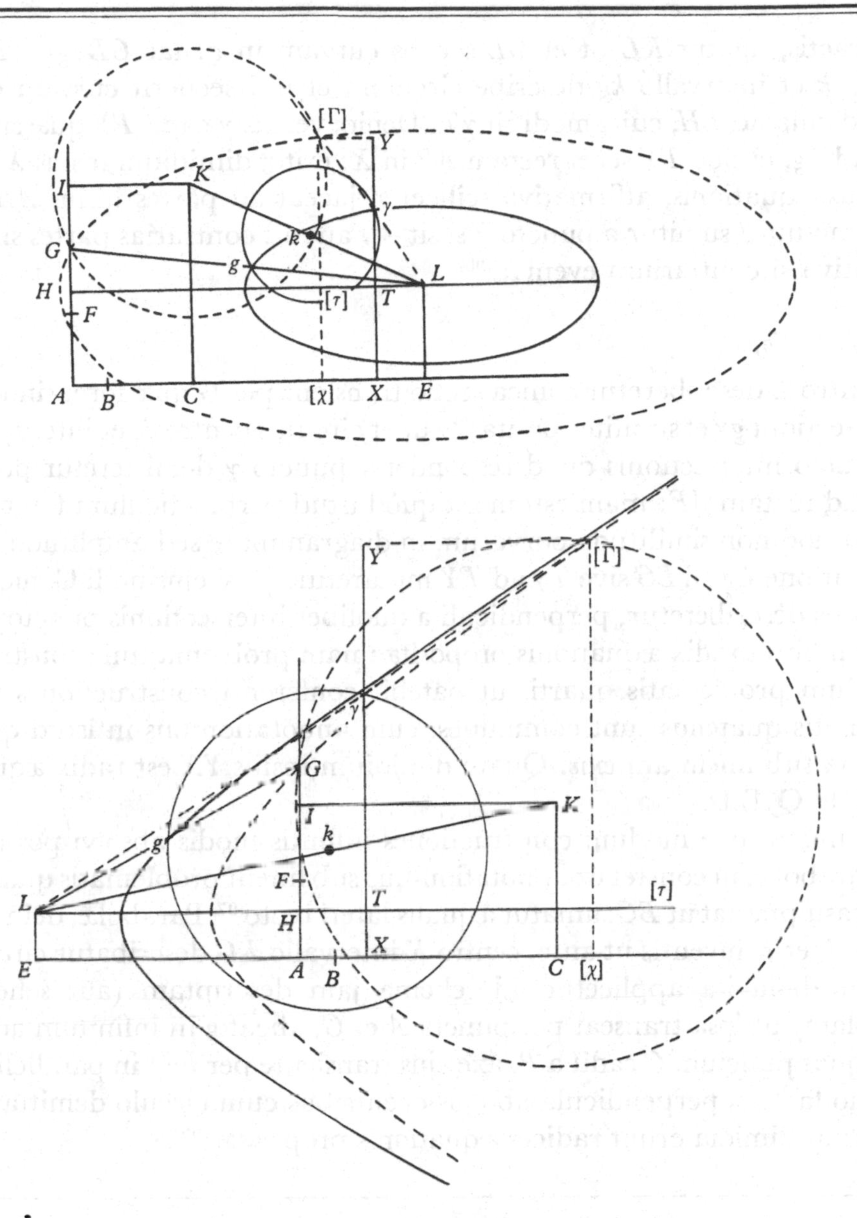

calling BC n, take $BA = q/n$, and that towards C if q be negative, otherwise away from it. Also, at the point A erect the perpendicular AI and in it take $AF = FG = p$; likewise, take $FI = r/n^2$ towards G if p and r have the same signs, otherwise towards A. Next make $BE:BC = FI:FH$, taking FH in the direction of I if the curve be an ellipse or in the opposite one if it be a hyperbola. Moreover, let the parallelograms $IACK$ and $HAEL$ be completed and let all these lines just now described be transferred to the given curve, or, what is the same thing, let the curve be superimposed so that its axis or transverse diameter coincides with the line LH and its centre with the point L. When this has been done, let KL

His peractis, agatur KL ut et GL secans curvam in g; fac $GL.gL::KL.kL$; centroq; k et intervallo kg describe circulum et ubi secuerit curvam demitte perpendicula ad LH, cujusmodi sit γT. Deniq; versus γ cape TY quæ sit ad $T\gamma$ ut LG ad Lg, et hoc TY secet rectam AB in X; eritq; dimidium ipsius XY una e radicibus æquationis, affirmativa scilicet si jaceat ad partes ipsius AB versus quas punctum I sumitur a puncto F si sit $+r$ aut ad contrarias partes si sit $-r$; et negativa si contrarium eveniat.[95]

Demonstratio.

Si centro L describeretur conica sectio transiens per G quæ foret similis datæ conicæ sectioni $g\gamma$ et similiter posita;[96] ut et circulus centro K et intervallo KG; et a puncto intersectionis quod responderet puncto γ demitteretur perpendiculum ad rectam AE: manifestum est quòd istud perpendiculum foret æquale YX, eo quod non similitudo curvarum in diagrammate sed amplitudo tantùm idq; in ratione Lg ad LG sive $T\gamma$ ad TY mutaretur. At si ejusmodi Conica sectio et circulus describeretur, perpendiculi a quolibet intersectionis puncto demissi dimidium foret radix æquationis propositæ: nam problema tunc construeretur ad modum problematis quarti, ut patebit conferenti constructiones utriusq; problematis quatenus sunt communes, cum annotationibus in istud quartum problema sub finem annexis. Quare dimidium ipsius YX est radix æquationis propositæ. Q.E.D.

Ad eundem fere modum constructiones infinitis modis absolvi possunt ope datæ Parabolæ, ut constet ex annotationibus sub finem problematis quarti: Sed in isto casu præstat ut BC sumatur æqualis lateri recto[97] Parabolæ, dein punctis A, F, G, I, et K inventis ut ante, centro K intervallo KG describatur circulus, et data Parabola ita applicetur ad schema jam descriptam (aut schema ad Parabolam) ut ipsa transeat per puncta A et G, abeatq; in infinitum ad partes versus quas punctum C cadit a B, axe ejus transiente per F[98] in parallelismo ad BC. Quo facto si perpendicula ab ejus occursibus cum circulo demittantur ad BC, eorum dimidia erunt radices æquationis propositæ.[99]

(95) An ingenious use of a homothety whose basic structure is outlined in the following 'Demonstratio'. Making full use of the two degrees of freedom implicit in the hitherto unrestricted choice of k and n, Newton fixes them by his two conditions that the constructing conic (Γ) be similar to the given conic (γ), and also be placed concentrically and in line with it round the point (and common centre) L. By suitable application of the scaling factor $GL/gL = KL/kL = \Gamma L/\gamma L (= YT/\gamma T)$ the construction of the preceding Problem 4 is reproduced in proportional facsimile on the given conic (γ) and its results then mapped back onto the true constructing conic (Γ). Newton's text-figures do not show either the constructing circle (K) or the conic (Γ) but we have added them in broken line to the equivalent diagrams reproduced with our English version so that the essence of the argument may be immediately clear.

(96) That is, sharing its centre L and the direction of its axes HL, EL.

18 19

Prob 5. Æquationem quamcunq cubicam $x^3=pxx.qx.r$ ope datæ cujuscunq sectionis Conicæ modis infinitis construere.

Constructio.

A puncto quovis B cape duas quascunq rectas BC, BE ad easdem plagas si data curva sit Ellipsis, vel ad contrarias si sit Hyperbola. Sit autem BC ad BE ut datæ sectionis conicæ latus rectum ad latus transversum. Et BC nominatâ n, cape $BA=\frac{q}{n}$ idq versus C si sit $-q$ aliter adversus. Et ad punctum A erige perpendiculum AG in eo cape $AF=FG=p$; item $FI=\frac{r}{nn}$ versus G cape si p et r habeant eadem signa, aliter versus A. Dein fac BE.BC::FI.FH et cape FH versus I si curva sit Ellipsis aut adversus I si sit Hyperbola. Porro compleantur parallelogramma IACK et HAEL, et hæ omnes jam descriptæ lineæ transferantur ad datam curvam, ut quod pars ejus et curva superponatur ita ut axis ejus sive transversa diameter cum recta LH conveniat et centrum cum puncto L. His peractis agatur KL ut et GL secans curvam in g; fac GL.gL::KL.kL; centroq k et intervallo kg describe circulum et ubi secuerit curvam demitte perpendicula ad LH, cujusmodi sit γT. Denique versus γ cape TY quæ sit ad Tγ ut LG ad Lg, et hoc TY secet rectam AB in X; eritq dimidium ipsius XY una e radicibus æquationis, affirmativa scilicet si jaceat ad partes ipsius AB versus quas punctum I sumitur a puncto F, si sit $+r$ aut ad contrarias partes si sit $-r$; et negativa si contrarium eveniat.

Demonstratio.

Si centro L describeretur conica sectio transiens per G quæ foret similis datæ conicæ sectioni gγ et similiter posita; ut et circulus centro K et intervallo KG; Et a puncto intersectionis quod responderet puncto γ demitteretur perpendiculum ad rectam AE: manifestum est quod istud perpendiculum foret æquale YX, eo quod non similitudo curvarum in diagrammate sed amplitudo tantum idq in ratione Lg ad LG sive Tγ ad TY mutaretur. At si ejusmodi Conica sectio et circulus describerentur, perpendiculi a quolibet

Plate IV. Geometrical construction of the general cubic as the meet of a conic with a circle (3, 2, §2).

be drawn and also GL intersecting the curve in g; then make $GL:gL = KL:kL$ and with centre k and radius kg describe a circle, dropping perpendiculars to LH (such as γT) where it intersects the curve. Finally, towards γ take TY which is to be to $T\gamma$ as LG to Lg and let this line TY meet the line AB in X: then will half XY be one of the roots of the equation, positive, namely, if it lie on the side of AB in the direction in which the point I is taken from the point F should r be positive, or on the opposite side if r be negative; and negative should the contrary happen.(95)

Demonstration

Should there be described with centre L a conic section passing through G so as to be similar to the given conic $g\gamma$ and similarly placed,(96) also a circle with centre K and radius KG; and should there be let fall from the point of intersection corresponding to the point γ a perpendicular to the line AE: it is manifest that that perpendicular would be equal to YX, seeing that not the similarity of the curves in the diagram but merely their size is changed, and that in the ratio of Lg to LG or $T\gamma$ to TY. But if a conic and circle of this kind were to be described, half the perpendicular let fall from any point of intersection would be a root of the propounded equation: for the problem would then be constructed after the manner of problem 4, as will be evident to anyone who compares the constructions of each problem, in so far as they are common to both, with the notes appended to that fourth problem at the end. Hence half of YX is a root of the propounded equation. As was to be demonstrated.

In much the same manner constructions may be performed in an infinity of ways with the aid of a given parabola, as will be evident from the notes at the end of problem 4. But in that case it will be an advantage to take BC equal to the latus rectum(97) of the parabola, then, when the points A, F, G, I and K have been found as before, to describe a circle with centre K, radius KG and to apply the given parabola to the scheme just now described (or the scheme to the parabola) so that it passes through the points A and G, and goes off to infinity in the direction in which the point C falls away from B, its axis passing through F parallel with BC. When that is done, if perpendiculars are let fall from its meets with the circle to BC, their halves will be roots of the propounded equation.(99)

---

(97) This should read (compare note (92)) 'semissi lateris recti' (half the latus rectum).

(98) That is, H (since $AF = p$ coincides with $AH = p + r/k^2$ when $k = \infty$).

(99) Newton perceives that since all parabolas with the same parameter are congruent, the given parabola (γ) will be identical with the constructing parabola (Γ) except in position and therefore may be used in its place straightforwardly in the manner of a template (by a suitable translation along their common axis).

Et nota quod cum secundus æquationis terminus deest et latus rectum Parabolæ existit 2,[100] hæc constructio evadit eadem cum illâ quam Cartesius attulit in Geometriâ, nisi quod lineamenta hic sunt illorum duplicia.[101]

Prob 6. Æquationem quatuor dimensionum, $x^4=px^3.\ qx^2.\ rx\ .\ s$ cujuscunq; sit formæ construere, idq; modis dupliciter infinitis.

A puncto quovis B cape duas quascunq; longitudines BC & BE ambas ad eandem plagam sitq; BD inter eas medium proportionale et BC dic n. Et ad eandem plagam cape etiam $B\beta=\frac{pp}{4n}$, ut et $\beta A=\frac{q}{n}$ si in æquatione habeatur $-q$, at si sit $+q$ cape βA ad plagam oppositam. Dein ad punctu[m] A erige perpendiculum AI inq; eo cape $AF=\frac{1}{2}p$, et $FG=\frac{r}{nn}$ adversus A si p et r habeant eadem signa, aliter versus A. Item fac $n\ .\ \frac{1}{2}p::AB\ .\ GI$, capiens GI ad partes ipsius G versus quas AF sumitur ab A si punctum B interjaceat punctis A et C, aliter ad partes contrarias. Insuper fac $BE\ .\ BC::FI\ .\ FH$, cape FH versus I et comple parallelogramma $IACK$ & $HAEL$. Et ultimò fac $BC\ .\ BD::AH\ .\ AM$, et subtensas ME, CI concipe ductas esse.

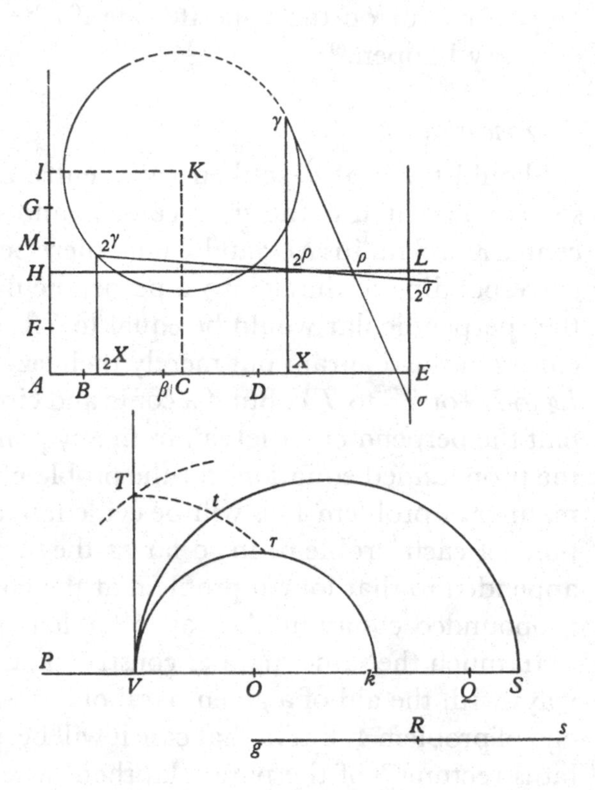

Jam in aliquâ lineâ PQ seorsim sume $VS=ME$, $Vk=IC$, $VQ=\frac{s}{nnl}$ et ad oppositam plagam $VP=4l$, ubi pro l quilibet numerus vel unitas ad arbitrium assumi debet.[102] Dein erige perpendiculum VT, et bisecto PQ in O inscribe $OT=OP$, et si habeatur $+s$ tunc centro K intervallo Tk describe circulū $\gamma_2\gamma$ et duc rectam $gs=TS$. Sed si habeatur $-s$, tunc circuli duo VtS & $V\tau k$ super diametris VS et Vk construendi sunt et in ipsis inscribendæ Vt, & $V\tau$ æquales VT, centroq; K et intervallo τk circulus describendus ac gs ducenda æqualis distantiæ tS.

Deniq; fac $BE\ .\ BD::gs\ .\ gR$ capiens gR ad utramvis partem puncti g, et hanc lineam gRs ita moveto ut punctum R super lineam HL et punctum s super LE continuò incedat donec tertium ejus punctum g occurrat circulo $\gamma_2\gamma$, quemad-

Note, too, that when the second term of the equation is lacking and the parabola's latus rectum is 2,(100) this construction comes out the same as that which Descartes presented in his *Geometrie* except that the present line-lengths are twice those of his.(101)

Problem 6. To construct an equation of four dimensions $x^4 = \pm px^3 \pm qx^2 \pm rx \pm s$ of any form whatever, and that in a doubly infinite manner.

From any point B take any two lengths BC and BE, both the same way, and let BD be the mean proportional between them. Call BC n, and the same way also take $B\beta = p^2/4n$ together with $\beta A = q/n$ if there be had $-q$ in the equation, but if it be $+q$ take βA the opposite way. Then at the point A erect the perpendicular AI and in it take $AF = \frac{1}{2}p$ and $FG = r/n^2$, away from A if p and r have same signs, otherwise towards it. Again, make $n:\frac{1}{2}p = AB:GI$, taking GI in that direction from G in which AF is assumed to be from A should the point B lie between the points A and C, otherwise in the contrary direction. In addition, make $BE:BC = FI:FH$, taking FH towards I, and complete the parallelograms $IACK$ and $HAEL$. And, lastly, make $BC:BD = AH:AM$ and suppose the subtended lines ME, CI to be drawn.

Now in some separate line PQ take $VS = ME$, $Vk = IC$, $VQ = s/n^2l$ and (the opposite way) $VP = 4l$, where in place of l must be assumed unity or any number at will.(102) Next erect the perpendicular VT and, having bisected PQ in O, inscribe $OT = OP$: then if $+s$ is supposed, with centre K and radius Tk describe the circle $\gamma_2\gamma$ and draw the straight line $gs = TS$. But if $-s$ be supposed, then two circles VtS and $V\tau k$ are to be constructed on diameters VS and Vk and within them are to be inscribed Vt and $V\tau$ equal to VT, then with centre K and radius τk is to be described a circle, while gs has to be drawn equal to the distance tS.

Finally, make $BE:BD = gs:gR$, taking gR on whichever side of the point g you wish, and then move this line gRs so that the point R travels uninterruptedly on HL and the point s on LE till its third point g meets the circle $\gamma_2\gamma$, as may be

(100) That is, when $p = 0$ and $n = 1$.

(101) See Descartes, [*Geometrie* (1637): 388–95 =] *Geometria* ($_2$1659): Liber III: 85–90: 'Modus generalis construendi omnia Problemata solida, reducta ad Æquationem trium, quatuórve dimensionum'; compare also I, 3, 3, §2: note (11). There Descartes constructed the reduced quartic $z^4 = pz - qz + r$ geometrically by the intersection of the parabola $z^2 = y + \frac{1}{2}(p+1)$ with the circle $(z+\frac{1}{2}q)^2 + y^2 = \frac{1}{4}[(p+1)^2 + q^2] + r$: he would therefore resolve Newton's present equation $x(x^3 - qx - r) = 0$ as the meets of the parabola $x^2 = y + \frac{1}{2}(q+1)$ with the circle $(x - \frac{1}{2}r)^2 + y^2 = \frac{1}{4}[(q+1)^2 + r^2]$. Newton's own construction is, as he says, but trivially different, for on setting $2x = X$, $2y = Y$ he makes corresponding use of the parabola $X^2 = 2Y$ and circle $(X-r)^2 + (Y-q-1)^2 = (q+1)^2 + r^2$.

(102) The parameter l is chosen to keep $PV \times VQ = 4s/n^2$ invariant while affording a convenient construction of the point T as the meet of the semicircle on diameter PQ with the normal at V.

modum videre est in positionibus $\gamma\rho\sigma$ et ${}_2\gamma_2\rho_2\sigma$. Et a punctis occursuum γ, ${}_2\gamma$ si demittantur perpendicula γX, ${}_2\gamma_2 X$ in lineam BC, eorum dimidia erunt radices æquationis propositæ.(103)

Et nota quod recta $\gamma\rho\sigma$ cum circulo $\gamma_2\gamma$ in tot punctis occurret quot sunt radices possibiles: et quod radices sunt affirmativæ quæ cadunt ad partes rectæ BC versus F si habeatur $+p$ et contra si $-p$. Aut si te[r]minus p desit tunc affirmativæ cadunt ad partes versus I si habeatur $+r$ et contra si $-r$.

Demonstratio.

Potest hæc constructio ad modum problematis quarti(104) demonstrari sed brevitatis et varietatis gratia mallem adhibere sequentem Analysin.

Concipiendo punctum γ esse intersectionem circuli et Ellipsis descriptæ per motum rectæ $\gamma\rho\sigma$, assumantur æquationes duæ $a+bx+cy-xx=yy$, et

$$a \begin{matrix} +b \\ -d \\ +ed \end{matrix} x \begin{matrix} +c \\ -n \\ +en \end{matrix} y - exx = yy,^{(105)}$$

quarum prior (ut ex terminis pateat) est ad circulum et posterior ad Ellipsin: adhibitis nempe a, b, c, d, e et n pro designandis quibuslibet cognitis quantitatibus, et x pro qualibet γX(106) perpendiculariter ordinatâ ad Basin AX sive y.(106) Jam ut a, b, c, d, e, et n ex hac assumptione determinentur, subduc priorem æquationem a posteriori et restabit $\begin{matrix} -d \\ +ed \end{matrix} x \begin{matrix} -n \\ +en \end{matrix} y \begin{matrix} -e \\ +1 \end{matrix} xx = 0$, divide per $e-1$ et oritur $dx+ny-xx=0$, sive $y=\dfrac{xx-dx}{n}$. Hunc valorem pro y in æquatione primâ substitue et orietur $\dfrac{x^4-2dx^3+ddxx}{nn}=a+bx\dfrac{+cxx-cdx}{n}-xx$.

(103) On taking $BC = n$ and $BD = k$ arbitrarily and setting $AF = \frac{1}{2}p$, $\beta A = q/n$, $FG = r/n^2$ there follows $BE = k^2/n$, $B\beta = p^2/4n$ and so

$$AH = AF+(n^2/k^2)(FG+(p/2n)AB) = (-p^3+4pq+4pk^2)/8k^2,$$
$$HL = AE = (-p^2+4q+4k^2)/4n,$$
$$AI = CK = AF+FG+(p/2n)AB = (-p^3+4pq+4pn^2)/8n^2$$

and $$IK = AC = (-p^2+4q+4n^2)/4n.$$

Further, $TV^2 = TO^2-VO^2 = PV\times VQ$ and so

$$\gamma\sigma^2 = gs^2 = TS^2 = VS^2 \text{ (or } ME^2 = AE^2+AM^2) + TV^2 = AE^2+(k^2/n^2)AH^2+PV\times VQ,$$

while $$\gamma\rho = gR = (n/k)gs = (n/k)\gamma\sigma.$$

Lastly $K\gamma^2 = Tk^2 = Vk^2$ (or IC^2) $+ TV^2 = AC^2+CK^2+PV\times VQ$. On taking the co-ordinates $AX = Y$ and $X\gamma = X$ the effect of Newton's construction is to define the family of ellipses $(\gamma) \equiv (k^2/n^2)\alpha+\beta = 0$ (k, n free) of centre L and semi-axes of length $\gamma\sigma$, $\gamma\rho$ which are all through the meets of the parabolas

$$\alpha \equiv X^2-pX-2nY = 0 \quad \text{and} \quad \beta \equiv Y^2-\lambda Y-\mu X-4s/n^2 = 0,$$

seen in the positions $\gamma\rho\sigma$ and ${}_2\gamma_2\rho_2\sigma$. If then from the meeting points γ, ${}_2\gamma$ be dropped perpendiculars γX, ${}_2\gamma_2 X$ onto the line BC, their halves will be roots of the propounded equation.(103)

And note that the straight line $\gamma\rho\sigma$ will meet the circle $\gamma_2\gamma$ in as many points as there are possible roots: and that those roots are positive which fall on the side of the line BC towards F if $+p$ is supposed, and the contrary if it is $-p$. Or if the term p is lacking, then the positive ones fall on the side towards I if there is had $+r$, and the contrary if $-r$.

Demonstration

This construction can be demonstrated in the manner of problem 4(104) but for the sake of brevity and variety I prefer to employ the following analysis.

By conceiving the point γ to be the intersection of the circle and an ellipse described by the movement of the straight line $\gamma\rho\sigma$, let there be assumed two equations, $a+bx+cy-x^2=y^2$ and $a+(b-d+ed)x+(c-n+en)y-ex^2=y^2$,(105) the first of which (as is evident by its terms) is that of a circle and the latter that of an ellipse: a, b, c, d, e and n are, of course, employed to designate any known quantities you please, while x is used for any line γX(106) ordinate normally to the base AX or y.(106) Now to determine a, b, c, d, e and n from this assumption take away the former equation from the latter and there will remain

$$(-d+ed)x+(-n+en)y+(-e+1)x^2=0;$$

divide this by $e-1$ and there arises $dx+ny-x^2=0$, or $y=(x^2-dx)/n$. Substitute this value for y in the first equation and there will arise

$$\frac{x^4-2dx^3+d^2x^2}{n^2}=a+bx+\frac{cx^2-cdx}{n}-x^2,$$

where $\lambda=(-p^2+4q)/2n$, $\mu=(-p^3+4pq+8r)/4n^2$. (Correctly, as we shall see (note (107)), the values of λ and μ should be $(+p^2+4q)/2n$ and $(+p^3+4pq+8r)/4n^2$, but this entails the trivial amendment to the construction that we set off $B\beta$ from B in the direction opposite to that indicated by Newton.) The circle on centre K is the member of the family, $\alpha+\beta=0$, for which $n=k$. Evidently this meets any of the ellipses in (up to) four points $\gamma(X, Y)$ such that $x^4=px^3+qx^3+rx+s$, where $x=\frac{1}{2}X$.

(104) That is, 'syntheticè' (by composition in classical style).

(105) Even in his present 'analysis' Newton is not completely open with his reader, for it is clear that he himself in the first instance (compare note (77)) assumed the two parabolic defining equations $\alpha\equiv x^2-dx-ny=0$ and $\beta\equiv y^2-(b-d)x-(c-n)y-a=0$, compounding the present expressions for the circle and ellipse as $\alpha+\beta$ and $e\alpha+\beta$ respectively. (The curious form of the latter's coefficients arise because Newton now sets the circle's equation as

$$x^2+y^2=a+bx+cy.)$$

(106) A first careless slip. Newton has forgotten that, to avoid numerical fractions in his constructional elements, he in fact made use of the 'doubled' co-ordinates $X\gamma=X=2x$ and $AX=Y=2y$. This is the reason for the numerical discrepancies between the values used in his preceding construction (note (103)) and those calculated in the sequel.

sive
$$x^4 = 2dx^3 + \begin{matrix} dd \\ -nn \\ +cn \end{matrix} xx \begin{matrix} +nnb \\ -ncd \end{matrix} x + nna.^{(107)}$$

Jam hujus terminos confer cum terminis æquationis $x^4 = px^3 + qxx + rx + s$, et prodibunt $\frac{1}{2}p = d$. $\frac{q - \frac{1}{4}pp}{n} = c - n$.(108) $\frac{r + \frac{1}{2}pq - \frac{1}{8}p^3}{nn} = b - d$.(108) & $\frac{s}{nn} = a$.(108) Unde si n pro arbitratu assumatur, dabuntur a, b, c, ac d, et etiamnum e restabit pro lubitu assumendum.(109) Ex hisce sic inventis determinantur æquationes duæ primò positæ, quæ si geometricè construantur emerget determinatio circuli et Ellipsis qualem in constructione problematis hujus assignavi. Quid plura?(110)

Cæterùm ex annotationibus in Prob 4 pateat hanc biquadraticarum æquationum constructionem non modò absolvi posse per intersectionem circuli et Ellipsis (quam concipe per motum rectæ $\gamma\rho\sigma$ describi:) sed etiam ad Parabolam vel Hyperbolam quamvis se extendere, prout punctum E vel ad infinitam distantiam vel ad contrarias partes puncti B assumitur.(111)

Quemadmodum si velis Hyperbolam cujuscunꝗ datæ speciei adhibere; debes assumere longitudinem BE ad partes contra BC quæ sit ad BC ut est Hyperbolæ latus transversum ad latus rectum ejus. Dein cætera peragenda sunt pro more præcedentis constructionis usꝗ ad descriptionem circuli $\gamma_2\gamma$ nisi quod FH capi debet ad partes contra FI. Postea longitudo rectæ gs paulò aliter determinanda est; nempe in præcedenti constructione fuit $gs = \surd : VS^q \pm VT^q$: sive $= \surd : AE^q + AM^q \pm VT^q$: sed hic debes efficere ut sit

$$gs = \surd : AE^q - AM^q \pm VT^q :$$

ubi VT^q habebit idem signum $+$ aut $-$ ac ultimus æquationis terminus s. Tum facto $gs.gR :: BE.BD$, erunt gs & gR semidiametri Hyperbolæ: nempe gs transversa semidiameter quæ poni debet in recta LH utrinꝗ a puncto L quod erit centrum curvæ, et altera semidiameter gR ad prioris extremitate[m] juxta verticem curvæ perpendiculariter erecta determinabit Asymptotos ejus.(112)

Quod si placet adhibere Parabolam, puncta D et E negligenda sunt ut et inventio punctorum H et M rectæꝗ gR, et linea per punctum F ducenda est

---

(107) A second careless slip: the first term in the coefficient of x^2 should be '$-dd$' ($-d^2$). The effect of this is to substitute $-p^2$ for each occurrence of p^2, and may be so corrected (note (103)).

(108) These are, of course, the coefficients of the parabola $\beta = 0$ (note (105)) and Newton should for clarity's sake have acknowledged this.

(109) In the preceding construction, in fact, Newton has set $e = k^2/n^2$.

(110) We may perhaps detect a hint of oncoming boredom with his construction in these words of Newton's.

(111) The position of E (with respect to B) depends on the sign of $BE = \pm k^2/n$ and so on

or $$x^4 = 2dx^3 + (d^2 - n^2 + cn)\,x^2 + (n^2b - ncd)\,x + n^2a.^{(107)}$$

Now compare the terms of this with those of the equation $x^4 = px^3 + qx^2 + rx + s$ and there will come out $\frac{1}{2}p = d$, $(q - \frac{1}{4}p^2)/n = c - n$,[108]

$$(r + \tfrac{1}{2}pq - \tfrac{1}{8}p^3)/n^2 = b - d^{(108)} \quad \text{and} \quad s/n^2 = a.^{(108)}$$

Hence if n be assumed arbitrarily, a, b, c and d will be given and even so e will still remain to be assumed at pleasure.[109] From the quantities found in this way are determined the two equations first supposed, and if these are constructed geometrically the determination of the circle and ellipse as I assigned it in the construction of this problem emerges. Why say more?[110]

For the rest, it should be evident from the notes on Problem 4 that this construction of quartic equations may not only be executed by the intersection of a circle and ellipse—conceive this to be described by the motion of the straight line $\gamma\rho\sigma$—but also extends to the parabola or any hyperbola at all, depending on whether the point E is assumed to be at an infinite distance or on the further side of the point B.[111]

For instance, if you wish to employ a hyperbola of any given species, you should assume the length BE to be on the contrary side of BC and in proportion to it as the hyperbola's main axis to its latus rectum. Next, everything else is to be performed in the style of the preceding construction up to the description of the circle $\gamma_2\gamma$, except that FH should be taken on the contrary side of FI. After that, the length of the straight line gs is to be determined somewhat differently: precisely, in the preceding construction it was $gs = \surd(VS^2 \pm VT^2)$ or

$$\surd(AE^2 + AM^2 \pm VT^2)$$

but here you should make $gs = \surd(AE^2 - AM^2 \pm VT^2)$, where VT^2 will have the same sign $+$ or $-$ as the equation's final term s. Then, having made

$$gs : gR = BE : BD,$$

gs and gR will be the semi-diameters of the hyperbola: precisely, gs will be the semi-diameter which should be set in the line LH either way from the point L (which will be the centre of the curve) and the second semi-diameter gR, when erected normally at the end-point of the former, on the vertex of the curve as it were, will determine its asymptotes.[112]

But should you desire to employ a parabola, the points D and E must be neglected and the finding of the points H and M and of the straight line gR also,

---

that of $\pm k^2$. In the terminology of note (103) the conic family $(k^2/n^2)\alpha + \beta = 0$ is one of ellipses, the family $(-k^2/n^2)\alpha + \beta = 0$ is one of hyperbolas, while the parabola

$$\alpha \equiv X^2 - pX - 2nY = 0$$

is the limit-case of both when $\pm k^2 = \infty$ or E passes to infinity.

(112) Newton's hyperbola is $(-k^2/n^2)\alpha + \beta = 0$, where α and β are defined as in note (103): that is, $(Y - AE)^2 - (k^2/n^2)(X - AH)^2 = AE^2 - (k^2/n^2)AH^2 + (4s/n^2 \text{ or})\, TV^2 = gs^2$. Evidently gs and $(n/k)gs = gR$ are the lengths of its semi-axes. Compare the figure in note (90).

parallela ad BC pro axe Parabolæ quæ describi debet cum latere recto BC ita ut transeat per punctum A, tendens in infinitum ad plagam versus quam punctum C sumitur a puncto B. Et quod ad circulum $\gamma_2\gamma$ attinet ille omninò determinatur ut in præcedentibus: Quemadmodum et perpendiculorum ab intersectionibus γ ad BC demissorum dimidia erunt radices æquationis propositæ.(113)

Et hæc omnia patebunt ex Analysi quam huic problemati adjeci, si vice $-e$ (quod designat rationem lateris transversi ad latus rectum) scribas $+e$ pro Hyperbola vel ponas e infinitum esse pro Parabola, id est si ponas $dx+ny=xx$ nam cæteri termini per infinitum e divisi evanescent.(114)

Præterea cum species conicæ sectionis hic sicut in Prob 4 sit ad arbitrium determinabilis sequentis etiam problematis per omnia similis problemati quinto constructionem similiter exhinc lucror.

Prob 7. Æquationem quamcunq; biquadraticam $x^4=px^3.\ qxx.\ rx.\ s$, ope datæ cujuscunq; sectionis conicæ modis infinitis construere.

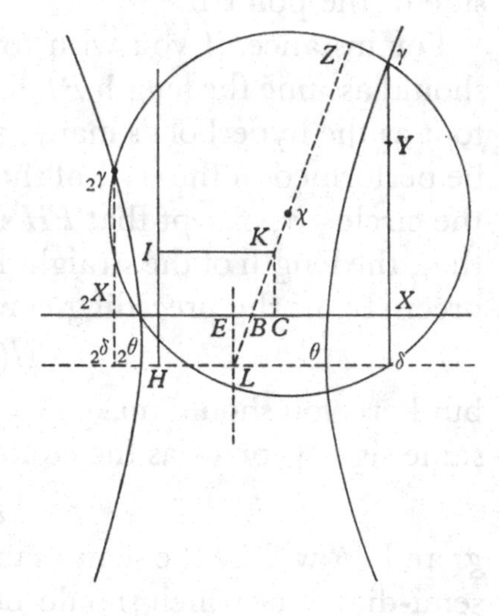

Ducantur quævis duæ rectæ BC[,] BE in ratione lateris recti datæ conicæ sectionis ad ejus latus transversum(115) idq; ad easdem partes si sit Ellipsis, aut ad diversas si Hyperbola. Dein prosequere constructionem problematis præcedentis aut animadversionum ei annexarum, usq; ad determinationē punctorum K et L, distantiæ kT vel $k\tau$, et longitudinis gs.

Sit jam data curva $\gamma\theta$ cujus centrū sit in L et latus transversum $\theta L[_2]\theta$ coincidens cum LH; Junge LK et fac $gs.L\theta::LK.L\chi::kT$ (vel $k\tau$) $.\chi Z$, centroq; χ et intervallo χZ describe circulum $Z\gamma$ et a punctis intersectionum cum curva demitte perpendicula $\gamma\delta$ ad rectam HL, et in illis versus γ cape δY in ratione ad $\delta\gamma$ ut est $L\theta$ ad gs[;] secent autem BC in X, et ipsorum XY dimidia erunt radices æquationis propositæ.(116)

---

(113) When $\pm k^2 = \infty$ the points D and E pass to infinity while F and H coincide and the conic locus reduces to $X^2 = pX+2nY$ (compare note (111)), a parabola with its axis through H (or F) parallel to BC. The circle $\alpha+\beta = 0$ remains, of course, unchanged.

(114) Compare notes (109) and (111).

(115) In the proportion $n^2:k^2$, that is.

(116) The justification of this construction follows, as Newton observes, from that of the preceding half of the problem in the same way that that of Problem 5 depends on Problem 4. Once again (compare note (95)) the elements of the constructing conic and circle (shown in broken line as (G) and G_2G respectively in the amplified diagram which accompanies our English version) are reproduced in the proportion $L\chi:LK = L\gamma:LG$ in the homocentric conic-circle pair drawn in Newton's figure (where for variety's sake a hyperbola exemplifies the conic locus); and, as before, when the intersections of the given conic $\gamma\theta$ with the circle

while the line through the point F must be drawn parallel to BC for the axis of the parabola, which ought to be described with latus rectum BC so as to pass through the point A, stretching to infinity in the direction in which the point C is taken from the point B. What relates to the circle $\gamma_2\gamma$ is determined wholly as in the preceding: and likewise, too, the halves of the perpendiculars let fall from the points of intersection γ to BC will be roots of the propounded equation.(113)

And all these things will be clear from the analysis which I have added to this problem if in place of $-e$ (which designates the ratio of the main axis to the latus rectum) you write $+e$ in the case of the hyperbola or suppose e to be infinite in that of the parabola, that is, if you set $dx+ny = x^2$ (for all the other terms divided by the infinity e will vanish).(114)

Furthermore, since here as in Problem 4 the species of conic section is determinable at choice, I hence gain in a similar way the construction of the following problem also, in every respect similar to Problem 5.

Problem 7. To construct any quartic equation $x^4 = \pm px^3 \pm qx^2 \pm rx \pm s$ with the help of any given conic section and in an infinity of ways.

Let there be drawn any two straight lines BC, BE in the ratio of the latus rectum of the given conic section to its main axis,(115) in the same direction if it be an ellipse or in opposite ones if a hyperbola. Next carry through the construction of the preceding problem, or of the observations appended to it, up to the determination of the points K and L, the distance kT or $k\tau$ and the length gs.

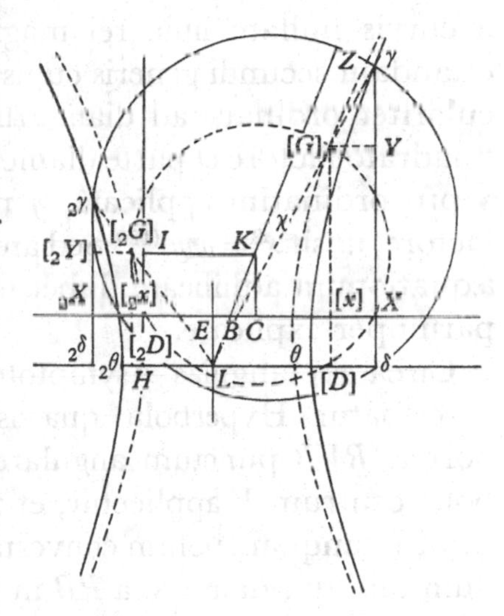

Let now the given curve be $\gamma\theta$, of centre L and main axis $\theta L_2\theta$ coincident with LH. Join LK and make $gs:L\theta = LK:L\chi = kT$ (or $k\tau$) $:\chi Z$, then with centre χ and radius χZ describe the circle $Z\gamma$ and from its points of intersection with the curve let fall perpendiculars $\gamma\delta$ to the line HL and in them in the direction of γ take δY in proportion to $\delta\gamma$ as $L\theta$ is to gs; let these intersect BC in X and the halves of those lines XY will be roots of the propounded equation.(116)

$\gamma_2\gamma$ (of centre χ) homothetic to the constructing circle G_2G (of centre K) have been found, they are then mapped back by the inverse proportion $LK:L\chi = LG:L\gamma$ onto the constructing conic-circle pair with respect to the similitude centre L. The point B may be chosen arbitrarily but the proportion $BE:BC = k^2:n^2$ is determined as being that of the conic's latus rectum to the latus transversum (or principal conic axis), identical in both given and constructing curves since they are similar.

Ad eundem fere modum si detur Parabola licitum est æquationes ejus beneficio infinitis modis construere assumendo *BC* ad arbitrium: sed præstat assumere æqualem lateri recto Parabolæ, ac dein pergere pro more ostenso in prob 6.(117)

Et horum demonstratio ex quinto problemate manifesta est.(118)

Hactenus de solidis problematibus. Restat ut de linearibus nonnulla adjiciam, ad quæ requiruntur motus sive curvarum descriptiones magis compositæ. Possunt equidem nonnulla per motus æque simplices ac solida problemata construi(119) sed id rarissimè evenit et ejusmodi casus sunt adeò restricti ut in exemplaria generaliter proponi nequeant. Et inde ad motus magis compositos necessariò recurrendum est, cui incommodo non melius remedium afferri posse videtur quàm si curva aliqua adhibeatur quæ per motus magis compositos semel descripta, generaliter inserviet constructioni quarumlibet æquationum. Nam ejusmodi curvam pro data licebit assumere et problematum difficultates fere ex solis motibus qui ad particularia requiruntur æstimare. Et e curvis nullam huic rei magis accommodatam invenio quàm Parabolam quandam secundi generis cujus hæc est proprietas ut rectæ cujusvis perpendiculariter ordinatæ ad diametrū cubus semper æquetur facto ex dato quovis quadrato factore et parte diametri ad datum punctū terminatæ. Hoc est, posito x pro ordinatim applicatâ, y pro interceptâ parte diametri, et nn pro dato factore, ut sit $x^3 = nny$.(120) Sed antequam hanc cubicam parabolam ad solutiones æquationum adhibeam, lubet naturam ejus parumper explicare.

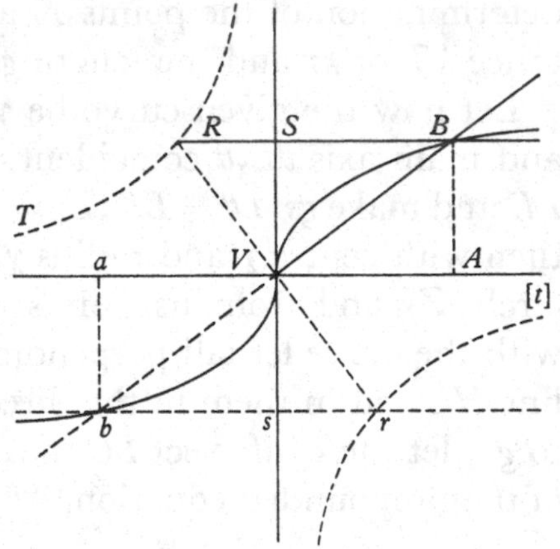

Circa rectangulas Asymptotos aVA, sVS describatur Hyperbola quævis RT. Dein normæ RVB punctum angulare ad Hyperbolæ centrum V applicetur, et norma circa istud V tanquam polum convertatur, interea dum alia quædam recta RB in parallelismo ad Asymptotorum alteram aVA sic moveatur ut continuò transeat per punctū R ubi normæ crus RV secat Hyperbolam. Et punctum B ubi secat alterum crus, describet

---

(117) All homothetic parabolas have the same parameter when, as here, the similitude centre L passes to infinity. They are therefore congruent, and so in the present case the given parabola may, by a suitable translation along their common axis, be used in place of the constructing parabola without change (or as its template).

(118) Compare notes (95) and (99).

In much the same way if a parabola be given it is allowed to construct equations with its aid in an infinity of ways by assuming *BC* arbitrarily: but it pays to assume it equal to the latus rectum of the parabola, and then proceed in the fashion shown in Problem 6.(117)

The demonstration of these things is manifest from Problem 5.(118)

So much for solid problems. It remains for me to add something on the subject of linear ones: for these more complex motions or descriptions of curves are required. Some may indeed be constructed(119) by motions just as simple as those used for solid problems but that very rarely happens and its cases are so restricted that they cannot be proposed generally as model ones. In consequence we must necessarily revert to more complex motions. For this inconvenience there appears to be no better remedy applicable than to employ some curve which, when once described by more complex motions, may serve generally to construct any equation you please; for it will then be permissible to assume such a curve as given and to reckon the difficulties of problems almost wholly in terms of the motions which are required in particular applications. Of curves I find none more suitable for this purpose than a certain parabola of second order whose property it is that the cube of any straight line ordinate normally to the diameter be equal to the product of any given square factor and that portion of the diameter terminated at the given point: in other words, on setting x for the ordinate, y for the intercepted portion of the diameter and n^2 for the given factor, that there be $x^3 = n^2y$.(120) But before I employ this cubic parabola in the resolution of equations, it is pleasant to spend a little while explaining its properties.

About the rectangular asymptotes *aVA*, *sVS* let there be described any hyperbola *RT*. Then let the angular point of the sector *RVB* be applied to the centre *V* of the hyperbola and let the sector rotate round that point *V* as pole while at the same time some other straight line *RB* moves parallel to one or other of the asymptotes *aVA* so as continually to pass through the point *R* in which the sector's leg *RV* intersects the hyperbola. The point *B* in which it intersects the other leg will describe the parabola *bVB* about which I am at present con-

---

(119) Newton first wrote 'solvi' (solved).

(120) This 'parabola quædam secundi generis' is, of course, the 'parabola of ye 2^d kind' in I, **3**, 3, §2, but the name 'parabola cubica' or 'parabola Wallisiana' (compare I, **1**, §3: note (94) above) was to become Newton's standard terminology. We may note that Jakob Bernoulli independently applied this 'vulgare paraboloides cubicale' to the geometrical construction of equations in his 'Animadversio in Geometriam Cartesianam, & Constructio quorundam Problematum Hypersolidorum' (*Acta Eruditorum* (June 1688): 323–30 = *Opera*, **1** (Geneva, 1744): N° XXXI: 343–51).

Parabolam bVB de qua jam agitur.[121] Sit enim S intersectio rectæ RB et Asymptoti sVS. Et quoniam ex naturâ Hyperbolæ rectangulum $VS \times SR$ sit dato æquale, pone istud datum esse nn et erit $RS = \frac{nn}{VS}$; et inde cùm sit $RS . VS :: VS . SB$ erit $SB = \frac{VS^{\text{cub}}}{nn}$. Sive demisso BA perpendiculo ad VA dictóq; $BA = x$ et $VA = y$, erit $\frac{x^3}{nn} = y$.

Ex his facile colligitur quod rectæ cujusvis BVb per punctum V ductæ partes VB, Vb hinc inde ad curvam terminatæ sint æquales quia portio Vb similiter per conjugatam Hyperbolam $[r]t$ describitur atq; portio VB per RT. Et hinc punctum V non impropriè dici potest Parabolæ centrum.[122]

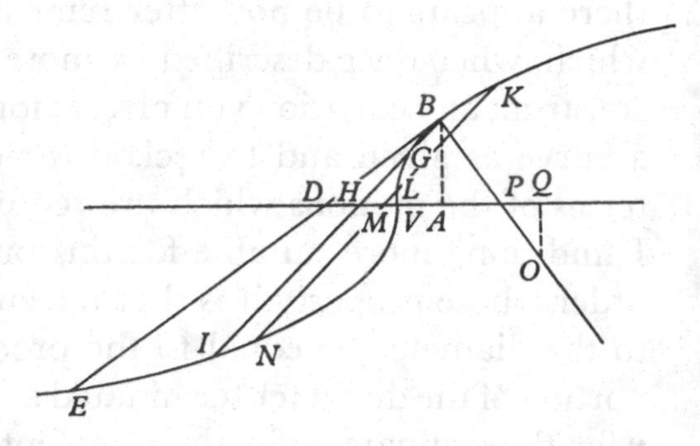

Porro ductâ quavis rectâ secante hanc Parabolam in tribus punctis I, G, B, et lineam AV in H: segmenta duo HG, HB ex unâ parte ejusdem AV ad curvam terminata erunt unà æquales tertio segmento HI ex alterâ parte ad curvam similiter terminato. Nam stantibus jam positis hoc est quod sit $VA = \frac{x^3}{nn}$, et insuper propter datam positionem rectæ HB posito $VH = c$, & $A[H] \left(c + \frac{x^3}{nn}\right) . [AB] (x) :: d . e$; erit $ec + \frac{ex^3}{nn} = dx$, sive $x^3 - \frac{dnn}{e}x + cn[n] = 0$. Cujus æquationis tres radices designant tria perpendicula a punctis B, G, et $[I]$ ad rectam AV demissa, et manifestum est ob defectum secundi termini quod eorum duo ex unâ ꝑte ejusdem rectæ AV sint æqualia tertio ex alterâ parte. Et proinde cùm HB, HG et HI sint inter se ut ista perpendicula, patet esse $HB + HG = HI$.[123] Hinc recta VA meritò dicetur Parabolæ hujus diameter quia rectas omnes ad curvam tripliciter terminatas dividit in ꝑtes hinc inde æquales, & proinde ejusmodi omnes rectæ lineæ (qualis BGI) dici possunt ordinatim applicatæ ad diametrum.[124] Et hanc denominationem retinere possunt etsi forte duæ e partibus sint imaginariæ.

Et patet non esse nisi unicam diametrum et ad eam rectas in quibuscunq; angulis ordinatim applicari posse,[125] contra quàm accidit in conicis sectionibus

(121) Note the identity of the present terminology with that used in the exposition of the apparatus for constructing curves organically in terms of given 'describing' ones (**1**, **3**, §2 above). The similarity is not coincidental, for his present construction is, in fact, an organic description of the cubic parabola (B) from the describing hyperbola (R) by means of the intersecting 'legs' VB, ωB and VR, ωR of the two 'sectors' $B\hat{V}R$ and $B\hat{\omega}R$ (of zero magnitude)

cerned.[121] For let S be the intersection of the line RB and the asymptote sVS. Then, since by the nature of the hyperbola the rectangle $VS \times SR$ is equal to a given quantity, suppose that given quantity to be n^2 and there will be

$$RS = n^2/VS;$$

and consequently, since $RS:VS = VS:SB$, it will be $SB = VS^3/n^2$. That is, when BA is let fall perpendicular to VA and BA is called x and VA y, then $x^3/n^2 = y$.

From this it is easily gathered that the parts VB, Vb of any straight line BVb drawn through the point V and terminated on either hand at the curve are equal because the portion Vb is described by the conjugate hyperbola rt in the same way as the portion VB by RT. The point V may hence not improperly be called the parabola's centre.[122]

Further, when any straight line is drawn intersecting this parabola in the three points I, G, B and the line AV in H, the two segments HG, HB terminated at the curve on one side of AV will together be equal to the third segment HI similarly terminated on the other. For, on continuing our suppositions (that is, that $VA = x^3/n^2$) and supposing in addition, on account of the given position of the line HB, that $VH = c$ and that AH (or $c+x^3/n^2$) $: AB$ (or x) $= d:e$, then $ec+ex^3/n^2 = dx$ or $x^3-(d/e)\,n^2x+cn^2 = 0$. The three roots of this equation designate the three perpendiculars let fall from the points B, G and I to the line AV, and it is manifest since the second term is lacking that two of these on one side of that same line AV are equal to the third on the other one. And therefore, because HB, HG and HI are to one another as those perpendiculars, it is evident that $HB+HG = HI$.[123] In consequence, the line VA will deservedly be called the diameter of this parabola since it divides all straight lines terminated in three points at the curve into parts equal on either side of it, and therefore all straight lines of the sort (such as BGI) may be called ordinates to that diameter.[124] And they may keep this title even though two of the parts happen to be imaginary.

It is evident that there is but a single diameter and that straight lines may be ordinately applied to it in any angle[125]—the reverse of what occurs in conic

---

rotating round the 'poles' V and ω, where ω is the point at infinity on VA. It is wholly typical of Newton that he should desire both to give an elementary construction of the cubical parabola (rather than merely present its Cartesian defining equation $x^3 = n^2y$) and yet conceal its theoretical basis.

(122) This is immediately obvious analytically since the parabola's Cartesian defining equation $x^3 = n^2y$ lacks even terms: compare **1**, 1, §3: note (94).

(123) Compare **1**, 2, §2: note (36) above. The cubic equation $x^3-(d/e)n^2x+cn^2 = 0$ is, of course, the eliminant of $x^3-n^2y = 0$ and $y-(d/e)x+c = 0$, the respective Cartesian defining equations of the cubic parabola and the line BH.

(124) For this terminology compare **1**, 2, §2.2: Theorem 1, Scholium. The following sentence is a late addition inserted in afterthought.

(125) Compare **1**, 2, §2.2: Theorem 1, Corollary 3.

ubi tot sunt diametri quot anguli in quibus rectæ ordinantur, si modò circulū demas ubi ordinatio semper fit in angulis rectis.

Patet etiam quo pacto centrum et diameter cùm datur hæc Parabola inveniri possunt, nempe duc duas quascunq; parallelas aut non parallelas rectas IGB et NLK secantes curvam in punctis I, G, B et N, L, K; easq; ita divide in H et M ut partes hinc inde sint æquales, videlicet $IH=HG+HB$, et $NM=ML+MK$ $\left(\text{hoc est cape } GH=\frac{GI-GB}{3} \text{ et } ML=\frac{LN-LK}{3},\right)$[126] et recta per H et M ducta erit quæsita diameter.

Est etiam manifestum quòd superficies utrinq; a diametro inter lineas parallelas terminatæ sint æquales, viz: $HMNI=HMLG+HMKB$, si modo BI et KN ponantur parallelæ.[127]

Porrò si recta BD ducatur quæ curvam alicubi in B tangent (id quod fit capiendo $VD=2VA$ ac jungendo BD) et si tangens ista producatur donec iterum curvæ occurrat in E; constat esse $DE=2BD$.[128]

Si deniq; ad idem quodvis curvæ punctum B ducatur perpendiculum BP et indefinitè producatur secans diametrum [in] P, et ab A versus P capiatur $AQ=\frac{1}{2}DP$ et ad Q normaliter erigatur QO secans præfatum perpendiculum in O[,] circulus centro O et intervallo OB descriptus erit ejusdem curvitatis cum hac Parabola in præfato ejus puncto B, quemadmodum alibi docetur.[129] Minima autem curvitas in illo punctorum B contingit esse ubi ponitur

$$AB=\frac{c}{\sqrt{qq:27:}}.^{(130)}$$

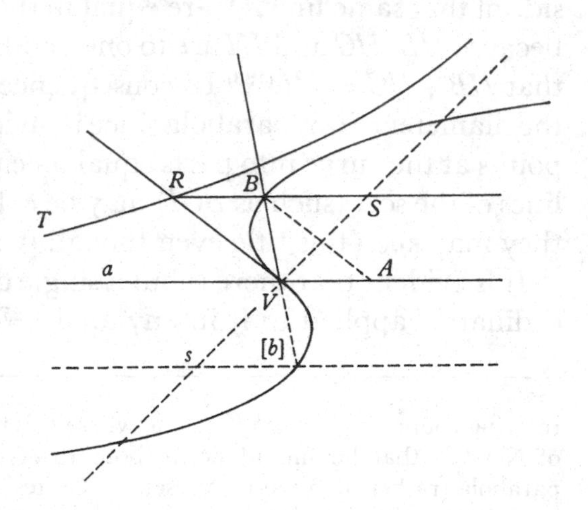

Cæterùm usui potest esse hanc curvam non modò ad rectos angulos describere sed et ita ut principales ordinatæ (id est quarum cubi æquantur facto ex data quantitate nn et parte diametri ad centrum[131] V terminatâ) diametrum in dato quovis angulo secent. Id verò sic præstabis. Sit datus angulus BAV et huic æqualem AVS. Tunc Asymptotis aVA et sVS in angulo

(126) For $GI = GH+HI = 3GH+GB$ and $LN = LM+MN = 3LM+LK$.

(127) See **1**, 2, §2.2: Theorem 1, Corollary 1.

(128) The limit-form of $IH = HG+HB$ when G coincides with B (and so H with D and I with E). Analytically, since $DA = x(dy/dx) = 3y = 3VA$ (or as, Newton remarks, $DV = 2VA$), then, where E has the co-ordinates (X, Y), $Y:y = X^3:x^3 = ED^3:DB^3 = (Y-2y)^3:(3y)^3$, or $Y^3-6yY^2-15y^2Y-8y^3 = 0 = (Y-8y)(Y+y)^2$, so that $Y = 8y$ and $ED:DB = 2:1$.

sections, where there are as many diameters as angles in which straight lines are ordinate, with the exception only of the circle where the ordinates are always applied at right angles.

It is evident also how the centre and diameter may be found when this parabola is given: precisely, draw any two straight lines IGB and NLK (parallel or non-parallel) intersecting the curve in the points I, G, B and N, L, K, and then divide them in H and M so that the parts on either hand are equal, namely, that $IH = HG+HB$ and $NM = ML+MK$ (that is, take $GH = \frac{1}{3}(GI-GB)$ and $ML = \frac{1}{3}(LN-LK)$)(126), and the straight line drawn through H and M will be the diameter sought.

It is further manifest that the surfaces terminated on either hand by the diameter between parallels are equal, namely, that

$$(HMNI) = (HMLG)+(HMKB)$$

provided BI and KN are supposed parallel.(127)

Moreover, if the straight line BD is drawn so as to touch the curve at some point B (which happens on taking $VD = 2VA$ and joining BD) and if that tangent be produced till it meets the curve again in E, then it is settled that $DE = 2BD$.(128)

If, finally, at the same arbitrary point B of the curve there be drawn the normal BP indefinitely produced and intersecting the diameter in P, and if from A in the direction of P there be taken $AQ = \frac{1}{2}DP$ and at Q is erected the perpendicular QO intersecting the previous normal in O, then a circle described with centre O and radius OB will be of the same curvature as this parabola at the above mentioned point B, as is explained elsewhere.(129) The least curvature, however, will occur at that point B for which AB is set equal to $c/\sqrt[4]{27}$.(130)

For the rest, it may be useful to describe this curve not only at right angles but also so that its principal ordinates (those, that is, whose cubes are equal to the product of the given quantity n^2 and the portion of the diameter terminated at the centre(131) V) intersect the diameter in any given angle. That, indeed, you will accomplish in this way. Let the given angle be $B\hat{A}V$ and $A\hat{V}S$ equal to it. Then with asymptotes aVA and sVS in the angle $S\hat{V}a$ which is their supplement

---

(129) By 'Problem 2d' of the October 1666 fluxional tract (I, 2, 7), where $\mathfrak{X} \equiv x^3-n^2y = 0$, then $AQ = (\cdot\mathfrak{X}^2\mathfrak{X}\cdot y^2+\mathfrak{X}\cdot^3x^2)/(-\cdot\mathfrak{X}^2\mathfrak{X}:+2\cdot\mathfrak{X}\mathfrak{X}\cdot\cdot\mathfrak{X}\cdot-\mathfrak{X}\cdot^2:\mathfrak{X})y$, in which

$$\cdot\mathfrak{X} = 3x^3, \quad \mathfrak{X}\cdot = -n^2y, \quad :\mathfrak{X} = 6x^3 \quad \text{and} \quad \cdot\mathfrak{X}\cdot = \mathfrak{X}: = 0;$$

in consequence, $AQ = (qy^2+x^2)/6y = \frac{1}{2}(3y+x^2/3y) = \frac{1}{2}DP$ since $DA = 3y$ and $AP = AB^2/DA$.

(130) Read '$\frac{n}{\sqrt{qq:45:}}$', since by the same 'Problem 2d' the radius of curvature BO at $B(x, y)$ on the cubic parabola $x^3 = n^2y$ is $(9x^4+n^4)^{\frac{3}{2}}/6n^4x$ and this is a minimum when $x = \pm 45^{-\frac{1}{4}}n$.

(131) Newton has cancelled 'verticem' (vertex).

SVa qui est deinceps, describe Hyperbolam aliquam *TR*, duasq; rectas *VR* et *VB* ad punctum *V* in angulo dato concurrentes fac gyrare circa istud *V* interea dum alia recta *RB* in parallelismo ad *VA* sic movetur ut continuò transeat per punctum *R* ubi recta *VR* secat Hyperbolam et intersectio ejus cum alterâ rectâ *VB* (quæ debet esse(132) ad partes *RV* versus quas *VA* jacet a *VS*) describet desideratam curvam. Nam cùm ex naturâ Hyperbolæ rectangulum $VS \times RS$ sit dato æquale, pono esse nn eritq; $SR = \frac{nn}{VS}$, sive $= \frac{nn}{AB}$. Item propter similia triangula *RVS*, *BVA*, erit $RS . VS(AB) :: AB . VA$, adeoq; $VA = \frac{AB^{\text{cub}}}{nn}$. Q.E.F.(133)

Potuit insuper ostendi quod ubicunq; polus *V* [(]circa quem angulus *RVB* convolvi debet) sumatur extra centrum Hyperbolæ, tamen hujusmodi curva describetur, dummodò angulus iste *RVB* æquetur residuo anguli a duobus rectis in quo Hyperbola describitur.(134) Sed(135) his & ejusmodi alijs minùs expeditis descriptionibus prætermissis propositum in sequente problemate comprehendo.

Prob: 8. Æquationem quamcunq; plusquam biquadraticam et secundo termino carentem, ope datæ cubicæ parabolæ construere.(136)

Species radicalis (sive radicem in æquatione designans) sit x et sit $x^3 = y$ (posito $n = 1$ et angulo *VAB* recto) æquatio designans datam Parabolam pro more jam explicato. Jam si æquationis propositæ numerus dimensionum non sit per numerum ternarium divisibilis duc in x vel xx ut talis evadat, et postea pro x^3 substitue y quoties id fieri potest et emerget æquatio designans aliquam curvam quæ si describatur et a punctis intersectionum cum præfata(137) Parabola, perpendiculares ad diametrū demittantur, ipsæ erunt radices æquationis propositæ.(138)

Instantiæ gratiâ, si proponatur æquatio $x^5 = px^3 + qxx + rx + s$, ubi p designat quantitatem cognitam tertij termini suis signis + et − adfectam, q 4ti, r 5ti, &c: Duco æquationem in x et fit $x^6 = px^4 + qx^3 + rxx + sx$. Tum pro x^3 ubiq; substi-

(132) 'sita' (situated) is cancelled.

(133) The same construction and proof as before (note (121)) except that now $A\hat{V}R$ is in general oblique.

(134) This is the condition, necessary and sufficient for the cubic (B) described in the organic construction from the hyperbola (R) with respect to poles V and ω (at infinity on VA) to be a Wallisian parabola, that the point at infinity on the hyperbola (R) correspond to a cusp at ω on (B), where the line at infinity is tangent to the cubic. (Compare note (121).) As before the point at infinity on the hyperbola (R) in the direction VA will determine the tangent at the cubic's centre V.

(135) A first cancelled continuation reads 'jam dicta sufficiunt proposito' (what has now been said suffices for our purpose).

describe some hyperbola *TR* and make two straight lines *VR* and *VB*, concurrent at the point *V* in the given angle, revolve round that point *V* while at the same time a second straight line *RB* moves parallel to *VA* so as continuously to pass through the point *R* in which the line *VR* intersects the hyperbola: its intersection with the second line *VB* (which should be(132) on that side of *RV* towards which *VA* lies from *VS*) will describe the desired curve. For since by the nature of the hyperbola the rectangle $VS \times RS$ is equal to a given quantity, I take it to be n^2 and there will be $SR = n^2/VS$, that is, n^2/AB. Again, on account of the similarity of the triangles *RVS* and *BVA*, there will be

$$RS : VS \text{ (or } AB) = AB : VA,$$

so that $VA = AB^3/n^2$. Which was to be done.(133)

It could have, in addition, been shown that, no matter where the pole *V* (round which the angle $R\hat{V}B$ is to be rotated) be taken outside the hyperbola's centre, none the less a curve of this sort will be described provided that angle $R\hat{V}B$ be equal to the supplement to two right angles of the angle in which the hyperbola is described.(134) But(135) I pass over these and other less ready descriptions of this sort and grasp my purpose in the following problem.

Problem 8. To construct any equation higher than a quartic and lacking its second term with the help of a given cubic parabola.(136)

Let the radical variable (the one, that is, which designates the root in the equation) be x and let $x^3 = y$ (where n is set equal to 1 and the angle $V\hat{A}B$ right) be the equation defining the given parabola in the manner just now explained. If now the number of dimensions in the propounded equation be not divisible by three, multiply it by x or x^2 to make it so, and afterwards for x^3 substitute y as many times as can be done, and there will emerge an equation defining some curve: if this be described and perpendiculars be let fall from its points of intersection with the aforesaid(137) parabola to the diameter, these will be roots of the propounded equation.(138)

For instance, if the equation $x^5 = px^3 + qx^2 + rx + s$ be proposed, where p designates the known quantity of the third term affected with its signs + and −, q that of the 4th, r that of the 5th, and so on: I multiply the equation by x and it becomes $x^6 = px^4 + qx^3 + rx^2 + sx$. Then, on substituting y everywhere for x^3

(136) A revised extended version of the sections 'A Generall rule wherby any Probleme may bee resolved' and 'Constructions performed by a Parabola of y^e 2^d kind' in Newton's May 1665 paper 'Of the construction of Problems' (I, 3, 3, §2).

(137) Newton first wrote 'data' (given).

(138) The condition that the second term of the given equation (of, or increased to become of, degree $3p$) in x be lacking is evidently necessary if it is to be the eliminant of the Cartesian defining equations $x^3 = y$ and $f(x, y) = 0$ (of degree p). The second term may, of course, always be removed without loss of generality by suitably increasing or decreasing the root.

tuendo y emergit $yy=pxy+qy+rxx+sx$. Est itaq; conica sectio describenda quam æquatio $yy=pxy+qy+rxx+sx$ designat, posito quòd y denotat eandem lineam AV ac in æquatione $x^3=y$ ad cubicā Parabolam, et quod x hic designat longitudinem perpendiculi Ab ad Conicam sectionem in b terminati sicut in alterâ æquatione $x^3=y$ designabat longitudinem ejusdem perpendiculi ad cubicam Parabolam similiter in B terminati. Et manifestum est quòd ubi hæ duæ curvæ concurrunt (quemadmodum in β) erunt ijdem valores de x in utrâq; æquatione $x^3=y$ et $yy=pxy+qy+rxx+sx$, et inde quod valores illi (hoc est perpendicula $\beta\alpha$ a punctis concursuum demissa) sunt radices æquationis propositæ, siquidem æquatio ista ex his vicissim supprimendo y produci potest.

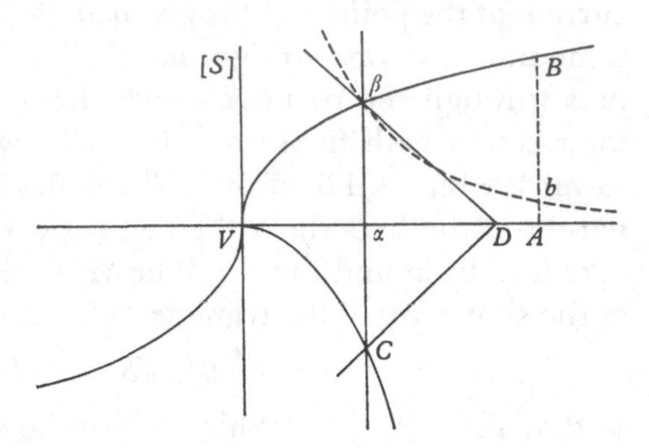

Ad eundem modum si proponatur æquatio cubo-cubica

$$x^6=px^4+qx^3+rxx+sx+t,$$

emerget $yy=pxy+qy+rxx+sx+t$ pro determinanda conica sectione quam describi oportet.

Et sic æquatio $x^7=px^5+qx^4+rx^3+sxx+tx+v$ dabit

$$xyy=pxxy+qxy+ry+sxx+tx+v.$$

Aut si ducas in xx juxta regulam, habebitur $x^9=px^7+qx^6+rx^5+sx^4+tx^3+vx^2$ et inde $y^3=pxyy+qyy+rxxy+sxy+ty+vxx$, ad determinationem curvæ describendæ. Et sic præterea. Ubi patet quod æquationes omnes quæ non transcendunt novem dimensiones solvuntur per descriptionem curvæ trium dimensionum, et quod illæ 12, 15, vel 18 dimensionum non nisi curvam 4, 5, vel 6 dimensionum requirunt, et sic porrò.(139)

Descriptio curvarum quæ per hujusmodi æquationes designantur si sint conicæ sectiones manifesta est ex aliorum scriptis præsertim Geometria Cartesij.(140) At si sint altioris generis præstat ut curva designetur per inventionem aliquot punctorum ejus, nisi fortè commoda aliqua descriptio per motum continuum occurrat. Quemadmodum si proponitur inventio sex mediarum continuè proportionalium inter a et b, posita x pro primâ emerget æquatio $x^7=a^6b$ et inde $xyy=a^6b$ designabit curvam describendam. Si itaq; Parabolam conicam in promptu descriptam habeas cujus latus rectum sit 1,

(139) See note (138). It continues to be understood that the equations in all cases lack their second term.

(140) In reducing the Greek 3/4 line problem analytically to a second degree Cartesian defining equation in co-ordinates x and y inclined to one another in a defined angle, and then

there emerges $y^2 = pxy+qy+rx^2+sx$. There has consequently to be described the conic section defined by the equation $y^2 = pxy+qy+rx^2+sx$, supposing that y denotes the same line AV as in the equation $x^3 = y$ to the cubic parabola, and that x here designates the length of the perpendicular Ab terminated at the conic section in b just as in the other equation, $x^3 = y$, it designated the length of the same perpendicular terminated similarly at the cubic parabola in B. And it is manifest that where these two curves meet (in β, for example) the values of x in either equation, $x^3 = y$ and $y^2 = pxy+qy+rx^2+sx$, will be the same and consequently that those values (in other terms, the perpendiculars $\beta\alpha$ let fall from the meeting points) are roots of the propounded equation, inasmuch as that former equation may be produced out of these in turn by suppressing y.

In much the same way if the sextic equation $x^6 = px^4+qx^3+rx^2+sx+t$ be propounded, there will emerge $y^2 = pxy+qy+rx^2+sx+t$ for determining the conic which must be described.

And thus the equation $x^7 = px^5+qx^4+rx^3+sx^2+tx+v$ will yield

$$xy^2 = px^2y+qxy+ry+sx^2+tx+v.$$

Or should you multiply by x^2 according to the rule there will be had

$$x^9 = px^7+qx^6+rx^5+sx^4+tx^3+vx^2$$

and thence $y^3 = pxy^2+qy^2+rx^2y+sxy+ty+vx^2$ to determine the curve to be described. And so forth. It is here evident that all equations which do not surpass nine dimensions are resolved by the description of a curve of three dimensions, while those of 12, 15 or 18 dimensions require a curve of but 4, 5 or 6 dimensions, and so on.(139)

The description of curves defined by equations of this sort is manifest, if they are conic sections, from the writings of others and particularly Descartes' *Geometrie*.(140) But if they are of higher order it is better that the curve be defined by finding some of its points unless some commodious description by a continuous motion chances to be met with. For example, if the finding of six mean continued proportionals between a and b be proposed, when x is set for the first there will emerge the equation $x^7 = a^6b$ and thence $xy^2 = a^6b$ will define the curve to be described. If, consequently, you have described in readiness a conic

transforming this locus by co-ordinate transformation into a set of canonical forms corresponding with the classical defining 'symptoms' of the conic, Descartes had in the second book of his *Geometrie* outlined, with some slight omissions later filled in by De Beaune and others, the identity of the conic with the geometrical locus whose co-ordinates are related analytically by an algebraic equation of second degree in two variables. The construction of a conic corresponding to any given equation of second degree is implicit in his general account and proceeds by reducing the latter to the form $(y+Ax+B)^2 = C^2+Dx+Ex^2$. (Compare *Geometrie*, (Leyden, 1637): 324–34 = *Geometria* ($_2$1659): 25–34.) For mechanical constructions of conics known to Newton compare I, **1**, 1, §1: passim.

pone verticem ejus ad punctum V et axem in rectâ VS. Dein regula $\beta\alpha C$ in parallelismo ad VS utcunq; moveatur et ad hanc norma βDC ita continuò applicetur ut ejus puncti angularis D in recta VA positi distantia αD a regula continuò sit $a\sqrt[3]{b}$, et crus DC transeat per intersectionem regulæ et parabolæ conicæ. Nam intersectio regulæ et cruris alterius $D\beta$ describet curvam quæsitam.(141) Quod si hæc descriptio minùs placeat, (ubi non opus est curvam delineare, sed regulam et normam in desideratâ positione tantùm sistere) curva designari potest quærendo aliquot ejus puncta beneficio æquationis $a^6b=xyy$.

Cæterùm ad hujus regulæ exemplar possunt aliæ componi. Sic enim curva jam explicata quam æquatio $1=xyy$ designat,(142) accommodatur ad solutiones æquationum penultimo termino carentium; et sic aliæ curvæ (ut $x^4=y$, vel $x^3y=1$(143)[)] usui esse possunt, etsi non ita generaliter. Sed de his satis. Restat tantùm ut in complementum hujus doctrinæ adjungam problema sequens.

Prob: 9. Quomodò Problemata solvenda sunt, ubi per intricatam terminorum complicationem non licet ad æquationes commodè pervenire.(144)

Conclusio.

Cùm præter generales jam traditas, regulas aliquæ occurrant quæ nonnumquam usui esse possunt, ipsas etiam in sequentibus breviter attingam.

1. In quavis recta KB assumptis ad arbitrium punctis [K], A, et B, biseca AB in C et ad C erigatur perpendiculū CX et agatur quævis AX. Jam ut inter has rectas CX et AX inscribatur linea EY ejusdem longitudinis cum AC, ita ut producta transeat per punctum K; sit(145) $KA=a$. $KB=b$, $CX=c$, $XY=x$, et $KE=y$. et erit $CK=\dfrac{a+b}{2}$, $CY=c+x$ & $YK=y-\dfrac{b-a}{2}$. Adeoq; per 47[,] 1 Elē est

$$\frac{aa+2ab+bb}{4}+cc+2cx+xx=yy-by+ay\frac{+bb-2ab+aa}{4}.$$

(141) The moving 'sector' βDC constructs $\alpha D = a\sqrt[3]{b}$ continuously as the mean between $\alpha\beta = x$ and $\alpha C = V\alpha^2 = y^2$. By the general phrase 'commoda aliqua descriptio per motum continuum' Newton may intend an oblique reference to his organic description of curves (1, 3, §2).

(142) Since by the transformation $x \to x^{-1}$ an equation lacking its last but one term is converted into one whose second term is missing while the curve $x^{-1} = y^2$ is converted into the parabola $x = y^2$ (and not, as he seems to require, $x = y^3$), Newton probably intends the curve '$1 = xy^3$'.

(143) Read perhaps '$x^4y = 1$' (compare note (142)).

parabola whose latus rectum is unity, set its vertex on the point V and its axis in the line VS. Then let the ruler $\beta\alpha C$ be moved in any way whatever parallel to VS and let the sector βDC be continuously applied to it so that the distance αD of its angular point D set in the line VA from the ruler be continuously $a\sqrt[3]{b}$, and that the leg DC pass through the intersection of the ruler and the conic parabola. For the intersection of the ruler and its other leg $D\beta$ will describe the curve sought.(141) But if this description does not please so much (where there is no need to delineate the curve but merely that the ruler and the sector should be fixed in the desired position), the curve may be defined by seeking several of its points with the aid of the equation $a^6b = xy^2$.

Still other rules may be composed on the pattern of this one. Thus, for instance, the curve just now deployed (which is defined by the equation $1 = xy^2$)(142) is suited to resolving equations lacking their last but one terms; and thus other curves (such as $x^4 = y$ or $x^3y = 1$)(143) may be of use, though not so generally. But enough of this. It merely remains for me to complement this doctrine by adjoining the following problem.

Problem 9. How problems are to be resolved when by reason of the involved complexity of their terms it is not permitted to arrive conveniently at equations.(144)

Conclusion

Since apart from the general ones now delivered there are some other rules to be met with which may sometimes be of use, I shall briefly touch on these also in the following.

1. Assuming the points K, A and B arbitrarily in any straight line KB, bisect AB in C and let there be erected at C the perpendicular CX and any line AX be drawn. Now to inscribe between these straight lines CX and AX a line EY of the same length as AC so as to pass, when produced, through the point K, let there be(145) $KA = a$, $KB = b$, $CX = c$, $XY = x$ and $KE = y$: then will there be $CK = \frac{1}{2}(a+b)$, $CY = c+x$ and $YK = y-\frac{1}{2}(b-a)$, so that by *Elements*, I, 47, $\frac{1}{4}(a^2+2ab+b^2)+c^2+2cx+x^2 = y^2-by+ay+\frac{1}{4}(b^2-2ab+a^2)$, or

---

(144) A page and a half following in the manuscript has been left blank for intended insertion of the resolution of the problem. What exactly Newton had in mind seems difficult to define. He may perhaps have envisaged the construction of problems whose exact algebraic formulation is complicated by some generalized 'mechanical' technique: the n-section of an angle, for example, when n is a large prime would lead to an impossibly large, irreducible equation of like degree algebraically, but would be effected at once by the introduction of such transcendental curves as the quadratrix, cycloid or Archimedean spiral. Only in a small minority of cases, however, would there be a significant gain in introducing a constructional apparatus which, in effect, resolved a higher-order algebraic or transcendental equation.

(145) Newton first wrote 'His ita constructis' (When these have thus been constructed).

Sive $yy\begin{smallmatrix}-b\\+a\end{smallmatrix}y\begin{smallmatrix}-ab\\-cc\end{smallmatrix}-2cx-xx=0$. Jam cùm per Lem 1 Prob 2 fuit $YX.AK::CX.KE$[,] hoc est $x.a::c.y$ scribo $\frac{ac}{y}$ pro x et oritur

$$yy\begin{smallmatrix}-b\\+a\end{smallmatrix}y\begin{smallmatrix}-ab\\-cc\end{smallmatrix}-\frac{2ac}{y}-\frac{aacc}{yy}=0.$$

Hoc per yy multiplico dividoq; per $y+a$[146] et fit $y^3-byy-ccy-acc=0$.[147]

Si rursus pro y scribatur $\frac{ac}{x}$, emerget $aac-abx-cxx-x^3=0$.[147] Et hæ duæ formulæ (signis $+$ et $-$ probè mutatis) inserviunt solutioni cubicarum æquationum ubi non sunt omnes radices affirmativæ vel omnes negativæ.[148]

2. Ut a puncto quovis K inter rectas quascunq; positione datas YX et EX inscribatur recta EY datæ longitudinis: junge KX, demitte perpendiculū KC ad rectarum alterutrā YX et eidem YX duc KF parallelā. Dein dic $KF=a$. $KX=b$. $CX=c$. $EY=d$. $XY=x$, et $EK=z$, eritq; $YK=z-d$, et (per 13, 2 Elem)

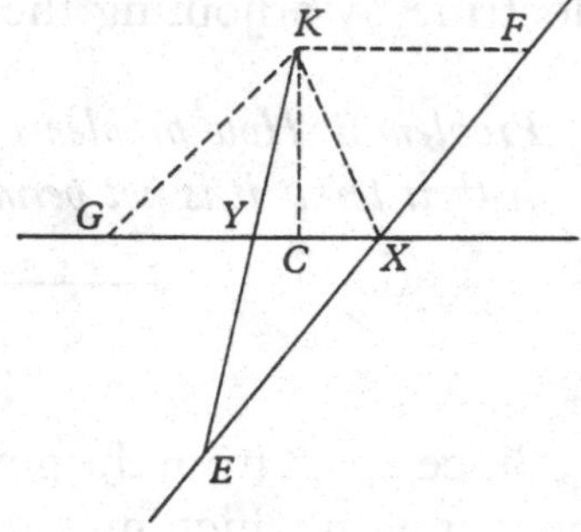

$$KX^q+XY^q=YK^q+2CXY_{[,]}$$

hoc est $bb+xx=zz-2dz+dd+2cx$. Quare cum insuper sit $EY.YX::EK.KF$ sive $d.x::z_{[.]}a$, substitue $\frac{ad}{x}$ pro z et orietur

$$bb+xx=\frac{aadd}{xx}-\frac{2add}{x}+dd+2cx.\quad \text{Sive}\quad x^4-2cx^3\begin{smallmatrix}+bb\\-dd\end{smallmatrix}xx+2addx-aadd=0.^{(149)}$$

Et rursus pro x si vicissim substituas $\frac{ad}{z}$ prodibit

$$aadd-2cadz\begin{smallmatrix}+bb\\-dd\end{smallmatrix}zz+2dz^3-z^4=0.^{(149)}$$

Et hæ duæ formulæ (signis $+$ et $-$ probè mutatis) inserviunt constructioni omnium biquadraticarum æquationum ubi diversa sunt signa primi et ultimi termini.[150] Et siquando impossibilitas in earum una contingat tunc altera debet adhiberi.[151]

(146) That is, $y+a=0$, the root when EY coincides with AC (and so XY with $XC=-a$).

(147) Much as in note (34) above, on setting $KX=\alpha$, $KC=\beta$, $CX=c$, $EY=AC=d$ and supposing $XY=x$, $KE=y$ we find $KE\times XY\times AC=EY\times CX\times KA$ or $xy=c(\beta-d)$ and also $y^3-(\beta+d)y^2+(\beta^2-\alpha^2)y-c^2(\beta-d)=0$. Hence, when $\hat{C}$ is taken to be right (or $\alpha^2=\beta^2+c^2$) and also $KA=\beta-d=a$, Newton's results are immediate consequences.

$$y^2+(a-b)\,y-ab-c^2-2cx-x^2=0.$$

Now since by Lemma 1 of Problem 2 it was $YX:AK = CX:KE$, that is, $x:a=c:y$, I write ac/y for x and there arises $y^2+(a-b)\,y-ab-c^2-2ac/y-a^2c^2/y^2=0$. This I multiply by y^2 and divide by $y+a$(146) and it becomes

$$y^3-by^2-c^2y-ac^2=0.\text{(147)}$$

If, on the other hand, for y there should be written ac/x, there will emerge $a^2c-abx-cx^2-x^3=0$.(147) And these two formulas (with the signs + and − appropriately changed) serve to resolve cubic equations in which not all roots are positive or all negative.(148)

2. To inscribe in line with K between any straight lines given in position YX and EX the straight line EY of given length: join KX, drop the perpendicular KC onto one of the two lines YX and to the same line YX draw KF parallel. Then call $KF=a$, $KX=b$, $CX=c$, $EY=d$, $XY=x$ and $EK=z$. It will then be $YK=z-d$ and (by *Elements*, II, 13) $KX^2+XY^2=YK^2+2CX\times XY$, that is, $b^2+x^2=z^2-2dz+d^2+2cx$. Hence, since in addition $EY:YX=EK:KF$ or $d:x=z:a$, substitute ad/x for z and there will arise

$$b^2+x^2=a^2d^2/x^2-2ad^2/x+d^2+2cx,$$

or $x^4-2cx^3+(b^2-d^2)\,x^2+2ad^2x-a^2d^2=0$.(149)

But, on the other hand, should you in turn for x substitute ad/z there will come $a^2d^2-2cadz+(b^2-d^2)\,z^2+2dz^3-z^4=0$.(149)

These two formulas (with the signs + and − appropriately changed) serve to construct all quartics in which the first and last terms are opposite in sign.(150) And if at any time an impossibility occurs in one of them then the other should be employed.(151)

---

(148) Geometrically, the points E which yield the roots $XY=x$ and $KE=y$ are the meet of the conchoid of pole K and sliding length $AC=CB$ through A and B with the line EX drawn through the 'lower' vertex A. In Newton's figure there can be only one real point of intersection E and so only one real root in either cubic, but all three roots may become real when AXE is rotated round A or K moved downwards along the axis ACB. The truth of Newton's concluding observation will be evident if all possible positions of the line AXE and pole K are considered.

(149) Alternatively, if we define the co-ordinates of the point $E(X, Y)$ with respect to origin C and axes KC and CG, then on taking $KC=\alpha$ (so that X and F are the points $(0, -c)$ and $(-\alpha, -a)$ respectively) the points E will be determined as the meets of the conchoid $X^2Y^2=(d^2-X^2)\,(X+\alpha)^2$ of pole K with the line EF, $(a-c)X-\alpha Y-\alpha c=0$, that is by the quartic $X^2((a-c)X-\alpha c)^2=\alpha^2(d^2-X^2)\,(X+\alpha)^2$. Newton's results follow by determining $X=\alpha x/(a-x)$ and $x=ad/z$ when for α^2 is set its equal b^2-c^2.

(150) Evidently so, since x^4 (or z^4) and $-a^2d^2$ can never have the same sign.

(151) This appears inexplicable. Since $xz=ad$, any complex roots x will have conjugate complex correspondents z, and conversely so, so that Newton cannot intend by 'impossibility' the complexity of the roots.

3. A centro C cujuscunq; circuli EB ad rectam AY positione datam demisso perpendiculo CA et in eo sumpto quovis puncto G; ut ab isto G inter dictum circulum et lineam rectam inscribatur alia recta linea EY datæ longitudinis: pone $GC=a$, $GA=b$. $CE=c$, $EY=d$, et $EG=x$, ac demitte perpendiculum EF; eritq;

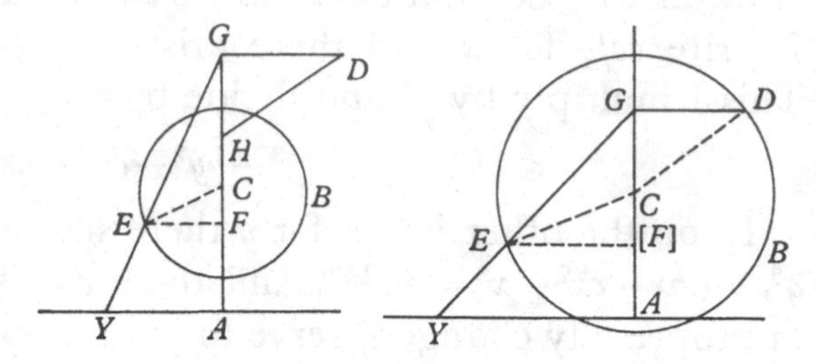

$$GY.GA::GE.GF,$$

adeoq; $GF=\frac{bx}{x+d}$. Item per 12, 2 Elem erit $\frac{GE^q+GC^q-CE^q}{2CG}$ $\left(\text{sive } \frac{xx+aa-cc}{2a}\right)=GF$. Positâ jam æqualitate inter valores GF, et factâ reductione, emerget $x^3+dxx \begin{matrix} +aa \\ -cc \\ -2ab \end{matrix} x \begin{matrix} +aad \\ -ccd \end{matrix}=0$, formula construendi cubicas æquationes.(152)

Potest autem ad hunc modum enunciari.

Æquationem cubicam $x^3=pxx+qx+r$ nullo (nisi fortè tertio) termino carentem infinitis modis solvere.

Cape ad arbitrium quamcunq; longitudinem n, sitq; ejus dimidium GC et ad punctum G erige perpendiculum $GD=\sqrt{\frac{r}{p}}$. Deinde si p et r habent eadem(153) signa centro C intervallo CD describe circulum. Sin eorum signa diversa(153) sunt, ad punctum D in angulo DGA inscribe $DH=GC$ et centro C radio GH describe circulum. Porrò cape $GA=\frac{q}{n}+\frac{r}{np}$ versus C si id (signis + & − terminorum q et r probè observatis) sit affirmativum, aliter cape adversus C et per A duc AY ad angulos rectos et inter hanc et circulum inscribe $EY=p$(154) ita ut per G transeat; et erit EG radix æquationis. Radices autem affirmativæ sunt ubi punctum E cadit inter G et Y si habetur $-p$, et contra si $+p$.(155)

4. Possem alia hujusmodi specimina afferre, sed mallem de constructionibus per intersectiones conicarum sectionum et circuli observare quanta sit earum

(152) Alternatively, if we suppose $AF = X$, $FE = Y$, the points $E(X, Y)$ may be constructed as the intersections of the conchoid (of pole G and sliding length EY)

$$X^2Y^2 = (d^2-X^2)(b^2-X^2)$$

with the circle (B), $(-a+b-X)^2+Y^2 = c^2$. When Y is eliminated there comes

$$2aX^3+(a^2-c^2-2ab+d^2)X^2-2bd^2X+b^2d = 0$$

and Newton's result follows on setting $X = bd/(x+d)$. Each value of X (and so of x) determines

3. Having from the centre C of any circle EB let fall to the straight line AY given in position the perpendicular CA and having taken in it any point G, to inscribe in line with G between this said circle and straight line another straight line EY of given length set $GC = a$, $GA = b$, $CE = c$, $EY = d$ and $EG = x$, and then let fall the perpendicular EF: there will then be $GY:GA = GE:GF$ and so $GF = bx/(x+d)$. Further, by *Elements*, II, 12,

$$\frac{GE^2+GC^2-CE^2}{2CG}\left(\text{or } \frac{x^2+a^2-c^2}{2a}\right) = GF.$$

Setting now the two values of GF equal to one another and carrying out reduction, there will emerge $x^3+dx^2+(a^2-c^2-2ab)\,x+a^2d-c^2d = 0$, a formula for constructing cubic equations.(152) It may, however, be expressed in this fashion:

To resolve the cubic equation $x^3 = px^2+qx+r$ lacking no term (except perhaps the third) in an infinity of ways.

Take any length n at will, let GC be its half and at the point G erect the perpendicular $GD = \sqrt{[r/p]}$. Then, if p and r have the same(153) sign, with centre C and radius CD describe a circle. But if their signs are opposite(153) at the point D in the angle DGA inscribe $DH = GC$ and with centre C, radius GH describe a circle. Further, take $GA = (q/n)+(r/np)$ towards C if it (with the signs + and − of the terms q and r duly observed) be positive, otherwise take it away from C, and through A draw AY at right angles, and between this and the circle inscribe $EY = p$(154) so as to pass through G: then will EG be a root of the equation. Positive roots, however, are those for which the point E falls between G and Y if there is had $-p$, and the contrary if $+p$.(155)

4. I might adduce other models of this kind, but I would prefer with regard to constructions by means of the intersections of conic sections and a circle to

a pair of corresponding values of $Y = \pm X^{-1}(b-X)\sqrt{[d^2-X^2]}$, which are, however, real only for $|X| \leqslant d$ or $|x+d| \geqslant b$. In consequence, the construction will not necessarily yield all the real roots of the cubic with which the present expression is identified.

(153) These two adjectives should be interchanged (note (155)).

(154) More precisely, '$-p$'.

(155) When the cubic $x^3-px^2-qx\mp r = 0$ is identified with

$$x^3+dx^2+(a^2-c^2-2ab)x+(a^2-c^2)d = 0$$

there comes $d = -p$, $a^2-c^2 = \pm r/p$ and so $2ab = q+(\pm r/p)$: that is, on setting $2a = n$ arbitrarily, $GC = a = \frac{1}{2}n$, $EY = d = -p$, $GD = \sqrt{[\pm r/p]}$, $GA = b = (q+GD^2)/n$ and $CE = \sqrt{[\frac{1}{4}n^2-r/p]} = \sqrt{[GC^2\mp GD^2]}$, which equals GH or CD according as p and r have the same or opposite signs. When $r = 0$ the point D will be on the circle. The value $p = 0$ is useless in the present construction since then the circle (B) would have infinite radius while the points E and Y coincide. Despite Newton's proviso to the contrary there seems no reason why q may not be zero.

varietas. Scilicet in præcedentibus[156] duæ fuerunt conditiones ad arbitrium determinandæ, sed ejusmodi constructiones componi possunt ubi non solùm duæ sed etiam tres[,] quatuor vel quinꝗ ejusmodi conditiones indeterminatæ relinquantur. Cujus rei cape hoc exemplum.[157]

Pro circulo designando fingatur æquatio $xx+yy=dxy\begin{matrix}+c\\+g\end{matrix}y\begin{matrix}+b\\+f\end{matrix}x\begin{matrix}+a\\+e\end{matrix}$ ubi x ad y tanquam ad diametrum ordinatur in angulo cujus complementi ad rectum sinus est $\frac{1}{2}d$ posito 1 pro radio, habitóꝗ respectu ad signum termini d. Et ad eundem modum sit $kxx+yy=dxy\begin{matrix}+c\\+gk\end{matrix}y\begin{matrix}+b\\+fk\end{matrix}x\begin{matrix}+a\\+ek\end{matrix}$ æquatio ad Ellipsin vel aliam conicam sectionem ubi y ad x similiter ordinatur. Jam subduc priorem æquationem e posteriori et restabit $\begin{matrix}k\\-1\end{matrix}xx=\begin{matrix}gk\\-g\end{matrix}y\begin{matrix}+fk\\-f\end{matrix}x\begin{matrix}+ek\\-e\end{matrix}$. Divide per $k-1$ et fit $xx=gy+fx+e$, sive $y=\frac{xx-fx-e}{g}$. Hunc valorem in æquatione prima pro y substitue et fiet $x^4\begin{matrix}-2f\\-dg\end{matrix}x^3\begin{matrix}-2e\\+ff\\+dgf\\-cg\end{matrix}xx\begin{matrix}+2ef\\+dge\\+cgf\\-ggb\end{matrix}x\begin{matrix}+ee\\+cge\\-gga\end{matrix}=0$.[158] Divide jam per $x-f$ et emerget $x^3\begin{matrix}-f\\-dg\end{matrix}xx\begin{matrix}-2e\\-cg\end{matrix}x\begin{matrix}+dge\\-ggb\end{matrix}=0$ si modò termini $dgef-ggbf+ee+[c]ge-gga$ sub finem divisionis possunt se mutuò destruere.[159] Jam hæc Æquatio

$$x^3=\begin{matrix}f\\+dg\end{matrix}xx\begin{matrix}+2e\\+cg\end{matrix}x\begin{matrix}-dge\\+ggb\end{matrix}$$

pro formula construendi quancunꝗ cubicam æquationem $x^3=pxx+qx+r$ inservire potest, et in eâ sunt octo quantitates a, b, c, d, e, f, g & k determinandæ, et tamen non nisi quatuor data (p, q, r, & æquatio $dgef-ggbf+ee+[c]ge-gga=0$) quorū ope determinentur; adeoꝗ ex octo quantitatibus restabunt quatuor pro lubitu assumendæ.[160]

De terminis $dgef-ggbf+ee+[c]ge-gga$ obiter notari potest quod se mutuò destruent si modò circulus et conica sectio describatur per idem quoddam commune punctum quod invenietur ponendo simul

$$x=f \text{ et } y\left(=\frac{xx-fx-e}{g}\right)=\frac{-e}{g}.^{(161)}$$

(156) In problems 2–7 above, that is.

(157) In generalization of problem 4 Newton obtains a further indeterminate condition by supposing the angle at which the co-ordinates are inclined to be generally oblique.

(158) With some slight disguise Newton considers the quartic in x obtained by eliminating y between $\alpha \equiv x^2-e-fx-gy=0$ and $\beta \equiv y^2-a-bx-cy-dxy=0$. With regard to Cartesian co-ordinates x and y inclined at the angle $\cos^{-1}\frac{1}{2}d$ these are a parabola and hyperbola whose meets are determined by that quartic eliminant, but as before (note (77)) he prefers to

observe the extent of their variety. Specifically, in the preceding[156] there were two conditions to be determined at will, but constructions of this kind may be framed in which not only two but even three, four or five indeterminate conditions are to be left. Of this circumstance take this as an example.[157]

For defining a circle let there be imagined the equation

$$x^2+y^2 = dxy+(c+g)\,y+(b+f)\,x+a+e,$$

where x is ordinate to y as diameter in an angle whose complementary sine is $\frac{1}{2}d$ (where unity is set for the radius and regard is paid to the sign of the term d). And in much the same manner let $kx^2+y^2 = dxy+(c+gk)\,y+(b+fk)\,x+a+ek$ be an equation to an ellipse or some other conic section, where y is similarly ordinate to x. Now take the former equation away from the latter and there will remain $(k-1)\,x^2 = (gk-g)\,y+(fk-f)\,x+ek-e$. Divide through by $k-1$ and it becomes $x^2 = gy+fx+e$, or $y = (x^2-fx-e)/g$. This value substitute in the first equation for y and there will come

$$x^4-(2f+dg)\,x^3+(-2e+f^2+dgf-cg)\,x^2 \\ +(2ef+dge+cgf-g^2b)\,x+e^2+cge-g^2a = 0.\text{[158]}$$

Divide now by $x-f$ and there will emerge

$$x^3-(f+dg)\,x_2-(2e+cg)\,x+dge-g^2b = 0$$

provided the terms $dgef-g^2bf+e^2+[c]ge-g^2a$ at the close of the division are able to destroy one another.[159] Now this equation

$$x^3 = (f+dg)\,x^2+(2e+g)\,x-dge+g^2b$$

can serve as a formula for constructing any cubic equation $x^3 = px^2+qx+r$: in it are eight quantities a, b, c, d, e, f, g and k to be determined but yet only four given elements (p, q, r and the equation $dgef-g^2bf+e^2+[c]\,ge-g^2a = 0$) by whose aid they are to be determined, and so of the eight quantities there will remain four to be assumed at pleasure.[160]

Concerning the terms $dgef-g^2bf+e^2+[c]ge-g^2a$ it may be noted incidentally that they will destroy one another should the circle and conic section be described through the same certain common point which will be found by setting simultaneously $x = f$ and $y = ([x^2-fx-e]/g) = -e/g$.[161]

---

construct these intersections as the common meets of the family of ellipses $k\alpha+\beta = 0$, one of which, $\alpha+\beta = 0$, (when $k = 1$) is a circle.

(159) Since the meets of the circle and ellipse have to satisfy $\alpha = \beta = 0$ (note (158)), therefore when $x-f = 0$ (or $x = f$ and so correspondingly $\alpha = 0$, $y = -e/g$) the hyperbola $\beta = 0$ must pass through the point $(f, -e/g)$, which is the present condition.

(160) The arbitrary inclination of the co-ordinates (at angle $\cos^{-1}\frac{1}{2}d$) allows an extra degree of freedom, certainly, implicit in the free choice of d, but since the constant e may, without noticeably affecting the geometrical construction, be absorbed into the co-ordinate y, its indeterminacy scarcely represents the gain of a fourth dimension of freedom.

(161) Compare note (159).

5. In præcedentibus si termini e et f ponantur nihil, habebitur

$$xx+yy=dxy\begin{smallmatrix}+c\\+g\end{smallmatrix}y+bx+a$$

æquatio ad circulum; et $kxx+yy=dxy\begin{smallmatrix}+c\\+gk\end{smallmatrix}y+bx+a$, æquatio ad conicam sectionem: et inde emerget $x^4=dgx^3+cgxx+bggx+agg$, æquatio cujus ope omnes æquationes tum cubicæ tum biquadraticæ construi possunt, idq; modis dupliciter infinitis. Ejus verò constructionem qua Lector possit sese exercere, prætermitto.(162)

(162) In the terminology of note (158), when there is set $\alpha \equiv x^2-gy$ and

$$\beta \equiv y^2-a-bx-cy-dxy,$$

the circle $\alpha+\beta = 0$ and the ellipse $k\alpha+\beta = 0$ intersect in the meets of the conics $\alpha = 0$ and $\beta = 0$, while the quartic in x is the result of eliminating y between them. Since identification of that eliminant with $x^4 = px^3+qx^2+rx+s$ yields only three conditions for fixing the five undetermined constants, a, b, c, d and g, two of them may be taken arbitrarily and so the resulting geometrical construction has two dimensions of freedom.

5. If in the preceding the terms e and f be set equal to zero, there will be had $x^2+y^2 = dxy+(c+g)\,y+bx+a$ as the equation to the circle, and

$$kx^2+y^2 = dxy+(c+gk)\,y+bx+a$$

as the equation to the conic section: and there will emerge therefrom

$$x^4 = dgx^3+cgx^2+bg^2x+ag^2,$$

an equation by whose help all equations both cubic and quartic may be constructed and that in a doubly infinite manner. Its construction I pass by, however, so that the reader may exercise himself with it.[162]

INDEX OF NAMES